PLASTICS

DESIGN

HANDBOOK

PLASTICS

DESIGN

HANDBOOK

BY

DOMINICK V. ROSATO

RHODE ISLAND SCHOOL OF DESIGN
PROVIDENCE, RI

DONALD V. ROSATO

SOCIETY OF PLASTICS ENGINEERS

MARLENE G. ROSATO

UNIVERSITY OF MASSACHUSETTS
LOWELL, MA

Kluwer Academic Publishers
Boston/Dordrecht/London

Distributors for North, Central and South America:
Kluwer Academic Publishers
101 Philip Drive
Assinippi Park
Norwell, Massachusetts 02061 USA
Telephone (781) 871-6600
Fax (781) 871-6528
E-Mail <kluwer@wkap.com>

Distributors for all other countries:
Kluwer Academic Publishers Group
Distribution Centre
Post Office Box 322
3300 AH Dordrecht, THE NETHERLANDS
Telephone 31 78 6392 392
Fax 31 78 6546 474
E-Mail <orderdept@wkap.nl>

 Electronic Services <http://www.wkap.nl>

Library of Congress Cataloging-in-Publication Data

Rosato, Dominick V.
  Plastics design handbook / by Dominick V. Rosato, Donald V. Rosato, Marlene G. Rosato.
    p.   cm.
  ISBN 0-7923-7980-2 (alk. paper)
    1. Plastics—Handbook, manuals, etc.,   2. Engineering design—Handbook, manuals, etc.
I. Rosato, M.G.   II. Rosato, Donald V.   III. Title.

TA455.P5 R66   2000
620.1′923—dc21

00-61054

Printed on acid-free paper.

Printed in the United States of America

Contents

Preface

This book provides a simplified and practical approach to designing with plastics that fundamentally relates to the load, temperature, time, and environment subjected to a product. It will provide the basic behaviors in what to consider when designing plastic products to meet performance and cost requirements. Important aspects are presented such as understanding the advantages of different shapes and how they influence designs. Information is concise, comprehensive, and practical.

Review includes designing with plastics based on material and process behaviors. As designing with any materials (plastic, steel, aluminum, wood, etc.) it is important to know their behaviors in order to maximize product performance-to-cost efficiency. Examples of many different designed products are reviewed. They range from toys to medical devices to cars to boats to underwater devices to containers to springs to pipes to buildings to aircraft to spacecraft. The reader's product to be designed can directly or indirectly be related to product design reviews in the book.

Important are behaviors associated and interrelated with plastic materials (thermoplastics, thermosets, elastomers, reinforced plastics, etc.) and fabricating processes (extrusion, injection molding, blow molding, forming, foaming, rotational molding, etc.). They are presented so that the technical or non-technical reader can readily understand the interrelationships.

This type of basic information has been reviewed for many centuries with different types of materials and more recently (in just over a century) with plastics. Recognize the design basics and/or fundamentals remain the same. Their interpretation and applicability improves with time. It is like saying $2+2=4$ for the many past centuries. Now we can say it with a computer where in the recent past we used an abacus, adding machine, slide rule, etc.

It has been prepared with the awareness that its usefulness will depend on its simplicity and its ability to provide essential information. Examples are provided of designing different plastic products and relating them to critical factors that range from meeting performance requirements in different environments to reducing costs and targeting for zero defects. Reviews range from small to large and simple to complex products.

As explained in the book many designs do not require the use of engineering equations since a practical approach can be used. The engineering equations needed for designs are plentiful and readily available. When using these equations in designs all that is required is to incorporate basically the load, temperature, time, and environment behavior of plastics. A limited amount of equations as well as plastics material properties and processing information presented are provided as comparative guides that relate to the many behavior patterns available to meet your design requirements. As reviewed and referenced in this book there are extensive resources available to obtain detailed worldwide engineering equations, material data, and processing techniques such as those reviewed in Appendix A: **PLASTICS DESIGN TOOLBOX** and references.

There is an endless amount of available data for many plastic materials worldwide. Unfortunately, as with other materials, there does not exist only one plastic material that will meet all performance requirements. However it can be stated that for practically any product requirement(s), particularly when not including cost, more so than with other materials, there is a plastic that can be used.

The data included provides examples of what are available. As an example static properties (tensile, flexural, etc.) and dynamic properties (creep, fatigue, impact, etc.) can range from near zero to extremely high values. They can be applied in different environments from below the surface of the earth, to over the earth, and into space.

Designing depends on being able to analyze many diverse, already existing products, such as those reviewed in this book. One important reason for studying these products and design approaches is that they provide a means to enhance the designers' skills. Design is interdisciplinary. It calls for the ability to recognize situations in which certain techniques may be used and to develop problem-solving methods to fit specific design situations.

The many problems that are reviewed in this book should not occur. They can be eliminated so that they do not effect the product performances when qualified people understand that the problems can exist. They are presented to reduce or eliminate costly pitfalls resulting in poor product performances or failures. With the potential problems or failures reviewed there are solutions presented. These failure/solution reviews will enhance the intuitive skills of those people who are already working in plastics.

This book provides the reader with useful reference of pertinent information readily available as summarized in the table of Contents and particularly the Index. From a pragmatic standpoint, any theoretical aspect that is presented has been prepared so that the practical person will understood it and put it to use. The theorist, for example, will gain an insight into the limitations that exist and relate to those that exist with other materials such as steel, wood, and so on.

Based on over a century of worldwide production of billions of plastic products, they can be designed and processed successfully, meeting high quality, consistency, long life, and profitability. All that is needed is understanding the behavior of plastics and properly applying these behaviors.

The information contained in this book is of value to even the most experienced designers and engineers, and provides a firm basis for the beginner. The intent is to provide a review of the many aspects of designing that goes from the practical elementary to the advanced or theoretical approach. In addition to the plastic designer, this book will be useful to different people where they can interrelate their interests that interface with designing. Included are the tool maker (mold, die, etc.), designer of other materials (metals, aluminum, glass, wood, etc.), fabricator, plant manager, material supplier, equipment supplier, testing and quality control personnel, cost estimator, accountant, sales and marketing personnel, new venture type, buyer, vendor, educator/trainer, workshop leader, librarian, industry information provider, lawyer, and consultant. People with different interests can focus on and interrelate across subjects that they have limited or no familiarity in the World of Plastics.

Patents or trademarks may cover information presented. No authorization to utilize these patents or trademarks is given or implied; they are discussed for information purposes only. The use of general descriptive names, proprietary names, trade names, commercial designations, or the like does not in any way imply that they may be used freely. While information presented represents useful information that can be studied or analyzed and is believed to be true and accurate, neither the authors nor the publisher can accept any legal responsibility for any errors, omissions, inaccuracies, or other factors.

In preparing this book and ensuring its completeness and the correctness of the subjects reviewed, use was made of the authors worldwide personal, industrial, and teaching experiences that total over a century, as well as worldwide information from industry and trade associations.

THE ROSATOS, YEAR 2001

1

Introduction

Overview

This book was written to serve as a useful reference source for the product designer new to plastics as well as providing an update for those with experience. It should also be of interest to non-designers and management personnel involved in plastic products that need a general overview of the concepts and critical issues related to plastic products in this World of Plastics. It highlights designing with plastics based on material & process behaviors. As with designing any materials (plastic, steel, aluminum, wood, ceramic, etc.) it is important to know their behaviors in order to maximize product performance-to-cost efficiency.

The mature plastics industry is a worldwide, multibillion-dollar industry in which a steady flow of new or improved plastic materials, new or improved production processes, and new or improved market demands has caused rapid and tremendous growth in the use of plastics. For over a century the World of Plastics product production, with over a billion products, continues to expand enormously with the passing of time. Manufacturers are introducing new products in record time. The ability to shrink time-to-market schedules continues to evolve through the more knowledgeable application and behavior or familiarity of the different plastic materials, processing techniques, and design approaches (Fig. 1-1).

During this time very successful and long lasting designed products have been in service that range from primary and secondary structures in toys to packages to computer hardware to electrical/electronic communication hardware to boats to aircraft to space craft and so on. There are those exposed to extreme environmental conditions that range from in the ocean, underground, earth's surface, and to outer space. Temperature exposures have been from below freezing to elevated conditions that extended to 2500°F (1370°C) for very short time periods. Products have performed meeting static and dynamic loads (creep and fatigue stresses, impact loads, and so on). An inherent characteristic that makes most plastics performance-to-cost efficient is the different available low to high production methods that results in relatively low processing cost (Fig. 1-2).

So the answer to the question of what is new in designing with plastics actually is that we continue to do it easier and quicker because new, improved, and more uniform behaviors of plastics and their processing capabilities are always developing. Designwise little is new conceptually. What changes with the passing of time is the level of sophistication that is applied to designing any products.

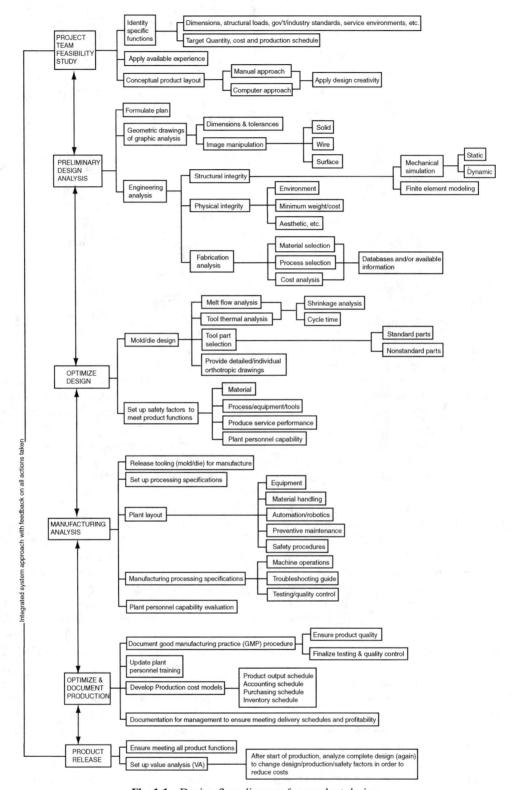

Fig. 1-1   Design flow diagram for product design.

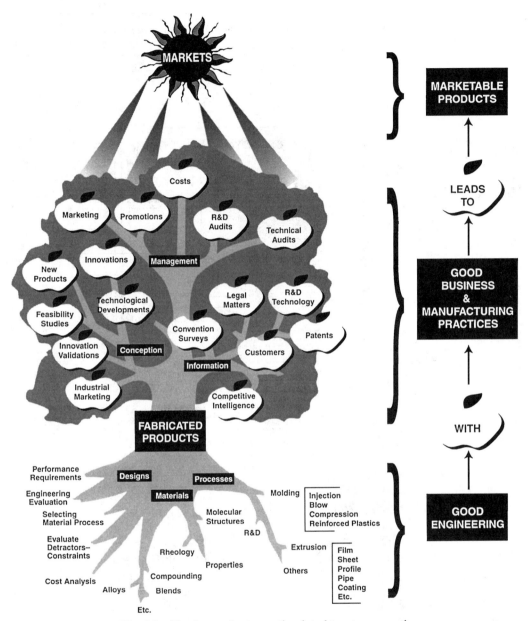

Fig. 1-2 Plastics product growth related to a tree growth.

Unfortunately there is no one plastic or process (as with other materials such as steel, wood, glass, etc.) that provide all types of performance requirements. However it can be stated that for practically any product requirement(s), particularly when not including cost, more so than with other materials, there is a plastic that can be used. The many different plastics meet different property and processing requirements that the de-

signer uses in selecting the most appropriate plastic that can be fabricated by the required process.

What has made these plastic products successful was that there were those that knew the behavior of plastics and how to properly apply this knowledge. Recognize they did not have the "tools" that make it easier for us to now design products. Now we are more knowledgeable and in the future it will

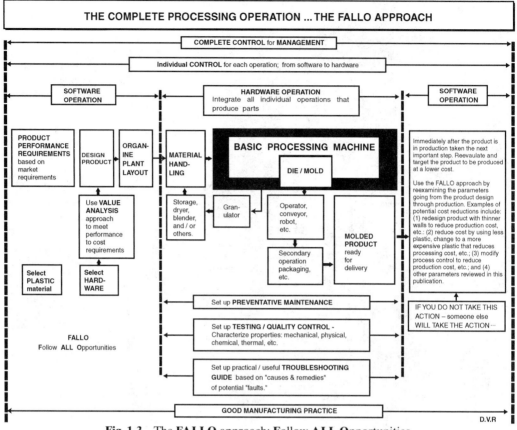

Fig. 1-3   The **FALLO** approach: Follow **ALL** Opportunities.

continue to be easier with new or improved materials and processing techniques ever present on the horizon. What is still needed, as usual, is to have a design plan conceived in the human mind and intended for subsequent fabricating execution by the proper method (Fig. 1-3).

Designers and processors to produce products at the lowest cost have unconsciously used the basic concept of the FALLO approach. This approach makes one aware that many steps are involved to be successful, all of which must be coordinated and interrelated. It starts with the design that involves specifying the plastic, and specifying the manufacturing process. The specific process (injection, extrusion, blow molding, thermoforming, and so forth) is an important part of the overall scheme and should not be problematic.

Following the product design is producing a tool (mold, die, and so forth) around

the product. Next is putting the proper fabricating process around the tool. This action requires setting up the necessary auxiliary and/or secondary equipment to interface in the complete fabricating line. Next is setting up completely integrated inline equipment controls to target for the goal of zero defects. Also in the FALLO process is that of purchasing equipment as well as materials, then properly warehousing the material and maintaining equipment. If processing is to be subcontracted ensure that the proper equipment is available and used efficiently. This interrelationship is different from that of most other materials, where the designer is usually limited to using specific prefabricated forms that are bonded, welded, bent, and so on.

The designer matches the end use requirements with the properties of the selected material using a practical or engineering technique (Fig. 1-4). Target is to achieve the basic three general requirements of design success:

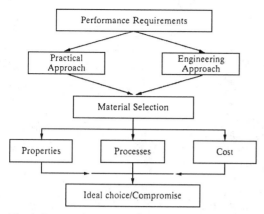

Fig. 1-4 Practical and engineering approaches to designing.

(1) economical, (2) functional, and (3) attractive in appearance. In turn the functional aspect relates to the product's three basic environment conditions of load, temperature, and time. The production method to be used will often set limitations on designs and vise versa. The way in which a product is manufactured has a profound influence on its design. There is a number of processing techniques from which to choose, each of which usually produces a different type of product.

Consider that the first step is a general product description with requirements such as what is it to do, how it is to be used, where it is to fit, etc. An example of a design program approach is reviewed in Table 1-1. Overall design or product conception can be initiated from many sources. The most obvious is the completely new product. Although such products are not in abundance compared to modifications, they offer designers the opportunity to utilize their abilities fully. Depending on its complexity, a new product requires several months to several years before commercial introduction. More commonly, overall design is using a practical approach such as a modification of an existing product. This may be initiated by a company's need to make the product more attractive or easier for consumers to use. Manufacturing may request a new design to simplify assembly or minimize breakage; or management may demand that costs be reduced.

In some cases, achieving these goals requires only a material substitution or a minor design change. But it could mandate that

the product be redesigned, different material used, and/or the components made and assembled using a different approach. After the product is defined, the functional requirements and the cost value are established, which is then followed by a preliminary design. After a preliminary design is completed and approved, different departments such as the manufacturer's engineering, marketing, manufacturing, and quality assurance departments should review it; or perhaps one person may have all these responsibilities. Inevitably, some changes will be required. If they are found to be practical (capable of achievement without compromising product cost or functionality in the intended use environment), the design project can proceed to its next stage.

The next step is to prepare a detailed design with drawings, Once the drawings are available, prototyping and testing can be initiated. Methods of prototyping vary greatly. In some cases, a painted model cut from polystyrene foam blocks will suffice. In other cases, a prototype must be made using the specified material and manufacturing process. Prototyping is essential in many designs, regardless of how it is done, in order to ensure that a product will perform properly prior to production (Chapter 3).

The aim of product design or redesign is to achieve the best possible product at the least practical cost. It is a dynamic procedure, with the key being communication. As the design project progresses and as more is learned, modifications may need to be made. Compromise may also be essential. For example, a superior design may cost more to produce than was originally estimated, but after an objective evaluation it may be determined that it is worth more and market acceptance will occur. Thus, product cost can be increased without jeopardizing its chances for success. Similarly, prototypes usually show where additional strength is required, where a product is over designed resulting in unnecessary cost, and so on.

A final note regarding overall product design procedure is that any design, no matter how good, can be improved. However, there comes a time when the design must be frozen and prototyping or production must begin. If

Table 1-1   Example of a design program approach

Design Category	Subcategories	Detailed Requirements
Establish functional and performance requirements	Estimate allowable size and shape	Product basic functions Aesthetics and marketing Shipping Available space Weight Standardization Strength and stiffness criteria Flexibility Process limitations

	Establish structural requirements	Loads:
		  Gravity
		  Dead—Own weight superimposed
		  Live—Occupancy
		    Snow
		    Misc.
		  Pressure
		  Fluid
		  Earth
		  Wind
		  Dynamic
		  Impact
		  Seismic
		  Handling and shipping
		  Cyclic
		Temperature:
		  Service range—$\begin{cases}\text{Interior}\\\text{Exterior}\end{cases}$
		  Gradient across component thickness
		  No. of cycles—high to low
		  No. of cycles—freeze–thaw
		  Solar gain, surface air flow
		Liquid, moisture, and/or vapor tightness
		Strength–weight ratios—relative significance

	Establish nonstructural requirements	Service environment:
		  Corrison resistance $\begin{cases}\text{Interior}\\\text{Exterior}\end{cases}$ $\begin{cases}\text{Chemical}\\\text{Soil}\\\text{Moisture}\\\text{Organic}\end{cases}$
		
		Weathering Moisture
		    $\begin{cases}\text{Wet–dry cycles}\\\text{Freeze–thaw cycles}\end{cases}$
		    UV exposure
		    Rain abrasion
		    Aging
		  Moisture
		  Temperature
		  Fire safety Incombustibility
		      Flame spread rate
		      Toxic gases
		      Fuel content

(*Continues*)

Table 1-1 (*Continued*)

Design Category	Subcategories	Detailed Requirements		
		Light transmission	Transparency Translucency Opaque	{ Control of sunlight and solar heat Color
		Surface texture		{Aesthetics
		Surface coatings		{Abrasion resistance
		Thermal insulation		{ Barrier Gradient
		Moisture and vapor penetration		{Condensation
		Electrical insulation		{Dielectric properties
	Establish cost targets	Examine economics for successful competition with similar products in conventional materials Consider total effect of new design on end product costs: materials, tooling, finishing, assembly, warehousing and inventory, quality control, packaging and shipping, and installation Consider effect on operating costs. Light weight is important in some applications		
	Establish production and marketing requirements	Number of identical pieces Minimum and maximum probable production rates Available plant Market locations Shipping costs Method of marketing Installation criteria, if applicable Cost restrictions imposed by competing products or technology. Prices can shift with short- and long-term changes in market conditions		
Preliminary design of component	Select size and general configuration	Consider end use and limitations of suitable plastics, efficient manufacturing processes, requirements for sufficient strength and stiffness with efficient use of materials, and cost		
	Select feasible plastics material or materials	Satisfy structural requirements with favorable cost ratios Satisfy nonstructural criteria with acceptable compromises and trade-offs where necessary Is efficient fabrication process available?		
	Select feasible manufacturing process or processes	Provide required size and configuration Tooling and plant capital costs to be appropriate for number of pieces and rate of production Compatible with available plant and marketing plan Provides required structural properties and quality control		
	Determine structural response based on approximate analysis	Develop suitably simplified concept of structural behavior to permit approximate determination of structural response—reactions, stress resultants, stability and stiffness requirements. Make appropriate assumptions within confines of laws of statics		

(*Continues*)

Table 1-1 (*Continued*)

Design Category	Subcategories	Detailed Requirements
	Establish design criteria for trial materials selected	Determine suitable allowable design strengths and stiffness, taking into account type and duration of load, service environments, process effects, quality expectations, etc.
	Proportion component for specific configuration and thickness	Determine trial shape of plates, shells, and ribs, depth of ribs and sandwiches, and wall thicknesses to meet strength, deflection, and stability criteria Review economics and suitability
	Develop significant details	Determine concept and principal details for shop and field connections, penetrations, and other subparts (if required) Determine materials for connections, coatings, subparts, etc.
Revise preliminary design of component	Evaluate preliminary design	Review economics and suitability of materials and process based on preliminary proportions. Consider overall compatibility and practicality of all materials and parts in component as a system Does it meet functional and performance requirements? Is it compatible with other components that may interact with it, relative to effects of expansion and contraction, structural support or movement, fire safety, etc.?
	Review performance and functional requirements	Determine if all original performance requirements are feasible within economic objectives, or whether compromises and tradeoffs should be considered
	Optimize design to reduce cost or satisfy functional and performance requirements	General configuration Configuration proportions such as rib depths, shell radii, fillet radii, etc. Material thickness Material alternatives—consider additives to tailor properties Process alternatives
Develop final design of component	Perform structural analysis of acceptable accuracy	Determine structural response—stresses, support reactions, deflections, and stability—based on a structural analysis of acceptable accuracy. Determine acceptable accuracy based on economic value of component, consequences of failure, state-of-the-art capability in stress and stability analysis, margin of safety, knowledge about loads and materials properties, conservatism of loads, provisions for further evaluation by prototype testing
	Establish final design criteria	Allowable stresses, strains, deflections Margins of safety against local and overall instability, vibrations, etc. Take into account type and duration of load, service environments, process effects, equality expectations

(*Continues*)

Table 1-1 (*Continued*)

Design Category	Subcategories	Detailed Requirements
	Evaluate proportions and design details; revise if necessary	Shape of plates, shells, ribs Depth of ribs and sandwiches Thickness of shells, flanges, and stiffeners Connections: Shop Field Edge conditions Penetrations Subparts, Inserts

	Prepare engineering drawings	Drawings are sometimes prepared in two stages:
		   Design drawings
		   Detail or fabrication drawings

	Prepare specifications for technical requirements of product and materials	Materials requirements including composition, quality standards, and minimum structural properties Fabrication requirements and standards, including dimensional tolerances, allowable defects, and minimum structural properties Requirements for prototype and quality control tests and procedures Shipping and handling Requirements for field assembly, installation, or erection
	Prepare manuals or instructions for maintenance and repair	Periodic maintenance, recoating Service conditions: temperature limits, chemical exposure limits Repair procedures
Evaluate design by prototype and materials tests	Develop practical full-scale prototype for structural tests	Develop practical test program to demonstrate components ability to meet structural and performance criteria. Extent of such test program, if any, depends on economic value of component, number of units to be produced, consequences of failure, accuracy of structural analysis and design, margins of safety used in design, knowledge about service loads and environments, and difficulty of duplicating service loads and conditions in test
	Test materials for structural properties and effect of service environment	Determine that materials produced in actual fabrication process will have the minimum structural properties and resistance to service environment assumed in the design. Extent of testing, if any, depends on available information about specific materials and processes to be used
	Revise design, if required	Correct design and detail problems, if any, revealed in tests Modify materials, or process, if production materials' properties not adequate Protect or modify materials if service environment causes excessive degradation of properties
Develop production and distribution system	Pattern design and drawings	
	Mold design and drawings	Take into account shape limitations and design rules that facilitate molding

(*Continues*)

Table 1-1 (*Continued*)

Design Category	Subcategories	Detailed Requirements
	Production process design and layouts	Take into account materials and configuration characteristics that simplify processes Automated processes are needed for high-volume production
	Develop any special equipment	
	Distribution and marketing plan	Production for inventory, or by special order Replacement part inventory Locate production facilities to optimize distribution
	Procedures for packaging, storing, handling, and shipping	Identify special requirements for protection in handling and shipment
	Installation requirements	Specify special requirements for assembly or installation

this is not done, the new design will remain on the board until competition beats you to the marketplace.

The fabricator places no limits on the design. There is a way to make the product if the price is justified. Any job can be done at a price. The truly limitations are factors such as usable tool size (mold, die, etc.), material shrinkage, substance finishing operation (that usually is not required), dimensional tolerance allowance, undercut, insert inclusion, parting line, fragile section, production rate, and the essential selling price. Note that there are companies with in-house fabricating capabilities that will replace existing equipment to produce a new product at a lower cost.

Table 1-2 provides estimates of the major types of plastics consumed yearly worldwide that now total 339,990 million lb (154 million tons). About 90% are thermoplastics (TPs) and 10% thermoset (TS) plastics. USA and Europe consumption's are each about one-third of the world total. There are well over 35,000 different type plastic materials worldwide. However, most of them are not used in large quantities; they have specific performance and/or cost capabilities generally for specific products by specific processes that principally include many thousands of products (Chapters 6 & 7).

The plastics industry is characterized by a wide variety of many different plastic materials and distinct processing methods that fabricate many different plastic materials into many different products. The following Fig. 1-5 provides a summary of the interrelations of plastics-to-processing-to-products. By following this type of a practical sequence of events permits fabricating products that meet performance and cost requirements used substantially in all industries. This is a "back to basic" approach that helps one to understand that there is a logical approach in producing products that range from the initial concept to the customer receiving the product.

Plastics are now among the nations and world's most widely used materials, having surpassed steel on a volume basis in 1983 (Fig. 1-6). By the beginning of this century, plastics surpassed steel even on a weight basis (Fig. 1-7). These figures do not include the two major and important materials consumed, namely wood and nonmetallic earthen (stone, clay, concrete, glass, etc.). Volumewise wood and construction materials each are possibly about 70 billion ft^3(2 billion m^3). Each represents about 45% of the total consumption of all materials. The remaining 10% include those shown in Figs. 1-6 & 1-7. Plastic materials and products cover the entire spectrum of the world's economy, so that its fortunes

Table 1-2 World plastic consumption (million 1b)

Plastic	United States	Canada	Mexico	Brazil	Other Latin America[1]	Western Europe[2]	Eastern Europe[3]	Japan	China	Other Asia-Pacific[4]	Africa & Middle East	Rest of World	Grand Total
LLDPE	8468	610	480	610	900	4093	1023	1439	2661	6121	580	367	27,352
LDPE	7748	1281	1176	1590	619	10,254	2563	1804	1938	3391	1210	457	34,031
HDPE	14,065	1136	952	1388	417	9178	2294	2158	1218	4967	1325	532	39,630
Urethane	5265	475	380	410	390	5481	1808	1512	1825	4850	390	310	23,096
PVC	14,698	1394	605	1456	1837	12,388	2477	3761	5338	11,633	1250	773	57,610
Polystyrene	6589	725	405	280	495	6180	1236	2175	2299	5275	190	352	26,201
Polypropylene	13,739	796	695	1366	1307	13,566	2713	5001	2667	11,176	1275	738	55,039
ABS	1409	145	175	365	290	1410	282	948	460	950	305	92	6,831
Acrylic	613	75	80	165	140	576	115	302	155	325	140	37	2,723
Unsaturated Polyester	1681	180	125	250	210	1036	207	1285	645	1175	215	95	7,104
Nylon	1267	110	90	170	145	1210	242	339	170	425	155	59	4,382
PET	4330	410	192	482	386	2464	493	1178	715	5441	390	224	16,705
Poly-carbonate	857	90	70	155	140	607	121	443	215	475	145	45	3,363
Thermoplastic Polyester	346	45	30	65	50	243	48	164	85	322	42	20	1,460
Acetal	389	40	30	70	45	317	63	212	106	235	63	21	1,591
Recycle Plastics	1800	166	121	195	162	1625	220	502	453	1254	165	88	6,751
Other Plastics[5]	8150	763	412	875	74	7250	1750	255	198	5310	740	344	26,121
Total	91,414	8,441	6,018	9,892	7,607	77,878	17,655	23,478	21,148	63,325	8,580	4,554	339,990

[1] Argentina, Chile, Columbia, Venezuela and all other.
[2] European Union plus Norway and Switzerland.
[3] Includes Russia and the Balkans.
[4] Australia, India, Indonesia, Malaysia, North Korea, Pakistan, South Korea, Taiwan, Thailand, Philippines, Singapore, Vietnam.
[5] High Performance, other thermosets, specialty elastomers, tailored blends and alloys.

Source: Concise Encyclopedia of Plastics: Fabrication & Industry, by Rosatos, Kluwer Publ., 2000; Injection Molding Handbook, 3ed Edition, by Rosatos, Kluwer Publ., 2000.

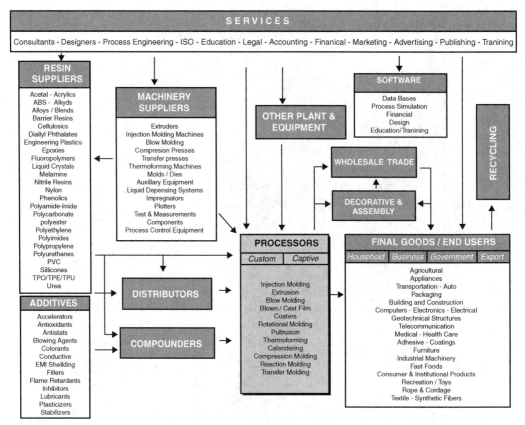

Fig. 1-5   The plastics industry.

are not tied to any particular business segment. Designers are in a good position to benefit in a wide variety of markets: packaging, building and construction, electronics and electrical, furniture, apparel, appliances, agriculture, housewares, luggage, transportation, medicine and health care, recreation, and so on (Chapter 4).

The effective exploitation of product design opportunities is the key to success. In turn, success hinges on other factors, such as the proper selection of materials and using the best available processing equipment. Because new materials and equipment continue to be more productive and produce better quality products, one should stay abreast of new material and equipment developments and evaluate them logically. With designing, there is an extremely vast area for improving profitability by ensuring that the best available material and equipment are used to meet specific design performance requirements. Recognize that there are occasions when a higher cost material can provide a lower processing cost that result in a lower cost product. To design, as to live is to change, and to aim for perfection is to have changed often (Chapter 4. **RISK, Perfection**).

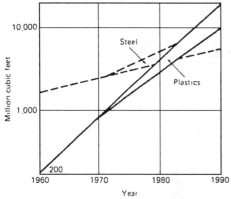

Fig. 1-6   World consumption of plastics by volume.

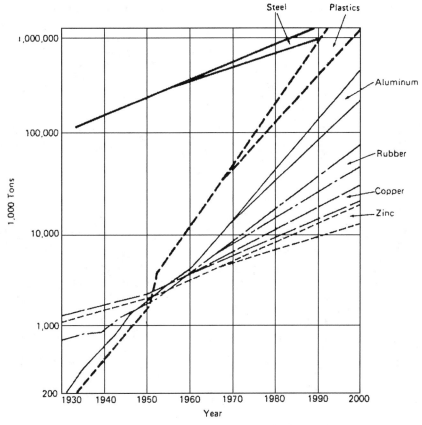

Fig. 1-7 World consumption of plastics by weight.

What this book provides is information on the extensive different properties (Figs. 1-8 & 1-9) and processing capabilities the many different plastics offer. It also provides facts such that most of the plastic products produced only have to meet the usual requirements we humans have to endure such as the environment (temperature, etc.). Thus there is no need for someone to identify that most plastics can not take heat like steels. For certain plastic products there are definite properties (temperature, chemical resistance, load, etc.) that have far better performance than steel and other materials. Recognize that most plastics in use also do not have a high modulus of elasticity or long creep and fatigue behaviors because they are not required in their respective product designs. However there are plastics with extremely high modulus and very long creep and fatigue behaviors (Chapter 2).

What this discussion identifies is that each material (specific plastic, specific steel, specific wood, etc.) behaves certain ways. If a product can be made from a specific steel rather than a specific plastic, that is the material to use. However, unfortunately for steel and other materials, plastics continue to expand its use where these other materials are not competitive propertywise and/or cost-wise.

This book will not provide extensive engineering equations since they are readily available from industry that are reviewed in Appendix A: **PLASTICS DESIGN TOOL-BOX** and references (3, 6, 10, 14, 20, 29, 31, 36, 37, 39, 43 to 125). Equations will be reviewed throughout this book where they are required to understand the behavior of plastics in order to meet different load requirements (static to dynamic). What this book provides is information on the behavior of

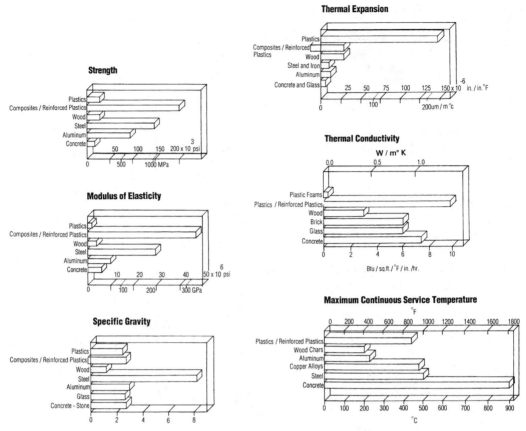

Fig. 1-8   General comparison of different materials.

plastics that permits the designer to properly design with plastics usually not requiring any engineering equations.

When one becomes familiar with plastics such as their viscoelastic behavior [that is a combination of viscous and elastic properties in a plastic with the relative contribution of each being dependent on time, temperature, stress, and strain rate (Chapter 2)], plastics can be properly applied to the equations, etc. This book is targeted to have you become both familiar and how to apply the behavior of plastics in any equations. Your book authors as well as many of the referenced authors have extensive experience in adapting the behavior of plastics to products with or without equations that have different shapes, decorations, etc. They include toys, packages, building panels, medical devices, electronic devices, lighting devices, chemical operations, building to bridge structures, small to very large pipe lines, boats, underwater devices, aircraft primary and sec-

ondary structures, missile components, reentry spacecraft protective shields, etc. These examples of a few of the products designed basically range from just requiring the proper aesthetics (practical approach) to those being subjected to extensive high dynamic creep and fatigue stress loads (engineering approach).

Generalization Justifiable

A short dissertation upon almost any extensive subject such as this subject is usually blessed by the reader's understanding that generalizations are not only justifiable but also mandatory in order to cover the scope of the subject. However, a learned treatise of ponderous bulk can be readily exempted from criticism for tedious passages devoted to details in that the authors are attempting to present a full and uncompromised assay of the subject. Somewhere in between lies this

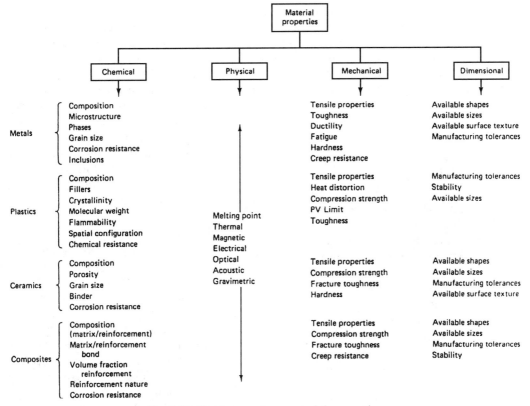

Fig. 1-9   Guide to various material properties.

book. For those desiring more details, appropriate references are provided.

Often the authors set their own ground rules in a probably futile attempt to satisfy the inquisitiveness of those from another technical discipline in an expeditious manner, and yet not to incur the criticism of those highly knowledgeable in the subject area. The writers have followed this important but perilous course.

Design Definition

The term "design" has many connotations. They can range from industrial designers to high structural load engineering designers. A few of these will be summarized in order to highlight that different designer skills are used to meet different product requirements. Essentially it is the process of devising a product that fulfills as completely as possible the total requirements of the user, and at the same time satisfies needs in terms of

cost-effectiveness or ROI (return on investment). It encompasses the important interrelationship practical factors such as shape, material selection (including unreinforced and reinforced, elastomers, foams, etc.), consolidation of subparts, fabricating selection, and others that provide low cost-to-performance products.

Product design is as much an art as a science. Recognize that a successful design is usually a compromise between the requirements of product function, productibility, and cost. Basically design is the mechanism whereby a requirement is converted to a meaningful plan. Design guidelines for plastics have existed for over a century.

With plastics to a greater extent than other materials, an opportunity exists to optimize product design by focusing on material composition and orientation to structural member geometry when required. The type of designer to produce a product depends on the product requirements. As an example in most cases an engineering designer is not needed because the product has no major load requirement. All that is needed is experience and/or a logical evaluation approach based on available material and processing data. This practical approach is the least consumer of time and least expensive (Fig. 1-4).

Design Technology

It is the prediction of performance in its broadest sense, including all the characteristics and properties of materials that are essential and relate to the processing of the plastic. To the designer, an example of a strict definition of a design property could be one that permits calculating product dimensions from a stress analysis. Such properties obviously are the most desirable upon which to base material selections.

However, like with metals, there are many stresses that cannot be accurately analyzed. Hence one is forced to rely on properties that correlate with performance requirements. Where the product has critical performance requirements, such as ensuring safety to people, production prototypes will have to be exposed to the requirements it is to meet in service.

In plastics, these correlative properties, together with those that can be used in design equations, generally are called engineering properties. They encompass a variety of situations over and above the basic static strength and rigidity requirements, such as impact, fatigue, flammability, chemical resistance, and temperature.

Industrial Designer

IDs are essential to all industries that relate to research, engineering, production, and marketing activities. They must exercise the creativity imagination that sets them apart from being a mere modifier of what the competition offers. There is a difference between IDs and other professions whose functions are sometimes confused with those of IDs. The true artist, for instance, produces a personal interpretation of what one feels and creates the final object alone. The IDs do not;

they help to provide directions by which others create the final product. They differ in their approach as an engineering designer.

The ID profession has embraced plastics with enthusiasm for several reasons. First, plastics provide enormous freedom of shape compared with traditional materials of design. They also permit product production that is faster and more consistent, and they can do it all at a fraction of the cost for making nonplastic products. This low product cost does not stem from the fact that plastics are low in cost. On a per-pound basis, they are actually more costly than many competing materials. But the processability and relatively low density of plastics (which translates into lower costs per volume) gives them a big economic advantage. The net result is that the ID can now achieve quality products at disposable price levels (216).

Colorability is another reason IDs select plastics for many products. Molding color into a product eliminates finishing and painting operations, thus reducing costs. Beyond cost, integral color also masks the nicks, chips and scratches that impair appearance during the life of the product. Color effects are almost limitless. Transparent, translucent, pearlescent, fluorescent, or marbleized colors are readily available for use in plastics.

Another design appeal of plastics is their ability to accept topical decoration. A permanently affixed multicolor label can be provided by means of heat transfer or hot stamping. When a more secure surface is required (for computer keycaps, containers, etc.) the markings, decorations, or labels can be placed directly in the mold cavities and subsequently molded into the product as it is formed. Two-color injection molding is another option. This is a process in which a product is first molded in line color and then (without demolding) a second cavity is placed over the part permitting a second color to be molded over a predetermined portion of it. Other in-mold decoration processes are available, including a selection of patterns that can be etched into the mold surface ranging from a very high polish usually reserved for lenses to a medium matte finish adequate to mask minor sink marks (Chapter 8).

Appearance means something different to each discipline involved in the development of a new product. Industrial designers, engineering designers, tool builders, and processors are each affected in a different way. Yet the cooperation of all is necessary in order to achieve the best possible appearance. Concern for appearance generally translates to more work for the designer. It would certainly be far easier to construct a rectangular box or a drafted cylinder with a few appropriately placed screws. But a world full of rectangular boxes and drafted cylinders usually would not be eye appealing. The principal problems associated with appearance factors are the development of contoured housings, space limitations, and assembly devices. Contoured housings are far more difficult to calculate than those with regular dimensions. In some cases, initially it is practically impossible to achieve complex molded shapes without creating wall thickness variations that can cause sink marks and warpage.

The ID function involves a great deal more than appearance design. The designer is often called on to create the very concept of the product. In doing so, they will consider the utility, cost, innovation and human engineering aspects of the proposed product that relates to its basic appeal to the end-user.

Human engineering. While the designer usually regards the problems of space limitations as being appearance related, they are most often the outcome of the ID's concern for human engineering, or the proper relationship of the product to the human body. For example, personal computers should be small enough to be carried by many people, coffee-cup handles should be comfortable to the hand, eyeglass frames should be easy on the ears, etc.

Human engineering requirements often dictate the size, weight, and form of a product. This translates to smaller, lighter and contoured products as the ID works from the outside to its interior. Often this results in conflict with the company's engineering designer, who works basically from the interior to its outside. Compromise becomes an important factor as they all bring their requirements to bear on the appearance and performance of the product.

Engineering Designer

It is the area of engineering that involves the application of graphic principles and practices to the solution of engineering loading equation problems. It is the systematic process by which a solution to a problem is created. A definition that contains the necessary ideas and speaks broadly is that engineering design is a decision-making activity whereby scientific and technological information is used to produce a product or system. It is different in some degree from what the designer knows to have been done before and that is meant to meet new needs.

Graphic Designer

It covers the principles of engineering drawings, computer graphics, descriptive geometry, and problem solving. The overall study of graphics involves the three basic aspects of terminology, skills, and theory.

Innovative Designer

A skilled designer blends a knowledge of materials, an understanding of manufacturing processes, and imagination of new or innovative designs. Recognizing the limits of design with traditional materials is the first step in exploring the possibilities for innovative design with plastics. Some designers operate by creating only the stylish outer appearance, allowing basic designer to work within that outside envelope. This approach is used very successfully such as in certain products or parts for furniture, etc. There are also the combination of designing appearance with engineering so that the stylish product incorporates the best combination with ease of processing when using a specific plastic, simplify assembly, provide capability of repair, streamlining quality control, and/or other conditions. The stylish envelope that eventually emerges will be a logical and aesthetic answer to the design challenge.

Material Optimization Designer

Designers can turn to materials as a means of dramatically improving their products' performancewise and costwise. Over 70% of product designs are geometric. With over 80,000 materials worldwide (including over 35,000 plastics) to choose from, the material software tools have become an asset to designers and engineers with or without experience in material familiarity. There is software that provides information on specific performance requirements so that only one or a few will be listed as the best material to meet the product performance requirements. These tools let designers consider materials as a variable in design to meet their specific product requirements. (Appendix A: **PLASTICS TOOLBOX**)

Maximum Diametrical Interference Designer

An example of an interference fit design is the maximum allowable interference for a particular hub and shaft. It depends on the types of materials used in the hub and shaft, and on the ratio of the shaft diameter to the hub outside diameter. It is determined to ensure that hoop stress in the interference fit does not exceed the allowable stress of materials used (Chapter 7).

Medical Device Designer

Designing medical devices is one of many others (buildings, chairs, jewelry, aircraft, toys, etc.) with each having their specific requirements. As an example in USA designing medical devices with developing controlled manufacturing environments are required to be submitted to FDA for approval. They are (necessary) time-consuming, costly activities for medical device manufacturers. Such activities generally have a target time line with a set completion date and budget. In the mean time, the daily operations of manufacturing in a controlled environment present continual challenge that vary as the regulations change and the cost of materials and manufacturer increases.

Different systems are used to aid people, including designers. An example is a system designed to ensure efficient contamination control operations. It is called PACT (prevention, assessment, corrective action, and training). PACT is designed to assist supervisor, managers, and engineers with contamination control management. It involves the continuous improvement principles of total quality management. Also involved is the quality system regulation (Appendix: **TERMINOLOGY, Quality system regulation**).

Design Features That Influence Performance

One of the earliest steps in product design is to establish the configuration of the product that will form the basis on which a suitable material is selected to meet performance requirements. During this phase certain design features have to be kept in mind to avoid problems such as reduction of properties. Such features are called detractors or constraints. Most of them are responsible for the unwanted internal stresses that can reduce the available stress level for load bearing purposes. Other features may be classified as precautionary measures that may influence the favorable performance of a product if they are properly incorporated.

For example, something as simple as a stiffening rib is different in size for a solid or foamed product even when both products use the same plastic. Familiarization of design constrains is a critical first step in design to eliminate product problems. A designer might have an expensive tool (mold or die) prepared based on a plastic's shrinkage value. It is discovered belatedly that the plastic did not meet some overlooked design constrain and the plastic required a much lower shrinkage value. The tool has too large a mold cavity or die opening requiring expensive modification or replacement of the tool.

This book contains information that can be used to setup a checklist on plastic capabilities and constrains based on performance requirements. A general guide of the design process anatomy can use the overlapping approach of chronological order, clarifying the task, conceptual design, embodiment of design, and detailed design. To do full justice to plastics, one should become familiar with the meaning of the data furnished by the material supplier, the long-term effect on properties when subjected to a load, the influence of surrounding environment, and other material requirements that, as a rule, may be insignificant with metals or some of the other materials (Chapter 5).

With this type of action, it will become rather routine to follow through with a design, to build a prototype, to test performance capability, and so forth. Design examples reviewed were selected for the purpose of demonstrating that the standard technical or engineering handbooks are in general of the same importance in conjunction with plastics as they are with steels, aluminum, wood, etc. Product reviews have been modified to eliminate strength detractors and to incorporating needed features in such a manner that overall product characteristics will be protected.

Before investigating any one or a group of any materials, product performance and environmental conditions are to be determined. This is a major area of product failure because, in most cases, a complete set of requirements is not properly determined. A few or many requirements may exist. Examples of a few of these requirements are reviewed.

1. Determine requirements under which the product will be used. Examples include color, temperature, moisture, ultraviolet exposure, exposure to fungus, flammability, chemical, electrical resistance, arc resistance, light transmission, stability or permanency, physical property, mechanical property, optical property, heat and/or electrical insulation, resistance to scratching (mar resistance), and special requirements such as self-lubrication, lightness, hinging property, spring property, time of exposure, etc. Also important will be meeting existing government and/or industry regulation.

2. Determine tolerance requirements that are expected in the performance of the product. Shrinkage characteristics of the selected plastic should be as small as possible so that tolerances can be anticipated with a reasonable degree of accuracy.

3. If required, determine the nature of the load to which the product will be exposed, such as impact, creep, deflection, stresses, bending, gliding, etc.

4. Color matching may be a factor.

5. Cost of plastic by volume and cost to fabricate.

Once the performance and environmental conditions have been defined, the selection of a suitable material can be made, and this in turn can be followed, if necessary, with the necessary engineering calculations to establish strength requirements. The basic data needed for calculations have to be collected and have to pertain to the specific grade of the selected material. The pertinent information required for making determinations for longevity of the product and obtaining a general concept of the character and behavior of the selected material should be supplied by the manufacturer of the raw material and/or obtained in-house or via a contractor.

Examples of information that could be required follows:

1. Data sheets of the specific grade of material containing the properties required,

2. Stress-strain curves at the conditions of product application. If applicable, this would usually indicate the toughness of material by sizing up the area under the curve (Chapter 2). It would also show the proportional limit, yield point, corresponding elongations, and other relevant data.

3. Curves showing change of tensile strength, flexural strength, and modulus with increasing temperatures or other environments.

4. Creep data for periods at 100 and 1000 hours (or more, if available) covering stress and temperature conditions closely comparable to those of product application.

5. The allowable working stress, based on successful performance at conditions of product usage.

6. Chemical and/or heat resistance at conditions in service.

7. Others (fatigue, etc.) (Chapter 2).

Not all product components are subjected to a load; in fact most are not subjected to loads requiring an engineering analysis via engineering equations, etc. Experience in the material behavior on similar products and/or similar performance requirements are all that is needed. In these products designers become involved in their processing features that will prevent or reduce internal stresses, with elements that will lead to consistent and economical production, with appearance and dimensional control, etc.

Products that are subjected to a load have to be analyzed carefully with respect to the type and duration of the load, the temperature conditions under which the load will be active, and the stress created by the load. A load can be defined as continuous when it remains constant for a period of 2 to 6 hours, whereas an intermittent load could be considered of up to two hours duration and is followed by an equal time for stress recovery. The temperature factor requires greater attention than would be the case with metals. The useful range of temperatures for plastic applications is relatively low and is of a magnitude that in metals is viewed as negligible.

Most design books continually report that plastics cannot take the heat of metal (steel, etc.) indicating that plastics cannot take heat. As reviewed, by far practically most plastic products do not have to take any more heat then the human body. Practically all plastics easily meet this heat requirement for these type products and in fact many types of these plastics meet the higher heat requirements of plastic products that exist under the engine hood of an automobile, in the trunk of an automobile (excellent user-environmental test), electrical/electronic devices, etc.

An important subject to introduce concerns the allowable working stress for a spe-

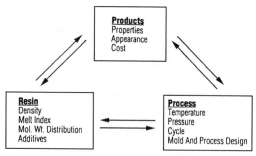

Fig. 1-10 Interrelating product-plastic-process performance.

cific plastic that relates to their viscoelasticity. Other important subjects include their static and dynamic loading capabilities as well as their creep to fatigue behaviors. Details on these subjects are presented in Chapter 2.

Interrelating Product-Plastic-Process Performance

In order to understand potential problems and solutions of design, it is helpful to consider the relationships of machine capabilities, plastics processing variables, and product performance (Fig. 1-10). A distinction has to be made here between machine conditions and processing variables. For example, machine conditions include the operating temperature and pressure, mold and die temperature, machine output rate, and so on. Processing variables are more specific, such as the melt condition in the mold or die, the flow rate vs. temperature, and so on (Chapter 8).

A distinction between machine conditions and processing variables must be made in order to avoid mistakes in using cause-effect relationships to advantage. It is the processing variables, properly defined and measured, not necessarily the machine setting, that can be correlated with product performance. There was a time when designers took little interest in the processing of the products they had designed. They simply sent the drawings to the processor in another department or company and expected perfect products to emerge, but design and processing are now so interrelated that this separation should not exist if products are to be consistently successful.

Those familiar with processing can detect and correct visible problems or readily measure factors such as color, surface condition, and dimension. However, less-apparent property changes are another matter. These may not show up until the products are in service, unless extensive testing and quality control are used.

As there are many different plastics, a number of techniques for defining and quantifying their characteristics exist. As an example molecular weight distribution (MWD) is an indication of the relative proportions of molecules of different weights and lengths. In turn MWD relates to processing characteristics that directly relate to product performances (Chapter 8).

Advantage and Disadvantage of Plastic

As a construction material, plastics provide practically unlimited benefits to the design of products, but unfortunately no one specific plastic exhibits all these positive characteristics. The successful application of their strengths and an understanding of their weaknesses (limitations) will allow designers to produce useful products. With any material (plastic, steel, etc.) products fail not because of its disadvantage(s). They failed because someone did not perform in the proper manner. It could be the designer even though the processor goofed and the designer was not aware that a goof could occur. When a situation exists that one person does not have the total responsibility, goofs can easily occur. If the designer does not have the overall responsibility, then his boss or someone else up the management line has the responsibility and is accountable for the goof.

There is a wide variation in properties among the over 35,0000 commercially available materials classified as plastics. They now represent an important, highly versatile group of commodity and engineering plastics. Like steel, wood, and other materials, specific groups of plastics can be characterized as having certain properties. As with other materi-

Become aware that for any gain there could be a loss not originally included in the design performance.

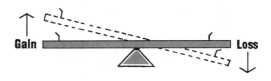

When you gain "something" there will be a loss .....does that loss influence product performance (for any material: plastic, wood, steel, glass, etc.).

Fig. 1-11 For any gain, there could be a loss not originally included in design performance using plastic or any other material.

als, for every advantage cited, a corresponding disadvantage can probably be found (Fig. 1-11).

In general, most plastics can easily be fabricated into all shapes and sizes. As reviewed throughout this book, if desired they can have highly intricate shapes held to tight tolerances by using certain plastics and processes suitable for either limited or mass production. Many plastics can be worked using common shop techniques. Other generalizations include, as reviewed, the fact that many consider plastics to be of low cost. In fact some are so expensive that their use is limited to the most sophisticated technology and applications. Regardless, if the cost of materials appears to be too high for an application, a look at the processing method to be used usually shows that it could be less expensive, based on the material-to-processing costs. There are designers who overlook this aspect. Many plastic products are very successful costwise because their fabricating costs are low.

Many plastics are typically not as strong or as stiff as metals, and they are prone to dimensional changes, especially under load and/or heat. Successful designs take these conditions into account when they influence design requirements. As will be seen, there are plastics that meet dimensional tight requirements, dimensional stabilities, and are stronger or stiffer based on product shapes than other materials.

Highly favorable conditions such as less density, strength through shape, good thermal insulation, high degree of mechanical dampening, high resistance to corrosion and chemical attack, and exceptional electric resistance exist for certain plastics. There are also those that will deteriorate when exposed to sunlight, weather, or ultraviolet light, but then there are those that resist such deterioration.

No matter what the material may be, there is always room for improvement, whether it is in plastics, metals, wood, design parameters, testing procedures, or any other category. To date designers are generally most familiar with metals and wood as well as their behavior under load and varying conditions of temperature and environment. For those designing in metals and the other materials that have been used for centuries there is extensive literature available (and much more continually being developed because they are needed). One can easily enter the field of designing with these materials and refer to the handbooks that tell one what to do similar to what this book provides when using plastics.

As an example, for room-temperature applications most metals can be considered to be truly elastic. When stresses beyond the yield point are permitted in the design, permanent deformation is considered to be a function only of applied load and can be determined directly from the stress-strain diagram. The behavior of most plastics is much more dependent on the time of application of the load, the past history of loading, the current and past temperature cycles, and the environmental conditions. Ignorance of these conditions has resulted in the appearance on the market of plastic products that were improperly designed. Fortunately, product performance has been greatly improved as the amount of technical information on the mechanical properties of plastics has increased in the past half century. More importantly, designers have become more familiar with the behavior of plastics rather than

just explaining that one cannot design with plastics.

Basics in Designing

Plastic materials are predominantly synthetic materials. Since their inception over a century ago they have enjoyed a growth that has been unequaled by any other group of materials. This demand continues to increase, and the facilities for meeting the new requirements are being expanded continuously. There have been good reasons for the phenomenal application of plastics in order to justify the large investments needed to produce the raw materials and to convert them into finished products.

Overall, it can be stated that plastic products meet the following criteria: their functional performance meets use requirements; they lend themselves to esthetic treatment at comparatively low cost; and, finally, the finished product is cost competitive. Examples of their desirable behaviors can start with providing high volume production. Plastic conversion into finished products for large volume needs has proven to be one of the most cost-effective methods. Combining bosses, ribs, and retaining means for assembly are easily attained in plastic products, resulting in manufacturing economies that are frequently used for cost reduction. It is a case where the art and technology of plastics has outperformed any other material in growth and prosperity.

Their average weight is roughly one-eighth that of steel. In the automotive industry, where lower weight means more miles per gallon of gasoline, the utilization of plastics is increasing with every model-year. For portable appliances and portable tools lower weight helps people to reduce their fatigue factor. Lower weight is beneficial in shipping and handling costwise, and as a safety factor to humans (no broken glass bottles, etc.).

The value of heat insulation is fully appreciated in the use of plastic drinking cups, plastic handles on cooking utensils, electric irons, and others where heat can cause discomfort or burning. As insulation in the walls of buildings, homes, etc. energy and cost advantages are obvious. In electrical devices the plastic material's application is extended to provide not only voltage insulation where needed, but also the housing that would protect the user against accidental electrical grounding. In industry the thermal and electrical uses of plastics are many, and these uses usually combine additional features that prove to be of overall benefit.

Corrosion resistance and color are extremely important in many products. Protective coatings for most plastics are not required owing to their inherent corrosion-resistant characteristics. The eroding effects of rust are well known with certain materials. Whereas certain plastics do not deteriorate offering distinct advantages. Colors for esthetic appearance are incorporated in the material compound and become an integral part of the plastic for the life of the product.

Those with transparency capabilities provide many different products that include toys, protective shields (high heat resistance, gunfire, etc.), transportation vehicle lighting, camera lenses, eyeglasses, contact lenses, etc. When transparency is needed in conjunction with toughness, plastic materials are the preferred candidates. Add to the capability of providing simple to very complex shapes.

Include plastics with coefficient of friction, chemical resistance, and others. Many plastic materials inherently have a low coefficient of friction. Other plastic materials can incorporate this property by compounding a suitable ingredient such as graphite powder into the base material. It is an important feature for moving products, which provides for self-lubrication. Chemical resistance is another characteristic that is inherent in most plastic materials; the range of this resistance varies among materials.

Materials that have all these favorable properties also have their limitations. As with other materials, every designer of plastic products has to be familiar with their advantages and limitations. It requires being cautious and providing attention to all details. Nothing new since this is what designers have

been doing for centuries with all kinds of materials if they want to be successful.

As reviewed throughout this book one must determine what measures were taken to evaluate the materials. Carefully study the test results and test methods that are employed in obtaining these properties and their interpretation for application purposes (Chapter 5), and finally determine the fine details of use conditions to establish the suitability of a plastic material for the intended product. It is easily accomplished to determine the plastic to be used but requires familiarity with the test results being evaluated and behavior of plastics to meet your performance requirements.

Design Approach

The highest skill a designer can possess consists of making full use of the properties of materials to create truly distinctive products. In the process the designer needs to know and explore the limits of design that can start with a feasibility study (Fig. 1-12). For example, limits are imposed by such factors as the manufacturing process limits on the material that will determine the shapes to which the material can economically be converted, the physical properties of the materials that will limit their applications and useful environments, and the designer's imagination in combining form and function. In theory, the imagination is limitless, but in practice the first two limitations affect a designer's ability to exercise it.

Design ideas in the industry using plastics continue to be extensive. New ideas and applications for meeting performance-to-cost requirements are seemingly end-less. They include the capability of being innovative to applying practical concepts. There are all sorts of novel approaches to design ideas with plastics. An innovative design exercise to produce automotive products that includes consolidations is reviewed in Fig. 1-13.

This design exercise relates to different products and their consolidation such as front-end parts (valance panel can be joined

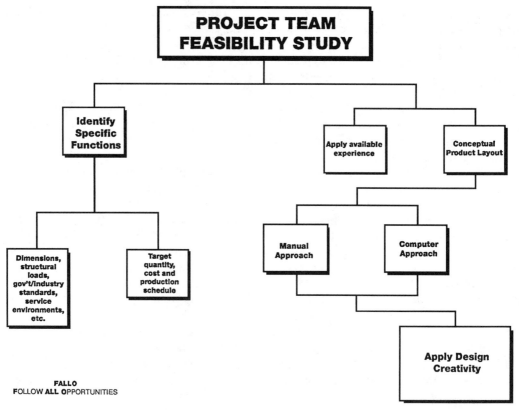

Fig. 1-12   Project team feasibility study.

to make a one-piece front end) and rear-end parts consolidation. The materials used include unreinforced and reinforced plastics. Products include fender extension, headlamp housing, taillamp housing, rear finish panel, front and upper grille panel, hoods, scoops, rear deck-lid, and spoilers. The future will probably include the mass production of one-piece plastic molded monocoque chassis and even incorporate other parts.

Most successful designers have the ability to develop products that will be instantly acceptable to a buyer. Their designs have a recognizable, functional improvement along with some visual appeal to set their products apart from conventional ones. Too many new product designs or redesigns are nothing more than slight improvements that anybody could make with a minimum of thought. Many companies inch there way to progress with just such a slight change every now and then. This is the easy way to give the appearance of improving a product line with

the least disruption of the manufacturing process and requires little adjustment by the sales staff, with the exception of printing new sales brochures.

Design Feature

As reviewed throughout this book and particularly in Chapter 3, **Design Concept**, there are many design features that keep expanding the use of plastics in different products. These features include shapes, sandwich constructions, shrinkages, tolerances, and processes.

One factor that has done a great deal to harm the reputation of plastics is that in many cases designers and engineers have, after deciding tentatively to try to introduce plastics, then lavishly copied the metal product it was to replace. Too much emphasis cannot be given to the general principle that plastics are to be used based on their behavior

An upper grille panel, and a valance panel could join to make a one-piece front end.

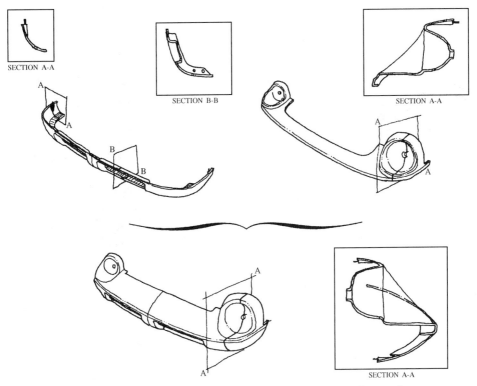

Fig. 1-13   An innovative design exercise to produce automotive products.

characteristics in order to eliminate failures. It is essential to cast aside all preconceived notions of design in metals and treat plastics on their own merit as one would with any other material.

A hard-and-fast rule to be followed by all intending to use plastics is to design for plastics. As an example, for the same-size cross-section the strength of conventional plastics (not the high-performance reinforced types) is considerably less than that of most metals. The designer will thus find it necessary to increase thickness, introduce stiffening webs, and/or possibly use design inserts of various types of threads to secure the proposed product. The process will in some instances also require modification to the shape of the equipment used to produce the product.

It will become obvious that what is considered good design practice insofar as metals are concerned will not necessarily be good practice for processing plastics. It is advisable

when in doubt to review this book, the referenced literature on the subject, and/or consult processing experts who know the behavior of plastics. Almost all current methods of design analysis are based on models of material behavior that are relevant to traditional metallic materials, as for example elasticity and plastic yield. These principles are embodied in design formulas' design sheets or charts and in more modem techniques, such as computer-aided design (CAD) using finite element analysis. The design analyst is merely required to supply appropriate elastic or plastic constants for the material as reviewed in Chapter 2.

Computer Use

The computer supports rather routine tasks of embodiment and detailed operation rather than the human creative activities of

The two-piece hood with outer panel and inner
support panel may give way in the future to a
one-piece, self-reinforced molded hood.

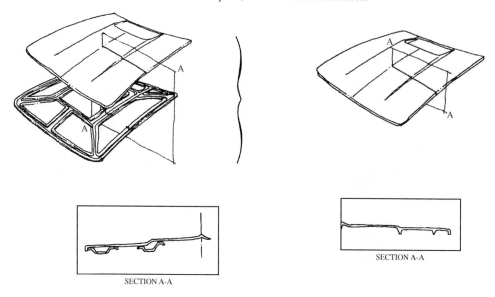

Fenders and adjoining parts in the future can be
molded in one piece.

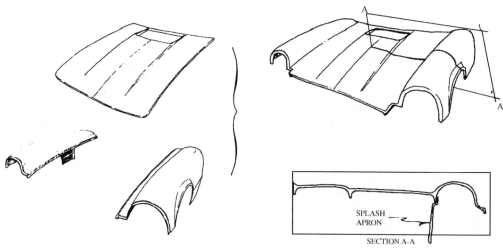

Fig. 1-13   (*Continued*)

conceptual human operation. Recognize that
if the computer can do the job of a designer,
there is no need for a designer. The computer is another tool for the designer to use.
It makes it easier if one is knowledgeable on
the computer's software capability in specific
areas of interest such as designing simple to
complex shapes, product design of combining

parts, material data, mold design, die design,
finite element analysis, etc. By using the computer tools properly, the results are a much
higher level of product designing and processing that will result in no myths.

Successful designed products require the
combination of various factors that includes sound judgment and knowledge of

Another step would join the hood-fender to the one-piece front end
--a single joint-free automobile front end assembly--as a
single molded component.

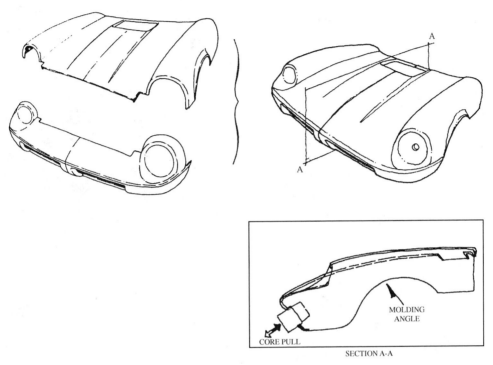

MOLDING
ANGLE

CORE PULL

SECTION A-A

Fig. 1-13   (*Continued*)

processing. Until the designer becomes familiar with processing, a fabricator must be taken into the designer's confidence early in development and consulted frequently. There are software programs that can provide some assistance in this area. It is particularly important during the early design phase when working with conditions such as shapes and sizes. There are certain features that have to be kept in mind to avoid degradation of plastic properties. Most of these detractors or constrains are responsible for the unwanted internal stresses that can reduce the available stress for load bearing purposes (Chapter 5).

Computers permeate all areas of the plastics industry from the concept of a product design, to raw material, to processing, to marketing, to sales, to recycling, to government and industry regulations, and so on. Computers have their place, but most important is the person involved with proper knowledge in using its software in order to operate and use them efficiently.

The industrial production process as practiced in today's business is based on a smooth interaction between regulation technology, industrial handling applications, and computer science. Particularly important is computer science because of the integrating functions it performs that includes the primary processing equipment, auxiliary equipment, material handling, and so forth up to business management itself. This means that CIM (computer-integrated manufacturing) is very realistic to maximize reproducibility that results in producing successful products.

The use of computers in design and related fields is widespread and will continue to expand. It is increasingly important for designers to keep up to date continually with the nature and prospects of new computer

Deck lids can take the same parts consolidation path as hoods.
The deck lid, inner, with spoiler and the deck lid, outer, could
be combined into a one-piece self-reinforced deck lid.

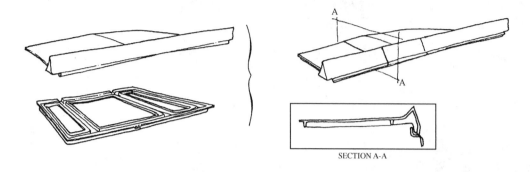

SECTION A-A

Rear finishing panels, rear valance panels and rear fender
extensions are realities. Tail lamp housings, both hidden and
with exterior finish are in common usage. A relatively simple
flight of imagination joins all these elements into a one-piece
rear end panel.

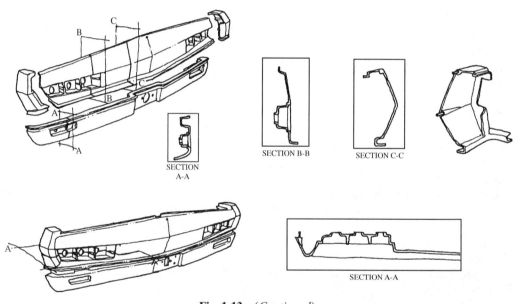

Fig. 1-13   (*Continued*)

hardware and software technologies. For example, plastic databases, accessible through computers, provide product designers with property data and information on materials and processes. To keep material selection accessible via computer terminal and a modem, there are design database that maintain graphic data on thermal expansion, specific heat, tensile stress and strain, creep, fatigue, programs for doing fast approximation of the stiffening effect of rib geometry, educational information and design assistance, and more (Appendix A: **PLASTICS DESIGN TOOL-BOX**).

Computer-Aided Design

CAD is the process of solving design problems with the aid of computers. This function includes the computer generation and modification of graphic images on a video display,

The one-piece deck lid, and the one-piece rear end panel, with the addition of a quarter panel section, could become a fully integrated rear end assembly in a single, joint-free molded component.

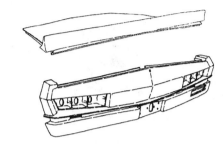

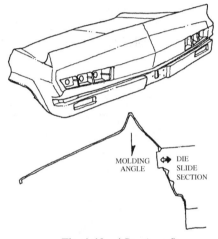

MOLDING ANGLE · DIE SLIDE SECTION

Fig. 1-13 (*Continued*)

printing these images as hard copy using a printer or plotter, analyzing the design data, and electronic storage and retrieval of design information. Many CAD systems perform these functions in an integrated fashion that can increase the designer's productivity.

It is important to recognize that the computer does not change the nature of the design process; it is simply a tool to improve efficiency and productivity. It is appropriate to view the designer and the CAD system together as a design team, with the designer providing knowledge, creativity, and control and the computer accurate, easily modifiable graphics and the capacity to perform complex design analysis at great speeds and store and recall design information. Occasionally, the computer can augment or replace many of the designer's other tools, but it is important to remember that this ability does not change

the fundamental role of the designer's innovation capability.

Computer-Aided Design Drafting

CADD, a part of CAD, is the computer-assisted generation of working drawings and other documents. The CADD user generates graphics by interactive communication with the computer. The graphics are displayed on a video terminal and can be converted into hard copy by a printer or plotter.

Computer-Aided Manufacturing

CAM describes a system that can take a CAD product, devise its essential production steps, and electronically communicate this information to manufacturing equipment such as robots. The CAD/CAM system has offered many advantages over past traditional manufacturing systems, including the need for less design effort through the use of CAD and CAD databases, more efficient material use, reduced lead time, greater accuracy, and improved inventory functions.

Computer-Integrated Manufacturing

CIM is the coordination of all stages of manufacturing, which enables the manufacturers to custom design products efficiently and economically, by a computer or a system of computers.

Computer-Aided Testing

In addition to computer-aided activities, CAT involves the testing that takes place in all stages of product development (Chapter 5). The advantage of CAT is that the output of sensors measuring the characteristics of the prototype or finished product can manipulate the product model to improve its accuracy or identify design modifications needed. In this way testing integrates design and fabrication into an ongoing, self-correcting development process.

It seems appropriate to close this design
guide, which featured tiny details of design,
with one designer's idea of how the first
all – plastic – automobile may
be assembled.

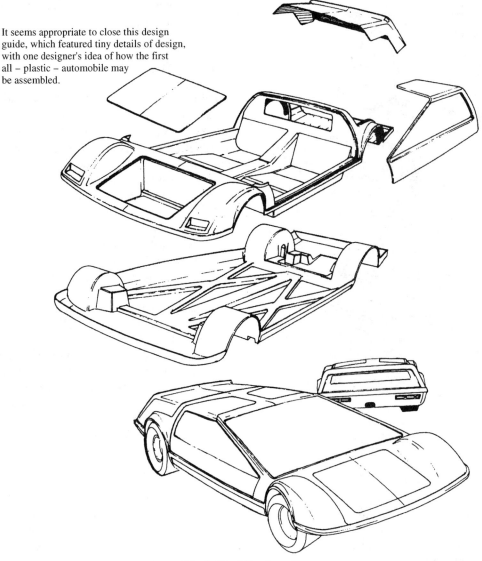

Fig. 1-13   (*Continued*)

Computer Software Program

Available are many software programs that
provide guides in simulating the different design and processing operations. These guides
provide a logical approach in training and
conducting research. There are programs that
allow fabricating of different designs using
different types of plastics. There are also simulated process controls that permit processing operators to make changes and see the effects that occur on a fabricated product such
as thickness or tolerance.

Basically a computer software is a set of
instruction guidelines that "tells and documents" what is to be done and how to
do it. There are many off-the-shelf software
instruction programs, with many more always on the horizon, in addition to some
operations developing there own. They include product design, mold design, die design, processing techniques, management,
storage, testing, quality control, cost analysis, and so on. The software tasks vary so
that if you need a particular program, one
should be available or close to it. Recognize

that if you are not successful in your selection, you probably did not set up the complete requirements. Also if your software can not easily accommodate change, then you have the wrong program. A few introductory examples of design software follow with more in Appendix A: **PLASTICS DESIGN TOOLBOX:**

1. MOLDEST: Provides product design, mold design, and injection molding process control by Fujitsu Ltd., Tokyo, Japan.

2. CAD[Plus] SOLID EDGE: Advanced mechanical simulation via finite element analysis by Algor, Inc. Pittsburgh, PA www.algor.com

3. DFMA: Design for Manufacture & Assembly provides determinants of costs associated with processes by Boothroyd Dewhurst Inc., Wakefield, RI www.dfma.com

4. Prospector: Examines and provides tabular, single-point (for preliminary material evaluation) and multi- point data (predict structural performance of a material under actual load conditions) for its 35,000 plastics by IDES Inc., Laramie, WY.

Remember we do not need people if the software does all the jobs of product design, mold or die design, material selection, processing setup, making financial profits, and so on. Software programs are useful tools and can perform certain functions. The key to success is the designer's capability in using what is available that includes understanding and putting to practical use software programs and recognizing their limitations.

As an example since the development of the first injection molding (IM) simulation program modules in the 1970s, they (rheology, thermal, mechanical, process control, quality control, statistical analysis, cost modeling, and so on) have become more and more powerful as well as compact. The predictive value of their computations much greater [but not perfect (Chapter 4, **RISKS, Perfection**)]. The software program systems also include areas of product applications to their performance characteristics. Simulation programs can be clearly differentiated partly by their user-friendliness and computational precision, and partly on theoretical grounds.

Thus a design can be evolved as a planar 2-D model or 3-D model.

Software and database. Many thousands of software and database programs expand design capabilities, simplifies design analysis, material processing properties, processing capability startups, training, and so on. As an example a worldwide software program that has been extremely useful, based on how it was organized, is the CAMPUS Database software. This Computer-Aided Material Selection by Uniform Standards of Testing Methods (CAMPUS) compares different plastics available from different material suppliers. Special CAMPUS pages are on their web sites, updated each time they finish further testing of present and new materials.

Its data can be directly merged into CAE programs. CAMPUS provides comparable property database on a uniform set of testing standards on materials along with processing information. The database contains single-point data for mechanical, thermal, rheological, electrical, flammability, and other properties. Multipoint data is also provided such as secant modulus vs. strain, tensile stress-strain over a wide range of temperatures, and viscosity vs. shear rate at multiple temperatures.

RAPRA free Internet search engine. The number of plastic-related web sites is increasing exponentially, yet searching for relevant information is often laborious and costly. During 1999 RAPRA Technology Ltd., the UK-based plastics and rubber consultancy, launched what is believed to be the first free Internet search engine focused exclusively in the plastics industry. It is called Polymer Search on the Internet (PSI). It is accessible at **www.polymersearch.com.** Companies involved in any plastic-related activity are invited to submit their web-site address for free inclusion on PSI. RAPRA Technology's USA office is in Charlotte, NC (tel. 704-571-4005).

Short and Long Term Performance

Product design starts by one visualizing a certain material, makes approximate calculations to see if the contemplated idea is practical to meet requirements that includes cost, and if the answer is favorable, proceeds to collect detailed data on a range of materials that may be considered for the new product. When plastics are the candidate materials, it must be recognized from the beginning that the available test data require understanding and proper interpretation before an attempt can be made to apply them to the product design. For this reason, an explanation of data sheets is required in order to avoid anticipating product characteristics that may not exist when merely applying data sheet information without knowing how such information was derived (Chapter 5).

The application of appropriate data to product design can mean the difference between the success and failure of manufactured products made from any material (plastic, steel, etc.). There are different sources of information on plastics. There is the data sheet compiled by a manufacturer of the material and derived from tests conducted in accordance with standardized specifications. Another source is the description of outstanding characteristics of each plastic, along with the listing of typical applications.

It is important for the designer to become familiar with all the information that is available for each plastic, especially that which is pertinent to the product design requirements. Designers, who are knowledgeable of the data derived from metal tests, could have a tendency to apply the plastic data sheet information in a manner similar to that used for metals. This could be understood because there is no warning that some of the data supplied by the manufacturer are applicable only when use and test conditions are nearly the same.

However, if suppliers' data were to be applied without a complete analysis of the test data for each property, the result could prove costly and embarrassing. The nature of plastic materials is such that an oversight of even a small detail in its properties or the method by which they were derived could result in problems and product failure.

Once it is recognized that there are certain reservations with some of the properties given on the data sheet, it becomes obvious that it is very important for the designer to have a good understanding of these properties. Thus the designer can interpret the test results in order to make the proper evaluation in selecting a material for a specific product.

Predicting Performance

Avoiding structural failure can depend in part on the ability to predict performance of materials. When required designers have developed sophisticated computer methods for calculating stresses in complex structures using different materials. These computational methods have replaced the oversimplified models of materials behavior relied upon previously. The result is early comprehensive analysis of the effects of temperature, loading rate, environment, and material defects on structural reliability. This information is supported by stress-strain behavior data collected in actual materials evaluations.

With computers the finite element analysis (FEA) method has greatly enhanced the capability of the structural analyst to calculate displacement, strain, and stress values in complicated plastic structures subjected to arbitrary loading conditions. Details on FEA are reviewed in Chapter 2, **Finite Element Analysis**.

Nondestructive testing (NDT) is used to assess a component or structure during its operational lifetime. Radiography, ultrasonics, eddy currents, acoustic emissions, and other methods are used to detect and monitor flaws that develop during operation (Chapter 7).

The selection of the evaluation method(s) depends on the specific type of plastic, the environment of the evaluation, the effectiveness of the evaluation method, the size of the structure, the fabricating process to be used, and the economic consequences of structural failure. Conventional evaluation methods are often adequate for baseline and acceptance inspections. However, there are increasing demands for more accurate characterization of the size and shape of defects that may require advanced techniques and procedures and involve the use of several methods.

A Changing World

It would be difficult to imagine the modern world without plastics. Today they are an integral part of everyone's life-style, with products varying from commonplace domestic articles to sophisticated scientific and medical instruments. Nowadays designers readily turn to plastics. Exceptional progress has been made in over the past century worldwide in all markets. As a matter of fact, many of the technical wonders we take for granted would be impossible without versatile, economical plastics. Yet some that are not mindful of the many benefits of plastics still carry negative feelings about them. Some examples of their creative use follow that in turn show the actions by creative designers.

Recreation

Because people everywhere tend to take their fun seriously, they spend freely on sports and recreational activities. The broad range of properties available from plastics has made them part of all types of sports and recreational equipment for land, water, and airborne activities. Roller-skate wheels are now abrasion- and wear-resistant polyurethane, tennis rackets are molded from specially reinforced plastics (using glass, aramid, graphite, or other fibers), skis are laminated with plastics, and so on.

Electronic

Most of the electrical equipment and electronic devices we use and enjoy today would not be practical, economical, or occasionally even possible without plastics.

Packaging

When packaging problems are tough, plastics often are the answer and sometimes the only answer. They can perform tasks no other materials can provide. As an example plastics have extended the life of vegetables after they are packaged. Packaging represents the largest consumer market for plastics.

Building and Construction

This market is the second-largest consumer of plastics, after packaging. Durable and easy to install, as well as cost effective, plastics continue to find more and more applications. This is a market where spectacular growth will occur when their performance is understood by the building industry (meeting their specifications, etc.) and/or the price is right. Recognize that if wood with its excellent record of performances for many centuries, based on present laws and regulations, could not be used. They burn, rot, etc. It would be ridiculous not to use wood.

Health Care

Plastics have made many major contributions to the contemporary scene. Health-care professionals depend on plastics for everything from intravenous bags to wheelchairs, disposable labware to silicone body parts, etc. The diversity of plastics allows them to serve in many ways, improving and prolonging lives, such as a braided, corrugated Dacron (Du Pont's polyester) aorta tube (24).

Another example of thousands is a biodegradable plastic developed at the Massachusetts Institute of Technology that may be saving lives in the form of a medical implant. This plastic is being tested nationwide to determine its effectiveness as a drug-releasing implant in brain cancer patients. These implants, roughly the size of a quarter, are being placed in patients' brains to release the chemotherapy drug BCNU (Carmustine). These biocompatible implants have been found to be safer than injections, which can cause the BCNU to enter the bone marrow or lungs, where the drug is toxic.

This plastic, known as polyanhydride, was designed shapewise so that water would trigger its degradation but would not allow a drug to be released all at once. The implant degrades from the outside, like a bar of soap, releasing the drug at a controlled rate, as it becomes smaller. The rate at which the drug is released is determined by the surface shaped area of the implant and the rate of plastic degradation, which can be customized to release drugs at rates varying from one day to many years. This design approach also holds promise for use with different drugs for various other-medical problems.

Transportation

For today's autos, trucks, busses, vans, etc. plastics offer a wide variety of benefits that include durability, light weight, corrosion resistance, safety, and fuel savings.

Aerospace

During the last half-century, aeronautics technology has soared, with plastics playing a major role. Lightweight durable plastics and reinforced plastics (RPs) save on fuel while standing up to forms of stress like creep and fatigue, in different environments.

Appliance

In this market plastics have been exceptionally beneficial. For example, in the early 1900s doing simple household tasks was a real chore. Washing, drying, and ironing clothes was a rigorous, two-day affair involving the filling of metal tubs, scrubbing by hand, hanging clothes to dry, and heating cast-iron flat irons on a stove. With new technology and "plastics" laundry rooms and kitchens worldwide are operating in relatively minutes and looking better than ever before.

As one of thousands of examples, in the fall of 1987, Milwaukee Electric Tool Corp. found itself on the short end of the age-old supply-and-demand equation. That is, it was unable to keep up with demand for its heavy-duty electric power tools. The problem was that their machining operations could not turn out enough aluminum die-cast motor housings to keep up with market demand. The firm briefly considered what would have been a long-term solution; a state-of-the-art machining center. But a feasibility study showed that capital costs for such a facility would run into hundreds of thousands of dollars, while resulting savings would amount to a few cents per part.

Fortunately, there was the other option of using plastic motor housings. Du Pont agreed to produce plastic prototypes of the housing in Zytel nylon 82G plastic. The prototypes were quickly assembled; then they endured demanding drop tests and other field tests that are standard for Milwaukee Electric tools. When the housings of impact-resistant Zytel passed the tests with no problem, the firm had a new, lower-cost solution to its machining problem; a plastic housing, produced from a production mold that required no machining.

The redesign presented several additional opportunities. Initial target was to replace aluminum die-casting, and thereby eliminate machining as well as deflashing, trimming, and spadoning (a surface treatment that imparts a matte finish). But they also wanted to eliminate as many parts as possible, simplify the assembly, and use a product that worked as well or better than aluminum. Achieving these goals produced some spectacular benefits; parts costs dropped by two-thirds while manufacturing throughput rates increased. Savings in labor, machining, and assembly operations were augmented by lower capital and maintenance costs. As many as one million plastic housings were injection molded without major tool repair or replacement, vs. 100,000 parts for the die-casting operation.

Six parts in the housing were eliminated. Because plastic used was not conductive, designers were able to do away with insulating parts, such as a coil shield that separated the electric brush holder area from the aluminum, and the cardboard insulating sleeve that went between the copper wiring of the field core and the housing. Removing the sleeve had the added benefit of creating better airflow inside the housing, so the motor ran cooler under load. Press-fitting a rear ball bearing into the housing and keeping the bearing securely in place proved to be a major obstacle. The solution was to use eight small ribs inside the rear-bearing pocket. The ribs increase the amount of interference that can be overcome when press-fitting the ball bearing, and keep the bearing in its pocket with a strong, uniform force.

Another concern was achieving overall perpendicularity of the housing face where it fitted with a mating gear case. The molder solved this problem by repeatedly adjusting molded housing dimensions by a few thousands of an inch. The key to this fine-tuning was to establish three adjustment spots; one at each screw hole location. Thus it was much easier to design mating parts so that they sat on the lands, specific points, rather than trying to align a complete surface. Accurately repeating such minute dimensions required batch-to-batch plastic consistency and process control.

Success by Design

A skilled designer blends a knowledge of materials, an understanding of manufacturing processes, and imagination into successful new designs. Recognizing the limits of design with traditional materials is the first step in exploring the possibilities for innovative design with plastics. What is important when analyzing plastic designs is the ease to incorporate ergonomics and empathy that results in products that truly answers the user's needs.

With designing there has always been the need to meet engineering, styling, and performance requirements at the lowest cost. To some there may appear to be a new era where ergonomics is concerned, but this is not true. What is always new is that there are continually easier methods on the horizon to simplify and meet all the specific requirements of a design. Some designers operate by creating only the stylish outer appearance, allowing basic engineers to work within that outside envelope. Perhaps this is all that is needed to be successful, but a more in-depth approach will work better. Beginning with a thorough understanding of the

user's needs and keeping an eye toward ease of manufacture and repair, designers should also work from the inside out. The envelope that eventually emerges will then be a logical and aesthetic answer to the design challenge (Fig. 1-14).

With new plastics and processing techniques always becoming available, the design challenge becomes easier, even when taking today's solid-waste problem into account. Today's plastics and processes allow designers to incorporate and interrelate all the aspects of success. In products such as electronics, medical devices, transportation controls, and many others where user-friendly design is required, it has to be obvious to all that plastics play an important role.

Responsibility

The responsibilities of designers encompass all aspects of design. Although functional design is of paramount importance, a design is not complete if it is functional but cannot easily be manufactured, or functional but not dependable, or if it has a good appearance but poor reliability, or the product will not fail but does not meet safety requirements. Designers have a broad responsibility to produce designs that meet all the objectives of function, durability, appearance, safety, and low cost. They should not contend that something is now designed and it is now the manufacturing engineer's job to figure out how to make it at a reasonable cost. The functional design and the production design are too closely interrelated to be handled separately.

Product designers must consider the conditions under which fabrication will take place, because these conditions affect product performance and cost. Such factors as production quantity, labor, and material cost are vital. Designers should also visualize how each product is to be fabricated. If they do not or cannot, their designs may not be satisfactory or even feasible from a production standpoint. One purpose of this book is to give designers sufficient information about manufacturing processes so that they

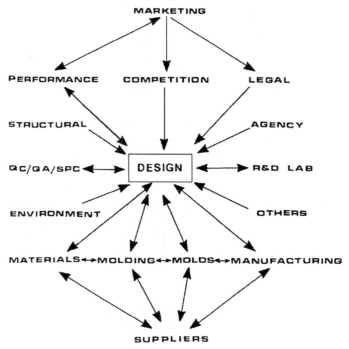

Fig. 1-14   The overall design challenge.

can design intelligently from a productivity standpoint.

Responsibility Commensurate with Ability

Designers have the responsibility of being committed to developing experience so that they can function properly. They design products to meet performance and other requirements such as those reviewed in Chapter 4, **DESIGNING AND LEGAL MATTER** and also **RISK**.

Recognize that people have certain capabilities; the law says that people have equal rights (so it reads that we were all equal since 1776) but some interpret it to mean equal capabilities. So it has been said via Sun Tzu, The Art of War, about 500 BC "Now the method of employing people is to use the avaricious and the stupid, the wise and the brave, and to give responsibilities to each in situations that suit the person. Do not charge people to do what they cannot do. Select them and give them responsibilities commensurate with their abilities."

Ethic

Although there is no substitute for individual action based on a firm philosophical and ethical foundation, designers have developed guidelines for professional conduct based on the experience of many of them who have had to wrestle with troublesome ethical questions and situations previously. These guidelines can be found in the published codes of ethics for designers and engineers of a number of industry and technical societies such as the Industrial Designer societies.

Terminology

Different terms are used when discussing the subject of designing with plastics. Many of them could be repeated in this book. To eliminate repeating the same definition the Appendix B: **TERMINOLOGY** has been prepared. Where it is important to describe a term in the text, its definition is only used in the text. The **INDEX** includes the definitions in the text.

2

Design Influencing Factor

Introduction

The basic information involved in designing with plastics concerns the load, temperature, time, and environment. As reviewed throughout this book there are other performance requirements that may exit such as aesthetics.

Design is essentially an exercise in predicting performance. The designer must therefore be knowledgeable in such behavioral responses of plastics as those ranging from short time static (tensile, flexural, etc.) to long time dynamic (creep, fatigue, impact, etc.) mechanical load performances in different environments. This chapter presents important basic concepts that concern this type information. Along with the other chapters, it provides the background needed to understand plastics different load performance behaviors in different environments.

Many plastic products seen in everyday life are not required to undergo sophisticated design analysis because they are not required to withstand extreme loading conditions such as creep and fatigue loads. Examples include containers; cups; toys; boxes; housings for computers, radios, televisions and the like; and nonstructural or secondary structural products of various kinds in buildings, aircraft, appliances, and electronic devices. These type products require reviewing the routine performance properties such as static strength and stiffness to impact that are primarily available in-house, material software, and/or from material suppliers.

In evaluating and comparing specific plastics to meet these requirements past experience and/or the material suppliers are sources of information. It is important to ensure that when making comparisons the data be available where the tests were performed using similar procedures (Chapter 5). Where information or data may not be available some type of testing can be performed by the designer's organization, outside laboratory (many around), and/or possible the material supplier if it warrants their participation (technicalwise and potential costwise). If little is known about the product or cannot be related to similar products prototype testing will definitely be required (Chapter 3).

Designing is, to a high degree, intuitive and creative, but at the same time empirical and technically mechanical. An inspired idea alone will not result in a successful design; experience plays an important part that can easily be developed. An understanding of one's materials and a ready acquaintance with the relevant processing technologies (Chapter 8) are essential for converting an idea to an actual product. In addition, certain basic tools are usually needed when conducting prototype evaluation, such as those for computation and measurement to ensure that the loads and forces the product is to absorb can be safely withstood.

When required plastics permit a greater amount of structural design freedom than any other material. Products can be small or large, simple or complex, rigid or flexible, solid or hollow, tough or brittle, transparent or opaque, black or virtually any color, chemical resist or biodegradable, etc. Materials can be blended to achieve any desired property or combination of properties (Chapters 6 and 7). The final product performance is affected by interrelating the plastic with its design and processing method. The designer's knowledge of all these variables can profoundly affect the ultimate success or failure of a consumer or industrial product.

For these reasons design is spoken of as having to be appropriate to the materials of its construction, its methods of manufacture, and the loads (stresses) involved in the product's environment. Where all these aspects can be closely interwoven, plastics are able to solve design problems efficiently in ways that are economically advantageous.

Material Behavior

An adequate description of material behavior is basic to all designing applications. Fortunately, many problems may be treated entirely within the framework of plastic's elastic material response. While even these problems may become quite complex because of geometrical and loading conditions, the linearity, reversibility, and rate independence generally applicable to elastic material description certainly eases the task of the analyst for static and dynamic loads that include conditions such as creep, fatigue, and impact.

However, we are increasingly confronted with practical problems that involve material response that is inelastic, hysteretic, and rate dependent combined with loading which is transient in nature. These problems include, for instance, structural response to moving or impulsive loads, all the areas of ballistics (internal, external, and terminal), contact stresses under high speed operations, high

speed fabricating processes (injection molding, extrusion, blow molding, thermoforming, etc.), shock attenuation structures, seismic wave propagation, and many others of equal importance.

As these problems were encountered in the past, it became evident that we did not have at hand the physical or mathematical description of the behavior of materials necessary to produce realistic solutions. Thus, during the past half century, there has been considerable effort expended toward the generation of both experimental data on the static and dynamic mechanical response of materials (steel, plastic, etc.) as well as the formulation of realistic constitutive theories (Appendix A: **PLASTICS DESIGN TOOLBOX**).

As a plastic is subjected to a fixed stress or strain, the deformation versus time curve will show an initial rapid deformation followed by a continuous action. Examples of the standard type tests are included in Fig. 2-1. Details on using these type specimens under static and dynamic loads will be reviewed throughout this chapter. (Review also Fig. 8-9 that relates elasticity to strain under different conditions.)

Dynamic loading in the present context is taken to include deformation rates above those achieved on the standard laboratory-testing machine (commonly designated as static or quasi-static). These slower tests may encounter minimal time-dependent effects, such as creep and stress-relaxation, and therefore are in a sense dynamic. Thus the terms static and dynamic can be overlapping.

Rheology and Viscoelasticity

Rheology is the science that deals with the deformation and flow of matter under various conditions. The rheology of plastics, particularly of TPs, is complex but understandable and manageable. These materials exhibit properties that combine those of an ideal viscous liquid (with pure shear deformations) with those of an ideal elastic solid (with pure elastic deformation). Thus, plastics are said to be viscoelastic.

Tensile load

Resistance
to
bending

Deflection,
rigidity

Buckling

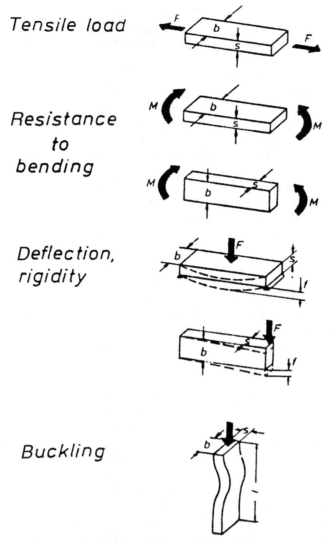

Fig. 2-1 Schematics of different type test specimens.

The mechanical behavior of plastics is dominated by such viscoelastic phenomena as tensile strength, elongation at breaks, stiffness, and rupture energy, which are often the controlling factors in a design. The viscous attributes of plastic melt flow are also important considerations in the fabrication of plastic products. (Chapter 8, **INFLUENCE ON PERFORMANCE, Viscoelasticity**).

Viscoelasticity Behavior

The viscoelasticity is a combination of viscous and elastic properties in a plastic with the relative contribution of each being dependent on temperature, load, and time. It relates to the important mechanical behavior of plastics in which there is a temperature and time dependent relationship between stress and strain. A material having this property is considered to combine the features of a perfectly elastic solid and a perfect fluid.

The viscoelastic nature of the material requires not merely the use of data sheet information for calculation purposes, but also the actual long-term performance experience gained that can be used as a guide. The allowable working stress is important for determining dimensions of the stressed area and also for predicting the amount of distortion and strength deterioration that will take place over the life-span of the product. This means that the allowable working stress for a constantly loaded product that is expected to perform satisfactorily over many years has to be established using creep characteristics of a material that has sufficient data with which a reliable long-term prediction of short-term test results can be made. (The book authors includes starting from 1943 in personally conducting extensive creep, fatigue, impact, etc. test in order to design and fabricated primary structural products such as the first all-plastic airplane.)

Pseudo-elastic design Since at least the 1940s there have been extensive research and developments in applying standard engineering equations for plastics based on the behavior of plastics. Many different plastic products have been designed using these equations, fabricated, and providing long service life of primary structures such as used in aircraft, buildings, boats, transportation vehicles, signs, deep space antennas, and many more.

For those not familiar with this type information recognize that the viscoelastic behavior of plastics shows that their deformations are dependent on such factors as the time under load and temperature conditions. Therefore, when structural (load bearing) plastic products are to be designed, it must be remembered that the standard equations that have been historically available for designing steel springs, beams, plates, cylinders, etc. have all been derived under the assumptions that: (1) the strains are small, (2) the modulus is constant, (3) the strains are independent of the loading rate or history and are immediately reversible, (4) the material is isotropic, and (5) the material behaves in the same way in tension and compression.

These assumptions are not justifiable when applied to plastics unless modified. The classical equations cannot be used indiscriminately. Each case must be considered on its own merits, with account being taken of such factors as the mode of deformation, service temperature and environment,

and fabrication method. (These factors can also apply to steels, etc.) (Appendix A: **PLASTICS DESIGN TOOLBOX**).

In particular, it should be noted that the past traditional equations that have been developed for other materials, principally steel, use the relationship that stress equals the modulus times strain, where the modulus is constant. Except for thermoset-reinforced plastics and certain engineering plastics, most plastics do not generally have a constant modulus of elasticity. Different approaches have been used for this non-constant situation, some are quiet accurate. The drawback is that most of these methods are quite complex, involving numerical techniques that are not attractive to the average designers.

One method that has been widely accepted is the so-called pseudo-elastic design method. In this method appropriate values of such time-dependent properties as the modulus are selected and substituted into the standard equations. It has been determined that this approach is sufficiently accurate in most cases if the value chosen for the modulus takes into account the projected service life of the product and the limiting strain of the plastic, assuming that the limiting strain for the material is known. Unfortunately this is not a straightforward value applicable to all plastics or even of one plastic in all applications. This value is often arbitrarily chosen, although several methods have been suggested for arriving at a suitable value.

For the past century one successful approach is to plot a secant modulus that is at 1% strain or 0.85% of the initial tangent modulus and noting where they intersect the stress-strain curve (Fig. 2-2). However for many plastics, particularly the crystalline thermoplastics, this method is too restrictive. So in most practical applications the limiting strain is decided based on experience and/or in consultation between the designer and the plastic material manufacturer. Once the limiting strain is known, design methods based on its creep curves become rather straightforward (additional information to follow).

Effect of strain rate Workers in the field of continuum mechanics have had occasion to

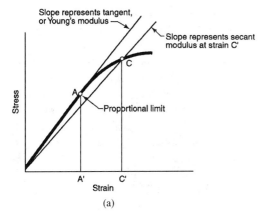

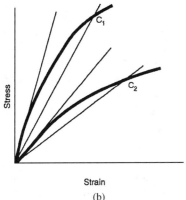

Fig. 2-2 (a) Example of the modulus of elasticity determined on the initial straight portion of the stress-strain curve and secant modulus and (b) secant modulus for two different plastics that are 85% of the initial tangent modulus.

some cases, they were unduly restrictive, but nonetheless they provided a basis from which viscoelastic behavior was better understood and moreover they led to the later development of mathematically more elegant expositions. In 1964 a paper was presented on Thermodynamics of Materials with Memory. This was a work of power and depth, such depth according to some people, that it gave "no status at all" to the other theories (144).

Correlating Rheological Parameter

Object in this section is to review how rheological knowledge combined with laboratory data can be used to predict stresses developed in plastics undergoing strains at different rates and at different temperatures. The procedure of using laboratory experimental data for the prediction of mechanical behavior under a prescribed use condition involves two principles that are familiar to rheologists; one is Boltzmann's superposition principle which enables one to utilize basic experimental data such as a stress relaxation modulus in predicting stresses under any strain history; the other is the principle of reduced variables which by a temperature-log time shift allows the time scale of such a prediction to be extended substantially beyond the limits of the time scale of the original experiment.

Ludwig Boltzmann (1844–1906) was born in Vienna. His work of importance in chemistry became of interest in plastics because of his development of the kinetic theory of gases and rules governing their viscosity and diffusion. They are known as the Boltzmann's Law and Principle, still regarded as one of the cornerstones of physical science.

Mechanical properties of plastics are invariably time-dependent. Rheology of plastics involves plastics in all possible states from the molten state to the glassy or crystalline state (Chapter 6). The rheology of solid plastics within a range of small strains, within the range of linear viscoelasticity, has shown that mechanical behavior has often been successfully related to molecular structure. Studies in this area can have two objectives: (1) mechanical characterization of

witness in the past, a significant evolution in the theory of irreversible thermodynamics of viscoelastic materials. Following work in the early 1930's that principally involved steels, there ensued an intense activity consisting in attempts to give a thermodynamic basis to the mechanical theory of small viscoelastic deformations in metals which, of course, constitute processes that may be regarded as small deviations from an equilibrium state. It is of historical interest that this activity left in its wake a divided opinion. Strong objections were voiced from the mathematical wing of "natural philosophers" who in a tours de force attacked such assumptions which, apparently, were arbitrary and, at most, of debatable validity.

These assumptions concerning steel were often based on physical intuition and in a plastic in order to predict its behavior in practical applications, and (2) rheological experimentation as a means for obtaining a greater structural understanding of the material. Much has been explored in these areas during the past many decades with the result that a great deal is known about the effect of structure on the properties of plastics, particularly in the case of amorphous plastics in a rubbery state.

In this approach the reviews concerned the rheology involving the linear viscoelastic behavior of plastics and how such behavior is affected by temperature. Next is to extend this knowledge to the complex behavior of crystalline plastics, and finally illustrate how experimental data were applied to a practical example of the long-time mechanical stability.

When a plastic material is subjected to an external force, a part of the work done is elastically stored and the rest is irreversibly (or viscously) dissipated; hence a viscoelastic material exists. The relative magnitudes of such elastic and viscous responses depend, among other things, on how fast the body is being deformed. It can be seen via tensile stress-strain curves that the faster the material is deformed, the greater will be the stress developed since less of the work done can be dissipated in the shorter time.

When the magnitude of deformation is not too great, viscoelastic behavior of plastics is often observed to be linear, i.e., the elastic part of the response is Hookean and the viscous part is Newtonian. Hookean response relates to the modulus of elasticity where the ratio of normal stress to corresponding strain occurs below the proportional limit of the material where it follows Hooke's law. Newtonian response is where the stress-strain curve is a straight line.

From such curves, however, it would not be possible to determine whether the viscoelasticity is in fact linear. An experiment is needed where the time effect can be isolated. Typical of such experiments is stress relaxation. In this test, the specimen is strained to a specified magnitude at the beginning of the test and held unchanged throughout the experiment, while the monotonically decay-ing stress is recorded against time. The condition of linear viscoelasticity is fulfilled here if the relaxation modulus is independent of the magnitude of the strain. It follows that a relaxation modulus is a function of time only.

There are several other comparable rheological experimental methods involving linear viscoelastic behavior. Among them are creep tests (constant stress), dynamic mechanical fatigue tests (forced periodic oscillation), and torsion pendulum tests (free oscillation). Viscoelastic data obtained from any of these techniques must be consistent data from the others.

If a body is subjected to a number of varying deformation cycles, a complex time dependent stress would result. If the viscoelastic behavior is linear, this complex stress-strain-time relation is reduced to a simple scheme by the superposition principle proposed by Boltzmann. This principle states in effect that the stress at any instant can be broken up into many parts, each of which has a corresponding part in the strain that the body has experienced. This is illustrated in Fig. 2-3, where the stress is shown to consist of two parts, each of which corresponds to the time axis as the temperature is changed. It implies that all viscoelastic functions, such as the relaxation modulus, can be shifted along the logarithmic time axis in the same manner by a suitable temperature change. Thus, it is possible to reduce two independent variables (temperature and time) to a single variable (reduced time at a given temperature). Through the use of this principle of reduced variables, it is thus possible to expand enormously the time range of a viscoelastic function to many years and many decades.

The relaxation modulus (or any other viscoelastic function) thus obtained is a mean's of characterizing a material. In fact relaxation spectra have been found very useful in understanding molecular motions of plastics. Much of the relation between the molecular structure and the overall behavior of amorphous plastics is now known.

Mechanical properties of crystalline plastics are much more complex than those of amorphous plastics (Chapter 6, **STRUCTURE AND MORPHOLOGY**). For

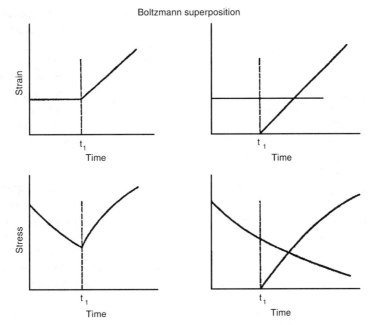

Boltzmann superposition

Fig. 2-3   Boltzmann superposition principle.

example, a simple temperature-time shift is respective strain. Viscoelastic data, at least in theory, can be utilized to predict mechanical performance of a material under any use conditions. However it is seldom practical to carry out the necessarily large number of tests for the long time periods involved. Such limitations can be largely overcome by utilizing the principle of reduced variables embodying a time-temperature shift.

While a plastic usually exhibits not one but many relaxation times, each relaxation time is affected by the temperature in exactly the same manner as another. That is the whole relaxation spectrum shifts in unison along the logarithmic no longer applicable in these materials, because the crystalline morphology changes with the temperature.

Mechanical Load

It is well known that mechanical loads on a structure induce stresses within the material such as those shown in Fig. 2-4. It is also well known that the magnitudes of these static and dynamic stresses depends on many factors, including forces, angle of loads, rate and point

of application of each load, geometry of the structure, manner in which that structure is supported, and time at temperature. The behavior of the material in response to these induced stresses determines the performance of the structure.

The behavior of materials (plastics, steels, etc.) under dynamic loads is important in certain mechanical analyses of design problems. Unfortunately, sometimes the engineering design is based on the static loading properties of the material rather than dynamic properties. Quite often this means overdesign at best and incorrect design resulting

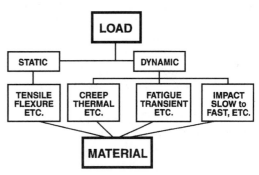

Fig. 2-4   Interaction of static and dynamic loads on materials.

in failure of the product in the worst case. The complex nature of the dynamic behavior problem can be seen from Fig. 2-4, which depicts a wide range of interaction of dynamic loads that occurs with various materials (metals, plastics, etc.). Ideally, it would be desirable to know the mechanical response to the full range of dynamic loads for each material. However, certain load-material interactions have more relative importance for engineering design, and significant work on them exists already. The mechanical engineers, civil engineers, and the metallurgical engineers have always found these materials to be most attractive to study. Even so, there is a great deal that we do not understand about them (includes steels, plastics, etc.) in spite of voluminous scientific literature existing in this area. Each type of load response, e.g., creep, fatigue/vibratory, or impact, is a major field in itself. Regardless one has been able to design products that are subjected to dynamic loads even though there is always a desire to obtain more data. The following review describes the types of material behavior that must be examined and evaluated in any structural design project involving plastics. They start with damping followed with short-term stress-strain behavior, long-term viscoelastic behavior (creep and stress relaxation), fatigue, load material interaction, and thermal expansion and contraction. Damping The dynamic mechanical behavior of plastics is of great interest and importance. For one thing, the dynamic modulus, or for that matter the modulus measured by any other technique, is one of the most basic of all mechanical properties, with its importance being well known in any structural application. The role of mechanical damping is, however, not as well known. Damping is to diminish progressively vibration or oscillation.

Damping is often the most sensitive indicator of all kinds of molecular motions going on in a material, even in its solid state. Aside from the purely scientific interest in understanding the molecular motions that can occur, analyzing these motions is of great practical importance in determining the mechanical behavior of plastics. For this reason, the absolute value of a given damping and the temperature and frequency at which the damping peaks occur can be of considerable interest and use.

High damping is sometimes an advantage, sometimes a disadvantage. For instance, in a car tire high damping tends to give better friction with the road surface, but at the same time it causes heat buildup, which makes tires degrade more rapidly.

Damping reduces mechanical and acoustical vibrations and prevents resonance vibrations from building up to dangerous amplitudes. However, the existence of high damping is generally an indication of reduced dimensional stability, which can be undesirable in structures carrying loads for long periods of time. Many other mechanical properties, including fatigue life, toughness and impact, and wear and the coefficient of friction are intimately related to damping.

Dynamic Mechanical Behavior

Dynamic mechanical tests measure the response or deformation of a material to periodic or varying forces. Generally an applied force and its resulting deformation both vary sinusoidally with time. From such tests it is possible to obtain simultaneously an elastic modulus and mechanical damping, the latter of which gives the amount of energy dissipated as heat during the deformation of the material.

The behavior of materials under dynamic load is of considerable importance and interest in most mechanical analyses of design problems where these loads exist. The complex workings of the dynamic behavior problem can best be appreciated by summarizing the range of interactions of dynamic loads that exist for all the different types of materials. Dynamic loads involve the interactions of creep and relaxation loads, vibratory and transient fatigue loads, low-velocity impacts measurable sometimes in milliseconds, high-velocity impacts measurable in microseconds, and hypervelocity impacts as summarized in Fig. 2-4.

Metals are unique under both static and dynamic loads that can be cited as outstanding cases. The continuum mechanical engineer and the metallurgical engineer have both found these materials to be most attractive to study. At the same time, metals, as compared to plastics, are easier to handle for analysis. Yet there is a great deal that is still not understood about metals, even in the voluminous scientific literature available. The importance of plastics and reinforced plastics (RPs) has been growing steadily, resulting in more dynamic mechanical behavior data becoming available. However more is required to meet new design challenges.

Material behavior have many classifications. Examples are: (1) creep, and relaxation behavior with a primary load environment of high or moderate temperatures; (2) fatigue, viscoelastic, and elastic range vibration or impact; (3) fluidlike flow, as a solid to a gas, which is a very high velocity or hypervelocity impact; and (4) crack propagation and environmental embrittlement, as well as ductile and brittle fractures.

Short-Term Load Behavior

The mechanical properties of plastics enable them to perform in a wide variety of end uses and environments, often at lower cost than other design materials such as metal or wood. This section reviews the static property tests. Chapter 5 provides more information on the meaning of these type data.

As reviewed thermoplastics (TPs) being viscoelastic materials respond to induced stress by two mechanisms: viscous flow and elastic deformation. Viscous flow ultimately dissipates the applied mechanical energy as frictional heat and results in permanent material deformation. Elastic deformation stores the applied mechanical energy as completely recoverable material deformation. The extent to which one or the other of these mechanisms dominates the overall response of the material is determined by the temperature and by the duration and magnitude of the stress or strain. The higher the temperature, the most freedom of movement of the individual plastic molecules that comprise the

TP and the more easily viscous flow can occur with lower mechanical performances.

Likewise, the longer the duration of material stress or strain, the more time for viscous flow to occur. Finally, the greater the material stress or strain, the greater the likelihood of viscous flow and significant permanent deformation. For example, when a TP product is loaded or deformed beyond a certain point, the material comprising it yields and immediate or eventually fails. Conversely, as the temperature or the duration or magnitude of material stress or strain decreases, viscous flow becomes less likely and less significant as a contributor to the overall response of the material; and the essentially instantaneous elastic deformation mechanism becomes predominant.

Consequently, changing the temperature or the strain rate of a TP may have a considerable effect on its observed stress-strain behavior. At lower temperatures or higher strain rates, the stress-strain curve of a TP may exhibit a steeper initial slope and a higher yield stress. In the extreme, the stress-strain curve may show the minor deviation from initial linearity and the lower failure strain characteristic of a brittle material.

At higher temperatures or lower strain rates, the stress-strain curve of the same material may exhibit a more gradual initial slope and a lower yield stress, as well as the drastic deviation from initial linearity and the higher failure stain characteristic of a ductile material.

There are a number of different modes of stress-strain that can be taken into account by the designer. They include tensile stress-strain, flexural stress-strain, compression stress-strain, and shear stress-strain.

Tensile Stress-Strain

One of the most informative properties of any material is their mechanical behavior specifically the determination of its stress-strain curve in tension (ASTM D 638). This is usually accomplished in a testing machine by measuring continuously the elongation (strain) in a test sample as it is stretched by an increasing pull (stress). Stress is defined as the force on a material divided by the cross sectional area over which it initially acts. If the area over which the force acts changes significantly because the material deforms, one can use that area to calculate its engineering stress. The usual method used is the stress with its prevailing initial area concept.

Strain is defined as the deformation of a material divided by a corresponding undeformed dimension. The units of strain are meter per meter (m/m) or inch per inch (in./in.). Since strain is often regarded as dimensionless, strain measurements are typically expressed either as a percentage deformation or in microstrain units. One microstrain is defined as 10^{-6} m/m or in./in.

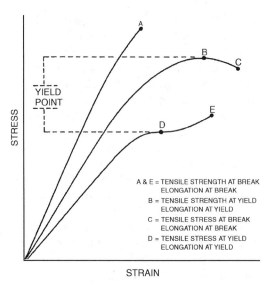

Fig. 2-5　Tensile designations according to ASTM D 638.

Tensile strength The maximum tensile stress sustained by a specimen during a tension test is its tensile strength. Figure 2-5 identifies the different types of tensile strengths. When a material's maximum stress occurs at its yield point this stress is designated its tensile strength at yield. When the maximum stress occurs at a break, the designation is its tensile strength at break. In practice these differences are frequently ignored. The tensile strength of different materials is shown in Fig. 2-6.

The generalized stress-strain curve for plastic shown in Fig. 2-7 serves to define several useful qualities that include the tensile strength, modulus (modulus of elasticity) or stiffness (initial straight line slope of the curve), yield stress, and the length of the elongation at the break point. The ultimate tensile strength is usually measured in megapascals (MPa) or pounds per square inch (psi). Values of tensile strength range from under 20 MPa (3000 psi) for low density polyethylene and some vinyls, to 76 MPa (11,000 psi) for Nylon 6, to more than 350 MPa (50,000 psi) for reinforced thermoset plastics (RTPs). The curves shown in Fig. 2-7 are typical for a plastic such as polyethylene. Figure 2-8 compares tensile curves for hard and soft steels with polycarbonate (top) and an extended scale for polycarbonates with specific behaviors usable in a design analysis.

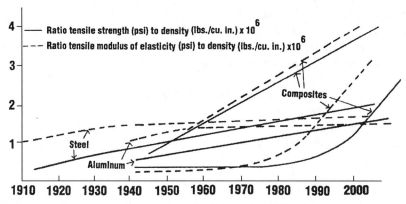

Fig. 2-6　The growth for structural properties of reinforced plastics, steel, and aluminum during the 20th century.

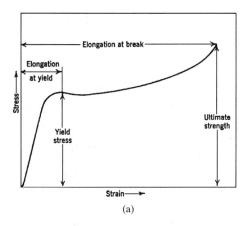

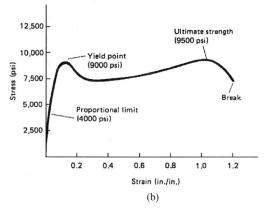

Fig. 2-7 (a) Generalized tensile stress-strain curve for plastics and (b) example of a commodity plastic's stress-strain diagram.

Area under the curve Generally, the area under the stress-strain curve is proportional to the energy required to break the specimen. It is thus sometimes referred to as the toughness of the plastic. Figure 2-9 shows tensile stress-strain curves for the usual different plastics that relate the area under the curve to their toughness or physical properties. However, there are types, particularly among the many fiber-reinforced TSs, that are very hard, strong, and tough, even though their area is extremely small.

Elongation The elongation is the stretch that a material will exhibit before break or deformation. It is usually identified as a percentage. Some materials like the styrenes and phenolics yield little before break, while others like polypropylene can be stretched many times their length before deformation.

These type curves can be related to the plastic's degree of flexibility as depicted by Fig. 2-10.

Yield It is the first point on a stress-strain curve at which an increase in strain occurs without any increase in stress. This yield point is also called yield strength or tensile strength at yield. Some materials may not have a yield point. Yield strength can in such cases be established by picking a stress level beyond the material's elastic limit. The yield strength is generally established by constructing a line to the curve where stress and strain is proportional at a specific offset strain, usually at 0.2% (Fig. 2-11). The stress at the point of intersection of the line with the stress-strain curve is called its yield strength at 0.2% offset.

Proportional limit A material's proportional limit is the greatest stress at which it is capable of sustaining an applied load without deviating from the proportionality of a stress-strain straight line (Fig. 2-2).

Elastic limit The elastic limit of a material is the greatest stress at which it is capable of sustaining an applied load without any permanent strain remaining, once stress is completely released.

Modulus of elasticity Most materials, including plastics and metals, have deformation proportional to their loads below the proportional limit. Since stress is proportional to load and strain to deformation, this implies that stress is proportional to strain. Hooke's Law, developed in 1676, follows that this straight line (Fig. 2-2) of proportionality is calculated as:

$$\text{Stress/Strain} = \text{Constant} \qquad (2\text{-}1)$$

The constant is called the modulus of elasticity (E) or Young's modulus (defined by Thomas Young in 1807 although the concept was used by others that included the Roman Empire and Chinese-BC), the elastic modulus, or just the modulus. This modulus is the straight line slope of the initial portion of the stress-strain curve, normally expressed in terms such as MPa or GPa (10^6 psi or Msi). A

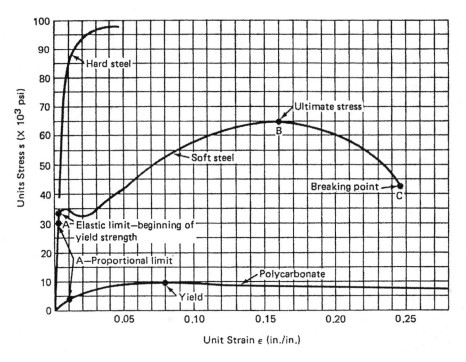

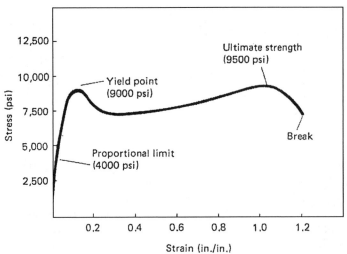

Fig. 2-8   Tensile stress-strain curves.

material not loaded past its proportional limit will return to its original shape once the load is removed. However, some elastic materials do not necessarily obey Hooke's law.

With certain plastics, particularly high performance RPs, there can be two or three moduli. Their stress-strain curve starts with a straight line that results in its highest E, followed by another straight line with a lower E, and so forth. To be conservative providing

a high safety factor the lowest E is used in a design however the highest E is used in certain designs where load requirements are not critical (Chapter 6, **REINFORCED PLASTIC, Basic Design Theory**).

In many plastics, particularly the unreinforced TPs, the straight region of the stress-strain curve is not linear or the straight region of this curve is too difficult to locate. It then becomes necessary to construct a

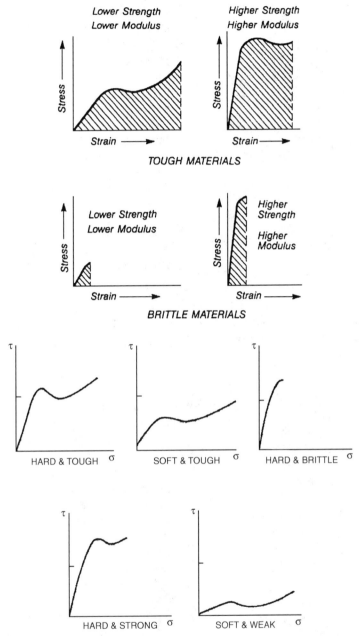

Fig. 2-9 Examples of areas under the tensile stress-strain curves.

straight-line tangent to the initial part of the curve to obtain a modulus called the initial modulus. Designwise, an initial modulus can be misleading, because of the nonlinear elasticity of the material. For this reason, a secant modulus (to be reviewed) is usually used to identify the material more accurately. Thus, a modulus could represent Young's modulus of elasticity, an initial modulus, or a secant modulus, each having its own meaning. The Young's modulus and secant modulus are extensively used in design equations.

Standard ASTM D 638 states that it is correct to apply the term modulus of elasticity to describe the stiffness or rigidity of a plastic where its stress-strain characteristics depend on such factors as the stress or strain rate, the temperature, and its previous history

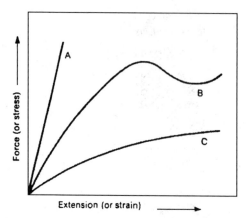

Fig. 2-10 Examples of elongation where usually A = hard to brittle, B = ductile, and C = rubbery.

as a specimen. However, D 638 still suggests that the modulus of elasticity can be a useful measure of the stress-strain relationship, if its arbitrary nature and dependence on load duration, temperature, and other factors are taken into account.

Secant modulus The secant modulus is the ratio of stress to the corresponding strain at any specific point on the stress-strain curve. As shown in Fig. 2-2(a), the secant modulus is the slope of the line joining the origin and a selected point C on the stress-strain curve; this could represent a vertical line at the usual 1% strain. The secant modulus line is plotted from the initial tangent modulus and where it intersects the stress-strain curve. The plotted line location is also based on the angle used in relation to the initial tangent line from the ab-

scissa (horizontal coordinate). Figure 2-2(b) shows curves for two different plastics each at 85% of their respective angles; for design purposes the 85% is usually used. However the 1% strain approach is preferred because it provides the E required and is easier to plot.

The secant modulus measurement is used during the designing of a product in place of a modulus of elasticity for materials where the stress-strain diagram does not demonstrate a linear proportionality of stress to strain or E is difficult to locate.

Hysteresis effect The hysteresis effect is a retardation of the strain when a material is subjected to a force or load. Figure 2-12 are examples of different hysteresis recovery rates.

The top view represents incomplete recovery of strain in a material subjected to a stress during its unloading cycle due to energy consumption. This energy is converted from mechanical to frictional energy (heat). It can represent the difference in a measurement signal for a given process property value when approached first from a zero load and then from a full scale. Middle view is an example of recovery to near zero strain. It shows that material can withstand stress beyond its proportional limit for a short time, resulting in different degrees of the hysteresis effect. Bottom view is a hysteresis loop applicable to dynamic mechanical measurement. The closed curve represents the successive stress-strain status of the material during a cycle deformation.

Poison's ratio It is the proportion of lateral strain to longitudinal strain under conditions of uniform longitudinal stress within the proportional or elastic limit. When the material's deformation is within the elastic range it results in a lateral to longitudinal strain that will always be constant. In mathematical terms, Poisson's ratio is the diameter of the test specimen before and after elongation divided by the length of the specimen before and after elongation. Poisson's ratio will have more than one value if the material is not isotropic

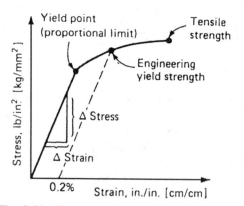

Fig. 2-11 Example of determining yield point and offset strain.

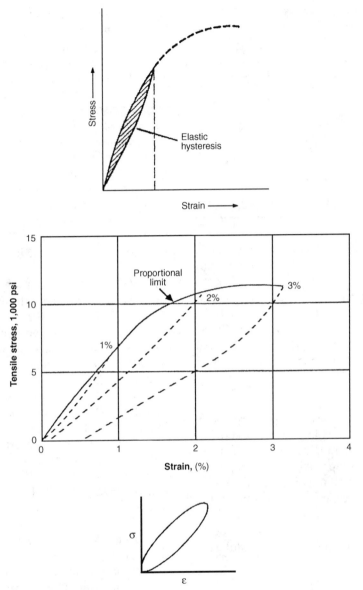

Fig. 2-12 Hysteresis effects on material.

(Chapter 8, **RP Directional Property, *Orientation of Reinforcement***).

Poisson's ratio always falls within the range of 0 to 0.5. A zero value indicates that the specimen would suffer no reduction in diameter or contraction laterally during elongation but would undergo a reduction in density. A value of 0.5 indicates that the specimen's volume would remain constant during elongation or as the diameter decreases. For most plastics the ratio lies between 0.10 and 0.40 (Tables 2-1 and 2-2).

Poisson's ratio is a constant in engineering design analysis for determining the stress and deflection properties of plastic, metal, and other structures such as beams, plates, shells, and rotating discs. When temperature changes, the magnitude of stresses and strains, and the direction of loading all have their effects on Poisson's ratio. However, these factors usually do not alter the typical range of values enough to affect most practical calculations, where this constant is frequently of only secondary importance.

Table 2-1 General range of Poison's ratio for different materials

Material	Range of Poisson's Ratio
Aluminum	0.33
Carbon steel	0.29
Rubber	0.50
Rigid thermoplastics	
Neat	0.20–0.40
Filled or reinforced	0.10–0.40
Structural foam	0.30–0.40
Rigid thermosets	
Neat	0.20–0.40
Filled or reinforced	0.20–0.40

The application of Poisson's ratio is frequently required in the design of structures that are markedly 2-D or 3-D, rather than one-dimensional like a beam. For example, it is needed to calculate the so-called plate constant for flat plates that will be subjected to bending loads in use. The higher Poisson's ratio, the greater the plate constant and the more rigid the plate.

Ductility A typical tensile stress-strain curve of many ductile plastics is shown in Fig. 2-13. As strain increases, stress initially increases approximately proportionately (from point 0 to point A). For this reason, point A is called the proportional limit of the material. From point 0 to point B, the behavior of the material is purely elastic; but beyond point B, the material exhibits an increasing degree of permanent deformation. Consequently, point B is called the elastic limit of the material.

The first point of zero slope on the curve (point C) is identified with material yielding and so its coordinates are called the yield strain and stress (strength) of the material. The yield strain and stress usually decrease as temperature increases or as strain rate decreases. The final point on the curve (point D) corresponds to specimen fracture. This represents the maximum elongation of the material specimen; its coordinates are called the ultimate, or failure strain and stress. Ultimate elongation usually decreases as temperature decreases or as strain rate increases.

Brittleness Brittle materials exhibit tensile stress-strain behavior different from that illustrated in Fig. 2-13. Specimens of such materials fracture without appreciable material yielding. Thus, the tensile stress-strain curves of brittle materials often show relatively little deviation from the initial linearity, relatively low strain at failure, and no point of zero slope. Different materials may exhibit significantly different tensile stress-strain behavior when exposed to different factors such as the same temperature and strain rate or at different temperatures. Tensile stress-strain data obtained per ASTM for several plastics at room temperature are shown in Table 2-3.

Crazing When tensile stress is applied to an amorphous glassy (Chapter 6) plastic such as polystyrene, crazing may occur before fracturing. Crazes are like cracks in that they are wedges shaped and formed perpendicular to the applied stress. However, they differentiate from cracks by containing plastic that is stretched in a highly oriented manner perpendicular to the plane of the craze, which is to say parallel to the applied stress's direction. Another major distinguishing feature is that unlike cracks, crazes are able to bear stress. Under static loading, the strain at which crazes start to form decreases as the applied stress decreases. In constant strain-rate testing crazes always start to form at a well-defined stress level.

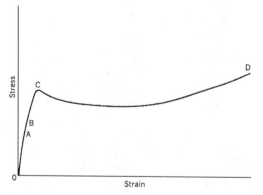

Fig. 2-13 Tensile stress-strain curve typical of many ductile plastics.

Table 2-2   Examples of specific room temperature shear stress-strain data and Poisson's ratio for several plastics and other materials

Generic Material Type	Shear Modulus, MPa	Shear Stress, MPa	At	Poisson's Ratio
ABS	960	51.0	Ultimate	0.35
	810	37.9	Ultimate	0.35
	810	32.9	Ultimate	0.35
	660	30.0	Ultimate	0.36
Acetal copolymer	1000	53	Ultimate	0.35
Acetal	1330	65.5	Ultimate	0.35
homopolymer	1330	68.9	Ultimate	0.35
Acrylic		44.6		
Nylon (DAM)		66.2	Ultimate	0.34–0.43
(0.2% moisture)		57.9	Ultimate	0.34–0.43
		59.3	Ultimate	0.34–0.43
		62.7	Ultimate	0.34–0.43
Nylon (50% RH)				0.35–0.50
(2.5% moisture)				0.35–0.50
		55.8	Ultimate	0.35–0.50
				0.35–0.50
Phenolic		82.7	Ultimate	
Polycarbonate	785	41.3	Yield	0.37
		68.9	Ultimate	
Phenylene ether		62.6	Ultimate	
copolymer		66.2	Ultimate	
Polysulfone	917	41.4	Yield	0.37
		62.1	Ultimate	
Steel, structural ASTM A7-61T	79200	120	Yield	0.27
Brass, naval	38000	280–310	Ultimate	
Aluminum,	30000	240	Yield	0.33
wrought 2014-T6		270	Ultimate	
Pine (southern long-leaf) (with grain)		10		
Oak (white) (with grain)		13.0		

Test rate and property The test rate or cross-head rate is the speed at which the movable cross-member of a testing machine moves in relation to the fixed cross-member. The speed of such tests is typically reported in cm/min. (in./min.). An increase in strain rate typically results in an increase yield point and ultimate strength. Figure 2-14 provides examples of the different test rates and temperatures on basic tensile stress-strain behaviors of plastics where: (a) is at different testing rates per ASTM D 638 for a polycarbonate, (b) is the effects of tensile test-

ing speeds on shapes of stress-strain diagrams, and (c) is a simplified version of the effects on curves of changes in test rates and temperatures.

For most rigid plastics the modulus (the initial tangent to the stress-strain curve) does not change significantly with the strain rate. For softer TPs, such as polyethylenes, the theoretical elastic or initial tangent modulus is usually independent of the strain rate. The significant time-dependent effects associated with such materials, and the practical difficulties of obtaining a true initial tangent

Table 2-3 Examples of room temperature tensile stress-strain data for several plastics and other materials

Generic Material Type	Modulus, MPa	Yield Stress, MPa	Elongation at Yield, %	Elongation at Break, %
ABS	2,600	52	2.5	25–75
	2,200	43	2.5	25–75
	2,200	40	2.5	25–75
	1,800	34	3.3	25–75
Acetal copolymer	2,800	61	12	75
	2,800	61	12	60
Acetal homopolymer	3,100	68.9	12	75
	3,100	68.9	12	40
	3,100	68.9	12	25
Acrylic	2,960	72		5.4
	2,239	48		
	1,720	38	5.0	35
Nylon (DAM) (0.2% moisture)		82.7	5	60
				60
		60.7	7	150
		51.0	20	290
Nylon (50% RH) (2.5% moisture)		58.6	25	>300
				210
		51.0	40	>300
		40.7	30	285
Phenolic	19,310	55.2		0.29
	10,340	58.6		0.57
	7,590	48[h]		0.63
Polycarbonate	2,380	62	6–8	110
	2,240	62	6–8	90
Polyethylene		30		900
		29		900
		24		900
		14		500
Phenylene ether copolymer	2,500	58	4–6	50–100
	2,500	55	4–6	50–100
Polypropylene	1,400	35.5	12	
	1,200	27.3	13	<300
	830	20.0	6.3	
Polystyrene	3,100	52		2.5
	2,070	31		30
	2,070	25		60
	1,930	25		50
Polysulfone	2,482	70.3	5–6	50–100
	2,482	70.3	5–6	50–100
	2,482	68.9	5–6	50–100
Steel, structural ASTM A7-61T	200,000	230		
Brass, naval	100,000	170–340		
Aluminum, wrought 2014-T6	73,000	410		
Pine (southern long-leaf)	13,700			
Oak (white)	11,200			

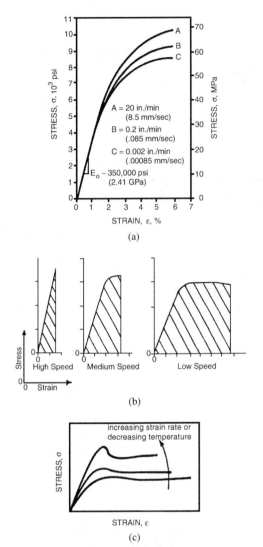

(a)

(b)

(c)

Fig. 2-14 Examples of using different tensile testing rates.

Viscoelasticity It is the plastics respond to stress with elastic strain. In the material, strain increases with longer loading times and higher temperatures.

Flexural Stress-Strain

Like tensile testing, flexural stress-strain testing according to ASTM D 790, determines the load necessary to generate a given level of strain on a specimen, typically using a three-point loading (Fig. 2-15). Testing is performed at a constant rate of crosshead movement, typically 0.05 in./min. for solids and 0. 1 in./min. for foamed samples.

Simple beam equations are used to determine the stresses on specimens at different levels of cross-head displacement. Using traditional beam equations and section properties, the following relationships can be derived where Y is the deflection at the load point (refer to Fig. 2-15):

Bending stress where $\sigma = 3FL/2bh^2$
$$(2\text{-}2)$$

Bending or flexural modulus where
$$E = FL^3/4bh^3Y \qquad (2\text{-}3)$$

Using these relationships, the flexural strength, also called the modulus of rupture,

modulus near the origin of a nonlinear stress-strain curve, render it difficult to resolve the true elastic modulus of the softer TPs in respect to actual data. Thus, the observed effect of increasing strain is to increase the slope of the early portions of the stress-strain curve [Fig. 2-10(c)], which differs from that at the origin. The elastic modulus and strength of both the rigid and the softer plastics each decrease with an increase in temperature. While in many respects the effects of a change in temperature are similar to those resulting from a change in the strain rate, the effects of temperature are relatively much greater.

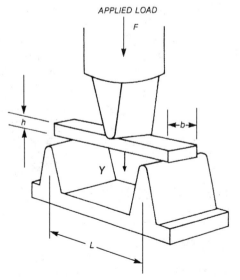

Fig. 2-15 Three-point flexural test schematic.

Table 2-4 Examples of flexural modulus of
elasticity for polypropylene compounds

Unreinforced (neat)	40% Glass Fiber*	40% Talc*
180,000 psi (1,240 MPa)	1,100,000 psi (7,600 MPa)	575,000 psi (3,970 MPa)

*Glass fiber and talc content are by weight.

and the flexural modulus of elasticity can be
determined. Table 2-4 provides examples of
the flexural modulus of elasticity for different formulations of polypropylene. The flexural modulus reported is usually the initial
modulus from the load-deflection curve. The
flexural data can be useful in product designs that involve such factors as bending
loads.

Significantly, a flexural specimen is not in a
state of uniform stress. When a simply supported specimen is loaded, the side of the
material opposite the loading undergoes the
greatest tensile loading. The side of the material being loaded experiences compressive
stress (Fig. 2-16). These stresses decrease linearly toward the center of the sample. Theoretically the center is a plane, called the neutral axis, experiences no stress.

Real differences between the tensile and
the compressive yield stresses of a material
may cause the stress distribution within the
test specimen to become very asymmetric at
high strain levels. This cause the neutral axis
to move from the center of the specimen toward the surface which is in compression.
This effect, along with specimen anisotropy
due to processing, may cause the shape of the
stress-strain curve obtained in flexure to dif-

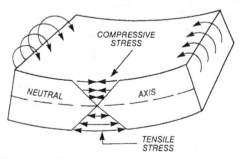

Fig. 2-16 Flexural specimen subjected to compressive and tensile stresses.

fer significantly from that of the tensile stress-
strain curve. Flexural stress-strain data obtained per ASTM for several plastics at room
temperature are shown in Table 2-5.

The stress-strain behavior of plastics in
flexure generally follows from the behavior
observed in tension and compression for either unreinforced or reinforced plastics. The
flexural modulus of elasticity is nominally the
average between the tension and compression moduli. The flexural yield point is generally that which is observed in tension, but
this is not easily discerned, because the strain
gradient in the flexural RP sample essentially
eliminates any abrupt change in the flexural
stress-strain relationship when the extreme
"fibers" start to yield.

The flexural strength for most plastics under standard ASTM bending tests is typically somewhat higher than their ultimate
tensile strength, but flexural strength itself
may be either higher or lower than compressive strength. Since most plastics exhibit
some yielding or nonlinearity in their tensile
stress-strain curve, there is a shift from triangular stress distribution toward rectangular distribution when the product is subject
to flexure (Fig. 2-17). This behavior is similar to that assumed for plastic design in steel
and for ultimate design strength in concrete.
Thus, the modulus of rupture reflects in part
nonlinearities in stress distribution caused by
plastification or viscoelastic nonlinearities in
the cross-section. Shifts in the neutral axis
resulting from differences in the yield strain
and post-yield behavior in tension and compression can also affect the correlation between the modulus of rupture and the uniaxial strength results.

Even plastics with fairly linear stress-strain
curves to failure, for example short-fiber reinforced TSs (RPs), usually display moduli of
rupture values that are higher than the tensile strength obtained in uniaxial tests; wood
behaves much the same. Qualitatively, this
can be explained from statistically considering flaws and fractures and the fracture energy available in flexural samples under a
constant rate of deflection as compared to
tensile samples under the same load conditions. These differences become less as the

Table 2-5 Room temperature flexural and compressive stress-strain data for several plastics and other materials

Generic Material Type	Flexural Modulus, MPa	Flexural Yield Stress, MPa	Compressive Modulus, MPa	Compressive Stress, MPa	At
ABS	2800	90	2600		
	2300	74	2200		
	2300	69	2200	42.4/45.1	10% yield
	1900	59	1800		
Acetal copolymer	2590	89.6		31/110	1%/10%
	2590	89.6		31/110	1%/10%
Acetal	2620	98.6	4600	35.9/124	1%/10%
homopolymer	2830	97.2	4600	35.9/124	1%/10%
	2960	96.5	4600	34.5/121	1%/10%
Acrylic	3170	110		117	Maximum
	2239	72		72	Maximum
	1720	62		41	Yield
Nylon (DAM)	2827			33.8	1%
(0.2% moisture)	1689			13.1	1%
	2034			16.6	1%
	1034				
Nylon (50% RH)	1207				
(2.5% moisture)	862				
	1241				
	745				
Phenolic		96.5		193.1	Ultimate
		89.6		206.9	Ultimate
		82^h		193.1	Ultimate
Polycarbonate	2340	93.0	2380	86.1	Yield
	2240	90.9	2240	86.1	Yield
Polyethylene	1100				
	1100				
	861				
	410				
Phenylene ether	2500	86	2500		
copolymer	2500	94	2500		
Polypropylene	1750	54.4			
	1295	48.2			
	1065	34.5			
Polystyrene		86.2			
Polysulfone	2689	106.2	2579	96/276	Yield/break
Steel, structural ASTM A7-61T				230	Yield
Aluminum, wrought 2014-T6				430	Yield
Pine (southern long-leaf) (with grain)		101		58.2	Ultimate
Oak (white) (with grain)		95.8		48.5	Ultimate

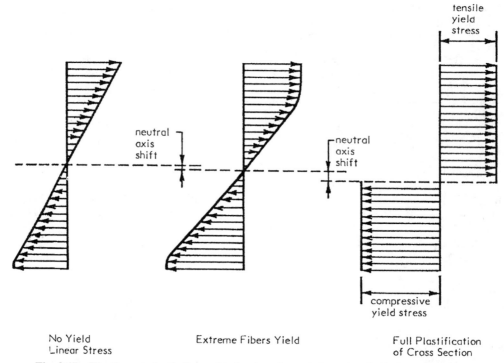

Fig. 2-17 Elastic and plastic flexural behavior of unreinforced and reinforced plastics.

thickness of the bending specimen increases, as would be expected by examining statistical considerations.

Another method of flexural testing that can be used is, for example, the cantilever beam method (Fig. 2-18), which is used to relate different beam designs. It provides an exam-ple of the effect of the modulus of elastic-ity on elastic deflection for different materi-als, using cantilever test specimens. All the test beams have the same lengths and cross-sections. It is used in creep and fatigue testing and for conducting testing in different envi-ronments.

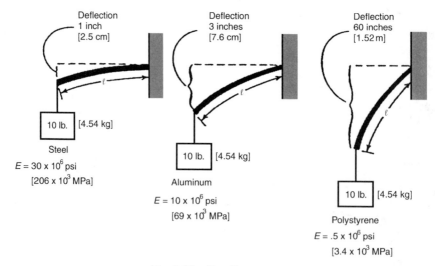

Fig. 2-18 Cantilever tests.

Compressive Stress-Strain

Data are generated by placing a test specimen between the two flat, parallel faces of a testing machine and then moving these faces together at a specified rate (ASTM D 695). A displacement transducer may be used to measure the compression of the specimen, while a load cell measures the compressive force exerted by the specimen on the testing machine. Stress and strain are computed from the measured compression load, and these are plotted as a compressive stress-versus-strain curve for the material at the temperature and strain rate employed for the test.

In general, the compressive strength of a non-reinforced plastic or a mat-based RP laminate is usually greater than its tensile strength. The compressive strength of a unidirectional fiber-reinforced plastic is usually slightly lower than its tensile strength. Room-temperature compressive stress-strain data obtained per ASTM for several plastics are shown in Table 2-5.

The majority of tests to evaluate the characteristics of plastics are performed in tension or flexure; hence, the compressive stress-strain behavior of many plastics is not well described. Generally, the behavior in compression is different from that in tension, but the stress-strain response in compression is usually close enough to that of tension so that possible differences can be neglected (Fig. 2-19). The compression modulus is not always reported, since defining a stress at a strain is equivalent to reporting a tensile secant modulus. However, if a compression modulus is reported, it will generally be an initial modulus.

As reviewed a general rule is that the compressive strength of plastics is greater than its tensile strength. However, this is not generally true for reinforced TSs (RTSs). Different results occur with different plastics. As an example the compression testing of foamed plastics provides the designer with the useful recovery rate. A compression test result (Fig. 2-20) for rigid foamed insulating polyurethane (3.9 lb/ft^3) resulted in almost one-half of its total strain recovered in one week.

Many of the procedures in compression stress-strain testing are the same as in tensile testing, but in compression testing particular care must be taken to specify the specimen's dimensions. If a sample is too long and narrow, for instance, buckling may cause premature failure. To avoid this, designers should test a specimen with a square cross-section and a longitudinal dimension twice as long as a side of the cross-section.

At higher stress levels, compressive strain is usually less than tensile strain. Unlike tensile loading, which usually results in failure, stressing in compression produces a slow, indefinite yielding that seldom leads to failure. Where a compressive failure does occur catastrophically, the designer should determine the material's strength in the same way as with tensile testing by dividing the maximum load the sample supported by its initial cross-sectional area in kPa (psi). When the material does not exhibit a distinct maximum load prior to failure, the designer should report the strength at a given level of strain (often 10%).

Since the ends of compression specimens usually tend to "flower" and not remain rigid, test results are usually very scattered requiring close examination as to what the results mean in reference to the behavior of the test specimens. Different clamping devices are used to eliminate the flowering action that could provide inaccurate readings that in turn influence results by usually making them stronger.

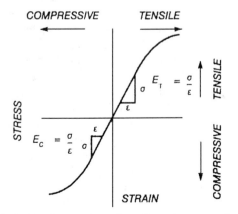

Fig. 2-19 Comparison of tensile and compression stress-strain behavior of TPs.

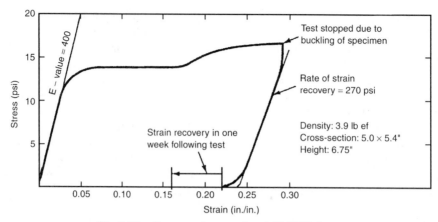

Fig. 2-20   Compression test for rigid PUR foam.

Shear Stress-Strain

The shear mode involves the application of a load to a material specimen in such a way that cubic volume elements of the material comprising the specimen become distorted, their volume remaining constant, but with opposite faces sliding sideways with respect to each other. Shear deformation occurs in structural elements subjected to torsional loads and in short beams subjected to transverse loads.

Shear stress-strain data can be generated by twisting (applying torque) a material specimen at a specified rate while measuring the angle of twist between the ends of the specimen and the torque load exerted by the specimen on the testing machine. Maximum shear stress at the surface of the specimen can be computed from the measured torque that is the maximum shear strain from the measured angle of twist.

The shear modulus of a material can be determined by a static torsion test or by a dynamic test employing a torsional pendulum or an oscillatory rheometer. The maximum short-term shear stress (strength) of a material can also be determined from a punch shear test.

Unlike the methods for tensile, flexural, or compressive testing, the typical procedure used for determining shear properties is intended only to determine the shear strength. It is not the shear modulus of a material that will be subjected to the usual type of

direct loading (ASTM D 732). Torsion pendulum and oscillatory rheometer techniques are used to determine the shear modulus. The shear strength values are obtained by such simple tests using single or double shear actions. In these tests the specimen to be tested is sheared between the hardened edges of the supporting block and the block to which the load is applied. The shearing strength is calculated as the load at separation divided by the total cross-sectional area being sheared.

The use of the word direct in these tests might seem to imply that this is the only stress being placed on the specimen. However, an inspection of the test fixtures in these test devices indicates that bending stresses do in fact exist and the stress cannot be considered as being purely that of shear. Therefore, the shearing stress calculated must be regarded as an average stress. This type of calculation is justified in analyzing bolts, rivets, and any other mechanical member whose bending moments are considered negligible.

Because strain measurements are difficult if not impossible to measure, few values of yield strength can be determined by testing. It is interesting to note that tests of bolts and rivets have shown that their strength in double shear can at times be as much as 20% below that for single shear. The values for the shear yield point (kPa or psi) are generally not available; however, the values that are listed are usually obtained by the torsional testing of round test specimens.

The data obtained using the test method above should be reported as direct shear strength. Designers are nevertheless cautioned to use the shear strength reported by this method only in similar direct-shear situations, because this is not a pure shear test. This test cannot be used to develop shear stress-strain curves or determine a shear modulus, because a portion of the load is transferred by bending or compression rather than pure shear. Also, the test results can depend on the susceptibility of the material to the sharpness of load faces. When analyzing plastics in a pure shear situation or when the maximum shear stress is being calculated in a complex stress environment, a shear strength equal to half the tensile strength or that given above is generally used, whichever is less.

Basically, shearing stresses are tangential stresses that act parallel to the planes they stress. For example, the shearing force in a beam provides shearing stresses on both the vertical and horizontal planes within the beam. The two vertical stresses must be equal in magnitude and opposite in direction to ensure vertical equilibrium. However, under the action of those two stresses alone the element would rotate. Clearly, this pair of stresses must be negated by another couple. If the small element is taken as a differential one, the magnitude of the horizontal stresses must have the value of the two vertical stresses. This principle is sometimes phrased as "cross-shears are equal." In other words, a shearing stress cannot exist on an element without a like stress being located 90 degrees around the corner.

The block diagram in Fig. 2-21 is subjected to a set of equal and opposite shearing forces (Q). The top view (a) represents a material with equal and opposite shearing forces and (b) is a schematic of infinitesimally thin layers subject to shear stress. If the material is imagined as an infinite number of infinitesimally thin layers, as shown at the bottom, then there is a tendency for one layer of the material to slide over another to produce a shear form of deformation or failure if the force is great enough. The shear stress will always be tangential to the area upon which it acts. The

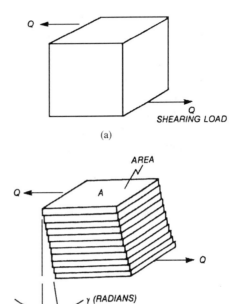

Fig. 2-21 Basic analysis of shear stress.

shearing strain is the angle of deformation γ as measured in radians. For materials that behave according to Hooke's Law, shear strain is proportional to the shear stress producing it.

The constant G, called the shear modulus, the modulus of rigidity, or the torsion modulus, is directly comparable to the modulus of elasticity used in direct-stress applications. Only two material constants are required to characterize a material if one assumes the material to be linearly elastic, homogeneous, and isotropic. However, three material constants exist: the tensile modulus of elasticity (E), Poisson's ratio (v), and the shear modulus (G). An equation relating these three constants, based on engineering's elasticity principles, follows:

$$E/G = 2(1 + v) \qquad (2\text{-}4)$$

This calculation, which holds true for most metals, is generally applicable to TPs. However, the designer is to be familiar with the inherently nonlinear, anisotropic nature of most plastics, particularly the fiber-reinforced and liquid crystal plastics (Chapter 6).

It is important to note material such as those plastics or wood that are weak in either tension or compression will also be basically weak in shear. For example, concrete is weak in shear because of its lack of strength in tension. Reinforced bars in the concrete are incorporated to prevent diagonal tension cracking and strengthen concrete beams. Similar action occurs with RPs using fiber filament structures.

Although no one has ever been able to determine accurately the resistance of concrete to pure shearing stress, the matter is not very important, because pure shearing stress is probably never encountered in concrete structures. Furthermore, according to engineering mechanics, if pure shear is produced in one member, a principal tensile stress of an equal magnitude will be produced on another plane. Because the tensile strength of concrete is less than its shearing strength, the concrete will fail in tension before reaching its shearing strength. This action also occurs with plastics.

Torsion property As noted, the shear modulus is usually obtained by using pendulum and oscillatory rheometer techniques. The torsional pendulum (ASTM D 2236: Dynamic Mechanical Properties of Plastics by Means of a Torsional Pendulum Test Procedure) is a popular test, since it is applicable to virtually all plastics and uses a simple specimen readily fabricated by all commercial processes or easily cut from fabricated products.

The moduli of elasticity, G for shear and E for tension, are ratios of stress to strain as measured within the proportional limits of the material. Thus the modulus is really a measure of the rigidity for shear of a material or its stiffness in tension and compression. For shear or torsion, the modulus analogous to that for tension is called the shear modulus or the modulus of rigidity, or sometimes the transverse modulus.

Applying Stress-Strain Data

The information presented is used in different load bearing equations such as those reviewed in the literature (3, 6, 10, 14, 20, 29, 31, 36, 37, 39, 43 to 125). As an example stress-strain data may guide the designer in the initial selection of a material. Such data also permit a designer to specify design stresses or strains either safely within the proportional/elastic limit of the material. On the other hand, if a vessel is being designed to fail at a specified internal pressure, the designer may choose to use the tensile yield stress of the material in the design calculations.

Designers of most structures specify material stresses and strains well within the proportional/elastic limit. Where required (with no or limited experience on a particular type product materialwise and/or processwise) this practice builds in a margin of safety to accommodate the effects of improper material processing conditions and/or unforeseen loads and environmental factors. This practice also allows the designer to use design equations based on the assumptions of small deformation and purely elastic material behavior. Other properties derived from stress-strain data that are used include modulus of elasticity and tensile strength.

Modulus of elasticity (E) is one of the two factors that determine the stiffness or rigidity (EI) of structures comprised of a material. The other is the moment of inertia (I) of the appropriate cross section, a purely geometric property of the structure. Figure 3-1 provides examples of moment of inertia (I). In identical products, the higher the modulus of elasticity of the material, the greater the rigidity; doubling the modulus of elasticity doubles the rigidity of the product. The greater the rigidity of a structure, the more force must be applied to produce a given deformation.

It is appropriate to use Young's modulus to determine the short-term rigidity of structures subjected to elongation, bending, or compression. It may be more appropriate to use the flexural modulus to determine the short-term rigidity of structures subjected to bending, particularly if the material comprising the structure is non-homogeneous, as foamed or fiber-reinforced materials tend to be. Also, if a reliable compressive modulus of elasticity is available, it can be used to determine short-term compressive rigidity, particularly if the material comprising a structure is fiber-reinforced. The room temperature moduli of elasticity for several plastics and some other materials are presented in Table 2-3.

Long-Term Load Behavior

Long time dynamic load involves behaviors such as creep, fatigue, and impact. Two of the most important types of long-term material behavior are more specifically viscoelastic creep and stress relaxation. Whereas stress-strain behavior usually occurs in less than one or two hours, creep and stress relaxation may continue over the entire life of the structure such as 100,000 hours or more.

Viscoelastic Creep

When a viscoelastic material is subjected to a constant stress, it undergoes a time-dependent increase in strain. This behavior is called creep. The viscoelastic creep behavior typical of many TPs is illustrated in Figs. 2-22 and 2-23. At time t_0 the material is suddenly subjected to a constant stress that is main-

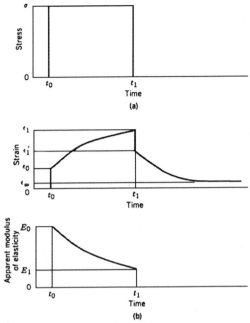

Fig. 2-23 Viscoelastic creep behavior typical of many TPs under short-term stress: (a) input stress vs. time profile and (b) output strain vs. time profile.

tained for a long period of time as shown in Fig. 2-22. The material responds by undergoing an immediate initial strain that increases to time t_r, when it fails. In Fig. 2-23 the constant stress is maintained for a shorter time. The material undergoes an immediate initial strain at t_0 which increases to t_1, at time t_1. When the stress is removed, the material immediately decreases in strain from ε_1 to ε_1^1 followed by a gradual decrease from ε_1, to a permanent residual strain.

Although the creep behavior of a material could be measured in any mode, such experiments are most often run in tension or flexure. In the first, a test specimen is subjected to a constant tensile load and its elongation is measured as a function of time. After a sufficiently long period of time, the specimen will fracture that is a phenomenon called tensile creep failure. In general, the higher the applied tensile stress, the shorter the time and the greater the total strain to specimen failure. Furthermore, as the stress level decreases, the fracture mode changes from ductile to brittle. With flexural, a test specimen

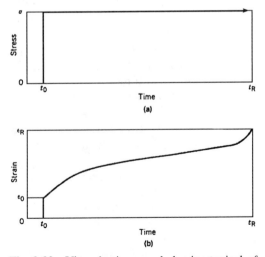

Fig. 2-22 Viscoelastic creep behavior typical of many TPs under long-term stress to rupture: (a) input stress vs. time profile and (b) output strain vs. time profile.

is subjected to a constant bending load and its deflection is measured as a function of time. Tests are conducted at different constant loads.

Viscoelastic creep data are usually presented in one of two ways. In the first, the total strain experienced by the material under the applied stress is plotted as a function of time. Families of such curves may be presented at each temperature of interest, each curve representing the creep behavior of the material at a different level of applied stress. Below a critical stress, viscoelastic materials may exhibit linear viscoelasticity; that is, the total strain at a given time is proportional to the applied stress. Above this critical stress, the creep rate becomes disproportionately faster. In the second, the apparent creep modulus is plotted as a function of time.

The viscoelastic creep modulus may be determined at a given temperature by dividing the constant applied stress by the total strain prevailing at a particular time. Since the creep strain increases with time, the viscoelastic creep modulus must decrease with time (Fig. 2-23). Below its critical stress for linear viscoelasticity, the viscoelastic creep modulus versus time curve for a material is independent of the applied stress. In other words, the family of strain versus time curves for a material at a given temperature and several levels of applied stress may be collapsed to a single viscoelastic creep-modulus-time-curve if the highest applied stress is less than the critical value.

Different viscoelastic materials may have considerably different creep behavior at the same temperature. A given viscoelastic material may have considerably different creep behavior at different temperatures. Viscoelastic creep data are necessary and extremely important in designing products that must bear long-term loads. It is inappropriate to use an instantaneous (short load) modulus of elasticity to design such structures because they do not reflect the effects of creep. Viscoelastic creep modulus, on the other hand, allows one to estimate the total material strain that will result from a given applied stress acting for a given time at the anticipated use temperature of the structure.

The viscoelastic creep modulus is particularly useful to the designer because it may be substituted for Young's modulus to predict the long-term rigidity of load bearing structures. Thus, creep data allows to design a structure so that the stress within the material comprising it will remain at or below the desired level. This testing procedure has been followed for almost a century in designing all kinds of products.

Stress Relaxation

When a viscoelastic material is subjected to a constant strain, the stress initially induced within it decays in a time-dependent manner. This behavior is called stress relaxation. The viscoelastic stress relaxation behavior is typical of many TPs. The material specimen is a system to which a strain-versus-time profile is applied as input and from which a stress-versus-time profile is obtained as an output. Initially the material is subjected to a constant strain that is maintained for a long period of time. An immediate initial stress gradually approaches zero as time passes. The material responds with an immediate initial stress that decreases with time. When the applied strain is removed, the material responds with an immediate decrease in stress that may result in a change from tensile to compressive stress. The residual stress then gradually approaches zero.

The stress-relaxation behavior of a material is normally determined in either the tensile or the flexural mode. In these experiments, a material specimen is rapidly elongated or compressed to produce a specified strain level and the load exerted by the specimen on the test apparatus is measured as a function of time. Specimens of certain plastics may fail during tensile or flexural stress-relaxation experiments.

Viscoelastic stress-relaxation data are usually presented in one of two ways. In the first, the stress manifested as a function of time. Families of such curves may be presented at each temperature of interest. Each curve representing the stress-relaxation behavior of the material at a different level of applied strain. Below a critical strain, viscoelastic materials may exhibit linear viscoelasticity; that is, the stress at a given time is proportional to the applied strain. Above this critical strain, the stress relaxation rate becomes disproportionately faster. In the second, the apparent stress relaxation modulus is plotted as a function of time. Apparent or viscoelastic stress relaxation modulus is a time- and temperature-dependent parameter that reflects the stress relaxation behavior of the material.

Although all viscoelastic materials undergo stress relaxation, the rate at a give temperature may differ considerably from material to material. A given viscoelastic material may have considerably different stress relaxation behavior at different temperatures. Such data are necessary in designing structures that will be subjected to long-term deformation, including gaskets, springs, and force-fit components. The viscoelastic stress relaxation modulus allows one to estimate the material stress that will result from a given applied strain after a given time at the anticipated use temperature of the structure. It is particularly useful to the designer because it may be substituted for Young's modulus (E) in the appropriate elastic design equations to predict the long-term resiliency of such structures.

Stress-relaxation data enables the design of a structure so that the strain of the material comprising it will remain at the desired level. Too high a strain level and the material may cease to be linearly viscoelastic, which may lead to a significantly higher rate of stress relaxation. Too low a level and the material stress may not be high enough to generate the required spring force.

Long-term Viscoelastic Behavior

The rate of creep and stress relaxation of TPs increases considerably with temperature; those of the TSs (thermoset plastics) remain relatively unaffected up to fairly high temperatures. The rate of viscoelastic creep and stress relaxation at a given temperature may also vary significantly from one TP to an-

other because of differences in the chemical structure and shape of the constituent plastic molecules (Chapter 6). These differences affect the way the plastic molecules interact with each other, and hence they're relative freedom of movement. For the sake of practicality, viscoelastic creep and stress relaxation experiments are normally terminated at 1000 hours. Time-temperature super-positioning is often used to extrapolate this 1,000 hours of data to approximately 100,000 hours ($\approx$12 years).

Creep Property

Basics Creep data can be very useful to the designer. In the interest of sound design-procedure, the necessary long-term creep information should be obtained on the perspective specific plastic, under the conditions of product usage (Chapter 5, **MECHANICAL PROPERTY, Long-Term Stress Relaxation/Creep**). In addition to the creep data, a stress-strain diagram under similar conditions should be obtained. The combined information will provide the basis for calculating the predictability of the plastic performance.

The factors that affect being able to design with creep data include a number of considerations.

1. The strain readings of a creep test can be more accessible to a designer if they are presented as a creep modulus. In a viscoelastic material, namely plastic, the strain continues to increase with time while the stress level remains constant. Since the creep modulus equals stress divided by strain, we thus have the appearance of a changing modulus.

2. The creep modulus, also known as the apparent modulus or viscous modulus when graphed on log-log paper, is normally a straight line and lends itself to extrapolation for longer periods of time.

3. Creep data applications are generally limited to the identical plastic, temperature, stress level, atmospheric conditions, and type of test. Data of a relatively short duration of 1000 h can be extrapolated to long

term needs. Reinforced thermoplastics and particularly reinforced thermosets display a much higher resistance to creep than do the unreinforced plastic.

An introduction is provided to analyzing creep data and determining guidelines on how they should be used. The viscoelastic nature of plastic reacts to a constant load over a long period of time by an ever-increasing strain. Since the modulus formula states that modulus = stress/strain and the stress is constant, while the strain is increasing, we see an ever decreasing modulus. This type of modulus is called an apparent modulus, and the data for it are collected from test observations for the purpose of predicting long-term behavior of plastics subjected to a constant stress at selected temperatures.

Creep modeling A stress-strain diagram is a significant source of data for a material. In metals, for example, most of the needed data for mechanical property considerations are obtained from a stress-strain diagram. In plastic, however, the viscoelasticity causes an initial deformation at a specific load and temperature and is followed by a continuous increase in strain under identical test conditions until the product is either dimensionally out of tolerance or fails in rupture as a result of excessive deformation. This type of an occurrence can be explained with the aid of the Maxwell model shown in Fig. 2-24.

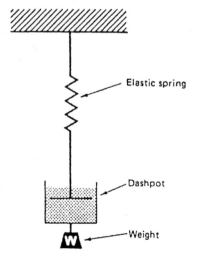

Fig. 2-24 Maxwell model used to explain viscoelastic behavior.

The Maxwell model is also called Maxwell fluid model. Briefly it is a mechanical model for simple linear viscoelastic behavior that consists of a spring of Young's modulus (E) in series with a dashpot of coefficient of viscosity (η). It is an isostress model (with stress δ), the strain (ε) being the sum of the individual strains in the spring and dashpot. This leads to a differential representation of linear viscoelasticity as $d\varepsilon/dt = (1/E)\, d\delta/dt + (\delta/\eta)$. This model is useful for the representation of stress relaxation and creep with Newtonian flow analysis.

When a load is applied to the system, shown diagrammatically, the spring will deform to a certain degree. The dashpot will first remain stationary under the applied load, but if the same load continues to be applied, the viscous fluid in the dashpot will slowly leak past the piston, causing the dashpot to move. Its movement corresponds to the strain or deformation of the plastic material.

When the stress is removed, the dashpot will not return to its original position, as the spring will return to its original position. Thus we can visualize a viscoelastic material as having dual actions: one of an elastic material, like the spring, and the other like the viscous liquid in the dashpot. The properties of the elastic phase are independent of time, but the properties of the viscous phase are very much a function of time, temperature, and stress. This phenomenon is further explained by looking at the dashpot again, where we can visualize that a thinner fluid resulting from increased temperature under a higher pressure (stress) will have a higher rate of leakage around the piston during the time that the conditions described prevail. Translated into plastic creep, this means that at higher use temperature and higher stress levels the strain will be higher, resulting in greater creep.

Visualizing the reaction to a load (without time) by such a dual-component interpretation is valuable to our understanding of the creep process but is basically meaningless for design purposes. For this reason the designer is interested in the actual deformation or product failure over a specific time span. Observations of the amount of strain at certain time intervals must be made, which will then make it possible to construct curves that can be extrapolated to longer time periods. The usual initial readings at 1, 2, 3, 5, 7, 10, and 20 hours, are followed by readings every 24 hours up to 500 hours, then readings every 48 hours up to 1,000 hours. The time segment for the creep test is common to all materials. Strains are recorded until the specimen ruptures or is no longer useful because it has yielded. In either case, a point of failure of the test specimen has been reached.

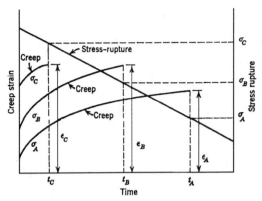

Fig. 2-26 Typical creep and stress-rupture curves.

Product performance data Products subjected to a given load develop a corresponding predictable deformation. If it continues to increase without any increase in load or stress, the material is said to be experiencing creep or cold flow. Creep in any product is defined as increasing strain over time in the presence of a constant stress (Figs. 2-25 and 26). The rate of creep for any given plastic, steel, wood, etc. material depends on the basic applied stress, time, and temperature.

Creep-test specimens may be loaded in tension or flexure (to a lesser degree in compression) in a constant temperature environment. With the load kept constant, deflection or strain is recorded at regular intervals of hours, days, weeks, months, or years. Generally, results are obtained at three or more stress levels.

Stress-strain-time data are usually presented as creep curves of strain versus log time. Sets of such curves, seen in Fig. 2-27, can be produced by smoothing and interpolating data on a computer. These data may also be presented in other ways, to facilitate the selection of information to meet specific design requirements. Sections may be taken

through creep curves at constant times to yield isochronous stress versus strain curves or at a constant strain, giving isometric stress versus log-time curves as in Fig. 2-27. Standard ASTM D 2990 provides details for different creep tests.

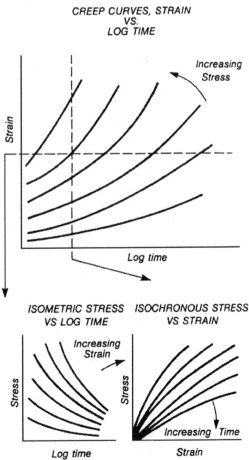

Fig. 2-27 Typical creep data vs. log time.

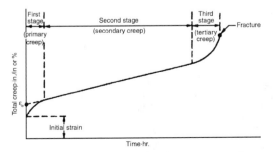

Fig. 2-25 Concept for evaluating creep-test data.

If a designer is faced with decisions concerning creep, the most reliable source of information is a test program run under simulated or actual conditions on the product itself or at least on test specimens. The expected operating life of most products designed to withstand creep is usually ten to twenty years. It is apparent that actual long-time testing is not likely to be undertaken, so available creep test-data must be used. The so-called long-time tests are undertaken for at least 1,000 hours, the recommended time specified in the ASTM standard based on extensive data accumulated since at least 1943.

The tests are performed under carefully controlled stress (load), temperature, time, and creep (elongation) conditions. To save time, tests for different constant loads are performed simultaneously on different specimens of the same material. Creep tests may be rather extensively conducted, as for example when developing creep data prior to the design and fabrication of the first all-plastic airplane (41). The usual procedure is to plot the creep versus time curve, but other combinations are possible.

The theoretically shaped curve in Fig. 2-25 provides the three typical stages for evaluation. An initial strain takes place almost immediately, consisting of the elastic strain plus a plastic strain, if the deformation extends beyond the yield point. This initial action in the first stage shows a decreasing rate of elongation because of strain hardening. The action most relevant to the designer concerns the second stage, which begins at a minimum strain rate and remains rather constant, because of the balancing effects of strain hardening and annealing. In the third stage a rapid increase in the creep rate is accompanied by severe necking (that is, thickness reduction) and ultimately rupture.

Designers are concerned with the second stage in the sense that their target is not to have the product being designed enter into the third stage. Thus, after plotting the creep versus time data of the 1,000 h test, the second stage can be extrapolated out to the number of hours of desired product life. This process is then followed for each of the creep curves. In making this extrapolation it is assumed that the 1,000-hour test has allowed the material to enter into the second stage. The material will have behaved similarly to that shown in the curve in Fig. 2-25.

Creep rupture. Creep-rupture data are obtained in the same way as creep data except that higher stresses are used and the time is measured to failure (Figs. 2-28 and 29). The strains are sometimes recorded, but this is not necessary for creep rupture. The results are generally plotted as the log stress versus log time to failure (110). In creep-rupture tests it is the material's behavior just prior to the rupture that is of primary interest. In these tests a number of samples are subjected to different levels of constant stress, with the time to failure being determined for each stress level. General technical literature and product data sheets seldom provide a complete description of a material's behavior prior to rupture. It should include the development of any crazing and stress whitening, its strain-time

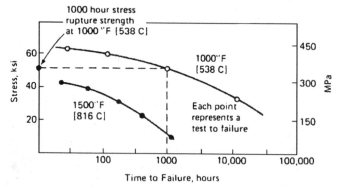

Fig. 2-28 Typical stress-rupture data vs. temperature.

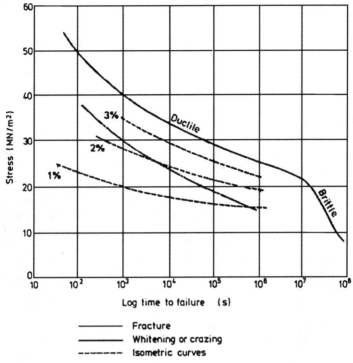

Fig. 2-29 Typical creep-rupture ductile-to-brittle behavior of TPs.

behavior, the nature of the fracture process, and describe yielding, necking, and brittleness.

The Fig. 2-30 shows the curves of a family of TPs describe as failure that is fairly typical of the behavior of certain TPs. The time-dependent strains resulting from several levels of sustained or creep stress are shown, together with the development of crazing and of stress whitening. The features that develop in the failure process follow a particular pattern:

Overall behavior. It is the time-dependent strain at which crazing, stress whitening, and rupture decreases with a

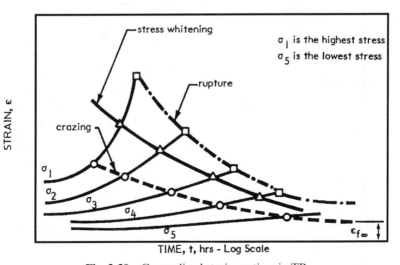

Fig. 2-30 Generalized strain vs. time in TP.

decreasing level of sustained stress. The time to development of these defects increases with a decreasing stress level.

Crazing. This develops in such amorphous plastics as acrylics, PVCs, PS, and PCs as creep deformation enters the rupture phase. Crazes start sooner under high stress levels. Crazing occurs in crystalline plastics, but in those its onset is not readily visible. It also occurs in most fiber-reinforced plastics, at the time-dependent knee in the stress-strain curve.

Stress whitening. This occurs in many types of plastics, including the amorphous ones like PVC and ABS, and in the crystalline types such as PE and PP. A stress-whitening zone may be a sign of crazing in some plastics where individual fine crazes may be difficult to detect. Stress whitening occurs fairly late in the rupture stage, just prior to yielding.

Rupture. Rupture strain decreases steadily with increases in the duration of stress. Alternately, the magnitude of stress needed to cause rupture decreases as the duration of stress increases. Figure 2-31 shows the development first of damage and then of yielding in a PVC compound as a function of its being under sustained stress. The decay at the onset of the first damage and of yield

strength with the increasing duration of sustained stress is also evident. In other words, a decrease in the magnitude of the sustained stress lengthens the time over which crazing, stress whitening, and yielding develop. Yielding is frequently taken as the failure criterion for plastics. However, some common types of standardized creep-rupture tests do not determine yielding, only the sustained stress and time to failure. One example of this is the ASTM D 1598 procedure for determining the time-dependent burst strength of plastic pressure pipes. Plotting the failure or bursting stress against the time to failure for a given material defines its strength regression relationship.

Comparisons have been made of the strength-regression characteristics of plastics with those of wood. The capacity of wood to resist sustained stress has been determined to decay at a rate of 8% for each decade of time change, that is, its capacity at the end of each decade is 92% of what it was at the start of the decade. The decay rates calculated from published strength-regression information on pressure-rated plastic pipe compounds are shown in Table 2-6. The decay rate for the specific plastics tested varies from 7 to 32% per decade, depending on the generic type of plastic and the specific compound within that type. The time-dependent strength behavior of some of these plastics is similar to that of wood.

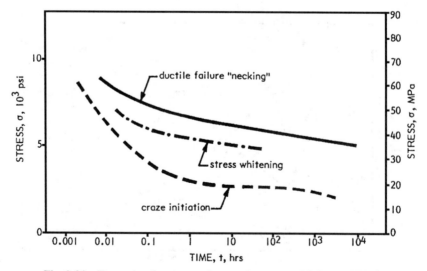

Fig. 2-31   Example of stress vs. time to damage and failure of PVC.

Table 2-6 Rate of strength decay for wood and TPs

Material	Range of Decay Rate,* % per Decade of Time
Wood	8
PVC	7–19
PE	8–13
ABS	12–32

* Change in strength under sustained stress from beginning to end of decade, or unit change in log time.

Apparent creep modulus. The concept of an apparent modulus is a convenient method for expressing creep, because it takes into account the initial strain for an applied stress plus the amount of deformation or strain that occurs over time. Thus, the apparent modulus E_a is calculated in a very simplified approach as:

$$E_a = \text{Stress/Initial strain} + \text{Creep}\quad(2\text{-}5)$$

Because products tend to deform in time at a decreasing rate, the acceptable strain based on the desired service life of the product is determined. The shorter the duration of load, the higher the apparent modulus and thus the higher the allowable stress. The apparent modulus is most easily explained with an example. As long as the stress level is below the elastic limit of the material, its modulus of elasticity E can be obtained from the usual equation:

$$E = \text{Stress/Strain}\quad(2\text{-}6)$$

For example, a compressive stress of 10,000 psi (69 MPa) gives a strain of 0.015 in./in. (0.038 cm/cm) for FEP plastic at 63°F (17°C). Then:

$$E = 10{,}000/0.015 = 667{,}000\,\text{psi}\,(4{,}600\,\text{MPa})$$
$$(2\text{-}7)$$

If the same stress level prevails for 200 hours, the total strain will be the sum of the initial strain plus the strain due to time. This total strain can be obtained from a creep-data curve. If, for example, the total deformation under a tension load for 200 hours is 0.02 in.

per in., then:

$$E = 10{,}000/0.015$$
$$= 5{,}000{,}000\,\text{psi}\,(3{,}500\,\text{MPa})\quad(2\text{-}8)$$

Similarly, E_a can then be determined for one year. Extrapolating from the creep-data curve, which is in fact a straight line, gives a deformation of 0.025 in. per in. Thus,

$$E = 10{,}000/0.025$$
$$= 400{,}000\,\text{psi}\,(2{,}800\,\text{MPa})\quad(2\text{-}9)$$

When plotted against time, these calculated values for the apparent modulus provide a simplified means of predicting creep at various stress levels (Fig. 2-32). For all practical purposes, curves of deformation versus time eventually tend to level off. Beyond a certain point, creep is small and may safely be neglected for many applications.

There have been continuing attempts made to create meaningful formulas for the apparent modulus change with respect to time. However the factors in the formulas that would fit all conditions are more cumbersome to use than presenting test data in a graph form and using it as the means for predicting the strain (elongation) at some distant point in time. The test data when plotted on log-log paper usually form a straight line and tend themselves to easy extrapolation.

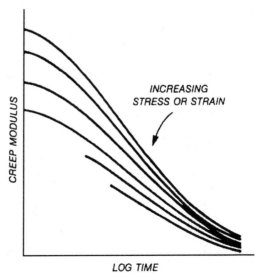

Fig. 2-32 Example of plotting the apparent creep modulus vs. log time.

The slope of the straight line indicating a decreasing modulus depends on the nature of a material (principally its rigidity and temperature of heat deflection), on the temperature of the environment in which the product is used, and on the amount of stress in relation to tensile strength.

Extensive amount of these type data has been plotted but unfortunately most of it is privately owned. Creep data available from material suppliers, college and government projects, etc. can provide guidelines. However where the product has to meet critical requirements that usually include safety of people and data from previous work does not exist, creep test have to be conducted and properly applied by the designer.

Stress relaxation. In a stress-relaxation test a plastic is deformed by a fixed amount and the stress required to maintain this deformation is measured over a period of time (Fig. 2-33) where: (a) recovery after creep, (b) strain increment caused by a stress step function, and (c) strain with stress applied (1) continuously and (2) intermittently. The maximum stress occurs as soon as the deformation takes place and decreases gradually with time from this value. From a practical standpoint, creep measurements are generally considered more important than stress-relaxation tests and are also easier to conduct.

Those interested in the theory of viscoelasticity and in the relationship of materials' properties to their molecular structure tend to concentrate more on stress-relaxation than

creep measurements. One reason may be that stress-relaxation figures are generally more easily interpreted in terms of viscoelastic theory than are creep data. Stress-relaxation data also provide practical information such as determining the stress needed to hold a metal insert in a plastic product, evaluating the additives needed such as antioxidants, choosing cantilever-type beams, and so on.

The stress-relaxation behavior of plastics is extremely temperature dependent, especially in the region of the plastics' glass transition temperature (T_g) (Chapter 7). Many unreinforced amorphous types of plastics at temperatures well below the T_g have a tensile modulus of elasticity of about 3×10^{10} dynes/cm^2 [300 Pa (0.04 psi)] at the beginning of a stress-relaxation test. The modulus decreases gradually with time, but it may take years for the stress to decrease to a value near zero. Not only is the stress-relaxation behavior of an amorphous plastic most sensitive to temperature in its transition region, but at a given temperature in that region the stress changes rapidly with time.

With crystalline plastics, the main effect of the crystallinity is to broaden the distribution of the relaxation times and extend the relaxation stress too much longer periods. This pattern holds true at both the higher and low extremes of crystallinity (Chapter 6). With some plastics, their degree of crystallinity can change during the course of a stress-relaxation test. This behavior tends to make the Boltzmann superposition principle difficult to apply.

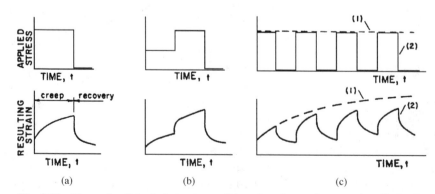

Fig. 2-33 Example of strain behavior under various intermittent and cyclic loads.

Many designs incorporate the phenomenon of stress-relaxation. For example, in many products, when plastics are assembled they are placed into a permanently deflected condition, as for instance press fits, bolted assemblies, and some plastic springs. In time, with the strain kept constant the stress level will decrease, from the same internal molecular movement that produces creep. This gradual decay in stress at a constant strain (stress-relaxation) becomes important in applications such as preloaded bolts and springs where there is concern for retaining the load. The amount of relaxation can be measured by applying a fixed strain to a sample and then measuring the load with time.

The resulting data can then be presented as a series of curves much like the isometric stress curves in Fig. 2-27. A relaxation modulus similar to the creep modulus can also be derived from the relaxation data. It has been shown that using the creep modulus calculated from creep curves can approximate the decrease in load from stress relaxation.

Plastic products with excessive fixed strains imposed on them for extended periods of time could fail. One example might be the eventual splitting of a plastic tube press fitted over a steel shaft. Unfortunately, there is no relaxation-rupture corollary to creep rupture. For developing initial design concepts, a strain limit of 20% of the strain at the yield point or of the yield strength is suggested for high-elongation plastics. Likewise, using 20% of the elongation at the break is suggested for low-elongation brittle materials without a yield point, as only a guideline for initial design. Prototype products should then be thoroughly tested at end-use conditions to confirm the design, or the available data on specific material of interest can provide more exacting limits.

Intermittent loading. The creep behavior of plastics that has been considered so far has assumed that the level of the applied stress will be constant. However, in service the material may be subjected to a complex pattern of loading and unloading cycles (Fig. 2-33). This variability can cause design

problems in that it would clearly not be feasible to obtain experimental data to cover all possible loading situations, yet to design on the basis of constant loading at maximum stress would not make efficient use of materials or be economical. In such cases it is useful to have methods for predicting the extent of the accumulated strain that will be recovered during the rest periods after a number of cycles of load changes (100).

Recovery is the strain response that occurs upon the removal of a stress or strain. The mechanics of the recovery process are illustrated in Fig. 2-34, using an idealized viscoelastic model. The extent of recovery is a function of the load's duration and time after load or strain release. In the example of recovery behavior shown in Fig. 2-34 for a polycarbonate at 23°C (73°F), samples were held under sustained stress for 1,000 hours, and then the stress was removed for the same amount of time. The creep and recovery strain measured for the duration of the test provided several significant points.

First the sample, that was loaded to about 20% of its short-term yield strength or 13.8 MPa (2,000 psi), recovered almost completely one hour after the release of the load, the net strain being 0.03%. Second, the sample loaded to 66% of its short-term yield strength, or 41.4 MPa (6,000 psi), retained a strain of 0.8% at 1,000 hours after the release of the load. The initial strain was 2.8%, the strain from the 1,000 hour creep an additional 1.7%. Thus only about one-half the creep strain was recovered. Visually extrapolating the recovery curve reveals that even after a year (10^4 hr.), about one-third of the creep strain (0.6%) will remain.

The first damage developed during creep or relaxation also affects recovery behavior. If limiting the magnitude and duration of the stress prevents the first damage, recovery will eventually be substantially complete for all practical purposes.

Conversely, at strains above the first damage limit recovery will be incomplete and permanent deformation should be expected and accounted for in the evaluation. This is true not only for plastics in general but also of reinforced plastics. When RPs are stressed

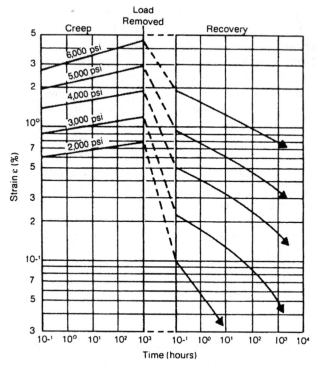

Fig. 2-34 Tensile creep and recovery during the intermittent loading.

beyond the knee in their stress-strain curve, recovery becomes incomplete and hysteresis is clearly evident.

Material and processing. As covered particularly in Chapters 6 and 7, the various material types and compositions as well as their processing methods, influence their properties including creep. When properly processed, in general crystalline materials have lower creep rates than the amorphous type, and RPs as a whole have significantly improved creep resistance.

Some examples of creep data presented in different formats are given in Figs. 2-35

to 2-40 and Table 2-7. The data show all kinds of creep behavior, including the effects of time and temperature on amorphous materials that basically have a curve spread over a much wider time scale than that of the crystalline types. If the temperature is well below the T_g, only the first part of the curve will be observed, for it might require years or even centuries to observe the complete curve. In the transition region nearly the complete curve can be observed in a period from a few seconds to a few hours. If

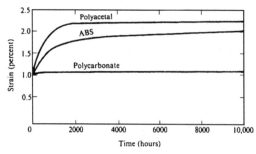

Fig. 2-35 Tensile creep curves for TPs.

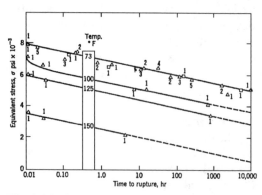

Fig. 2-36 Stress-rupture data for rigid 2 in. diameter PVC pipe as a function of temperature.

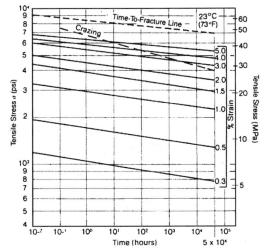

Fig. 2-37 Tensile stress-strain-time correlation resulting from creep for PC.

the temperature is well above the T_g, only the upswing in the curve (that is, an increase in the creep) will be observed, unless measurements can be made in a fraction of a second.

Not only do the creep properties of crystalline polymers change rapidly with temperature, but in some cases at a given temperature a crystalline type will creep more with time than will the rigid amorphous or cross-linked (TS) types. However, a crystalline type above its T_g creeps very little, compared to the others. Thus, crystalline types tend to

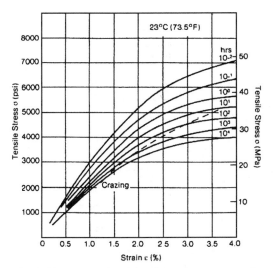

Fig. 2-38 Example of isochronous stress-strain curves for PCs resulting from stress relaxation.

have an even broader distribution of retardation times than do the amorphous types (the term crystalline refers to plastics that are actually semicrystalline).

At small loads the compliance of most materials at a given time is independent of stress. For example, doubling the load doubles the deformation. At higher loads, especially those approaching that which is required to break the plastic, compliance at any given time increases with the load. This effect is generally most pronounced with the crystalline types, the tough polyblends, and the amorphous types in the transition region or above it. However, the rigid types like polystyrene and the highly cross-linked phenol-formaldehyde plastics also show creep elongation, which increases at a rate greater than the first power of the stress at high loads. As a result, doubling the stress more than doubles the amount of elongation.

The load or stress has another effect on the creep behavior of most plastics. The volume of isotropic or amorphous plastic increases as it is stretched unless it has a Poisson ratio of 0.50. At least part of this increase in volume manifests itself as an increase in free volume and a simultaneous decrease in viscosity. This decrease in turn shifts the retardation times to being shorter.

A creep test can be carried out with an imposed stress, then after a time have its stress suddenly changed to a new value and have the test continued. This type of change in loading allows the creep curve to be predicted. The simple law referred to earlier as the Boltzmann superposition principle, hold for most materials, so that their creep curves can thus be predicted.

The first assumption involved in using the Boltzmann superposition principle is that elongation is proportional to stress, that is, compliance is independent of stress. The second assumption is that the elongation created by a given load is independent of the elongation caused by any previous load. Therefore, deformation resulting from a complex loading history is obtained as the sum of the deformations that can be attributed to each separate load.

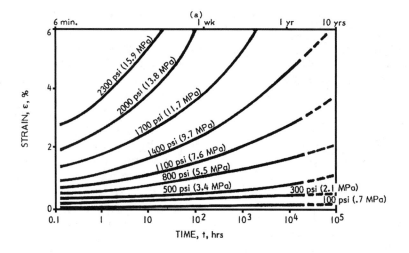

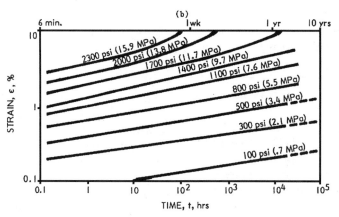

Fig. 2-39　Tensile-creep behavior of PP; top on semilog scale and bottom on log-log scale.

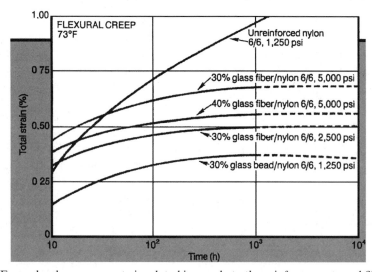

Fig. 2-40　Example where creep rate is related inversely to the reinforcements and filler content.

Table 2-7   Flexural creep data of reinforced plastics at 23°C

Base Resin	Filler Type and Content (%)	Stress (psi)*	Strain (%) Hours			Apparent Modulus (10³ psi)* Hours		
			10	100	1,000	10	100	1,000
Nylon 6/6	Glass fiber 15,	2,500	0.555	0.623	0.709	450	401	353
	mineral 25	5,000	0.823	0.967	1.140	607	517	439
Polyester	Glass fiber 15,	2,500	0.452	0.470	0.482	553	532	519
(PBT)	mineral 25	5,000	0.693	0.742	0.819	721	674	610
Nylon 6/10	Ferrite 83	2,500	0.463	0.507	0.568	540	493	440
		5,000	0.638	0.732	0.952	784	683	525
Polypropylene	Carbon powder	2,500	1.100	1.140	1.970	114	87	63
		5,000	6.230	6.920	8.660	40	36	29
Nylon 6/6	Glass fiber 15,	2,500	2.160	2.400	2.510	116	104	100
	carbon powder	5,000						
Nylon 6	Glass beads 30	1,250	0.140	0.320	0.368	893	391	340
		5,000	0.290	0.650	0.750	862	385	333

* To convert psi to pascals (Pa), multiply by 6.895×10^3.

Designing with creep data. The factors that affect being able to design with creep data include a number of considerations. First, the strain readings of a creep test can be more accessible to a designer if they are presented as a creep modulus. In a viscoelastic material the strain continues to increase with time while the stress level remains constant. Since the creep modulus equals stress divided by strain, we thus have the appearance of a changing modulus.

Second, the creep modulus, also known as the apparent modulus or viscous modulus when graphed on log-log paper, is normally a straight line and lends itself to extrapolation for longer periods of time. The apparent modulus should be differentiated from the modulus given in the data sheets, which is an instantaneous or static value derived from the testing machine, per ASTM D 638.

Third, creep data application is generally limited to the identical material, temperature use, stress level, atmospheric conditions, and type of test (that is tensile, flexural, or compressive) with a tolerance of ±10%. Only rarely do product requirement conditions coincide with those of a test or, for that matter, are creep data available for all the grades of materials that may be selected by a designer. In such cases a creep test of relatively short duration, say 1,000 hours, can be instigated, and the information be extrapolated to long-term needs. In evaluating plastics it should be noted that reinforced thermoplastics and thermosets display a much higher resistance to creep than do unreinforced plastics.

Finally, there have been numerous attempts to develop formulas that could be used to predict creep information under varying usage conditions. In practically all cases the suggestions have been made that the calculated data be verified by actual test performance. Furthermore, numerous factors have been introduced to apply such data to reliable predictions of product behavior.

Creep data can be very useful to the designer. The data in Fig. 2-41 have been plotted from material available from or published by material manufacturers. It shows the apparent modulus versus time at 23°C (73°F) for (A) Merlon polycarbonate at 13.8 MPa (2,000 psi); (B) an extrapolation of (A) beyond 10^7 hrs.; (C) Noryl 731 modified PPO at 13.8 MPa (2,000 psi); (D) Delrin 500 acetal at 6.9 MPa (1,000 psi); and (E) Zytel 109 nylon at 50% relative humidity and 6.9 MPa (1,000 psi). The broken lines represent extrapolated values, the circles are actual test-reading points. The log-log graph sheets are 9 in. × 15 in. and contain 3 × 5 cycles. The end of the first time cycle of 10^3 represents 1,000 hours.

The first point is the 100 h time interval. The data for shorter intervals do not as a rule

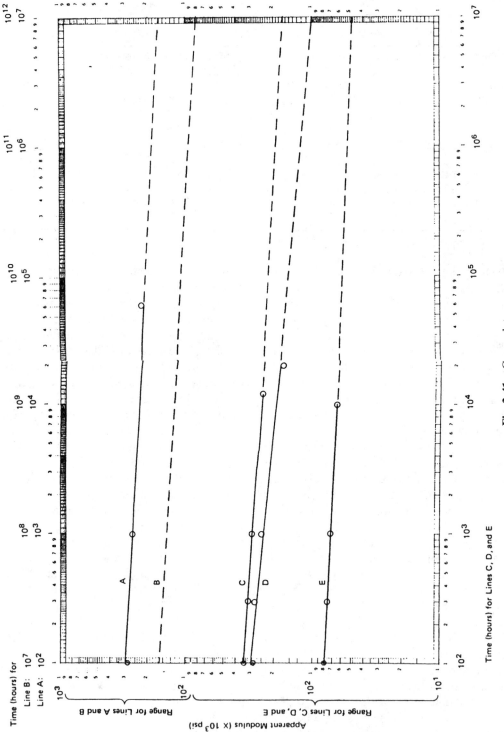

Fig. 2-41 Creep data.

fit the straight-line configuration that exists on log-log charts for the long-term duration beyond the first 100 h test period. The circled points are the 100 h, 300 h, and 1,000 h test periods, and other observed values, and a straight line is fitted either through the circles or tangent to them to give the line a slope for long-term evaluation.

From this line can be estimated the time at which the strain will be such as to cause tolerance problems in product performance. Or by using the elongation at yield as the point at which the material has attained the limit of its useful life, we can estimate the time at which this limit will be reached.

The equation "modulus (apparent) equal stress/strain" enables us to locate the modulus that corresponds to the test stress and strain (the strain being obtained by using the dimensional change or elongation limit) where it intersects the straight line leading to an appropriate time value. The polycarbonate creep line shows that a limit of 0.010 in elongation is reached at the end of 10^5 hours (apparent modulus = 200,000 psi) and an elongation (yield) of 0.06 is arrived at after 10^6 hours, or indefinitely if the 0.010 limitation does not exist.

As reviewed in the interest of sound design-procedure, the necessary creep information should be procured on the prospective material, under the conditions of product usage. In addition to the creep data, a stress-strain diagram, also at the conditions of product usage, should be obtained. The combined information will provide the basis for calculating the predictability of material performance in the designed product.

Allowable working stress. The viscoelastic nature of the material requires not merely the use of data sheet information for calculation purposes but also the actual long-term performance experience gained from it which can be used as a guide. The allowable working stress is important for determining dimensions of the stressed area and for predicting the amount of distortion and strength deterioration that will take place over the life span of the product. The allowable working

stress, for a constantly loaded product that is expected to perform satisfactorily over many years has to be established using creep characteristics for a material with enough data to make reliable long-term predictions of short-term test results.

Creep test data when plotted on log-log paper usually form a straight line and tend themselves to extrapolation. The slope of the straight line, which indicates a decreasing modulus, depends on the nature of the material (principally its rigidity and temperature of heat deflection), the temperature of the environment in which the product is used, and the amount of stress in relation to tensile strength.

Certain conclusions can now be developed, based on creep-data test results: First, for practical design purposes, the data accumulated for up to 100 hours of creep are of no real benefit. There is usually too much variation during this test period, which is of a relatively short duration.

Next, the apparent modulus values, starting with a test period of 100 hours and continuing up to 1,000 hours, form a straight line when plotted on log-log paper. This line may be continued for longer periods on the same slope for interpolation purposes. This action can be taken (based on a guide) provided the stress level is one quarter to one fifth that of the ultimate strength and the test temperature is no greater than two thirds of the difference between room temperature and the heat deflection temperature at 264 psi (1,800 kPa). When these limitations are exceeded, there is a potential sharp decrease in the apparent modulus after 1,000 hours, with indications that failure from creep is approaching (that is, the material has attained the limit of its usefulness).

Since the designer will be expected to plot curves to suit requirements, some examples will be cited that can serve as a guide for potential needs (Fig. 2-42).

This example for ABS uses creep data for 1,000 psi stress at 23°C. When the line is extended to 10^5 h, the apparent modulus is 140,000 psi. If the product is designed for the duration of 10^5 h and calculations are made for product dimensions, the modulus

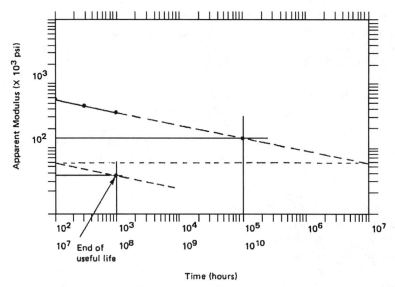

Fig. 2-42   Creep data for ABS.

of 140,000 psi should be inserted into any formula in which the modulus appears as a factor. At 10^5 h the total strain is:

$$E = \text{Stress/Strain} \qquad (2\text{-}10)$$
$$140{,}000 = 1{,}000/\text{Strain} \qquad (2\text{-}11)$$
$$\text{Strain} = 1{,}000/140{,}000 = 0.007 \text{ or } 7\%$$
$$(2\text{-}12)$$

Based on this calculation, if the product can tolerate this type of strain without affecting performance, then the dimensional requirements are met.

The elongation at yield for this particular ABS is 0.0275, which could be considered the end of the useful strength of the material. The apparent modulus corresponding to this strain at 1,000 psi and 23°C is:

$$E = 1{,}000/0.0275 = 36{,}364 \text{ psi} \qquad (2\text{-}13)$$

In the lower part of the graph in Fig. 2-42, draw at the point of 56×10^3 on the left side a line parallel to the original creep line and find that it intersects the apparent modulus line at a time of $10^9 \times 0.5$ h. The product would fail at this time, owing to its loss of strength, even if dimensional changes permitted satisfactory functioning of the product.

Some charts show creep test data beyond the 1,000 h duration, and in fact under most conditions the straight line between the 100

and 1,000 h points is continued into the 10,000 and 20,000 h range. Even in such charts a deviation from the straight line occurs occasionally, which should not be considered unreasonable, because of all the variables that enter into the test data.

Selecting an allowable continuous working stress at the required temperature must be a procedure that allows for making an estimation of the elongation at the end of the product's life. For example, if a product will be stressed to 1,700 psi at a temperature of 66°C (150°F), and data are available for 2,000 psi stress at 71°C (160°F), this information plotted on log-log paper should allow to extrapolate the long-term behavior of the material.

Isometric and isochronous graph   Creep curves are a common method of displaying the interdependence of stress-strain-time. However, there are other methods that may also be useful in particular applications, specifically isometric and isochronous graphs. An isometric graph is obtained by taking a constant strain section through the creep curves and replotting this as stress versus time (Figs. 2-33 and 2-34). It is an indication of the relaxation of stress in the plastic when strain is kept constant. These data are often used as a good approximation of stress relaxation in a plastic. In addition, if the vertical (stress) axis is divided by the strain, one obtains a graph of the modulus against time (Figs. 2-28 and 2-29). These graphs provide a good illustration of the time-dependent variation of the modulus.

An isochronous graph may be obtained by taking a constant time section through the creep curves and then plotting stress versus strain as shown in Fig. 2-38. It can also be obtained experimentally by performing a series of brief creep and recovery tests on a plastic. In this procedure a stress is applied to a plastic test piece and the strain is recorded after a specified time, typically 100 seconds. The stress is then removed and the plastic allowed to recover, normally for a period of 4 (4 × 100 see.). A larger stress is then applied to the same specimen, after recording the strain at the 100 s. time period; then this stress is removed and the material allowed to recover. This procedure is repeated until enough points have been obtained to let an isochronous graph to be plotted.

Isochronous data are usually presented in log-log scales. One reason for doing so is that on linear scales any slight, but possibly important, nonlinearity between stress and strain may go unnoticed. Whereas the use of log-log scales will usually produce a relatively straight-line graph the slope of which gives an indication of the linearity of the material. If the material is perfectly linear, the slope will be at 45 degrees, but if it is nonlinear the slope will be less than 45 degrees.

Isochronous graphs are particularly valuable when obtained experimentally, because they are less time consuming and require less specimen preparation than creep curves. Such graphs at several time intervals can also be used to build up creep curves and indicate areas where the main experimental creep program could be most profitable. They are also popular as means of evaluating deformational behavior, because their method of data presentation is similar to the conventional tensile test data.

Deformation or fracture. The failure of a plastic product in the performance of its normal long-time function is usually caused by one of two factors: excessive deformation or fracture. For plastics it is more often than not found that excessive creep deformation is the limiting factor. However, if fracture occurs, it can have more catastrophic results. Therefore, it is essential that designers recognize the factors that are likely to initiate fracture, so that steps can be taken to avoid them.

Fractures are usually classified as either brittle or ductile. Although any type of fracture is serious. Brittle fractures are potentially more dangerous, because there is no observable deformation of the material prior to or during breakage. When the failure is ductile, however, large nonrecoverable deformations become evident, which serve as a warning that all is not well. Plastic fractures are ductile or brittle depending on such variables as their polymer structure, additives, processing conditions, strain rate, and temperature with its stress system. The principal external causes of fracture are a prolonged steady stress (creep rupture), the continuous application of a cyclically varying stress (fatigue), and the application of a stress (creep rupture). In all these cases the fracture processes can be accelerated if the plastic is in an aggressive environment.

Creep guideline. Here is a summation of the factors to consider when reviewing creep properties:

1. Predictions can be made on creep behavior based on creep and relaxation data.

2. There is generally a less-pronounced curvature when creep and relaxation data are plotted log-log. This facilitates extrapolation and is commonly practiced, particularly with creep modulus and creep-rupture data.

3. Increasing the load on a product increases its creep rate.

4. Increasing the level of reinforcement in a composite increases its resistance to creep.

5. Particulate fillers provide better creep resistance than unfilled plastics but are less effective than fibrous reinforcements.

6. Glass-fiber-reinforced amorphous TP RPs generally have greater creep resistance than glass-fiber-reinforced crystalline

TP RPs containing the same amount of glass fiber.

7. Carbon-fiber reinforcement is more effective in resisting creep than glass-fiber reinforcement.

8. The effect of a flame-retardant additive on the flexural modulus provides an indication of its effect on long-time creep.

9. Over the past century, many plastic products have been successfully designed for long-time creep performance based on the information and test data then available, but much more exists now and will in the future.

Fatigue Property

Introduction   Fatigue is the phenomenon of having materials under cyclic loads at levels of stress below their static yield strength. Fatigue data are used so the designer can predict the performance of a material under cyclic loads. The fatigue test, analogous to static long-term creep tests, provides information on the failure of materials under repeated stresses. This fatigue behavior is by no means a new problem. The term was applied to the failure of a wooden mast by hoisting too many sails too often in the pre-Christian era.

As plastics replaced metals and other materials in many critical structural applications, fatigue tests became important. Examples of products subject to fatigue when they are stressed repeatedly or in some defined cyclic manner are a snap-action plastic latch that is constantly opened and closed, a reciprocating mechanical part on a machine, a gear tooth, a bearing, and any structural component subjected to vibration, such as an aircraft wing or any product that will be subjected to repeated impacts. Such cyclic loading can cause mechanical deterioration and progressive fracturing of the material, leading to its ultimate failure.

Under a repeated applied cyclic load, fatigue cracks begin somewhere in the product and extend during the cycling. Eventually the crack will expand to such an extent that the remaining material can no longer support the stress, at which point the product will fail suddenly. However, failure for different service conditions may be defined differently than just as the separation of two parts. ASTM D 671 defines failure as occurring also when the elastic modulus has decreased to 70% of its original value.

The failure effect is generally a loss of toughness, lowered impact strength, and lowered tensile elongation. Failure includes the melting of any part of a product, excessive change of dimensions or the warping of the product, and the crazing, cracking, or formation of internal voids or deformation markings. These types of defects all may seriously affect performance strength.

Plastics are susceptible to brittle crack-growth fractures as a result of cyclic stresses in much the same way as metals. In addition, because of their high damping and low thermal conductivity, plastics are prone to thermal softening if the cyclic stress or cyclic rate is high. Examples of the TPs with the best fatigue resistance include PP and ethylene-propylene copolymers.

Fatigue data are normally presented as a plot of the stress (S) versus the number of cycles (N) that cause failure at that stress; the data plotted defined as an S-N curve (Fig. 2-43). The use of an S-N curve is used to establish a fatigue endurance limit strength. The curve asymptotically approaches a parallel to the abscissa, thus indicating the endurance limit as the value that will produce failure. Below this limit the material is less susceptible to fatigue failure.

Examples of fatigue curves for unreinforced (top) and reinforced (bottom) plastics are shown in Fig. 2-44. The values for stress amplitude and the number of load cycles to failure are plotted on a diagram with logarithmically divided abscissa and English or metrically divided ordinates.

The fatigue behavior of a material is normally measured in a flexural but also in a tensile mode. Specimens may be deliberately cracked or notched prior to testing, to localize fatigue damage and permit measuring the crack-propagation rate. In constant-deflection amplitude testing a specimen is

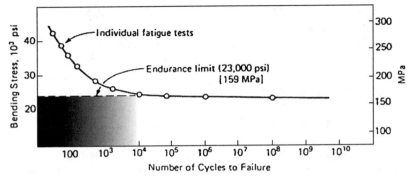

Fig. 2-43 S-N curve establishes fatigue endurance limit strength.

repeatedly bent to a specific outer surface strain level. The number of cycles to failure is then recorded. In constant flexural load amplitude testing a bending load is repeatedly applied to the specimen. This load causes a specified outer-surface stress level. The number of cycles to failure is then recorded. Both modes of flexural fatigue testing can be related to the performance of real structures, one to those that are flexed repeatedly to a constant deflection and the other to those

that are repeatedly flexed with a constant load.

Since fatigue cracks often start at a random surface imperfection, considerable scatter occurs in fatigue data, increasing with the increasing lifetime wherever crack initiation occupies most of the fatigue life of a specimen. When a line of the best fit is drawn from the available data points on an S-N curve, this represents the mean life expected at any given stress level or the stress that would cause, say, 50% of the product failures in a given number of cycles.

If sufficient data are available, much more information can be provided when different curves for various percentages of failure are plotted. Where such data are available, reasonable design criteria would be based on some probability for failure, depending on how critical the effects of failure occur. If a large, expensive repair of a complex mechanism would result from the fatigue failure of one product, then a 10 or even 1% probability of failure would be a more likely design criterion than the 50% suggested above.

The fatigue strength of most TPs is about 20 to 30% of the ultimate tensile strength determined in the short-term test but higher for RPs. It decreases with increases in temperature and stress-cycle frequency and with the presence of stress concentration peaks, as in notched components.

ASTM Special Technical Publication No. 91 discusses in detail the important ramifications to be considered in the various statistical aspects of fatigue testing. Most often, the fatigue curves as well as the tabulated values

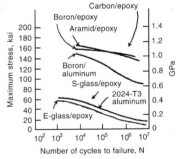

Fig. 2-44 Examples of room temperature fatigue curves.

of endurance strengths and endurance limits are based on the 50% probability curve. As a result, designers do not resort to using scatterband curves unless they are involved with a design that takes a statistical approach. The designer requiring information on the highest order of reliability should always contact the plastic manufacturer a nd/or run tests.

Testing mode Basically material fatigue failure is the result of damage caused by repeated loading or deformation of a structure. The magnitudes of the stresses and strains induced by this repeated loading or deformation are typically so low that they would not be expected to cause failure if they were applied only once.

Constant deflection amplitude fatigue testing is probably the less demanding of the two techniques, because any decay in the modulus of elasticity of the material due to hysteretic heating would lead to lower material stress at the fixed maximum specimen deflection. In the constant load amplitude tests, maximum material stress is fixed, regardless of any decay in the modulus of elasticity of the material.

The test frequency used during fatigue evaluation of plastics is typically 30 Hz, and test temperature is typically 23°C (73°F). The behavior of viscoelastic materials is very temperature and strain rate dependent. Consequently, both test frequency and test temperature has a significant effect upon the observed fatigue behavior. Material fatigue data are normally presented in one of two ways: constant stress amplitude or constant strain amplitude plotted versus the number of cycles (N) to specimen failure to produce a fatigue endurance S-N curve for the material. The fatigue testing of TPs is normally terminated at ten million (10^7) cycles.

Two conclusions can be drawn from an inspection of the S-N curve: (1) the higher the applied material stress or strain, the fewer cycles the specimen can survive; and (2) the curve gradually approaches a stress or strain level called the fatigue endurance limit below which the material is much less susceptible to fatigue failure. Different materials may

show different fatigue behavior at the same test temperature and test frequency.

The standard way to deal with fatigue effects is by a statistical approach. A curve of stress to failure vs. the number of cycles to this stress level to cause failure is made by testing a large number of representative samples of the material under cyclical stress, each one at a progressively lowered stress level. This S-N curve is used in designing for fatigue failure by determining the allowable stress level for a number of stress cycles anticipated for the product. In the case of materials such as metals, this approach is relatively not complicated. Unfortunately, in the case of plastics the loading rate, the repetition rate, and the temperature all have a substantial effect on the S-N curve, and it is important that the appropriate data be used.

Endurance limit To develop S-N curves the fatigue specimen is loaded until, for example, the maximum stress in the sample is 275 MPa (40 ksi) (Fig. 2-43). At this stress level it may fail in only 10 cycles. These data are recorded and the stress level is then reduced to 206 MPa (30 ksi). This specimen may not break until after 1,000 stress cycles at this rather low stress level.

This procedure is repeated until a stress level is determined below which failure does not occur. In this example of a relatively high fatigue performance material develops a flat portion of the S-N curve at a stress level of 159 MPa (23 ksi). Test duration of 10^7 stress cycles is usually considered infinite life. This type of testing is expensive, principally because it involves a large number of samples and much statistical evaluation. The end result, determining the fatigue endurance limit of a material, is an extremely important design property. This property should be used in determining the allowable stresses in products, rather than just the short-term yield strength, any time a product will see cyclic loading in service.

Cyclic loading significantly reduces the amount of allowable stress a material can withstand. If data are not available on the endurance limit of a material being considered for use, a percentage of its tensile strength can be used. This percentage varies with the different material systems. For engineering plastics the endurance limit could be about 50% of its tensile strength, as with metals. Taking this 50% approach requires the designer to become familiar with fatigue testing results on plastics and other materials, so that the proper evaluation can be applied. However, to design correctly, requires obtaining reliable S-N curves with the required endurance limit, as in Fig. 2-43. Plastics are subject to fatigue, with a wide range of performance, and efforts should be made to arrive at endurance limit information if a fail-safe design is desired.

Heat generation Since plastics are viscoelastic, there is the potential for having a large amount of internal friction generated within the plastics during mechanical deformation, as in fatigue. This action involves the accumulation of hysteretic energy generated during each loading cycle. Examples of products that behave in this manner include coil or leaf springs and shoe soles.

Because this energy is dissipated mainly in the form of heat, the material experiences an associated temperature increase. When heating takes place the dynamic modulus decreases, which results in a greater degree of heat generation under conditions of constant stress. The greater the loss modulus of the material, the greater the amount of heat generated that can be dissipated. Plastics for fatigue applications can therefore have low losses.

If the plastic's (TP's) surface area is insufficient to permit the heat to be dissipated, the specimen will become hot enough to soften and melt. The possibility of adversely affecting its mechanical properties by heat generation during cyclic loading must therefore always be considered. The heat generated during cyclic loading can be calculated from the loss modulus or loss tangent of the plastics.

The rate dependence of fatigue strength demands careful consideration of the potential for heat buildup in both the fatigue test and in service. Generally, since the buildup is a function of the viscous component of the material, the materials that tend toward viscous behavior will also display sensitivity to cyclic load frequency. Thus TPS, particularly the crystalline polymers like polyethylene that are above their glass-transition temperatures, are expected to be more sensitive to the cyclic load rate, and highly cross-linked plastics or glass-reinforced TS plastics are less sensitive to the frequency of load.

From this review it should be obvious that care must be taken in the use of the type of accelerated fatigue testing that is common for metals. For example, a frequency of 30 Hz is not uncommon for metal tests. However, significant change in the fatigue life of PMMA occurs as measured by excessive thermal softening at frequencies well below 30 Hz. Depending on the type of plastic, testing at frequencies of a few Hz or less is required, to avoid such softening. In contrast, if the component is to be subjected to high-frequency loads in service, the test should be performed at similar frequencies. High-frequency loadings may show no significant heat buildup, provided stresses are small, particularly when the product is to be cooled.

Fatigue results in a shift from ductile to brittle failure with the increased number of load cycles. Strength-regression behavior obtained under sustained stress with the regression under a 0.5 Hz cyclic stress can be applied in a square waveform. The curves for an equal duration of tensile stress, is represented by either the time under sustained load or the cumulative time under stress during the square-wave loading of the fatigue test. Compared to the static loading, the fatigue loading results in both a pronounced shift from ductile to brittle fracturing and a marked decrease in the time to failure at a given stress.

Fracture mechanic The fracture mechanics theory developed for metals is also adaptable for use with plastics. The basic concepts remain the same, but since metals and plastics are different they require different techniques to describe their fatigue-failure behaviors. Some of the comments made about crack and fracture influences on fatigue performance relate to the theory of fracture mechanics. The fracture mechanics theory method, along with readily

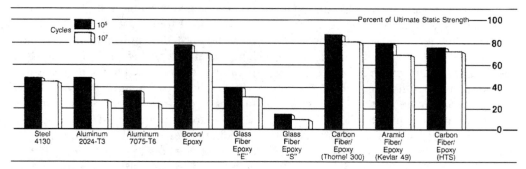

Fig. 2-45 Summary of high-performance fatigue properties of different materials based on their percent of ultimate static tensile strength.

measured material properties, component geometry, and loading information, can be used to design against fatigue failure. The fracture mechanics model also gives insight into materials' development by showing how their resistance to crack propagation depends on both molecular and structural factors.

Service failures in plastics can be caused by fatigue. When time is the critical factor, this type of failure is called static fatigue or creep rupture. For mechanical load reversal or the number of cycles controls failure, the term employed is cyclic fatigue. Interaction between the material and an environment capable of damaging it can lead to stress cracking in the static case and fatigue in the cycle one. An additional failure mode is thermal degradation, in which the temperature increases within the sample from hysteretic energy dissipation.

Traditional fatigue testing produces the familiar S-N diagram. In this type of testing the crack initiation phase usually represents a large fraction of a product's life. However, the crack propagation phase reveals a material's inherent fracture resistance under fracture mechanics testing. The mechanical description of a fracture is usually divided into three stages: crack initiation, stable or incremental fatigue crack propagation, and rapid or catastrophic fracture.

Reinforced plastic In common with metals and unreinforced plastics, RPs also is susceptible to fatigue. However, they provide high performance when compared to un-

reinforced plastics and many other materials (Fig. 2-45). If the matrix is a TP, there is a possibility of thermal softening failures at high stresses or high frequencies. However, in general the presence of fibers reduces the hysteretic heating effect, with a reduced tendency toward thermal softening failures. When conditions are chosen to avoid thermal softening, the normal fatigue process takes places as a progressive weakening of the material from crack initiation and propagation.

Plastics reinforced with carbon, graphite, boron, and aramid are stiffer than the glass-reinforced plastics (GRP) and are less vulnerable to fatigue. E-glass is the most popular type used; S-glass improves both short- and long-term properties (10). In short-fiber GRPs cracks tend to develop easily in the matrix, particularly at the interface close to the ends of the fibers. It is not uncommon for cracks to propagate through a TS matrix and destroy the material's integrity before fracturing of the fabricated product occurs (Fig. 2-46). With short-fiber composites fatigue life can be prolonged if the fiber aspect ratio of its length to its diameter is large, such as at least a factor of five, with ten or better for maximum performance.

In most GRPs debonding can occur after even a small number of cycles, even at modest levels. If the material is translucent, the buildup of fatigue damage can be observed. The first signs (for example, with glass-fiber TS polyester) are that the material becomes opaque each time the load is applied. Subsequently, the opacity becomes permanent and more pronounced, as can occur in

(i) Crack initiation

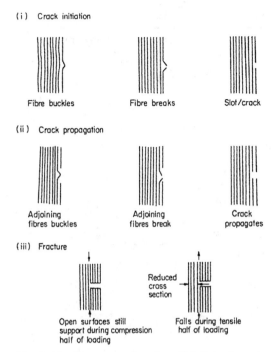

Fibre buckles   Fibre breaks   Slot/crack

(ii) Crack propagation

Adjoining   Adjoining   Crack
fibres buckles   fibres break   propagates

(iii) Fracture

Reduced
cross
section

Open surfaces still   Falls during tensile
support during compression half of loading
half of loading

Fig. 2-46 Sequence of push-pull fatigue in unidirectional glass-fiber/TS polyester RPs by microbuckling processes.

corrugated RP translucent roofing panels. Eventually, plastic cracks will become visible, but the product will still be capable of bearing the applied load until localized intense damage causes separation in the component. However, the first appearance of matrix cracks may cause sufficient concern, whether for safety or aesthetic reasons, to limit the useful life of the product. Unlike most other materials, GRPs give visual warning of their fatigue failure.

Since GRPs can tend not to exhibit a fatigue limit, it is necessary to design for a specific endurance, with safety factors in the region of three to four being commonly used. Higher fatigue performance is achieved when the data are for tensile loading, with zero mean stress. In other modes of loading, such as flexural, compression, or torsion, the fatigue behavior can be worse than that in tension due to potential abrasion action between fibers if debonding of fiber and matrix occurs. This is generally thought to be caused by the setting up of shear stresses in sections of the matrix that are unprotected

by some method such as having properly aligned fibers that can be applied in certain designs. Another technique, which has been used successfully in products such as high-performance RP aircraft wing structures, incorporates a very thin, high-heat-resistant film such as Mylar between layers of glass fibers. With GRPs this construction significantly reduces the self-destructive action of glass-to-glass abrasion and significantly increases the fatigue endurance limit.

Designing with fatigue data   The ranking of fatigue behavior among various plastics should be conducted after an analysis is made of the application and the testing method to be used or being considered. It is necessary to also identify whether the product will be subjected to stress or strain loads. Plastics that exhibit considerable damping may possess low fatigue strength under constant stress amplitude but exhibit a considerably higher ranking in constant deflection amplitude and strain testing. Also needed consideration is the volume of material under stress in the product and its surface area-to-volume ratio. Because plastics are viscoelastic, this ratio is critical in that it influences the temperature that will be reached. At the same stress level, the ratio of stressed volume to area may well be the difference between a thermal short-life failure and a brittle long-life failure, particularly with TPs.

Another factor is whether the product will be in an isothermal or adiabatic heat condition or its thermodynamic behavior. This heat condition is strongly dependent on the loading rate and environmental influences such as temperature, water, solvents, ultraviolet light, air speed, and others. There are different design approaches to eliminating the basically hysteretic heating. For example, using a plastic with a low viscous response to mechanical stress minimizes heat generation because this material is usually very stiff. Heat-transfer conditions can be improved by increasing the flow of air or other coolant (water, gases, etc.) across the surface of the product. The product's design can also be altered to decrease mechanical energy input, slow the cyclic loading rate, increase the

Table 2-8   Elevated temperature properties of glass fiber/nylon 6/6 RPs

Property	Short Fiber		Long Fiber	
	30%	50%	30%	50%
At 300°F				
Tensile strength (10^3 psi)	12.8	13.8	14.3	19.2
Elongation (%)	9.3	7.8	5.3	5.6
Flexural strength (10^3 psi)	13.8	14.3	17.4	23.7
Flexural modulus (10^5 psi)	4.64	5.27	5.50	8.90
At 400°F				
Tensile strength (10^3 psi)	6.3	7.3	7.8	8.3
Elongation (%)	8.6	9.5	6.2	6.8
Flexural strength (10^3 psi)	6.9	7.4	8.9	10.0
Flexural modulus (10^5 psi)	3.96	4.80	5.19	7.51

Data on long-fiber glass-reinforced grades are for Verton compounds.

* To convert psi to pascals (Pa), multiply by 6.895×10^3.

surface area for dissipating heat through fins and the like, and other alterations.

As usual with plastics and other materials, sharp comers or abrupt changes in their cross-sectional geometry or wall thickness should be avoided because they can result in weakened, high-stress areas. The areas of high loading where fatigue requirements are high need more generous radii, combined with optimal material distribution. Radii of ten to twenty times are suggested for extruded parts, and one quarter to one half the wall thickness may be necessary for moldings to distribute stress more uniformly over large areas. In evaluating plastics for a particular cyclic loading condition, the type of material and the fabrication variables are quite important. Remember that the many plastics perform differently, whether they are TPs, TSs, unreinforced and reinforced plastics.

The basic rules to providing fatigue endurance can be summarized. Fiber reinforcement provides significant improvements in fatigue with carbon fibers and graphite and aramid fibers being higher than glass fibers (Fig. 2-45). The effects of moisture in the service environment should also be considered, whenever hygroscopic plastics such as nylon, PCs, and others are to be used. For service involving a large number of fatigue cycles in TPs, crystalline-types offer the potential of more predictable results than those based on amorphous types, because the crystalline ones usually have definite fatigue endurance. Also, for optimum fatigue life in service involving both high-stress and fatigue loading, the reinforced high-temperature performance plastics like PEEK, PES, and PI are recommended (Table 2-8).

High Speed Property

For the most part, many of the behavioral characteristics discussed are valid for a wide range of loading rates. There may be significant shifts in behavior, however, at load or strain durations that are much shorter than those discussed, usually take about a second or less to perform (Figs. 2-47 and 2-48). This section deals with loading rates significantly faster than those covered so far, namely rapid and impact loading.

An example of the type, significance, and character of the data obtainable under rapid loads is shown in Fig. 2-49. The data shown describe behavior over a spectrum of elongation rates up to several orders of magnitude higher than those obtained in standard tests. These data illustrate trends for the specific materials examined. (a) Tensile strength usually increases with higher strain rate, for all plastics. (b) The elongation at break decreases with the strain rate. (c) Energy to failure, as determined from the area under the stress-deformation plot, generally

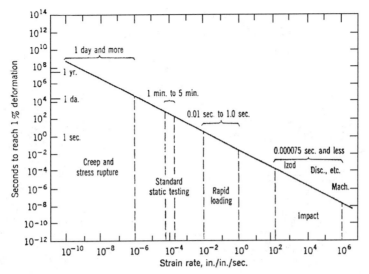

Fig. 2-47 Relation of rapid loading strain rates to those developed in other methods of testing.

decreases or remains the same with an increasing test rate. Moreover, different plastics show a markedly different rate of decay of failure energy with increased test speed.

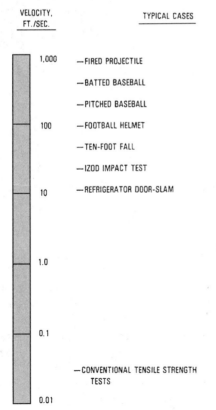

Fig. 2-48 Typical velocities that refer to rapid loading.

Designers with a background in using other materials will recognize both the similarities and the differences in the behavior of the plastics discussed. As an example, impact resistance has been a continuing issue with engineering materials, particularly certain metals with similarities to many of the phenomena observed in plastics.

The concept of a ductile-to-brittle transition temperature in plastics is likewise well known in metals, notched metal products being more prone to brittle failure than unnotched specimens. Of course there are major differences, such as the short time moduli of many plastics compared with those in steel, that may be 30×10^6 psi (207×10^6 kPa). Although the ductile metals often undergo local necking during a tensile test, followed by failure in the neck, many ductile plastics exhibit the phenomenon called a propagating neck. These different engineering characteristics also have important effects on certain aspects of impact resistance.

There are a number of basic forms of energy loads or impingement on products to which plastics react in a manner different from other materials. These dynamic stresses include loading due to impact, impulse, puncture, frictional, hydrostatic, and erosion. They have a difference in response and degree of response to other forms of stress. Analyzing these differences provides

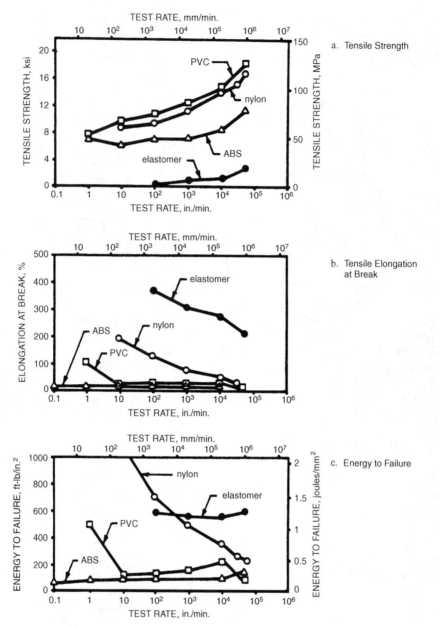

Fig. 2-49 Examples of rapid loading or high speed tests.

Impact loading

Whenever a product is loaded rapidly, it can be said to be subjected to impact loading. Any product that is moving has kinetic energy. When this motion is somehow stopped the designer with information applicable to product performances. because of a collision, its energy must be dissipated. The ability of a plastic product to absorb energy is determined by such factors as its shape, size, thickness, type of material, method of processing, and environmental conditions of temperature, moisture, and so on. Although the impact strengths of plastics are widely reported, these properties have no particular design value. However, they are important, because they can be used to provide an initial comparison of the relative responses of materials. Impact strength can pick up a discriminatory response to notch sensitivity. A better value via impact tensile values, is unfortunately not generally reported.

With limitations, the impact value of a material can broadly separate those that can withstand shock loading from those that fare poorly in this response. Of great importance is that they can be compared to the impact performance on the fabricated products. The results provide guidelines that will be more meaningful and empirical to the designer. To eliminate broad generalizations, the target is to conduct impact tests on the final product or, if possible, at least on its components. In conducting impact tests on products the usual problem that has to be resolved as well as possible is how it should be conducted. The real test is after the product has been in service and field reports are returned for evaluation. Regardless, the usual impact tests conducted on test samples can be useful if they are properly coordinated with product requirements.

Design feature The overall impact resistance of a structure is defined as its ability to absorb and dissipate the energy delivered to it during relatively high speed collisions with other objects without sustaining damage that would jeopardize its intended function. Several design features affect impact resistance. For example, rigidizing elements such as ribs may decrease a part's impact resistance, while less-rigid sections may absorb more impact energy without damage by deflecting elastically.

Likewise, dead sharp corners or notches subjected to tensile loads during impact may decrease the impact resistance of a product by acting as stress concentrators, whereas generous radii in these areas may distribute the tensile load and enhance the impact resistance. This point is particularly important for products comprised of materials whose intrinsic impact resistance is a strong function of a notch radius. Such notch sensitive materials are characterized by an impact resistance that decreases drastically with notch

radius. Wall thickness may also affect impact resistance. Some materials have a critical thickness above that the intrinsic impact resistance decreases dramatically.

Impact loads are a particularly important kind of load for plastics. While many materials such as PE and nylon have good impact strength, other plastics such as crystal PS and some grades of PVC have low impact strength. Many of the tests for impact strength have been based on tests for steel and other metals and the applicability of such tests to plastics has always been questionable. For example PVC which rates low in notched Izod impact tests performs well in normal applications that involve impact loading. However, some grades of rubber-modified high impact styrenes that show up well in the Izod test break on impact under field test conditions. These results have led to continual reexamination of the tests used to determine the toughness of plastics.

Methods employed to determine the impact resistance of plastics include: pendulum methods (Izod, Charpy, tensile impact, falling dart, Gardner, Dynatup, etc.) and instrumented techniques. In the case of the Izod test, what is measured is the energy required to break a test specimen transversely struck (the test can be done either with the specimen notched or unnotched). The tensile impact test has a bar loaded in tension and the striking force tends to elongate the bar (Chapter 5, **Impact Strength**).

There are plastics that tend to be very notch sensitive on impact. This is apparent from the molecular structure of the materials that consist of random arrangements of plastic chains. If the material exists in the glassy state at room temperature the notch effect is to cut the chains locally and increase the stress on the adjacent molecular chains which will scission and propagate the effect through the material. At the high loading rate encountered in impact loading the only form of molecular response is the chain bending (spring) reaction which is limited in extent and generally low in magnitude compared to the viscoelastic response which responds at longer loading times.

There are several ways in which the impact properties of plastics can be improved if the material selected does not have sufficient impact strength. One method is by altering the composition of the material so that it is no longer a glassy plastic at the operating temperature of the product (Chapter 6). In the case of PVC this is done by the addition of an impact modifier which can be a compatible plastic such as an acrylic or a nitrile rubber. The addition of such a material lowers the glass transition temperature and the material becomes a rubbery viscoelastic plastic with much improved impact properties. This is one of the methods in which PVC materials are made to exhibit superior impact properties.

Another way in which to improve impact properties is by orienting the material. Nylon has a fair impact strength but oriented nylon has a very high transverse impact strength. The intrinsic impact strength of the nylon comes from the polar structure of the material and the fact that the polymer is crystalline. The substantial increase in impact strength as a result of the orientation results from the molecular chains being aligned. This makes them very difficult to break and, in addition, the alignment improves the polar interaction between the chains so that even when there is a chain break the adjacent chains hold the broken chain and resist parting of the structure. The crystalline nature of the nylon material also means that there is a larger stress capability at rapid loading since the crystalline areas react much more elastically than the amorphous glassy materials.

Other methods in which impact strength can be substantially improved is by the use of fibrous fillers. These materials act as a stress transfer agent around the region that is highly stressed by the impact load. Since most of the fibrous fillers such as glass and asbestos have high elastic moduli, they are capable of responding elastically at the high loading rates encountered in impact loading (39).

One general method of improving the performance of plastic products in impact loading is to prevent, by design and handling, the formation of notched areas which act as stress risers. Especially under impact conditions the possibility of localized stress intensification can lead to product failure. In almost every case the notched strength is substantially less than the unnotched strength.

It is important to recognize that impact strength is sensitive to temperature conditions. The impact strength of plastics is reduced drastically at low temperatures with the exception of fibrous filled materials that improve in impact strength at low temperature. The reduction in impact strength is especially severe if the material undergoes a glass transition where the reduction in impact strength is usually an order of magnitude.

Impulse Loading

A related form of stress to impact is impulse loading that differs in two ways from impact loading. Impact loading implies striking the object and consequently there is a severe surface stress condition present before the stress is transferred to the bulk of the material. In addition, in impact loading the load is applied instantly, limited in straining rate only by the elastic constants of the material being struck. A significant portion of the energy of impact is converted to heat at the point of impact and complicates any analytically exact treatment of the mechanics of impact.

In the case of impulse loading the load is applied at very high rates of speed limited by the member applying the load. However, the loading is not generally localized and the heat effects are similar to conventional dynamic loading in that the hysteresis characteristics of the material determines the extent of heating and the effects can be analyzed with reasonable accuracy. For example, the load of two billiard balls striking is definitely an impact condition. The load applied to a brake shoe when the brake is applied or the load applied to a fishing line when a strike is made is an impulse load. The time constants are short but not as short as the impact load and the entire structural element is subjected to the stress.

Plastics generally behave in a much different manner under impulse loading than they do under loading at normal straining rates. Some of the same conditions occur as under impact loading where the primary response to load is an elastic one because there is not sufficient time for the viscoelastic elements to operate. The primary structural response in the polymer is by chain bending and by stressing of the crystalline areas of crystalline polymers. The response to loading is almost completely elastic for most materials, particularly when the time of loading is of the order of milliseconds.

Since the entire load is applied to the elastic elements in the structure and the long-range strain adaptation is precluded, the material will exhibit a high elastic modulus and much lower strain to rupture. It is difficult to generalize as to whether the material is stronger under impulse loading than under normal loading. For example, PMMA and rigid PVC materials that appear to be brittle under normal loading conditions, exhibit high strength under impulse loading conditions. Rubbery materials such as thermoplastic urethane elastomers and some other elastomers behave like brittle materials under impulse loading. This is an apparently unexpected result that upon analysis is obvious because the elastomeric rubbery response is a long time constant response and the rigid connecting polymer segments which are brittle are the ones that respond at high loading rates.

The comments regarding improvements made with respect to impact loading for structures apply equally well to impulse loading conditions. Fibrous fillers improve impulse loading strength. Oriented materials withstand impulse loading much better than unoriented materials. As an example fibrous forms of materials are used in rope because they take impulse loading well. Crystalline polymers generally perform well under impulse loading, especially polar materials with high interchain coupling.

Using plastics under impulse loading conditions requires a careful design approach. Test data taken with high-speed testing machines are essential before using a plastic for these applications since it is difficult to predict the response of the material from the available data. High-speed testing machine are used to determine the response of materials at millisecond loading rates. In the absence of such test data, the only first sorting evaluation that can be done is from the results of the tensile impact test. The test should be done with a series of loads below break load, through the break load, and then estimating the energy of impact under the non-break conditions as well as the tensile impact break energy. As indicated above, apparently brittle materials perform well and rubbery materials that would seem to be a natural for impulse loading behave in a brittle manner.

Puncture Loading

Resistance to puncture is another type of loading. It is of particular interest in applications involving sheet and film as well as thin-walled tubing or molding and other membrane type loaded structures. The surface skins of sandwich panels are another area where it is important. A localized force is applied by a relatively sharp object perpendicular to the plane of the sheet of material being stressed. If the material is thick compared to the area of application of the stress, it is effectively a localized compression stress with some shear effects as the material is deformed below the surface of the sheet.

In the case of a thin sheet or film the stresses cause the material to be displaced completely away from the plane of the sheet and the restraint is by tensile stress in the sheet and by hoop stress around the puncturing member. Most cases fall somewhere between these extremes, but the most important conditions in practice involve the second condition to a larger degree than the first condition.

To analyze the second condition take the material at the point where the puncturing object has almost pierced the membrane but has not broken through. At this point one can see the nature of the forces which are resisting the puncture and qualitatively relate them to the primary physical characteristics of the material so that we can indicate which materials are suitable for resistance to this type of stress and how to improve the resistance to puncture.

There are three principal stresses that result from the puncture forces through relatively thin material. They are a compressive stress under the point of the puncturing member, a tensile stress caused by the stretching of the material under the penetrating force, and a hoop stress caused by the material being displaced around the penetrating member. Part of the hoop stress is compressive adjacent to the point which changes to tensile stress to contain the displacing forces. It is evident that anisotropic materials will have a more complicated force pattern and, in fact, uniaxially oriented materials will split rather than puncture under this type of loading. To improve the puncture resistance materials are needed with high tensile strength. This is evident as required to have both the stretching load and the hoop stress. In addition, the material should have a high compression modulus to resist the point penetration into the material. Resistance to notch loading is also important.

Based on this analysis it is evident that materials which are biaxially oriented will have good puncture resistance. Highly polar polymers would be resistant to puncture failure because of their tendency to increase in strength when stretched. The addition of randomly dispersed fibrous filler will also add resistance to puncture loads. From some examples such as oriented polyethylene glycol terephthalate (Mylar), vulcanized fiber, and oriented nylon, it is evident that these materials meet one or more of the conditions reviewed. Products and plastics that meet with puncture loading conditions in applications can be reinforced against this type of stress by use of a surface layer of plastic with good puncture resistance. Resistance of the surface layer to puncture will protect the product from puncture loads. An example of this type of application is the addition of an oriented PS layer to foam cups to improve their performance.

Frictional Loading

The frictional properties of plastics are of particular importance to applications in machine products and in sliding applications such as belting and structural units such as sliding doors. The range of friction properties are rather extensive. The relationship between the normal force and the friction force is used to define the coefficient of static friction.

Friction coefficients will vary for a particular material from the value just as motion starts to the value it attains in motion. The coefficient depends on the surface of the material, whether rough or smooth, as well as the composition of the material. Frequently the surface of a particular plastics will exhibit significantly different friction characteristics from that of a cut surface of the same smoothness. These variations and others that are reviewed make it necessary to do careful testing for an application which relies on the friction characteristics of plastics. Once the friction characteristics are defined, however, they are stable for a particular material fabricated in a stated manner.

The molecular level characteristics that create friction forces are the intermolecular attraction forces of adhesion (2). If the two materials that make up the sliding surfaces in contact have a high degree of attraction for each other, the coefficient of friction is generally high. This effect is modified by surface conditions and the mechanical properties of the materials. If the material is rough there is a mechanical locking interaction that adds to the friction effect. Sliding under these conditions actually breaks off material and the shear strength of the material is an important factor in the friction properties. If the surface is not rough, but smoothly polished, the governing factor induced by the surface conditions is the amount of area in contact between the surfaces. In a condition of large area contact and good adhesion, the coefficient of friction is high. In the case of smoothly polished surfaces and adhesion forces the coefficient is very high since there is intimate surface contact.

Several other factors affect the frictional forces. If one or both of the contacting surfaces have a relatively low compression modulus it is possible to make intimate contact between the surfaces which will lead to high friction forces in the case of plastics having good adhesion. It can add to the friction forces in another way. The displacement of material in front of the moving object adds a mechanical element to the friction forces.

All sliding friction forces are dramatically affected by surface contamination. If the surface is covered with a material that prevents the adhesive forces from acting, the coefficient is reduced. If the material is a liquid which has low shear viscosity the condition exists of lubricated sliding where the characteristics of the liquid control the friction rather than the surface friction characteristics of the materials. It is possible by the addition of surface materials that have high adhesion to increase the coefficient of friction.

The use of plastics for gears and bearings is the area in which friction characteristics have been examined most carefully. As an example highly polar polymers such as nylons and the TP polyesters have, as a result of the surface forces on the material, relatively low adhesion for themselves and such sliding surfaces as steel. Laminated plastics also make excellent bearings. The physical properties of these materials make them a good choice for both bearing and gear materials. The typical coefficient of friction for such materials is 0.1 to 0.2.

In the injection molded condition the skin formed when the plastic cools against the mold tends to be harder and slicker than a cut surface so that the molded product exhibit lower sliding friction and are excellent for this type of application. Good design for this type of application is to make the surfaces as smooth as possible without making them glass smooth which tends to increase the intimacy of contact and to increase the friction above that of a fine surface. The problems in this type of application related to friction are heat effects due to the rubbing surfaces. For successful design the heat generated by the friction must be dissipated as fast as it is generated to avoid overheating and failure.

Obviously the addition of appropriate lubricants will lower the friction and help to remove the heat. There are several other ways in which the friction can be reduced. One is by the incorporation of fillers. The fillers can be used to increase the thermal conductivity of the material such as glass and metal fibers. The filter can be a material like TFE plastic that has a much lower coefficient of friction and the surface exposed material will reduce the friction. Another approach that is used is the incorporation of slightly incompatible materials such as silicone oil into the molding material. After molding the material migrates to the surface of the product and acts as a renewable source of lubricant for the product. In the case of bearings it is carried still further by making the bearing material porous and filling it with a lubricating material in a manner similar to sintered metal bearings, graphite and molybdenum sulfide are also incorporated as solid lubricants.

A different type of low friction or low drag application is encountered with sliding doors or conveyor belts sliding on support surfaces. In applications like this the normal forces are generally quite small and the friction load problems are of the sticking variety. Some plastics exhibit excellent track surfaces for this type of application. TFEs have the lowest coefficient of any solid material and represent one of the most slippery surfaces known. The major problem with TFE is that its abrasion resistance is low so that most of the applications utilize filled compositions with ceramic filler materials to improve the abrasion resistance.

There is a whole field of applications for TFE in reducing friction using solid materials as well as films and coatings. Another material with excellent properties for surface sliding is ultra high molecular weight polyethylene (UHMWPE). Polyethylene and the polyolefins in general have low surface friction, especially against metallic surfaces. The UHMWPE has an added advantage in that it has much better abrasion resistance and is preferred for conveyor applications and applications involving materials sliding over the product. In the textile industry loom products also use this material extensively because it can handle the effects of the thread and fiber passing over the surface with low friction and relatively low wear.

The specific friction characteristics of plastics at the high friction end are also an area of significant applications. Some plastics, notably polyurethanes and some plasticized vinyl compositions, have very high friction coefficients. These materials make excellent traction surfaces for products ranging from power belts to drive rollers where the plastics either drives, or is driven, by another member. Conveyor belts made of oriented nylon and woven fabrics are coated with polyurethane elastomer compounds to supply both the driving traction and to move the objects being conveyed up fairly steep inclines because of the high friction generated. Drive rollers for moving paper through printing presses and business machines are frequently covered with either urethane or vinyl to act as the driver members with minimum slippage. The materials are also used as the torque surfaces in clutches and brakes.

In all of the friction applications suggested as well as in many others, there are two areas where the design effort is introduced. The first is in material selection and modification to provide either high or low friction as required by the application. The other is in determining the required geometry to supply the frictional force level needed by controlling contact area and surface quality to provide friction level. A controlling factor limiting any particular friction force application is heat dissipation. This is true if the application of the friction loads is either a continuous process or a repetitive process with a high duty cycle. The use of cooling structures either incorporated into the products or by the use of external cooling devices such as coolants or air flow should be a design consideration.

Hydrostatic Loading

This is another behavior to be considered in this type of loading. The surface properties of the material are quite significant. If the water does not wet the surface, the tendency will be to have the droplets that do not impact close to the perpendicular direction bounce off the surface with considerably less energy transfer to the surface. Non-wetting coatings reduce the effect of wind and rain erosion.

Impact of air-carried solid particulate matter is more closely analogous to straight impact loading since the particles do not become disrupted by the impact. The main characteristic required of the material, in addition to not becoming brittle under high rate loading, is resistance to notch fracture. The ability to absorb energy by hysteresis effects is also important as is the case with the water. In many cases the best type of surface is an elastomer with good damping properties and good surface abrasion resistance. An example is polyurethane coatings and products that are excellent for both water and particulate matter that is air-driven. Besides such applications as vehicles, these materials are used in the interior of sand and shot blast cabinets where they are constantly exposed to this type of stress. These materials are fabricated into liners in hoses for carrying pneumatically conveyed materials such as sand blasting hoses and for conveyor hose for a wide variety of materials such as sand, grain, and plastics pellets.

In general when the surface impact loading by fluid-borne particulate matter, liquid or solid, or cavitation loading is encountered, the method of minimizing the effects of erosion produced are by material selection and modification. The plastics used should be ductile at impulse loading rates and capable of absorbing the impulse energy and dissipating it as heat by hysteresis effects. The surface characteristics of the materials in terms of wettability by the fluid and frictional interaction with the solids also play a role. In this type of application the general data available for materials should be supplemented by that obtained under simulated use conditions since the properties needed to perform are not readily predictable from the usually available data.

Another loading condition in underwater applications is the application of external hydrostatic stress to plastic structures (also steel, etc.). Internal pressure applications such as those encountered in pipe and tubing or in pressure vessels such as aerosol containers are easily treated using tensile stress and creep properties of the plastic with the appropriate relationships for hoop and membrane stresses (108). The application of external pressure, especially high static pressure, has a rather unique effect on plastics. The stress analysis for thick walled spherical and tubular structures under external pressure are available.

The interesting aspect that plastics have in this situation is that the high compressive stresses increase the resistance of plastic materials to failure. Glassy plastics under conditions of very high hydrostatic stress behave in some ways like a compressible fluid. The density of the material increases and the compressive strength is increased. In addition, the material undergoes sufficient internal flow to distribute the stresses uniformly throughout the product. As a consequence, the plastic products produced from such materials as PMMA and PC make excellent view windows for undersea vehicles that operate at extreme depths where the external pressures are 1000 psi (7MPa) and more.

Erosion Loading

This subject is emphasizing a specialty high speed loading that is part of the previous section on hydrostatic loading. It is the effect of erosion forces such as wind driven sand or water, underwater flows of solids past plastic surfaces and even the effects of high velocity flows causing cavitation effects on material surfaces. One major area for the utilization of plastics is on the outside of moving objects that range from the front of automobiles to boats, aircraft, missiles, and submarine craft. In each case the impact effects of the velocity driven particulate matter can cause surface damage to plastics. Stationary objects such as radomes and buildings exposed to the weather in regions with high and frequent winds are also exposed to this type of effect.

A type of wind erosion analysis that has been extensively studied is the effect of water drop erosion on rapidly moving missile parts.

Aircraft radomes have also been extensively studied for the effects of wind-driven water and solids. The erosion effects are very dramatic and the surfaces are usually protected with elastomeric materials that have good resistance to this type of stress.

To determine the type of physical properties materials used in this environment should have, it is necessary to examine the mechanics of the impact of the particulate matter on the surfaces. The high kinetic energy of the droplet is dissipated by shattering the drop, by indenting the surface, and by frictional heating effects. The loading rate is high as in impact and impulse loading, but it is neither as localized as the impact load nor as generalized as the impulse load, Material that can dissipate the locally high stresses through the bulk of the material will respond well under this type of load. The plastic should not exhibit brittle behavior at high loading rates.

In addition, it should exhibit a fairly high hysteresis level that would have the effect of dissipating the sharp mechanical impulse loads as heat. The material will develop heat due to the stress under cyclical load. Materials used are the elastomeric plastics used in the products or as a coating on products.

Cavitation erosion With increasing ship speeds, the development of high-speed hydraulic equipment, and the variety of modern fluid-flow applications to which metal materials are being subjected, the problem of cavitation erosion becomes ever more important. Erosion may occur in either internal-flow systems, such as piping, pumps, and turbines, or in external ones like ships' propellers (36).

Osborne Reynolds identified the phenomenon of cavitation as early as 1873. By the turn of the century it had been called by its present name by R. E. Froude, the director of the British Admiralty Ship Model Testing Laboratories.

Cavitation occurs in a rapidly moving fluid when there is a decrease in pressure in the fluid below its vapor pressure and the presence of such nucleating sources as minute foreign particles or definite gas bubbles. As a result, vapor bubble forms that continues to grow until it reaches a region of pressure higher than its own vapor pressure, when it collapses. When these bubbles collapse near a boundary, the high-intensity shock waves that are produced radiate to the boundary, resulting in mechanical damage to the material. The force of the shock wave or of the impinging may still be sufficient to cause a plastic flow or fatigue failure in a material after a number of cycles, depending on the properties of the material, the existing hydrodynamic conditions, and the foil-design parameters.

The behavior of materials, particularly steel, in cavitating fluids results in an erosion mechanism, including mechanical erosion and electrochemical corrosion. The straightforward way to fight cavitation is to use hardened materials, chromium, chrome-nickel compounds, or elastomeric plastics. Other cures are to reduce the vapor pressure with additives, reduce the turbulence, change the liquid's temperature, or add air to act as a cushion for the collapsing bubbles.

Rain erosion   One that walks through a gentle spring rain seldom considers that raindrops can be small destructive "bullets" when they strike high-speed aircraft. These bulletlike raindrops can erode paint coatings, plastic products, and even steel, magnesium, or aluminum leading edges to such an extent that the surfaces may appear to have been sandblasted. Even the structural integrity of the aircraft may be affected after several hours of flight through rain. This problem is of special interest to aircraft engaged in all weather flying.

It affects commercial aircraft, missiles, high-speed vehicles on the ground, spacecraft before and after a flight when rain is encountered, and even buildings or structures that undergo high-speed rainstorms. The critical situations exist in flight vehicles, since flight performance can be affected to the extent that a vehicle can be destroyed. Research and development concerning rain erosion on aircraft has been extensive since the 1940s.

Erosion by rain of the exterior of so called (at that time) high-speed aircraft during flight was observed during World War 11 on all-weather fighter airplanes capable then of only flying at 400 mph. The aluminum edges of wings and particularly of the glass-fiber-reinforced TP polyester-nose radomes (particularly the long Eagle Wing on B-29s flying over the Pacific) were particularly susceptible to this form of degradation. The problem continues to exist as can be seen on the front of commercial and military airplanes with their neoprene protective coated RP radomes; the paint coating over the rain erosion elastomeric plastic erodes and then is repainted prior to the rain erosion elastomeric coating is affected.

Actual flight tests to determine the severity of this phenomenon of rain erosion carried out in 1943 established that aluminum and RP leading edges of airfoil shapes exhibited serious erosion after exposure to rainfall of only moderate intensity. Inasmuch as this problem originally arose with military aircraft, the U.S. Air Force initiated research studies at the Wright-Patterson Development Center's Materials Laboratory in Dayton, Ohio. Based on the results of a young girl physicist (worked for DVR), it resulted in applying an elastomeric neoprene coating adhesively bonded to RP radomes. The usual 5-mil coating of elastomeric material used literally bounces off raindrops, even from a supersonic airplane traveling through rain. There is a slight loss of radar transmission of about 1% per mil of the plastic thickness, but this is better than losing 100% when the radome is destroyed.

Thermal Expansion and Contraction

Unconstrained specimens of almost all materials respond to temperature increases by expanding and to temperature decreases by contracting. The coefficient of linear thermal expansion (CLTE) of a material is determined by varying the temperature of a representative test specimen. Measurement is made of its length as a function of temperature over the desired range, computing the total change in specimen length over that range, and then dividing that change in length by both the specimen length at the reference temperature and the total temperature excursion. In determining the coefficient of linear thermal expansion of plastics per ASTM D696, the temperature range is −30°C to +30°C (−22 to +86°F); and the reference temperature is 23°C (73°F).

There are plastics that have equal or less than those of other materials of construction (metals, glass, or wood). In fact with certain additives such as graphite powder contraction can occur rather than the expected expansion with the application of heat. However many plastics typically have coefficients that are considerably higher than those of other materials of construction such as metals. This difference may amount to a factor of 10 to 30. Also available are plastics, particularly TS-RPs, with practically no change.

Obviously, the designer must take thermal expansion and contraction into account if critical dimensions and clearances are to be maintained during use where material is in a restricted design. Less obvious is the fact that products may develop high stresses when they are constrained from freely expanding or contracting in response to temperature changes. These temperature-induced stresses can cause material failure.

Plastic products are often constrained from freely expanding or contracting by rigidly attaching them to another structure made of a material (plastic, metal, etc.) with a lower coefficient of linear thermal expansion. When such composite structures are heated, the plastic component is placed in a state of compression and may buckle, etc. When such composite structures are cooled, the plastic component is placed in a state of tension, which may cause the material to yield or crack. The precise level of stress in the plastic depends on the relative compliance of the component to which it is attached, and on assembly stress.

To minimize the stresses induced by differential thermal expansion/contraction one must: (1) employ fastening techniques that allow relative movement between the component parts of the composite structure, (2) minimize the difference in coefficient of linear thermal expansion between the materials comprising the structure; and/or (3) minimize the temperature differences the structure will experience during use or shipment. Examples of proper fastening methods include the use of screws, bolts, spring clips, etc. with oversize holes, slots, or compliant bushings.

Hysteresis Effect

The hysteresis heating failure occurs more commonly in plastic members subject to dynamic loading. It would be well to point out what comprises dynamic loading and what types of stresses are encountered. One example commonly encountered is a plastic gear. In the course of operation the gear teeth are periodically, once per revolution, subjected to a bending load that transmits the power from one gear to another. Another example is a link that is used to move a paper sheet in a copier or in an accounting machine from one operation to the next. The load may be simple tensile or compressive stresses, but more commonly it is a bending load.

There are some less obvious but quite important dynamic stress situations that illustrate the importance of dynamic loading. A belt that is used to drive a pulley is subject to repeated bending and tensile stresses during operation. A tire on a vehicle is subjected to a complicated combination of bending, bearing, and compressive stresses during the movement of the vehicle that it supports. The keys on a computer or printer are subjected to repeated impulse loading during use even though the action is not strictly cyclical in nature. Add to this the effects of vibration induced by vehicle motion or machine action that is an induced cyclical stress in products that are attached to the vibration-inducing object and it becomes apparent that cyclical loading is a widely encountered type of loading. In many instances the dynamic stress exists in conjunction with static stresses and with other longer-term periodic loads.

An example will be given to show how dynamic loading can lead to product failure by hysteresis heating. When this condition exists the failure will be catastrophic rather than gradual. This is not generally true of creep failure or of normal fatigue failure. The example is a link that is loaded in tension and is the connecting link that drives a flywheel type of unit by means of a linear actuating force. The primary load on the product is alternate tension and compression loading. The compression loading is low since it represents the flywheel driving the link back against friction forces. The tensile force is the driving force and it varies from zero to a maximum that is determined by the torque load on the flywheel member.

The relationship of force to time is determined using engineering equations (Appendix A: **PLASTICS DESIGN TOOLBOX**). Result is the force function per revolution and this divided by the time of a revolution will give the force as a function of time. Increasing temperature increases heat transfer from the product to the surroundings and if the rate of heat transfer equals the rate of heat generation at a temperature below the softening temperature of the material, the process will stabilize and the product will not fail. It is apparent that the major factor in hysteresis failure is not the ability of the product to dissipate the heat generated to its surroundings (190).

The designer can use several approaches to prevent hysteresis failure. The first is material selection. The stiffer the material is, the smaller the strain is for a given stress level and the lower the hysteresis loss per cycle. Some materials are additionally fairly linear in stress-strain characteristics and have smaller hysteresis loops. These would be preferred in dynamic loading applications.

Another approach is to improve the heat transfer conditions from the product. This can be accomplished in several ways. One way is to operate in a coolant medium that would also act as the lubricant for the system. The heat transfer to a liquid is usually much better than to air, and the liquid can be cooled by passing it through a heat exchanger device. A second approach is to improve the heat transfer to air. This can be done by increasing the surface area of the product by means of fins or other surface projections. The larger area will increase the heat flow

out of the product substantially. The third approach is the use of air circulating techniques which can be areas added to the stressed unit such as air deflector sections or the use of fan cooling as part of the system. An approach that may fit some applications is the use of metal heat-sink elements buried in the plastic that conducts the heat into other parts of the complete machine to dissipate it to the surroundings.

Basically, anything that can be done to reduce the temperature of the product by removal of heat generated by the cyclical stress will improve the possibilities of surviving the cyclical stress. If the heat transfer capability is limited, then the only alternative is to use stiff materials and low stress levels on the product compared with the strength capability of the material. The heavier products that result will be relatively inefficient in the use of material. In some cases when the load applied is an inertial load (such as an impeller on a pump) it may be that only a trade-off of weight for low stress level can cause failure.

Energy and Motion Control

There are plastics such as TP elastomers that are frequently subjected to dynamic loads where heat energy and motion control systems are required. One of the serious dynamic loading problems frequently encountered in machines and vehicles is vibration-induced deflection (Chapter 4, **DYNAMIC LOAD ISOLATOR**).

Such effects can be highly destructive, particularly if a product resonates at one of the driving vibration frequencies. One of the best ways to reduce and in many cases, eliminate vibration problems is by the use of these viscoelastic plastics. Some materials such as silicone elastomers, flexible vinyl compounds of specific formulations, polyurethane plastics, and a number of others have very large hysteresis effects. By designing them into the structure it is possible to have the viscoelastic material absorb enough of the vibration inducing energy and convert it to heat so that the structure is highly damped and will not vibrate. In each case the viscoelastic material is arranged in such a way that movement or flexing of the product results in large deflections of the viscoelastic materials so that a large hysteresis curve is generated with a large amount of energy dissipated per cycle.

By calculating the energy to heat it is possible to determine the vibration levels to which the structure can be exposed and still exhibit critical damping. There is one area that must be evaluated. Plastics exhibit a spectrum of response to stress and there are certain straining rates that the material will react to almost elastically. If this characteristic response corresponds to a frequency to which the structure is exposed the damping effect is minimal and the structure may be destroyed. In order to avoid the possibility of this occurring, it is desirable to have a curve of energy absorption vs. frequency for the material that will be used.

Viscoelastic damping The same approach can be used in designing power transmitting units such as belts. In most applications it is desirable that the belts be elastic and stiff enough to minimize heat buildup and to minimize power loss in the belts. In the case of a driver which might be called "noisy" in that there are a lot of erratic pulse driven forces present, such as an impulse operated drive, it is desirable to remove this noise by damping out the impulse and get a smooth power curve.

This is easily done using a viscoelastic belt that will absorb the high rate load pulses. The same approach can be used by making one gear in a gear train or one link in a linear drive mechanism an energy absorber. The viscoelastic damping can be a valuable tool for the designer to cope with impulse loading that is undesirable and potentially destructive to the product.

There is another type of application where the damping effect of plastic structures can be used to advantage. It has a long although not obvious history. The early airplanes used doped fabric as the covering for wings and other aerodynamic surfaces. The dope was cellulose nitrate and later cellulose acetate that is a damping type of plastic. Conse-

quently, surface flutter was a rare occurrence. It became a serious problem when aluminum replaced the fabric because of the high elasticity of the metal surfaces. The aerodynamic forces acting on the thin metal coverings can easily induce flutter and this was a difficult design problem that was corrected for minimizing the effect.

Weathering/Environment

Plastics have been used, and used very successfully, in applications where continuous or intermittent exposure to weather and different environments has been involved. The successes, however, have been the result of thoughtful and discriminating sifting of the weatherability data, rather than blind choice or wishful thinking. Due to their extreme versatility, plastics are used in almost every imaginable type of media. Not only must they resist weather and temperature extremes ranging from cryogenic to above 1,370°C (2,500°F) heat generated in rocket motors, but also extreme conditions of corrosion, irradiation, fluid degradation as typified by liquid rocket propellants, and mechanical energy such as abrasion.

Like other materials with which designer's work, different plastics can be sensitive to the environments. Even ordinary exposure from sunlight or to household cleaning agents can change the properties of certain plastics. Whereas rust, corrosion, and loss of its properties can plague metals, the cracking, crazing, and loss of its properties can affect certain plastics in the presence of different environments.

The age-old problem of predicting what will happen to any material after it is subjected to service also exists with plastics. Different data on plastics are available, but typical of so-called progress, there is never sufficient or adequate useful information to predict the service life of products being designed. It is suggested that rather than assume that a lack of data exists, one should determine what is logically available and apply it most efficiently. A potential example of improper design with plastics concerns toys.

Toys can be made of plastics that will resist mechanical destruction even though some people consider all toys "immediately destructible."

Designing and developing plastics for use in different environments requires the usual practical or design approach for any material; namely, knowledge of its behavior and capability. Since insufficient data will continually exist for some materials, particularly the newer ones or those being subjected to different environments, different methods of evaluation and developing data can be used. These include both experimental and theoretical approaches. Experimental techniques include static or dynamic specimen or product testing, simulated service testing of specimen, or product and field-testing indoors and/or outdoors. In order to shorten the time cycle required during testing, theoretical tests can be conducted such as studies in rheology (science dealing with the deformation and flow of matter) and aging (Chapter 5, **WEATHERING**).

Temperature Review

As reviewed throughout this book, certain plastics can be affected in different ways by temperature. Among other things, it can influence short- and long-time static and dynamic mechanical properties (Fig. 2-50), aesthetics, dimensions, electronic properties, and other characteristics. Some plastics cannot take boiling water, most others can operate up to at least 150°C (300°F), and the so-called high-temperature types can take various degrees of continuous use way above 150°C and there are plastics that reach at least 538°C (1,000°F). Then there are the reinforced plastics used as heat-shield ablative materials on the nose cones of space vehicles that reach temperatures up to 1,370°C (2,500°F) for fractions of a second upon reentering the atmosphere. As reviewed practically all plastics can take heat up to at least what the human body can endure, which is one important reason they are extensively used.

TPs soften to varying degrees at elevated temperatures, but TSs are much less affected. The maximum temperatures under which plastics can be employed are generally higher than the temperatures found in buildings, including walls and roofs, but some such as LDPE are marginal and others cannot carry appreciable stresses at moderately elevated temperatures without undergoing noticeable creep. Many plastics can take shipping conditions that are more severe than their service conditions, as in an automobile trunk or railroad boxcar that can reach at least 52°C (126°F).

The response of a plastic to an applied stress depends on the temperature and the time at that temperature to a much greater extent than does that of a metal or ceramic. The variation of an amorphous TP over an extended temperature range can be exemplified by the behavior of its elastic modulus as a function of temperature.

With a temperature change the short-term static strength, the elastic modulus, and the elongation behavior of a material will be similar for it's tensile, compressive, flexural, and shear properties. A material's strength and modulus will decrease and its elongation increase with increasing temperature at constant strain. Curves for creep isochronous stress and isometric stress are usually produced from measurements at a fixed temperature. Complete sets of these curves are sometimes available at temperatures other than the ambient. It is common, for instance, to find creep rupture or apparent modulus curves plotted against log time, with temperature as a parameter. Figure 2-51 shows time-temperature shifting of apparent modulus curves shifting to estimate the extended time values at lower temperatures.

These curves suggest that it would be reasonable to estimate moduli at somewhat longer times than the data available from the lower temperatures. However, a set of creep-rupture curves from various temperatures, as in Fig. 2-52, would suggest that projecting the lowest-temperature curves to longer times as a straight line could produce a dangerously high prediction of rupture strength, so

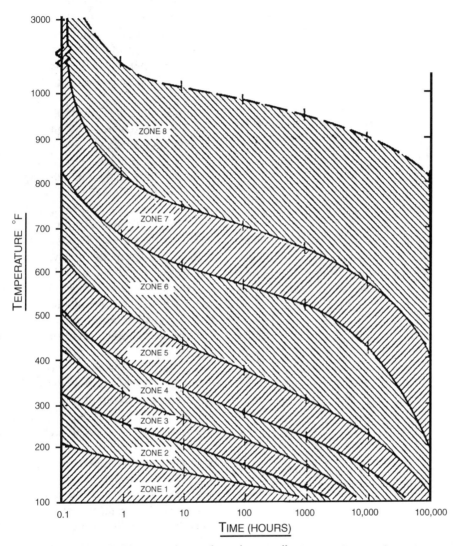

Fig. 2-50   Guide to maximum short-time tensile stress vs. temperature.

this approach is not recommended. As previously reviewed one advantage of conducting complete creep-rupture testing at elevated temperatures is that although such testing for endurance requires long times, the strength levels of the plastic at different temperatures can be developed in a relatively short time, usually just 1,000 to 2,000 h. The Underwriters Laboratories and other such organizations have employed such a system for many years.

Testing different impact properties at various temperatures produces a plot that looks very much like an elongation vs. temperature curve. As temperatures drop significantly below the ambient temperatures, most TPs lose much of their room-temperature impact strength. A few, however, are on the

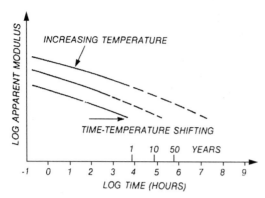

Fig. 2-51 Apparent modulus at different temperatures.

lower, almost horizontal portion of the curve at room temperature and thus show only a gradual decrease in impact properties with decreases in temperature. One major exception is provided by the glass fiber RPs, which have relatively high Izod impact values, down to at least $-40°C$ ($-40°F$). The S-N (fatigue) curves for TPs at various temperatures show a decrease in strength values with increases in temperature. However the TSs, specifically the TS RPs, in comparison can have insignificant or very low losses in strength.

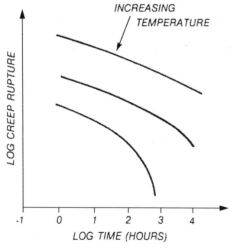

Fig. 2-52 Creep-rupture curves indicating the danger of making linear projections to longer times at lower temperatures.

Stress Cracking and Crazing

Environmental stress cracking is the cracking of certain plastic products that becomes exposed to a chemical agent while it is under stress. This effect may be caused by exposure to such agents as cleaners or solvents. The susceptibility of affected plastics to stress cracking by a particular chemical agent varies considerably among plastics, particularly the TPs.

The resistance of a given plastic to attack may be evaluated by using either constant-deflection or constant-stress tests in which specimens are usually coated with the chemical or be immersed in the chemical agent. After a specified time the degree of chemical attack is assessed by measuring such properties as those of tensile, flexural, and impact (Figs. 2-53 and 2-54). The results are then compared to specimens not yet exposed to the chemical. In addition to chemical agents and the environment for testing may also require such other factors as thermal or other energy-intensive conditions.

A classic example illustrating the effects of stress cracking is the case of the PE milk bottle from the 1950s. A PE plastic and a process to blow mold the bottles were successfully integrated to the point where the lactic acid in the milk would not cause a premature split in the highly stressed neck area of the bottle. As noted, stress cracking is intensified by an increase in temperature. As an example, the results from testing HDPE pressure-pipe specimens in water at $82°C$ ($180°F$) show results in a life span of just a few hundred hours but when the water temperature is at $23°C$ ($74°F$) the life expectancy becomes fifty years. In both tests, water was moving through the pipes.

It is possible with solvents of a particular composition to determine quantitatively the level of stress existing in certain TP products where undesirable or limited fabricated-in stresses exist. The stresses can be residual (internal) stresses resulting from the molding, extrusion, or other process that was used to fabricate the plastic product. Stresses can also be applied such as bending the product. As it has been done for over a half century, the

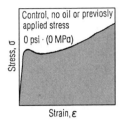

Control, no oil or previosly applied stress

0 psi · (0 MPa)

Stress, σ

Strain, ε

Previously no stress or applied stress lasting 16 hours with sample coated with vegetable oil prior to testing for the short-term stress-strain behavior shown.

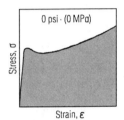
0 psi · (0 MPa)

Stress, σ

Strain, ε

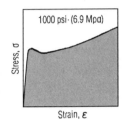
1000 psi · (6.9 Mpa)

Stress, σ

Strain, ε

2000 psi · (13.8 MPa)

Stress, σ

Strain, ε

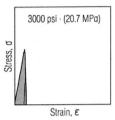
3000 psi · (20.7 MPa)

Stress, σ

Strain, ε

Fig. 2-53 Example of the influence of tensile stress-strain curves subjected to an environment that influences the ductility of a specific plastic.

product is immersed in the solution that attacks the plastic for various time periods. Any initial cracks or surface imperfections provides information that stresses exist. Other tests conducted can be related to the stress-time information. Information on the solvent mixtures suitable for this type of test and how to interrupt them are available from plastic material suppliers or determining from industry test data which show solvents that effect the specific plastic to be evaluated.

TP cracking develops under certain conditions of stress and environment sometimes on a microscale. Because there are no fib-

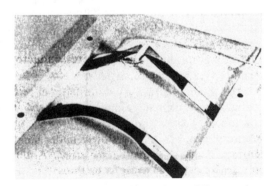

Fig. 2-54 Tensile test bars of two different plastics under the same stress were sprayed with acetone. The top one cracked quickly, but the other did not fail.

rils to connect surfaces in the fracture plane (except possibly at the crack tip), cracks do not transmit stress across their plane. Cracks result from embrittlement, which is promoted by sustained elevated temperatures and ultraviolet, thermal, chemical, and other environments.

For the designer it is not important whether cracking develops upon exposure to a benign or an aggressive medium. The important considerations are the embrittlement itself and the fact that apparently benign environments can cause serious brittle fractures when imposed on a product that is under sustained stress and strain, which is true of certain plastics.

Crazing or stress whiting is damage that can occur when a TP is stretched near its yield point. The surface takes on a whitish appearance in regions that are under high stress. Crazing is usually associated with yielding. For practical purposes stress whiting is the result of the formation of microcracks or crazes, which is another form of damage. Crazes are not true fractures, because they contain strings of highly oriented plastic that connect the two flat faces of the crack. These fibrils are surrounded by air voids. Because they are filled with highly oriented fibrils, crazes are capable of carrying stress, unlike true fractures. As a result, a heavily crazed product can still carry significant stress, even though it may appear to be fractured.

It is important to note that crazes, micro-cracking, and stress whiting represent irre-versible first damage to a material, which could ultimately cause failure. This damage usually lowers the impact strength and other properties of a material compared to those of undamaged plastics. One reason is that it exposes the interior of the plastic to attack and subsequent deterioration by aggressive fluids. In the total design evaluation, the for-mation of stress cracking or crazing damage should be a criterion for failure, based on the stress applied.

Weather Resistance

Ultraviolet rays and the heat from solar ra-diation degrade the natural molecular struc-ture of certain plastics. Acrylics, PCs, PPO, TFE, silicone, and TS polyester are exam-ples of plastics that have outstanding dura-bility under UV exposure. The resistance to sunlight of those that degrade can become weather resistant by using chemical heat sta-bilizers and/or various fillers that can screen and protect the plastic from radiation such as is done with acrylics, polypropylene, etc. Weather resistant paints and coatings can also protect plastics from UV damage. These chemical heat stabilizers are also used to en-sure no damage occurs to the plastic dur-ing heat processing such as during extrusion or injection molding. If the processing heat is increased above its normal requirement (to reduce cycle time, etc.), the stabilizer that should have remained to provide weather protection is consumed causing the product to be destroyed in outdoor service. This situ-ation actually occurred when the first IM PP outdoor stadium seats were installed many moons ago.

The effects of UV radiation on degradable plastics are usually confined to the exposed surface layers. The general effect is one of embrittlement. Tensile strength may either increase or decrease, but the elongation upon breaking is always reduced. A loss of impact strength is the usual measure of UV degra-dation. The creep rupture strength will also be reduced dramatically, and the onset of the knee in the stress-strain curve of certain plas-tics such as PE will be accelerated. UV degra-dation is aggravated by stresses or strains, and the plastic may stress crack or craze after de-terioration has occurred. The secondary ef-fect of UV degradation is usually a yellowing or browning of certain plastic.

Other elements of weather and outdoor exposure can interact with UV radiation to accelerate degradation in degradable types of plastics. They include humidity, salt spray, wind, industrial pollutants, and atmospheric impurities such as ozone, biological agents, and temperature. The wavelengths that have the most effect on plastics range from 290 to 400 nm (2,900 to 4,000 A).

One of the insidious disadvantages of cer-tain plastics is their tendency to absorb mois-ture from ambient air and then change their size and properties. There are protective measures that can be taken with these plas-tics such as coatings, chemical treatments, ad-ditives, and so on. To be practical, the best way to circumvent problems of this type is to select a plastic with the lowest possible ab-sorption rate.

Sterilization-Irradiation

Fundamentally, radiation is the emission of energy in such forms as light and heat or the transfer of energy through space by electromagnetic waves. Irradiation basically identifies the radiant energy per unit of inter-cepting area. The effect of these energies on degrading certain plastics and in changing or improving their properties is measurable. Most nontechnical people consider only that radiation results in degradation, but the irra-diation of plastics is an important science for plastic packaging sterilized medical products, curing TS plastics, converting certain TPs to TSs, and so on.

Sterilization is an important process that involves a major market for the use of plas-tics in packaging. The most common methods of sterilization are those using heat, steam (autoclaving), radiation, and gas (EtO-ethylene oxide). Unfortunately, each of these methods has its limitations. There are, however, plastics that do meet performance requirements based on the various different processes, including radiation.

Harmful Weather Component

Weather is a complex all-embracing term that includes many components. However, these elements can be listed and would seem to be amenable to analysis, since the recognized segments are not too great in number. Among the portions that affect the properties of plastics are such things as solar radiation, temperature, oxygen, humidity, precipitation, wind, biological agents, and atmospheric impurities. Different aspects complicate the picture because factors such as their concentrations and degree of two or more components influence results. Normally these components reduce performances. However there are data in the literature that show that some TS plastics grow stronger for periods of 2-4 years due to postcuring when exposed to the elements.

One of the most common impurities in coastal areas which acts in a chemical manner rather than a physical one is salt water. However, with the ever-increasing spread of the chemical industries, and the stepped-up use of gasoline powered vehicles, the problem of chemical degradation are also of interest particularly in inland areas. While plastics in general are corrosion resistant, the multiplicity of chemical agents which can be in the air in industrial atmospheres, plus the chemical nature of the various plastics indicates that it cannot be assumed that all plastics are chemically resistant to all atmospheres.

Assessing Weathering Effect

In assessing the effects of weathering, some change in property is measured. The relative rankings of the various plastics (and to a large extent the degree of correlation between artificial tests and outdoor exposure) will depend upon the property chosen for measuring. When consideration is given to the various types of properties and the number of possible choices within each type, the list becomes legion.

Depending upon the interests of the person conducting the program, and the potential uses foreseen, the test may be of a mechanical, optical (including appearance), electrical, or thermal nature. In some cases, support of a biological culture or change in dimension could be the criterion of success or failure of the material to resist exposure. Within each of the above general groups, there are still many choices.

The ASTM Book of Standards lists many different tests such as over 25 mechanical tests applicable to plastics, more than a dozen thermal tests, etc. (128). These tests will differ in sensitivity and in applicability to a potential use. For instance, a test that establishes a value dependent upon volume properties might very well rank materials in an order differing from a test dependent upon surface properties. In a similar fashion, if the material will not be extended in use, a measurement of tensile elongation loses significance. The multiplicity of available test methods has led to the paradoxical situation wherein extensive data are available but knowledge as to the performance of a specific material in a specific use must be based in many cases upon educated guesses rather than substantial data. This free choice is highly defensible and can lead to a disturbing dilemma.

In many cases, even the method of conditioning prior to test will influence the ratings. For example, flexural tests run on standard conditioned specimens (50% relative humidity and 73.5°F) may rank materials differently from tests conducted on specimens which have been immersed in water or which have been heated to some elevated temperature after outdoor exposure.

No one can afford to run all the tests even on one material. Yet the test or tests chosen may not be related to a contemplated use. This has lead to an enormous amount of duplication of time-consuming exposures. Consideration has been given to means of eliminating this duplication by the different testing organizations (particularly ASTM, UL, and ISO). To date no entirely satisfactory answer has been found. One means of obtaining comparable data would be to have all investigators perform the same test by a standard procedure on the same type of specimens prepared in the same fashion. This would still not tell how weather would affect some other property that might be of legitimate interest. Another argument against this would be the reluctance of a manufacturer to report data that would make their product appear inferior.

As might be expected from some of the previous discussion, the results of any weathering tests will be largely influenced by the exposure method used. While it is obvious that the general environment will have a bearing on the ranking assigned, it is normally not enough to know that the climate is hot, dry, and rural, or cold, wet, and industrial. In addition, the local conditions specific to a location are important. Such relatively minor aspects as angle of exposure, height above the background and the nature of the background and the season of the year when exposure is initiated all have a bearing on the results obtained.

Due to the unpredictable scheduling and high dollar costs of all weather natural testing, much of the environmental testing has been brought into laboratories or other such testing centers. Artificial conditions are provided to simulate various environmental phenomena and thereby aid in the evaluation of the test item before it goes into service under natural environments. This environmental simulation and testing does require extensive preparation and planning. It is generally desirable to obtain generalizations and comparisons from a few basic tests to avoid prolonged testing and retesting. The type and number of tests to be conducted, natural or simulated, as usual are dependent on such factors as end item performance requirements, time and cost limitations, past history, performance safety factors, shape of specimens, available testing facilities, and the environment. Specifications, such as ASTMs', provide guidelines.

When considering environment it generally becomes difficult since actual service conditions are most of the time unpredictable. As an example, there is a systematic difference in the frequency distributions of liquid water content in rain. It appears that the areas most likely to have high values of liquid water are where there is a plentiful supply of moisture and a high instability in the atmosphere. The lowest values of liquid water are obtained from the climatic areas of light continuous rains such as that found along the northwest coast of the United States.

Outer Space

The space environment, seen as beginning in the center of the earth, extends to infinity. In the past few decades outer space has been penetrated. These initial successful steps depended on a number of factors, one of which was the use of plastics. As in terrestrial uses, plastics have their place in space.

Plastics will continue to be required in space applications from rockets to vehicles for landing on other planets. The space structures, reentry vehicles, and equipment such as antennas, sensors, and an astronaut's personal communication equipment that must operate outside the confines of a spaceship will encounter bizarre environments. Temperature extremes, thermal stresses, micrometeorites, and solar radiation are sample conditions that are being encountered successfully that include the use of plastics.

Perhaps the most striking phenomenon encountered in outer space is the wide variation in temperature that can be experienced on spacecraft surfaces and externally located equipment. Temperatures and temperature gradients not ordinarily encountered in the operation of ground or airborne structures and equipment are ambient conditions for spacecraft equipment. On such hardware, not suitably protected externally or housed deep within the space vehicle in a controlled environment, these temperature extremes can wreak destruction. Designers of earthbound electronics must fight temperatures that will produce system degradation, but spacecraft electronic designers may be fighting temperatures that will cause their equipment to melt.

On either ends of the temperature scale, the ground- or airborne-equipment designer has a simpler environment problem. In addition, the space designer has a temperature paradox to consider. A black box cannot simply be placed in a superinsulated enclosure anymore than a human being can. All other factors aside, both would rapidly be destroyed because of self-generated heat. The equipment must therefore be exposed to its environment in some manner, but it also needs a great deal of protection. The problem is not as simple as putting on or taking off a sweater, depending on whether the temperature is 21 or 70°F (6 or 21°C). The problem is to put something on and keep it on, regardless of whether the temperature is −250 or +250°F (−155 or +121°C). Many factors give rise the temperature extremes encountered.

Ocean

Plastics are already vital for operation on top and within the sea, even though comparatively little is known about the sea. To develop more knowledge, radically new basic ideas and approaches were needed, such as consideration of plastic structural hulls for deep submersibles, or elastomeric plastics for undersea housing and storage.

Oceans occupy 70.8% or 125 million square miles of the surface of the earth. Within or beneath this "inner space" are foods, fuels, and minerals. Thus interest in the sea is obvious. At least 4/5 of all life on earth exists in saltwater. It is predicted that of the oil and gas demand in future years will come from oil at 2,000 ft. depths operated by manned submarines and marine robots. All the equipment needed to collect and store oil or gas will be installed and operated on the sea floor. Underwater housing and decompression chambers will be required. The sea bottom is also reported to include trillions of tons of copper, nickel, cobalt, iron, and other important minerals.

From ships to submarines to mining the sea floor, certain plastics can survive sea environments, which are considered more hostile than those on earth or in space. For water-surface vehicles many different plastic products have been designed and used successfully in both fresh and the more hostile seawater. Figure 2-55 is an example where extensive use is made using unreinforced and reinforced plastics meeting structural and nonstructural product requirements. Included are compartments, electronic scanners, radomes, optically transparent devices, food storage and dispensing containers, medical products, buoyant devices, temperature insulators, and many more.

Boats have been designed and built up to at least 37 × 9 m (120 × 30 ft.) in RP. Plastics have become vital for operating within the sea. In 1965 extensive test were conducted by the U.S. Navy' Sealab 11 to assay man's ability to live and work in ocean depths for long periods of time (Fig. 2-56). For forty-five days three groups of ten men each lived fifteen-day periods in a 57 ft. × 12 ft. habitat at a depth of 188 m (205 ft.) one half mile off La Jolla, Calif. Plastic parts as well as other materials were used to provide a highly successful experiment.

This frontier's practical opportunities were first developed with submarines, which until the nuclear ones were limited to depths of only a few hundred feet. Many thousands of feet can now be navigated. The crushing pressures below the surface, which increase at a rate of about 1/2 psi per foot of depth, make corrosion a major threat to the operation and durability of many materials. For example, the life of uncoated magnesium bolts in contact with steel nuts is less than seventy-two hours, aluminum buoys will corrode and pit after only eleven months at just four hundred feet, and low-carbon steel corroded at a rate one-third greater than in surface waters.

Tests on plastics in deep water have been extremely successful. As an example filament-wound RP cylinders and PVC buoys retained their strength. PVC washers and the silicone-seating compound used in steel-to-aluminum joints helped prevent their corrosion. Black twisted nylon and polypropylene

Fig. 2-55   Extensive uses made in military, commercial, and pleasure boats.

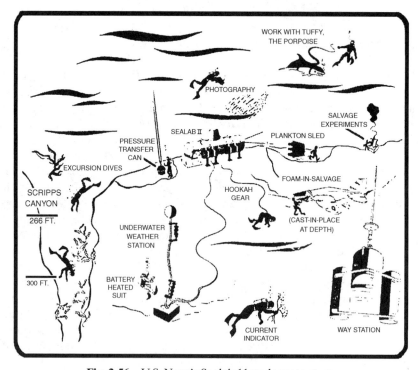

Fig. 2-56   U.S. Navy's Sealab 11 underwater test.

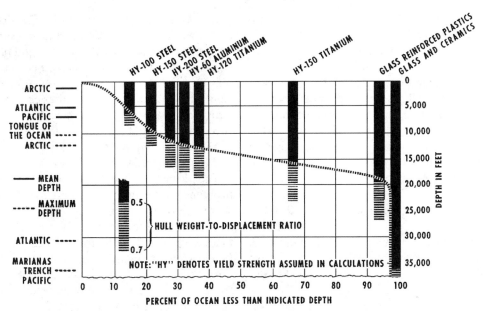

Fig. 2-57   Example of boat hull materials subjected to seawater depths.

ropes used to rig and retrieve test platforms are unaffected. Grappling lines attached to platforms, made of steel wire jacketed with extruded HDPF prevented corrosion of the steel.

PE is also used to protect submerged telephone cables for over a half century. Plastic primers such as epoxy are used to prevent antifouling plastic paints from corroding metals. These paints generally use cuprous oxide to prevent the growth of barnacles. Plastics are used successfully in instruments to determine depth, velocity of currents, temperature, and as echo sounders. Products operating to depths of 4,500 m (15,000 ft.) include molded polystyrene rotors, neutrally buoyant polyethylene control vanes, PVC buoy supports, 0-ring seals, PE flotation, and watertight electrical connectors using PVC, polyurethane, and DAP. Plastics such as PE, PP, and PUR are used to develop the shapes and provide different combinations of desirable characteristics such as ease of wrapping around standard winches, resistance to the water environment and abrasion, good electrical insulation in wire-conducting cables, and ease of fabrication and repair.

The materials studied for deep-submergence hulls are generally limited to steel (Hy 170), aluminum, titanium, reinforced plastics, and glass. Figure 2-57 shows the depth limitations of various hull materials in near-perfect spheres, superimposed on the familiar distribution curve of ocean depths. To place the materials in their proper perspective, the common factor relating their strength-to-weight characteristics to a geometric configuration for a specified design depth is the ratio showing the weight of the pressure hull to the weight of the seawater displaced by the submerged hull, a factor referred to as the weight displacement (W/D) ratio. The portions of the bars above the depth-distribution curve correspond to hulls having a 0.5 W/D ratio, the portion beneath showing the depth attainable by heavier hulls with a 0.7 W/D.

The ratio of 0.5 and 0.7 is not arbitrary, as it may appear, for small vehicles can normally be designed with W/D ratios of 0.5 or less, and vehicle displacements can become quite large as their W/D ratio approach 0.7. Using these values permits making meaningful comparisons of the depth potential of various hull materials. An examination of the data reveals that for all the metallic pressure-hull materials taken into consideration, the best results would permit operation to a depth of about 18,288 m (20,000 ft.) only at the expense of increased displacement. The nonmetallic materials of reinforced plastics (those with just glass-fiber TS polyester) and glass alone would permit operation to 20,000 ft. or more with minimum-displacement vehicles.

The submergence materials show the variation of the collapse depth of spherical hulls with the weight displacement of these materials. All these materials initially would permit building the hull of a rescue vehicle operating at 1,800 m (6,000 ft.) with a collapse depth of 2,700 m (9,000 ft.).

For a search vehicle operating at 6,000 m (20,000 ft.) with collapse depth of 9,000 m (30,000 ft.), the only materials that appear suitable are solid glass and RP. No metals can be used, because they potentially do not have sufficient strength-to-weight values. One of the drawbacks to using glass in hulls is its lack of toughness. The inside of the glass hull would need protection from impacts, etc.: thus an elastomeric plastic would be used to cover the glass. Another serious problem is the difficulty in designing penetrations and hatches in a glass hull. A solution to these problems could be a filament winding around the glass or using a tough plastic skin.

These glass structural problems show that the RP hull is very attractive on a weight-displacement ratio, strength-weight ratio, and for its fabrication capability. A significant advantage of RP over solid glass is that it is available today and the technology of fabricating large, thick-wall structures already exists. Also, with an increased modulus of elasticity in fibers other than glass additional gains are obtained beyond what is presently available in conventional RPs.

RPs have already been used in different structural applications, to replace conventional metal in seawater-compressed air surfacing ballast tanks in the Alvin depth vehicle. This vehicle, a first-generation deep research vehicle, also used RP in its outer hull construction to enclose the pressure tanks and aluminum frame. In the unmanned acoustical research vehicle of the Ordnance Research Laboratory called Divar, an RP cylinder with a 16 in. OD, 3/4 in. wall thickness, $12\frac{1}{2}$ in. ID with nine ribs, a 60 in. length and weight of 180 pounds went to depths of 950 m (6,500 ft.).

In addition to developing solid RP structures, work has been conducted on sandwich structures such as filament-wound plastic skins with low-density foamed core or a plastic honeycomb core to develop more efficient strength-to-weight structures. Sandwich structures using a syntactic core have been successfully tested so that failures occurred at prescribed high-hydrostatic pressures of 28 MPa (4,000 psi).

The design of a hull is a very complex problem. Under varying submergence depths there can be significant working of the hull structure, resulting in movement of the attached piping and foundations. These deflections, however slight, set up high stresses in the attached members. Hence, the extent of such strain loads must be considered in designing attached components.

Buoyancy in some form is employed in nearly all categories of underwater and surface systems to support them above the ocean bottom or to minimize their submerged weight. The buoyant material can assume many different structural forms utilizing a wide variety of densities. The choice of materials is severely restricted by operational requirements, since different environmental conditions exist. For example, lighter, buoyant liquids can be more volatile than heavier liquids. This factor can have a deleterious effect on a steel structure by accelerating stress corrosion or increasing permeability in reinforced plastics.

The typical syntactic foam used for buoyancy in many vehicles is made of hollow glass, ceramic, or plastic microspheres of 30 to 300 micron size, uniformly dispersed in a plastic such as epoxy. The navy, desiring to develop a material to replace the more conventional gasoline flotation one, produced an excellent syntactic foam. Strict processing and quality control in producing the foam can develop a static hydrostatic pressure of 10,000 psig and fatigue testing of 1,000 cycles.

In the Woods Hole Oceanographic Institute's three-man 1,800 m (6,000 ft.) depth vehicles, approximately 5,000 lb. of syntactic foam were used to provide buoyancy. With a specific gravity of 0.68, it required three pounds of material to gain one pound of buoyant effect. However, its main attributes were that of being able to tailor it to fit the available space and being useful to at least a 5,000 psi (35 MPa) load.

Time-Dependent Data

Different developments and theoretical approaches are always being considered for predicting expected service of plastics. Since there appears to be an endless list of new materials being developed, or improvements continually are developed in the science of plastics, the problem of obtaining immediate 5 to 20 year service data sets up problems. Even though this situation exists, recognize that there are plastics with 10 to over 50 years service; i.e., polyvinyl chloride, acrylic, silicone, phenolics, glass fiber-TS polyester RPs, etc. So from a truly designers, engineering, or scientific approach one has to be realistic that certain plastics exist with actual service data. In regard to those who desire long time data on plastics that have not been subjected to field tests, or accelerated tests, other approaches such as the following evaluations exist.

Viscoelastic and rate theory   To aid the designer the viscoelastic and rate theories can be used to predict long-term mechanical behavior from short-term creep and relaxation data. Plastic properties are generally affected by relatively small temperature changes or changes in the rate of loading application.

Time dependence   Viscoelastic deformation is a transition type behavior that is characterized by the occurrence of both elastic strain and time-dependent flow. It is the time dependence of the mechanical properties of plastics that makes the behavior of these materials difficult to analyze by mathematical theory.

Creep behavior   Creep is the deformation that occurs over a long period of time in a material subjected to a continuous load, and stress relaxation is the reduction in stress with time that occurs in a material when it is de-

formed to some specific strain which is maintained constant.

Failure can be considered as an actual rupture (stress-rupture) or an excessive creep deformation. Correlation of stress relaxation and creep data has been covered as well as a brief treatment of the equivalent elastic problem. The method of the equivalent elastic problem is of major assistance to designers of plastic products since, by knowing the elastic solution to a problem, the viscoelastic solution can be readily deduced by simply replacing elastic physical constants with viscoelastic constants.

Linear viscoelasticity   Linear viscoelastic theory and its application to static stress analysis is now developed. According to this theory, material is linearly viscoelastic if, when it is stressed below some limiting stress (about half the short-time yield stress), small strains are at any time almost linearly proportional to the imposed stresses. Portions of the creep data typify such behavior and furnish the basis for fairly accurate predictions concerning the deformation of plastics when subjected to loads over long periods of time. It should be noted that linear behavior, as defined, does not always persist throughout the time span over which the data are acquired; i.e., the theory is not valid in nonlinear regions and other prediction methods must be used in such cases.

The basic viscoelastic theory assumes a timewise linear relationship between stress and strain. Based on this assumption and using mechanical models thought to represent the behavior of a plastic material, it can be shown that the stress, at any time t, in a plastic held at a constant strain (relaxation test), is given by:

$$\sigma = \sigma_0 e^{-t/\gamma} \qquad (2\text{-}14)$$

where σ is the stress at any time t, γ is the relaxation time, σ_0 is the initial stress, and e is the natural logarithmic base number.

Using the same mechanical models and assumptions, it cam also be shown that the total deformation experienced in a creep process (with the same under constant stress σ)

is given by:

$$\varepsilon = (\sigma/E_0) + (\sigma/E)(1 - e^{-t/\gamma}) + (\sigma t/\eta)$$

$$(2\text{-}15)$$

where ε is the total deformation, E_0 is the initial modulus of the sample, E is the modulus after time t, and η is the viscosity of the plastic. Excluding the permanent set or deformation and considering only the creep involved, Eq. 2-15 may be stated as:

$$\varepsilon = (\sigma/E) + (\sigma/E)(1 - e^{-t/\gamma}) \qquad (2\text{-}16)$$

Note that the term γ in Eqs. 2-15 and 2-16 has a different significance than that in Eq. 2-14. In the first equation it is based on a concept of relaxation and in the others on the basis of creep. In the literature, these terms are respectively referred to as a relaxation time and a retardation time, leading for infinite elements in the deformation models to complex quantities known as relaxation and retardation functions. One of the principal accomplishments of viscoelastic theory is the correlation of these quantities analytically so that creep deformation can be predicted from relaxation data and relaxation data from creep deformation data.

Using viscoelastic theory, it is possible to demonstrate that:

$$(\sigma_0/\sigma)\ \text{relaxation} = (\varepsilon/\varepsilon_0)\ \text{creep} \qquad (2\text{-}17)$$

Thus, by determining values of (σ_0/σ) from a relaxation test, creep strains can be calculated using Eq. 2-17 in the form of:

$$\varepsilon = \varepsilon_0(\sigma_0/\sigma) = (\sigma_0/\sigma)(\sigma_0/E_0)(\sigma_0/\sigma)$$

$$(2\text{-}18)$$

where $(1/E_0)(\sigma_0/\sigma)$ may be thought of as a time-modified modulus, i.e., equal to $1/E$, from which the modulus at any time t, is:

$$E = E_0(\sigma_0/\sigma) \qquad (2\text{-}19)$$

that is the value to replace E in the conventional elastic soultions to mechanical problems. Where Poisson's ration, γ, appears in the elastic solution, it is replaced in the viscoelastic solution by:

$$\gamma = (3B - E)6B \qquad (2\text{-}20)$$

where B is the bulk modulus, a value that remains almost constant throughout deformations.

Creep and stress relation   Creep and stress relaxation behavior for plastics are closely related to each other and one can be predicted from knowledge of the other. Therefore, such deformations in plastics can be predicted by the use of standard elastic stress analysis formulas where the elastic constants E and γ can be replaced by their viscoelastic equivalents given in Eqs. 2-19 and 2-20.

If data are not available on the effects of time, temperature, and strain rate on modulus, creep tests can be performed at various stress levels as a function of temperature over a reasonable period of time. In this regard, reasonable is a relative term. For applications like rockets and missiles, data obtained over a time period of 4–5 sec to an hour provide the essential information. For structural applications, such as pipelines, data over a period of years are required.

This is the one serious limitation in plastic design problems. Even if the designer did wait for data on one material, chances are the final design might be switched to another plastic or formulation. Thus, as a compromise, data from relatively short-term tests are extrapolated by means of theory to long-term problems. However, when this is done, the limitations inherent in the procedure should be kept in mind.

Rate theory   An alternate method available involves the manipulation of the rate theory based on the Arrhenius equation. This procedure requires considerable test data but the indications are that considerably more latitude is obtained and more materials obey the rate theory. The method can also be used to predict stress-rupture of plastics as well as the creep characteristics of a material, which is a strong plus for the method.

If it is assumed that the physical and chemical properties of the material are the same before and after rupture (so that the concentration of material undergoing deformation is related to the rate constant, K, by $x = Kt$, where t is time) then it can be shown, as in the following equation, that for plastics:

$$A/R = K^1 = [TT_0/(T_0 - T)](20 + \log t)$$

$$(2\text{-}21)$$

where A is the activation energy for the process, R is the gas constant, T is the absolute temperature of the process, T_0 is the absolute temperature at which the material has no strength, K^1 is a constant, and t is time.

For some materials, rupture curves can be computed for all values of T related to the magnitude of the stress applied. For design purposes, if the required time and operating temperature are specified, K^1 can be computed and the value of stress required to cause rupture at that time and temperature can be read off charts.

Creep deformations are calculated by dividing the stress by the modulus of the material. The deformation observed in a short-term tensile test at an elevated temperature is related to the deformation that takes place at a lower temperature over a longer period of time. The short-term data thus obtained can be used to obtain long-term modulus data through the development of a master modulus curve. Being able to determine the modulus at any time t and knowing the constant value of stress to which a material is subjected, it is then possible to predict the creep which will have been experienced at time t by simply dividing the stress by the modulus using conventional elastic stress analysis relationships.

Designing plastic Basically the general design criteria applicable to plastics are the same as those for metals at elevated temperature; that is, design is based on (1) a deformation limit, and (2) a stress limit (for stress-rupture failure). There are, of course, cases where weight is a limiting factor and other cases where short-term properties are important.

In computing ordinary short-term characteristics of plastics, the standard stress analysis formulas may be used. For predicting creep and stress-rupture behavior, the method will vary according to circumstances. In viscoelastic materials, relaxation data can be used in Eqs. 2-16 to 2-20 to predict creep deformations. In other cases the rate theory may be used.

Molecular Weight and Aging

MW and aging may each be cause and/or effect on plastics. Reactivity with oxygen, ozone, moisture, and UV light sensitization via outdoor weathering and/or high temperature all become important with aging particularly the NEAT plastics. Different additives are used with different plastics to provide long-time aging. Certain plastics will improve with aging based on actual service tests and extensive creep tests. However, certain plastics have limited endurance. This action is somewhat related to MW where low MW materials tend to degrade and the higher MWs become stronger through cross-linking (Chapter 8).

Arrhenius Plot Theory

Another technique known and available for evaluating and predicting performance in special applications concerns the Arrhenius plot.

Why materials age Aging involves both chemical and physical changes, although many of the latter are just the visible manifestation of the former. While most of these should be obvious to the chemist and engineer, it is important that they be reviewed as background to the basic approach to be taken.

The chemical changes are due to reactions between materials of different energies, and may involve some complex reaction kinetics. Whenever two materials are combined, many products may be formed. Thus, oxygen may be absorbed and oxidation of the materials result. The exact reaction depends on the presence of catalytic agents. The oxidation may cause the materials to cross-link and cause hardening or shrinkage that may show up as stress cracking. It may degrade the molecules to lower molecular weight or volatile products, causing a volume decrease or stress cracking. Or it may lead to acidic products which will discolor the material or bring about further degradations.

Other degradative reactions can occur in the absence of oxygen. The presence of water may bring about hydrolitic reactions and, if oxygen is also present, may increase oxidation reactions. All of these may be going on simultaneously and in the most complicated manner. The lower molecular weight units can cause changes in flow, shape, strength, or other desired properties.

Physical changes may result from the chemical changes, but even without a chemical reaction the low molecular weight materials may be lost by volatilization, and the material may be undergoing a stress relaxation. Volatilization is a slow process that depends on the partial vapor pressure of the lower molecular weight volatile constituents, unless they are polar or have electrostatic attraction for other molecules. It may result in hardening from the loss of plasticizers identified as such and also from the loss of low molecular weight materials, or even water, which can act as plasticizers. Stress relaxation may occur due to mechanical or thermal stress. Differential expansion of organic and inorganic parts often results in cracking. These effects are particularly possible at temperatures near those where changes or property transitions occur (for example, softening or melting points or brittle points of materials).

Rate of aging process   It has been known for a long time as an empirical fact that many reactions approximately double or treble their rates with a $10°C$ rise in temperature. A more quantitative relation is given by the classical Arrhenius modified equation:

$$\log k = (E/2.303R)(1/T) + C \qquad (2\text{-}22)$$

where k = specific reaction rate; E = activation energy for the reaction; R = gas constant per gram molecular weight; and T = absolute temperature, $°K$.

A straight line is produced when the logarithm of a specific reaction rate is plotted against the reciprocal of the absolute temperature. Temperature has a marked influence on the reaction rates, but the range between reactions that are too slow or too fast to measure is really quite narrow.

Similarly, the rate of evaporation of materials depends on the vapor pressure, p, of the volatile constituent, which in turn varies directly as its molar concentration and the temperature:

$$\log p = (M/T) + C \qquad (2\text{-}23)$$

Volatilization is also affected by the ventilation rate over the surface of the material, but when this is constant, a straight-line result from the plot of the logarithm of the vapor pressure against the reciprocal of the absolute temperature.

Diffusion of a reactive component or a volatile constituent into or out of a material is also a temperature dependent rate phenomenon, as:

$$\log D = K - (E/RT) \qquad (2\text{-}24)$$

where D is the diffusion rate and K is a force factor dependent on the velocity of motion of the molecule and the frictional resistance to this motion.

Thus, whether the changes in the material are due to chemical reactions, volatilization, or diffusion, one can expect a linear relationship between the logarithm of life (i.e., time to failure) and the reciprocal of absolute temperature. But there is no sound basis for extrapolating the effect of changing the concentration of the environmental exposure medium or the physical functions.

It is possible, in some situations, that two different phenomena which proceed at different rates with different temperature coefficients or activation energies will affect the physical properties. In such complex cases, it is not expect to obtain a linear relation between the logarithm of life and reciprocal absolute temperature. If one obtains a nonlinear curve, however, it may he possible to identify the reaction causing the nonlinearity and correct for it. When one can make such a correction, one obtains a linear relationship.

Using the Arrhenius equation to predict performance   It has been shown that temperature alone is a sufficient accelerating means. Now one must consider how best to apply this criterion to a prediction of performance. The conditions of an accelerated aging test should correspond as closely as possible to the conditions encountered in actual service, and the material must be tested in the form and in the function in which one wants to evaluate its performance.

The accelerated aging test should take into account the associated materials as well as the atmosphere that will be encountered in actual use, since they are also controlling factors. It is helpful to include materials of known performance against which to rate the new material, since this allows a check of controlling factors and further validates the extrapolation. Thus, existing data from long term tests may be of considerable value.

For a criterion of failure, life tests should measure the time required for a material to deteriorate to a condition where it is no longer capable of performing its intended function. Careful analysis and testing are required to determine the most important condition or property at the time of failure or the point when a material becomes inadequate to its intended function.

Test temperature. Tests should be run at a minimum of three temperatures and preferably four to confirm the linear relation between the logarithm of life and the reciprocal of absolute temperature. Several samples are necessary to plot the results of changes occurring versus time for each testing temperature. The time available and the accuracy of the extrapolation desired determine the lowest test temperature. Usually the most desirable lowest temperature is one that will give results in about 1,000–2,000 h (6–12 weeks).

Normal oxidative degradation. There is seldom any reason to go below a 75% or a 50% retention of properties, as the material has usually changed rather drastically at these levels.

Cross-linking. In cross-linking the materials often improve before they begin to degrade; otherwise, the results are the same as in the oxidative degradation.

Catalytic degradation. There are materials that show no change for a period of time, then degrade rapidly. As an example the degradation of certain polypropylenes, which initiates at a site and then propagates by chain scissor, would give this effect, resulting from the depletion of a protective ingredient such as a heat stabilizer or antioxidant. In this case, the only significant end point is the very early loss of properties.

Arrhenius plot. In an Arrhenius plot the ordinate is the log of the material life. The abscissa is the reciprocal of the absolute temperature. The linear curves obtained with the Arrhenius plot overcome the deficiency of most of the standard tests, which provide only one point and indicate no direction in which to extrapolate. Moreover, any change in any aspect of the material or the environment could alter the slopes of there curves. Therein lies the value of this method.

There is supporting evidence in the literature for the validity of this method; two cases in particular substantiate it. In one, tests were made on plastics heated in the pressure of air. Differential infrared spectroscopy was used to determine the chemical changes at three temperatures, in the functional groups of a TP acrylonitrile, and a variety of TS phenolic plastics. The technique uses a film of un-aged plastic in the reference beam and the aged sample in the sample beam. Thus, the difference between the reference and the aged sample is a measure of the chemical changes.

The results showed that the rate of change was markedly temperature dependent and that the degradation at each of the three temperatures was identical. In other words, the amount of degradation that occurred in 4 h at 200°C (392°F) was identical to the amount in 725 h at 100°C (212°F). The same tests in a vacuum showed no degradation, indicating that the initiating step in the degradation process in air must be the attack of oxygen. With acid catalyzed phenolic plastic, similar results were obtained. These results are important; they lend some authenticity to this technique.

The second case concerns the results, published by du Pont, of actual service life tests on Zytel 101 and 103, which were run for more than 12 years. Some of these data were plotted on an Arrhenius plot. It is not surprising that the heat stabilized Zytel 103 is markedly more stable to heat oxidation than the Zytel 101, since the heat stabilizers were developed by high temperature evaluations until a stabilizing additive was found. What may not have been readily apparent, however, is that the two materials would have about the same life expectancy at room temperature.

Data was collected over a two-year period on the effect of water on DuPont's Zytel 101. In an Arrhenius plot of this data the failure point was the time when the elongation and impact strength started to decrease. This is not a chemical degradation, but rather a permeation or diffusion rate phenomenon. It shows that high temperature water tests can be used to predict normal temperature exposure results.

Usefulness of thermal evaluation technique
The following list includes examples where the technique can be used.

Fundamental Studies

Guiding research in tailor-making plastics

Fast evaluations and comparisons will show if desired improvements are being made

Curing studies-pot life

Engineering data for design

Stress relaxation at low loads under different exposures

Long term strength retention of glass fiber in wet service

Fatigue testing-flexure, rotation or fold

Product or end use evaluations

Selecting or screening materials specifications

Control testing

Predicting end use applications

Water immersion

Vacuum or space conditions

Pipe burst

Storage effects on solid propellant binders

Resistance to chemicals

Radiation

Soil burial

Weathering-UV exposure

Ozone

Permeation through films

Detergent effects on blown bottles

Discoloration

High Temperature

Plastics have found numerous uses in specialty areas such as hypersonic atmospheric flight and chemical propulsion exhaust systems. The particular plastic employed in these applications is based on the inherent properties of the plastics or the ability to combine it with another component material to obtain a balance of properties uncommon to either component. Some of the compositions and important properties of plastics are given in Tables 2-9 and 2-10 that have been developed over the years for use in flight vehicles and propulsion systems that are dependent upon chemical, mechanical, electrical, nuclear, and solar means for accelerating the working fluid by high temperatures.

Since 1950, plastics have been development for uses in very high temperature environments. By 1954, it was demonstrated that plastic materials were suitable for thermally protecting structures during intense propulsion heating. This discovery, at that time, became one of the greatest achievements of modern times, because it essentially initially eliminated the "thermal barrier" to hypersonic atmospheric flight as well as many of the internal heating problems associated with chemical propulsion systems.

Only chemical propulsion will be further discussed, and in particular, that associated with liquid, solid, and hybrid motors and engines. These motors and engines are uniquely different from other chemical propulsion systems in that they carry on board the necessary propellants, as contrasted to jet engines that rely on atmospheric oxygen for combustion of the fuel.

Table 2-9 Typical ablative compositions

Plastic-Base	Ceramic-Base	Elastomer-Base	Metal-Base
Polytetrafluoroethylene	Porous oxide (silica) matrix infiltrated with phenolic resin	Silicone rubber filled with microspheres and reinforced with a plastic honeycomb	Porous refractory (tungsten) infiltrated with a low melting point metal (silver)
Epoxy-polyamide resin with a powdered oxide filler	Porous filament wound composite of oxide fibers and an inorganic adhesive, impregnated with an organic resin	Polybutadiene-acrylonitrile elastomer modified phenolic resin with a subliming powder	Hot-pressed refractory metal containing an oxide filler
Phenolic resin with an organic (nylon), inorganic (silica), or refractory (carbon) reinforcement	Hot pressed oxide, carbide, or nitride in a metal honeycomb		
Precharred epoxy impregnated with a noncharring resin			

Hypersonic Atmospheric Flight

Progress in aeronautics and astronautics within the past decades has been remarkable because people have learned to master the difficult feat of hypervelocity flight. A variety of manned and unmanned aircraft have been developed for faster transportation from one point on earth to another. Similarly, aerospace vehicles have been constructed for further exploration of the vast depths of space and the neighboring planets in the solar system.

All bodies traveling in a fluid experience dynamic heating, the magnitude of which depends upon the body characteristics and the environmental parameters. Modern supersonic aircraft, for example, experience appreciable heating. This incident flux is accommodated by the use of an insulated metallic structure, which provides a near balance between the incident thermal pulse and the heat dissipated by surface radiation. Hence, only a small amount of heat has to be absorbed by mechanisms other than radiation.

Table 2-10 Plastics for propulsion environments

Major Property of Interest	Type of Polymer	Propulsion System Application
Ablative	Phenol-formaldehyde	Charring resin for rocket nozzle
Chemical resistance	Fluorosilicone	Seals, gaskets, hose linings for liquid fuels
Cryogenic	Polyurethane	Insulative foam for cryogenic tankage
Adhesion	Epoxy	Bonding reinforcements on external surface of combustion chamber
Dieletric	Silicone	Wire and cable electrical insulation
Elastomeric	Polybutadiene-acrylonitrile	Solid propellant binder
Power transmission	Diesters	Hydraulic fluid
Specific strength	Epoxy-novolac	Resin matrix for filament wound motor case
Thermally nonconductive	Polyamides	Resin modifier for plastic thrust chamber
Absorptivity:emissivity ratio	Alkyd silicone	Thermal control coating
Gelling agent	Poly(vinyl chloride)	Thixotrophic liquid propellant

As flight speeds increased to about 8,000 fps heating increases to a point where some added form of thermal protection was necessary to prevent thermostructural failure. In a somewhat similar manner, hypervelocity vehicles transcending through a planetary atmosphere also encounter gas-dynamic heating. The magnitude of heating is very large, however, and the heating period is much shorter. This latter type of thermal problem is frequently referred to as the "reentry heating" problem, and it posed one of the most difficult engineering problems of the twentieth century.

The intended mission of a hypervelocity vehicle will dictate its flight velocity and trajectory. This point is illustrated in Fig. 2-58, which presents the general altitude-velocity. The severity of the gas-dynamic heating problem increases with flight velocity. It becomes particularly acute when a vehicle must be slowed from a very high speed to a much lower impact or landing speed. In other words, hypervelocity vehicles possess a tremendous amount of kinetic and potential energy that must be dissipated during deceleration.

As shown in Fig. 2-58, a body entering the earth's atmosphere at 25,000 fps has a kinetic energy equivalent to 12,500 Btu/lb of vehicle mass. Assuming the vehicle weighs a ton, it possesses a thermal energy equivalent to 25,000,000 Btu. This magnitude of energy greatly exceeds that required too completely vaporize the entire vehicle. Fortunately, only a very small fraction of the kinetic energy converted to heat reaches the body while the remainder is dissipated in the gas surrounding the vehicle.

Hyperenvironment Materials performance during hypersonic atmospheric flight depends upon certain environmental parameters. These thermal, mechanical, and chemical variables differ greatly in magnitude and with body position. In general, they are concerned with temperatures from about 2,000 to over 20,000°F (1,100–11,000°C), gas enthalpies up to 40,000 Btu/lb, convective/radiative heating from 10 to over 10,000 Btu/ft^2/sec, stagnation pressures less than 1 to over 100 atm., surface shear stresses up to about 900 psf, heating times from a few to several thousand seconds, and gaseous compositions involving molecular, dissociated, and ionized species. The type and magnitude of effects produced by these environmental parameters are dependent upon the local aerothermo-chemical and gas-dynamic state of the flow field.

Thermal protection The design of vehicles for hypersonic atmospheric flight represents a compromise between the intended

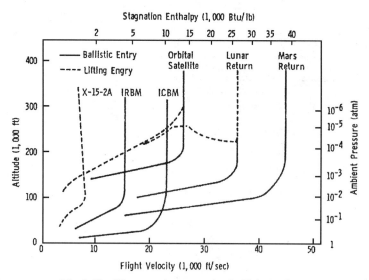

Fig. 2-58 Hypersonic atmospheric flight regime.

mission, the thermostructural aspects of the environment, the rate and magnitude of vehicle deceleration permitted, and the amount of lift necessary for flight control and landing at a predetermined point on some planet. The heating problem associated with high performance vehicles has been solved by a variety of design techniques. These include radiative cooling, heat sinks, transpiration cooling, ablation, and combinations thereof. Each thermal protective scheme is applicable to a particular portion of the flight regime, with reduced efficiency or no utility at other flight conditions.

Ablation The most common design approach for handling intense heating and extremely high temperatures is ablation. In this process, surface material is physically removed or a temperature-sensitive component of a composite is preferentially removed. The injected vapors alter the chemical composition, transport properties, and temperature profile of the boundary layer, thus reducing the beat transfer to the material surface. At high ablation rates, the heat transfer to the surface may be only 15% of the thermal flux to a non-ablating surface. Up to tens of thousands of Btu's of heat can be absorbed, dissipated and blocked per pound of ablative material through the sensible heat capacity, chemical reactions, phase changes, surface radiation and boundary layer cooling of the ablator (Fig. 2-59).

Ablative systems are not limited by the heating rate or environmental temperature, but rather by the total heat load. In spite of this limitation, however, the versatility of ablation has permitted it to be used on various hypervelocity atmospheric vehicles. No single, universally acceptable ablative material has been developed. Nevertheless, the interdisciplinary efforts of materials scientists and engineers have resulted in obtaining a wide variety of ablative compositions and constructions. These thermally protective materials have been arbitrarily categorized by their matrix composition, and typical materials are given in Table 2-9.

Plastic-base composites that employ an organic matrix, is the most widely used class of plastic ablative heat protective materials. They respond to a hyperthermal environment in a variety of ways, such as depolymerization-vaporization (polytetrafluoroethylene), pyrolysis-vaporization (phenolic, epoxy) and decomposition-melting-vaporization (nylon fiber reinforced plastic). The principal advantages of plastic-base ablators are their high heat shielding capability and low thermal conductivity. The major limitations are high erosion rates during exposure to very high gas-dynamic shear forces and, limited capability to accommodate very high heat loads.

Elastomeric-base materials represent a second major class of ablators. They thermally decompose by such processes as

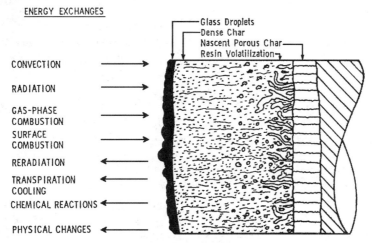

Fig. 2-59 Surface heat balance of an ablating glass fiber reinforced phenolic plastic.

depolymerization, pyrolysis, and vaporization. Most of the interest to date has been focused on the silicone plastics because of their low thermal conductivity, high thermal efficiency at low to moderate heat fluxes, low temperature properties, elongation of several hundred percent at failure, oxidative resistance, low density, and compatibility with other structural materials. They are generally limited by the amount of structural quality of char formed during ablation, that restricts their use in hyperthermal environments of relatively low mechanical forces.

Thermoplastic and elastomeric plastics tend to thermally degrade into simple monomeric units with the formation of considerable liquid and a lesser amount of gaseous species. Little or no solid residue generally remains on the ablating surface. On the contrary, most thermoset plastics and highly cross-linked plastics (especially those with aromatic ring structures) form a hard surface residue of porous carbon. The amount of char formed depends upon various factors. (1) The carbon-to-hydrogen ratio present in the original plastic structure. (2) Degree of cross-linking and tendency to further cross-link during heating. (3) Presence of foreign elements like the halogens, asymmetry and aromaticity of the polymer structure. (4) Degree of vapor pyrolysis of the ablative hydrocarbon species percolating through the char layer. (5) Type of elemental bonding.

With the formation of a carbonaceous layer, the primary region of pyrolysis gradually shifts from the surface to a substrate zone beneath the char layer. The newly formed char structure is attached to the virgin substrate material and remains thereon for at least a short period of time. Meanwhile, its refractory nature serves to protect the temperature-sensitive substrate from the environment. Gaseous products formed in the substrate pass through the porous char layer, undergo partial vapor phase cracking, and deposit pyrolytic carbon (or graphite) onto the walls of the pores.

As the organic plastic or its residual char are removed by the ablative aspects of the hyper-environment, the reinforcing fibers or particle fillers are left exposed and unsup-

ported. Being vitreous in composition, they undergo melting. The resultant molten material covers the surface as liquid droplets, irregular globules, and/or a thin film. Continued addition of heat to the surface causes the melt to be vaporized. A fraction of the melt may be splattered by internal pressure forces, or sloughed away when acted upon by external pressure and shear forces of the dynamic environment.

Chemical Propulsion Exhaust

The basic purpose of a propulsion system is to convert the thermal energy of a chemical reaction into useful kinetic energy by directing the flow of the resultant products. In other words, the propulsion system is to provide thrust for the movement of a vehicle. Expulsion of material is the essence of thrust production, and without material to expel no thrust can be produced, regardless of how much energy is available. The amount of thrust generated is equal to the rate of propellant consumption multiplied by the exhaust gas velocity. In order to maximize the exhaust velocity, it is necessary to have the combustion process take place at the highest possible temperature and pressure. Energy is released in the process, with a major fraction appearing as thermal (heat) energy.

The combustion process is carried out in a thrust chamber or a motor case, and the reaction products are momentarily contained therein. The newly formed species are heterogeneous in composition and involve a wide variety of low molecular weight products. The temperature of these products is generally high, and it ranges from about $2,000°F$ ($1,100°C$) in gas generators to well over $8,000°F$ in advanced liquid propellant engines. The combustion products leave the chamber and are directed and expanded in a nozzle to obtain velocities from about 5,000 to 14,000 ft/sec.

Cooling techniques Various methods have been developed to cope with high temperature and heating problems. They are based on absorptive, dissipative, and mass transfer cooling systems. More specifically, they include regenerative cooling, inert or endothermic heat sinks, ablation, and combinations of the preceding techniques.

A form of cooling, and the one of prime interest, concerns ablative cooling. It is essentially a heat and mass transfer process in which mass is expended to achieve thermal dissipation, absorption, and blocking. The process is passive in nature, serves to control the surface temperature, and greatly restricts the flow of heat into the material substrate. As a result of these desirable attributes, ablative cooling (includes use of plastic compositions) has been widely used for thermal protection of solid propellant motors and less extensively in liquid propellant motors.

Flammability

When plastics are used, their behavior in fire must be considered. Ease of ignition, the rate of flame spread and of heat release, smoke release, toxicity of products of combustion, and other factors must be taken into account. Some plastics bum readily, others only with difficulty, and still others do not support their own combustion A plastic's behavior in fire depends upon the nature and scale of the fire as well as the surrounding conditions. Fire is a highly complex, variable phenomenon, and the behavior of plastics in a fire is equally complex and variable (Chapter 5, **FIRE**).

Early in the past century it was thought that the matter of fire hazard was simple enough: does the material bum, or not? Wood burns; steel does not. Although these statements are certainly true, they are almost irrelevant to the relative fire risk of the two materials. Compare fires in two different buildings, one framed of heavy timbers (or plastic bonded-laminated wood arches, RP-TS plastics, etc.) and the other of steel framing. The steel frame could collapse after only a few minutes of exposure to fire, but it may require a fire of very long duration (days) to bring down the timber-framed building (and other materials). Fire reaches 1,370°C (2,500°F). Steel basically takes only

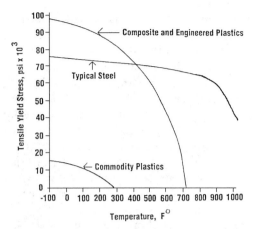

Fig. 2-60 Strength vs. temperature of steels and plastics.

up to about 538°C (1,000°F), making it collapse like a pretzel (Fig. 2-60). Wood, like certain plastics, can take the heat, and it takes a rather long time to self-destruct, thus giving time for people to leave the scene of the fire and so on.

Fire tests of plastics, like fire tests generally, are frequently highly specific, with the results being specific to the tests. The results of one type of test do not in fact often correlate directly with those of another and may bear little relationship to actual fires. Some tests are intended mainly for screening purposes during research and development, whereas others, such as large-scale product tests, are designed to more nearly approximate actual fires. Consequently, such often-used terms as self-extinguishing, nonburning, flame spread, and toxicity must be understood in the context of the specific tests with which they are used.

Some materials may bum quite slowly but may propagate a flame rapidly over their surfaces. Thin wood paneling will burn readily, yet a heavy timber post will sustain a fire on its surface until it is charred, then smolder at a remarkably slow rate of burning. Bituminous materials may spread a fire by softening and running down a wall. Steel of course does not burn, but is catastrophically weakened by the elevated temperatures of a fire. PVC does not bum, but it softens at relatively low temperatures. Other plastics may not burn readily but still emit copious amounts of smoke. And some flammable plastics, such as polyurethane, may be made flame retardant (FR) by incorporating in them additives such as antimony oxide. Other plastics basically do not bum, such as silicone and fluorine.

The principles of good design for fire safety are as applicable to plastics as to other materials. The specific design must be carefully considered the properties of the materials taken into account, and good designer judgment applied. When evaluating the fire risk that exists with plastic products it is always best to perform appropriate tests on the end items (Chapter 4, **RISK**). However, it is often helpful to select plastic materials for specific applications by first evaluating the flammability of the plastics under consideration in laboratory tests if the data is not available. These tests, often used for specifying materials, fall into the category either of small-scale or large-scale tests. Of course, as in evaluating any properties, having prior knowledge or obtaining reliable data applicable to fire or other requirements is the ideal situation.

Smoke   Toxic smoke and fumes have became generally recognized as the major cause of fire deaths, making the combustion products released by burning plastics and other materials particularly important. Smoke is recognized by firefighters as being in many ways more dangerous than actual flames because (1) it obscures vision, making it impossible to find safe means of egress, thus often leading to panic, (2) it makes helping or rescuing victims difficult if not impossible; and (3) it leads to physiological reactions such as choking and tearing. Smoke from plastics, wood, gasoline, coal, oil, and other materials usually contains toxic gases such as carbon monoxide (CO), which has no odor, often accompanied by noxious gases that may lead to nausea and other debilitating effects as well as panic warning the fire victim of danger. With only CO the victim would die whereas the start of a fire with noxious gases could alert a person that a fire has started and leave the area.

Whether a plastic gives off light or heavy smoke and toxic or noxious gases depends on the basic plastic used, its composition of additives and fillers, and the conditions under which its burning occurs. Some plastics burn with a relatively clean flame, but some may give off dense smoke while smoldering. Others are inherently smoke producing (Fig. 2-61).

In a particular application, therefore, careful consideration should be given to the relative importance of smoke and flame, including creating designs favoring the rapid elimination of smoke by venting, for fending off smoke, and other approaches.

Different regulations, such as those of the Federal Aviation Administration, Department of Transportation, and local building codes, mandate that the designs of certain products comply with specific flammability test requirements. Flame-retardancy requirements generally include limits on flame spread, burning time, dripping, and smoke emission. A multitude of flammability tests has been developed, with more than 100 known just in the United States. Different organizations are actively involved providing information, software, etc.; they include UL, ASTM, ISO, and NIST (Chapter 5).

One of the most stringent and most widely accepted test is UL 94 that concerns electrical devices. This test, which involves burning a specimen, is the one used for most flame-retardant plastics. In this test the best rating is UL 94 V-0, which identifies a flame with a duration of 0 to 5 s, an afterglow of 0 to 25 s, and the presence of no flaming drips to ignite a sample of dry, absorbent cotton located below the specimen. The ratings go from V-0, V-1, V-2, and V-5 to HB, based on specific specimen thicknesses.

Intumescence coating   Historically, smoke and the resultant toxic fumes caused by the burning of a flammable substrate were part of any fire, regardless of whether or not a fire retardant treatment was applied. What was needed was to smother the fire and thus stop the generation of toxic smoke and save the flammable substrate from further damage. Intumescence coatings were developed over a half century ago via the US Navy R&D programs for use on board ships and thereafter industry projects developed different types of water resistant intumescent coatings.

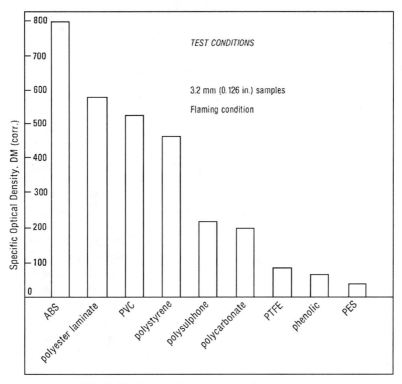

Fig. 2-61 Example of smoke emission results.

These intumescent coatings, when subjected to fire, form a char between the substrate and the fire source. The basic product coated becomes flameproof (36, 159).

Instability Behavior

Shrinkage/Tolerance

Among the inherent characteristics of each plastics material is its tendency to shrink or expand. While the problems of shrinkage is of major concern to the designer of tools (molds, dies, etc.) and in turn the fabricators of the products, it should also be understood, as reviewed in this and other Chapters, that the product designer has a major role in setting the required and proper product dimensions. By setting up the required shrinkage and tolerance requirements, the product designer can significantly influence eliminating or reducing problems in production that directly increases costs. Where tight tolerances are not required, they may

be desired to reduce material and fabricating costs (Chapter 8, **INFLUENCE ON PERFORMANCE, Tolerance and Dimensional Control**).

Shrinkage is defined as the difference between corresponding linear dimensions of the product and fabricating tool (under controlled atmospheric conditions). Though the shrinkage characteristics of plastics are known and compensated for by a tool design, the complex and overlapping nature of the factors affecting shrinkage prevents exact scientific prediction. These factors include the temperature range to which the plastic is exposed during the fabricating operation, the material's coefficient of thermal expansion, and the degree to which the plastic has been compressed during the fabricating operation.

Shrinkage can influence product performances such as mechanical properties. Anisotropy directional property can be used when referring to the way a material shrinks during processing, such as in injection molding (Fig. 2-62) and extrusion. Shrinkage is an important consideration when fabricating

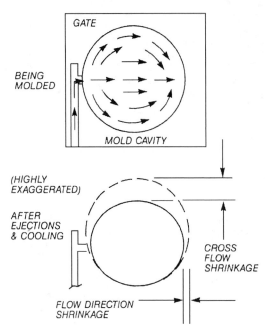

Fig. 2-62 Example of directional shrinkage in an injection molded product.

plastics, particularly crystalline TPs or those with glass fibers. The plastic melt flow direction can have more shrinkage than the cross-flow direction. The control of shrinkage is made to meet design requirements by factors such as the design of the mold or die shape, the processing-machine controls, the change of product shape, and the type of plastics.

Heat Generation

This chapter has covered the reaction of plastics to varying loads and environments. The heating of plastics by hysteresis and the methods of designing to avoid frictional heat failures of such products has been discussed. Fatigue effects on plastics subject to periodic loading as a failure mechanism was reviewed with emphasis on the design engineering approach to avoid fatigue failure. The treatment is concluded with a discussion of the use of viscoelastic effects to damp vibration and transient loads to reduce flutter and, in general, to absorb mechanical energy and prevent damage from induced mechanical vibration. By using the energy absorbing

properties of plastics the range of useful components is increased.

Annealing

Also called hardening, tempering, physical aging, and heat treatment. The annealing of plastics can be define as a heat-treatment process directed at improving performance by removal of stresses or strains set up in the material during its fabrication. These unwanted stresses in the fabricated product can cause service instability such as warping and cracking. Depending on the plastic used, it is brought up to a required temperature for a definite time period, and then liquid (usually water; also use oils and waxes) and/or air-cooled (quenched) to room temperature at a controlled rate.

Basically the temperature is usually near, but below the melting point. At the specified temperature the molecules have enough mobility to allow them to orient to a configuration removing or reducing residual stress. The objective is to permit stress relaxation without distortion of its shape and obtain maximum performances and/or dimensional control. For certain plastics and product configurations, a holding frame is used during the annealing process to eliminate unwanted warpage.

Annealing is generally restricted to thermoplastics, either amorphous or crystalline. Result is increasing density, thereby improving the plastics heat resistance and dimensional stability when exposed to elevated temperatures. It frequently improves the impact strength and prevents crazing and cracking of excessively stressed products. The magnitude of these changes depends on the nature of the plastic, the annealing conditions, and the product's geometry.

The most desirable annealing temperatures for amorphous plastics, certain blends, and block copolymers is just above their glass transition temperature (T_g) where the relaxation of stress and orientation is the most rapid. However, the required temperatures may cause excessive distortion and warping.

The plastic is heated to the highest temperature at which dimensional changes owing to strain are released. This temperature can be determined by placing the plastic product in an air oven or water liquid bath and gradually raising the temperature by intervals of 3° to 5°C until the maximum allowable change in shape or dimension occurs. This distortion temperature is dictated by the thermomechanical processing history, geometry, thickness, and size. Usually the annealing temperature is set about 5°C lower using careful quality control procedures.

Rigid, amorphous plastics such as PS and PMMA are frequently annealed for stress relief. Annealing crystalline plastics, in addition to the usual stress relief, may also bring about significant changes in the nature of their crystalline state. The nature of the crystal structure, degree of crystallinity, size and number of spherulites, and orientation control it. In cases when proper temperature and pressure are maintained during processing, the induced internal stresses may be insignificant, and annealing is not required.

Plastic blends and block copolymers typically contain other low and intermediate molecular weight additives such as plasticizers, flame-retardants, and UV or thermal stabilizers. During annealing, phase and micro-phase separation may be enhanced and bleeding of the additives may be observed. The morphologies of blends and block copolymers can be affected by processing and quenching conditions. If their melt viscosities are not matched, compositional layering perpendicular to the direction of flow may occur. As in the case of crystalline plastics, the skin may be different both in morphology and composition. Annealing may cause more significant changes in the skin than in the interior.

Plastic Material and Equipment Variable

The subject of plastic materials and equipment variables are important to understand so products can be properly designed and processed. Details on **MATERIAL VARIABLE** are in Chapter 6 and on **EQUIPMENT/PROCESSING VARIABLE** are in Chapter 8.

Finite Element Analysis

Introduction

Designing products is usually performed based on experience since most products only require a practical approach (Fig. 1-4). Experience is also used in producing new and complex shaped products usually with the required analytical evaluation that involves stress-strain characteristics of the plastic materials. Testing of prototypes and/or preliminary production products to meet performance requirements is a very viable approach used by many.

Classical equations and formulas from engineering handbooks for stress-strain static and dynamic loads are utilized. Computer-aided design analysis such as the use of finite element analysis (FEA) can be used. FEA can help a designer to take full advantage of the unique properties of plastics by making products lighter, yet stronger while at the same time also saving processing time and money. The use of FEA has expanded rapidly over the past decades. Unlike metals, plastics are nonlinear, so they require different software for analysis, The early software programs were difficult and complex, but gradually the software for plastics has become easier to use. Graphic displays are better, they consume less time, and are easier to understand (162).

One uses FEA to shorten the lead-time to less than half. FEA helps to reduce material costs while lessening the expense of building prototypes and remachining tools. By eliminating excess material, it can save weight. It can simulate what will happen, allowing immediate redesign to prevent premature failure, The designer does less guessing and more engineering, thereby improving reliability. Accuracy is increased.

With the computer hundreds of simultaneous equations can be solved that would take literally years to solve without a computer. FEA can be defined as a numerical technique involving breaking a complex problem down into small subproblems via computer models that can be solved by a computer. The key to effective FEA modeling is to concentrate element details at areas of highest stress. This approach produces maximum accuracy at the lowest cost.

The first step in applying FEA is the construction of a model that breaks a component into simple standardized shapes or (usual term) elements located in space by a common coordinate grid system. The coordinate points of the element corners, or nodes, are the locations in the model where output data are provided. In some cases, special elements can also be used that provide additional nodes along their length or sides. Nodal stiffness properties are identified, arranged into matrices, and loaded into a computer where they are processed with certain applied loads and boundary conditions to calculate displacements and strains imposed by the loads (Appendix A: **PLASTICS DESIGN TOOLBOX**).

In this method, the modeling technique is critical because it establishes the structural locations where stresses will be evaluated. If a component is modeled inadequately for a given problem, the resulting computer analysis could be quite misleading in its prediction of areas of maximum strain and maximum deflection values. An inadequate model could be quite expensive in terms of computer time.

A cost-effective model concentrates on the smallest elements at areas of highest stress. This configuration provides greater detail in areas of major stress and distortion, and minimizes computer time in analyzing regions of the component where stresses and local distortions are smaller.

Unfortunately modeling can be a stumbling block because the process of separating a component into elements is not essentially straightforward. Some degree of insight, along with an understanding of how materials behave under strain, is required to determine the best way to model a component for FEA. The procedure can be made

easier by setting up a few ground rules before attempting to construct the model.

For example, in the case of plane stress analysis, quadrilateral elements should be used wherever possible. These elements provide better accuracy than corresponding triangular elements without adding significantly to calculation time. Also, 2-D and 3-D elements should have corners that are approximately right-angled, and should resemble squares and cubes as much as possible in regions of high strain gradient. Generally, element size should be in inverse proportion to the anticipated strain gradient with the smallest elements in regions of highest stress.

Fundamental

In its most fundamental form, FEA is limited to static, linear elastic analyses. However, there are advanced finite element computer software programs that can treat highly nonlinear dynamic problems efficiently. Important features of these programs include their ability to handle sliding interfaces between contacting bodies and the ability to model elastic-plastic material properties. These program features have made possible the analysis of impact problems that only a few years ago had to be handled with very approximate techniques. FEA have made these analyses much more precise, resulting in better and more optimum designs.

Operational Approach

The opportunity for creative design by viewing many imaginative variations would be blunted if each variation introduced a new set of doubts as to its ability to withstand whatever stress might be applied. From this point of view the development of computer graphics has to be accompanied by an analysis technique capable of determining stress levels, regardless of the shape of the product. This need is met by FEA.

The FEA computer-based technique determines the stresses and deflections in a structure. Essentially, this method divides a structure into small elements with defined stress and deflection characteristics. The method is based on manipulating arrays of large matrix equations that can be realistically solved only by computer. Most often, FEA is performed with commercial programs. In many cases these programs require that the user know only how to properly prepare the program input.

FEA is applicable in several types of analyses. The most common one is static analysis to solve for deflections, strains, and stresses in a structure that is under a constant set of applied loads. In FEA material is generally assumed to be linear elastic, but nonlinear behavior such as plastic deformation, creep, and large deflections also are capable of being analyzed. The designer must be aware that as the degree of anisotropy increases the number of constants or moduli required to describe the material increases.

Uncertainty about a material's properties, along with a questionable applicability of the simple analysis techniques generally used, provides justification for extensive end use testing of plastic products before approving them in a particular application. As the use of more FEA methods becomes common in plastic design, the ability of FEAs will be simplified in understanding the behavior and the nature of plastics.

FEA does not replace prototype testing; rather, the two are complementary in nature. Testing supplies only one basic answer about a design that either passed or failed. It does not quantify results, because it is not possible to know from testing alone how close to the point of passing or failing a design actually exists. FEA does, however, provide information with which to quantify performance.

Safety Factors

In order to take uncertainties into account in a product's design, there exist what is familiarly called the safety factory (SF) or sometimes the "factor of ignorance." Observe that the safety factors have been omitted from most calculations, because different designers working on varying products use the appropriate criteria for choosing SFs. In general, a SF used based on experience is 1.5 to 2.5, as is commonly used with metals.

Many designers have already used or calculated a safety factor on material, perhaps without recognizing it such as deciding what approach is used in determining the tensile secant modulus. The process appears to be simple and straightforward, but unfortunately things are never quite that simple.

The designer must be fully aware of what one means when one calculates such a factor or bases a design on it. Improper use of a presumed safety factor may in some case result in a needless waste of material or in other cases even product failure. Thus, one must define what is meant when using a safety factor.

Designers unfamiliar with plastic products can use the suggested preliminary safety factor guidelines in Table 2-11. They provide for extreme safety. Any product designed with these guidelines in mind should conduct tests on the products themselves to relate the guidelines to actual performance (Chapter 4, **RP PIPES, Stress-Strain Curves**). With more experience, more-appropriate values will be developed targeting to use 1.5 to 2.5. After field service of

Table 2-11   Safety factors[a] for preliminary guidelines

Type of Load	Factor[b]
Static short-term loads	1.5–2.5
Static long-term loads	2.0–5.0
Variable of changing loads	4–10
Repeated loads	5–15
Fatigue or load reversal	5–15
Impact loads	10–20

[a]The material strength determined is the minimum required, not the average or maximum, which is what is normally provided on manufactures' published data sheets.

Low-range values represent situations where failure is not critical; the higher values are for where failure is critical.

Note: These values are intended for preliminary design analysis only and are not to be used in place of thorough product design.

the preliminary designed products have been obtained, action should be taken to evaluate reducing your SF in order to reduce costs. Structures are designed where the SF approaches 1.2 such as in aircraft because extreme close controls are made that range from the raw materials to their final assembly with initial flight tests.

There are no hard-and-fast rules to follow in setting safety factors for any given material unless experience exists. The most important consideration is of course the probable consequences of failure. For example, a little extra deflection in an outside wall or a hairline crack in one of six internal screw bosses might not cause concern, but the failure of a pressure vessel or aircraft wing might have serious safety or product-liability implications.

Before putting any product onto the market, prototype tests should be run at their most extreme operating conditions. For instance, the maximum working load should be applied at the maximum temperature and in the presence of any chemicals that might be encountered in the end use. Furthermore, the loads, temperatures, and chemicals to which a product will be exposed prior to reaching its end use must not be overlooked. Impact loading should be applied at the lowest temperature expected, including what occurs during shipping and assembly. The effects of variations in plastic lots and molding conditions must also be considered. The results should be to provide more logical SFs pertinent to the product and the materials used in it.

Many situations discovered during the testing of preproduction products can be corrected with a selective use of increased thickness in walls, ribs, and gussets or by eliminating stress concentrations. Changing a material to another grade of the same plastic or to a different plastic with a suitable mechanical property profile might also be the solution.

Uncertainty

In addition to the basic uncertainties of graphic design, a designer may also have to consider additional conditions such as:

1. *Variations in material properties.* Because no two plastic (or steel, for that matter) melts are exactly alike, some may have inclusions and so on, the strength properties given in materials tables are usually average values. If the value stated is a manufacturer's value, it probably is the minimum value, which can significantly reduce or eliminate its uncertainty.

2. *Effect of size in stating material strength properties.* Property tables, unless otherwise stated for plastics, metals, and so on, list strength values based on a specified size, yet larger components generally fail at a lower stress than a similar smaller component made of the same material.

3. *Type of loading.* A simple static load is relatively easy to recognize, but there are cases that fail between impact and suddenly applied loads. One thus takes into account infrequently applied fatigue loading mixed with some shock loads, as for example cams, links, or feeding devices.

4. *Effect of processes.* The fabricating operations for plastics, steel, glass, and so forth may, and usually do, introduce unwanted stress concentrations and residual stresses if not properly processed.

5. *Overall concern for human safety.* All design must consider safety of the user who may be near or in contact with the product. Unexpected overloads or other situations may cause breakage and considerable bodily harm.

3

Product Design Feature

Introduction

Design problems with the other conventional materials of construction are usually solved with the aid of textbooks or handbooks that refer the reader to data sheets where the characteristics of a specific material are listed. However, products designed with plastics involve some special considerations when using these textbooks or handbooks as reviewed in Chapter 2.

Plastic material suppliers provide material data sheets for each grade they produce. At first glance, there could be a tendency to apply the plastic information in a similar fashion to that of other materials. If such a procedure were to be followed, the result would not only lead to disappointment but also perhaps even to failure for many products. The reason for the difference in treating the plastic data sheets from those of other materials is the behavior of plastics under load and under varying environmental conditions, which normally are not factors with other materials such as steel (Chapter 2).

The reaction of plastics under test conditions to their behaviors is explained in other Chapters, particularly Chapters 2, 5 to 7. Until this phase of the information is properly understood, it is best not to apply the numbers from the data sheet, for they can be a source of misinterpreted information. A con-siderable segment of the data is usually only usable for preliminary comparative evaluation of various grades of materials. Even in this case one must be sure that the test procedure and the test conditions were the same.

In many cases, product use conditions are very different from material suppliers data sheet test conditions; therefore, it would not be safe to attempt interpolating the available information. The needed data should be obtained under conditions simulating factors such as the same specific test procedure, fabricating, and end use requirements. The reader should recognize that knowledge of plastic material tests is a prerequisite to understanding the meaning and value of data for design purposes as presented in the suppliers' sheets (Chapter 5).

Another factor to consider in the early stages of design is material selection in relation to cost per volume rather than by weight. This subject volume vs. weight will be reviewed latter in this chapter entitled **Analysis Method**. Since the material value in a plastic product is usually over one-half of its overall cost, it becomes important to select a candidate material with extraordinary care.

After a material type and grade that will fulfill performance requirements has been decided upon, steps should be taken to ensure that degrading features such as inside sharp comers, nonuniform wall thicknesses, etc., are eliminated or reduce as much as possible from the design. There are features that can degrade properties such as those reviewed latter in this Chapter entitled **FEATURES INFLUENCING PERFORMANCE**.

Design Analysis

Overview

In structural applications for plastics, which generally include those in which the product has to resist substantial static and/or dynamic loads, it may appear that one of the problem design areas for many plastics is their low modulus of elasticity. The moduli of unfilled plastics are usually under 1×10^6 psi (6.9×10^3 MPa) as compared to materials such as metals and ceramics where the range is usually 10 to 40×10^6 psi (6.9 to 28×10^4 MPa). However with reinforced plastics (RPs) the high moduli of metals are reached and even surpassed as summarized in Fig. 2-6.

Since shape integrity under load is a major consideration for structural products, low modulus plastic products are designed shape-wise for efficient use of the material to afford maximum stiffness and overcome their low modulus. These type plastics and products represent most of the plastic products produced worldwide (1, 3, 6, 9, 10, 14, 20, 28, 35, 36, 62, 64).

Pseudo-Elastic Design Method

Throughout this book as the viscoelastic behavior of plastics has been described it has been shown that deformations are dependent on such factors as the time under load and the temperature. Therefore, when structural components are to be designed using plastics it must be remembered that the standard equations that are available (such as in Figs. 3-1 and 3-2) for designing springs, beams, plates, and cylinders, and so on have all been derived under the assumptions that (1) the strains are small, (2) the modulus is constant, (3) the strains are independent of the loading rate or history and are immediately re-

versible, (4) the material is isotropic, and (5) the material behaves in the same way in tension and compression.

Since these assumptions are not always justifiable when applied to plastics, the classic equations cannot be used indiscriminately. Each case must be considered on its merits, with account being taken of such factors as the time under load, the mode of deformation, the service conditions, the fabrication method, the environment, and others. In particular, it should be noted that the traditional equations are derived using the relationship that stress equals modulus times strain, where the modulus is a constant. From the review in Chapter 2 it should be clear that the modulus of a plastic is generally not a constant. Several approaches have been used to allow for this condition. The drawback is that these methods can be quite complex, involving numerical techniques that are not attractive to designers. However, one method has been widely accepted, the so-called pseudo-elastic design method.

In this method appropriate values of such time-dependent properties as the modulus are selected and substituted into the standard equations. It has been found that this approach is sufficiently accurate if the value chosen for the modulus takes into account the projected service life of the product and/or the limiting strain of the plastic, assuming that the limiting strain for the material is known. Unfortunately, this is not just a straightforward value applicable to all plastics or even to one plastic in all its applications. This type of evaluation takes into consideration the value to use as a safety factor. If no history exist a high value will be required. In time with service condition inputs, the SF can be reduced if justified.

This modulus value is often arbitrarily chosen, although several methods have been suggested for arriving at a suitable value. One is to plot a secant modulus based on 1% strain or that is 0.85% of the initial tangent modulus (Chapter 2, **SHORT-TERM LOAD BEHAVIOR**). However, for many plastics, particularly the crystalline TPs, this method is too restrictive, so in most practical situations the limiting strain is decided in consultation

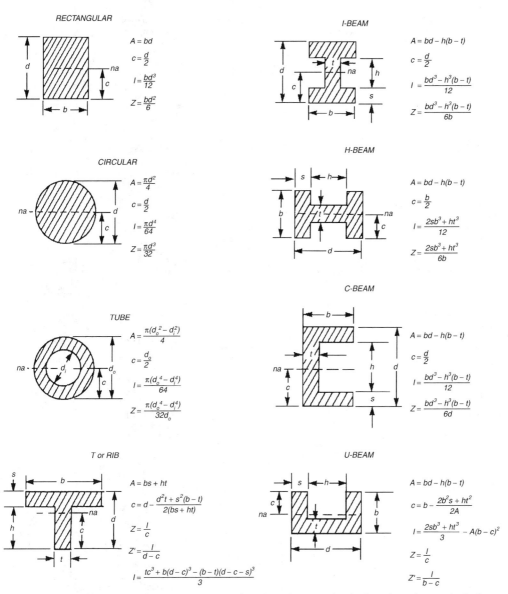

Fig. 3-1   Properties of some common cross-sections based on mechanical engineering analysis (*na* = neutral axis).

between the designer and the plastic material's manufacturer. Once the limiting strain is known, design methods based on its static and/or dynamic load becomes rather straightforward.

Analysis Method

Plastics have some mechanical characteristics that differ significantly from those of the familiar metals. Consequently, design analysts may have less confidence in these materials and in their own ability to design with them. Materials selection thus may tend to confine itself to familiar materials, or else products may be overdesigned, and failures may even occur in service due to faulty design.

Also, the statistics available on materials are often presented so as to favor a particular bias, which complicates the process of

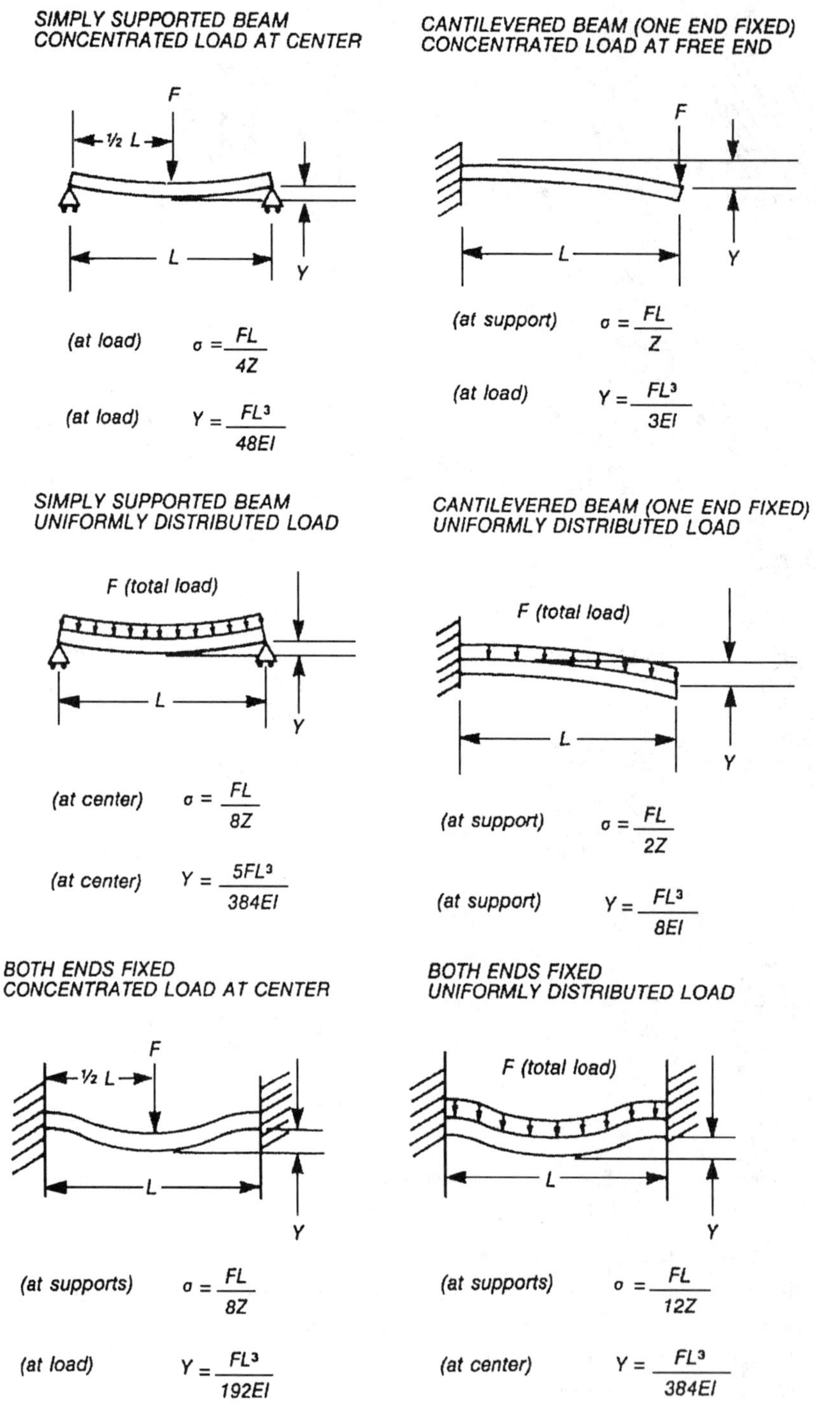
SIMPLY SUPPORTED BEAM
CONCENTRATED LOAD AT CENTER
F
½ L
L
Y
(at load) σ = FL/4Z
(at load) Y = FL³/48EI
CANTILEVERED BEAM (ONE END FIXED)
CONCENTRATED LOAD AT FREE END
F
L
Y
(at support) σ = FL/Z
(at load) Y = FL³/3EI
SIMPLY SUPPORTED BEAM
UNIFORMLY DISTRIBUTED LOAD
F (total load)
L
Y
(at center) σ = FL/8Z
(at center) Y = 5FL³/384EI
CANTILEVERED BEAM (ONE END FIXED)
UNIFORMLY DISTRIBUTED LOAD
F (total load)
L
Y
(at support) σ = FL/2Z
(at support) Y = FL³/8EI
BOTH ENDS FIXED
CONCENTRATED LOAD AT CENTER
F
½ L
L
Y
(at supports) σ = FL/8Z
(at load) Y = FL³/192EI
BOTH ENDS FIXED
UNIFORMLY DISTRIBUTED LOAD
F (total load)
L
Y
(at supports) σ = FL/12Z
(at center) Y = FL³/384EI

Fig. 3-2 Maximum stress and deflection equations for selected beams.

assessing their relative merit and adds to the confusion. Essentially, what the design analyst requires is relevant and credible design data, together with valid methods for calculating, predicting, and optimizing a product's performance. This type information has been available for over a half century. These methods may involve design formulas and charts

Table 3-1   Examples of the mechanical properties of metal and plastic materials

Property	Aluminum	Mild Steel	Polypropylene (PP)	Glass-fiber Reinforced Plastics (GRP)
Tensile modulus (E) 10^6 GN/m^2 (psi)	70 (10)	210 (30)	1.5 (0.21)	15 (2.2)
Tensile strength (σ) 10^3 MN/m^2 (psi)	400 (58)	450 (65)	40 (5.8)	280 (40.5)
Specific gravity (S)	2.7	7.8	0.9	1.6

GN/m^2 kPa.

such as those included in computer-aided designs (CADs) that provide an opportunity for plastics to be handled on a basis equivalent to that of other materials. The CAD's software includes applicable static and dynamic data.

In the past almost all the methods for the design analysis of plastics were base on models of material behavior relevant to traditional metals, as for example elasticity and plastic yield. These principles were embodied in design formulas, design sheets and charts, and in the modern techniques such as those of CAD using finite element analysis (FEA). The design analyst was required only to supply appropriate elastic or plastic constants for the material, and not question the validity of the design methods. Traditional design analysis is thus based on accepted methods and familiar materials, and as a result many designers have little, if any, experience with such other materials as plastics, wood, and glass.

Under these circumstances it is both tempting and common practice for designers to treat plastics as though they were traditional materials and to apply familiar design methods with what seem appropriate materials constants. It must be admitted that this pragmatic approach does often yield acceptable results. However, it should also be recognized that the mechanical characteristics of plastics are different from those of metals, and the validity of this pragmatic approach is often fortuitous and usually uncertain.

It would be more acceptable for the design analysis to be based on methods developed specifically for the materials, but this action will require the designer of metals to accept new ideas. Obviously, this acceptance becomes easier to the degree that the new

methods are presented as far as possible in the form of limitations or modifications to the existing methods discussed in this book.

Table 3-1 gives typical mechanical property data for four materials, the exact values of which are unimportant for this discussion. Aluminum and mild steel have been used as representative metals and polypropylene (PP) and glass fiber-TS polyester reinforced plastics (GRP) as representative plastics. Higher-performance types could have been selected for both the metals and plastics, but those in this table offer a fair comparison for the explanation being presented.

Also, it appears from the data that these metals are much stiffer and significantly stronger than the plastics. This approach to evaluation could eliminate the use of plastics in many potential applications, but in practice it is recognized by those familiar with the behavior of plastics that it is the stiffness and strength of the product that is important, not its material properties.

To illustrate the correct approach, consider applications in which a material is used in sheet form, as in automotive body panels, and suppose that the service requirements are for stiffness and strength in flexure. First imagine four panels with identical dimensions that were manufactured from the four materials given in Table 3-1. Their flexural stiffnesses and strengths depend directly on the respective material's modulus and strength. All the other factors are shared in common with the other materials, there being no significantly different Poisson ratios. Thus, the relative panel properties are identical with the relative material properties illustrated in Fig. 3-3. Obviously, the metal panels will be stiffer and

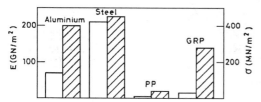

Fig. 3-3 The relative stiffness (open bars) and strength (shaded bars) of sheets made from the materials listed in Table 3-1.

significantly stronger than the plastic ones, based on the identical panel dimensions resulting in the use of equal volumes of materials.

Obviously, the lower densities of plastics allow them to be used in thicker sections than metals, which will have a significant influence on panel stiffness and strength. For example, assume that the four panels have equal weights and therefore different thickness (t). When the panels are loaded in flexure, their stiffnesses depend on (Et^3) and their strength on (σt^2) where E and σ are the material's modulus and its strength. For panels of equal weight it follows that their relative stiffnesses are governed by (E/s^3) and their relative strengths by (σ/s^2) where s denotes specific gravity. These relative panel properties illustrated in Fig. 3-4, show that the plastics now appear in a much more favorable light. Figures 3-3 and 3-4 present the same basic data from Table 3-1, but in two different forms and the superficial use of either form can be misleading.

In practice, metals and plastics usually do not have to compete under either of the extreme conditions of equal volume or equal weight, and their positioning between these extremes will depend on the requirements of the particular application. For vehicle body panels, plastics may be used with thicker sections than their metals counterparts (that is, not of equal volume), but the desire to save weight will ensure that they are not used to their extremes. Thus the designer has the opportunity to balance out the requirements for stiffness, strength, and weight saving.

For the materials data given in Table 3-1 a GRP panel having 2.4 times the thickness of a steel panel has the same flexural stiffness but 3.6 times its flexural strength and only half its weight. The tensile strength of the GRP panel would be 50% greater than that of the steel panel, but its tensile stiffness is only 17% that of the steel panel. The designer's interest in this GRP panel would then depend in this context on whether tensile stiffness was what was required.

No general conclusions should be drawn on the relative merits of various materials based on this description alone. These examples have been presented merely to illustrate the dangers of superficially interpreting property data and of making dogmatic or generalized statements about the relative merits of various classes of materials. Similar remarks could be made with respect to various materials' costs and energy contents, which can also be specified per unit of volume or weight. If these factors are to be treated properly, they too must relate to final product values, including the method of fabrication, expected lifetime, repair record, in-service use, and so on.

One important conclusion illustrated by the example given is that plastic products are often stiffness critical, whereas metal products are usually strength critical. Consequently, metal products are often made stiffer than required by their service conditions, to avoid failure, whereas plastic products are often made stronger than necessary, for adequate stiffness. Thus, in replacing a component in one material with a similar product in another material is not usually necessary to have the same product stiffness and strength. It follows that general statements about energy content or cost per unit of stiffness or strength, as well as other factors, should be treated with caution and applied only where relevant.

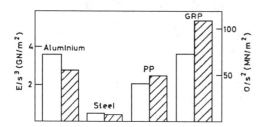

Fig. 3-4 The relative plate stiffness (open bars) and strength (shaded bars) in flexure for panels of equal weight.

This review identifies the need for using design analysis methods appropriate for plastics. It also indicates the uncertainty of using with plastics methods derived from metals, and demonstrates the dangers of making generalized statements about the relative merits of different classes of materials. The designer who has basically no familiarity with plastics needs to be receptive to the different methods of handling them.

It is necessary to keep an open mind when designing products with plastics, rather than limiting the design to being an exact replica of the metal product. Let us assume from this point on that this approach is accepted, so a more-detailed examination of the needs of the design analysis methods can follow.

Analysis Requirement

It should be evident that the full spectrum of the possible materials and applications in load-bearing situations involves many factors that may have to be taken into account. Fortunately, most products involve only a few factors, and others will not be significant or relevant. Regardless, the methods of design analysis must be made available to handle any possible combinations of such factors as the materials' characteristics, the product's shape, the loading mode, the loading type, and other service factors and design criteria.

Material Characteristic

The wide choice available in plastics makes it necessary to select not only between TPs, TSs, reinforced plastics (RPs), and elastomers, but also between individual materials within each family of plastic types (Chapters 6 and 7). This selection requires having data suitable for making comparisons which, apart from the availability of data, depends on defining and recognizing the relevant plastics behavior characteristics. There can be, for instance, isotropic (homogeneous) plastics and plastics that can have different directional properties that run from the isotropic to anisotropic. Here, as an example, certain

engineering plastics and RPs that are injection molded can be used advantageously to provide extra stiffness and strength in pre-designed directions.

Reinforced plastic   It can generally be claimed that fiber based RPs offer good potential for achieving high structural efficiency coupled with a weight saving in products, fuel efficiency in manufacturing, and cost effectiveness during service life. Conversely, special problems can arise from the use of RPs, due to the extreme anisotropy of some of them, the fact that the strength of certain constituent fibers is intrinsically variable, and because the test methods for measuring RPs' performance need special consideration if they are to provide meaningful values.

Some of the advantages, in terms of high strength-to-weight ratios and high stiffness-to-weight ratios, can be seen in Figs. 3-5 and 2-6, which shows that some RPs can outperform steel and aluminum in their ordinary forms. If bonding to the matrix is good, then fibers augment mechanical strength by accepting strain transferred from the matrix, which otherwise would break. This occurs until catastrophic debonding occurs. Particularly effective here are combinations of fibers with plastic matrices, which often complement one another's properties, yielding products with acceptable toughness, reduced thermal expansion, low ductility, and a high modulus (10, 62, 92).

As a further advantage, RPs make effective use of some materials that are otherwise

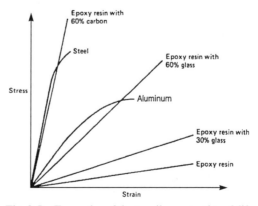

Fig. 3-5   Examples of the tensile properties of different materials.

unable to stand alone. When incorporated into plastics, in particular those such as TS unsaturated polyesters or phenolics, particles (fibers, flakes, etc.) can reduce manufacturing shrinkage and yield a more usable product. In service, zero thermal expansion coefficients can be achieved by a suitable choice of starting materials.

Design Concept

Design analysis is required to convert applied loads and other external constraints into stress and strain distributions within a product and calculate the associated deformations. The nature and complexity of these calculations will be strongly influenced by the product's shape. The designing will be simplified if the product approximates a simple engineering form like a plate or shell, beam or tube, or some combination of idealized forms such as a box structure. In such cases standard design formulas can be used, with appropriate parameters relating to the factors being reviewed: short- and long-time loadings, creep, fatigue, impact, and so on using viscoelastic materials (Chapter 2).

There are of course products whose shapes do not approximate a simple standard form or where more detailed analysis is required, such as a hole, boss, or attachment point in a section of a product. With such shapes the component's geometry complicates the design analysis for plastics, glass, metal, or other material and may make it necessary to carry out a direct analysis, possibly using finite element analysis (FEA) followed with prototype testing. Examples of design concepts are presented.

Loading Mode

Loads applied on products induce tension, compression, flexure, torsion, and/or shear, as well as distributing the loading modes. The product's particular shape will control the type of materials data required for analyzing it. The location and magnitude of the applied loads in regard to the position and nature of such other constraints as holes, attachment

points, and ribs will determine its shape. Also influencing the design decision will be the method of fabricating the product. The load's magnitude and distribution can be difficult to specify, especially in a system composed of several interacting components of the product (37).

Load

In a simplified approach the first step in analyzing any product is to determine the loads to which it will be subjected. These loads will generally fall into one of two categories, directly applied loads and strain-induced loads. Directly applied loads are usually easy to understand. They are defined loads that are applied to defined areas of the product, whether they are concentrated at a point, line, or boundary or distributed over an area. The magnitude and direction of these loads are known or can easily be determined from the service conditions. Figure 3-2 shows examples of directly applied loads.

Frequently, a product becomes loaded when it is subjected to a defined deflection. The actual load then is a result of the structural reaction of the product to the applied strain. Unlike directly applied loads, strain-induced loads are dependent on the modulus of elasticity and, with TPs, will generally decrease in magnitude over time. Many assembly and thermal stresses could be the result of strain-induced loads. They include metal insert press fits in the plastic and clamping or screw attachments.

When a load is applied, if the product is to remain in equilibrium there must be equal force acting in the opposite direction. These balancing forces, as an example, are the reactions at the supports. For purposes of structural analysis there are several supports conditions that have been defined. The free (unsupported), simply supported, and fixed supports are the most frequently encountered. The free (unsupported) condition occurs where the edge of a body is totally free to translate or rotate in any direction. The fixed (clamped or built-in) support condition at the end of a beam or plate prevents transverse displacement and rotation. The condition can be thought of as its ends support firmly embedded into fixed solid walls. In practice this condition rarely exists in its pure form, especially with plastic, since the mounting points of the products usually have some give via factors such as relaxation.

The other support conditions include a guided system that is similar to the free end except that its edge is prevented from rotating. In a simple support condition transverse displacement in one direction is restricted. There is the held (pinned) situation that is similar to the simply supported one except that here only rotations are allowed.

Loading Type

The mechanical behavior of plastics on time-dependent applied loading can cause different important effects on materials viscoelasticity. Loads applied for short times and at normal rates (Chapter 2) causes material response that is essentially elastic in character. However, under sustained load plastics, particularly TPs, tend to creep, a factor that is included in the design analysis.

Products can also be subjected to intermittent loading involving successive creep and recovery over relatively long time scales. It is not unusual, for instance, for creep deformation arising during one loading phase to be only partly recovered in the unloading cycle, leading to a progressive accumulation of creep strain (Figs. 3-6 and 3-7) and possibly resulting in creep rupture. An analogue of creep behavior is the stress-relaxation cycle that can occur under constant strain.

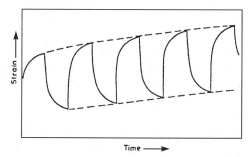

Fig. 3-6 An example of intermittent loading involving successive creep strain and recovery.

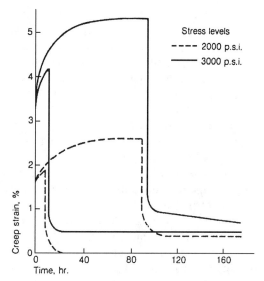

Fig. 3-7 Examples are shown of elasticity changes for engineering TPs involving one cycle of loading and unloading. The curves show effects of stress and time under load and strain recovery after loading.

This behavior is particularly relevant with push-fit assemblies and bolted joints that rely on maintaining their load under constant strain. Special design features or analysis may be required to counteract excessive stress-relaxation.

In many applications, intermittent or dynamic loads arise over much shorter time scales. Examples of such products include chair seats, panels that vibrate and transmit noise, engine mounts and other antivibration products, and road surface-induced loads carried to wheels and suspension systems. Plastics' relevant properties in this regard are material stiffness and internal damping, the latter of which can often be used to advantage in design (Chapter 2). Both properties depend on the frequency of the applied loads or vibrations, a dependence that must be allowed for in the design analysis. The possibility of fatigue damage and failure must also be considered.

Load-Bearing Product

A fundamental concept in structural analysis is that the structure as a whole and each of its elements together are in a state of equilibrium. This means that there are no un-balanced forces of tension, compression, flex, or shear acting on the structure at any point. All the forces counteract one another, which results in equilibrium. When all the forces acting on a given element in the same direction are summed up algebraically, the net effect is zero, with no acceleration. However, the object does respond to the various forces internally. It can be pushed or pulled and otherwise deformed, with internal stresses of varying types and magnitudes accompanying these deformations.

Basically, designing a load-bearing product with any material involves first selecting a suitable material and then specifying the shape into which it is to be formed or assembled. One important aspect of shape is its effect on internal stress. As the cross-sectional area of a product increases for a given load, the stresses are reduced. Design is concerned with determining the stresses for a given or hypothetical shape and subsequently adjusting the shape until the stresses are neither high enough to risk fracture nor low enough to suggest that material is being wasted.

Stress analysis involves using the descriptions of the product's geometry, the applied loads and displacements, and the material's properties to obtain closed-form or numerical expressions for internal stresses as a function of the stress's position within the product and perhaps as a function of time as well. The term engineering formulas refers primarily to those equations reviewed previously and given in engineering handbooks by which the stress analysis can be accomplished.

Multiaxial Stresses and Mohr's Circle

Sophisticated design engineers unfamiliar with plastics' behavior will be able to apply the information contained in this and other chapters to applicable sophisticated equations that involve such analysis as multiple and complex stress concentrations. The various machine-design texts and mechanical engineering handbooks listed in the Appendix A: **PLASTICS TOOLBOX** and **REF-**

ERENCES section at the end of this book provide detailed analysis of these stress-concentration factors and other load-bearing parameters.

Many structural products are stressed in a manner that is more complex than simple tension, compression, flex, or shear. Because yielding will also occur under complex stress conditions, a yield criterion must be specified that will apply in all stress states. Any complex stress state can be resolved into three normal components acting along three mutually perpendicular (X, Y, Z) axes and into three shear components along the three planes of those axes. Then, by making a proper choice it is possible to find a set of three axes along which the shear stresses will be zero. These are the principal axes, with the normal stresses along them being called the principal stresses. Determining these principal stresses in a complex loaded member is the responsibility of the designer, a task normally performed by using approaches such as Mohr's circle and its associated relationships.

Design Criteria

The nature of design analysis obviously depends on having product-performance requirements. The product's level of technical sophistication and the consequent level of analysis that can be justified costwise basically relate to control of these requirements. The analysis also depends on the design criteria for a particular product. If the design is strength limited, to avoid component failure or damage, or to satisfy safety requirements, it is possible to confine the design analysis simply to a stress analysis. However, if a plastic product is stiffness limited, to avoid excessive deformation from buckling, a full stress-strain analysis will likely be required.

Even though many potential factors can influence a design analysis, each application fortunately usually involves only a few factors. For example, TPs' properties are dominated by the viscoelasticity relevant to the applied load. Anisotropy usually dominates the behavior of long-fiber RPs.

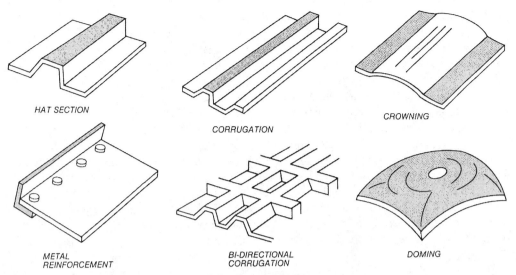

HAT SECTION

CORRUGATION

CROWNING

METAL REINFORCEMENT

BI-DIRECTIONAL CORRUGATION

DOMING

Fig. 3-8 Geometric shapes.

Geometrical Shape

There are different techniques that have been used for over a century to increase the modulus of elasticity of plastics. Orientation or the use of fillers and/or reinforcements such as RPs can modify the plastic. There is also the popular and extensively used approach of using geometrical design shapes that makes the best use of materials to improve stiffness even though it has a low modulus. Structural shapes that are applicable to all materials include shells, sandwich structures, and folded plate structures (Fig. 3-8). These widely used shapes employed include other shapes such as dimple sheet surfaces. They improve the flexural stiffness in one or more directions.

EI theory In each case displacing material from the neutral plane makes the improvement in flexural stiffness. This increases the EI product that is the geometry material index that determines resistance to flexure. The EI theory applies to all materials (plastics, metals, wood, etc.). It is the elementary mechanical engineering theory that demonstrates some shapes resist deformation from external loads.

This phenomenon stems from the basic physical fact that deformation in beam or sheet sections depends upon the mathemat-

ical product of the modulus of elasticity (E) and the moment of inertia (I), commonly expressed as EI. This theory has been applied to many different plastic constructions including solid and different sandwich structures.

In the case of plastics, emphasis is on the way plastics can be used in these structures and why they are preferred over other materials. In many cases plastics can lend themselves to a particular field of application only in the form of a sophisticated lightweight stiff structure and the requirements are such that the structure must be of plastics. In other instances, the economics of fabrication and erection of a plastics lightweight structure and the intrinsic appearance and other desirable properties make it preferable to other materials.

When compared to other materials, formability into almost any conceivable shape is one of plastics' design advantages. It is important for designers to appreciate this important characteristic. Both the plastic materials and different ways to manufacture products provide this rather endless capability (Chapter 8). Shape, which can be almost infinitely varied in the early design stage, is capable for a given volume of materials to provide a whole spectrum of strength properties, especially in the most desirable areas of stiffness and bending resistance. With shell structures, plastics can be either singly or doubly curved via the different processes and so on.

There are different design approaches to consider as reviewed in different engineering textbooks concerning specific products (Appendix A: **PLASTICS TOOLBOX**). They range from designing a drinking cup to a complex shape such as the roof of a house to a wing structure. As an example shapewise consider a house to stand up to the forces of a catastrophic hurricane. Low pitch roofs are less vulnerable than steeper roofs because the same aerodynamic factors that make an airplane fly can lift the roof off the house. Also, the roof requires being properly attached to the building structure.

Other examples include the advantage of basic beam structures as well as hollow channel, I-shape, T-shape, etc. They are used to provide more efficient strength-to-weight products and so forth. While this construction may not be as efficient as the sandwich panel, it does have the advantage that it can be molded, extruded, etc. directly in the required configuration at a low cost and the relative proportions be designed to meet the flexural, etc. requirements. One of the potential limitations is that generally it imparts increased stiffness in one direction much more than in the other. However processing techniques can be used to develop bidirectional or any other directional properties such as combining extrusion with filament winding.

In most cases, plastic products can take advantage of a basic beam structure in their design. Hollow-channel, I-, and T-shapes designed with generous radii (and other basic plastic flow considerations) rather than sharp corners are more efficient on a weight basis in plastics because they use less material, thus provide a high moment of inertia, etc. The moments of inertia of such simple sections, and hence their stresses and deflections, can be fairly easily calculated, using simple engineering equations (Fig. 3-1).

Rib

In the discussion of uniform wall thickness, ribbing was one of the suggested remedies. Ribs are also used to increase load-bearing requirements when calculations indicate wall thicknesses are above recommended values. They are provided for spacing purposes, for supporting components, etc. The first step in designing a rib is to determine dimensional limitations followed by establishing what shape the rib is to have in order to realize a product with good strength and satisfactory appearance that can be produced economically.

Unless the reinforcing ribs are added in the correct engineered proportions, some additional material may be used and placed so that it creates high stresses, actually decreasing the loads that cause yielding or fracture. Lengthy equations for the moment of inertia and for deflection and stress are normally required to determine the effect of ribs on stress. However, nondimensional curves have been developed to allow quick determination of proper rib proportions and a corresponding program for a pocket calculator or computer will allow for obtaining greater precision when required (Fig. 3-9).

If performance calculations indicate wall thicknesses well above those recommended for a particular material, one of the solutions to the problem is to find equivalent cross-sectional properties by ribbing. Heavy walls can be responsible for reduction in properties due to poor heat conductivity during fabrication, thus creating temperature gradients throughout the cross section, and thereby causing residual stresses. Cycle times are usually longer, thus adding another potential cause for stresses when using too short a cycle time. Also, close tolerance dimensions are more difficult to maintain, material is wasted, quality is degraded, and material and processing costs are increased.

Solid plastic wall thicknesses for most materials should be targeted to be below 0.2 in. and preferably around 0.125 in. in the interest of avoiding the above pitfalls. In most cases ribbing will provide a satisfactory solution; in other cases sandwich structures or reinforced materials may have to be considered. As reviewed elsewhere when presenting the ideal target to meet the best design such as the thinner wall just reviewed, does not mean that a thicker wall can not be processed, etc. The thicker wall can be processed requiring closer process controls (Chapter 8).

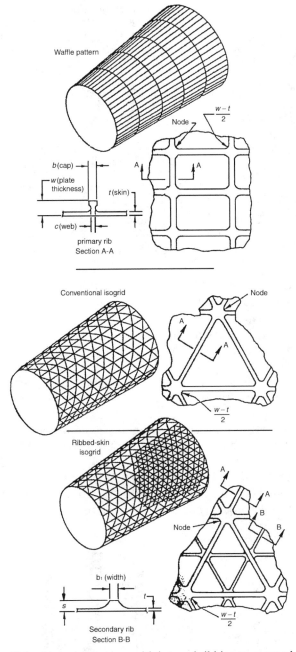

Fig. 3-9 Thin-skinned structures with integral ribbing to carry edge loads.

Rib design An example of how ribbing will provide the necessary equivalent moment of inertia and section modulus will be given. A flat plastic bar of $1\frac{1}{2}$ in. $\times\, 3/8$ in. thick and 10 in. long, supported at both ends and loaded at the center, was calculated to provide a specified deflection and stress level under a given load. The favorable material thickness of this plastic is 0.150 in. Using judgment as a guide, it would appear that the $1\frac{1}{2}$ in. width would require about two ribs. So, as a starting point, calculate the equivalent cross-sectional data as if we were dealing with two "T" sections.

According to the handbooks under "Stress and Deflections in Beams" and "Moments of Inertia," etc., the moment of inertia and resistance to deflection expresses the resistance

Case	Shape	Change	Moment Of Inertia	Increase In I	Increase In Weight	Ratio $\frac{I}{Wt.}$
1		Base 2" x ¼"	.0026			
2	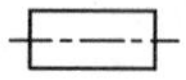	Double Height	.0208	700%	100%	7
3		Add ⅛" wx ¼" H Rib	.0048	85%	6.25%	14
4	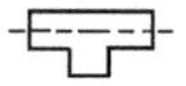	Add ¼" wx ¼" H Rib	.0064	146%	12.5%	12
5	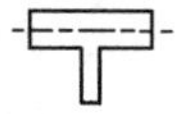	Add ⅛" wx ½" H Rib	.0118	354%	12.5%	28
6	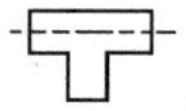	Add ¼" wx ½" H Rib	.0194	646%	25%	26

Fig. 3-10   Examples of ways in using ribs to increase rigidity and reduce weight.

to stress by the section modulus. By finding a cross section with the two equivalent factors, we will ensure equal or better performance. Summarizing this subject, the moment of inertia can be changed substantially by adding ribs or gussets or some combination of them.

As shown in Figs. 3-10 and 3-11 and Table 3-2, there is a better way to achieve this result and still keep weight at a minimum by using ribbing, if space exists for it. The views include sections of equal stiffness. Adding ribs to a part maintains its thin walls and thus allows faster fabricating cycles. Summarizing this subject, it is possible to reduce the cross-sectional area of a product and con-

sequently reduce the amount of material used in it, with a corresponding weight reduction.

Beam

In simple beam-bending theory a number of assumptions must be made, namely that (1) the beam is initially straight, unstressed, and symmetrical; (2) its proportional limit is not exceeded; (3) Young's modulus for the material is the same in both tension and compression; and (4) all deflections are small so that planar cross-sections remain planar before and after bending. The maximum stress

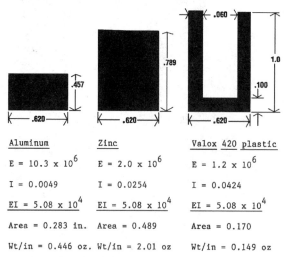

Aluminum	Zinc	Valox 420 plastic
$E = 10.3 \times 10^6$	$E = 2.0 \times 10^6$	$E = 1.2 \times 10^6$
$I = 0.0049$	$I = 0.0254$	$I = 0.0424$
$EI = 5.08 \times 10^4$	$EI = 5.08 \times 10^4$	$EI = 5.08 \times 10^4$
Area = 0.283 in.	Area = 0.489	Area = 0.170
Wt/in = 0.446 oz.	Wt/in = 2.01 oz	Wt/in = 0.149 oz

Fig. 3-11   Different cross-sectional TP polyester profiles with equivalent stiffness in bending.

Table 3-2 Design examples to obtain the same part rigidity for a section 1 ft. × 2 ft.

Property	Steel	Solid Plastic	Structural Foam	Ribbed Solid[a]
Thickness (in.)	0.040	0.182	0.196	0.125
E (psi)	3×10^7	3.2×10^5	2.56×10^5	3.2×10^5
I (in.⁴)	0.000064	0.006	0.0075	0.006
E × I (rigidity)	1,920	1,920	1,920	1,920
Weight (lbs.)	3.24	1.98	1.78	1.60

[a] Rib height = 0.270 in., thickness = 0.065 in., rib spacing = 2.0 in.

occurs at the surface of the beam farthest from the neutral surface, as given by the following equation:

$$\sigma = Mc/I = M/Z \qquad (3\text{-}1)$$

where M = the bending moment in in./lbs., c = the distance from the neutral axis to the outer surface where the maximum stress occurs in inches, I = the moment of inertia in in.⁴, and $Z = I/c-$, the section modulus in in.³.

Observe that this is a geometric property, not to be confused with the modulus of the material, which is a material property. I, c, Z, and the cross-sectional areas of some common cross-sections are given in Fig. 3-1, and the mechanical engineering handbooks provide many more. The maximum stress and defection equations for some common beam-loading and support geometries are given in Fig. 3-2. Note that for the T- and U-shaped sections in Fig. 3-1 the distance from the neutral surface is not the same for the top and bottom of the beam. It may occasionally be desirable to determine the maximum stress on the other nonneutral surface, particularly if it is in tension. For this reason, Z is provided for these two sections.

Beam Bending and Spring Stress

To illustrate how traditional materials such as metals limit the design process, consider a spring. The manufacturing process in metals limits the options available in producing a variety of shapes in this material. As a result, steel springs are produced in basically only three shapes: the torsion bar, the helical coil, and the flat-shaped leaf spring. By comparison, TPs and TSs can be easily fabricated into a variety of shapes to meet different product requirements. Switching from metal to plastic thus lets the designer overcome configuration barriers and environmental operations to new spring designs.

As an example plastic replaced a metal pump in a PVC plastic bag containing blood. The plastic spring hand-operating pump did not contaminate the blood.

Figure 3-12 is an example of a TP spring action with a different shape. It is an injection molded Du Pont Delrin acetal plastic stapler illustrating spring design with the body and curved spring section molded in a single part. This complex shape could not have been achieved in a single operation in steel. The designer has taken advantage of molding's versatility to reinforce the curved, frequently stressed back section. When the stapler is depressed, the outer curved shape is in tension and the ribbed center section is put into compression. When the pressure is released, the tension and compression forces are in turn released and the molded stapler returns to its original position. With this type of plastic having these inherently desirable properties as well as other desirable properties, this repeated spring action has a virtually unlimited life span.

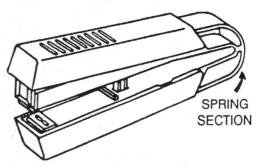

Fig. 3-12 TP spring action.

Leaf springs constructed of unidirectional fiber-reinforced plastics have come to be recognized as viable replacements for steel springs in truck and automotive suspension applications and have been used in aircraft landing systems since the early 1940s taking advantage of weight savings. Because of the material's high specific strain energy storage capability as compared to steel, a direct replacement of multileaf steel springs by monoleaf composite springs can be justified on a weight-saving basis. Such springs have in fact been in use since the 1960s in ground transportation vehicles. Further advantages of RP springs accrue from the ability with them to design and fabricate spring leaves having continuously variable widths and thicknesses along their lengths (3, 10, 62).

Such design features will no doubt lead to new suspension arrangements in which leaf springs will serve multiple functions, thereby providing a consolidation of parts and simplification of suspension systems. One distinction between steel and plastic is that complete knowledge of shear stresses is not important in a steel part undergoing flexure, whereas with RP design shear stresses, rather than normal stress components, usually control the design. Procedures have thus been developed for evaluating design stresses because of simple flexure as well as secondary loads like axle windup.

Developments with RP leaf springs have highlighted the need to reassess standards for testing and evaluating them. Because of the anisotropic properties of RPs, the standards previously developed for steel components in the laboratory and on the proving ground can give misleading test results in plastics. The concept of spring design has been well documented in various SAE and ASTM-STP design manuals from the 1970s on. These also give the equations for evaluating design parameters, which are simply derived from geometric and material considerations. Further information enables the calculation of windup (that is, accelerating and braking) and roll stiffnesses for springs as a check against the design requirements. (See, for instance, SAE J788A, Oct. 1970, and STP 376, Jan. 1973.)

The design of any RP product is unique and rather difficult, because the stress conditions within a given structure depend on its manufacturing methods, not just its shape. Programs have therefore been developed on the basis of the strain balance within the spring to enable suitable design criteria to be met. Stress levels are then calculated, after which the design and manufacture of RP springs become feasible.

The cantilever spring can be employed to provide a simple format from a design standpoint. Cantilever springs, which absorb energy by bending, may be treated as beams, with their deflections and stresses being calculated as short-term beam-bending stresses. The calculations arrived at for multiple-cantilever springs (that is, two or more beams joined in a zigzag configuration, as in Fig. 3-13) are similar to, but may not be as accurate as, those for a single-beam spring.

A zigzag configuration may be seen as a number of separate beams each with one end fixed. The top beam is loaded (F) either along its entire length or at a fixed point. This load gives rise to deflection y at its free end and moment M at the fixed end. The second beam is then loaded by moment M (upward) and load F (the effective portion of load F, as determined by the various angles) at its free end. This moment results in deflection y_2 at the free end and moment M_2 at the fixed end (that is, the free end of the next beam). The

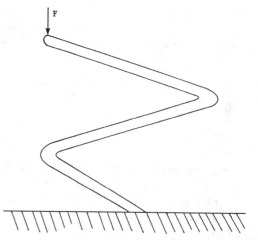

Fig. 3-13 Zigzag configured multiple-cantilever beam spring.

third beam is then loaded by M_2 (downward) and force F_2 (the effective portion of F_1), and so on.

The total deflection, y, is the sum of the deflections of the individual beams. The bending stress, deflection, and moment at each point can be calculated by using standard equations. To reduce stress concentration, all corners should be fully radiused. This type of spring is often favored because of its greater design flexibility over the single-beam spring. The relative lengths, angles, and cross-sectional areas can be varied to give the desired spring rate F/y in the available space. Thus, the total energy stored in a cantilever spring is equal to:

$$E_c = 1/2 F_y \qquad (3\text{-}2)$$

where F = total load in lb, y = deflection in., and E = energy absorbed by the cantilever spring, in-lbs.

Torsional Beam Spring

A torsional beam spring absorbs energy by twisting through an angle 0 (Fig. 3-14) and may thus be treated as a shaft in torsion.

A shaft subject to torque is generally considered to have failed when the strength of the material in shear is exceeded. For a torsional load the shear strength used in design should be the published value or one half the tensile strength, whichever is less. The maximum shear stress on a shaft in torsion is given by the following equation:

$$\tau = T_c / J \qquad (3\text{-}3)$$

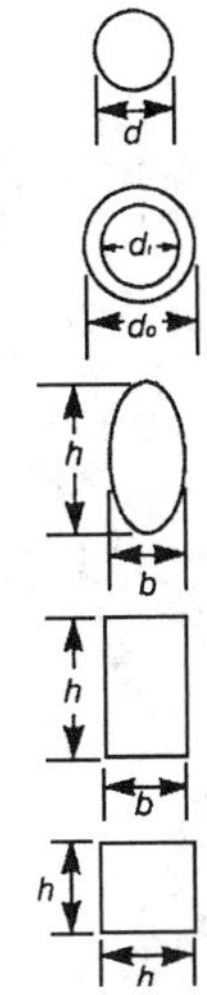

CROSS SECTION	POLAR MOMENT OF INERTIA J	LOCATION OF MAX. SHEAR STRESS c
circle, d	$\dfrac{\pi d^4}{32}$	$\dfrac{d}{2}$
annulus, d_i, d_o	$\dfrac{\pi(d_o{}^4 - d_i{}^4)}{32}$	$\dfrac{d_o}{2}$
ellipse, h, b	$\dfrac{\pi b^3 h}{32}$	$\dfrac{b}{2}$
rectangle, h, b	$\dfrac{b^3 h}{9}$	$\dfrac{b}{2}$
square, h, h	$\dfrac{h^4}{9}$	$\dfrac{h}{2}$

Fig. 3-15 Polar moments of inertia for common cross-sections.

where T = applied torque in in-lb, c = the distance from the center of the shaft to the location on the outer surface of shaft where the maximum shear stress occurs, in. (Fig. 3-15), and J = the polar moment of inertia, in.[4] (Fig. 3-15).

The angular rotation of the shaft is caused by torque is given by:

$$\theta = TL/GJ \qquad (3\text{-}4)$$

where L = length of shaft, in., G = shear modulus, psi = $E/2(1 + v)$, E = tensile modulus of elasticity, psi, and v = Poisson's ratio.

The energy absorbed by a torsional spring deflected through angle θ equals:

$$E_t = 1/2 M_\theta \times \theta \qquad (3\text{-}5)$$

where M_θ = the torque required for deflection θ at the free end of the spring, in-lb.

Folded Plate

The methods of analysis and design presented for beams can be applied to the more-complex products such as folded plate structures, which range from bottles to roofing to

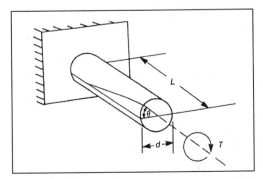

Fig. 3-14 Shaft with diameter d and length L under torque T undergoing angular deformation.

outer-space structures. They are basically assemblies of rectangular, triangular, spherical, or other shapes that behave much like beams, portal frames, arches, or shells.

The stresses in some folded structures can be determined with acceptable accuracy by applying elementary beam theory to the overall cross-sections of the plate assemblies. When assemblies are plates whose lengths are large relative to their cross-sectional dimensions (thin-wall beam sections, ribbed panels, and so on) and are in large plates whose fold lines deflect identically, such as the interior bays of roofs, they can be analyzed as beams. More elaborate procedures must be used to determine the transverse bending stresses in assemblies of large plates and longitudinal stresses in structures with "pinned" connections along folded lines that do not deflect identically.

There are also bellows-style collapsible plastic containers such as blow molded bottles (jars) that are foldable. As shown in Figs. 3-16 and 3-17, the technology of foldable

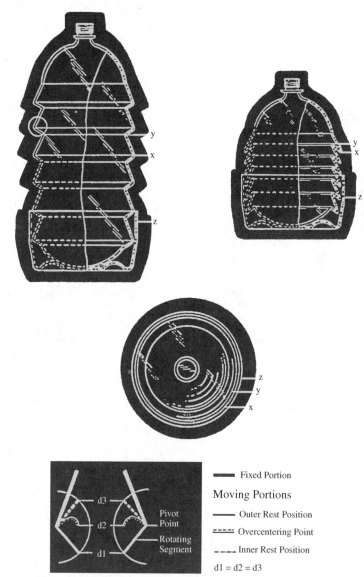

Fig. 3-16 Theory and operation of the collapsible bottle: (a) an uncallapsed bottle, (b) a collapsed bottle, (c) top view of the bottle, and (d) definitions.

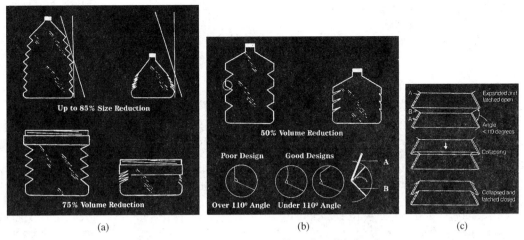

Fig. 3-17   Design ideas for the collapsible bottle.

containers in contrast to that of the usual "passive" bottles provides advantages and conveniences such as reduced storage, transportation, and disposal space; prolonged product freshness by reducing oxidation and loss of carbon dioxide; and provides continuous surface access to foods like mayonnaise and jams.

The bellows of collapsible containers overlap and fold to retain their folded condition without external assistance, thus providing a self-latching feature. This latching is the result of bringing together under pressure two adjacent conical sections of unequal proportions and different angulations to the bottle axis. On a more technical analytical level the latching is created by the swing action of one conical section around a fixed pivot point, from an outer to an inner, resting position. The two symmetrically opposed pivot points and rotating segments keep a near-constant diameter as they travel along the bottle axis. This action explains the bowing action of the smaller, conical section as it approaches the overcentering point.

After fabrication an initial collapsing of such a bottle should occur as soon as possible (no later than one to two hours after manufacture; the sooner the better). Additional pressure is needed for this first-time collapse in order to create permanent fold rings and completely orient the plastic molecules (Chapter 8, **PROCESSING AND PROP-**

ERTIES, Orientation). The subsequent collapsing and expansion of such bottles before filling them can be performed at the recommended ambient temperature of 20°C (68°F) or higher depending on the type of plastic used. In most disposable applications these bottles would undergo three changes of volume: (1) an initial collapsing of the container before shipment and storage; (2) expansion of the container at its destination, before or during filling; and (3) finally collapsing the bottle for reuse or disposal.

The fold rings designed on the bellows for such containers have proven to be durable and sturdy. Prototype bottles made of PETG and 75 durometer PVC were able to withstand dozens of collapses and still pass their stress tests. A guide is proved as shown in Fig. 3-17. The two adjacent conical sectors providing the latching should not exceed an angle of 110° to make a sharp fold ring. The size of rotating conical section B should not exceed 80% of conical section A, to prevent confusion and wobbling as the bottle is being collapsed.

Product labeling can be accomplished by attaching a floating sleeve to the neck or shoulder section of the bottle. This cylindrical sleeve then accommodates the bellows as they fold from the bottom up and contains them within the sleeve as the jar is collapsed. The maximum length of the sleeve is limited by the collapsed dimension of the bottle. An extended cap can also be used to hold the label to the side of the bottle.

Sandwich Construction

Increasing the stiffness capability of any material (plastic, metal, etc.) can be accomplished by thickening the product. Result is use of more materials and more processing costs. As reviewed, other approaches include the use of ribs, corrugations, and sandwich constructions.

Each have their place with plastic sandwiches being very popular when space is available providing high performance, lightweight, noise suppression, heat insulation, permit fabricating special shapes, and for certain products provide cost reductions (Fig. 3-18). They are similar to the I-beam shape in which its facings (plastics unreinforced or reinforced) correspond to the flanges and the cores (plastic foams; honeycomb structure of plastics, etc.) representing the webs. The facings, also called skins, resist axial loads and provide stiffness. The sandwich core stabilizes the facings against buckling or wrinkling under axial compression, and provides resistance to shear in bending (109).

A sandwich panel performs by improving loading characteristics such as in bending in the direction perpendicular to the plane of the panel. It basically exhibits no improvement in performance in other directions such as parallel to the plane of the sandwich (unless high strength facings are used). It is, in fact, subject to failure under lower load conditions under edge loading because of the susceptibility to failure of the skins by buckling. Despite these potential limitations structural sandwich panels are a very efficient lightweight structural element widely used in buildings, aircraft, surface vehicles, spacecrafts, and many commercial and industrial applications.

Different core materials are used. They include foam, honeycomb core (plastic, paper, aluminum, etc.), ribs, balsa wood, filler spacers, corrugated sheet spacers, etc. Materials such as polyurethane foam, cellulosic foams, and polystyrene foams are widely used as core materials. Plastics, such as glass-reinforced polyester, are frequently used as the skins for panels. Different skin materials are used such as metallic skins alone or in conjunction with plastic skins.

They range from structural foam molded products (which come from the mold as completed molded products) incorporating low density cores and high density skins of the same materials to products vacuum formed of a plastics material, the core of which becomes cellular during the heating process (Chapter 8). RP translucent structural panels for curtain wall building construction using

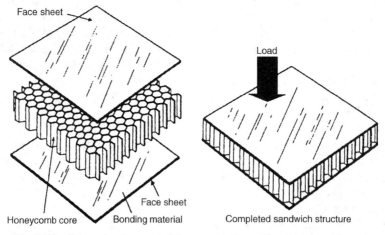

Fig. 3-18   Example of a sandwich construction with a honeycomb core.

honeycomb cores or other decorative cores are examples of the widespread use of the panel concept. Plated plastics products covered with substantial layers of metal are another use of reinforcing skins to improve the stiffness of plastics products as well as to provide a different appearance. Extruded cellular plastic shapes with applications that range from molding substitutes for wood to structural shelving are other examples where the sandwich panel stiffening principle is applied.

In the usual building and construction practice a structural sandwich construction is a special case of a laminate with flat, curved, or otherwise two thin facings. The facings are of relatively stiff, hard, dense, strong material that are bonded to a relatively thick core of a lightweight material that is considerably less dense, stiff, and strong than the facings. Structural sandwiches can be all plastics, all metals, or combination of plastic and metal, etc.

With this geometry and relationship of mechanical properties, facings are subjected to almost all the stresses in transverse bending or axial loading. The geometry of the arrangement provides for high stiffness combined with lightness, because the stiff facings are at a maximum distance from the neutral axis, similar to the flanges of an I-beam (Fig. 3-1). The continuous core takes the place of the web of an I-beam or box beam, absorbs most of the shear, and stabilizes the relatively thin facings against buckling or wrinkling under compressive stresses. The bond between the core and its facings must resist shear and any transverse tensile stresses set up as the facings tend to wrinkle or pull away from the core.

Stiffness   For an isotropic material with a modulus of elasticity E, the bending stiffness factor (EI) of a rectangular beam b wide and h deep stiffness is:

$$EI = E(bh^3/12) \qquad (3\text{-}6)$$

In a rectangular structural sandwich with the same dimensions just given whose facings and core have moduli of elasticity E_f and E_c, respectively, and a core thickness C,

the bending stiffness factor EI is:

$$EI = (E_f b/12)(h^3 - c^3) + (E_c b/12)(C^3) \qquad (3\text{-}7)$$

This equation is exact if the facings are of equal thickness, and approximate if they are not, but the approximation is close if the facings are thin relative to the core. If, as is usually the case, E is much smaller than E_f, the last term in the equation can be ignored. For asymmetrical sandwiches with different materials or different thicknesses in their facings or both, a more general equation may be used (109).

In many isotropic materials the shear modulus G is high compared to the elastic modulus E, and the shear distortion of a transversely loaded beam is so small that it can be neglected in calculating deflection. In a structural sandwich the core shear modulus G, is usually so much smaller than E_f of the facings that the shear distortion of the core may be large and therefore contribute significantly to the deflection of a transversely loaded beam. The total deflection of a beam is thus composed of two factors: the deflection caused by the bending moment alone, and the deflection caused by shear, that is, $\delta = \delta_m + \delta_s$, where δ = total deflection, δ_m = moment deflection, and δ_s = shear deflection.

Under transverse loading, bending moment deflection is proportional to the load and the cube of the span and inversely proportional to the stiffness factor, EI. Shear deflection is proportional to the load and span and inversely proportional to shear stiffness factor N, whose value for symmetrical sandwiches is:

$$N = [(h + c)b/2]G_c \qquad (3\text{-}8)$$

where G_c = the core shear modulus.

The total deflection may therefore be written:

$$\delta = (K_m W L^3)/EI + (K_s W L/N) \qquad (3\text{-}9)$$

The values of K_m and K_s depend on the type of load. Examples of these values are given in Table 3-3.

Table 3-3 Typical loading conditions

Loading	Beam Ends	Deflections at	K_m	K_s
Uniformly distributed	Both simply supported	Midspan	5/384	1/8
Uniformly distributed	Both clamped	Midspan	1/384	1/8
Concentrated at midspan	Both simply supported	Midspan	1/48	1/4
Concentrated at midspan	Both clamped	Midspan	1/192	1/4
Concentrated at outer quarter points	Both simply supported	Midspan	11/768	1/8
Concentrated at outer quarter points	Both simply supported	Load point	1/96	1/8
Uniformly distributed	Cantilever, 1 free, 1 clamped	Free end	1/8	1/2
Concentrated at free end	Cantilever, 1 free, 1 clamped	Free end	1/3	1

Reinforced Plastic Directional Property

The term reinforced plastic (RP), also called composites (more accurately plastic composites), refers to combinations of plastic (matrix) and reinforcing materials that predominantly come in fiber forms such as chopped, continuous, woven and nonwoven fabrics, etc.; also in other forms such as powder, flake, etc. They provide significant oriented property and/or cost improvements than the individual components (10, 14, 35, 38, 39–43, 62).

Primary benefits include high strength and modulus, lightweight, high strength-to-weight ratio, high dielectric strength and corrosion resistance, long term durability, and particularly oriented strength or controlled directional properties. Their exists with many unreinforced and reinforced plastics directional properties requirements. Included are materials such as oriented film, pressure pipe, support beams, laminates, reinforced plastics, and so on. (Chapter 8, **PROCESSING AND PROPERTIES, Orientation**).

Both thermoset (TS) and thermoplastic (TP) are used. Included in the RTPs (thermoplastic RPs) are stampable reinforced thermoplastics (SRTPs). At least 90wt% use glass fiber and about 40% of all RPs use TS polyester plastic of the RP products fabricated. Improved understanding and control of processes continue to increase performance and reduce variability (Chapter 8). Fiber strengths have risen to the degree that 2-D and 3-D RPs can be used producing very high strength and stiff RP products having long service lives of over a half century. In-

cluded in these RTPs are stampable reinforced thermoplastics (SRTPs).

Thermoplastic RPs (RTPs), even with their relatively lower properties when compared to thermoset RPs (RTSs), consume about 55wt% of all RP products. Practically all of the RTPs are injection molded with very fast cycles using short glass fiber producing highly automated and high performance products.

Isotropic material   In an isotropic material the properties at a given point are the same, independent of the direction in which they are measured (Fig. 3-19). The term isotropic means uniform. As one moves from point-to-point in this type of homogeneous plastic the material's composition remains constant. Also, the smallest samples of material

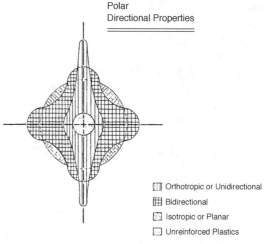

Fig. 3-19 Examples of the performance of RPs with different orientations of their fiber reinforcements.

cut from any location has the same properties. A cast, unfilled plastic is a good example of a reasonably homogeneous material. With RPs and other similar materials, isotropic refers only to the plane of the fiber layup; that is, it is only two-directional (2-D), rather than a complete isotropic behavior in three planes. However there are 3-D RPs; they use a 3-D fabric layup.

Anisotropic material In an anisotropic material the properties vary, depending on the direction in which they are measured. There are various degrees of anisotropy, using different terms such as orthotropic or unidirectional, bidirectional, heterogeneous, and so on (Fig. 3-19). For example, cast plastics or metals tend to be reasonably isotropic. However, plastics that are extruded, injection molded, and rolled plastics and metals tend to develop an orientation in the processing flow direction (machined direction). Thus, they have different properties in the machine and transverse directions, particularly in the case of extruded or rolled materials (plastics, steels, etc.).

Wood is anisotropic with distinct different properties in three directions. Its highest mechanical properties are in the growth (fiber) direction, with the perpendicular (or second plane) direction having lower properties and the other perpendicular (or third plane) direction having much lower properties.

During World War 11RTPs were developed. These glass fiber-TS polyester plastics were used in many high-performance, structurally loaded products in aircraft, ground vehicles, to ships. The RPs used many different glass-fiber nonwoven and woven constructions to produce the required directional properties for the different products. The design equations and engineering technology approaches used were based on the technology and engineering knowledge of the anisotropic wood performance (based on centuries of wood applications in buildings, bridges, etc.) that the Forest Products Laboratory (FPL) in Madison, WI used. The Materials Laboratory's Plastic Section at the Wright-Patterson Air Force Laboratories in Dayton, Ohio, with the help of FPL provided the original engineering equations and technical approach to designing with RPs that latter were applied to unreinforced plastics (14, 41, 43, 106).

Monocoque Structure

Plastics provides a means to producing monocoque constructions such as has been done in different applications that include toys to automotive body, motor truck, railroad car, aircraft fuselage and wings, and houses. Its construction is one in which the outer covering "skin" carries all or a major part of the stresses. The structure can integrate its body and chassis into a single structure. Unreinforced and RPs are used in these constructions (14, 34).

Integral Hinge

Although integral hinges are feasible with a number of TPs, the concept is generally associated with injection or blow molding polypropylene. This section discusses the integral hinge as it is generally applicable to polypropylene. There are various techniques used to fabricate integral hinges; molded-in (by injection or blow molding), cold worked, extruded, and coining. These so-called "living hinges" take advantage of molecular orientation to provide the bending action in the plastic hinge. As an example an integral hinge can be molded by conventional processing techniques providing certain factors are observed. The required molecular orientation runs transverse to the hinge axis. This can best be achieved by a proper fast melt flow through a thin hinge section, using a proper high melt temperature (Fig. 3-20).

The main concern in integral hinge molding is to avoid conditions that can lead to delamination in the hinge section. These include filling the mold too slow, having too low a melt temperature, having a nonuniform flow front through the hinge section, suffering material contamination as from pigment agglomerates, and running excessively high mold temperatures near the hinge area.

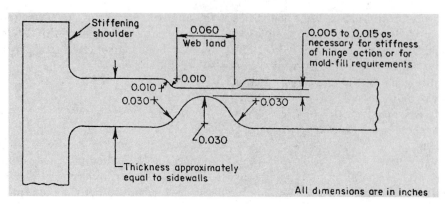

Fig. 3-20   Example of a PP living hinge.

Immediate post-mold flexing while it is still hot is usually required to ensure its proper operation.

Conventional coining mechanical operation is used where the plastic at the hinge section is compressed to the desired thickness using matching bars that produce the required shape of the hinge (Fig. 3-21). Pressure applied to properly cut metal plates, heated or unheated depending on type plastic used, produces the hinge. The plastic in the hinge section is stressed beyond its yield point after creating a necking down effect that causes stretching or orienting of its molecules perpendicular to the hinge folding direction.

A press, a homemade toggle job, or a hot-stamping machine can be used to perform the cold working operation. When heat is required (usual requirement) the male forming die should be about 132 to 138°C (270 to 290°F). Pressure should be maintained for about 10 seconds. This time can be reduced if the product still retains residual molding heat or is preheated. The recommended preheating temperature is from 80 to 110°C (175 to 230°F).

The die backings may be either hard or flexible. With a hard backing, such as steel, the softened polypropylene is die formed into the desired hinge contour. Using a flexible backing like stiff rubber usually makes thinner hinges. The deformation of this type of backing produces the hinge contour by stretching the softened plastic and generally results in thinner cross-sections.

Hinge dimensions of lids, boxes, and many other products made of TPs, particularly polypropylene, have been well established (Fig. 3-22). The successful operation of such hinges depends not only on processing technique, but determining the proper dimensions based on the type plastic used. The dimensions in Figs. 3-20 and 3-21 are typical. Recognize dimensions can differ if the hinge is to move 45° or 180°. If the web land length is too short for the 180° it will self-destruct. Also during injection molding of large size products, there may be a tendency to place a mold gate at the center of the box and another at the center of the lid. The result is that the flow patterns are not an ideal combination to creating a favorable hinge strength, in fact a weld line can develop in the hinge area.

Forming the hinge cross-section by using an extruder die results in a hinge with poor flex life. Because hinges are formed in the direction of the polymer flow, they cannot be sufficiently oriented when flexed. However, if an extruded hinge is formed by the take-off mechanism while the polypropylene

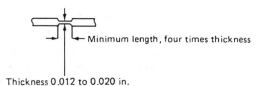

Fig. 3-21   Coining dimensions.

(1) Self-hinged spray closure   (2) Box lid hinge   (3) Single flap cal hinge closure
                                                                    with snap spud

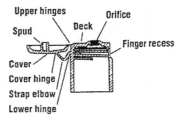

Thickness

Land Length

Land Length
To Thickness Ratio
At Least 3 To 1

Upper hinges   Orifice
Spud   Deck
Cover   Finger recess
Cover hinge
Strap elbow
Lower hinge

(4) Action of package design combining a hinge with snap fits.

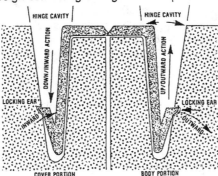

HINGE CAVITY   HINGE CAVITY

DOWN/INWARD ACTION   UP/OUTWARD ACTION

LOCKING EAR   LOCKING EAR

INWARD   OUTWARD

COVER PORTION   BODY PORTION

Fig. 3-22   Examples of living hinges.

still retains internal heat, the hinge will have properties approaching those of cold working.

Snap Joint

A snap joint is economical in two respects: it allows the structural member to be molded simultaneously with the molded product, and it allows rationalizing the assembly, compared with such other joining processes as screws. Table 3-4 provides a comparison of its advantages and disadvantages. Some examples of the various types and their design considerations are shown in Figs. 3-23 to 3-25.

The geometry for snap joints should be chosen in such a manner that excessive increases in stress do not occur (Table 4-3). The arrangement of the undercut should be chosen in such a manner that deformations of the molded product from shrinkage, distortion, unilateral heating, and loading do not

disturb its functioning. The following guidelines are recommended regarding the position of the snap joint to the injection molded gate and the choice of the wall thicknesses in the area of flow to the place of joining: (1) there should be no binding seams at critical points; (2) avoid binding seams created by stagnation of the melt during filling; (3) the plastic molecules and the filler should be oriented in the direction of stress; and (4) any uneven distribution of the filler should not occur at high-stress points (Chapter 4, **JOINING AND ASSEMBLING. Snap Fit**).

Product Size and Shape

Product size is limited to available equipment that can handle the size and pressure as well as other processing requirements. Also involved are factors such as packaging and shipment to the customer. The ability to achieve specific shapes and design details is dependent on the way the process operates.

Table 3-4   Advantages and disadvantages of snap joints

Advantages	Disadvantages
Can be easily integrated into the structural member	The fixing of the joined parts is weaker than in welding, bonding, and screw joining
Compact, space-saving form	The conduct of force at the joining place is lesser than in areal joining (bonding, welding)
Takes over other functions like bearing, spring cushioning, fixing	Effects of processing on the properties of the snap joints (orientation of the molecules and of the filler, distribution of the filler, binding seams, shrinkage, surface, roughness and structure)
Higher forces can also be transmitted with proper designing	
Small number of individual parts	Narrow tolerances are required in complicated applications (in plastics, this is associated in some cases with considerable expenditure)
Assembly of a construction system with little expenditure of production facilities and time	Influence of environmental effects (for example, distortion due to temperature differences) on the functioning
	Difficulties with a continuous loading of the snap joint

Generally the lower the process pressure, the larger the product that can be produced. With most labor-intensive fabricating methods, such as RP hand lay-up with TS plastic, relatively slow process curing reaction time of the plastic can be used so that there is virtually no limit on size (Fig. 8-63).

A general guide to practical processing thickness limitations follows based on heat transfer capability through plastics is in inches: injection molding 0.02 to 0.5; extrusion 0.001 to 1.0; blow molding 0.003 to 0.2; thermoforming 0.002 to 1.0; compression molding 0.05 to 4.0; and foam injection molding 0.1 to 5.0. However with the proper process controls on materials and equipment, products are produced that range below these figures (Chapter 8).

Although there is no limit theoretically to the shapes that can be created, practical considerations must be met. These relate not only to product design but also to mold or die design, since these must be considered one entity in the total creation of a usable, economically feasible product. In the sections that follow, various phases considered important in the creation of such products are examined for their contribution to and effect on design and function.

Prior to designing a product, the designer should understand such basic factors as those summarized in Fig. 3-26 and Table 3-5. Success with plastics, or any other material for that matter, is directly related to observing design details. For example, something as simple as a stiffening rib is different for an injection molded or structural foam product, even though both may be molded from the same plastic (Fig. 3-26). However, a stiffening rib that is to be molded in a low-mold-shrink, amorphous TP will differ from a high-mold-shrinkage, crystalline TP rib, even though both plastics are injection molded. Ribs molded in RP have their own distinct requirements. Hollow stiffening ribs of the type produced by thermoforming, blow, or rotational molding have the same function, but they are designed to be totally different.

The important factors to consider in designing can be categorized as follows: product thicknesses, tolerances, ribs, bosses and studs, radii and fillets, drafts or tapers, holes, threads, colors, surface finishes and gloss levels, decorating operations, parting lines, gate locations, shrinkages, assembly techniques, mold or die designs, production volumes, tooling and other equipment amortization periods, as well as the plastic and process selected. As previously reviewed (Fig. 1-11) with the gains obtain by these type factors, there could be losses such as surface imperfections, sink marks, or voids that could occur.

Preparing a complete list of design constraints is a crucial first step in the design; failure to take this step can lead to costly errors. For example, a designer might have an

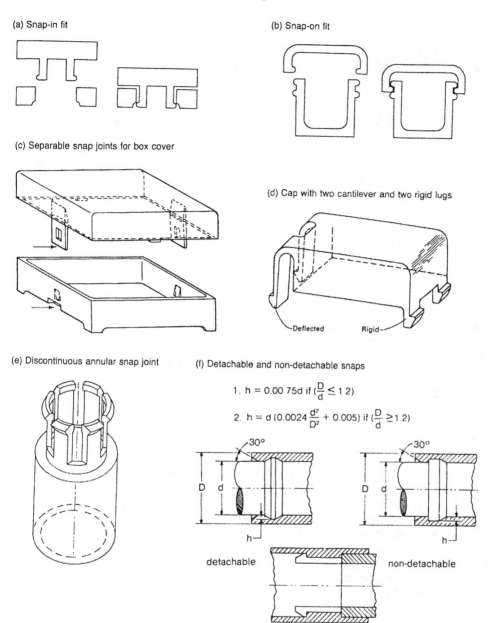

(a) Snap-in fit

(b) Snap-on fit

(c) Separable snap joints for box cover

(d) Cap with two cantilever and two rigid lugs

Deflected   Rigid

(e) Discontinuous annular snap joint

(f) Detachable and non-detachable snaps

1. $h = 0.00\,75d$ if $(\frac{D}{d} \leq 1\,2)$

2. $h = d\,(0.0024\,\frac{d^2}{D^2} + 0.005)$ if $(\frac{D}{d} \geq 1.2)$

30°   30°

D   d   D   d

h   h

detachable   non-detachable

Fig. 3-23   Different snap-fit designs.

expensive injection mold prepared, designed for a specific material's shrink value to meet specific product dimensions, only to discover belatedly that the initial material chosen did not meet some overlooked design requirement or constraint.

The designer may have the difficult if not impossible task of finding a plastic that does meet all the design constraints, including the important appropriate shrink value for the existing mold cavity(s) otherwise expensive mold modifications may be required, if not replacing the complete mold. Such desperation in the last stages of a design project can and should be avoided. As emphasized from one end of this book to the other, it is vital to set up a complete checklist of product requirements, to preclude the possibility that a critical requirement may be overlooked initially. Fortunately there are occasions where

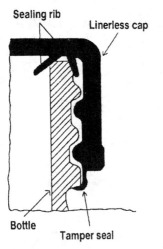

Fig. 3-24 Section of a linerless cap with a tamperproof snap seal.

changes in process control during fabrication can be used to produce the required dimensional product and meet product performance properties without any or excessive cost.

Basic Feature

Design involves establishing the configuration of the products that will form the basis on which suitable material selected for anticipated requirements and processes can be applied. During the drawing of shapes and cross sections, there are certain design features with plastic materials that have to be kept in order to avoid failure or degrada-

tion of properties during fabrication and/or in service.

Such features may be called property detractors since most of them are responsible for internal stresses that can result in products cracking, surface crazing, reducing the available stress level for load-bearing purposes, etc. Other features may be classified as precautionary measures that may influence the favorable performance if not properly incorporated. Examples of these property detractors along with other features suggest means of circumventing their potential negative effects are presented. This subject will be reviewed in detail latter in this chapter.

Tolerance

Tight tolerances can be met. However, the specific dimensions that can be obtained on a finished fabricated plastic product basically starts with the design configuration. In turn it depends on the performance and control of the fabrication process, the plastic material, and, in many cases, upon properly integrating the materials with the process. A number of variable characteristics exist with designing, materials, and processes as described at the end of this chapter and Chapters 6 and 8. Unfortunately, many designers tend to consider dimensional tolerances on plastic products to be complex, unpredictable, and not controllable. This is simply not true since there is a logical approach to controlling and operating within tolerances that can be met.

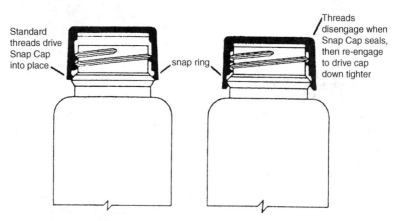

Fig. 3-25 Nonbackoff snap cap provides liquid-tight closure.

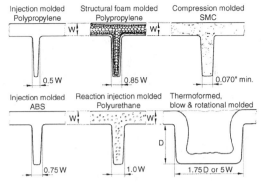

Fig. 3-26 Examples of how different plastics and processes affect the design details of a stiffening rib.

Plastics are no different in this respect than other materials. If steel, aluminum, and ceramics were to be made into a different complex shapes and no prior history on their behavior for that processing shape existed, a period of trial and error would be required to ensure their meeting the required measurements. If relevant processing information or experience did exist, it would be possible for these metallic (or plastic) products to meet the requirements with the first product produced. Experience on new steel shapes always took trial and error time that included different shaped high pressure hydraulic steel cylinders that failed in service when used in a new injection molding hydraulically operating machine (author's experience).

This same situation exists with plastics. To be successful with plastics requires experience with their melt behavior, melt-flow behavior during processing, and the process controls needed to ensure meeting the dimensions that can be achieved in a complete processing operation. Based on the plastic to be used and the equipment available for processing, certain combinations will make it possible to meet extremely tight tolerances.

Fortunately, there are many different types of plastics that can provide all kinds of properties, including specific dimensional tolerances. It can thus be said that the real problem is not with the different plastics or processes but rather with the designer, who requires knowledge and experience to create products to meet the desired requirements. The designer with no knowledge or experience

has to become familiar with the plastic design concepts expressed throughout this book and work initially with capable people such as qualified and independent consultants, suppliers of plastic materials, and/or the processors.

Some plastics, such as the TSs and in particular the TS-RPs (RTSs), can produce products with exceptionally tight tolerances that practically meet zero tolerances. In injection or compression molding of relatively thin to thick and complex shapes, tolerances can be held to less than 0.001 in. or to even zero, as can also be done using hand layup RP fabricating techniques.

At the other extreme are the unfilled, unreinforced extruded TPs. Generally, unless a very thin uniform wall is to be extruded, it is impossible to hold the tight tolerances just given. The thicker and more complex an extruded shape is, the more difficult it becomes to meet tight tolerances. However extruded products meet satisfactory tolerances such as those required in products such as window frames, rain gutters electrical/electronic devices, and medical device (6, 242). What is important to the designer is to determine the tolerances that can be met and then design using these tolerances.

To maximize control in setting tolerances there is usually a minimum and a maximum limit on thickness, based on the process to be used such as those in Tables 3-6 to 3-9. Each plastic has its own range that depends on its chemical structure, composition (additives, etc.), and melt-processing characteristics. Any dimensions and tolerances are theoretically possible, but they could result in requiring special processing equipment, which usually becomes expensive. There are of course products that require and use special equipment such as polycarbonate compact discs (CDs) to meet extremely tight tolerances.

An influence on dimensions and tolerances involves the coefficient of linear thermal expansion or contraction. This CLTE value has to be determined at the product's operating temperature (Chapter 2, **THERMAL EXPANSION AND CONTRACTION**). Plastics can provide all extremes in CLTEs. As an

Table 3-5   Design guides for processes vs. product requirements

		Compression Molding			Injection Molding (Thermo-plastics)	Cold Press Molding	Spray-up and Hand Lay-up
		Sheet Molding Compound	Bulk Molding Compound	Preform Molding			
Minimum inside radius, in. (mm)		1/16″ (1.59)	1/16″ (1.59)	1/8″ (3.18)	1/16″ (1.59)	1/4″ (6.35)	1/4″ (6.35)
Molded-in holes		Yes	Yes	Yes	Yes	No	Large
Trimmed in mold		Yes	Yes	Yes	No	Yes	No
Core pull & slides		Yes	Yes	No	Yes	No	No
Undercuts		Yes	Yes	No	Yes	No	Yes
Minimum recommended draft, in./°		1/4″ to 6″ (6.35–152 mm) depth: 1° to 3° 6″ (152 mm) depth and over: 3°or as required				2° 3°	0°
Minimum practical thickness, in. (mm)		.050″ (1.3)	.060″ (1.5)	.030″ (0.76)	0.35″ (0.89)	.080″ (2.0)	.060″ (1.5)
Maximum practical thickness, in. (mm)		1″ (25.4)	1″ (25.4)	.250″ (6.35)	.500″ (12.7)	.500″ (12.7)	No limit
Normal thickness variation, in. (mm)		±.005 (±0.1)	±.005 (±0.1)	±.008 (±0.2)	±.005 (±0.1)	±.010″ (±0.25)	±.020″ (±0.51)
Maximum thickness buildup, heavy buildup and increased cycle		As req'd.	As req'd.	2-to-1 max.	As req'd.	2-to-1 max.	As req'd.
Corrugated sections		Yes	Yes	Yes	Yes	Yes	Yes
Metal inserts		Yes	Yes	Not recommended	Yes	No	Yes
Bosses		Yes	Yes	Yes	Yes	Not recommended	Yes
Ribs		As req'd	Yes	Not recommended	Yes	Not recommended	Yes
Molded-in labels		Yes	Yes	Yes	No	Yes	Yes
Raised numbers		Yes	Yes	Yes	Yes	Yes	Yes
Finished surfaces (reproduces mold surface)		Two	Two	Two	Two	Two	One

Table 3-6 Examples of dimensional tolerances for injection molded TP products

Dimensions, in	ABS		Acetal		Acrylic		Nylon		Polycarbonate	
	Commercial	Fine	Commercial	Fine	Commercial	Fine	Commercial	Fine	Commercial	Fine
To 1.000	0.005	0.003	0.006	0.004	0.005	0.003	0.004	0.002	0.004	0.0025
1.000–2.000	0.006	0.004	0.008	0.005	0.006	0.004	0.006	0.003	0.005	0.003
2.000–3.000	0.008	0.005	0.009	0.006	0.007	0.005	0.007	0.005	0.006	0.004
3.000–4.000	0.009	0.006	0.011	0.007	0.008	0.006	0.009	0.006	0.007	0.005
4.000–5.000	0.011	0.007	0.013	0.008	0.009	0.007	0.010	0.007	0.008	0.005
5.000–6.000	0.012	0.008	0.014	0.009	0.011	0.008	0.012	0.008	0.009	0.006
6.000–12.000, for each additional inch	0.003	0.002	0.004	0.002	0.003	0.002	0.003	0.002	0.003	0.015
············	0.004	0.002	0.004	0.002	0.003	0.003	0.004	0.003	0.003	0.002
············	0.003	0.002	0.004	0.002	0.005	0.003	0.005	0.003	0.003	0.002
0.000–0.125	0.002	0.001	0.002	0.001	0.003	0.001	0.002	0.001	0.002	0.001
0.125–0.250	0.002	0.001	0.003	0.002	0.003	0.002	0.003	0.002	0.002	0.015
0.250–0.500	0.003	0.002	0.004	0.002	0.004	0.002	0.003	0.002	0.003	0.002
0.500 and over	0.004	0.002	0.006	0.003	0.005	0.003	0.005	0.003	0.003	0.002
0.000–0.250	0.003	0.002	0.004	0.002	0.004	0.002	0.004	0.002	0.002	0.002
0.250–0.500	0.004	0.002	0.005	0.003	0.004	0.002	0.004	0.003	0.003	0.002
0.500–1.000	0.005	0.003	0.006	0.004	0.006	0.003	0.005	0.004	0.004	0.003
0.000–3.000	0.015	0.010	0.011	0.006	0.010	0.007	0.010	0.004	0.005	0.003
3.000–6.000	0.030	0.020	0.020	0.010	0.015	0.010	0.015	0.007	0.007	0.004
TIR	0.009	0.005	0.010	0.006	0.010	0.006	0.010	0.006	0.005	0.003

(*Continues*)

Table 3-6   (*Continued*)

Dimensions, in	Polyethylene, high-density		Polyethylene, low-density		Polypropylene		Polystyrene		Vinyl, flexible		Vinyl, rigid	
	Commercial	Fine	Commercial	Fine	Commercial	Fine	Commercial	Fine	Commercial	Fine	Commercial	Fine
To 1.000	0.008	0.006	0.007	0.004	0.007	0.004	0.004	0.0025	0.011	0.007	0.008	0.0045
1.000–2.000	0.010	0.008	0.010	0.006	0.009	0.005	0.005	0.003	0.012	0.008	0.009	0.005
2.000–3.000	0.013	0.011	0.012	0.008	0.011	0.007	0.007	0.004	0.014	0.009	0.010	0.006
3.000–4.000	0.015	0.013	0.015	0.010	0.013	0.008	0.008	0.005	0.015	0.011	0.012	0.007
4.000–5.000	0.018	0.016	0.017	0.011	0.015	0.009	0.010	0.006	0.017	0.012	0.013	0.008
5.000–6.000	0.020	0.018	0.020	0.013	0.018	0.011	0.011	0.007	0.018	0.013	0.014	0.009
6.000–12.000, for each additional inch add	0.006	0.003	0.005	0.004	0.005	0.003	0.004	0.002	0.005	0.003	0.005	0.003
....	0.006	0.004	0.005	0.004	0.006	0.003	0.0055	0.003	0.007	0.003	0.007	0.003
....	0.006	0.004	0.005	0.004	0.006	0.003	0.007	0.0035	0.007	0.003	0.007	0.003
0.000–0.125	0.003	0.002	0.003	0.002	0.003	0.002	0.002	0.001	0.004	0.003	0.004	0.003
0.125–0.250	0.005	0.003	0.004	0.003	0.004	0.003	0.002	0.001	0.005	0.004	0.004	0.003
0.250–0.500	0.006	0.004	0.005	0.004	0.005	0.004	0.002	0.0015	0.006	0.005	0.005	0.004
0.500 and over	0.008	0.005	0.006	0.005	0.008	0.006	0.0035	0.002	0.008	0.006	0.006	0.005
0.000–0.250	0.005	0.003	0.003	0.003	0.005	0.003	0.0035	0.002	0.004	0.003	0.004	0.003
0.250–0.500	0.007	0.004	0.004	0.004	0.006	0.004	0.004	0.002	0.005	0.004	0.005	0.004
0.500–1.000	0.009	0.006	0.006	0.005	0.009	0.006	0.005	0.003	0.006	0.005	0.006	0.005
0.000–3.000	0.023	0.015	0.020	0.015	0.021	0.014	0.007	0.004	0.010	0.007	0.015	0.010
3.000–6.000	0.037	0.022	0.030	0.020	0.035	0.021	0.013	0.005	0.020	0.015	0.020	0.015
TIR	0.027	0.010	0.010	0.008	0.016	0.013	0.010	0.008	0.015	0.010	0.010	0.005

Table 3-7 Guide on allowable undercut tolerances TPs

Material	Average Maximum Strippable Undercut [mm (in)]
Acrylic	1.5 (0.060)
Acrylonitrile butadiene styrene	1.8 (0.070)
Nylon	1.5 (0.060)
Polycarbonate	1.0 (0.040)
Polyethylene	2.0 (0.080)
Polypropylene	1.5 (0.060)
Polystyrene	1.0 (0.040)
Polysulfone	1.0 (0.040)
Vinyl, flexible	2.5 (0.100)

example graphite-filled molding compounds could work in reverse. Upon heating, they contract rather than expand, and vice-versa.

To assist the designer the Society of the Plastics Industry (SPI) issued Standards and Practices of Plastics Molders that contains tolerance ranges for various plastics as an initial guide. These ranges theoretically encompass the accuracy involved in moldmaking, shrink variations, and molding variations. Each material supplier converts these data to suit their specific plastics. Tables 3-10 to 3-14 are examples of this information. This type of information is intended to give the designer a guide for tolerances that are to be shown on the drawings; these tolerances include varia-

tions in product manufacture and some degree of variation in the tooling for TPs and TSs.

These SPI tables can also be used as the basis for establishing standards for molded products between the designer, molder, and customer. Users will find that two separate sets of values are represented. Commercial values represent common production tolerances that can be achieved at the most economical level. Fine values represent closer tolerances that can be held, but at a greater cost. The selection of one or the other will depend on the application under consideration and the economics involved.

Refer to the hypothetical molded article and its cross-section, illustrated in the tables. Then using the applicable code number (such as **A** that represents the diameter) in the first column of the table and the exact dimensions indicated in the second column, one can find the recommended tolerances either in the chart at the top of the table or in the two columns underneath. Note that the typical article shown in cross-section in the table may be round or rectangular or some other shape. Thus, dimensions **A** and **B** may be either diameters or lengths.

Tight tolerances on dimensions should be specified only where absolutely necessary. Too many drawings show limits of sizes where other means of attaining desired results would be more constructive or the tolerances

Table 3-8 Guide for wall thicknesses of TS molding materials

	Minimum Thickness in. (mm)	Average Thickness in. (mm)	Maximum Thickness in. (mm)
Alkyd—glass filled	.040 (1.0)	.125 (3.2)	.500 (13)
Alkyd—mineral filled	.040 (1.0)	.187 (4.7)	.375 (9.5)
Diallyl phthalate	.040 (1.0)	.187 (4.7)	.375 (9.5)
Epoxy-glass filled	.030 (0.76)	.125 (3.2)	1.000 (25.4)
Melamine—cellulose filled	.035 (0.89)	.100 (2.5)	.187 (4.7)
Urea—cellulose filled	.035 (0.89)	.100 (2.5)	.187 (4.7)
Phenolic—general purpose	.050 (1.3)	.125 (3.2)	1.000 (25.4)
Phenolic—flock filled	.050 (1.3)	.125 (3.2)	1.000 (25.4)
Phenolic—glass filled	.030 (0.76)	.093 (2.4)	.750 (19)
Phenolic—fabric filled	.062 (1.6)	.187 (4.7)	.375 (9.5)
Phenolic—mineral filled	.125 (3.2)	.187 (4.7)	1.000 (25.4)
Silicone glass	.050 (1.3)	.125 (3.2)	.250 (6.4)
Polyester premix	.040 (1.0)	.070 (1.8)	1.000 (25.4)

Table 3-9   Guide to tolerances of TP extrusion profiles

	HIPS	PC, ABS	PP	PVC Rigid	PVC Flex.	LDPE
Wall thickness (%, =)	8	8	8	8	10	10
Angles (Deg., =)	2	3	3	2	5	5
Profile dimensions (in., ±)						
To 0.125	0.007	0.010	0.010	0.007	0.010	0.012
0.125 to .500	0.012	0.020	0.015	0.010	0.015	0.025
.500 to 1	0.017	0.025	0.020	0.015	0.020	0.030
1 to 1.5	0.025	0.027	0.027	0.020	0.030	0.035
1.5 to 2	0.030	0.035	0.035	0.025	0.035	0.040
2 to 3	0.035	0.037	0.037	0.030	0.040	0.045
3 to 4	0.050	0.050	0.050	0.045	0.065	0.065
4 to 5	0.065	0.065	0.065	0.060	0.093	0.093
5 to 7	0.093	0.093	0.093	0.075	0.125	0.125
7 to 10	0.125	0.125	0.125	0.093	0.150	0.150

where not sufficiently specific. For example, if the outside dimensions of an electric drill housing halves were to have a tolerance of ±0.003 in., this would be a tight limit and can be met. And yet if half of the housing were to be on the minimum side and the other on the maximum side, there would be a resulting step that would be uncomfortable to the feel of the hand while gripping the drill.

A realistic specification would call for matching of halves that would provide a smooth joint between them, and the highest step should not exceed 0.002 in. The point is that limits should be specified in a way that those responsible for the manufacture of a product will understand the goal that is to be attained. Thus we may indicate dimensions for gear centers, holes as bearing openings for shafts, guides for cams, etc. This type of designation would alert a mold maker as well as the molder to the significance of the tolerances in some areas and the need for matching products in other places and clearance for assembly in still other locations.

Most of the engineering plastics reproduce faithfully and easily conform to the mold configuration, and when processing parameters are appropriately controlled, they will repeat with excellent accuracy tolerancewise, etc. As an example for the past many decades, we see plastic gears and other precision products made of acetal, nylon, polycarbonate,

etc. Their tooth contour and other precision areas are made with a limit of 0.0002 in., and the spacing of the teeth is extremely uniform to meet the most exacting requirements.

The problem with any precision-type product is to recognize what steps are needed to reach the objective and follow through in a detailed manner every phase of the process to safeguard the end product. Generally speaking, if we segregate the tolerances we should come up with feasible tolerances that will be reasonable and useful. The segregation can be into (a) functional need, such as running fit, sliding fit, gear tooth contour, etc.; (b) assembly requirements that are to accommodate products with their own tolerances; and (c) matching parts for appearance or utility. This approach would be more productive than trying to apply tolerances strictly on a dimensional basis.

Adaptation of metal tolerances to plastics is not advisable. With plastics reaction to moisture and heat, for example, is drastically different from metals, so that pilot testing under extreme use conditions is almost mandatory for establishing adequate tolerance requirements. Also important to control cost is that close tolerances should be indicated only where needed, carefully analyzed for their magnitude, and proven out as to their usefulness.

Table 3-10   High density polyethylene (HDPE)

Drawing Code	Dimensions (Inches)		Plus or Minus in Thousands of an Inch

Drawing Code	Dimensions	Comm. ±	Fine ±
A = Diameter (See note #1) B = Depth (See note #3) C = Height (See note #3)	6.000 to 12.000 for each additional inch add (inches)	0.006	0.003
D = Bottom Wall	(See note #3)	0.006	0.004
E = Side Wall	(See note #4)	0.006	0.004
F = Hole Size Diameter (See note #1)	0.000 to 0.125	0.003	0.002
	0.126 to 0.250	0.004	0.002
	0.251 to 0.500	0.006	0.004
	0.501 & over	0.008	0.005
G = Hole Size Depth (See note #5)	0.000 to 0.250	0.005	0.003
	0.251 to 0.500	0.007	0.004
	0.501 to 1.000	0.009	0.006
H = Corners, Ribs, Fillets	(See note #6)	0.025	0.010
Flatness (See note #4)	0.000 to 3.000	0.023	0.015
	3.001 to 6.000	0.037	0.022
Thread Size (Class)	Internal	1	2
	External	1	2
Concentricity	(See note #4) (F.I.M.)	0.027	0.010
Draft Allowance Per Side	(See note #5)	2.0°	0.75°
Surface Finish	(See note #7)		
Color Stability	(See note #7)		

REFERENCE NOTES

1. These tolerances do not include allowance for aging characteristics of material.

2. Tolerances are based on 0.125 inch wall section.

3. Parting line must be taken into consideration.

4. Part design should maintain a wall thickness as nearly constant as possible. Complete uniformity in this dimension is sometimes impossible to achieve. Walls of non-uniform thickness should be gradually blended from thick to thin.

5. Care must be taken that the ratio of the depth of a cored hole to its diameter does not reach a point that will result in excessive pin damage.

6. These values should be increased whenever compatible with desired design and good molding techniques.

7. Customer-Molder understanding is necessary prior to tooling.

Shrinkage

One factor associated with tolerance is shrinkage. Generally, shrinkage is the difference between the dimensions of a fabricated product at room temperature and after cooling, checked usually twelve to twenty-four hours after fabrication. Having an elapsed time is necessary for many plastics, particularly the commodity TPs, to allow products to complete their inherent shrinkage behavior after processing. The extent of this postshrinkage can be near zero for certain plastics or may vary considerably.

Shrinkage can also be dependent on such climatic conditions as temperature and

Table 3-11   Polypropylene (PP)

Drawing Code	Dimensions (Inches)	Comm. ±	Fine ±
A = Diameter (See note #1) B = Depth (See note #3) C = Height (See note #3)	0.000 / 0.500 / 1.000 / 2.000 / 3.000 / 4.000 / 5.000 / 6.000 *(graph — Plus or Minus in Thousands of an Inch: 5, 10, 15, 20, 25; curves labeled "Commercial" and "Fine")*		
6.000 to 12.000 for each additional inch add (inches)		0.005	0.003
D = Bottom Wall	(See note #3)	0.006	0.003
E = Side Wall	(See note #4)	0.006	0.003
F = Hole Size Diameter (See note #1)	0.000 to 0.125	0.003	0.002
	0.126 to 0.250	0.004	0.003
	0.251 to 0.500	0.005	0.004
	0.501 & over	0.008	0.006
G = Hole Size Depth (See note #5)	0.000 to 0.250	0.005	0.003
	0.251 to 0.500	0.006	0.004
	0.501 to 1.000	0.009	0.006
H = Corners, Ribs, Fillets	(See note #6)	0.029	0.016
Flatness	0.000 to 3.000	0.022	0.014
(See note #4)	3.001 to 6.000	0.036	0.021
Thread Size (Class)	Internal	1	2
	External	1	2
Concentricity	(See note #4) (F.I.M.)	0.015	0.012
Draft Allowance Per Side	(See note #5)	1.5°	0.5°
Surface Finish	(See note #7)		
Color Stability	(See note #7)		

REFERENCE NOTES

1. These tolerances do not include allowance for aging characteristics of material.

2. Tolerances are based on 0.125 inch wall section.

3. Parting line must be taken into consideration.

4. Part design should maintain a wall thickness as nearly constant as possible. Complete uniformity in this dimension is sometimes impossible to achieve. Walls of non-uniform thickness should be gradually blended from thick to thin.

5. Care must be taken that the ratio of the depth of a cored hole to its diameter does not reach a point that will result in excessive pin damage.

6. These values should be increased whenever compatible with desired design and good molding techniques.

7. Customer-Molder understanding is necessary prior to tooling.

humidity, under which the product will exist in service, as well as its conditions of storage. If proper stipulations are not in the fabricators job order, molded products could be delivered at the time the ideal climatic conditions exist to meet the customer's tolerance requirements. As ridiculous as this may appear, it has happened unfortunately to customers.

Plastic suppliers can provide the initial information on shrinkage that has to be added to the design shape and will influence its processing. The shrinkage and postshrinkage will depend on the types of plastics and their

Table 3-12 Polyvinyl chloride (PVC)

Drawing Code	Dimensions (Inches)	Comm. ±	Fine ±
	Plus or Minus in Thousands of an Inch		

Drawing Code	Dimensions (Inches)	Comm. ±	Fine ±
A = Diameter (See note #1) B = Depth (See note #3) C = Height (See note #3)	0.000 / 0.500 / 1.000 / 2.000 / 3.000 / 4.000 / 5.000 / 6.000		
	6.000 to 12.000 for each additional inch add (inches)	0.005	0.003
D = Bottom Wall	(See note #3)	0.007	0.003
E = Side Wall	(See note #4)	0.007	0.003
F = Hole Size Diameter (See note #1)	0.000 to 0.125	0.004	0.003
	0.126 to 0.250	0.005	0.004
	0.251 to 0.500	0.006	0.005
	0.501 & over	0.008	0.006
G = Hole Size Depth (See note #5)	0.000 to 0.250	0.004	0.003
	0.251 to 0.500	0.005	0.004
	0.501 to 1.000	0.006	0.005
H = Corners, Ribs, Fillets	(See note #6)	0.030	0.010
Flatness	0.000 to 3.000	0.010	0.007
(See note #4)	3.001 to 6.000	0.020	0.015
Thread Size (Class)	Internal		
	External		
Concentricity	(See note #4) (F.I.M.)	0.015	0.010
Draft Allowance Per Side	(See note #5)	1.5°	1.0°
Surface Finish	(See note #7)		
Color Stability	(See note #7)		

REFERENCE NOTES

1. These tolerances do not include allowance for aging characteristics of material.

2. Tolerances are based on 0.125 inch wall section.

3. Parting line must be taken into consideration.

4. Part design should maintain a wall thickness as nearly constant as possible. Complete uniformity in this dimension is sometimes impossible to achieve. Walls of non-uniform thickness should be gradually blended from thick to thin.

5. Care must be taken that the ratio of the depth of a cored hole to its diameter does not reach a point that will result in excessive pin damage.

6. These values should be increased whenever compatible with desired design and good molding techniques.

7. Customer-Molder understanding is necessary prior to tooling.

additives/fillers that interrelate to the processing conditions. The type and amount of filler, such as its reinforcement, can significantly reduce shrinkage and tolerances where it could be at zero change. If a plastic product is free to expand and contract (shrink) with temperature change, then its thermal expansion property is usually of little significance. However, if it is attached to another material having a lower or different thermal expansion, then movement of the product will be restricted. Temperature change will then result in the development of thermal stresses in the product. The magnitude of the stresses will

Table 3-13 Nylon (polyamide) (PA)

Drawing Code	Dimensions (Inches)	Plus or Minus in Thousands of an Inch	
A = Diameter (See note #1) B = Depth (See note #3) C = Height (See note #3)	0.000 — 0.500 — 1.000 — 2.000 — 3.000 — 4.000 — 5.000 — 6.000 (graph: Fine, Commercial lines; scale 5, 10, 15, 20, 25)		

	Dimensions (Inches)	Comm. ±	Fine ±
	6.000 to 12.000 for each additional inch add (inches)	0.003	0.002
D = Bottom Wall	(See note #3)	0.004	0.003
E = Side Wall	(See note #4)	0.005	0.003
F = Hole Size Diameter (See note #1)	0.000 to 0.125	0.002	0.001
	0.126 to 0.250	0.003	0.002
	0.251 to 0.500	0.003	0.002
	0.501 & over	0.005	0.003
G = Hole Size Depth (See note #5)	0.000 to 0.250	0.004	0.002
	0.251 to 0.500	0.004	0.003
	0.501 to 1.000	0.005	0.004
H = Corners, Ribs, Fillets	(See note #6)	0.021	0.013
Flatness (See note #4)	0.000 to 3.000	0.010	0.004
	3.001 to 6.000	0.015	0.007
Thread Size (Class)	Internal	1	2
	External	1	2
Concentricity	(See note #4) (F.I.M.)	0.005	0.003
Draft Allowance Per Side		1.5°	0.5°
Surface Finish	(See note #7)		
Color Stability	(See note #7)		

REFERENCE NOTES

1. These tolerances do not include allowance for aging characteristics of material.

2. Tolerances are based on 0.125 inch wall section.

3. Parting line must be taken into consideration.

4. Part design should maintain a wall thickness as nearly constant as possible. Complete uniformity in this dimension is sometimes impossible to achieve. Walls of non-uniform thickness should be gradually blended from thick to thin.

5. Care must be taken that the ratio of the depth of a cored hole to its diameter does not reach a point that will result in excessive pin damage.

6. These values should be increased whenever compatible with desired design and good molding techniques.

7. Customer-Molder understanding is necessary prior to tooling.

depend on the temperature change, method of attachment, and relative expansion and modulus characteristics of the two materials at the exposed heat.

Expansion or contraction can be controlled in the plastic by orientation, cross-linking, adding fillers and/or reinforcements, etc. Any cross-linking has a substantial beneficial effect on TPs. With the amorphous type, expansion is reduced. In a crystalline TP, however, the decreased expansion may be partially offset by the loss of crystallinity. A compounded plastic can be made to match those of the attached material (plastic, steel, etc.). With certain additives, such as graphite filler, the thermal change could be zero or near zero;

Table 3-14 Phenol-formaldehyde (PF)

Drawing Code	Dimensions (Inches)	Plus or Minus in Thousands of an Inch	
A = Diameter (See note #1) B = Depth (See note #3) C = Height (See note #3)	0.000 / 0.500 / 1.000 / 2.000 / 3.000 / 4.000 / 5.000 / 6.000		
	6.000 to 12.000 for each additional inch add (inches)	Comm. ±	Fine ±
		0.003	0.002
D = Bottom Wall	(See note #3)	0.006	0.004
E = Side Wall	(See note #4)	0.004	0.003
F = Hole Size Diameter (See note #1)	0.000 to 0.125	0.002	0.001
	0.126 to 0.250	0.003	0.002
	0.251 to 0.500	0.004	0.003
	0.501 & over	0.005	0.003
G = Hole Size Depth (See note #5)	0.000 to 0.250	0.004	0.002
	0.251 to 0.500	0.005	0.003
	0.501 to 1.000	0.007	0.004
H = Corners, Ribs, Fillets	(See note #6)	0.030	0.015
Flatness (See note #4)	0.000 to 3.000	0.014	0.008
	3.001 to 6.000	0.021	0.014
Thread Size (Class)	Internal	1	2
	External	1	2
Concentricity	(See note #4) (F.I.M.)	0.007	0.004
Draft Allowance Per Side	(See note #5)	1.0°	0.5°
Surface Finish	(See note #7)		
Color Stability	(See note #7)		

REFERENCE NOTES

1. These tolerances do not include allowance for aging characteristics of material.
2. Tolerances are based on 0.125 inch wall section.
3. Parting line must be taken into consideration.
4. Part design should maintain a wall thickness as nearly constant as possible. Complete uniformity in this dimension is sometimes impossible to achieve. Walls of non-uniform thickness should be gradually blended from thick to thin.
5. Care must be taken that the ratio of the depth of a cored hole to its diameter does not reach a point that will result in excessive pin damage.
6. These values should be increased whenever compatible with desired design and good molding techniques.
7. Customer-Molder understanding is necessary prior to tooling.

in fact during a temperature rise, the plastic could contract rather than expand.

The condition of anisotropy can be used when referring to the way a material shrinks during processing, such as in injection molding (Fig. 2-62) and extrusion. Shrinkage is an important consideration when fabricating plastics, particularly crystalline TPs where the flow direction can have more shrinkage than the cross-flow direction. The control of shrinkage can be made to meet design requirements by factors such as the design of the mold with its gate locations or die shape, the processing machine controls, the change of product shape, and the type of plastics.

If it has been determined in advance that a product must be postcured, stress relieved, or baked, allowance must be made for probable additional shrinkage. These requirements must be specified on the initial drawings. Especially for long runs, mold or die design is an important factor. The metals that will be used, particularly in mold cavities, and the forces required will largely be determined by the complexity of the product design. This complexity will, of course, dictate in turn the intricacy of the tool design that will eventually be used (Chapter 8, **TOOLING**). In general, pack hardening, oil hardening, and prehardened steels are used, with materials such as beryllium copper and electroformed cavities finding use in applications for specialized purposes (2–7, 10, 20).

Processing and Tolerance/Shrinkage

Processing is extremely important in regard to tolerance control; in certain cases it is the most influential factor. The dimensional accuracy of the finished product relates to the process, the machining accuracy of mold or die, and the process controls, as well as the shrinkage behavior of the plastic.

The mold or die should also be recognized as one of the most important pieces of production equipment in the plant. These controllable, complex devices must be an efficient heat exchanger and provide the product's shape. The mold or die designer thus has to have the experience or training and knowledge of how to produce the tooling needed for the product and to meet required tolerances with the plastic to be processed.

Adequate process control and its associated instrumentation are essential to have product quality control. In some cases the goal is precise adherence to a control point. In others it is simply to maintain the temperature within a comparatively narrow range.

A knowledge of processing methods will be useful to the designer to help determine what tolerances can be obtained. With such high-pressure methods as injection and compression molding of 2,000 to 30,000 psi

(13.8 to 206.9 MPa) it is possible to develop tighter tolerances, but there is also a tendency to develop undesirable stresses (orientations, etc.). The low-pressure or no pressure processes, including RP contact, casting, and rotational molding, usually do not permit meeting tight tolerances. There are exceptions, such as certain RPs that are processed at contact pressures resulting in meeting tight tolerances. Regardless of the process used, exercising the required and proper control over it will maximize obtaining and repeating of tolerances that are achievable.

Table 3-15 reviews factors affecting tolerances. Many plastics change dimensions after molding principally because they're molecular orientations or molecules are not relaxed (Chapter 2). To ease or eliminate the problem, one can change the processing cycle so that the plastic is "stress relieved," even though that may extend the cycle time. Also applicable is annealing according to one's experience or the plastic supplier's suggestions.

An easy method for estimating shrink allowance for injection molding is as follows:

$$SD_1 = FL(1 + SR) \qquad (3\text{-}10)$$

where $SD_1 = $ mold dimension, $SR = $ plastic's shrinkage (in/in or mm/mm), and $FL = $ product dimension.

If the products are small and have thin walls, this estimate is the best guide. If they are large (>10 in. or 25 cm) or use rather high-shrink plastics, consider using the following method of analysis (214):

$$SD_2 = FL(1 - SR) \qquad (3\text{-}11)$$

where $SD_2 = $ the mold dimension as determined by the corrected equation.

The error, *ER*, would simply be the difference between the SD_1 and SD_2 equations. To be more accurate for calculating mold dimensions where the product size and shrink rate increase, this error value should be considered or Table 3-16 is be used. This table shows, as one example, which in the low shrink (0.008 mil/in. or less) materials, the products must be larger than 15 in.

Table 3-15   Parameters that influence product performances

PART DESIGN:	Part configuration (size/shape). Relate shape to flow of melt in mold to meet performance requirements that should at least include tolerances.
MATERIAL:	Chemical structure, molecular weight, amount and type of fillers/additives, heat history, storage, handling.
MOLD DESIGN:	Number of cavities, layout and size of cavities/runners/gates/cooling lines/side actions/knockout pins/etc. Relate layout to maximize proper performance of melt and cooling flow patterns to meet part performance requirements; preengineer design to minimize wear and deformation of mold (use proper steels); lay out cooling lines to meet temperature to time cooling rate of plastics (particularly crystalline types).
MACHINE CAPABILITY:	Accuracy and repeatability of temperature/time/velocity/pressure controls of injection unit, accuracy and repeatability of clamping force, flatness and parallelism of platens, even distribution of clamping on all tie rods, repeatability of controlling pressure and temperature of oil, oil temperature variation minimized, no oil contamination (by the time you see oil contamination damage to the hydraulic system could have already occurred), machine properly leveled.
MOLDING CYCLE:	Set up the complete molding cycle to repeatedly meet performance requirements at the lowest cost by interrelating material/machine/mold controls.

before an error of 0.001 in. will be realized. The allowable error will depend upon each product's particular application. In some cases it will be important to ensure proper mold-size calculations. In others, changing the calculation method will be purely academic.

Experience is still a basic requirement for mold design with regard to determining cavity dimensions. The costs for changing mold cavities are high, even when similar moldings are to be produced. Until now, theoretical efforts to forecast linear shrinkage have been limited because of the number of existing variables. One way to solve this problem is to simplify the mathematical relationship, leading to an estimated but still acceptable assessment. This means, however, that the number of necessary processing changes will also be reduced (3).

As a first approximation, a superposition method has been used that can provide a guide to predicting mold shrinkage (Fig. 3-27). However, problems arise in measuring the influencing variables, because they are often interrelated, such as variations in the pressure course in a mold with a varying wall thickness.

The parameters of the injection process must be provided. They can either be estimated or, to be more exact, taken from the thermal and rheological layout. The position of a length with respect to flow direction is in practice an important influence. This is used primarily for glass-filled material but can also be used for unfilled TPs.

Regarding this relationship, when designing the mold it is necessary to know the flow direction. To obtain this information, a simple flow pattern construction can be used (Fig. 3-28) via computer analysis. However, the flow direction is not constant. In some cases the flow direction in the filling phase differs from that in the holding phase. Here the question arises of whether this must be considered using superposition.

In order to get the flow direction at the end of the filling phase and the beginning of the holding phase (representing the onset of shrinkage), an analogous model can

Table 3-16 Error in mold size as a result of using incorrect shrinkage equation

Part Size Inches	Plastic Shrink Rate (inches/inch)																
	0.004	0.008	0.012	0.016	0.020	0.030	0.040	0.050	0.060	0.070	0.080	0.090	0.100	0.200	0.300	0.400	0.500
1.0	-0.02	-0.06	-0.1	-0.3	-0.4	-0.9	-1.7	-3	-4	-5	-7	-9	-11	-50	-129	-267	-500
3.0	-0.05	-0.19	-0.4	-0.8	-1.2	-2.8	-5.0	-8	-11	-16	-21	-27	-33	-150	-386	-800	-1500
5.0	-0.08	-0.32	-0.7	-1.3	-2.0	-4.6	-8.3	-13	-19	-26	-35	-45	-56	-250	-643	-1333	-2500
7.0	-0.11	-0.45	-1.0	-1.8	-2.9	-6.5	-11.7	-18	-27	-37	-49	-62	-78	-350	-900	-1867	-3500
9.0	-0.14	-0.58	-1.3	-2.3	-3.7	-8.4	-15.0	-24	-34	-47	-63	-80	-100	-450	-1157	-2400	-4500
11.0	-0.18	-0.71	-1.6	-2.9	-4.5	-10.2	-18.3	-29	-42	-58	-77	-98	-122	-550	-1414	-2933	-5500
13.0	-0.21	-0.84	-1.9	-3.4	-5.3	-12.1	-21.7	-34	-50	-68	-90	-116	-144	-650	-1671	-3467	-6500
15.0	-0.24	-0.97	-2.2	-3.9	-6.1	-13.9	-25.0	-39	-57	-79	-100	-134	-167	-750	-1929	-4000	-7500
17.0	-0.27	-1.10	-2.5	-4.4	-6.9	-15.8	-28.3	-45	-65	-90	-118	-151	-189	-850	-2186	-4533	-8500
19.0	-0.31	-1.23	-2.8	-4.9	-7.8	-17.6	-31.7	-50	-73	-100	-132	-169	-211	-950	-2443	-5067	-9500
21.0	-0.34	-1.35	-3.1	-5.5	-8.6	-19.5	-35.0	-55	-80	-111	-146	-187	-233	-1050	-2700	-5600	-10500
23.0	-0.37	-1.48	-3.4	-6.0	-9.4	-21.3	-38.3	-61	-88	-121	-160	-205	-256	-1150	-2957	-6133	-11500
25.0	-0.40	-1.61	-3.6	-6.5	-10.2	-23.2	-41.7	-66	-96	-132	-174	-223	-278	-1250	-3214	-6667	-12500
27.0	-0.43	-1.74	-3.9	-7.0	-11.0	-25.1	-45.0	-71	-103	-142	-188	-240	-300	-1350	-3471	-7200	-13500
29.0	-0.47	-1.87	-4.2	-7.5	-11.8	-26.9	-48.3	-76	-111	-153	-202	-258	-322	-1450	-3729	-7733	-14500
31.0	-0.50	-2.00	-4.5	-8.1	-12.7	-28.8	-51.7	-82	-119	-163	-216	-276	-344	-1550	-3986	-8267	-15500
33.0	-0.53	-2.13	-4.8	-8.6	-13.5	-30.6	-55.0	-87	-126	-174	-230	-294	-367	-1650	-4243	-8800	-16500
35.0	-0.56	-2.26	-5.1	-9.1	-14.3	-32.5	-58.3	-92	-134	-184	-243	-312	-389	-1750	-4500	-9333	-17500
37.0	-0.59	-2.39	-5.4	-9.6	-15.1	-34.3	-61.7	-97	-142	-195	-257	-329	-411	-1850	-4757	-9867	-18500
39.0	-0.63	-2.52	-5.7	-10.1	-15.9	-36.2	-65.0	-103	-149	-205	-271	-347	-433	-1950	-5014	-10400	-19500
41.0	-0.66	-2.65	-6.0	-10.7	-16.7	-38.0	-68.3	-108	-157	-216	-285	-365	-456	-2050	-5271	-10933	-20500
43.0	-0.69	-2.77	-6.3	-11.2	-17.6	-39.9	-71.7	-113	-165	-227	-299	-383	-478	-2150	-5529	-11467	-21500
45.0	-0.72	-2.90	-6.6	-11.7	-18.4	-41.8	-75.0	-118	-172	-237	-313	-401	-500	-2250	-5786	-12000	-22500
47.0	-0.76	-3.03	-6.9	-12.2	-19.2	-43.6	-78.3	-124	-180	-248	-327	-418	-522	-2350	-6043	-12533	-23500
49.0	-0.79	-3.16	-7.1	-12.7	-20.0	-45.5	-81.7	-129	-188	-258	-341	-436	-544	-2450	-6300	-13067	-24500

*Error values in table are in mil (0.001 inch); thus, for shrink rate of 0.050 in/in and part size of 11 in, the error is 29 mil (0.029 in).

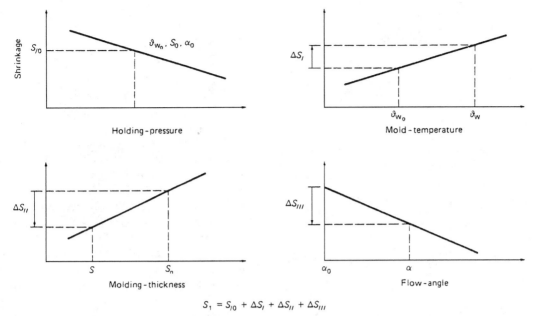

$$S_1 = S_{l0} + \Delta S_l + \Delta S_{ll} + \Delta S_{lll}$$

Fig. 3-27 Superposition approaches to determine shrinkage.

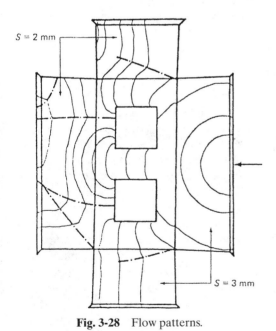

Fig. 3-28 Flow patterns.

be developed that provides the flow direction at the end of the filling phase. To control flow with respect to the orientation direction, color studies are used.

Cost advantage with tight tolerance Economical production requires that tolerances

not be specified tighter than necessary. However, after a production target is met, one should mold "tighter", if possible, for greater profit by using less material and/or reducing cycle time. Sometimes tight tolerances are specified when they are not needed. This action is usually taken when the designer is in a hurry or at a lost to set up the correct values. If relatively wide tolerances are specified consider specifying that tighter tolerances are desirable.

In fact if a cost savings occurs due to less plastic being used and/or cycle time is reduced also consider a financial reward (split the savings) with the fabricator. True the processor usually takes this action to reduce costs particularly after prototypes are accepted and production starts. If a large amount of plastic is consumed for the production run, the fabricating specification could include the weight of the product. Thus if a weight reduction occurs, it will be obvious. If tighter tolerances develop during molding and no reduction in weight occurs, the molder may be overpacking the cavity(s) providing you with possibly unwanted residual stresses or other problems that did not exist on the prototypes. To eliminate production process changes after a product is approved, the fabricated should follow their fabricating process control setting. This action would follow what FDA requires plastic medical device fabricators to meet called quality system regulation (QSR).

Blowing agent and tolerance For example, certain injection molded products can be molded to extremely close tolerances of less than a thousandth of an inch, or down to 0.0 percent, particularly when TPs are filled with additives or TS compounds are used. To practically eliminate shrinkage and provide a smooth surface, one should consider using a small amount of a chemical blowing agent (<0.5wt%) and a regular packing molding procedure. For conventional molding, tolerances can be met of ±5% for a product 0.020 in. thick, ±1% for 0.050 in., ±0.5% for 1.000 in., ±0.25% for 5.000 in., and so on. Thermosets generally are more suitable than TPs for meeting the tightest tolerances.

Impact Load

As reviewed in Chapter 2, loads are often applied abruptly, resulting in significant stress and strain increases. However, the elasticity of most TPs lets recovery usually be complete. Therefore, the steady-state stress and deflection of plastic can be considered identical to that of a product that is loaded gradually. However, when impact becomes severe, failure can result from it.

Many high impact resistant plastics can survive large deflections or strains during impact without suffering the permanent deformation or failure one might expect from the stress-strain curves of the plastics as measured at the standard loading rates. Therefore, the calculated impact stress of successful products will often appear to be unreasonably high. Recall that stress-strain behaviors are very different under rapid loading as compared to slow, steady loading conditions. Many plastics tend to have an exceptional capability of dissipating large amounts of mechanical energy when subject to these impact conditions.

Thermal Stress

When materials with different coefficients of linear thermal expansion (CLTE) are bolted, riveted, bonded, crimped, pressed, welded, or fastened together by any method that prevents relative movement between the products, there is the potential for thermal stress. Most plastics, such as the unfilled commodity TPs, may have ten times the expansion rates of many nonplastic materials. However there are plastics with practically no expansion. Details are reviewed in Chapter 2, **THERMAL EXPANSION AND CONTRACTION**.

In many assemblies the clearances around fasteners, the degrees of failure or yield in adhesives, and warpage or creep will tend to relieve the thermal stress. As with metal-to-metal attachments having different CLTEs, proper design allows for such temperature changes, especially with large parts that might be subject to wide temperature variations.

Film

Plastic films represent the largest worldwide market for plastics with practically all extruded (6). They are used to meet different performance requirements particularly for its major packaging market. Worldwide just for biaxial oriented (Chapter 8) polypropylene consumption is about $5\frac{1}{2}$ billion lb. Their use includes tape, food, tobacco, and confectionery. Thermoforming film (and extruded sheets) is a major processing technique producing all kinds of products.

History from Kodak relates to a product design feature approach that is applicable to the subject of developing film features. As reported by Kodak as of 1879 they were coating its photographic emulsions only on glass plates (187). Since 1920 cellulose triacetate safety plastic was used for the movie film and since then there has been relatively little change in the film base material. By the late 1950s Kodak was using acetate as the support across a range of products. But it was determined that polyethylene terephthalate (PET) had better dimensional stability and tensile strength. The dimensional stability made it a better material than acetate for microphotographic applications, especially for four-color work.

The tensile strength of PET provided characteristics that were important in X-ray applications. The modulus of acetate is half that of PET therefore PET was adopted in X-ray film so that images could be handled and displayed more easily, and in microphotographics for its greater accuracy.

In the early 1990s interest developed in the packaging material polyethylene naphthalate (PEN), a close cousin of PET. PEN has thermal stability 20°C higher than PET. Kodak had samples of the material sifting in its labs from the makers Teijin in Japan, as early as the beginning of the 1970s. Teijin is still the premier producer of PEN film and plastic today, and is involved with DuPont on the film manufacturing side.

The Kodak Advantix format film utilizes PEN. The material has two advantages in that context. First the Advantix film has smaller apertures for the film advance sprockets and PEN locates more accurately in the camera. Secondly its relaxation characteristics are better than PET. When film comes off the cylinder core it has a tendency to curl which can complicate both in-camera use and photofinishing. With PEN, which is a little stiffer than PET, the film relaxes into a flatter profile more quickly.

All consumer films and professional stock continue to be offered on cellulose triacetate. All its X-ray and micro photographic film is supported on PET. Only Advantix is on PEN. The primary issue is the cost of PEN, which may be why it has not been adopted widely in packaging. Its transport characteristics for bottles is excellent because of its physical characteristics. Its cost is high but a premium price in a premium product is more acceptable than it would be in Kodak's mainstream product range.

Motion film has been moving from acetate to PET because the tensile strength is better and the product is more durable. Going to PET has no cost penalty in comparison with acetate. Kodak uses a huge volume of PET and manufactures its own in Rochester, NY, USA as well as at its polyester plants in Colorado/USA, Chalon, France and London, UK.

In terms of trend the company is looking at a number of materials with interest. Photo manufacturers are studying sindiotactic polystyrene because it has very low moisture take-up. All its products need dimensional stability so materials which resist the effects of temperature and humidity changes are highly desirable.

Weld Line

Weld lines are also called knit lines. During processing, such as by injection molding and extrusion, weld lines can occur. They can form during molding when hot melts meet in a cavity because of flow patterns caused by the cavity configuration or when there are two or more gates. With extrusion dies, such as those with "spiders" that hold a center metal core, as in certain pipe dies, the hot melt that is separated momentarily produces a weld line in the direction of the extrudate and machine direction. The results of these weld lines could be a poor bond at the weld lines, dimensional changes, aesthetic damages, a reduction of mechanical properties, and other such conditions.

To illustrate the influence of processing on mechanical properties, the test specimens in Fig. 3-29 can be analyzed and related to what can happen in a fabricated product. It shows three sets of injection-molded specimens where the same plastic is processed in all specimens. There are three sets of similar specimens: a tensile one on top, a notched Izod impact one on the right side, and a flexural one on the left. The top set has a single gate for each specimen, the center set has double gates that are opposite each other for each specimen, and the bottom set has fan gates on the side of each specimen. The highest mechanical properties come with the top set of specimens, because of its melt orientation being in the most beneficial direction. The bottom set of specimens, with its flow direction being limited insofar as the test method

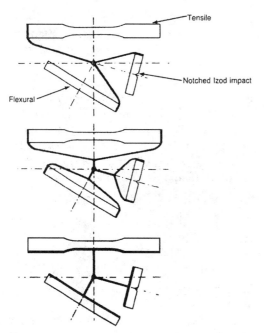

Tensile

Notched Izod impact

Flexural

Fig. 3-29 Injection molded test specimens that include weld lines.

is concerned, results in lower test data performance. With the double-gated specimens (the center set) weld lines develop in the critical testing area that usually results in this set's having the potential lowest performance of any of the specimens in this diagram.

Fabricating techniques can be used to reduce this problem in a product. However, the approach used in designing the product, particularly its mold (relocate gates), is most important to eliminate unwanted orientation or weld lines. This approach is no different from that of designing with other materials like steel, aluminum, or glass.

Meld Line

A meld line is similar to a weld line except the flow fronts move in parallel rather than meet head. Usually the meld line is identified as a weld line.

External Thread

Threads can be molded or tapped into a plastic. Molded internal threads usually require some type of unscrewing or collaps-

ing mechanism (3). Either splitting the mold halves or using unscrewing devices in the mold can produce external threads; split mold produces a parting line on the threads. With a split lower cost mold, it is basically easier to design the mold and easier to remove the threaded part from the mold during processing (Chapter 8, **TOOLING**). The proper design of the thread shape is required to prevent excessive shear, resulting in stripping the threads when torque is applied, and also to limit hoop stresses, which can result in tensile failure. Although the mechanics of stress analysis for screw threads are readily available, the equations for them can be rather complicated.

Coating

Coatings for all types of products (plastics, steels, etc.) are essential to meet all kinds of environmental requirement for certain plastic products but more with other materials such as steel and wood. Plastics continue to be the backbone in the coating industry since practically all coatings are composed of some type of plastics. The most widely used include acrylics, alkyds, vinyls, urea-melamine, styrenes, epoxies, and phenolics. Growth has been steady and reliable so that rational and economic paint production can no longer be regarded, as was the case until comparatively recent, as an art or craft based solely on empirical experiences. Although color matching tends to still be more of an art or craft.

Coatings are generally identified as paints, varnishes, and lacquers. Other nomenclature includes enamels, hot melts, plastisols, organosols, aerospace coatings, masonry water repellents, polishes, magnetic tape coatings, and overlays. They each have their performance characteristic (2). There are 100% plastic coatings such as vinyl-coated fabrics or polyurethane floor coverings. The more popular, and the largest user of plastics, are the paints. Almost all the binders in paints, varnishes and lacquers are made up principally of plastics.

The properties of the coating industry are essentially for the protection and decoration of the majority of manufactured products that characterize our complex material civilization. The protective function includes resistance to air, water, organic liquids and aggressive chemicals such as acids and alkalis, and together with improved superficial mechanical properties such as greater hardness and abrasion resistance. The decorative effect may be obtained through color, gloss or texture, or combinations of these techniques.

In the case of many surfaces such as walls or floors, or objects such as interior fittings, furniture and other articles, the surface coating can also fulfill hygienic requirements. The surface should not be prone to collect dirt, bacteria and other impurities. It should be easy to clean with common cleaning agents. In certain cases special qualities are required of the surface coatings, for example, in road-marking paints, in safety-marking paints, in factory floors, and in paints which make the surface either a good or poor electrical conductor.

Substrates protected from different environmental conditions basically include the metals (steel, zinc, aluminum and copper), inorganic materials (plaster, concrete and asbestos), and organic materials (wood, wallboard, wallpaper and plastics). Metals may be surface coated to improve their workability in mechanical processing.

Different technical developments have occurred in the coating industry which permit the use of a variety of raw materials. It is possible to formulate surface coatings that are suitable for each and every kind of material. In many cases a number of different coating systems may come into consideration for painting a particular substrate.

Functional Surface and Lettering

Surfaces of plastics may be provided with designs that can give a good grip or that can simulate wood, leather, etc. These types of surfaces should be specified in a manner that will not create undercuts to the withdrawal action from a mold. The undercut effect can be responsible for stresses and marring. A similar condition applies to lettering, and the location of such lettering should conform to smooth withdrawal requirements.

Fiber Reinforcement

Fiber behavior in reinforced plastics usually occurs at strains of only 1–3%. Designers that are accustomed to designing to yield with built-in safety factor margin of at least 10% strain might hesitate in using these materials. However, designing with these strains is logical and has been used since at least the 1940s. Data has been developed and used that includes variability in properties, creep, fatigue, static and dynamic sustained loading, etc. Both short and long fibers are used (106).

The conclusion that short stable fibers will not produce maximum physical properties is not theoretically correct. Both experiment and theory have concluded that with proper adhesion or bond between fibers and plastic matrix, maximum properties can basically be achieved by using relatively short stable fibers rather than continuous filament construction (39). To date the higher performances is overwhelming achieved with the continuous fibers. Also, the fibers used in RPs have the important potential of reaching values that are far superior (7, 10).

Process

There are conditions during the fabrication of plastic products that ensure meeting product performance requirements. However there are also constraints as reviewed in Chapter 8.

Prototype

The basic approach in designing any product made from any material (steel, aluminum, wood, plastic, etc.) involves knowing the behaviors and characteristics of the materials and manufacturing influences on the materials. In turn this knowledge is to be correctly applied such as using, when required, the processed material's static and/or dynamic properties. Should a need arise for data at conditions different from those at which test data are available, with few exceptions, it would not be too difficult or costly to obtain.

Depending on product performance requirements, that could involve the safety of people, the more costly engineering approach such as finite-element analysis may be required (Chapter 2). Finally, in addition to taking into account all of the relevant elements that are targeted to ensure a sound product design, it must be kept in mind that prototype testing to verify performance is usually the most important step in the overall design process of any product (Chapter 8, **MODEL/PROTOTYPING BUILDING**). These products would include highly load types to those basically not exposed to loads. The no load product may be one that has to meet certain requirements (abusive handling to meeting safety requirements) or have a long production run.

A prototype is a 3-D model suitable for use in the preliminary testing and evaluation of a product (also used for modeling a die, mold and other tool). It provides a means to evaluate the product's performances before going into production. The ideal situation is for the prototype to be the actual product made in production. However machining stock material and using rapid prototype techniques can make prototypes (Chapter 4, **BOOK SHELVES**).

Conventional machining operations are used preferably from the same plastic to be used in the product (Chapter 8, **SECONDARY EQUIPMENT**). Different casting techniques are used that provide low cost even though they are usually labor intensive. The casting of unfilled or filled/reinforced plastic used include TS polyurethane, epoxy, structural foam, and RTV silicone. Also used are die cast metals.

These materials are reviewed elsewhere in this book except RTV. The RTV (room temperature vulcanization) silicone plastic is a very popular type. It solidifies by vulcanization or curing at room temperature by chemical reaction, made up of two-part components of silicones and other elastomers/rubbers. RTV are used to withstand temperatures as high as 290°C (550°F) and as low as −160°C (−250°F) without losing their strength. Their rapid curing makes them

useful in different applications such as prototypes, prototype molds, etc.

Rapid Prototyping and Tooling

Rapid prototyping (RP) is technology used for building physical models and prototype products from 3-D computer-aided design data. Rapid tooling is any method or technology that enables one to produce tooling quickly. The term "rapid tooling" is derived from rapid prototyping technology and its application. It refers to RP-driven tooling. Even though these systems are more expensive than the past usual methods, they provide the desirable end result to the industry that is a much quicker way to obtain prototypes (hours instead of days). These systems are continuously being up dated and expanding their capabilities (165, 174).

Methods are used to produce the more costly rapid prototypes include those that produce models within a few hours. They include photopolymerization, laser tooling, and their modifications. The laser sintering process uses powdered TP rather than chemically reactive liquid photopolymer used in stereolithography. Models are usually made from certain types of plastics. Also used in the different processes are metals (steel, hard alloys, copper-based alloys, and powdered metals). With powder metal molds, they can be used as inserts in a mold ready to produce prototype products. These systems enable having precise control over the process and constructing products with complex geometries.

As an example stereolithography is a 3-D rapid process that produces automatically simple to very complex shaped models in plastic. Basically it is a method of building successive layers across sections of photopolymerized plastics on top of each other until all the thin printed layers can be joined together to form a whole product. The chemical key to the process, photopolymerization, is a well established technology in which a photo initiator absorbs UV energy to form free radicals that then initiate the polymerization of the liquid monomers. The degree of polymerization is dependent upon the total amount of light energy absorbed.

This process uses a moving laser beam, directed by a computer, to prepare the model. The model is made up of layers having thicknesses about 0.005–0.020 in. (0.012–0.50 mm) that are polymerized into a solid product. Advanced techniques also provides fast manufacturing of precision molds (152). An example is the MIT three-dimensional printing (3DP) in which a 3-D metal mold (die, etc.) is created layer by layer using powdered metal (300- or 400-series stainless steel, tool steel, bronze, nickel alloys, titanium, etc.). Each layer is inkjet-printed with a plastic binder. The print head generates and deposits micron-sized droplets of a proprietary water-based plastic that binds the powder together.

Once the lay-up is completed, product is removed and placed in a sintering oven. It goes through three cycles where plastic is burned off, metal powder is sintered together, and the product is solidified by infiltrating with another material to fill the voids such as lower melting point metal or a plastic (epoxy, etc.). Total time is 50 h. Shape is accurate within 0.005 in., plus 0.002 in./in. (0.0127 cm, plus 0.0051 cm) and may be acceptable for prototyping. The tool can be machined to tighter tolerances and polishing. This process permits creation of any type of internal voids such as cooling lines that conform to the part shape (160).

Features Influencing Performance

This section provides more detail on this important basic subject of design detractors and constrains. They represent conditions that can usually be incorporated in a design but it is to show that there is an easier way to fabricate the product so that you have a target to meet if the design permits it. Even though some of the analyses here will pertain to a specific process, many will relate to other processes, so it is best to review them all. Product designers should have some idea of where problems can develop that includes how a tool (mold or die) is designed and manufactured (Chapter 4, **JOINING**

AND ASSEMBLING and Chapter 8, **PROCESSING BEHAVIORS** and **PROCESSING AND PROPERTY**).

One of the earliest steps in product design is to establish the configuration that will form the basis on which strength calculations will be made and a suitable material selected to meet the anticipated requirements. During the sketching and drawing phase of working with shapes and cross-sections there are certain design features with plastics that have to be kept in mind to obtain the best cost-performances and avoid degradation of the properties. As previously reviewed, such features may be called property detractors or constraints. Prior to designing a product, the designer should understand such basic factors as those summarized in Table 3-5 and Fig. 3-26. Success with plastics, or any other material for that matter, is directly related to observing design details.

The important factors to consider in designing can be categorized as follows: part thickness, tolerances, ribs, bosses and studs, radii and fillets, drafts or tapers, holes, threads, colors, surface finishes and gloss levels, decorating operations, parting lines, gate locations, shrinkages, assembly techniques, production volumes, mold or die designs, tooling and other equipment amortization periods, as well as the plastic and process selections. The order that these factors follow can vary, depending on the product to be designed and the designer's familiarity with particular materials and processes.

Residual Stress

Such processing-induced residual stresses that influence properties as mechanical, physical, environmental, and aesthetic factors (which also exist in other materials like metals and ceramics) can have favorable or unfavorable effects, depending on the application of the load with respect to the direction of the stresses or orientation.

Residual stresses and molecular orientation play an important role in the toughness enhancement of plastics, because toughness is primarily based on the mechanics of craze formation and shear band (crazes and flaws) formation. The shear bands determine the fracture mode and toughness of a plastic when subjected to impact loads. The amount of energy dissipated will depend on whether the material surrounding the flaws deforms plastically. For toughness enhancement the residual stresses play an important role in the suppression of craze formation, by avoiding the stress state that promotes brittle fracture.

The term residual stress identifies the system of stresses that are in effect locked into a product, even without external forces acting on it. For instance, minute stresses may be induced in a material by nonuniform heating and cooling during processing. The production of residual stresses is usually the result of nonhomogeneous plastic deformation occurring during thermal and mechanical actions, arising from changes in either volume or shape. Thermal treatments like quenching (rapid cooling) and annealing (slow cooling) introduce changes in physical and mechanical properties. For example, with sheet plastic the stresses created by quenching are the result of uneven cooling, when the surfaces cool faster than the core. This produces nonuniform volume changes and properties throughout the thickness. The compressive stresses on the surfaces of the quenched plastic produce tensile stresses in the core, which maintain the equilibrium of the forces.

Cold Working

Generally, a variety of mechanical deformation processes cause the nonuniform deformation that results in the formation of residual stresses. This nonhomogeneous deformation in a material is produced by the material's parameters, largely its process parameters such as the tool geometry and frictional characteristics. For example, the rolling of a strip can be accomplished by using relatively cold squeeze rolls. In the rolling process, parameters with a small roll diameter and little reduction produce deformation penetration that is shallow and close to the surface, whereas the interior of the strip remains almost undeformed. After the removal of the deformation forces and a complete

elastic recovery, this condition produces compressive residual stresses at the surface and tensile residual stresses in the core.

The logic of this situation is that the surface material is forced to elongate more than the relatively rigid core permits. When there is a large diameter and much reduction, the deformation penetration occurs deeper in the core and there is a tendency for the plastic to lag at the surface, the result of friction at the material-to-tool interface occurs. Thus, the cold working processes like rolling, drawing, extrusion, and forging produce residual stresses along with their molecular orientation.

Generally, the more discussed and technically reviewed residual stresses are in injection molded products. It usually occurs because melt in the cavity next to the cavity is cold and does not properly flow. Cause could be the melt and/or the cavity surface was cold. Their presence can often be detected by (1) the product's performances being reduced or changed such as dimensions, or (2) qualitatively by immersing TP products in appropriate stress-cracking solvents for a short time, then observing the crazing caused by surface tensile residual stresses (Chapter 5, **STRESS ANALYSIS**). Such methods are ineffective for a product with compressive or insufficient tensile stresses on its surface. To determine their magnitude includes the layer removal technique of microtoming (Chapter 5, **FLAW DETECTION**).

Stress Concentration

Sharp corners should always be avoided in designing. Although sharp-cornered designs are common with certain sheet metal and machined products, good design practice in any material dictates the use of generous radii, to reduce stress concentrations. RPs and metal products will often tolerate sharp corners, because the stresses at their corners are low compared to the strength of the material or because localized yielding redistributes the load. However, neither of these factors should be relied upon in TP products. Sharp corners, particularly the inside comers, can cause severe molded-in stresses as a material shrinks onto the corner, as well as poor flow patterns, reduced mechanical properties, increased tool wear, and so on.

The elementary formulas used in design are based on structural members having a more or less constant cross-section, or at least only a gradual change of contour, but these conditions are seldom found in practice. The presence of shoulders, bosses, grooves, holes, threads, and corners result in modifying the simple stress distribution so there are no localized, high stresses. This localization, known as the stress concentration factor, is defined as $K =$ maximum stress divided by nominal stress. Localized high stresses must in most cases be determined experimentally rather than theoretically. The photoelastic technique is one of the more effective methods used to do this. To interpret a photoelastic diagram qualitatively it is sufficient to know that the number of fringes (the density of lines) is proportional to the absolute stress level (Chapter 5, **STRESS ANALYSIS**).

Basically, in the vicinity of a sharp comer all fringes converge toward the apex. Having a high density of lines at this point indicates the presence of high stress level. At a rounded corner there will be considerably less concentration. Besides the molding problems, sharp corners often cause premature failure because of the stress concentration. To avoid these problems, inside comer radii should be equal to one-half the nominal wall thickness, with a 0.020 in. radius considered as a minimum for products subjected to stress and a 0.005 in. minimum for the stress-free regions. Having inside radii less than 0.005 in. is not recommended for most materials. Outside comers should have a radius equal to the inside corner plus the wall's thickness.

Injection Molding

Design concept In designing a totally new product or redesigning an existing one to improve the product, bring about cost savings, or some combination of these or other reasons, consideration should be given to the key advantages of IM. These advantages include the ability to produce finished, multifunctional, or complex molded products accurately and repeatedly in a single, highly automated operation (Chapter 8). While keeping this in mind during the initial planning stage, one should also be aware of the general design considerations presented in this section (3).

Many injection molded products will influence the final product's performance, dimensions, and other characteristics. The mold includes the cavity shape, gating, parting line, vents, undercuts, ribs, hinges, and so on (Table 3-17). The mold designer must take all these factors into account to eliminate problems. At times, to provide the best design

Table 3-17 Functions of an injection mold

Mold Component	Function Performed
Mold base	Hold cavity (cavities) in fixed, correct position relative to machine nozzle
Guide pins	Maintain proper alignment of the two halves of a mold
Sprue bushing (sprue)	Provide means of entry into mold interior
Runners	Convey molten plastic from sprue to cavities
Gates	Control flow into cavities
Cavity (female) and force (male)	Control size, shape, and surface texture of molded article
Water channels	Control temperature of mold surfaces, to chill plastic to rigid state
Side (actuated by cams. gears, or hydraulic cylinders)	From side holes, slots, undercuts, threaded sections
Vents	Allow escape of trapped air and gas
Ejector mechanism (pins, blades, stripper plate)	Eject rigid molded article from cavity or force
Ejector return pins	Return ejector pins to retraced position as mold closes for next cycle

the product designer, processor, and mold designer may want to jointly review where compromises can be made to simplify meeting product requirements. With all this interaction, it should be clear why it takes a certain amount of time to ready a mold for production.

The subject of who is responsible for designing the mold can become confusing. One might say it is not the product designer responsibility. The product designer may provide a design that the mold maker can produce but the products do not meet performance requirements, etc. because the original design was not complete or accurate. It permitted the mold designer liberty to use whatever approach that made it easier to meet the design configuration.

Stretching this point let us assume that an optical lens was designed with no mention that a gate could not be located in the middle of the lens. The mold was made with a gate in the center of the mold cavity. The responsibility is on the product designer who should have known better and specified no gate on the lens' surface or without experience contacted a knowledgeable and cooperative person such as the mold designer that would explain the options available to meet the mold design requirements. In this example we have to assume that product designer was not familiar with gate location problems. However one of the requirements should have included that the lens' surface not be marred, meet certain optics (index of refraction, etc.), and so on. If these requirements had been listed and translated to the mold designer the gate would not have been located in the center of the lens.

Thus, in the design of any IM product there are certain desirable goals that the designer should use. In meeting them, problems can unfortunately develop. For example, the most common mold design errors of a sort that can be eliminated usually occur in the following areas: (1) thick/thin section transitions, (2) multiple gates resulting in distorted products or weld lines, (3) wrong gate locations, (4) inadequate provision for cavity air venting entrapping microscopic voids, (5) products too thin to mold properly for the plastic being used, (6) products too thick to mold properly for the plastic being used, (7) plastic flow path too long and tortuous, (8) runners too small, (9) gates too small, (10) poor temperature controls, (11) runner too long, (12) product symmetry vs. gate symmetry clashes, (13) orientations of plastic melt in flow direction, (14) hiding gate stubs, (15) stress relief for interference fits, (16) living hinges, and (17) thread inserts.

As reviewed in other chapters, different plastics have different melt and flow characteristics. What is used in a mold design for a specific material may thus require a completely different type of mold for another material. These two materials might, for instance, be of the same plastic but use different proportions of additives and reinforcements. This situation is no different than that of other materials like steel, ceramics, and aluminum. Each material will require its own cavity shapes and possibly have its own runner system.

What follows is a general summary of how to reduce problems to tolerable limits. They represent conditions that can be molded but it is to show that there is an easier way to mold so that you have a target to meet if the design permits it.

First as reviewed, inside comers should normally not be shown as two intersecting straight lines with a sharp corner. Corners are stress-concentration areas, quite similar to a notch in a test bar. The Izod impact strength of notched and unnotched test bars shows that the relative impact strength of each material at these two conditions that has a relationship to a radius vs. a sharp corner. Thus, for example, polycarbonate has an impact strength of the notched 1/8 in. test bar of 12 to 16 ft.-lb./in., whereas the same bar unnotched does not fail the test. Polypropylene has an impact strength 30 times greater in the unnotched than the notched bar. Nylon shows a drastic increase in impact strength as the radius increases from sharpness to 3/64 in. A similar trend exists for most other materials.

These examples point out that brittleness increases with the decreasing of a radius in a comer. Visually, a radius of 0.020 in. on a product may be considered sharp, with an influence on strength that is much more favorable than a radius of 0.004 in. To the moldmaker, a sharp corner is usually easier to produce, but in the plastic product it is a source of brittleness and, in most cases is highly undesirable. Inside sharp corners on plastic-part drawings are a frequent occurrence.

Second, varying wall thicknesses from thick to thin sections can lead to problems in molding. Having a uniform wall throughout a part gives it good strength and appearance. Thick and thin sections could have molded-in stresses, different rates of shrinkage (causing warpage), and possibly void formation in the thick portion. Since the molding solidify from their outer surfaces toward the center, sinks will tend to form on the surface of a thick portion. When thick (3/16 in. and over) and thin (1/8 in. or less) portions are unavoidable, the transition should be gradual and coring should be utilized whenever possible. Influencing these behaviors can be related to the locations of mold gates. The usual approach is gating at the thin section.

Third, sinks are not only the result of the causes listed above but also can occur whenever supporting or reinforcing ribs, flanges, or similar features are used in an attempt to provide functional service without changing the basic wall thickness of a product. If the appearance of a sink on the surface is objectionable, the ribs and transition radius should be proportioned so that their contribution to the sink is minimal. Sinks can usually be eliminated by changing the process controls that usually results in the unwanted longer cycle times.

Fourth, molded-in metal should be avoided whenever alternate methods will accomplish the desired objective. If it is essential to incorporate such inserts, they should be shaped so that they will present no sharp inside corners to the plastic. The effect of the sharp edges of a metal insert would be the same as explained in the first point above, namely, brittleness and stress concentration can occur. The cross-section that surrounds a metal insert should be heavy enough that it will not crack upon cooling. A method of minimizing cracking around the insert is to heat the metal insert prior to mold insertion to a temperature of 250 to 300°F (121 to 149°C) so that it will tend to form the plastic into its finished shape. The thickness of the plastic enclosure will vary from material to material. A reasonable guide is to have the thickness 1.75 to 2 times the size of the insert diameter.

Fifth, plastic threads have a very limited strength and may be further degraded if the thread form is not properly shaped. The V-shaped portion at the outside of a female thread will present a sharp inside corner that will act as a stress concentrator and thereby weaken the threaded cross-section. A rounded form that can be readily incorporated in a molding insert will appreciably improve the strength over a V-shaped form. When self-tapping, thread-cutting, or thread-form screws are used, their holding power can be increased if either the screws or plastics are heated to a temperature of 180 to 200°F (82 to 93°C) at joining time. This will provide forming action to some degree and keep the stress level caused by the joining action at a low point. More on this subject is reviewed latter under ***Internal plastic thread***.

These possible sources of problems in a molded part should be marked on the product drawing and explained to the mold designer for corrective action or creating an awareness of possible product defects. This is a necessary step in the chain of events in which the aim is to produce a tool that will provide useful products. Even if the mold's design, workmanship, and operation are carried out to the highest degree of quality, they cannot overcome a built-in weakness due to the product design.

Sharp corner As reviewed, and never to many times, when a drawing does not show a radius, the tendency is for the toolmaker while manufacturing a mold to leave the intersecting machined or ground surfaces as they are generated by the machine tool. The result is a sharp corner on the molded product. Such sharp corners on the insides of products are the most frequent property detractors.

Sharp corners become stress concentrators. The stress-concentration factor increases as the ratio of the radius R to the part thickness T decreases. An R/T of 0.6 is favorable, and an increase in this value will be of some limited benefit. The ASTM Izod impact strength value of nylon with various notch radii change. With a radius of 0.005 in. the impact strength is about 1.3 ft-lb./in., with an R of 0.020 in. it is 4.5 ft-lb./in., and with an R of 0.040 in. it is 12 ft-lb./in. In most cases a radius of 0.020 in. can be considered a sharp corner as far as end use is concerned, a size that is a decided improvement over a 0 to 5 mil radius; therefore, it should be considered a minimum requirement and be so specified.

The recommended radius not only reduces the brittleness effect but also provides a streamlined flow path for the plastic melt in the mold cavity. The radiused corner of the metal in the mold reduces the possibility of its breakdown and thus eliminates a potential repair need. Too large a radius is also undesirable because it wastes material, may cause sink marks, and may even contribute to stresses from having excessive variations in thickness.

Uniform wall thickness Wall requirements are usually governed by the load, the support needs for other components, attachment bosses, and other protruding sections. Designing a product to meet all these requirements while still producing a reasonably uniform wall will greatly benefit its durability. A uniform wall thickness will minimize stresses, differences in shrinkage, possible void formation, and sinks on the surface; it also usually contributes to material saving and economy in production.

Most of the features for which heavy sections are intended can be modified by means of ribbing, coring, and shaping of the cross-section to provide equivalent strength, rigidity, and performance. Top of Fig. 3-30 shows a small gear manufactured from metal bar stock. The same gear converted to a molded plastic would be designed as shown in the bottom of Fig. 3-30. This plastic gear design compared to the metal gear saves material, eliminates stresses from having thick and thin sections, provides uniform shrinkage in teeth

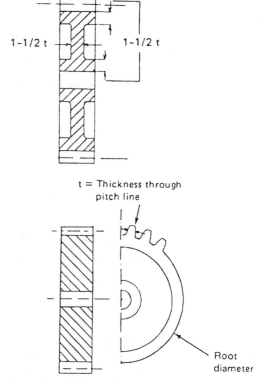

Fig. 3-30 Design of the solid steel gear (top) is redesigned using plastic (bottom).

and the remainder of the gear, avoids the danger of warpage, with its thin web and tooth base prevents bubble formation and potential weak spots, and, having no sink in the middle of the thickness, provides a full load-carrying capacity for the teeth.

Wall thickness tolerance When relatively deep products are being designed, a tolerance for the wall thickness on the order of ±0.005 in. is usually given. What this tolerance should mean is that a product will be acceptable when made with this tolerance, but that the wall thickness must be uniform throughout the circumference.

Let us analyze the molding condition of such a product and assume that one side is made to minimum specifications and the opposite to maximum specifications where the gate is unfortunately located. Result is that the resistance to plastic flow decreases with the third power of the thickness, which means that the thick side will be filled first, while the thin side will fill from all sides.

This type of filling can create a pocket on the thin side and compresses cavity air and gases to such a point that the rising temperature caused by compression results in the material to be charred while the pocket is filling up.

The charred plastic will create porosity, a weak area, and an electronically defective surface. Furthermore, the filling of the thick side ahead of the thin side creates a pressure imbalance generated by the usual at least 5 to 10 tons/sq. in. (69 to 138 MPa) injection pressure that can cause the core to deflect toward the thin side, further aggravating the difference in wall thickness. This pressure imbalance could eventually contribute to mold damage and make production of products difficult if not impossible. It can be conclude that the wall uniformity throughout the circumference must be within narrow limits, such as ±0.002 in., whereas the thickness in general may vary from the specified value by ±0.005 in. Logical corrective action is to gate from the thin section of the cavity.

Flow pattern Ultimately, product quality can be considered a direct outcome of a plastic melt's flow behavior in its mold cavity or cavities. Excessive restrictions and obstructions to the flow of material spell trouble in injection molding.

Parting line PLs on the surface of a molded product, which are produced by the parting line of the mold, when required can often be concealed on a thin, inconspicuous edge of the product. Doing so preserves the good appearance of the molding and in most cases eliminates the need for any finishing.

Gate size and location Because of high melt pressure, the area near a gate is highly stressed, both by the frictional heat generated at the gate and the high velocities of the flowing material. Using a small gate is desirable for separating the product from the feed line, but not for a product with low stresses. Gate openings are usually two thirds of a product's thickness. If they are that large or larger it will reduce frictional heat, permit lower velocities, and allow the application of higher pressures for increasing the product density

of the material in the cavity resulting in better thickness tolerance control. Temperature controlled valve gates are used to eliminate these type problems as well as other problems that can develop such as overpacking the melt near the gate.

The product designer should caution the tool designer to keep the gate area away from load-bearing surfaces and to make the gate size such that it will improve the quality of the product. It so happens that the product wall in the gate area develops the minimum tolerance due to the high melt pressure in that area.

Taper of draft angle It is desirable for any vertical wall of a molded product to have an amount of draft that will permit its easy removal from a mold. The amount of draft may vary from 1/8 degree up to several degrees depending on what the circumstances permit and behavior of the plastic. A fair average may be from 1/2 degree to 1 degree. The possibility of having voids close to the base is avoided, and increased cycle time in manufacturing is minimized.

The vertical cavity surfaces, particularly with a draft of 1/8 degree, will demand a much higher surface finish, with polishing lines in the direction of product withdrawal. On shallow walls the draft angle can be considerably larger, since the influence of the drawbacks will be minor. The designer should be cognizant of the need for drafts on vertical walls. If problems are encountered during the removal of products, stresses can result, the shape of the product can be distorted and surface imperfections be introduced.

If a vertical wall is required with no taper, it can be accomplished. However cost of mold is significantly increased since more action will be required in the mold such as moving its sidewalls to release the molding and higher ejection pressure mechanisms are required.

Weld line With moldings that include openings (holes), problems can develop. In the process of filling a cavity the flowing melt is obstructed by the core, splits its stream, and surrounds the core. The split stream then reunites and continues flowing until the cavity is filled. The rejoining of the split streams forms a weld line. It lacks the strength properties that exist in an area without a weld line because the flowing material tends to wipe air, moisture, and/or lubricant into the area where the joining of the stream takes place and introduces foreign substances into the welding surface. Furthermore, since the plastic material has lost some of its heat, the temperature for self-welding is not conducive to the most favorable results.

A surface that is to be subjected to load bearing should be targeted not to contain weld lines. If this is not possible, the allowable working stress should be reduced by at least 15%. Under the ideal molding conditions up to about 85% of available strength in the solidified plastic can be developed. At the other extreme where poor process controls exists the weld line could approach zero strength. In fact the two melt fronts could just meet and not blend so that there is relatively a microscopic space. Other problems occur such as influencing aesthetics. Some examples of different aspects pertaining to weld lines are shown in Fig. 3-31.

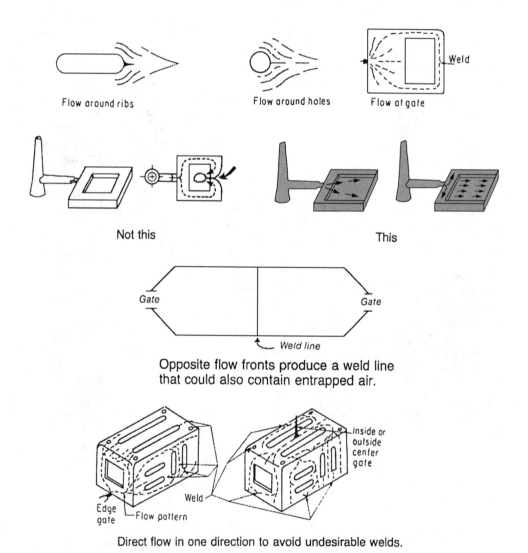

Fig. 3-31 Examples of melt flow patterns to consider during the design stage to eliminate or at least minimize weld lines to obtain maximum strength.

Undercut Whether external or internal, undercuts should be avoided if possible to reduce mold cost (by about 25 to 30%) and simplify melt flow during molding. However many molds use external and/or internal undercuts. In cases where it is essential to incorporate them in a design, appropriate mold design is required. The mold will include action such as sliding components on tapered surfaces, split cavity cam actions to produce the needed undercut, etc. (Chapter 8, **TOOLING**).

Some conditions will, however, permit incorporating undercuts with conventional stripping of the product from the mold. Certain precautions are necessary in order to attain satisfactory results. First, the protruding depth of the undercut should be two thirds of the wall thickness or less. Second, the edge of the mold against which the product is ejected should be radiused to prevent shearing action. Finally, the product being removed should be hot enough to permit easy stretching and return to its original shape after removal from the mold.

With particularly flexible type materials their elasticity and springback can simplify removal. As an example certain threaded plastic caps are stripped from the cores instead of being unscrewed. Coarse threads with the crest of the core thread rounded and a material with good elongation and ability to spring back make it more feasible to apply conventional stripping. The undercut problem can be solved by the cooperation of the designer, moldmaker, and processor, since each product configuration presents different possibilities.

Blind hole In regard to molding products that include holes, it is important to ensure that sufficient material surrounds the holes and melt flows property. A core pin forming blind holes is subjected to the bending forces that exist in the cavity due to the high melt pressures. Calculations can be made for each case by establishing the core pin diameter, its length, and the anticipated pressure conditions in the cavity (3).

From engineering handbooks we know that a pin supported on one end only will deflect up to forty-eight times as much as one supported on both ends. This suggests that the depth of hole in relation to diameter should be small in order to maintain a straight hole. Sometimes a deep, small-diameter hole is needed, as in pen and pencil bodies. In this case the plastic flow is arranged to contact the free end of the core, as an example, from four to six evenly spaced gates. This design will cause a centering action, and the plastic will continue flowing over the diameter in an umbrella like pattern to balance the pressure forces on the core.

When this type of flow pattern is impractical, an alternative may be a through hole or tube formation combined with a postmolding sealing or closing operation by spinning or ultrasonic welding. At the other extreme, consider a 1/4 in. (0.6 cm) diameter core exposed to a pressure of 4,000 psi (28 MPa) with an allowance for deflection of 0.0001 in. (0.00025 cm) and determine how deep a blind hole can be molded under these conditions.

Boss Bosses and other projections from the nominal wall are commonly found in IM products. These often serve as mounting or fastening points. As with rib design, avoiding overly thick wall sections is important, to minimize the chance of appearance or molding problems. When bosses are designed to accommodate self-tapping screws, the inside diameter and wall thickness must be controlled to avoid excessive buildup of hoop stresses in the boss. Ribs are frequently used in conjunction with bosses when lateral forces are expected. Special care must be used with tapered pipe threads, since they can create a wedging action on the boss. If there is a choice, the male rather than the female pipe thread should be the one molded into the plastic.

Coring The term coring in IM refers to the addition of steel to the mold for the purpose of eliminating plastic material in that area. Usually, coring is necessary to create a pocket or opening in the product, or simply for the purpose of reducing an overly heavy wall section. For simplicity and economy in injection molds, cores should be parallel to the line of draw of the mold. Cores placed in any other direction usually create the need for some type of side action (such as a mechanical cam or hydraulic cylinder) or manually loaded and unloaded loose cores.

Blind holes in molded plastics are created by a core supported by only one side of the mold. The length of the core and depth of the hole are limited by the ability of the core to withstand the bending forces produced by the flowing plastic without excessive deflection. For this reason, the depth of a blind hole should not exceed three times its diameter or minimum cross-sectional dimension. For small blind holes with a minimum dimension below 1/4 in., the L/D ratio should be kept to two. With through holes the cores can be longer, since the opposite side of the mold cavity supports them (3).

Sometimes the cores can be split between the two sides and interlocked when the mold is closed, allowing for the creation of long through holes. With through holes, the overall length of a given-size core can generally be twice as long as that of a blind hole. Some-

times, even longer cores are necessary. The tool can be designed to balance the pressure on the core pin, thus limiting the deflection.

Press fit   Products or components of any material in a press-fit assembly are assembled to a plastic product using an interference fit to maintain the assembly. The main advantage of this system is that the tooling is kept relatively simple. This method can, however, create very high stresses in the plastic. Amount of stress will depend on factors such as temperature during and after assembly, modulus of the mating material, type of stress, usage environment, and the type of material being used. Some materials will creep or stress relax, while others will fracture or craze if the strain is too high. Except for light press-fits, this type of assembly can be damaging due to the hoop stress in the boss that might already be weakened by a knit-line. Figures 3-32 and 33 provide hoop stress equations for two typical press fit and alternate methods of designing press-fits that result in lower risk of failure.

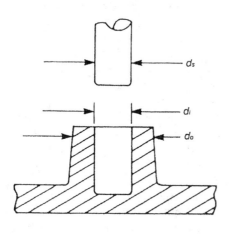

GEOMETRY FACTOR

$$\Gamma = \frac{1 + \left(\dfrac{d_s}{d_o}\right)^2}{1 - \left(\dfrac{d_s}{d_o}\right)^2}$$

E_p = MODULUS OF ELASTICITY OF PLASTIC
E_m = MODULUS OF ELASTICITY OF METAL
v_p = POISSON'S RATIO OF PLASTIC
σ_a = ALLOWABLE DESIGN STRESS FOR PLASTIC
$i = d_a - d_i$ = DIAMETRAL INTERFERENCE
i_a = ALLOWABLE INTERFERENCE

CASE A
SHAFT AND HUB ARE BOTH THE SAME OR ESSENTIALLY SIMILAR MATERIALS
HOOP STRESS GIVEN ''i'' IS

$$\sigma = \frac{i}{d_s} E_p \frac{\Gamma}{\Gamma + 1}$$

OR, THE ALLOWABLE INTERFERENCE IS

$$i_a = d_s \cdot \frac{\sigma_a}{E_p} \frac{\Gamma + 1}{\Gamma}$$

CASE B
SHAFT IS METAL, HUB IS PLASTIC
HOOP STRESS GIVEN ''i'' IS

$$\sigma = \frac{i}{d_s} E_p \frac{\Gamma}{\Gamma + v_p}$$

OR, THE ALLOWABLE INTERFERENCE IS

$$i_a = d_s \frac{\sigma_a}{E_p} \frac{\Gamma + v_p}{\Gamma}$$

Fig. 3-32   Determining press fit stresses for two typical situations.

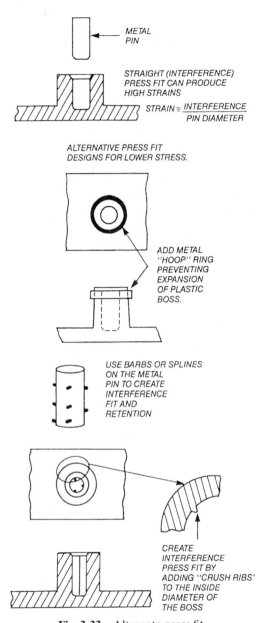

METAL
PIN

STRAIGHT (INTERFERENCE)
PRESS FIT CAN PRODUCE
HIGH STRAINS

$$STRAIN \cong \frac{INTERFERENCE}{PIN\ DIAMETER}$$

ALTERNATIVE PRESS FIT
DESIGNS FOR LOWER STRESS.

ADD METAL
"HOOP" RING
PREVENTING
EXPANSION
OF PLASTIC
BOSS.

USE BARBS OR SPLINES
ON THE METAL
PIN TO CREATE
INTERFERENCE
FIT AND
RETENTION

CREATE
INTERFERENCE
PRESS FIT BY
ADDING "CRUSH RIBS"
TO THE INSIDE
DIAMETER OF
THE BOSS

Fig. 3-33   Alternate press fit.

A common use is with a plastic hub or boss accepting either a plastic or metal insert. The press fit operation tends to expand the hub creating a tensile or hoop stress. If the interference is too great, a very high strain and stress will develop. The plastic product will (1) fail immediately by developing a crack parallel to the axis of the hub to relieve the stress, a typical hoop stress failure, (2) survive assembly but fail prematurely when the product is in use for a variety of reasons related to the high induced stress levels, or (3) undergo stress relaxation sufficient to reduce the stress to a lower level which can be maintained.

A simpler, although less accurate, method of evaluating these press fits is to assume that the shaft will not deform when pressed into the plastic. This is reasonably accurate when a metal shaft is used in a plastic hub. The hoop strain developed that is reasonably accurate in the hub is then given by the equation:

$$\varepsilon = i/d_I \qquad (3\text{-}12)$$

The hoop stress can then be obtained by multiplying by the appropriate modulus. For high strains, the secant modulus will give the initial stress (Chapter 2). The apparent or creep modulus should be used for the longer time stresses. The main point is that the maximum strain or stress must be below that value which produces creep rupture in the material. It is to be noted that there is usually a weld line present in the hub that can significantly effect the creep rupture strength of most plastic materials. An additional frequent complication with press fits is that a round hub or boss if often difficult to mold. There is a tendency for the hub to be slightly elliptical in cross section increasing the stresses on the product. In view of the above, all press fits must be given prototype life testing under actual operating conditions to assure product reliability (84).

Internal plastic thread   The strength of plastic threads is limited, and when molded in a product involving either an unscrewing device or a rounded shape of thread similar to bottle-cap threads, they can be stripped from the core. Screw threads, when needed, should be of the coarse type and have the outside of the thread rounded so as not to present a sharp V to the plastic that can produce a notch effect.

If a self-threading screw can be substituted, it will not only appreciably decrease mold maintenance and mold cost but most likely, with proper type selection, also give better holding power. A screw that has a thin thread with relatively deep flights can give high holding power. If the screw or plastic is preheated to about 121°C (250°F), a condition of forming in combination with material displacement will exist, thereby improving the holding power. When male plastic threads are being considered, the coarser threads are again preferred, and the root of the thread should be rounded to prevent the notch effect.

Molded-in insert If metal inserts are to be molded into a plastic product consideration should be given to wall thickness around the insert (Table 3-18) and their shape. The shape should present no sharp edges to the plastic, since the effect of the edges would be similar to that of a notch. A knurled insert should have the sharp point smoothed, again to avoid the notch effect.

The practice of molding inserts in place is usually employed to provide good holding power for plastic products, but there are drawbacks to this method. It normally takes a pin to support the insert, and since this pin is small in relation to the cored hole for the insert, it is easily bent or sheared under the influence of injection pressure. Should the insert fall out of position, there is danger of mold damage. Also, the hand placement of inserts contributes to cycle variation and with it potentially product quality degradation. Some of these problems can be overcome by higher mold expenditures,

Table 3-18   Suggested minimum wall thicknesses for inserts of various diameters [in. (mm)]

Plastic Material	Diameter of Inserts, in.					
	.125 (3.17)	.250 (6.35)	.375 (9.52)	.500 (12.7)	.750 (19.0)	1.00 (25.4)
ABS	.125 (3.17)	.250 (6.35)	.375 (9.52)	.500 (12.7)	.750 (19.0)	1.00 (25.4)
Acetal	.062 (1.57)	.125 (3.17)	.187 (4.75)	.250 (6.35)	.375 (9.52)	.500 (12.7)
Acrylics	.093 (2.36)	.125 (3.17)	.187 (4.75)	.250 (6.35)	.375 (9.52)	.500 (12.7)
Cellulosics	.125 (3.17)	.250 (6.35)	.375 (9.52)	.500 (12.7)	.750 (19.0)	1.00 (25.4)
Ethylene vinyl acetate	.040 (1.02)	.085 (2.16)	N.R.	N.R.	N.R.	N.R.
F.E.P. (fluorocarbon)	.025 (0.64)	.060 (1.52)	N.R.	N.R.	N.R.	N.R.
Nylon	.125 (3.17)	.250 (6.35)	.375 (9.52)	.500 (12.7)	.750 (19.0)	1.00 (25.4)
Noryl (modified PPO)	.062 (1.57)	.125 (3.17)	.187 (4.75)	.250 (6.35)	.375 (9.52)	.500 (12.7)
Polyallomers	.125 (3.17)	.250 (6.35)	.375 (9.52)	.500 (12.7)	.750 (19.0)	1.00 (25.4)
Polycarbonate	.062 (1.57)	.125 (3.17)	.187 (4.75)	.250 (6.35)	.375 (9.52)	.500 (12.7)
Polyethylene (H.D.)	.125 (3.17)	.250 (6.35)	.375 (9.52)	.500 (12.7)	.750 (19.0)	1.00 (25.4)
Polypropylene	.125 (3.17)	.250 (6.35)	.375 (9.52)	.500 (12.7)	.750 (19.0)	1.00 (25.4)
Polystyrene	Not Recommended					
Polysulfone	Not Recommended					
Surlyn (ionomer)	.062 (1.57)	.093 (2.36)	.125 (3.17)	.187 (4.75)	.250 (6.35)	.312 (7.92)
Phenolic G.P.	.093 (2.36)	.156 (3.96)	.187 (4.75)	.218 (5.53)	.312 (7.92)	.343 (8.71)
Phenolic (medium impact)	.078 (1.98)	.140 (3.56)	.156 (3.96)	.203 (5.16)	.281 (7.14)	.312 (7.92)
Phenolic (high impact)	.062 (1.57)	.125 (3.17)	.140 (3.56)	.187 (4.75)	.250 (6.35)	.281 (7.13)
Urea	.093 (2.36)	.156 (3.96)	.187 (4.75)	.218 (5.53)	.312 (7.92)	.343 (8.71)
Melamine	.125 (3.17)	.187 (4.75)	.218 (5.53)	.312 (7.92)	.343 (8.71)	.375 (9.52)
Epoxy	.020 (0.51)	.030 (0.76)	.040 (1.02)	.050 (1.27)	.060 (1.52)	.070 (1.78)
Alkyd	.125 (3.17)	.187 (4.75)	.187 (4.75)	.312 (7.92)	.343 (8.71)	.375 (9.52)
Diallyl phthalate	.125 (3.17)	.187 (4.75)	.250 (6.35)	.312 (7.92)	.343 (8.71)	.375 (9.52)
Polyester (premix)	.093 (2.36)	.125 (3.17)	.140 (3.56)	.187 (4.75)	.250 (6.35)	.281 (7.14)
Polyester T.P.	.062 (1.57)	.125 (3.17)	.187 (4.75)	.250 (6.35)	.375 (9.52)	.375 (9.52)

as for example shuttling cavities (Chapter 4, **JOINING AND ASSEMBLY**).

However the desired results in fastening can be attained by other means. One example is by coring holes in the molding that will permit ultrasonic welding of inserts in place. Coring a hole in the product will be of a size when the product is removed from the mold that will permit a slight press fit plus a gain in the holding power from postmolding shrinkage. Also is the approach of coring a hole in the product that will permit dropping the insert and providing a retaining shoulder by spinning or ultrasonic forming.

All these type assembly methods usually require the same time to perform as placing inserts in the mold, but they also lower IMM time. There are several other means of accomplishing the desired result that depend on the circumstances at hand. In any event, conventional molded-in inserts usually prove costlier. There are highly automatic injection molding machines available designed just for insert molding that will reverse the cost.

Screw Screw threads can be molded or tapped into a plastic. Molded internal threads, that can be produced meeting tighter and performance requirements, usually require some type of unscrewing or collapsing mechanism. External threads can be molded either by splitting the mold halves or parting the line across the thread if parting the line on the threads is permitted. With a split mold it is basically easier to design the mold and remove the threaded part from the mold during processing.

The design of the threads requires control, to prevent excessive shear, resulting in stripping the threads when torqued, and also to limit hoop stresses that can result in tensile failure. Although the mechanics of stress analysis for screw threads are readily available, the equations for them can be rather complicated.

For mechanical assemblies using screws can be detached indefinitely, with the exception of self-tapping screws, which can be loosened and retightened only a limited number of times. The best guideline for the designer is to prefer any assembly design that converts eventual tensile loads to compression loads. Those plastics generally are subject to crazing or stress cracking. Compression loads tend to reduce this problem.

When feasible, use metal-to-metal force-locking connections, particularly with many of the TPs, to release plastics from stresses. The forces that can be applied with a single small screw can be surprisingly high. As in metals, consider the use of torque-limiting wrenches in designs where the degree of loading is critical.

External and internal threads can be molded economically in plastic parts. Screw threads produced by the mold itself using rotating cores, split inserts, or collapsible cores will eliminate the normally expensive postmolding threading operations (Chapter 8, **TOOLING**). Coarse threads can be molded easier than fine ones, so threads less than 32-pitch should be avoided. American Standard screw threads should be designed and molded carefully. If the thread end form notches, a reduction in impact strength and ultimate elongation under tensile stress can be significant, depending on the type of plastic used. With certain applications and materials, trapezoidal and knuckle threads are better.

Generally, the length of thread used should be more than 1.5 times the diameter, the section thickness around the hole more than 0.6 times the diameter. Avoid having feather edges, and limit tightening with the bolt shoulder.

Bottle caps made from different plastics are extensively used. Some closures are of the simple cork snap type design, but most are of the screw type. Strong, accurate threads can be molded, which represent undercuts. Simple designs should be used when permitted, such as wide-pitch threads. The thread should be designed to start about 1/32 in. (0.08 cm) from the end of the face perpendicular to the axis of the thread. It is usually practical to mold up to 32 threads per in.; more than this number can give certain molders trouble.

Self-threading screws are an economical means of securing separable plastic joints. They can be thread cuttings. To select the correct self-threading screw, the designer should know which plastic will be used and what its mechanical properties are, particularly its modulus of elasticity. These self-threading screws are driven into the molded product, eliminating the need for a molded-in thread or a secondary tapping operation. They differ in their thread spacing and body design. Thread-forming screws, which provide the highest stripping torques, have less tendency to damage threads in repeated assembly operations than do other types. The thread-forming screw displaces material as it is being installed in the receiving hole.

This type of screw induces high stress levels in the plastic part, so it is not recommended for use with certain plastics such as those with a low modulus, unless careful procedures are used in forming the threads.

Screws or threaded bolts with nuts require through-going holes to provide an easy assembly system. Washers are recommended to distribute the load upon a larger area, wherever feasible or required. If a screw is tightened too far, excessive bending or tensile stresses will easily be created, possibly causing cracking based on stress-to-failure data curves. A change in design or the use of a spacer can convert tensile into compressive stresses. Different screw- and bolt-heads can be used, but flat-underside types of heads are best.

In regard to screw threads, certain observations should be considered. First, the torque values are based on the coefficients of friction of the mating parts and can thus vary significantly. The use of any compatible lubricant that reduces friction will increase the shear and hoop stresses if the torque remains the same. Therefore, with lubricants, reduce the amount of allowable torque.

Second, having high assembly torque to prevent vibrational loosening is frequently ineffective, since creep in the plastic will reduce the effective assembly torque even if the fastener does not rotate. Using vibrationproof screws, lockwashers, locknuts, and thread-locking adhesives are usually a better alternative when loosening is considered a problem.

Finally, self-tapping screws require additional torque to cut or form their threads. This torque can usually be added to the allowable safe-assembly torque, but for the first assembly only. The appropriate hole design for self-tapping screws is quite dependent on the material and screw design.

Rib As previously reviewed if there is sufficient space, the use of ribs is a practical, economic means of increasing the structural integrity of plastic products without creating thick walls. Ribs are provided for spacing purposes, to support components, and for other uses. Table 3-19 shows a summary of the results of using a rib design. Although the use of ribs gives the designer great latitude in efficiently tailoring the structural response of a plastic product, ribbing can result in warping and appearance problems (sink marks, etc.). In general, experienced designers do not use ribs if there is doubt as to whether they are structurally necessary. Adding ribs after the tool is built is usually simple and relatively inexpensive since it involves removing steel in the mold.

There are certain basic rib-design guidelines that should be followed (Fig. 3-9). The most general is to make the rib thickness at its base equal to one-half the adjacent wall's thickness. With ribs opposite appearance areas, the width should be kept as thin as possible. In areas where structure is more important than appearance, or with very low shrinkage materials, ribs are often 75 or even 100 percent of the outside wall's thickness. As can be seen in Fig. 3-10, a goal in rib design is to prevent the formation of a heavy mass of material that can result in a sink, void, distortion, long cycle time, or any combination of these problems.

Extrusion

Basically, the size of the die orifice initially controls the thickness, width, and shape of any extruded product dimension. In general, it is developed oversize to allow for the drawing and shrinkage that occur during conveyor pulling and cooling operations (Chapter 8, **TOOLING** and **EXTRUSION**). The rate of takeoff has significant influences

Table 3-19   Example of the effect of rib and cross-section changes

Geometry	Cross Section Area, square inches (mm^2)	Maximum Stress, psi (mPa)	Maximum Deflection, inches (mm)
(19.1 mm) ← 0 75 → ORIGINAL SECTION 0 080 (2.0)	0.0600 (38.7)	6800 (46.9)	0.694 (17.6)
ORIGINAL SECTION WITH RIB 0 040 0 400	0.0615 (39.7)	2258 (15.6)	0.026 (0.66)
← - 0.75 → THICK SECTION 0.239	0.1793 (115.7)	2258 (15.6)	0.026 (0.66)

on dimensions and shapes. This action, called drawdown, can also influence keeping the melt extrudate straight and properly shaped, as well as permitting size adjustments. The drawdown ratio is the ratio of orifice die size at the exit to the final profile size (6).

The range of extrudable profiles is practically unlimited, but to realize a full, practical design with economic potential, particular attention must be given to factors like wall thickness, hollows and cores, legs and projections, corners and radii, and so on. The most important consideration in profile design is the balancing of various wall thicknesses. A profile with a uniform wall thickness throughout its cross-section is the easiest to produce. Having uneven walls will cause material flow variations between the large and small portions of the profile. Also, thinner sections cool faster, causing bowing or warpage toward the heavy side. To compensate, it is necessary to provide external cooling for the bulkier sections and, usually, some

special orifice die design in which the land lengths (distances along metal surfaces) are changed significantly in respect to their cross-sectional openings. This usually requires additional costs, equipment, and reduced extrusion speed, resulting in higher production costs. Most important requirement is experience.

Tolerance   The penalty for having an unbalanced wall is the reduction of tolerance control. Tolerance limits are usually at least doubled. Also, with certain plastics it is more difficult to process them, such as those with low melt strength. Although the balanced wall is the ideal, having it is not always possible. Recognize that the unbalanced wall can be extruded with proper die design and control of the extruder line from upstream to downstream equipment.

As discussed in the previous section on injection molding, a sink mark almost always occurs in extrusion on a flat surface that is

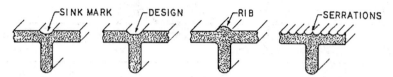

Fig. 3-34 Sink marks can be eliminated by creating a design, rib, or serration.

opposite to and adjoining a leg or rib, because of unbalanced heat removal or similar factors. As with IM, sink marks can be practically eliminated or eliminated by slowing down the extrusion line permitting more uniform cooling action. A popular method is to conceal the marks by adding a design feature, such as a series of serrations on the area where they occur (Fig. 3-34).

Figure 3-35 provides design features in extrusion dies that influence product performance. The guiding principle should be to keep it simple whenever possible. (a) The method of balancing flow to produce this shape requires having a short land where the thin leg is extruded. This design provides the same rate of flow for the thin section as for the heavy one. (b) This die for making square

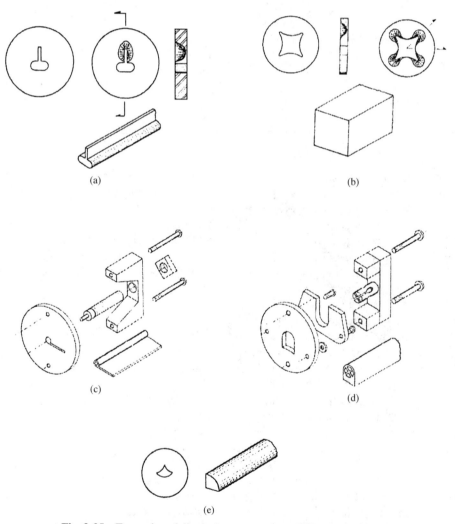

(a)

(b)

(c)

(d)

(e)

Fig. 3-35 Examples of die designs to produce different profiles.

extrusions uses convex sides on the die opening so that straight sides are formed upon melt exiting; the comers have a slight radius to help obtain smooth comers. The rear and sectional views show how part of the die has been machined away to provide short lands at the corners to balance the melt flow. (c) In this die for a P shape, a pin mounted on the die bridge forms the hole in the P. The rate of flow in thick and thin sections is balanced by the shoulder dam behind the small-diameter section of the pin. The pin can be positioned along its axis to adjust the rate of flow to meet the melt characteristics. (d) In this die to extrude a rather complicated, nonuniform shape, a dam or baffle plate restricts the flow at the heavy section of the extrudate to obtain uniform flow for all sections. The melt flows between the die plate and the dam to fill the heavy section. The clearance between the dam and the die plate can be adjusted as required for different plastics with different melt behaviors. (e) In this die for extruding a quarter-round profile the die opening has convex sides to give straight sides on the right-angled portion, and the comers have a slight radius to aid in obtaining smooth comers on the extrusion.

Figure 3-36 provides design tips for coextrusion. (a) A dual extrusion for a modular cabinet wall panel. (b) If the flexible sealing portion wears out from abrasion, a replacement flexible insert can be slid into the slot in a rigid portion. (c) A cross-section of a dual extrusion (a ball-return trough for a billiard table). (d) A bowling-ball return trough made from a 6 in. (15 cm) diameter extruded tube with one or more layers. The tube is slit while still workable and guided over a forming die. (e) Typical dual extrusions of rigid and flexible PVCs. (f) Typical extrusions of rigid and flexible PVCs showing different applications. (g) A cross-section of a window frame with a metal embedment. (h) Keying or fitting can join nonbondable plastic. (i) noncircular hollows are easier to form if each part of the surrounding wall is made from the same family of plastic: (A) the rigid PVC base will remain flat and not bulge and (B) the air pressure inside the hollow will cause the flexible base section to bulge. (j)

Different applications for metal-embedment extrusions.

Blow Molding

Blow molding, provides designers with the capability to make products ranging from the simple to rather complex 3-D shapes (Chapter 8, **BLOW MOLDING**). Designers should become aware of the potentials BM offers since intricate and complex shapes can be fabricated. The BM process is especially amenable to the designer's goal of consolidating as much function as possible into a single product. Some of the features that can be incorporated include threads, inserts, fasteners, hinges, and others somewhat similar to those covered under injection molding. Hinges include the different mechanical types as well as integral hinges (9, 20).

Hinge In addition to the information concerning Fig. 3-20 on the general approach to molding living hinges, this review specifically concerns BM. To produce a hinge during extrusion BM (Figs. 3-37 and 3-38) the hinge is formed perpendicular to the parison flow from the die. With injection blow molding, the hinge is perpendicular to the melt flow in the preform mold (Chapter 8, **BLOW MOLDING**).

Here are some guidelines to mold living hinges with polypropylene (other plastics are similar but may require their own specific dimensions dependent upon their particular viscoelastic behavior):

1. The land of the hinge should be at least 1.5 mm (0.06 in.) wide for a proper flow pattern and at least wide enough so that when the product is bent in service it will not develop strains. Too short a land length will cause the hinge to have limited flex life.

2. The minimum plastic thickness or pinch-off gap at the center of the hinge should be 0.25 to 0.38 mm thick (0.010 to 0.015 in.) and 0.5 mm (0.020 in.) wide.

3. When the plastic melt flows across the small hinge gap, frictional heat will be

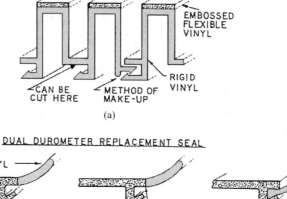

(a)

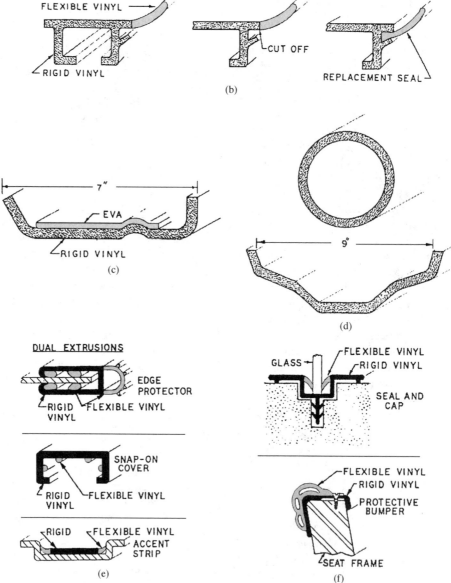

DUAL DUROMETER REPLACEMENT SEAL

(b)

(c)

(d)

DUAL EXTRUSIONS

(e)

(f)

Fig. 3-36   Designs for coextrusion.

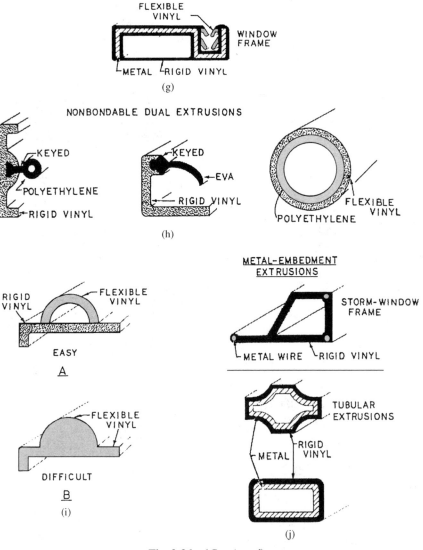

Fig. 3-36   (*Continued*)

generated. There should be sufficient cooling of the mold around the hinge area.

4. With injection blow, the hinge gap is a difficult area to flow across. Therefore, the gate should be placed so that the molten plastic can flow perpendicular to the hinge, to ensure a good fill. If the melt flows along the length of the hinge, there is bound to be a short shot or cold weld at the hinge.

5. Shoulders and lips should be included in the two mating parts to help alignment.

6. The finished product should be flexed immediately upon its ejection from the mold, while the heat from the mold is still in it. De-

pending on its service operation flex angle between 90° to 180° is recommended. The flexing action can stretch the hinge area by 200% or more; thus, the initial 0.25 to 0.38 mm thickness (0.10 to 0.015 in.) will be thinned down to less than 0.13 mm (0.005 in.). This elongation helps align the plastic's molecules and increases its tensile strength from 34×10^6 to 552×10^6 Pa (5,000 to 80,000 psi).

An extrusion-blown hinge can be compared to a stamping (Fig. 3-21). Hot- and cold-stamped hinges are made by compressing a sheet of material down to the desired thickness [12 to 20 in. (3 to 5 mm)]. Stamped

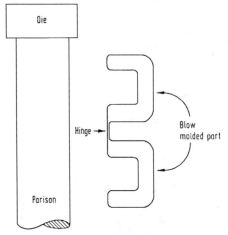

Fig. 3-37 Example of an extrusion blow molded container with a living hinge.

hinges are less durable in flexing but strong in tearing.

Consolidation Even though blow molding provides the capability of providing multiple products combined into one, by including hinges the designer has an added feature. The ability to produce many articulated products in one shot has always opened new design

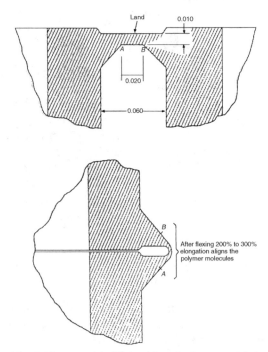

Fig. 3-38 A molded living hinge in the as-molded position (a) and a flexed position (b).

possibilities. Because this hinge is practically free, the cost of these products is just the cost of the material and the molding.

Summarizing this subject for a designer concerns becoming familiar with the possibilities in designing BM products, particularly the complex shapes. Gradually designers are becoming more familiar with the capabilities inherent in BM.

Thermoforming

Designers should follow the inherent nature of thermoforming that basically uses flat panels (film or sheet) instead of the solid, enclosed, boxlike, cylindrical, rodlike, or structural shapes of other processes (Chapter 8, **THERMOFORMING**). They should be aware of and observe the material's depth-of-draw limitations. It can vary depending on the type of TP, the thickness tolerance of the material, and the degree of pinhole freedom the material enjoys. Generally, for straight vacuum forming into a female mold, the depth-to-width ratio should not exceed 0.5/1. For drape forming over a male mold, this ratio should normally not exceed 1/1. For products to be used with the plug-assist, slip-ring, or one of the reverse-draw methods, the ratio can exceed 1/1 and perhaps even reach 2/1 under normal circumstances. However, shallow drafts are in general more readily formed than deep ones and result in more uniform wall thickness (1).

Undercuts and reentrant shapes are possible in many designs. They require movable or collapsible mold members, but with small undercuts they can often be sprung from a female mold while the formed product is still warm. This type of action works best when the plastic has some flexibility, as do the TPEs, or the material is very thin. Guidelines for the maximum amounts of undercutting that can be stripped from a mold are as follows: 0.04 in. (0.1 cm) for acrylics, PCs and other rigid plastics; 0.060 in. (0.15 cm) for PEs, ABSs, and PAs; 0.100 in. (0.25 cm) for flexible plastics such as the PVCs.

When female tooling is split to permit the removal of products with undercuts, a parting

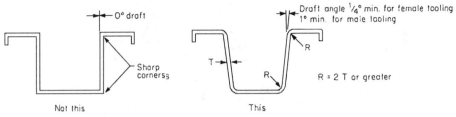

Fig. 3-39 Example of the draft on thermoformed side walls.

line of the split halves becomes visible on the formed part. If this is objectionable, the designer can sometimes incorporate the parting line in the decoration of the product or at some natural line on the product. Sharp corners generally should never be specified, since they hamper the flow of material into the mold's comers. This results in excessive thinning of the materials and causes concentrations of stress. A minimum radius of two times the stock's thickness is recommended. It is also more desirable from several standpoints to have large, flowing curves in a thermoformed product than to have squared comers or rectangular shapes. However there are thermoforming techniques that can produce thicker corners (1).

The best products have smooth, natural curves and drawn sections that are spherical or nearly so in shape. Their walls will be more uniform, they will be more rigid, their surfaces will be less apt to show tool marks, and their tooling and molds will be lower in cost. Notches or square holes should be avoided when punching formed products. Round holes are preferred to oval ones for minimizing stress buildup.

Some draft is required in side walls to facilitate the easy removal of the product from the mold. Female molds require less draft since products tend to pull away from mold walls as they shrink during cooling. With female or male tooling, for most plastics the draft on each side wall should be at least 1 degree (Fig. 3-39).

Metal inserts are usually not feasible, because thin walls are not sufficiently strong to hold inserts, particularly if thermal expansion and contraction of the product takes place in service. Figure 3-40 shows a method of holding metal fittings. It may be desirable to in-

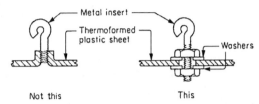

Fig. 3-40 Recommended method for holding metal fittings in thermoformed products.

crease the stiffness of thermoformed products. Many are panel shaped and made of thin walls, so they may lack rigidity. Corrugations, which if used are preferable in two directions, or an embossed pattern can add to their rigidity. With short-run production it may be more economical just to use thicker sheet plastic to gain stiffness. If the function of the product permits, use curved, dished, or domed surfaces to gain stiffness.

When thermoformed products such as caps are stacked, without controlled spacing they will jam together (as with other similar shaped processed products), which could cause sufficient stress to cause products to split. To avoid jamming and control the space between parts, a stacking boss or shoulder system can be used (Fig. 3-41). Within this stacking area the plastic must be sufficiently rigid to prevent the deflection of bosses that would cause jamming. The height of the bosses is generally greater than their vertical cross-sections at the point of least taper; otherwise the tapered walls will interfere before the stacking sections can engage. There are also other designs that can be used to eliminate jamming.

Tolerance Thermoformed products lack the dimensional accuracy of processes such as injection and compression molded products.

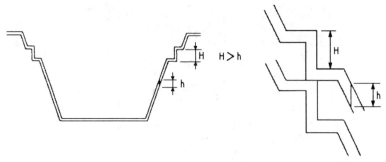

Fig. 3-41 Example for stacking thermoformed products to avoid jamming.

With its low pressure, thermoforming reduces the degree to which the sheet or film being formed is forced to conform to the mold. Material variations, mainly in their thickness and degree of existing pinholes, affect the final accuracy of the product. This is particularly true because tooling is generally one sided where the product is pushed against a female or male mold cavity. However there is the thermoforming operation where a male and female mold are used together which is a take-off of compression molding. The objective should be to use material with tight thickness controls that is pinhole free, rather than just to determine its weight. Some fabricators buy by the lower-cost method where weight is the controlling factor because they do not have to meet any tolerance. However it is possible that by buying by thickness amount of material consumption per product will provide a significant material cost saving. Other factors that effects cost savings also could include uniform heat absorption of the sheet/film that could result in more uniform thickness tolerances and quicker cycle times.

Products are affected dimensionally by the difference between their forming temperature and their product-use temperature. Thus, a plastic's coefficient of thermal expansion and contraction has a significant effect on service conditions. The thermoforming pressure, time, and temperature variations that can exist will affect the final dimensions. Of these factors, evenness in heating throughout the sheet thickness before forming is usually the most important control. Type of heater has a direct effect on obtaining uniform heat

(1). An allowance must also be made for post-forming shrinkage. Molds should be designed oversize so that when shrinkage is complete the product dimensions will be correct to within the design tolerances.

The dimensional tolerances with the more conventional single-mold system are generally ±0.6% (±0.35% for close tolerances). With female molds ±0.5% (±0.3% close) with male molds under 3 ft., ±0.8% (±0.4% close) and with male molds over 3 ft., ±30% (±10% close) for wall thicknesses.

Rotational Molding

RM, which uses single or multiple arms to hold the molds, is appropriate to different sizes and shapes of products such as tanks or containers ranging from small squeeze bulbs of vinyl plastisol to large storage tanks (Chapter 8, **ROTATIONAL MOLDING**). This technique can produce uniform wall thicknesses even when the product has a deep draw off the parting line or small radii. The liquid or powdered plastic used in this method flows freely into corners or other deep draws upon the mold's being rotated and is then fused by heat passing through the mold's wall.

Mold This process is particularly suited economically to producing small production runs and large-sized products, because molds are not subjected to relatively any pressure during molding and inexpensive thin sheet metal molds can thus be used in many applications. Lightweight cast aluminum and electroformed or vaporformed nickel molds, which are light in weight and low in cost, can also be used. Large RM machines can be purchased or built economically, because they can use relatively inexpensive gas-fired, hot-air ovens with relatively lightweight mold-rotating mechanism.

Cost When it is necessary to equal the production rates of other processes, the mold cost with RM may exceed that of other processes such as flow molding. The plastics used in RM are generally more expensive than the pelleted plastics used in many other processes, because they must be more finely and evenly powdered, such as to a 35 mesh. However, this process generates low levels of regrind or scrap, even when it is operating poorly. Products can have no flash at all if properly designed molds are used.

The molding of two or more different types of plastics in a single product may be accomplished to combine their specific properties and/or a better performing or lower-cost product. This process, called corotation, is similar to coextrusion or coinjection in terms of the performance of the designed product. An expensive plastic may be backed with a less costly material (recycled, etc.), and a skin surface layer can be backed with a foamed plastic molded in one operation. The dissimilar molding powders, which may have different softening temperatures, can be molded simultaneously or separately, depending on the processing conditions and the end product's requirements. Any greater than normal thickness must usually be designed to form multilayered products, especially if a foam component is to be included.

Some combinations of materials are not feasible with this method. For instance, after molding the first layer against the mold wall, the second material cannot have a higher melt temperature, which, of course, would melt the first layer, probably causing them to mix.

With RM one inherent overall disadvantage exists. It is that the complete cycle for a single mold is significantly longer than it is for many other processes. However, in many cases it is possible to run multiple molds on each arm or arms, to offset the effect of having slower cycles. Also in many applications total cost (mold, operator, etc.) is lower than the other processes.

The preferred contour for any parting line is the straightest path possible. By this means, mold construction costs can be reduced and demolding will be the easiest means possible. When two products like a container and its lid are to be molded together, as in blow molding, they may be separated after the molding by employing a removable cutter or annular wedge at the parting line. Another technique is by molding it oversize to provide a resting flange, then cutting it to separate the products.

Wall thickness/surface The wall thickness in a mold can be changed just by increasing the amount of plastic put into the mold, because the wall is basically produced by a coating or plating process that operates on the inside surface of the mold. However, changes in heating time would be necessary to fuse the plastic properly. Thus, adjustments to product's wall thicknesses can be made to increase rigidity, impact strength, or load-carrying capacity. A maximum thickness does exist, based on the type of material used and the material chosen to construct the mold as well as the heat source. These factors all influence the rate of heat transfer through the plastic. Because in this process the plastic is deposited on the mold without pressure, the finished part is generally stress-free.

Products in this process can have deep sections and relatively sharp corners. However RM flat, particularly large relatively uniform wall thickness surfaces are difficult if not impossible to produce. This process can be used to mold complex products that may require three or four split molds. Also, different finished surfaces are obtained. For example, the products' surface finish is dictated by the inside surface of the mold. This makes it easy to obtain smooth as well as textured surfaces on the product. Raised or depressed letters, fluting, and other decorative inscriptions may also be molded.

Processing technique The inside surfaces are influenced by the type of plastic used and may be made smoother by selecting an easily flowing melt with a high melt index. Because such plastics are sometimes chemically or mechanically inferior, the better plastic may be made smoother by resorting to higher molding temperatures and longer cycle times, short of damaging the plastic. In-mold decorating methods, such as decals that are deposited on the mold surface, are used that can become part of the finished product's surface and can be designed to provide increased structural performance, as also in injection molding, blow molding, and other processes.

Ample draft is suggested on sidewalls to facilitate product removal. A recommended minimum for most plastics is 1 degree. The lower-shrinkage plastics like PC and PMMA will require 1 1/2 to 2 degrees. Undercuts are possible, but they should be kept to a minimum. Making provisions for undercuts usually requires higher mold costs, because of having to use some type of action such as core pulls or splitting a mold to allow separation parallel to the undercut groove. Undercuts may also require extra time for unloading molds.

Inside or outside comers should use large radiuses, not sharp ones even though sharp corners can be molded. By doing so any potential cracking, molded-in stresses, and undesirable thickening will be prevented. A useful guide to the smallest allowable inside radius is 1/16 in. (0.16 cm) with 1/4 in. (0.66 cm) for optimal filling conditions. The goal should be to have a radius equal to the wall thickness for easier melt flow. Although RM produces uniform wall thicknesses, comers in it can have greater variation than the rest of the product. A sharp inside comer tends to heat at a slower rate, causing the plastic to flow away from it, thus making it thinner. Conversely, sharp outside comers heat at a faster rate and tend to hold the plastic longer, thus building up more thickness. If required techniques are used to reduce or eliminate these type problems by controlling the heat input at these corners.

It is usually difficult to produce internal or external bosses and T sections, because they are not conducive to producing uniform walls. It is possible to produce interior extensions, by placing a metallic screen in contact with the inner mold wall. This screen heats up, attracts plastic, and becomes covered remaining in place after molding. By using this method a hollow product can be molded to have two or more separate chambers, with the screen being extended entirely across the inside of the mold.

A hole can be formed by molding a dome and cutting it after molding. One technique that can be used for this design is to mount securely on the inside mold wall a fluorocarbon (PTFE) plug to prevent plastic from adhering to the mold at that location. Another method involves inserting machined brass plugs, pins, or tubing through the mold wall. During molding, the heat passes from the mold to the insert, causing plastic to form around it. Care must be taken to select inserts that will heat easily and a plastic that will not crack, because stresses are created as the plastic shrinks around the insert upon cooling.

Moldable holes and inserts can complicate molding and may require extra postmolding operations. Thus, the most economical designs are those that minimize the number of holes. With many plastics, both external and internal threads can be molded, but sharp V threads should be avoided because they can cause the plastic to bridge, resulting in incomplete thread fill. Rounded or modified buttress threads will allow improved thread fill.

The stiffening of solid ribs or projections is possible and easily moldable if the requirement of maintaining uniform wall thickness is followed. A narrowed rib will not fill and will leave inside stringers. It also can prevent the melt from reaching the bottom before fusing. A small, shallow, narrow rib will fill completely but have limited strengthening effect. The correct rib design requires a wide gap to form the rib, with a generous draft so that the melt is allowed to fill uniformly, without bridging. As a guide, deep ribs that are four times the wall's thickness generally require at least five times the wall's thickness between the parallel sides of the rib to prevent the plastic from bridging as it flows into the rib before fusing to the mold's wall.

Design Failure Analysis

The process of analyzing designs includes the modes of failure analysis. At an early stage the designer should try to anticipate how and where a design is most likely to fail. A few examples of potential problems due to loading conditions on products are reviewed.

The most common conditions of possible failure are elastic deflection, inelastic deformation, and fracture. During elastic deflection a product fails because the loads applied produce too large a deflection. In deformation, if it is too great it may cause other parts of an assembly to become misaligned or overstressed. Dynamic deflection can produce unacceptable vibration and noise. When a stable structure is required, the amount of deflection can set the limit for buckling loads or fractures.

Because many plastics are relatively flexible, analysis should consider how much deflection might result from the loadings and elevated temperatures the products might see in service. The equations for predicting such deflections should use the modulus of the material; its tensile strength is not pertinent. Usually, the most effective way to reduce deflection is to stiffen a product's wall by changing its cross-section.

Inelastic deformation can cause product failure arising out of a massive realignment of the plastic's molecular structure. A product undergoing inelastic deformation does not return to its original state when its load is removed. It should be remembered that there are plastics that are sensitive to this situation and others that are not.

The existence of an elevated temperature, with or without long-term or continuous loading, would suggest the possibility that a material might exceed its elastic limits. As explained in Chapter 2 concerning momentary loading, the properties to consider are the proportional limit and the maximum shear stress. The presence of fracture reflects a load that exceeds the strength of the design. The load may occur suddenly, such as upon impact, or at a low temperature, which will reduce the elongation of the material. A failure may develop slowly, from a steady, high load applied over a long time (creep rupture) or from the gradual growth of a crack from fatigue. If fracture is the expected mode of failure, analysis should examine the greatest principal stresses involved (Chapter 2, **HIGH SPEED PROPERTIES**).

4

Designing Plastic Product

Introduction

This chapter reviews the design of different current and potential applications for plastics. Plastics are used uniquely in these applications because of factors such as availability of their extensive capability in material modifications to meet specific material and processing requirements. By incorporating some innate behaviors of the materials and adapting them to operate in unusual environments, low cost products are produced as well as some of the most significant and sophisticated problems of man-kind are being solved by the use of plastics.

A product under consideration must have some utility; it must fulfill a need which may be aesthetic or functional and, generally, both. To proceed with the design we must know what function it is to perform. We also need to know the context or surroundings of the product to determine what effect they will have on its function. A careful definition of the function will simplify the design and permit the widest latitude of alternatives possible in the design without compromising the function of the product.

Book Shelve

A good example is designing a shelf. The function served is a support that can hold several objects in a desired location for storage or for display. Unless a more specific function is defined, the shelf can take on a wide range of possible shapes, structures, and materials. It would be necessary to define the shelf as a bookshelf, for example, compatible with the environment found in a library, suitable for holding five or more books per foot of shelf, and for constant reference use rather than for storage (190).

This information will sufficiently define the function so that a design can be started. It defines the environment, sets the load level and the type of loading situation, and gives some idea of the shape requirements, as well as the possible aesthetics of the unit. It still permits a wide range of design choices as to material, structure, and shape but they would be limited to those normally used in a library environment. The more accurately and completely the function is defined, the more restricted are the design possibilities and the more detailed the specifications for the function.

Size is the next factor to be considered. A product has to fit its function within the confines of the space in which it is used. Continuing with the example of the shelf, it is obvious that we must know the length of the shelf, either by deciding how many books it will hold or by stating the size of the supporting rack that will be used. The size can then be decided either by burden or by space restrictions. In most cases, one or the other of these considerations will apply and in a typical case both may apply to some extent. In the example given, the width of the shelf would be determined by the width of a book, which ranges from 6 to 11 in. (15 to 28 cm) The typical bookshelf is supported at 3 ft (0.9 m) intervals so that the shelf would be this wide to fit a typical book rack. The shelf will hold about 5 books per foot (5 per 0.3 m) with an average weight of about 2 lb (0.91 kg) each to make a maximum load on the shelf of 30 lb (13.6 kg). If the shelf were completely filled, it could be considered a distributed load, or it can be considered as a set of discreet loads.

Material

The type of shelf design is the next consideration. The shelf can be a solid plate of plastic material, an inverted pan-like structure with reinforcing ribs, a sandwich-type structure with two skins and an expanded core, or even a lattice type sheet that has a series of openings. The choice between these is dictated by a number of factors. One is appearance or aesthetics.

The lattice-type shelf is functionally as good as the others, but it may not look appropriate for a book shelf in the context of a library. A second consideration is a combination of physical requirements and appearance. A simple plastic beam that will function adequately in terms of strength and stiffness may be rather thin. A shelf of this type can look flimsy even if it is functional. This impression is useful to the designer since the solid plate is probably an uneconomical use of material. A requirement was added that the design should look like a wood shelf since this is the context in which it is to be used. To produce the desired thickness appearance either a lipped pan with internal reinforcement can be used or, alternatively, a sandwich-type structure with two skins and a separator core. In either case the displacement of the material from the plane of bending will improve the stiffness efficiency of the product. The appropriate procedure is to

leave both as possibilities and do some trial designs.

The next step in the design procedure is to select the materials. The considerations are the physical properties, tensile and compressive strength, impact properties, temperature resistance, differential expansion environmental resistance, stiffness, and the dynamic properties. In this example, the only factor of major concern is the long-term stiffness since this is a statically loaded product with minimum heat and environmental exposure. While some degree of impact strength is desirable to take occasional abuse, it is not really subjected to any significant impacts.

Using several materials such as PP, glass-filled PS, and PS molded structural foam that is a natural sandwich panel material, the design procedure follows to determine the deflection and stress limitations of the material in each of the several designs.

There are two criteria to use as the basis for evaluation. The design life of the shelf is determined by deciding what the product will tolerate in deflection and still be useful. This is combined with the cost effectiveness value the product must meet. For example, we can say that if it costs X the life must be A months, if it costs Y it must last B months, and if it costs Z, it must last C months. This can be presented as a table or it can be graphed as the criteria range that it must meet.

Using the several materials selected and the basic design possibilities, products should be designed to meet the criteria as far as deflection is concerned, and the cost of manufacture estimated. If the designer does not have this type experience it is best done with the assistance of the fabricator who will fabricate the product. In addition to the material cost and the production labor costs, the amortization costs of the tools are to be included.

The various designs and costs can be tabulated and the ones that are the most economical can be determined. At this point, it may become evident that the design life can be long and the cost of increasing the design life small, or, alternatively, it may be that the cost of small increments in the design life are quite costly. In the latter case, the design life should be limited to the acceptable minimum.

A value judgment must be made as to the product quality requirements and the final design made to meet these requirements. By leaving the options open until this conclusion is reached, the decision as to what is best in terms of product can be based on more than a single valued solution, with the probable result that a more economical and practical product will result.

The next step is the performance evaluation. In order to do this the product must be completely designed as to the dimensions necessary to fit any surrounding products, as well as to the necessary cross-section thickness for strength and stiffness. The material, color and manufacturing process must be selected. The product must include in the design the necessary features to make for proper processability and whatever design features are necessary to improve the performance in the service environment. The latter may include reinforced areas, coating or plating, inserts, etc.

Prototype

Next is to make sample prototype tooling and sample prototype products for the test. Samples made by machining or other simplified model making techniques do not have the same properties as the product made by molding or extrusion or whatever process is to be used (Chapter 3, **PROTOTYPES**). A product made this way is a sample rather than a testable prototype. Simplified prototypes may reduce trial mold cost and produce adequate test data in some cases. Its main value is appearance and feel to determine whether the aesthetics are correct. Any testing has to be done with considerable reservation and caution.

If prototype tooling is not made, then production tooling must be made to provide samples for evaluation testing. This is justified if the product does not represent a substantial departure from previously made units whose performance is known in similar applications. If there is no prior history and as an example

there will be a very long production run on producing the shelves, it mandates the use of a production type prototype tooling.

In the situation where a similar application existed, the risk that the tools may have to be scrapped or drastically altered as a result of the testing is not high and is justified. The other reason that a production tool is made with no prototype tooling is because of the lack of lead time. Here the risk is usually not justified and the shelves of processors are littered with tools that were the result of bad guesses made under severe time pressure.

By proper organization of time and tool design, the sample tooling can be made to fit into the normal tooling cycle with a minimum of added time and expense. Sometimes the prototype tooling can be made part of an existing master chase or holder or it can be made of soft materials such as epoxy or kirksite that can be worked rapidly. In any event, the information obtained in the prototype tooling stage can have a major effect on the final design and tooling to give a successful and economical product while prejudgments may result in a poor compromise as a result of drastically reworked tooling.

Testing

After obtaining the prototypes, tests must be made to determine the utility. Generally these include a short time destructive test to determine the strength and to check out the basic design. Another test that is done is to use the product in the projected environment with stress levels increased in a rational manner to make for an accelerated life test. Other tests may include consumer acceptance tests to determine what instructions in proper use are required, tests for potential safety hazards, electrical tests, self-extinguishing tests, and any others that the product requires. In the case of high risk products, the test program is continued even after the product enters service.

The remainder of the test program requires the generation of quality control (QC) monitoring tests. For example, in the case of an injection molded product such as the library shelf, the quality control monitoring might include critical dimension checks and an oven test to see if the product has a tendency to warp or self destruct, indicating improper molding conditions. A solvent test for monitoring proper process conditions might be included to check for stress crack resistance.

The last step in the design cycle is the end-use testing, or field trials. This can be done using selected individuals or applications that are closely monitored or controlled. This information will indicate whether or not the product performed as the designer anticipated because of some unexpected situation. For example, the library shelves are often cleaned with lemon oil that disintegrates the polystyrene (this requirement was not previously known). Some of the product can be test marketed to uncontrolled users and their reactions sampled with a standard response form generated by the advertising people. Sample responses from a larger group of people frequently produces what might be referred to as the "idiot response." It is unbelievable what some users can do to a product simply because they did not understand its limitations.

The results of the field testing must define the basis for labeling and the instructions to be used with the product. Unless there has been a serious misjudgment by the product development designers, the field tests do not lead to redesign. There may be the need for a more durable unit for the serious abuse situations, as well as the need for proper instructions, but the main need is for the labeling and instructions. The designer can supply the necessary data for the do's and don'ts and the instructions will at least minimize liability on the product of the producer for failure due to abuse. The most obvious case in which this is not true is when the abuse may occur easily and result in a personal hazard to the user. In this case, loss of the instructions is not an adequate defense against responsibility by the user. In all cases, the product must be designed to prevent danger to the user. Potential failures that can cause personal injury must be avoided in all product designs

(see near end of this Chapter, **LAWS AND REGULATIONS**).

Summary

A table can be prepared summarizing the steps in the design of a plastic product. It would indicate the different types of options that can be used at different stages in the design sequence. It also would indicate the areas of potential high risk failure and the alternative approaches to use under these conditions. Example of a design program approach follows:

1. Define the function of the product with life requirements.

2. State space and load limitations of the product.

3. Define all of the environmental stresses that the product will be exposed to in its intended function.

4. Select several materials that appear to meet the required environmental requirements and strength behaviors.

5. Do several trial designs using different materials and geometries to perform the required function.

6. Evaluate the trial designs on a cost effectiveness basis. Determine several levels of performance and the specific costs associated with each to the extent that it can be done with available data.

7. Determine the appropriate manufacturing process for each design.

8. Based on the preliminary evaluation select the best apparent choices and do a detailed design of the product.

9. Based on the detailed design select the probable product design, material, and process.

10. Make model if necessary to test the effectiveness of the product.

11. Build prototype tooling.

12. Make prototype products and test products to determine if they meet the required function.

13. Redesign the product if necessary based on the prototype testing.

14. Retest.

15. Make field tests.

16. Add instructions for use.

Pipe

A major and important market for plastics is in producing pipe (tube) for use such as on the ground, underground, in water, and electrical conduit. The largest use is in transporting water, gas, waste matter, industrial mining, etc. Use of extruded thermoplastic, such as HDPE, PVC, and PP, provide most of the world markets. The other major product is reinforced plastic principally using glass fiber with TS polyester plastic that use fabricating methods such as bag molding and filament winding fabricating techniques (Chapter 8, **REINFORCED PLASTIC**).

Since the 1930s, the TP pipe industry continues to expand its use worldwide. It now represents over 30% of the dollar share compared to other materials (iron/steel at 45%, copper at 12%, concrete at 8%, aluminum at 4%, etc.). Although RP TS pipe represents a small portion of the market, it is a product of choice for many special high performance applications. Corrosion resistance, toughness, and strength contribute to its growing acceptance.

A common pressure vessel application for pipe is with internal pressure. In selecting the wall thickness of the tube, it is convenient to use the usual engineered thin-wall-tube hoop-stress equation (top view of Fig. 4-1). It is useful in determining an approximate wall thickness, even when condition ($t < d/10$) is not met. After the thin-wall stress equation is applied, the thick-wall stress equation given in Fig. 4-1 (bottom view) can be used to verify the design (Appendix A: **PLASTICS DESIGN TOOLBOX**).

RP Pipe

RP pipes are used in different applications. The following review concerns the use and acceptance of buried large-diameter glass fiber,

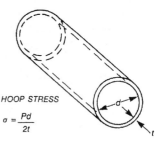

UNIFORM INTERNAL PRESSURE, P

HOOP STRESS

$$\sigma = \frac{Pd}{2t}$$

This equation is reasonably accurate for $t < d/10$. As the wall thickness increases the error becomes quite large.

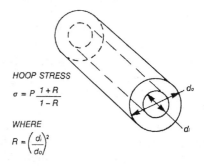

UNIFORM INTERNAL PRESSURE, P

HOOP STRESS

$$\sigma = P\frac{1+R}{1-R}$$

WHERE

$$R = \left(\frac{d_i}{d_o}\right)^2$$

This equation is for the maximum hoop stress which occurs on the surface of the inside wall of the tube.

Fig. 4-1 Cylindrical pressure pipe of thin-wall construction (top view) and cylindrical pressure pipe of thick-wall construction (bottom view).

plastic reinforced, filament-wound pipe that has increased steadily since the 1950s. Such RP was selected for its superior corrosion-resistance characteristics and installation-cost savings. ASTM standards use the term reinforced thermoset resin pipe (RTR pipe or RTRP); the general plastic industry uses the term reinforced thermoset plastic pipe (RTP pipe). Filament-wound pipe with a double helical angle of continuous-glass reinforcement (Chapter 8, **REINFORCED PLASTIC**) is but one of several types of RTP pipe constructions. Since at least 1944 pipe design equations have been used that specifically provide useful information to meet internal and/or external pressure loads (37).

Attempts have been made to utilize performance standards based upon internal pressures and pipes' stiffness, but other factors must be carefully considered in designing buried piping systems, especially the longitudinal effects of internal pressure, temperature gradients, and pipe bridging. Failing to recognize these factors incurs the risk of under designing a system. Because of its plastic-glass fiber construction, the physical characteristics of RTR pipe and therefore the design techniques needed for it differ considerably from those of older, traditional pipe materials.

It is true that RTR pipe design does to a degree parallel the design philosophy for steel pipe, but there is a point where the steel and RTR pipe design approaches part company, even though steel and RTR pipe are by definition both flexible conduits. In other words, both kinds of pipe can bend and deflect after burial, within certain limitations, without suffering structural or functional failure. In this regard they both differ from concrete pipe, which is a rigid conduit that cannot tolerate bending or deflection to the same extent as RTR pipe. Since an appreciation of the differences between flexible and rigid conduit is essential to a better understanding of RTR pipe design, let us examine these differences.

Load Testing

The diagrams in Fig. 4-2(a) illustrate the results of actual load testing on both types of conduit by the Roadway Committee of the American Railway Engineering Association. Both the flexible and the rigid pipes were buried under thirty-five feet of identical fill material. Obviously, specific pressures vary from installation to installation, but the relationship in the way the two kinds of pipe react to the same burial condition generally remains constants.

Let us start by examining a rigid pipe. Because of its rigid, inflexible characteristics, surface load intensifies at the crown of a rigid pipe and is transmitted through the pipe directly to the bed of the trench in which the pipe rests. This is not true with flexible conduit. Because a flexible conduit deflects under covering load of earth, this deflection transfers portions of the load to the surrounding envelope of soil. This is true of both steel and RTR pipe. The result is that the support of the surrounding earth actually increases the strength of the flexible conduit. Therefore, analyzing the type and consolidation of backfill materials must be considered an integral part of the design process.

Two additional observations can be made. First, because a rigid pipe transmits almost all the load of the earth cover to the trench bed, someone will occasionally be heard to say that rigid pipe, such as concrete pipe, does not require side support. This is not true. Second, because of the difference in the ways rigid and flexible conduit distribute the load of their earth cover, flexible piping materials are often said to require less bedding bearing strength, because they impose less of a load

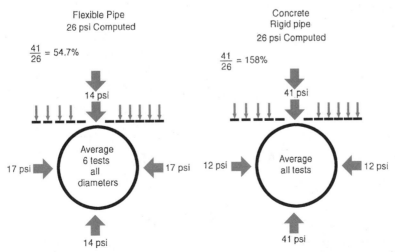

Fig. 4-2(a) Load-testing profile of flexible and rigid underground pipes.

on the trench bed. This is indeed true. In fact, it is one of several factors that help to reduce the installed cost of RTR large-diameter flexible pipe.

Directional Property

Given these differences between rigid and flexible conduit, let us examine the differences between steel and RTR pipe, both of which are, of course, flexible conduits. First, steel pipe is by definition constructed from a material, steel, that for our purposes is a homogeneous isotropic substance. Therefore, steel pipe can be considered to have the same material properties in all directions; that is, it is equally strong in both the hoop and longitudinal directions [Fig. 4-2(b)].

RTR filament-wound pipe is, however, an anisotropic material. That is, its material properties, such as its modulus of elasticity and ultimate strength, are different in each of the principal directions of hoop and longitude. It is here where the design approaches for steel and RTR pipe part company [Fig. 4-2(c)]. This behavior is a result of the construction of filament-wound RTR pipe.

Filament Wound Structure

Its manufacturing is done by winding continuous strands of plastic-impregnated glass fiber around a steel mandrel at a precisely controlled helix angle, under controlled tension. A cross-sectional view of an RP layup is shown in Fig. 4-2(d). As seen, the structural wall of the pipe is made up of con-

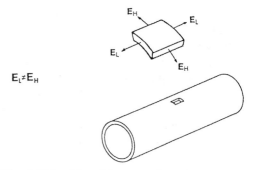

Fig. 4-2(c)   Material properties of anisotropic RTR pipe.

tinuous strands of fiber glass embedded in a plastic matrix, plus an internal corrosion barrier liner. The liner can be constructed from a number of different plastics and reinforcement materials, depending on what will eventually be put through the pipe. Incidentally, the thicknesses of the liner are not considered during design analysis, except for calculating buckling and pipe deflection (Chapter 8, **REINFORCED PLASTICS, Filament Wound Structure**).

Broadly speaking, three factors control the physical properties of RTR pipe. These are the amount of continuous-glass filament used to construct the pipe wall, the prescribed dual-helix angle at which the glass is wound around the mandrel, and the type and amount of plastic matrix used to bind the glass filaments together. Controlling the strength of the pipe in the hoop and longitudinal directions is done by selecting the winding angle and ratio of glass to plastic content. The winding angle for the structural wall is usually from 55 to 65 degrees to the horizontal, and the glass-fiber content is not less than 45 wt%. The final material composition of the pipe is determined by calculating the longitudinal

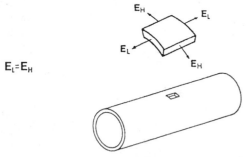

Fig. 4-2(b)   Material properties of relatively homogeneous isotropic steel pipe.

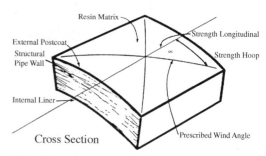

Fig. 4-2(d)   Cross-sectional view of RTR pipe.

and hoop strengths needed to meet installation requirements demanded by the project.

By now it should be apparent that, while both steel and RTR pipe are by definition flexible conduit, they are also quite different and therefore require different design approaches, even though initially at least their design considerations are identical. As with steel pipe, the RTR pipe designer must concern oneself with both pipe deflection and buckling analysis. Unlike the steel pipe designer, however, the RTR pipe designer must also examine a third area of concern.

The third factor is a combined strain analysis in both the hoop and longitudinal directions. This analysis demands a thorough examination of such important considerations as diametrical bending, internal pressure, temperature gradients, and the ability of the pipe to bridge voids in bedding. In such a design system the pipe is seen as part of a buried pipe system in the ground.

Stiffness and Strength

In simplest terms the design goal is to select the correct RTR pipe configuration for a specific application. In other words, we want to design a pipe wall structure of sufficient stiffness and strength to meet the combined loads that the pipe will experience over the long term. There are two primary ways to achieve correct pipe stiffness. One is to design a straight-wall pipe in which the wall thickness controls the stiffness of the pipe [Fig. 4-2(e)].

Another way is to design a rib-wall pipe on which reinforcement ribs of a specific shape and dimension are wound around the circumference of the pipe at precisely calculated intervals. The advantage of rib-wall pipe is that the nominal wall thickness of the pipe can be reduced while maintaining or even increasing its overall strength-to-weight ratio. Generally, a rib-wall pipe design is selected for applications where burial conditions are extreme or for difficult underwater installations. The ability to increase or maintain pipe stiffness by means of reinforcement ribs also provides the ability to design an RTR pipe system to fit the economic as well as mechanical parameters of a project.

Deflection Load

The next step in design is to determine the pipe deflection requirements, based on the equation shown in Fig. 4-2(f). The accepted maximum allowable pipe deflection should be no more than 5%.

This value is the basic standard that AWWA M-II specifies for steel conduit and pipe, as do the ASTM and ASME. As is obvious, there are a number of factors that contribute to pipe deflection. These are the external loads that will be imposed on the pipe, both the dead load of the overburden as well as the live loads of such things as wheel and rail traffic. The factors affecting RTR pipe deflection can be summarized as follows:

1. Design pipe deflection

2. Dead load-trench shape, overburden weight, depth of cover

3. Live load-wheel load, spacing surcharge

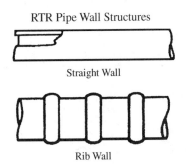

RTR Pipe Wall Structures

Straight Wall

Rib Wall

Fig. 4-2(e)   Example of RTR pipe wall structures.

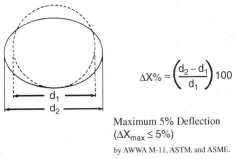

$$\Delta X\% = \left(\frac{d_2 - d_1}{d_1}\right) 100$$

Maximum 5% Deflection
($\Delta X_{max} \leq 5\%$)

by AWWA M-11, ASTM, and ASME.

Fig. 4-2(f)   How to calculate maximum allowable pipe deflection.

4. Modulus of soil reaction-native soils, type of backfill, differential soil stress and consolidation

5. Deflection lag factor

6. Bedding shape

7. Pipe stiffness (El)

8. Pipe radius

In terms of dead loads, the shape of the trench in which the pipe will be buried is also a factor. Generally speaking, a narrow trench with vertical sidewalls will impose less of a load on the pipe than will a wider trench with sloping side walls. It is necessary also to know the modulus of soil reaction (E), which is dependent on the type or classification of the native soil, the backfill material that is contemplated, and the desired consolidation of the backfill material. Soil consolidation is important, because it contributes to the strength of a flexible conduit in a buried pipe system.

If the designer is to do the job properly, it is important to have accurate data on which to base calculations. That is why test borings and proper laboratory analysis to determine the E value of the soil sample are essential. An arbitrary textbook selection of a soil modulus should always be avoided. However, if a pipe is to be buried deeper than the sampling zone that underwent laboratory testing to determine E and if the test bore shows the deeper material to be equal or better, then the designer may increase the E value proportionally to the square root of the differential soil stress.

Assuming that all the necessary data are available, determining the necessary pipe stiffness for the maximum allowable pipe deflection is relatively simple. The Spangler-Iowa equation provides a useful, reasonable determination of what wall structure will be needed (111).

Stiffness and Buckling

Assume for the purposes of this project that the calculations indicate that this pipe stiffness of 0.0365×10^6 lb-in^2/linear in. is in fact required to meet this deflection-design

criterion of no more than 5% deflection. Theoretically, we could choose a pipe-wall structure to meet this required stiffness of either a straight-wall pipe with a thickness of approximately 1.3 cm (0.50 in.), or a rib-wall pipe that would provide the same stiffness.

But would the wall structure selected be of sufficient stiffness to resist the buckling pressures of burial, or superimposed longitudinal loads? At this point we do not really know. To find out, we must know a few more things, one of being the amount of resistance to buckling that is wanted in the pipe. The ASME Section III Standard of a four-to-one safety factor on critical buckling, based on many years of field experience, should be used. To calculate the stiffness or wall thickness capable of meeting that design criterion one must know what anticipated external loads will occur (Fig. 4-2(g)). This time, in addition to the dead loads one must also consider the effects of possible flooding on both an empty and full pipe, as well as the vacuum load it is expected to carry.

The analysis should include the modulus of soil reaction, because in a buried RTR piping system the elastic medium surrounding the pipe helps increase the pipe's resistance to buckling. The formula into which all these factors can be inserted to determine the critical buckling pressure of the pipe is called the Luscher & Hoeg formula (44, 56).

It has been determined that, with burial depths greater than two thirds the radius of the pipe, this equation provides a means of determining the required pipe stiffness for critical buckling. To make the equation easier to use, it can be rewritten by substituting certain values and solving for the required stiffness for buckling. Suppose that this Luscher & Hoeg equation says we will require a pipe stiffness of 0.123×10^6 lb-in^2/lineal in. to

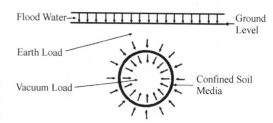

Fig. 4-2(g)   Buckling analysis.

meet a four-to-one critical buckling pressure safety factor.

This is a straight-wall thickness of approximately 1.9 cm (0.75 in.). But remember that we earlier calculated that a 1.3 cm (0.50 in.) thick wall would be sufficient to withstand the anticipated deflection pressure. Which of these two wall thicknesses is correct? Quite logically, it is the larger one of the 0.75 in. thickness, or a rib-wall pipe of equivalent stiffness. To put it another way, after carefully completing both a deflection and a buckling analysis, always select the pipe stiffness that is greater. Now if we were designing in steel pipe, the work would be about over. But since the design is a large diameter RTR buried piping system, we are not. From experience it has been learned that the final choice of an RTR pipe configuration cannot be made until the effects of strains in the longitudinal and hoop directions have been carefully investigated.

Anisotropic Behavior

The reason is obvious since the material, continuous glass-reinforced TS polyester plastic, is anisotropic. Unlike a homogeneous isotropic material, such as steel, the strength of RTR pipe in its longitudinal and hoop directions is not equal (Fig. 3-19). The effects of this unequal strength in the two directions must therefore be seriously considered during design if an RTR piping system is to meet the long-term operating requirements of the system being designed.

Therefore, before a final wall structure can be selected, it is necessary to conduct a combined strain analysis in both the longitudinal and hoop directions. This analysis will consider thermal contraction strains, the internal pressure, and the pipe's ability to bridge soft spots in the trench's bedding. In order to do this we must know more about the inherent properties of the material we are dealing with; that is a structure made up of successive layers of continuous filament-wound fiberglass strands embedded within a plastic matrix. We must know the modulus of the material in the longitudinal direction and the

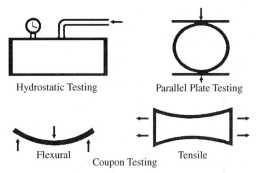

Fig. 4-2(h) Examples of using strain gauges to develop stress-strain curves.

hoop direction, plus the material's allowable strain.

These values are determinable through standard ASTM-type tests, including those for hydrostatic testing, parallel plate loading, coupon tests, and accelerated aging tests [Fig. 4-2(h)]. The next step is to examine Fig. 4-2(i), which shows the tensile stress-strain curve for typical steel-pipe materials. In the steel pipe business, designing is based on the curve's yield point. As noted previously, the yield point on a stress-strain curve beyond which steel pipe will enter into the range of plastic deformation could lead to a total collapse of the pipe. Generally to provide a safety factor, steel-pipe designers select an allowable design strain of approximately two thirds of the yield point.

Stress-Strain Curve

RTR pipe designers also use a stress-strain curve similar to that used by steel pipe

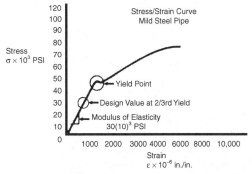

Fig. 4-2(i) Example of a tensile stress-strain curve for mild steel pipe material.

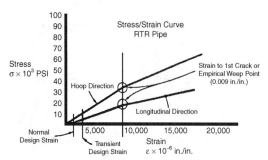

Fig. 4-2(j) Example of a tensile stress-strain curve for RTR.

designers. However, instead of a yield point, they use what is called an empirical weep point, or the point of first crack [Fig. 4-2(j)]. It is determined by either coupon or hydrostatic tests. The weep point is the point at which the matrix becomes excessively strained so that minute fractures begin to appear in the structural wall.

At this point it is probable that in time even a more elastic liner will be damaged and allow water or whatever else the pipe is carrying to ooze or weep through the wall. As is the case with the yield point of steel pipe, reaching the weep point is not cataclysmic. The pipe can still continue to withstand quite a bit of additional load before it reaches the point of ultimate strain and failure. Recognize that a more substantial, stronger liner can easily extend the weep point.

Weep Point

The weep point or strain-to-first-crack in a wall for filament-wound pipe constructed using isophthalic plastic is currently found to be not less than 0.009 in./in. This has been repeatedly demonstrated by careful coupon testing and burst testing of pipes with strain gauge instrumentation attached.

Thus, design values are based on the strain-to-first-crack or the empirical weep point. For normal design conditions a strain of 0.0018 in./in. is used, which provides a five-to-one safety factor. For transient design conditions a strain of 0.0030 in./in. is used, for a safety factor of approximately three to one. To those familiar with the design safety fac-

tors of other pipe manufacturers following (NBS/NIST) Voluntary Product Standard PS 15-69, these safety factors may seem modest. However, PS 15-69 is based on the ultimate tensile strength of the material (Chapter 2, **SAFETY FACTOR**).

The next step is to proceed with a strain or stress analysis in the longitudinal and hoop directions. When conducting this analysis the designer has the option to work in terms of either stress or strain. Strain is generally used, since it can be accurately measured using reliable strain gauges, whereas stresses have to be calculated. From a practical standpoint both the longitudinal and the hoop analysis determine the minimum structural wall thickness of the pipe. However, since the longitudinal strength of RTR pipe is less than it is in the hoop direction, it is wise to approach longitudinal analysis first.

In doing so there are three major factors to consider: the effects of internal pressure, the expected temperature gradients, and the ability of the pipe to bridge voids in the bedding. Analyzing these factors requires that several equations be superimposed, one on another. Even though from a practical standpoint all these longitudinal design conditions are solved simultaneously, it is interesting to examine each individually.

Poisson's Effect

Some tend to disregard the effects of internal pressure in the longitudinal direction on buried pipe because they theorize that the longitudinal load is cancelled out by the earth surrounding the pipe. Or they may assume that in a gasketed joined pipeline the gasket joints will allow the pipe to move freely, so that no longitudinal load will exist. However, this situation is not necessarily true. Poisson's ratio can have an influence (Chapter 2). This effect occurs when an open-ended cylinder is subjected to internal pressure. As the cylinder expands diametrically, it also attempts to shorten longitudinally. These movements will not be visible to the naked eye in all cases but can be easily measured with strain gauges. Or the movements can be observed in the shortening of pressurized pipe where a test fixture absorbs the pressure thrust.

Since a buried pipe movement is resisted by the surrounding soil, a tensile load is produced within the pipe. The internal longitudinal pressure load in the pipe is independent of the length of the pipe. Thus, Poisson's effect must be considered when designing any length of pipe, whether long or short that is part of a buried pipe system. Buried pipes are influenced by friction with their surrounding media.

Several equations can be used to calculate the result of Poisson's effect on the pipe in the longitudinal direction in terms of stress or strain. Equation provides a solution for a straight run of pipe in terms of strain. However, where there is a change in direction and thrust blocks are eliminated through the use of harness-welded joints, a different analysis is necessary. This is so because, compared to in straight runs of pipe, the longitudinal load imposed on either side of an elbow is greater. This increased load is the result of internal pressure, a temperature gradient, and/or a change in momentum of the fluid. Because of this increased load, the pipe joint and elbow thickness may have to be increased to avoid overstraining. There is a special equation, shown in Fig. 4-2(k) to calculate the longitudinal strain in pipe at harnessed elbows. For the sake of simplicity the effects of internal pressure, temperature gradient, and change in momentum of the fluid have been combined into one equation.

After examining the effects of internal pressure in the longitudinal direction, the next step is to investigate the longitudinal tensile loads generated by a temperature gradient in the piping system. The goal is to determine the extent of the tensile forces imposed on the pipe because of cooling. When an open-ended cylinder cools, it attempts to shorten longitudinally. A tensile load is then imposed by the resistance of the surrounding soil. As a matter of fact, any temperature change in the surrounding soil or medium that the pipe may be carrying can produce a tensile load. The effects of temperature gradient on pipe can be written in terms of strain.

In this analysis the designer must consider two conditions and base the pipe design on the one that is worse. One condition is where the temperature differential is one half the difference between the maximum temperature and minimum temperature. The second condition considers the temperature differential between the maximum pipeline temperature at installation and the minimum design temperature.

The next step in longitudinal analysis is to examine the bridging. Bridging can occur, and if so must be considered, wherever the bedding grade's elevation or the trench bed's bearing strength varies, when a pipe projects from a headwall, or, of course, in all subaqueous installations. It is a good practice to design the pipe to be strong enough to support the weight of its contents, itself, and its overburden while spanning a void of two pipe diameters [Fig. 4-2(l)].

Conservative Approach

For simplicity, the condition considers the conservative case where the pipe acts simply as a support. The normal practice is to solve all these equations simultaneously, then determine the minimum wall thickness that has strains equal to or less than the allowable design strain. Thus, the minimum structural wall thickness is dictated by the longitudinal tensile load.

The importance of combining longitudinal strain analyses is that it often provides the designer with a minimum wall thickness on which to base the ultimate choice of pipe configuration. For instance, assume that the combined longitudinal analysis indicates a

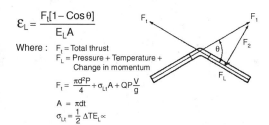

$$\varepsilon_L = \frac{F_t[1 - \cos\theta]}{E_L A}$$

Where : F_t = Total thrust
F_L = Pressure + Temperature + Change in momentum

$$F_t = \frac{\pi d^2 P}{4} + \sigma_{Lt} A + Q P \frac{V}{g}$$

$A = \pi dt$

$\sigma_{Lt} = \frac{1}{2} \Delta T E_L \propto$

Fig. 4-2(k) Example of longitudinal tensile strain with harnessed elbows.

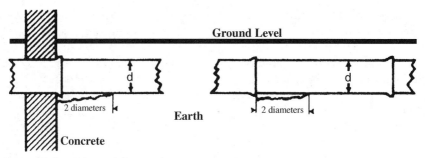

Fig. 4-2(l)   Example of the longitudinal tensile load on a pipe bridging a void of two pipe diameters.

minimum of 5/8 in. (1.59 cm) of wall thickness. However, deflection analysis calls for a 1/2 in. (1.27 cm) wall, and the buckling analysis says we need a 3/4 in. (1.9 cm) wall. We had already decided that the most likely candidate was the 3/4-in. wall. Now longitudinal analysis says that a 5/8 in. wall is enough to handle the longitudinal strains likely to be encountered. Which wall thickness, or what pipe configuration, straight wall or ribbed wall, do we now settle on? Economic considerations would have to be weighed, but the experienced designer would most likely choose the 3/4 in. straight-wall pipe. This is, of course, if the design analysis is complete, but it is not since there still remains strain analysis in the hoop direction. The target is to determine if the combined loads of internal pressure and diametrical bending deflection will exceed the allowable design strain [Fig. 4-2(m)]. This entails investigating the effects of rerounding or decreasing the diametrical deflection that occurs because of internal pressure [Fig. 4-2(n)].

There is a method that can be used for this analysis. It is extremely complex so it requires using a computer. In general, equations are generated to determine the moment and thrust created in the invert area of the deflected pipe, where a pressure term is superimposed. This analysis must examine the strains in the outer and innermost fibers of the pipe to verify that its wall structure is adequate and not overstrained. During this analysis the pipe must be examined under conditions of no pressure, minimum pressure, and maximum pressure.

Although this analysis should be conducted for both straight-walled and rib-walled pipe, it is particularly important in the case of rib walled. That is, because the rib is often thicker than the structural wall of the pipe, by several times the wall's thickness. Strains along the ribs may be higher than along the straight-walled sections, particularly at the top of the rib. For the sake of this discussion, assume that strain analysis in the hoop direction has confirmed that

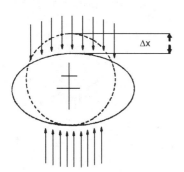

Fig. 4-2(m)   Example of strain analysis in the hoop direction with external load only; no internal pressure rerounding.

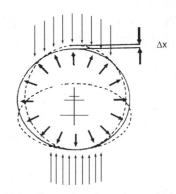

Fig. 4-2(n)   Example of strain analysis in the hoop direction with external load only plus internal pressure rerounding.

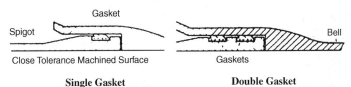

Single Gasket   **Double Gasket**

Fig. 4-2(o)   Example of using elastomeric bell-and-spigot seals.

a structural wall thickness of 3/4 in. is satisfactory. Does that mean the pipe design has been completed? Not yet. There's more to a piping system than just the pipe walls.

Pipe Joint

We must still design the joints to connect the straight lengths of pipe together. The designing of joints perhaps tends to be one of the most overlooked areas in any piping-system design. Since the performance of the whole piping system is directly related to the performance of the joints, this subject deserves serious attention.

For example, use a bell-and-spigot joint with an elastomeric seal [Fig. 4-2(o)]. This type of joint permits rapid assembly of a piping system and thus offers an economic advantage in terms of installed cost. It should be used as much as possible for connecting straight runs of pipe, especially at points where the pipe projects from a rigid structure. In terms of flexibility, the joint should be able to rotate at least two degrees without a loss of integrity. Thus, the threat of failure from unanticipated pipe subsidence is substantially reduced, and changes in the grade line during installation can be the more easily accommodated. The joint must also be designed with corresponding bell and spigot stiffnesses. And the spigot should have a special control ridge to ensure proper gasket seal, even when the pipes on either side of the joint are not uniformly supported.

The next type of joint is weld overlays, which are often utilized to eliminate the need for costly thrust blocks [Fig. 4-2(p)]. In designing the pipe an analysis was made to ensure that it possessed sufficient longitudinal strength. It makes sense, then, to make the weld joints be at least as strong as the longitudinal strength of the pipe rather than just as an internal pressure-seal pipe.

This can be done by noting the pipe's structural wall thickness and the strength relationship between the pipe and the overlay weld. These equations show one way of relating the structural wall thickness of a pipe to its longitudinal design's allowable values first in regard to the longitudinal strength of the pipe and its overlay laminate, to determine the proper thickness of the joint, and second to the longitudinal pipe strength and the weld overlay's bond strength.

Bearing

Similarly to gears, some plastic bearings have a long history of successful performance. The injection molded TPs that are considered for bearing applications are usually for relatively small, light duty work such as food mixers, adding machines, and similar devices. However small to large bearings also go into the heavy duty applications. Some TPs have inherent lubricating characteristics that can be enhanced by additives such as PTFE, molybdenum disulfide, and others. Other TPs, as well as TSs,

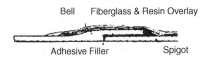

External Harness-Welded Joint   **Internal Harness-Welded Joint**

Fig. 4-2(p)   Joints with overlay on either side.

by the addition of PTFE and/or molybdenum disulfide, become excellent candidates for bearing materials (Chapter 4, **OTHER BEHAVIOR, Friction, Wear, and Hardness Property**).

The laminated (RP) fabric, bonded with phenolic plastic incorporating antifriction ingredients and cured under heat and pressure, gives excellent service when properly applied in various applications. This group of bearings has a low coefficient of friction, antiscoring properties, and adequate strength for use in steel mills and other heavy-duty applications and is well established in the industry.

PV Factor

When designing a plastic bearing, one must recognize that the primary cause for favorable performance will be keeping the frictional heat to a low value and having prevailing conditions that lead to dissipation of such generated heat. It has been established that the basic factors in a bearing system contributing to frictional heat are the pressure P exerted on the projected area of the bearing and the velocity V or the speed of the rotating member. This is known as the PV factor, and maintaining the values within the prescribed limits of each material will lead to a successful bearing.

The overall elements that contribute to the limiting of the PV factor are magnitude of pressure, speed of rotation, coefficient of friction of mating materials, lubrication, clearance between bearing and shaft, surrounding temperature, and surface finish, as well as hardness of the mating materials. Bearing wall thickness is also an element in the PV factor since it determines the heat dissipation.

By treating individually each of the above elements, the following occurs. The limit of the PV factor for each material or the internally lubricated materials for the constant wear of bearing is usually available from the supplier of the plastic. Neither the pressure nor the velocity should exceed a value of 1000. Thus, if the PV limit of acetal is 3000, the PV factor could be 1000 ft/min (300 m/min)

times 3 pounds or 1000 pounds times 3 ft/min at the extreme, provided heat conditions resulted in uniform rate of wear. The coefficient of friction data, available from suppliers, can also provide guidelines to the efficiency in comparing the different materials.

Lubrication whether incorporated as an RP material or provided by feeding the lubricant to the bearing will raise the PV limit 2.5 or more times over the dry system. The possibility of a rusting shaft should be guarded against in order to prevent excessive bearing wear.

Clearance between shaft and bearing will be large in comparison with, for example, a bronze bearing, mainly due to the high coefficient of expansion of these plastics-roughly up to 10 times that of steel. The average allowed clearance is 0.004 to 0.005 in./in. (0.010 to 0.013 cm/cm) of bearing diameter. This is needed in order to counteract the tendency to close in the bearing ID when temperatures rise or assembly conditions cause a decrease in the shaft hole.

The surface roughness of the shaft can play a large part in heat generation in the bearing. The protruding ridges from machining on the shaft act as minute cutting tools and disturb the smooth surface of the bearing thereby creating heat due to interference with the displaced material. A surface reading of 5 microinches is almost mandatory on the bearing portion of the shaft in order to ensure good life. Surface hardness of at least 300 Brinell on the shaft is another favorable feature for smooth operation. It prevents the pick-up of loose particles that can abrade the plastic material.

Wall thickness of the bearing should be on the low side to facilitate heat transfer into the housing, and yet it should be of sufficient value to facilitate effective manufacturing that will be conducive to a quality product. The thickness should be in proportion to shaft diameter, starting with 0.05 in. (1.27 cm) on the low side and ending with 0.2 in. (0.51 cm) on the high side. For filled materials, these thicknesses can be 50% higher. Bearing length should be equal to or less than 1.5 diameters. Concluding the remarks on plastic bearings, one must appreciate the numerous variables that enter into their design, and for that reason a prototype test is very much in order.

Gear

Gear design is one of the more complicated areas for designing with plastics (but understood with the extensive past half century experiences), because the bending, shear, rolling, and sliding stresses all act upon a mechanism whose purpose is to transmit uniform motion and power. In this age of lightweight and quieter operation, plastic gears have become increasingly important as a means of cutting cost, weight, and noise without reducing performance.

Because plastics are not as strong as steel, they must often perform far closer to their design limits than do metal gears (134). Although many plastic gear designs are derived from metal-gear technology, plastics demand special consideration, for instance to deal with heat buildup from hysteresis (Chapter 2).

Load Requirement

The basic difference between metal and plastic in gear design is that designs for metal are based on the strength of a single tooth, whereas plastic shares the load among the various gear teeth to spread it out. Thus, in plastics the allowable stress for a specific number of cycles to failure increases as the tooth size decreases to a pitch of about 48. Very little increase is seen above a 48 pitch, because of the effects of size and other considerations. The following guidelines for good gear design with TPs should be observed: (1) determine the gears' conditions of service, such as temperature, load, velocity, space, and environment; (2) establish the short-term plastic properties as against the initial performance requirements; (3) compare the long-term property retention factor as opposed to the life of the gear; (4) using physical property data, calculate the stress levels caused by the various loads and speeds; and (5) then compare these calculated values with the allowable stress levels and redesign as needed to provide an adequate safety factor.

Plastic gears fail for many of the same reasons as metal gears that include wear, scoring, plastic flow, pitting, fracture, creep, and fatigue. The causes of these failures are essentially the same. If a gear is lubricated, bending stress will be the most important parameter. Because nonlubricated gears may wear out before a tooth fails, contact stress is the prime factor in their design. Plastic gears usually have a full fillet radius at the tooth root, so they are not as prone to stress concentrations as are metal gears. The bending stress in engineering TPs is based on fatigue tests run at specific pitch-line velocities. A velocity factor should be used if the operating pitch-line velocity exceeds the test speed. Continuous lubrication can increase the allowable bending stress by a factor of at least 1.5.

As with bending stresses, calculating surface-contact stress requires using a number of correction factors. For example, a velocity factor is used when the pitch-line velocity exceeds the test velocity. A correction factor is also used to account for changes in operating temperature, gear materials, and the pressure angle. Stall torque, another important factor, could be considerably more than the normal loading torque.

Hysteresis Effect

At high speeds, plastic gears are also subject to hysteresis heating, which may be severe enough that they actually melt. Avoid this failure by designing the gear drive so that there is favorable thermal balance between the heat that is generated and that, which is removed by an inherent cooling process.

Reducing the rate of heat generation or increasing the rate of heat transfer will stabilize the gear's temperature so that they will run indefinitely until stopped by genuine fatigue failure. In such cases the wear resistance and durability of plastic gears makes them quite useful. Using unfilled engineering plastics usually gives them a fatigue life on an order of magnitude higher than metal gears.

Hysteresis heating in plastics can be reduced by several methods, the usual one being to reduce the peak stress by increasing the tooth root area available for torque transmission. Another way to reduce stress on the teeth is by increasing the gear's diameter. Peak stress can also be reduced by geometrically repositioning it using various conventional published gear theories.

Using stiffer plastics to reduce hysteresis provides other improvements. For example, the higher crystalline TPs like acetal and nylon (types of plastics extensively used) can be further increased by compounding techniques that can reinforce their stiffness by 25 to 50%. The most effective way to improve stiffness is through the use of fillers and reinforcements, particularly the high-stiffness fibers. Fillers and reinforcements are available that will also significantly increase heat transfer. The surrounding fluid, whether liquid or air, can have substantial cooling effects. A fluid like oil is at least ten times better at cooling than air. Agitating these mediums increases their cooling rates, particularly when employing a cooling heat exchanger.

Processing

When discussing plastic gears, one has to recognize that one is dealing with two basic materials. One type is a gear hobbed (cut) in a conventional manner from sheet blanks (the same as steel gears). The other type that represents practically all produced are injection molded into the required shapes. This type of gear has been in use for over at least a half century in electric power tools, instruments, meters, registers, windshield wiper mechanisms, steel mills, heavy duty machinery, or wherever the advantages of smoother operation, longer life, lower noise levels, lightweight, toughness, and lower costs existed. Timing gears for automobiles, for example, have been made on the same basic principle in very large volume over a period of many decades. This type of gear has established its reliability and is considered as a proven candidate for many applications.

Use is made of unfilled and filled or reinforced TPs.

Careful consideration of plastic material characteristics and its processing requirements can lead to gears with the enumerated benefits and, above all, to a more successful product performance. Considerable attention is given to the moldability of gears. One can make all the perfect calculations and insert the necessary values for plastic gears, but if molding conditions and molding materials are not compensated for to obtain a high-quality gear, one may end up with mediocre or even unsatisfactory results.

One of the first questions the designer faces in connection with a gear drive is to determine the pitch diameter of the pinion, and number of teeth required. Designs have to take into consideration backlash and working clearance. Backlash is defined as the measurement by which the space between teeth exceeds the thickness of the engaging tooth on the pitch circle. Backlash is necessary to prevent simultaneous contact on the two sides of the space and thus eliminates the possibility of binding. Backlash, tip relief, and similar arrangements are the means of providing satisfactory working clearance and thus minimizing excessive wear and noise.

The factors that cause a gear to mesh tightly are: (1) tolerance on concentricity of shaft hole with pitch diameter, (2) tolerance on center distance, (3) tolerance on quality, (4) coefficient of thermal expansion, and (5) change in dimensions due to moisture absorption, which is a consideration/detriment in some materials. The first three apply to gears of any material. Item 4, and in some cases Item 5, for plastic gears deserves special consideration.

The subject of transmitting motion and power by means of gears, their construction, and detail requirements are fully covered in textbooks, technical handbooks, and industrial literature of gear suppliers. The knowledge of gear fundamentals is a prerequisite for the understanding of where and how to insert the appropriate plastic behavioral information into the gear formulas so that the application results in favorable operation.

Gasket and Seal

Different plastics are used to fabricate gaskets and seats. With many applications, usually the chemical or heat resistance will suggest the choice of the plastic, but often it can be below the optimum in stress relaxation. As an example PTFE is used for being virtually inert and having outstanding high-temperature performance. PTFE is extremely vulnerable to creep and stress relaxation. However the many different filled grades permit eliminating or improving these type limitations; its use includes being subjected to severe environments.

There is a great range of plastics and elastomers to meet widely varying service requirements and all the types of geometric shapes and stress relaxation characteristics. Different tests have been set up to develop industry standards and test evaluations that should be directly useful in their applications. For example, a gasket relaxometer applies a compressive stress to a flat annular specimen, similar to the way many gaskets are stressed in service (ASTM F 38). This device is simple to operate, inexpensive, and capable of measuring the effects of such pertinent variables as stress relaxation in regard to time and environment.

The ability of a gasket to seal against leakage resulting from the pressure of a confined fluid is directly related to retained stress. High initial stresses often are required to be able to handle high pressures. In this example, however, high stresses serve to increase the tendency of a gasket to creep, thus requiring stronger and more expensive construction. The usefulness of stress relaxation data to the designer is that they provide a guide in arriving at the usually required suitable design compromise, without overdesigning. These data show that the thinner the gasket, the less stress relaxation occurs. In some material evaluations, stress relaxation can be correlated with geometric variables by means of a shape factor, as follows:

Shape factor

$$= \text{Annular area/total lateral area}$$

$$= (OD - ID)/4t \qquad (4\text{-}1)$$

where OD = outside diameter, ID = inside diameter, and t = thickness.

The trend of this factor is generally consistent with plastics' behaviors. However, stress-relaxation information has to be interrelated with the individual behavior of the plastics, as derived from the relaxation-test data (Chapter 2).

Grommet and Noise

To quiet a noise-generating mechanism, the first impulse is often to enclose it. Sometimes an enclosure is in fact the best solution, but not always. If it can be determined what is causing the noise, appropriate action can be taken to be more specific and provide a cost-effective fix. In some cases the problem is caused by a component such as a stepper motor or gear set that does not produce objectionable noise by itself. The trouble typically develops because a small noise is transmitted to a metal frame or cabinet that then serves to amplify the sound; using a plastic cabinet can isolate the noise problem.

Figure 4-3 is an example where in addition to reducing noise, the injection molded polyurethane (PUR) grommet (right) replaces five individual parts and saves time in assembling the lever linkage. During assembly it is snapped into a hole in the steel lever, then a grooved rod is inserted into the grommet. Intended to isolate vibration as well as connect metal parts, such a PUR grommet eliminates the hardening and cracking that use to shorten the life of the old assembly. This design might appear to be mechanically weaker than the cotter pin assembly, but it is at least as strong. The 1 in. OD grommets can withstand a 200 lb. (90 kg) pull on the rod without undergoing pull-out. In addition, the assembly can withstand a 100 lb. (45 kg) cyclic load (about 5,000 cycles at 60 cycles/min.) applied at 60 degrees off the rod axis at 300°F (149°C).

A cabinet that resonates can be quieted by damping its large flat areas so that they do not act like loudspeakers. Different approaches can be used, such as applying plastic foam

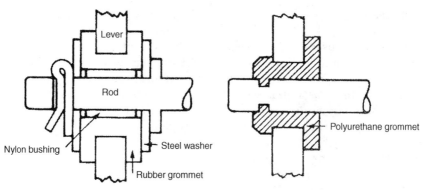

Fig. 4-3 Grommet designs.

sound insulator or, as reviewed, plastic panels, which have low damping characteristics. Various plastics have helped alleviate problems in all types of noisemakers, including rotating systems and hammer actions. One popular approach is to use plastic grommets where applicable.

In the past residential trash compactors were objectionably noisy. They were reduced to acceptable noise levels by redesigning them. Sound-absorbing grommets were used on the motors' bolt attachments and plastic gears were employed. Testing and all other types of equipment can take advantage of grommets or be redesigned to use plastic. Grommets provide their greatest noise reduction through damping in the octave frequency bands above 500 Hz where the ear is most sensitive and sound most annoying.

Electrical/Electronic Product

The early development of modern plastic materials (over a century) can be related to the electrical industry. The electronic and electrical industry continues to be not only one of the major areas for plastic applications, they are a necessity in many applications worldwide (2, 190). The main reasons is that plastic designed products are generally basically inexpensive, easily shaped, fast production dielectric materials with variable but controllable electrical properties, and in most cases the plastics are used because they are good insulators (Chapter 5, **ELECTRICAL PROPERTY**).

The function that plastics serve in most applications is that of a dielectric or insulator that separates two conductors with an electrical field between them. The field can be a steady direct current (DC) field or an alternating current (AC) field and the frequency range may vary such as from 0 to 10^{10} Hz.

The fact that plastics are good insulators does not mean that plastics are inert in an electrical field. They can in fact conduct electricity using certain plastics but more so by the addition of fillers such as carbon black and metallic flake (Fig. 4-4). The type and

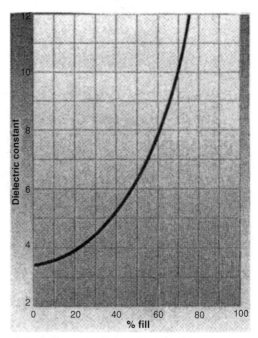

Fig. 4-4 Example of how additives or fillers provide a wide dielectric constant range.

degree of interaction depends on the polarity of the basic plastic material and the ability of an electrical field to produce ions that will cause current flows. In most applications for plastics, the intrinsic properties of the plastics are related to the performance under specific test conditions. The properties of interest are the dielectric strength, the dielectric constant at a range of frequencies (Fig. 4-4), the dielectric loss factor at a range of frequencies, the volume resistivity, the surface resistivity, and the arc resistance. The last three are sensitive to moisture content that may exist in certain materials. These properties are determined by the use of standardized tests such as those described by ASTM or UL. The properties of the plastics are temperature and/or moisture dependent as are many of their other properties. Temperature and/or moisture dependence must be recognized to avoid problems in electrical products made of plastics.

Examples of ASTM plastics insulation for wire and cable specifications are presented that relate to type plastic and specific field of application.

D 1047: Polyvinyl Chloride Jacket for Wire & Cable.

D 1351: Polyethylene Insulated Wire & Cable.

D 2219: Vinyl Chloride Plastic Insulation for Wire & Cable; 60 C Operation.

D 2220: Vinyl Chloride Plastic Insulation for Wire & Cable, 75 C Operation.

D 2308: Polyethylene Jacket for Electrical Insulated Wire & Cable.

D 2633: Thermoplastic Insulated and Jacketed Wire & Cable.

D 2770: Ozone-Resist Ethylene-Propylene Rubber Integral Insulation & Jacket for Wire & Cable.

D 2802: Ozone-Resistant Ethylene-Propylene Rubber Insulation for Wire & Cable.

Property

Electric currents can vary from fractions of a volt such as in communications signals to millions of volts in power systems. The currents carried by the conductor range from micro-amperes to millions of amperes. With this wide range of electrical conditions the types of plastic that can be used are different; no one plastic meets the different operating conditions. The selection of the materials and the configuration of the dielectric to perform under the different voltage, current, and frequency stresses are the primary design problem in electrical applications for plastics.

The primary function served by the dielectric or insulator is to separate the field-carrying conductors. This function can be served by air or vacuum, but these media do not offer any mechanical sup- port to the conductors. From this, the second function of the plastic insulator is derived. Since it is a mechanical support for the field-carrying conductors, the mechanical properties of the material are important.

The dielectric materials interact with the electrical fields and alter the characteristics of the electrical field. In some cases this is desirable and in others it is deleterious to the operation of the system and must be minimized. This is done by both the selection of the material and the configuration of the dielectric. To see how these concepts are applied, an example is presented of one of the major applications of plastics materials, i.e., to insulate wires, and show how a dielectric is designed to meet the service requirements. The specific requirements on a standard wire are:

1. The voltage between the conductors.

2. The current-carrying capacity.

3. The maximum operating temperature.

4. The frequency of the electric field.

5. The mechanical requirements on bending, etc.

6. Flame retardance.

The simplest wire configuration is a solid conductor with a sheath of insulation that might be flexible PVC or PE. If the wire is rated for 600 volts power frequency AC, the wall thickness would be about 0.020 to 0.030 in. (0.051 to 0.076 cm). The dielectric strength of the PVC would be about 300 volts per mil and the PE about 600 volts per mil so that the insulation would be more than adequate at room temperature.

Insulation Since the insulation value drops sharply with temperature, the wire would be limited in service temperature to 140°F (60°C), where both of these materials soften. The additional wall thickness above the theoretical minimum is used to give some mechanical strength to the insulation as well as to improve the resistance to cut through and bending. Since each of the conductors can handle 600 volts, it is possible to use two of the wires to handle 1200 volts. This is usually not done because of the possibility of grounding one of the conductors that would expose the other one to the full field.

The current-carrying capacity of the wire is not directly related to the dielectric. This is determined by the conductor resistance and the heating effect that it produces in the wire. The required current-carrying capacity determines the size of the wire and thus the size of the insulator. The temperature rise caused by the current flow determines the type of insulation to be used. If the wire is limited to 140°F (60°C) service, the insulation can be one of those discussed above. If the wire is to operate at 300°F (150°C), another specification for plastic wire with better heat resistance such as TP polyester or PTFE is used.

If the wire is to be used to carry much higher frequency currents, the design problem in geometry and plastic selection becomes more complicated. The dielectric constant and dielectric loss values for the plastics become important in the design. At a frequency of one megahertz the effect of the dielectric on the power transmission behavior of the wire is substantial and, even at frequencies of 10 to 100 kilohertz, the insulation on the wire must be considered in the design as a major electrical element in the circuit. More on the subject of insulation will be following this section.

Leakage resistance When dealing with low value currents, the leakage resistance of the insulation is also a major problem in the application of the wire. Such wire is used primarily in communications applications. The leakage of current from the wire is related to the volume resistivity of the dielectric material. In most plastics, the volume resistivity is high and in the case of the plastic most used in commercial communications wire, PE, the leakage is so low it causes no problems. When there is appreciable current leakage, the signal strength in the wire is reduced and noise from the environment is conducted into the wire to add to the loss of signal content (signal to noise ratio).

Dielectric constant/loss The value of the dielectric constant is important in the wire because of the effect that it has in coupling currents in one set of wires into another set of wires. The higher the dielectric constant, the higher the value capacitor that is formed between two wires. The capacitor thus formed is a signal carrying device at the frequencies used in communications and a signal can be capacitively coupled from one circuit to another. PE is the preferred choice for insulation of communication wire because of its low dielectric constant that minimizes the inter-circuit coupling effect usually referred to as cross-talk.

Dielectric loss The dielectric loss factor represents energy that is lost to the insulator as a result of its being subjected to alternating current (AC) fields. The effect is caused by the rotation of dipoles in the plastic structure and by the displacement effects in the plastic chain caused by the electrical fields. The frictional effects cause energy absorption and the effect is analogous to the mechanical hysteresis effects except that the motion of the material is field induced instead of mechanically induced.

Materials that have highly polar structures, permitting the field to have strong coupling to the plastic structure, have high loss factors particularly if they exhibit large viscoelastic behavior (Chapter 2). Materials with low polarity structures have a minor effect; combining a crystalline or crosslinked rigid molecular structure, show a low dielectric loss.

In signal processing applications the dielectric loss represents an attenuation of the signal. Where there are large amounts of power generated, such as in a radio transmitter, the dielectric loss represents a sufficient power drain that it will heat the material of the insulator and possibly destroy it. In both cases, it is important to minimize the amount of power dissipated into the dielectric material. This is done primarily by the use of plastics that have low dielectric loss factors and generally low dielectric constants. There is a relationship of dielectric loss factor as a function of frequency at two temperatures. The higher the temperature the higher the loss factor at all frequencies. This is due to the greater mobility of the plastic structure at higher temperatures permitting increased movement by the electrical fields.

The greater amount of frictional energy generated by the greater excursions possible at the higher temperature increases the dielectric loss. It should be pointed out that in the case of high power applications, this tendency produces an effect similar to that in the dynamic mechanical loading (Chapter 2) in that the heating produces an increase in the ability to be heated so that, if the heat is not dissipated, the material proceeds to catastrophic destruction.

The other approach to the reduction of the power loss to the dielectric material is by reducing the amount used. This is done by replacing part of the dielectric by air, an inert gas, or by vacuum. As examples there are three cable constructions in common use which employ these approaches to minimize dielectric loss. The first is the use of a foamed dielectric PS plastic that is commonly used in either twin lead transmission lines or in coaxial cables used for antenna lead-in wires in the UHF-TV antenna applications. The second system, which is illustrative of several sectional spacers, is used widely in communications cables of the coaxial type to minimize losses to the dielectric by reducing the amount of dielectric material in the cable.

Regarding the third system, the cable design must be modified to take into account the lower dielectric constant of the air that tends to increase the diameter of the cable so that it is not a simple replacement situation. The additional diameter will tend to increase the amount of plastic required so that an optimum must be reached in terms of the geometry to reduce the material to a minimum and still have a mechanically stable cable structure. The third scheme uses bead-like spacers at intervals along the cable (3). This type of cable is frequently evacuated to improve the dielectric performance of the cable.

Connector

The second major area for the use of plastics in electrical applications is at the terminations of the conductors. The connectors that are used to tie the wires into the equipment using the power, or used to connect the wires to the power source, are rigid members with spaced contacts. These are designed to connect with a mating unit and to the extension wires. The other type of wire termination is terminal boards where there are means to secure the ends of the wire leading to the equipment and the internal wiring in the equipment. These termination units require:

1. Adequate dielectric strength to resist the electric field between the conductors.

2. Good surface resistivity to prevent leakage of current across the surface of the material of the connector.

3. Good arc resistance to prevent permanent damage to the surface of the unit in case of an accidental arc over.

4. Good mechanical properties to permit accurate alignment of the connector elements so that the connectors can be mated properly.

If the connectors are to be used in high frequency applications, they must be made of plastics with low dielectric loss to avoid either damage to the part or signal loss in the circuit.

The design of a connector is a fairly straightforward process. It is easily illustrated with the example of a two-element connector of the type that may be used for either a power connection or to plug in an audio system. However multiple-element (32 and over) are extensively used. The plastic selected will be one that has the required dielectric strength at the maximum operating temperatures and at the frequency of intended use. For a power connector TP polyester, semi-rigid PVC, or phenolic are examples of what might be used. The anticipated voltage is used to calculate the probable leakage current which should be less that 0.1% of the magnitude of the current in the conductor.

The arc resistance of the plastic would not be critical for an appliance plug, since this condition rarely occurs in this application. For connectors used for industrial power connections, the plastic should have good arc resistance because of the possibility of flashover. The remainder of the problem would be to make sure that the connector is stiff enough to hold the contact members in alignment when the connector is inserted into the mating receptacle. The receptacle design is essentially the same as the plug except that it has the opposite set of contacting elements.

When the connector is used to make connection in an audio circuit, the configuration can be essentially the same. The additional consideration is that the material have low loss dielectric at the frequencies to be transmitted and that the spacing of the contact elements be determined by the transmission line characteristics required. Spacing is an electrical design function and its determination requires knowledge of the desired transmission line characteristics of the circuit. This part of the design is usually done by the electrical engineer and is an operating parameter for the plastic designer. The voltage resistance and other design factors are based on the data usually supplied for the material by the manufacturer.

In many electronic and electrical applications the internal wiring of the systems is done by high mechanical strength printed circuit methods. The substrates on which the printed wiring is done are usually plastics. Commonly used materials are epoxy-glass fiber RP laminates; chopped glass fiber-TS polyester injection molded boards, also to a lesser degree paper-based phenolic laminates. These internal wiring assemblies introduce a special design area in this application because of

basically good electrical properties and generally good chemical resistance to the chemicals and solvents used in processing the printed circuitry. They cannot be used in applications for high frequency circuitry because of the dielectric loss of the two materials. The phenolic is poor even at low RF frequencies and the epoxy has high loss factors at higher RF frequencies.

For these applications the printed circuit board materials used have included TS polyester-glass fiber, silicone-glass fiber, PTFE-glass fiber, different polyolefin-glass fiber, and glass-bonded mica. Materials such as glass-filled TS polyester plastic have good electrical and mechanical properties and with the contemporary wave soldering techniques it is possible to solder the boards without distortion.

The past introduction of TP materials into the area of printed circuit substrates has led to a broader type of application for the circuits. The products can be injection molded and the circuit applied to a molded part which will have a molded-in connector structure used to interconnect the device to the rest of the system. By combining the connector and substrate functions, it is possible to make very compact printed circuit units.

Such units are used in the watches with electronic drives instead of the traditional mechanical spring driven drives. Another area where the combination function is being used is in large-scale integrated circuit unit supports where the complex interconnection requirements make a combination circuit support and printed circuit unit an attractive way to achieve high packing efficiency such as computer hardware systems.

When using molded plastics parts such as the connectors and the circuit supports, it is important to make sure that the moldings, in addition to being made to close tolerances, be made under molding conditions that make for stable products. Electrical products are subjected to strong electrical fields in addition to the usual environmental abuse. Distortion of the product can lead to serious electrical malfunction by changing spacings that will alter electrical characteristics and if extreme enough, result in short circuits with serious results.

Designs for electrical products made of plastics should take into account the effect of the processing on the performance. In addition to possible distortion from heat, improper molding conditions can lead to premature failure from the effect of chemical agents if the product is used in an etched circuit application or from the effect of corona degradation if any product is used in a high voltage application. Conductor-to-insulator must be tight to eliminate corona.

Insulation

There are two areas of application for electrical insulation where the effects of long time exposure to electrical fields produces a fatigue effect that can be compared to the creep that occurs under static stress or the fatigue effects that occur under sustained dynamic loading (Chapter 2). The first effect is encountered in high voltage DC cables such as are used in X-ray equipment, some industrial and research equipment, and in the new high voltage DC distribution systems.

Dielectric break down and mechanical creep   Under sustained DC fields the plastic moves internally so that the dipoles align themselves in the field in response to the continuing voltage stress. As this continues, the dielectric constant of the material increases and the dipoles begin to break loose and migrate through the material. In doing so they disrupt the structure of the material, reducing the dielectric strength. After an extended period of time, usually several years, the dielectric will spontaneously break down and the system will arc over. This effect is similar to mechanical creep (Chapter 2) since the same sort of field based diffusion effects are at work to produce structural changes that take place.

The second effect is caused by the operation of alternating electrical fields on the dielectric in the system. This can happen to insulation at power frequencies as well as higher frequencies. This is not the dielectric

heating effect mentioned earlier, but an actual disruption of the plastic structure caused by the alternating stresses imposed by the alternating field. Some regions of the structure develop higher than average stresses and the plastic structure is broken at these points.

Dielectric break down and S-N analysis   As the number of defects grow with time, the structure becomes electrically less resistant to the imposed fields and the dielectric strength decreases to the point where the field arcs over. These effects occur after a period of years, usually in electrical insulation that is operating near the limit of its dielectric strength. The same type of S-N analysis that is used in mechanical fatigue is used in predicting this type of electrical fatigue (Chapter 2). It is essential that any materials used for this type of service be carefully evaluated for fatigue and are resistant to this effect.

Environment

In terms of environmental exposure, water and humidity must be carefully evaluated in electrical applications. In general, if a plastic absorbs a significant amount of water, the electrical resistivity drops. As examples this is the case for nylons and phenolic. Care must be used in selecting a dielectric to insure that the electrical properties such as the insulation resistance and dielectric strength, as well as other electrical properties are adequate under the conditions of field use, particularly if this involves exposure to high humidity conditions. Temperature also causes changes in most electrical products.

There is another type of condition that results from exposure to high humidity. The alteration in electrical properties caused by moisture absorption in nylon and phenolics is reversible. When the moisture content is decreased, the properties of the materials recover to close to the original values. In some instances the exposure to moisture and electrical fields can cause irreversible damage that can lead to failure.

One case is that of the TS polyester materials such as the alkyd molding compounds.

These materials, when exposed to continuous high humidity, especially in the presence of an electrical field, hydrolyze into the acid and alcohol precursors from which they are made. The acid plus water present make a conductive material that will cause the material to short the electrical circuit. The process by which the decomposition of the TS polyester takes place is very gradual at first and then accelerates so that extended testing of the material is necessary to be sure that the particular polyester composition used is resistant to hydrolytic degradation.

One other long-term condition that takes place with relatively low level DC fields in the presence of moisture is the migration of the metal of the conductor into the plastic. This was discovered to be a common thing in the past with silver conductors and phenolic insulators. The first instance of field failures were discovered in telephone equipment. The problem can occur with other metals with phenolic and also conceivably with other plastics that are moisture sensitive and can have a solvating action on the conductor metals that they contact. Most of these type plastics should be avoided inside hermetically sealed containers with movable contacts. Vapors released from the organic plastic deposit on the contacts to produce an insulation layer leading to contact failure.

Other peculiarities have occurred. As an example of a potential problem that can occur is when a silicone release agent is used when injection molding electrical connectors, etc. It can behave like the metal migration just reviewed.

Different Behavior

Capacitor There are several applications for plastics in electrical devices that use the intrinsic characteristics of the plastics for the effect on the electrical circuit. The most obvious of these is the use of plastics particularly in the form of thin films as the dielectric in capacitors. TP polyester films such as Mylar are especially useful for this type of application because of the high dielectric strength in conjunction with a good dielectric constant.

Mylar has the additional desirable feature that it is available in very thin films down to at least 2.5 microns.

Since the value of a capacitor is directly proportional to the area and inversely proportional to the spacing of the conductive plates, the thinner materials permit high values of capacitance in small size units. There are other materials that make good capacitors such as polyvinylidene fluoride that has a very high dielectric constant and good dielectric strength, oriented PS which makes a good capacitor for high frequencies because of its low dielectric loss constant, and others.

Electret Another application for plastics which uses the intrinsic properties is in electrets (a dielectric body in which a permanent state of electric polarization has been set up). Some materials such as highly polar plastics can be cooled from the melt under an intense electrical field and develop a permanent electrical field that is constantly on or constantly renewable.

These electret materials find a wide range of applications that vary from uses in electrostatic printing processes, to supplying static fields for electronic devices, to some specialized medical applications where it has been found that the field inhibits clotting in vivo. One application for the electret material is in a microphone that has a high degree of sensitivity and the electrical waves are produced by the field variations caused by the change in spacing of an electrode to an electret.

Structural binder A wide range of applications in electronics makes use of the plastics as a structural binder to hold active materials. For example, a plastic such as polyvinylidene fluoride is filled with an electroluminescent phosphor to form the dielectric element in electroluminescent lamps. Plastics are loaded with barium titanate and other high dielectric powders to make slugs for high K capacitors. The cores in high frequency transformers are made using iron and iron oxide powders bonded with a plastic and molded to form the magnetic core.

Magnetic recording tape, in addition to using plastic films as a support for the recording surface, also use polyvinyl alcohol and urethane plastics as binders for the magnetic oxides that form the recording medium. The range of special behavior materials that can be made from plastics is broad.

Electro-optic The liquid crystal plastics exhibit some of the properties of crystalline solids and still flow easily as liquids (Chapter 6). One group of these materials is based on low polymers with strong field interacting side chains. Using these materials, there has developed a field of electro-optic devices whose characteristics can be changed sharply by the application of an electric field.

Summary

A number of areas in which plastics are used in electrical and electronic design have been covered; there are many more. Examples include fiber optics, computer hardware and software, radomes for radar transmitters, sound transmitters, and appliances. Reviewed were the basic use and behavior for plastics as an insulator or as a dielectric material and applying design parameters. The effect of field intensity, frequency, environmental effects, temperature, and time were reviewed as part of the design process. Several special applications for plastics based on intrinsic properties of plastics materials were also reviewed.

Other areas such as static electricity and its use and control were not discussed since they represent a different type of application (2). As new materials became available and the electrical art continued to develop, the uses for plastics in electrical applications has increase both in the basic application as a dielectric and in special applications using the special intrinsic properties of the plastics.

Toy and Game

Extensive use is made in using different flexible to rigid plastics and processes to produce all kinds of toys and games. We see them all around us particularly through the advertising media (stores, TV, Internet, etc.). Extensive use is made in using different flexible to rigid plastics. They are designed to take a "beating" and survive within certain time periods.

Toys-Electronic

For the electronic component industry, different types of plastics and processes are extensively used. Not too evident is the high powered action of electronics in the plastic toy industry. The digital revolution has opened up a variety of new applications in "smart" microprocessor-based toys that use technology in innovative ways. Foremost player is the MIT Media Laboratory's Toys of Tomorrow (TOT) consortium that was organized in April 1998.

Members include Acer, Bandai America, Deutsche Telekom, Energizer, Intel, Disney, LEGO, Mattel, Motorola, Polar Electro Oy, TOMY, and the International Olympic Committee. This action is being taken since toys of the future may give birth to technologies that eventually end up in the workplace. They will be the first devices that carry new forms of networking into the home. They report that toys will lead the way to bring a home networking technology infrastructure faster than anything else.

Transparent and Optical Product

Overview

The use of plastics in certain transparent or optical applications is marked by selective but significant advantages of plastics over glass. Plastics weigh less and in many cases cost less, yet provide higher performance, such as impact strength and safety. They also have many more configuration possibilities to simplify assembly. There are the more expensive plastics with added performance features related to chemical resistance, heat resistance, high tensile and flexural strengths, and others that are used in specialty products (Chapter 5, **OPTICAL PROPERTY**).

The factor of configuration flexibility is particularly useful in systems that use aspheric or curved surfaces to simplify design and reducing the product count, weight, and cost. Moreover, there are light-transmission abilities of plastic optics that are comparable to those of high-grade crown glass. And from a safety standpoint, when plastics break they do not splinter, like glass, and thus are not hazardous or less hazardous.

Plastic's main disadvantages are its lower scratch resistance and, in some systems, comparative intolerance to severe temperature fluctuations. Even if plastic does have less temperature tolerance than glass, most optical systems do not operate in ambient temperatures beyond the thermal limits of plastics or the human body.

The processing advantages of plastics that exist, such as injection molding with multicavity molds, allows low-cost manufacturing to be combined with comparatively inexpensive materials resulting in low cost products used in automobiles, cameras, etc. By carefully sizing a mold for the required production volume, plastics' breakeven cost, compared to glass, will be very low. Another advantage of plastics fabrication is that in the mounting and assembly features like brackets, holes, and flanges, they can be molded integrally with the optical element to result in a single-piece design eliminating mounting hardware and simplifying alignment. Multiple elements can thus easily be combined and molded in unique optical configurations.

Property, Performance, and Product

There are plastics that are transparent and translucent in the unpigmented state. They have a range of optical properties that make them interesting for a wide spectrum of optical applications that extends from windows to lens systems to sophisticated applications involving action via polarized light. Used for over a half century are aircraft canopies (thermoformed) and windows in many different structures.

The application for plastics most widely known, based on the transparency and clarity of plastics, is their use as a window or cover. There are a number of relatively inexpensive plastics in the families of acrylics, polystyrene, cellulosics, and vinyls that have been widely used to make boxes for displaying merchandise, for windows in instruments, for glazing applications for buildings, outdoor signs, and so on. The primary optical property used in these applications is the transparency and lack of haze or light scatter.

A primary requisite for these materials is their mechanical, physical, and aging properties. Design of a box, a light, or a display unit will involve the requirements for static and/or dynamic loading that may be encountered in the end-use application. The only change is that the product is designed with a clear transparent material. (Recognize that one can say that glass has the highest compression strength of any material. However any other load, including a very minor load other than direct compression, will basically destroy it.)

One of the applications for plastics in optics is in refracting and reflecting elements where they are used as glass in lenses, prisms, mirror supports, and other refracting and reflecting units. The range of refractive index for plastics is generally in the same range as that for optical glasses. As a result, lenses having the same general properties as glass can be made from plastics such as the acrylics and polystyrene.

The ophthalmic applications for plastics lenses include contact lenses which are now made of acrylic plastics. Another material for this application is a special hydrophillic acrylic polymer used in soft contact lenses. These lenses are much more comfortable than rigid contact lenses.

Important applicable definitions that concern this subject follows:

1. Refractive index is the ratio of the velocity of light in free space to the velocity of light in the medium.

2. Light scattering is the ratio of the velocity of light in free space to the velocity of light in the medium.

3. Birefringence is the property of an anisotropic optical media that causes polarized light with one orientation to travel with

a different velocity than polarized light with another orientation.

4. Polarized light is light that has the electric field vector of all of the energy vibrating in the same plane. Looking into the end of a beam of polarized light one would see the electric field vectors as parallel or coincident lines.

5. Dichroism is a property of an optical material that causes light of some wave lengths to be absorbed when the incident light has its electric field vector in a particular orientation and not absorbed when the electric field vector has other orientations.

6. Light transmissability is the ratio of the light exiting from an optical material to the light entering the material.

7. Haze is the cloudy appearance in a plastic caused by inclusions that produce light scattering.

8. Color is the sum effect of the wavelengths of light transmitted by or reflected from a material.

9. Dispersion is a property of an optical material which causes some wavelengths of light to be transmitted through the material at different velocities and the velocity is a function of the wavelength. This causes each wavelength of light to have a different refractive index.

Lens

There are differences between plastic and glass lenses that the designer must considered. The first is that the plastics have a much greater change of dimension with temperature and a much greater change of optical constants with temperature. The other major difference is that, while the plastics are much more resistant to impact than glass, their resistance to scratching and to deformation is much lower than that of glass. However there are coatings and surface treatments that have been used for over a half century that significantly improves scratch resistance. In fact on the automotive design drawing boards, future cars are targeting to replace windows with these type plastics. As

a result of these limitations, most of the applications for plastics in optical elements are for low precision optics. Those products happen to be the major market for optical products.

In addition to the differences reviewed, plastics are different in another optical property from optical glass or crystalline optical materials. The degree of dispersion of light is much greater for plastics than for glass. Dispersion is the difference in refractive index for the different wavelengths of light and it is greater for plastics. As a result, a plastic prism will separate the different colors of the spectrum much more than a glass prism. This characteristic makes it more difficult to make lenses of plastics without fringing colors. Despite this limitation, good camera lenses corrected enough to use for color photography have been made out of plastics for low cost cameras. It is possible to design around the limitations of the plastic materials when the cost advantage justifies the additional design effort.

Plastics are the preferred optical materials used in lenses for controlling the light in warning lamps such as emergency lights, stop lights on cars, and retro-reflective lenses used on cars, on high way signs to show the presence of an obstacle, etc. These lenses are usually made with a large number of specially shaped lenticulations that are used to direct the incident or transmitted light in a direction where it can be readily seen. The lenticulations may be pyramidal or they may be spherical sections. Both can be designed as excellent light directors. The taillight lens on automobiles probably represents one of the largest applications of optical plastics and the retro-reflector element used both on cars. Highway posts is another major use of optical plastics.

Fresnel Lens

There are other groups of optical elements that use plastics in very fine patterns to make special optical elements. One of these is the Fresnel lens that is a collapsed lens structure that has the effect of a strong magnifier but is essentially a flat sheet. The lens is made by a special molding technique from a carefully machined master. It is used as a focusing lens for light sources, as an intensity-leveling unit in reflex camera viewers, and as a coarse view magnifier of simple objects. (Compact discs molding of polycarbonates are another example of precision molding of grooves.)

Large Fresnel lenses are used in solar furnaces to gather large areas of sunlight and to focus it at a point to achieve high temperatures. The other fine pattern application for plastics is in replica diffraction gratings. When a pattern of lines is made with a count of 50 to 500 lines per millimeter it acts as a diffraction grating which can break light into its components by selective interference. Diffraction gratings are made by ruling lines on a metal or glass plate by means of a ruling engine that is a tedious and expensive process.

The gratings can be replicated by using a plastic material applied to the surface of the grating that takes the pattern of the grating and reproduces it in the plastic. This is usually done with a curable plastic solution and at low temperature to avoid damage to the grating. Using this technique, it is possible to make low cost gratings that can be used for light analysis or for displays, or even to make an interesting form of iridescent jewelry.

Lenticular

A lenticular is a tiny lens or a groove on a screen. The term lenticular image refers to a specially constructed graphic viewed through a plastic sheet extruded with a series of mathematically calculated grooves or lenticules, that give the image depth, movement, or both. To achieve a 3-D effect, for example, several angles of one image are recorded. This is usually done by a digital camera or created right on a computer.

Interlacing then takes place, a precise digital merging by computer of the angles into a master image. When printed on a lenticular sheet such as Eastman Chemical's Spectar PETG, the grooves of the sheet force the

eye to view different sections of the images at the same time, thus rendering a 3-D effect (235).

To get high resolution the sheets are extruded with a precalculated number of lenticules per inch, or LPI. This varies by format. A hand held graphic needs high resolution and is usually extruded with 75 LPI. A large display such as 4×8 ft (1.2×2.4 cm) viewed from a distance will require as few as 15 LPI. The extruded lenticular formats are predetermined and designed for various imaging and viewing applications. To create the sheet, different pattern-roll cylinders are mounted on the extrusion line, each capable of engraving the precise LPI required for an application.

The extrusion line must be capable of running the sheet in a highly controlled and precise environment, to make sure the lenticules are accurately placed (6). Extrusion must also be a clean process in order to prevent image-altering contamination. Sheets are usually extruded in a thickness range of 0.015 to 0.100 in. (0.038 to 0.254 cm).

Piping Light

The high degree of clarity and low haze of some plastics, particularly acrylics, makes possible the use of these materials in applications using the light piping effect. Any light that enters the end of a long rod of a clear transparent refractive medium at angles less than the critical angle (defined as the angle at which light is refracted parallel to the surface) is trapped in the rod and transmitted down the length of the rod by multiple internal reflection. The trapped light is reemitted at the end of the rod or plate, or at any place along the length of the "light pipe," where the surface is changed in angle or where the surface is roughened to form a light diffusing area.

Use is made of materials such as acrylic that are very clear. They have very little light absorption in the visible spectrum, and have a very low haze level to scatter the light and change direction. The light can be piped over distances of the order of three to four meters with a minimum of light attenuation.

This effect has been used in several areas of plastics product design. One application is in the illumination of instrument dials and similar indicia where it is impractical to use lamps close to the indicia. The lamps can be placed in a convenient location and the light piped to the indicia where the surfaces are shaped to release the light. Another application is for lights that must be inserted in confined spaces where suitable bright light sources either do not fit or are too hot to use. An example of such an application is the light used in medical devices for examination of a patient's throat, etc.

The effect is widely used in signs and display devices to make them self illuminated. Edge lighted signs and panels are widely used in offices, automobiles, aircraft, etc. A sheet of acrylic material has light introduced into one edge from suitable lamps. The light is carried across the sheet. The indicia that are to be displayed are cut into the surface opposite the side from which the sign is to be viewed. The indicia can be either polished angle cuts or an area that is roughened to be a diffusing surface. In either case, the light piped through the sheet is altered in direction, emitted from the sheet toward the viewer, and appears as a self-illuminated sign.

Fiber Optic

These devices are also used in the transmission of light to form images as well as to transmit small areas of light. An optic fiber consists of a small diameter monofilament of a clear plastic such as acrylic that has a thin coaxial layer of another clear material of lower refractive index. The monofilaments are in the order of 10 to 50 microns in diameter and the coating is usually 5 to 10% of the diameter. These fibers are very efficient light piping elements as a result of the coatings. A bundle of optic fibers can transmit light over distances of at least 10 to 20 meters with a low degree of attenuation.

Random bundles of fiber optics make a good medium for bringing light to a specific region for illumination. A coherent bundle of fibers (one in which the fibers are aligned so that they occupy the same position everywhere along the length of the bundle) can be used to transmit images over long distances, around corners, and past other obstacles. The image can be viewed directly by looking into the end of the bundle. When the object is properly illuminated it can be seen directly. In other cases the object is imaged on one end of the fiber optic bundle and observed at the other end as a clear image as if it were the focal plane of the lens.

This feature of the fiber optics is used in a number of optical systems for remote viewing. One major application has been for cyctoscopes in medical diagnosis and for borescopes used to examine inaccessible areas in machinery such as plasticator barrels. Coherent bundles of fibers, properly transposed, are used as an encoding and decoding device to handle confidential image information. Short bundles of the fibers are used in conjunction with cathode ray tubes and other self-illuminated displays to improve visual contrast by minimizing the effect of ambient light on the display. Other applications for fiber optics are for decorative effects and for optical and photographic applications where the ability to transmit an image or to alter it in a predescribed manner simplifies the system.

In using fiber optics the designer is mainly concerned with a standardized material which has specific characteristics in terms of optical performance. Fiber optics made of plastics can be affected by exposure to the environment with deterioration of performance. Heat is an important environmental factor and the most likely cause of damage in optical applications. The heat can be generated by the light sources used. Some of the infrared generated by light sources can be removed with the use of appropriate filters.

Polarized Lighting

There are a number of applications for plastics in optics which involve the interaction with polarized light. Polarized light is distinguished from ordinary light that is called incoherent light. Incoherent light has wavelengths varying over a range of values, the phase of the individual wave trains do not have any phase relationship with each other, and the plane of the electric and magnetic vectors of the individual wave trains have random orientation with respect to each other. Looking into a beam of incoherent light, the electric vector can have any angle.

When the light is plane polarized, all of the light which passes through the polarizing device has the electric vectors parallel and the magnetic vectors parallel to each other. The effect of the polarizer can be likened to a picket fence through which you can attempt to make wave trains with a rope or string. The only waves that will come through are those with the plane of vibration parallel to the fence pickets.

There are several ways in which light can be plane polarized. In some light sources because of the effect of strong electric and/or magnetic field present when the light is generated, it is polarized (A). When light is reflected at low angles from a dielectric medium which is transparent, the reflected light is polarized parallel to the surface and the transmitted light is polarized perpendicular to the surface (C). Naturally birefringent (double refracting) materials can be made into prisms which pass light polarized in one plane and reflect out of the prism light polarized in the plane at right angles (D). The method most generally used for generating polarized light is by passing incoherent light through a polarizing filter having the property of absorbing light which is polarized in the principal absorbing plane of the material and passing through polarized light whose electric vector plane is perpendicular to the absorbing plane (B).

The dichroic polarizer widely utilized to produce polarized light employs plastics in the dichroic materials with the polarizing capability. It consists of a plastic with the molecules oriented strongly in the direction desired for polarization (Chapter 8, **PROCESSING AND PROPERTIES, Orientation**). The plastic has attached color absorbing structures along its length. When light passes across the plastic molecules perpendicular to the length of the molecule, there is a minimum of interaction with the color ab-

sorbing centers. When the light passes across the plastic chain parallel to the chain, there is a high degree of light absorption. With a substantial thickness of the oriented material, the light that passes through is plane polarized in the plane perpendicular to the orientation direction.

Examples of such dichroic polarizers are polyvinyl alcohol that is oriented. It is oriented with iodine absorbed on the alcohol side groups and polyvinylene which is made from oriented polyvinyl alcohol by heating to a temperature which causes splitting off of water to form unsaturation along the plastic chain. There is also multiple conjugate unsaturation in an organic molecule produces a light absorbing structure. In addition to the two examples given, it is possible to make absorption type dichroic polarizers by other techniques such as attaching dichroic dyes on to oriented plastic chains which have an affinity for the dye. Most dichroic polarizers pass about 45% of the incident energy through as plane polarized light.

Application One of the simplest and most interesting applications for polarizers in optical applications is the use of two polarizers to control the amount of light passing through the pair. If two polarizers are used serially on a light beam, the orientation of the second polarizer to the first will determine the amount of light passed. If the planes of polarization are parallel, then all of the light passed by the first polarizer will be passed by the second polarizer. If the planes of polarization are at right angles for the two polarizers then no light is passed. At angles between, differing amounts of light are passed based on the angular relation of the two polarizers to each other.

Two crossed polarizers are frequently used to inspect transparent materials placed between them for optical activity, either for birefringence or for optical rotary effects. Birefringence effects are produced by materials with a regular ordered structure that allows light to pass through at one orientation at a higher velocity than at another orientation. As a result of this, the two wave trains generated by the different velocities cause the phase angle of the light beam to change so that the light exits from the birefringent material with a different optical orientation than the entering light. Usually this is at an angle where part of the light can pass through the second polarizer. In many cases the emerging light is elliptically or circularly polarized instead of plane polarized so that part of the beam is absorbed in the second polarizer. Light which is elliptically or circularly polarized has the electric vector varying in value and angle along the wave train and it behaves similarly to incoherent light on passage through a plane polarizer (Chapter 5, **STRESS ANALYSIS**).

Optical rotary effects are also referred to as optical activity and are caused by optically assymetric groups in the structure of the material. A typical optically active group would be a carbon atom which has a different organic group attached at each position on the molecule. Cellulose and sugar have this type of structure. In this case the beam of polarized light has its plane rotated as it passes through the structure and the light emerges as plane polarized light with a different plane of polarization. By rotating the second polarizer's plane of polarization it is possible to find the exit angle and pass all of the light.

The crossed polarizer effects of both types are used in analysis work. The concentration of optically active organic materials is determined by the degree of rotation. In plastic processing the residual strains in molded materials as well as the degree of orientation of polymers is determined by the effect on polarized light. Crossed polarizers are used with special wave plates to control the amount of light that passes through an optical system.

Another application for the crossed polarizers is in electrically modulating the strength of a light beam. Electric fields have the effect of making certain substances variably birefringent. The most important of these materials are the liquid crystal materials (Chapter 6). Many liquid crystals are low polymers with highly polar side chains. By varying the field on the material the birefringence is varied and the light transmission is controlled. This effect is used in optical devices and has application in communications

systems, especially those using lasers as a signal source.

Laser Lighting

Plastics such as acrylic are used to make one type of laser material. When the appropriate luminescent dyes are incorporated into the material, the acrylics can be made into large laser units. One advantage of using acrylics is that very large clear castings can be made. As a result, large amounts of laser light can be produced at low volume densities of light. Consequently, heating effects are at a minimum. Optical systems can be used to concentrate the light from the large laser elements.

Color Filter

Plastics are suitable for most optical applications that utilize transparent materials, including color carriers. Color filters have all types of standard transmission characteristics that can be made and, because of the uniqueness of the plastic structure, a large number of dichroic and trichroic materials are possible that have different colors when viewed from different angles. One application for this is in polarizing filters.

An interesting application is in sunglasses where the tinting effect is combined with the polarizing effect to get sunglasses that are particularly effective against low angle glare. The lens materials are polarized in the vertical plane and the low angle ground reflections are polarized in the horizontal plane so that the glare light is strongly absorbed in the lenses. An antifogging coating on a plastic lens soaks up condensed moisture.

Processing

One of the major advantages of plastics optical elements is that they can me made by different processes such as injection molding, casting, or extrusion to good accuracy at low cost. Some glass lenses are pressed from hot glass but the majority of the lenses by far are ground and polished from rough blanks. Glass lenses are not molded because it is difficult to get good surfaces and because the residual stress left in the glass as a result of the pressing operation affect the optical properties of the lenses. In the case of plastics it is possible to get excellent surface quality with no stress using precision molds and appropriate molding methods and fabricating procedures.

The plastics may have stresses generated (less than glass) in the materials as well as flow orientation caused by the molding process. In the case of plastics the effect on these conditions are easily corrected by process controls (Chapter 8). When required fabricating processes such as injection-compression or injection molding a blank followed with a separate compression molding into the final lens (Chapter 8). Consequently, there are a large number of magnifiers, lenses, prisms, opthalmic, and others that are injection molded or cast from plastics.

Lenses for safety glasses are made from highly impact resistant plastics such as modified acrylics and polycarbonate. They will resist puncture from flying objects and offer the exceptional eye protection. They can be molded to prescription requirements.

Allyl diglycol carbonate (CR-39) is the most highly scratch resistant of the transparent plastics. Unlike most of the other transparent plastics that are TP, CR-39 is a TS plastic that has been in use for over a half century in applications such as bullet-proof shields, high temperature steel blast furnace eye and face guards, aircraft window side panels, etc. Processing methods include injection molding and casting.

To be successful, molded optical elements of plastics must be produced with careful control of the fabricating process. In the case of these optical products it is particularly important that the molding conditions be carefully controlled to minimize molded-in stress. In addition to these stresses reducing the dimensional stability of the products leading to distorted images, the stresses themselves affect the quality of the image. This is a result of the fact that the stresses/strained areas have a different refractive index from that of the unstrained areas with resultant distortion of the image.

Most optical products are injection molded in special molds in a dry atmosphere (including dried plastic material) that sustain packing pressure on the product as it cools to give good surface quality. In addition, the process controls on the molding machines are of the best type to insure close control over the melt temperature and pressure. Static mixers at the end of the IMM plasticator can be employed to eliminate thermal gradients in the material (3). Schemes for quality control use the optical image quality as a test, and use polarized light inspection methods to check for residual molding stress and orientation (Chapter 8, **INJECTION MOLDING**).

Designing

The following is a summary to the design information presented. The design of optical elements from plastics follows conventional optical design procedures that are covered in many texts on lens and prism design. The basic principles of design are based on ray tracing to determine the focal point of a lens or the image distance for a lens system. For the geometry of a suitable optical element it is necessary to use a text on optical design literature such as handbook of optics (101) and/or the services of an optical designer. The aspects of the design are first the selection of the appropriate plastic and then the design of the product and process to make an accurate stress free product.

Within the limitations on the physical properties which generally restrict plastics to low precision optics, plastics materials have found wide applications in optical products that range from lights to binders for electroluminescent phosphors to fiber optics and lasers. They represent an easily worked material with a wide range of desirable optical properties in simple to complex shapes. In this review the discussion has been limited to the differences between plastics and optical glass materials and to some of the unique design possibilities that are especially important for plastics. Using the optical arts and the special properties of plastics, unique products are attainable. The breakage resistance of the plastics has placed them into many common functional applications such as instrument windows and street lighting globes.

Packaging

The packaging industry and its technology is the major outlet for plastics where it consumes about 30 wt% of all plastics with sales at about $40 billion. Saleswise about 30% is HDPE, 16% LLDPE, 14% TP polyester, 13% PP, 11% LDPE, and 16% others. Different products and processing techniques are used to produce many different packaging designed products. These different products show how innovative designs have created different packages based on plastic behaviors and they're processing capabilities. Most of these products are extruded film and sheet. Other processes are used with thermoforming, injection molding and blow molding being the other principal types used (Chapter 8). The following information is examples of different designing packaging products with performance requirements.

Aseptic

In food processing, it is a process condition that renders a processed food product essentially free of microorganisms capable of growing in the food in un-refrigerated distribution and storage conditions. The aseptic food packaging include film pouches and pre-sterilized molded containers that are filled with aseptic foods, then hermetically sealed in a commercially sterile atmosphere.

Bag-in-Box

BIB refers to a sealed, sprouted (small too large, with or without dispensing valves) plastic film bag inside a molded rigid container, generally for packaging liquid products. Outer box may be made of disposable corrugated cardboard, disposable or reusable

plastic or wood, etc. The BIBs offer space and cost efficiencies. The empty BIBs in a warehouse may occupy as little as 20% of the space required for the equivalent volume in glass or rigid plastic, reduce shipping weight, etc.

Beverage Can

While aluminum cans dominate the USA market for soft drink containers with about 70% of the market, PET and glass are in second place. Note that most aluminum cans have an inside coating, usually epoxy, to protect its contents from the aluminum.

Biological Substance

Many of these substances are classified as hazardous requiring specialty packaging where plastics play an important role to meet strict requirements.

Blister

Also called blister carded packaging. It is a package in which thin plastics film or sheet is formed so that a product is placed in the blister, backed up by a material (plastic, paper, aluminum, etc.), and sealed.

Bubble Pack

Very popular is plastic cushioning material used in packaging, usually laminated thermoplastic films that incorporate air bubble pockets.

Clasp

Plastic molded clasps are commonly used to secure tops and lids on different type packages. Their holding power can be obtained from simple friction between the joint surfaces or from positive mechanical engagement. Plastic flexing capability is important because the products must flex in order to release the clasp and spring back to their original position.

Contour

Contour packaging is also called skin packaging. It is package in which thin plastic film or sheet is formed over a product and usually simultaneously sealed to a paperboard or plastic backing. The product serves as a mold.

Dual-Ovenable Tray

DOT is used for frozen foods. They have been a major outlet for plastics. In the past thermoset plastic materials were used but it practically all went to thermoplastics such as CPET.

Electronic

Plastic ease of processing with low cost have given them a wide application in solving problems in electronic packaging. They range from inexpensive consumer devices to sophisticated expensive computer systems and cellular phones.

The demand for antistatic plastics in electronics is of major importance. The size reduction of components and the higher density packing of components on computer chips make the devices more susceptible to static damage. In many cases the environment can be adjusted through increased humidity, antistatic mats, or ionizing sprays to control the problem. During transportation and storage static protection is critical. Components must be protected from stray electric fields such as electric motors and discharges that can destroy microcircuits. Method of protection is to pack them in antistatic plastic film bags, etc.

Film Breathable

A major factor contributing to the growth of this category is refinements of breath-

able films that are identified as controlled-atmospheric packaging (CAP). They substantially extend the shelf life of perishables by regulating oxygen, CO_2, and moisture permeability. This type packaging has extended the global food trade. There are different types for use in food, horticulture, medicine, etc. It was introduced in 1994 by researchers in Brunel University, Oxbridge, UK. It has basically a two-plastic layer structure containing small holes that open and close as the temperature changes.

The principle is the same as that used in bimetallic strips (thermocouples, etc.) taking advantage of differences in coefficients of linear thermal expansion. As the temperature rises, the edges of the holes peel away. Typically PE laminated to TP polyester with an acrylic adhesive is ideal for packaging vegetables and fruits. Structure can be modified to meet a wide range of respiratory rates. Other applications include medical dressings for burns, drug controlled release-delivery systems, variable vapor barriers for shoes and clothing, temperature sensitive warning labels, etc.

Food

If plastic packaging were not used, the amount of packaging contents (food, etc.) discarded from USA households would more than double. Plastics are the most efficient packaging materials due to their higher product-to-package ratio as compared to other materials. One ounce of plastic packaging can hold about 34 ounces of product. A comparison of product delivered per ounce of packaging material shows 34.0 plastics, 21.7 aluminum, 6.9 paper, 5.6 steel, and 1.8 glass. USA food container (annual $14 billion business) consumption by major materials is 38wt% paperboard, 28% plastic, 26% metal, and 8% glass.

Most packaged foods require a barrier against gases, flavors, or odors to maintain product quality and provide acceptable shelf life. Baked foods usually need moisture protection, while fresh meats and vegetables require low or controlled exposure to oxygen to maximize shelf life and consumer appeal. Polyethylene films, both single and multi-layer, are widely used to package such products. Certain key properties and processing conditions affect the permeability of PE. As an example with blown film factors such as the blow-up ratio, frost line height, and die gap width can be used to control permeability (Chapter 8).

Food, Oxygen Scavenger

It is impregnated plastics with chemically reactive additives that absorb oxygen, ethyl, and other agents of spoilage inside the package once it has been sealed. Food package types include flexible pouches, PET bottles, microwave trays, and cartons (coat plastic and paper). When such films are combined with conventional rigid or flexible barrier systems, they can allow food packagers to greatly retard spoilage and prolong shelf life.

Grocery Bag

Polyethylene sack or T-shirt bags are stronger than paper bags, take up 30% less storage space, provide water and puncture resistance, recyclable, etc.

Hot Fill

Thin wall plastics are used to hot fill (injection and blow molded bottles, thermoformed containers, etc.) without sagging during filling and maintaining mechanical properties such as impact strength and stiffness in temperatures from at least $-40°F$ to $250°F$ ($-40°C$ to $120°C$). Plastic used includes special grades of PEN, PET, PP, PS, and PVC.

Loose Fill

They use principally polyethylene and polystyrene foam in different shapes such as peanuts and pretzels.

Modified-Atmosphere

Modified atmosphere packaging (MAT) is a packaging method that uses special mixtures of gases (carbon dioxide, nitrogen, oxygen, or their combinations) and polyolefin blown or cast film barrier films to change the ambient atmosphere influencing its food content. Film construction includes polyolefin plastomers (POP, EVOH/PVDC, ULDPE, LLDPE, PP, SB, LDPE/PA/EVOH, and PVDC/coated PET/LDPE). The hermetically sealed MAT extends the shelf life of red meat, skinless turkey breast, chicken, half-baked bread, pizza's crust, bagels, etc. and allows them to be presented in a more palatable manner.

As an example, microbial growth of red meat is retarded maintaining a deoxygenated blue or gray coloring until the meat is placed on display. Ground beef normally may last 3 to 4 days but with MAT can go to 14 days before the sell-by date. This gas-flushed MAT permits grocers to sell uncooked fresh meats, marinated varieties, and ready-made meals for quick preparation for round-the-clock sales. This more expensive (but with reduced costs in service) and sophisticated packaging concept of using a gas flushed barrier film is not new compared to the basic traditional method.

Peelable Film

Peelable film in case-ready ground beef package add color and shelf life. As an example Cryovac Div. of W. R. GRACE & CO uses a peelable barrier lid and foam tray system. It is two packages in one; there is an oxygen barrier structure, which is peeled off, leaving an oxygen-permeable film over the meat.

*Pouch Heat-Sealed, Wrap,
and Reusable Container*

These packaging systems help keep food fresh and free of contamination. Thus the resources that went into producing the food was not wasted.

Retortable Pouch

Also called retortable container or flexible packaging. It has superior flavor retention and longer shelf life are principal advantages of single or laminated plastic (such as PP, PC, PEI, CPET, and EVOH), with or without aluminum foil, paper, etc. They are made impervious to light, air, and most other gases, microbial organisms, water, and most other liquids, etc. Many applications require them to be capable of withstanding temperatures of at least 250°F (121°C) for at least 20 minutes. Different processing methods are used such as blow molding and thermoforming. The USA converted flexible packaging industry consumes about 7 billion lb (900 kg) of which 75% is plastics, 22% is paper, and 3% is aluminum foil.

Shrink Wrap Tunnel

An oven in the form of a tunnel mounted over or containing a continuous conveyor belt is used to shrink oriented films in the shrink packaging process. Also use a hot air blower on the film to provide the heat required in specific areas.

Container Content Misrepresentation

When you use metric, such as in USA, be sure they are correct. Unwitting errors with upper and lower case letters can be dramatically misrepresenting package contents. As an example, there is a drastic difference between 100 ML and 100 ml or 100 million liters and 100 milliliters. Some designers prefer using capital letters for content designations on the label. Thus, with metric usage fluid ounces are replaced by milliliters; quarts are replaced by liters; pounds by dry ounces are replaced by grams or kilograms, and so on. Metric unit short forms are called symbols, not abbreviations; therefore they never should be followed by a period unless the symbol is the last word in a sentence. There must always be a space between digits and symbols. Metric symbols are always used in singular form, without adding an "s" to indicate quantity.

Permeability

The ability of a plastic to protect and preserve products in storage and distribution depends in part upon the diffusion (i.e., transport) of gases, vapors, and other low-molecular-weight species through the materials. A substance's tendency to diffuse through the plastic bulk phase is its diffusivity or diffusion coefficient. The rate of diffusion is related to the resistance, within the plastic wall, to the movement of gases and vapors. Two important aspects of the transport process are permeability and the migration of additives. Possible migrants from plastics can include residual monomers, low molecular-weight polymers, catalyst residues, plasticizers, antioxidants, antistatic agents, chain transfer agents, light stabilizers, FR (fire resistant) agents, polymerization inhibitors, reaction products, decomposition products, lubricants and slip agents, colorants, blowing agents, residual solvents, and others.

An important selection of materials to packaging, particularly food, is based on the permeability of the materials to oxygen, water vapor, and, in the case of packaging bananas, to ethylene gas that is used to artificially ripen the bananas. Selective permeability provides chemical separations, one of the most interesting of which is the use of PTFE materials to separate the hexafluorides of the different isotopes of uranium.

There are a number of industrial gas separation systems that use the selective permeability of plastics to separate the constituents. In design problems relating to such applications, the designer must consider the environmental conditions to determine whether the materials having the desired properties will withstand the temperatures and physical and chemical stresses of the application. Frequently the application will call for elevated temperatures and pressures. In the case of uranium separation, the extreme corrosivity of the fluorine compounds precluded the use of any material but PTFE. The PTFE material, however, requires careful design to make a sturdy membrane because of the poor mechanical properties of the PTFE plastics.

Basics The driving force for gases and vapors penetrating or diffusing through, for example, permeable packages is the concentration difference between environments inside and outside the package. A diffusing substance's transmission rate is expressed by mathematical equations commonly called Fick's First and Second Laws of Diffusion:

$$F = -D(dC/dX) \qquad (4\text{-}2)$$

$$dC/dt = D(d^2C/dX^2) \qquad (4\text{-}3)$$

where F = flux (the rate of transfer of a diffusing substance per unit area), D = diffusion coefficient, C = concentration of diffusing substance, t = time, and X = space coordinate measured normal to the section.

To measure gas and water vapor permeability, a film sample is mounted between two chambers of a permeability cell. One chamber holds the gas or vapor to be used as the permeant. The permeant then diffuses through the film into a second chamber, where a detection method such as infrared spectroscopy, a manometric, gravimetric, or coulometric method; isotopic counting; or gas-liquid chromatography provides a quantitative measurement (2). The measurement depends on the specific permeant and the sensitivity required.

Three general test procedures used to measure the permeability of plastic films are the absolute pressure method, the isostatic method, and the quasi-isostatic method. The absolute pressure method (ASTM D 1434, Gas Transmission Rate of Plastic Film and Sheeting) is used when no gas other than the permeant in question is present. Between the two chambers a pressure differential provides the driving force for permeation. Here the change in pressure on the volume of the low-pressure chamber measures the permeation rate.

With the isostatic method, the pressure in each chamber is held constant by keeping both chambers at atmospheric pressure. In the case of gas permeability measurement, there must again be a difference in permeant partial pressure or a concentration gradient between the two cell chambers. The gas that has permeated through the film into the lower-concentration chamber is then conveyed to a gas-specific sensor or detector by a carrier gas, for quantitation. Commercially available isostatic testing equipment has been used extensively for measuring the oxygen and carbon dioxide permeability of both plastic films and complete packages.

The quasi-isostatic method is a variation of the isostatic method. In this case at least one chamber is completely closed, and there is no connection with atmospheric pressure. However, there must be a difference in penetrant partial pressure or a concentration gradient between the two cell chambers. The concentration of permeant gas or vapor that has permeated through into the lower-concentration chamber can be quantified by a technique such as gas chromatography (2).

Three related methods based on the quasi-isostatic method are used to measure permeability. The most commonly used technique allows the permeant gas or vapor to flow continuously through one chamber of the permeability cell. The gas or vapor permeates through the sample and is accumulated in the lower-concentration chamber. At predetermined time intervals, aliquots are withdrawn from the lower cell chamber for analysis. The total quantity of accumulated permeant is then determined and plotted as a function of time. The slope of the linear portion of the transmission-rate profile is related to the sample's permeability.

Permeability and barrier resistance In the past, the usual materials used to contain food, gasoline, chemicals, perfumes, medication, and many other items that keep them from permeating or being contaminated were metal and glass. For over a century, however, plastic containers have been entering the arena of packaging. At first only certain plastics could be used, which were usually rather thick or heavy compared with what is used today. There have been various plastics that could provide permeability protection.

With the growth of plastic use in containers and packages, requirements to make them more compatible or useful resulted in new developments occuring and continue to occur. The two major approaches for providing permeability resistance in plastic containers involve chemically modifying the plastics' surfaces and, more important from a marketing standpoint, the use of barrier plastics with nonbarrier types to meet cost-to-performance requirements. This is achieved through coextrusion, coinjection, corotation, and other such processes (Chapter 8).

Chemically modifying a plastic's surface during or after fabrication permits controlling the permeation behavior of such products as diaphragms, film, and containers. These techniques are becoming increasingly important. There is an endless search for better barrier materials for packaging applications. As an example in blow-molded gasoline containers/tanks, the amount of gasoline permeation through HDPE even though it is very low, is still excessive, thus has required some type of barrier. Including a barrier in a multilayer construction can create such a barrier. Another approach is functionalized PE formed on the inside of the container wall by a chemical reaction, mostly sulfonation or fluorination.

There is also oxifluorination that is a process in which fluorine gas is thinned with nitrogen to which several percent of oxygen by volume have been added. Subjecting PE to fluorine and oxygen at the same time leads to functionalization of the PE, making it impermeable. This technique permits substantially reducing the required amount of fluorine, resulting in a cost-to-performance improvement.

Barrier plastics using oxifluorination are widely used for foods. With these, barriers are needed to protect them against spoilage from oxidation, moisture loss or gain, and changes or losses in favor, aroma, or color. Most plastics can be considered barrier types to some degree, but as barrier properties are maximized in one area (as the gases such as O_2, N_2, or CO_2), such other properties as permeability and moisture resistance diminish.

Product

Practically all markets use some type of packaging. Examples include the following:

1. Packaging (food, medical devices, egg cartons, dairy containers, meat and produce trays, electronic devices, tools, dinnerware, picnic dishes, drinking cups, lids, etc.)

2. Refrigeration (inner door liners, food compartments, crisper trays, etc.)

3. Appliances (housings for electrical, mechanical, chemical, etc.)

4. Signs and displays (point of purchase, interior and external, etc.)

5. Automotive (interior and exterior parts, air ducts, crash pads, arm rests, etc.)

6. Industrial (tote boxes, many different shaped machine and other device housings, etc.)

7. Military (aircraft canopies, contour maps, etc.)

Others include building and construction, cosmetics, dental, drugs, electrical and electronics, furniture, aerospace, agriculture, horticulture, industrial, mechanical, medical, public transportation, recreation, toys, and so on.

Building

Overview

The usually reported second largest market for plastics is building and construction consuming about 20 wt%. However, the amount of plastics is only about 5% of all materials consumed in building and construction so that a large growth area exists for plastics when the price is right since their properties provide durability, performances, insulation, cosmetics, etc. (Fig. 4-5). Different plastics are used that include PVCs, PEs, PMMA, PSs, phenolics, TS polyesters, and many more. Examples of products are listed in Table 4-1.

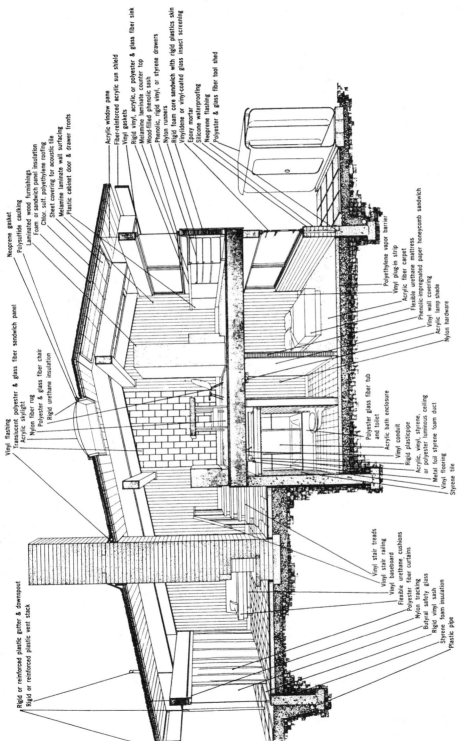

Fig. 4-5 Concept of plastics in buildings continues to expand.

Table 4-1   Applications of plastics in buildings

Exterior	
Adhesives	Topping-walk,
Air support	driveways
structures	Vent stacks
Air vents	Water proofings
Cables	Weather strippings
Caulkings	Window panes
Coating-metal, wood	Window sash (prime
Concrete forms	and storm)
Concrete mixes	Wire insulations
Curtain walls	
Doors (prime and	**Interior**
storm)	Acoustical panels
Expansion joints	Adhesives
Facings	Baseboards
Flashings	Cabinets
Gaskets	Ceilings
Glazings	Conduits
Grilles	Counter tops
Hardwares	Coverings
Illuminating panels	Decorative panels
Lighting fixtures	Drawers
Louvers	Ducts
Moisture barriers	Electrical fixtures
Mortar mixes	Floorings
Paints	Gaskets
Panels	Graphic arts
Pipes	Grilles
Railings	Hardwares
Rain system-gutters,	Insulations
downspout, etc.	Light diffusers
Roof edging, panels	Molding, trims
Safety and thermal	Paints
glasses	Panelings
Screens	Partitions
Sealants	Pipe fittings
Sheathings	Plaster backings
Shingles	Plumbing fixtures
Shutters	Railings
Sidings	Sealants
Signs	Shower stalls
Skylights	Stair treads
Stuccos	Tanks
Sun shields	Tile-floor, wall,
Swimming pools	ceilings
Tapes	Vapor barriers
Tool sheds	Wall coverings
	Wire insulations

Application and the Environment

The present and growing large market for plastics in building construction is principally due to its suitability in different environ-ments. The versatility of different plastics to exist in different environments perhaps may be related to another characteristic; namely, ability to be maintenance-free when compared to the more conventional and older materials. This section will review the different parameters that are important in building construction and are related to different environments.

From a practical review, perhaps it can be stated that buildings and construction materials are exposed to the most severe environments on earth, particularity when the long time factor is included. The environments include such conditions as temperature, ultraviolet, wind, snow, corrosion, hail, wear and tear, etc. Basically the following inherent potentials continue to be realized in different plastics: ease of maintenance, light weight, flexibility of component design, combine with other materials, corrosion/abrasion/weather resistance, variety of colors and decorative appearance, multiplicity of form, ease of fabrication by mass production techniques, and total cost advantages (combinations of base materials, manufacture and installation).

Success in applying plastics has been based on a combination of factors; such as, adequate testing, keeping up-to-date on customer problems, product identifications, quality control establishment of engineering standards, approval of regulatory agencies, supervise installations, accurate cost and time estimations, organizational responsibilities defined, meeting delivery schedules, development of proper marketing and sales approaches, resolution of profit potential based on careful selection of applications, and acknowledge competition exists.

The functional attributes that permits its growth at an accelerated rate are reliability, acceptability, feasibility and economics. Field installations of the new products are now providing more of the necessary reliable long time data. The field tests continue to be the best approach in demonstrating acceptance.

The obstacles, limitations or disadvantages confronting acceptance of plastics are:

1. Service life versus legal risk: The architect and builder can appreciate the limited

10 to 20 year service test results but rarely can appreciate the more abundant zero to two year or accelerated weathering test results. Their thinking is that plastic companies inherit their portion of risk (latter in this Chapter review **RISK** and **DESIGNING AND LEGAL MATTER**).

2. Properties: Fire safety continues to be a major performance requirement. Creep and heat distortion are other important properties to be considered. A major deterrent for the architect and builder is lack of common knowledge about plastics physical properties.

3. Cost: In no business is there more resistance to increase cost even if it represents true increase value.

4. Codes: The code problems are usually over emphasized. It is a recognized fact that obstacles do exist and many "heated" debates are already on the books and will continue to be on the books. The codes are important to society and must be recognized in plans and development programs for plastic building products. There are many examples of approvals such as pipe since 1965 as well as previous acceptance of paneling in 1959 by FHA on Sinclair-Koppers Company expandable polystyrene beads (EPSs) faced with asbestos cement or plywood. Recognize that practically the rest of the world accepted and used plastic pipe and other plastic building materials since the late 1940s.

In USA the major obstacle to using plastics in the past has been to convert standards and codes to include the use of plastics. There were standards and codes where plastic would meet all their requirements with "flying" colors except when they specifically stated that the material to be used had to be iron, steel, or other material (not plastics). Eventually (years passed) and plastics were included. So plastics could not be used until other materials such as plastics would be included or no specific material was specified. Outside USA the changes in most cases were immediately particularly immediately after 1945.

5. Competition: In line with the so-called sportsmanship approach or competitive business behavior, the entrenched steel, wood,

concrete and other industries would logically continue to resist plastics acceptance and use all humanly available resources to fight (restrict or eliminate) plastics. This situation existed for almost a century when plastics were entering into a competitive product. Eventually things changed including where the competing (old time) companies usually enter the plastics industry to make the products.

6. Aesthetics: The trend is to resist change in appearance so that plastic originally had to look like something else. With time passing the beauty of plastic was accepted and at the same time usually at lower costs and more benefits developed.

7. Identification: To the nonplastic user and even certain plastic users, identification of over 35,000 plastics tends to be either misleading or confusing even though this situation should never exist. As explained in this book, certain specific plastics meet certain product requirements.

8. Standards: Most buildings have never been subjected to thorough engineering analyses. Traditional precedent and judgment as reviewed under Codes have (logically) controlled the standards. In fact if wood had never been used, it would be difficult to approve (burns, rots, etc.); of course it is ridiculous to assume that wood would not be approved.

9. Performance data: Rather than make available principally sales type of data, the architect should review realistic and understandable technical data. They should learn that the behavior of certain plastics provide for more cost-efficient products rather than just be contented with only past history on the materials of construction.

10. Consumer: General demands for traditional materials continue to exist but it is gradually changing since more plastic products are all around them.

11. Labor: They generally sets-up problems but education on use of new plastics has been helpful. An example is when plastic pipes were approved for use in buildings where labor initially was against their use because it interfered by reducing labor hours, etc.

In the meantime ideas for using plastic in building continues to vary in all proportions (Table 4-1). Designers produced spray-foamed homes that were reviewed during the October 1965 annual National Decorative and Design Show in New York City. It was described that entire rooms and furnishings molded-in-place would be both practical, appealing, and survive the environment. During the 1950s different military groups built different foam structures, etc. During this time period all structures (wall, roof, etc.) of buildings were successfully built from extruded PVC hollowed sandwich-ribbed panel structures that were unfilled and filled with insulation material (PUR foam, etc.), concrete, or other materials (34–37, 151). Also building structures were made from extruded PS foam logs that were heat-bonded wrapped in dome shaped structures (14), RP filament wound room structures, and so on.

The Architect Approach

Breakdown in communications between the building industry and plastic manufacturer probably accounts for a large amount of lost motion and dollars. Perhaps another major cause is the pure sales approach within any industry that, in many cases, can delay technical progress.

Architects and builders desire factual data on products. However, their standards or codes in many cases only identify the composition of the end item, with no performance data. In most cases, the plastics were never subjected to engineering analysis but approved based on past performances. This situation is not new to the engineering community, where patience, time and/or money resolve the problem.

It has been stated that architecture as a profession has often stood in the way of progress. Being a learned profession, it generally looked to the past for knowledge and inspiration. Although most people have an inherent resistance to change, there are always enlightened architects, builders, and consumers who have the courage to overcome traditional beliefs and the foresight to anticipate new developmental trends (14).

Architects and builders do exist who foresee plastics as a major building material and represent a means to provide extending the building block or modular construction concept. The versatility of plastics permits developing single units containing water piping, electrical conduit, heating elements and other services. This building block approach has been used successfully in other industries, but, in most cases, developed due to government or military requirements. Many architects also see that plastics make possible the housewife's dream of a dust-proof construction.

The architect continues to look for products that can he multifunctional. As an example in roofing, the product could perform a part or all of the functions. The roof has to provide structural integrity, temperature and sound insulation, vapor and moisture control, weather resistance, elastic qualities for change in weather, fire protection, aesthetic appeal, and so on.

What the architect looks for in any new material has been expressed. The first interest concerns the continuing search for practical, aesthetically pleasing, and economically priced buildings with long service life. The materials or products are to provide new or better solutions to the myriad problems plaguing the construction industry. The second is that a complete and accurate account of each new material exists. The material is to fit within the manufacturing and installation practices of the present construction industry. The third is the assurance that architects are not left holding the bag when new materials do not perform satisfactorily. The economic situation as to who should be responsible represents a major problem area since legal suits can occur and architects lose prestige.

House of the Future

One of the first all plastic house was the Monsanto House of the Future erected in Disneyland, CA, USA in 1957 (Fig. 4-6). The key structural components were four

(a)

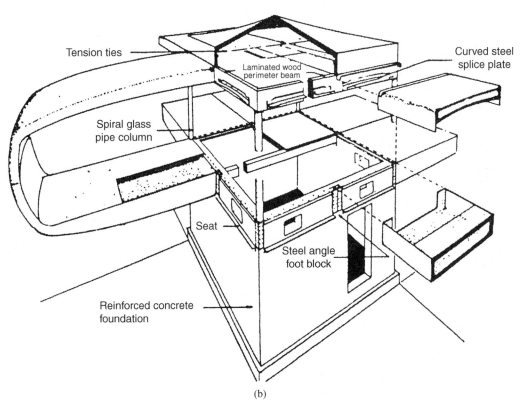

Tension ties

Laminated wood
perimeter beam

Curved steel
splice plate

Spiral glass
pipe column

Seat

Steel angle
foot block

Reinforced concrete
foundation

(b)

Fig. 4-6   House of the future structure: (a) view of one section, (b) design layout, and (c) cantilever support beam.

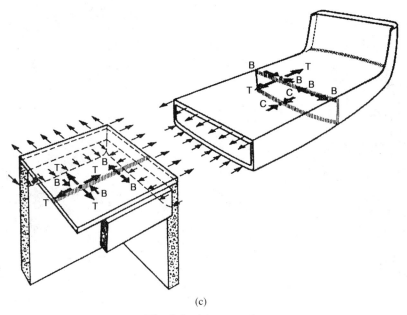

(c)

Fig. 4-6   *(Continued)*

16 ft (4.9 m) U-shaped cantilever (mono-coque box girders) reinforced plastic designs by MIT. Different plastics were used throughout the house including different plastic sandwich panels.

When this house was to be removed to provide a different scene (a main attraction for two decades), it had suffered almost no change in deflection. It was estimated to have been subjected to winds, earthquakes, subjected to families using it to the equivalent of centuries based on all the people that passed through it, etc. Destruction by conventional techniques (wrecking ball, etc.) was impossible without first cutting sections, etc.

Designing a Structure

The following example provides information on designing of plastic structural products to take static loads. It is a structural problem common to a number of different structures to show how the different structural requirements will affect the choice architectural designers has to make. The design problem will be a roof section which may be used for anything from a work shed,

to a house, to a vehicle, or even to a simple weather shelter (190).

The analysis begins with a definition of the function that a roof performs. A roof is the overhead product of a structure intended to protect the occupants and/or contents of the structure from the outside environment. It involves rain, snow, wind, sun, falling objects, hurricanes, and the other elements that make up the outside or surrounding environment. In order to perform this function the roof must be capable of supporting its own weight and the weight of snow or any other possible accumulations on the roof. It must be resistant to wind loads that are quite severe in some regions. The roof must also support loads imposed by people walking on it, usually for maintenance. In some instances the roof may double as a deck and the traffic may be constant.

The roof must be able to shed water that falls on it, although it need not be waterproof in the sense of being a waterproof membrane structure. The roof surface is exposed to sun, wind and driven debris and must be resistant to erosion by the action of sunlight and the abrasive action of wind driven debris. In most cases the roof is insulated thermally to prevent heat loss in cold weather and heat input during warm weather. Obviously, not all these roof requirements apply to all roof situations, but most of them do so you can set up your own requirements.

The major load applied to a roof is the static load of the roof structure itself. Since roofs come in a wide variety of types the self load will depend on the basic roof design. The simplest is the corrugated RP panel structure. This type of structural element is widely used for roofs on industrial buildings to admit daylight, porch and patio roofs, shelter roofs such as those used at bus stops, and a variety of similar applications. Variations of this simple roof are used for roof sections on transportation vehicles such as buses and trains. Since this section is one of the easier ones on which to describe loading conditions, it will be used to illustrate the design procedure. Other roof sections such as the domes, arches, geodesics, and paraboloids involve complicated stress analysis and the results would not be particularly useful in a general analysis of a static structure.

The corrugated materials are available in sheets which vary from 4 ft × 8 ft to as large as 10 ft × 20 ft. A typical material is 0.100 in. thick with 2 in. corrugations, and a corrugation depth of 1 in. The RP material from which they are made is glass fiber mat as the reinforcement and a weather-resistant TS polyester plastic. In general, the sheet material is nailed or screwed to wooden supports (could be pultruded RP supports if the price was right) at proper intervals (Chapter 8). In some cases the roof section is made in one piece with spars of TS polyester-glass material molded into the product to provide the stiffening support needed. In this case the only requirement for installation involves anchoring the edge of the section to the structure.

This type of design problem is somewhat different from others in that the unit is made from standardized sections that have specific physical properties and are available in only a limited number of thicknesses and configurations. The design problem now consists of trying the available materials in the structure with the supports that can be used and then determining if the material will perform. The self load is easily determined from the weight of the materials. The snow load is a design value available from experience obtained in the area where the structure is to be used. Similarly, the maximum wind load and people load can be determined from experience factors that are generally known.

The problem is worked out using several different sheet types and different support spacings in an environment that would be typical of a city in the Midwestern part of the USA. The indicated solution is that the material selected will take the required loads without severe sagging for a 15 year period with no danger that the structure will collapse due to excessive stress on the material. If a standard material had not been suitable, it would have been possible to use one specifically molded for the application, or by the use of several layers of the material. One typical way in which excessive loading for a single section is handled is to bond two layers of the corrugated panel together with the corrugations crossed. This results in a very stiff section capable of substantially greater weight bearing than a single sheet and it will meet the necessary requirements. The double sheet material also provides significant thermal insulation because of the trapped air space between the sheets particularly if they are edged sealed.

The roof section was designed to meet the static load requirements. However, it is necessary to consider transient loads such as people walking on the roof and fluctuating wind loads. The localized loads represented by people walking on the roof can be solved by assuming concentrated loads at various locations and by doing a short time solution to the bending problem and the extreme fiber stress condition. The local bearing loads and the localized shear should also be examined since they may cause possible local damage to the structure.

Stresses from varying winds are general alternating stress loads and occur over wide areas of the structure. When the wind changes direction, the stress frequently changes direction, and the tendency is for the roof to lift away from the structure. The main point of stress caused by the wind is at the anchorage points of the roof to the rest of the structure. They should be designed to take lifting forces as well as bearing forces; the lower the angle of roof, the less wind lifting force. Proper anchorage of the support structure to the ground is also essential. Certain local fire and building codes impose additional restrictions.

A large area of plastics, such as described here, has a change in dimension with temperature. Surprisingly, very few of the traditional building materials, including wood, have significant expansion under normal temperature shifts. The RP materials generally are not a problem since they have low linear thermal coefficients of expansion and the corrugated shape tends to flex and accommodate the changes caused by heating and cooling. In the case of materials such as vinyl siding, the expansion factor becomes significant and is an important consideration in the fastening system.

The effects of the environment on the performance of the material must be considered. Using the initial physical properties of the materials, the structure is sound. Exposure to weather, which includes water and sunlight, has a significant effect on the physical properties of the materials and this must be taken into account in the design. This type data is available from the reliable panel producers. Let us assume that there is a 50% or more drop in the physical properties in 5 years; actually far less. This can be due to surface damage and to changes in the bulk of the material. In general, this type of loss of physical properties levels off to a low rate of deterioration in suitable materials so that any potential failure can be anticipated. This loss of properties can be compensated for by increasing the strength requirements by a suitable safety factor (Chapter 2), probably about three in this case, and by using a protective coating on the sheet material to minimize the effects of weathering. The preferred type of coating would be a fluorocarbon material that has the best resistance to sunlight and other weathering factors of all of the plastics. If this type of surfacing is used, the material will retain its surface integrity for at least 20 years.

The example of the roof structure represents the simplest type of problem in static loading in that the loads are clearly long term and well defined. Creep effects can be easily predicted and the structure can be designed with a sufficiently large safety factor to avoid the probability of failure.

Chair

A seating application is a more complicated static load problem than the building example just reviewed because of the loading situation. The self load on a chair seat is a small fraction of the normal load and can be neglected in the design. The loads are applied for relatively short periods of time of the order of 1 to 5 hours, and the economics of the application requires that the product be carefully designed with a small safety factor.

A different design approach is used in this case. Instead of assuming an apparent modulus of elasticity using a constant creep situation covering the life of the chair, it is better to determine the actual creep deflection over a typical stress cycle, the creep recovery over a non-use cycle, and so on until the creep is determined after a series of what might be considered typical hard usage cycles for the chair. The accumulated creep after a period of two weeks can be assumed to represent the base line for an apparent modulus of elasticity to determine the design life of the chair.

Load Requirement

With this basic approach in mind, let us do a design on a typical molded chair seat. The load will be assumed to be a 250 lb person and the load cycle which includes loading times of 4 to 6 hours two or three times in 24 hours and a relaxation period of 1 to 2 hours during the day and of 10 hours during the night. The curve of loading is a random collection of these cycles over a 2 week period. The first step in the design is to select a section for the chair seat that will have the required strength to prevent breakage with the stress calculated from the extreme fiber formula.

The next step is to see that the seat does not deflect more than a given amount to be able to continue to function as a seat. An arbitrary deflection of 2 in. in the length of a seat 16 in. long will be assumed since consumer comfort testing usually arrives at such values. It might be noted that in some chair designs where the creep did not result in failure of the chair, the fact that the seat was too resilient and gave a feeling of insecurity led to poor consumer acceptance. In many cases the "feet" of a product is important to its success and the feeling of solidity is important in furniture applications.

Within the limits set above the design can vary widely. The seat can be attached to the rest of the chair frame by leg supports at the four corners, or it can be cantilevered from the back with a floor pad support, or, in another version, from the front. The seat construction can range from a formed sheet in two or three dimensions to one with rolled edges for reinforcement. It can have structural ribs molded in or it can be a sandwich panel construction made up of two molded parts bonded together. It can also be a structural foam molding. In each of the configurations there are tradeoffs of stiffness and strength that may make one more effective then the others in meeting the seating requirements.

Form and Dimension

In this case the designer has freedom of choice of both form and dimension as well as in the selection of the materials. Given this freedom, it would be desirable to examine several of the alternatives to see which would provide the best seating at the lowest cost. Obviously, there is no point in doing all of the possibilities so a selection should be made on the basis of anticipated use as well as style requirements. Three types will be analyzed. They are the single curve sheet cantilever mounted from the back, the molded pan supported on four legs, and the structural foam molding which is front supported.

In order to simplify the analytical exercise, a particular material was selected for each. The single curved sheet is made of TS polyester fiber glass molded to the shape. The corner supported pan is molded from ABS plastics. The structural foam unit is molded from PP with glass fiber filler.

From inspection of the three designs it is apparent that the main stress of the loading will be at the support point for the seat. This will be assumed to be sufficiently strengthened to prevent failure, either by excessive stress or bending at the support point. The analysis will be concerned with the fact that the seat itself will not break as a result of the load and will not sag excessively after continued use. For this example the impulse load caused by dropping into a chair will be ignored.

In each case the section is designed to keep the deflection to less than 2 in. in 16 in. for a design life of 5 years and the extreme fiber stress is kept to a value less than the yield strength of the material. The first step in the analysis is to determine the necessary section to resist the bending load using the short-term tensile and compressive strength and modulus values. The extreme fiber stress is calculated for these sections to determine that the chair will not break when deflected.

A time dependent modulus is then calculated using the extreme fiber stress level for each of the materials at the initial stress value level using the loading-time curve developed. If the deflection at the desired life is excessive, the section is increased in size and the deflection recalculated. By iteration the second can be made such that the creep and load deflection is equal to the maximum allowed at the design life of the chair. This calculation can be programmed for a computer solution.

Stiffness

The selection from the possible designs is made on a cost effectiveness basis. The least costly construction would be the best unless there is an inherently more useful construction for aesthetic or other reasons. In most design cases this will be an aesthetic consideration or, in this case, it may be a user consideration such as initial stiffness which leads to a better feeling of security. This may lead to the selection of something other than the lowest cost design. This value can be added to the consideration by examining the products' stiffness (El) for each of the constructions at the initial short-term loading (Chapter 3, **Geometrical Shape**, *EI theory*). The higher the El product, the less the seat will flex under load, and the more secure it will feel. With the materials and constructions examined, the initially stiffest construction may have a higher creep level and require a heavier and more costly construction to meet the creep criteria. This may be justified by the better feel of the seat.

As always, the designer must be concerned with the utility of the product because this is justification for designing it in the first place. The most important part of the design is that it satisfy the need for which it is made. The technical qualities must be such that they result in good acceptance. A technical success that leaves the user unsatisfied is not a product design success.

Environment

As with any design, the environmental and end-use requirements must be considered. The chair will be exposed to cleaning agents, children and dogs climbing on it, possible abuse in storage and shipment, etc. Such conditions should be considered as part of the material selection and design procedure. Of the three materials suggested, PP has the best chemical resistance, ABS has the best abrasion resistance, and the RP normally would have medium properties in both areas. All three are tough materials that would take rough handling, and the ABS would be best in terms of exposure to sunlight. In this particular design there would generally be no reason to choose any one of the materials over the other unless it is anticipated that the chair will have substantial exposure to conditions other than those typical of the home environment.

In the other design procedures we would tend to follow a minimum test sequence for acceptance testing in use to see that the design is functional, and set a reasonable quality testing procedure to insure that the processing is under control. While the use of plastics in a chair represents a situation where there is substantial personal injury risk it is one that is anticipatible and the normal design procedure would anticipate and eliminate the possibility of premature failure.

Deterioration of the chair with age should be examined to see if environmental exposure would lead to a shortened life. The indoor environment where a chair is normally used does not produce severe damage. Indoor sunlight is much less severe than outdoor exposure and room temperatures do not vary excessively. The only source of possible damage is in the use of cleaning and debugging agents that may attack the plastics. This can be controlled by following proper instructions. If difficulty is expected, a chemical resistant material is indicated. Abuse by dropping and impact should also be considered. This may cause surface or structural damage. Public seating is subject to much abuse.

As examples we have examined different types of common statically loaded structural units. In all case they represent long-life expectancy units which will carry substantial loads. Since the failure of the product would involve substantial risk of personal injury, the designs must be done with caution. In one case, strong considerations are present and in the other traditional performance requirements must be considered. The loadings and the basis for making the design judgments are different. These examples do not exhaust the possible combinations of conditions the designer will face, but they indicate what might be expected.

The approach to the problem is to make the best analysis of the product requirements, including what at first appear to be intangible requirements, and then to determine what are the important elements in the design. Using these as the guide, several types of structural possibilities are examined with different materials to see if they meet the performance requirements of the application. The loads, the duration of the loads, the environment, and the intangible use factors will favor one design over the other, particularly if the economics are made the final basis for choice. The final design is made and tested for performance and sent to production with suitable quality control tests indicated.

The process of design for static loads involves a great deal more than the mechanical operation of the stress-strain data to determine the performance of a section. The results obtained from the stress analysis are used to determine the functionality of the product and then, combined with the other factors involved to decide on a suitable design.

Prototype

Different tests are used on prototypes. As an example industry has a test where a load may be double that of a heavy person. Its two rear legs are positioned in front of an anchored board. The top of the chair has a rope or chain extending backwards to an oscillating device. The top of the chair will be pulled back to the point of almost failing backward and then released. The loaded chair will bounce on its two front legs. This cycle is repeated thousands of times. The industry test has requirements so if the chair is to be used in commercial environment its number of cycles will be many more than a noncommercial chair.

Using the approach suggested, designers can be guided in the design of static structure for performance in any environment from space, to aircraft, to land applications, to subsea use. Defining the requirements and using the data available or generated for the application, the end result can be made predictable to a sufficient extent that successful products result with minimum cost.

Automobile

For today's and tomorrow's transportation vehicles (automobiles, trucks, motorcycles, boats, airplanes, etc.) plastics offer a wide variety of benefits. Plastics play a very im-

portant role in these vital areas of transportation technology by providing special design considerations, process freedom, novel opportunities, economy, aesthetics, durability, corrosion resistance, lightweight, fuel savings, recyclability, safety, and so on (14, 153, 179, 234). Designs include lightweight and low cost principally injection molded thermoplastic car body to totally eliminate metal structure to support the body panels (Figs. 1-13 and 4-7). Example include pick-up trucks that use 100 lb thermoformed cargo-bed liners or RP boxes, RP bus interiors and bodies, injection molded TP fenders, etc. With more fuel-efficiency regulation new developments in lightweight vehicles is occurring with plastics. Plastics used include ABS, TPO, PC, PC/ABS, PP, PA, PVC, PVC/ABS, PUR, and RPs.

Different cars have and are being designed worldwide to produces low cost purchases, light weights, reduce fuel costs, reduce contaminating emissions, etc. Extensive use has been made by using unreinforced or reinforced TPs and/or TS plastics. A few of the many plastic products used follows. (1) Chrysler Corp.'s light-weight Composite Concept Vehicle (CCV) includes TP structural body panels with only a limited amount of metal underneath; it is an all-plastic body requiring a very large mold. (2) Ford Motor Co. has an all-aluminum body with plastic in some of the other parts. (3) GM focusing some plastics in their electric vehicle. (4) Asha/Taisun of Singapore producing taxi cabs for China with thermoformed body panels mounted on a tubular stainless steel space frame. (5) NA Bus Industries of Phoenix is delivering buses in USA and Europe with all RP bodies. (6) Brunswick Tech. Inc. of Brunswick, ME produces 30 ft RP buses except for the metallic engine (209).

There is Europe's plastic skin, 2-seat coupe, called the Smart car with molded-in color that virtually eliminates painting. The idea was to eliminate the need for three coats of paint and reducing both cost and emission problems, Project started in 1994 via a joint venture of Daimler-Benz in Stuttgart, Germany, then known as Mercedes-Benz, and the Swiss watchmaker SMH AG in Biel. They created a

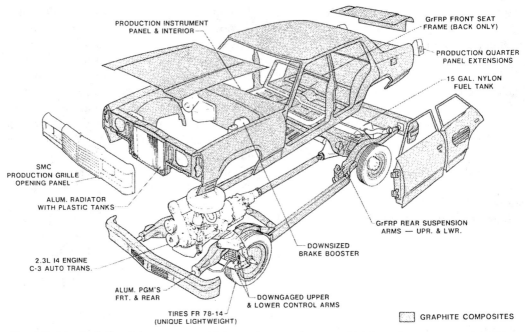

PRODUCTION INSTRUMENT
PANEL & INTERIOR

GrFRP FRONT SEAT
FRAME (BACK ONLY)

PRODUCTION QUARTER
PANEL EXTENSIONS

15 GAL. NYLON
FUEL TANK

SMC
PRODUCTION GRILLE
OPENING PANEL

ALUM. RADIATOR
WITH PLASTIC TANKS

GrFRP REAR SUSPENSION
ARMS — UPR. & LWR.

DOWNSIZED
BRAKE BOOSTER

2.3L I4 ENGINE
C-3 AUTO TRANS.

ALUM. PGM'S
FRT. & REAR

DOWNGAGED UPPER
& LOWER CONTROL ARMS

TIRES FR 78-14
(UNIQUE LIGHTWEIGHT)

▨ GRAPHITE COMPOSITES

Fig. 4-7   Ford's lightweight concept vehicle made extensive use of high performance graphite fiber RPs.

new company called Micro Compact Car AG, or MCC. The first of the cars had plastic injection molded outer body panels using GE's PC/TP polyester blend (Xenoy). Its unitized TP body that tied together the front fender, outer door panels, front panels, rear valence panels, and wheel arch in one wrap-around package. The entire car weighs 1,440 lb (650 kg), about 600 lb (270 kg) less than most steel-body compacts.

World's First All-Plastic Car Body

From Sichuan Huatong Motors Group, Chengdu, China is the all-plastic car called Paradigm, a 4-door/5-passenger midsize vehicle. Its features include glass fiber-TS polyester RP sandwich chassis, thermoformed coextruded ABS body panels with molded-in color, adhesively bonded body, and, for the high-end model, a coextruded acrylic (ASA) cap cover providing high gloss. Automotive Design & Composites Ltd., San Antonio, TX, USA has served as the primary contractor for developing the concept. Chassis features a single thermoformed lower

tub and an upper skeleton X-brace roof. It employs a monocoque structure where body panels are stitched-bonded to the chassis, forming a unitized structure.

Thermoformed chassis and body panels are featured on the car. The products were made initially in the USA for assembly in China. The car will weigh less than 2000 lb (900 kg). Automotive Design & Composites, Inc of San Antonio, TX, designed the vehicle to have body panels and trunk formed from coextruded sheet of ABS with an ASA cap layer that will hang on a pultruded composite frame. Ceramic tooling is used to thermoform plastic products.

The chassis is made from a 1/4 in. sheet of either ABS or TPO vacuum formed into a tub and reinforced with reinforced pultruded glass fiber-TS polyester plastic tubing. The hood and other products are being made from a 20 mm thick sandwich of thermoformed PPO-alloy skins, glass fabric infused with thermosetting vinyl ester, and a urethane foam core. The bumper and front fascia is thermoformed from a polyolefin elastomer sheet with an UV-resistant cap layer of DuPont's Tediar PVF film. The dash and inner console is thermoformed from soft vinyl over ABS.

Quadraxially oriented (four directional layer) glass fabric-TS vinyl ester polyester RP sheet panels with a foam core and gel coating are used. Most of the panels are 3 mm thick with molded-in rib structure supports. Body skins are bonded to the chassis with a double-stick acrylic tape developed by 3M Co. as well as mechanical fasteners. Unlike most steel designs, no B-pillar structural component between the front and rear doors is required thus providing more interior space and easy entry since doors open in opposite directions.

The bumper to bumper measures 4.6 m (15.18 ft); weighs 815 kg (1793 lb) that includes 1200 lb (550 kg) of plastics; has a gas/electric hybrid power system, air bags, neon tube tail-lamps, etc.; and gets 132 km (60 miles) per gallon of fuel. Huatong and the Chinese government have funded $100 million in this global project.

Evaluation was made on four prototype cars, and China's Huatong Motors expects to assemble 5000 units at its Sichuan plant in the first year. Once on the market, the product will be a moving example of the capability of plastics engineering that includes using principally thermoforming technology. The all-plastic car will be using joining techniques that have been used for more than a half of a century. Many examples where adhesives are used exist. A prime example where adhesive joints have been used since the 1950s is on military aircraft where joints are subjected to all types of environments as well as static and dynamic loads, fatigue loads, weather changes, and so on.

Aircraft

Since the 1940s extensive use is made externally and internally of light weight, durable, and high performance plastic in commercial and military aircraft. Included are unreinforced and reinforced plastics as well as specialty plastics such as anti-icing coating (Figs. 4-8 and 4-9). Different RP parts (wing fairings, floor beams, rudder, elevators, engine cowl, etc.) are used on the Boeing 777 (Fig. 4-10) (Appendix A: **PLASTICS DESIGN TOOLBOX**).

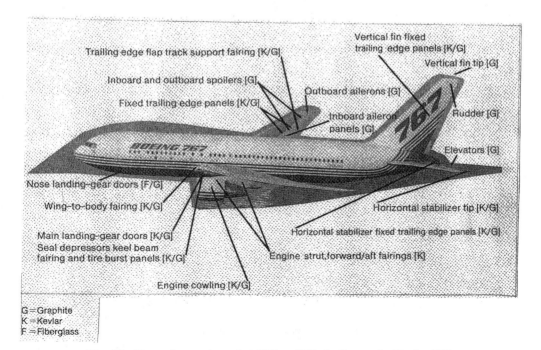

Fig. 4-8 Extensive use is made of TP and TS plastics on the Boeing 767.

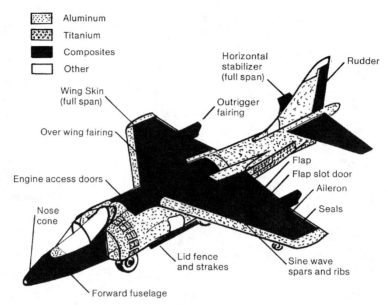

Fig. 4-9   Over 26% of this McDonald-Douglas AV-8B Harrier aircraft's weight uses carbon fiber-epoxy reinforced plastic; other plastics also used extensively.

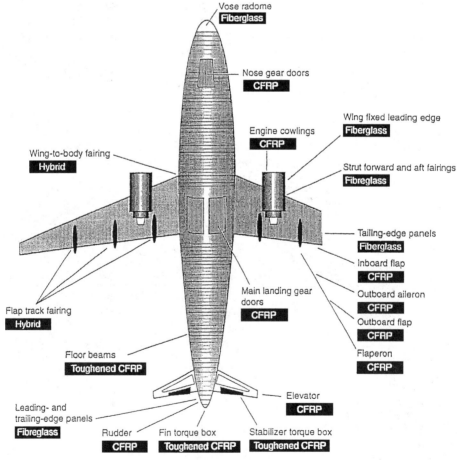

Fig. 4-10   RP products on the Boeing 777.

During 1944 at U.S. Air Force, Wright-Patterson AFB, Dayton, OH the first successful all plastic airplane (primary and secondary structures) was designed, fabricated, and flight tested (Fig. 4-11). It used glass fiber-TS polyester hand lay-up RP that included the use of the lost-wax process sandwich constructions for the monocoque fuselage, wings, vertical stabilizer, etc. (Chapter 8, **REINFORCED PLASTIC**).

(a)

(b)

Fig. 4-11 (a) BT-15 in flight, (b) sandwich wing section using RP skins and cellular cellulose acetate foam, and (c) section of the monocoque fuselage, and (d) example of a design flexibility in RP construction.

(c)

- Modulus, strength, and thickness can be varied
  through design and material choices

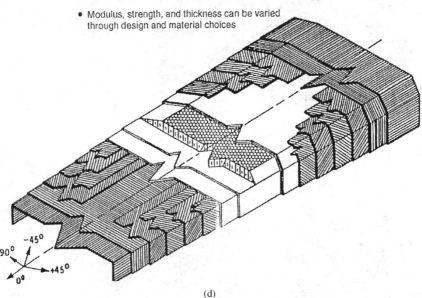

(d)

Fig. 4-11   (*Continued*)

Over the years innovations in aircraft have given rise to more new plastic developments and have kept the plastics industry profits at a higher level than any other major market principally since they can meet different environmental and load conditions. Virtually all plastics have received the benefit of the aircraft industry's uplifting influence. Practically all conceivable top quality plastics are used to provide cost advantages and improvements on flight system performance. At least 5% of commercial plane's weight is plastics. This percentage is expected to double or perhaps triple in the future. Military aircraft include those with up to at least 50 wt% of primary and secondary structures.

The chief challenge to the plastics industry is not in pounds of plastics needed but rather in the translation of plastic development technology into production line know-how. The aircraft industry is geared to pay high prices for plastics with exceptional properties; as high as hundreds of dollars per pound with up to ten-fold cost increases for fabricated products plus additional dollars to conduct continual testing and evaluation to insure safety of aircraft operation.

Medical Product

Plastics continue to make inroads regarding medical applications. Certain plastics have been found to exist in the environment of living tissues. In addition to being of direct aid in medicine, they are also important in medical devices and packaging medical items.

In effect doctors are designers. They are constantly helping design, test and evolve new materials and equipment to augment new and proposed methods of treatment and surgery. Their long, intensive training as basic scientists uniquely suits them to the designer capacity.

Bioplastic

New plastic applications and plastics are continually in use to help in the field of medicine (also biological systems, etc.). They include both mechanical and chemical applications and show the makeup of the field that could be called bioplastics (186). The heart valve that is often used in surgery to correct heart deficiencies was a spectacular contribution to medicine. In order for it to be successful it required first, ingenuity in designing a product that would function as a replacement for the mitral valve and to perform as well as the one replaced long enough to justify the risk involved in the operation. Second, it also required using a material that would function in the highly complex environment of the human circulatory system without being degraded and without causing harm to the circulatory system.

The use of special elastomer plastics (polyurethanes, silicones, etc.) designed ex-

pressly for the purpose and tested for years to determine possible adverse in-vivo effects combined with good design resulted in a mechanical device that has been a major lifesaver (29). The role of the plastic designer in an area such as this is one of knowing material limitations, processing problems and of devising test procedures to monitor the performance of the product in the patient as well as in continuous laboratory testing.

Other surgical implants are essentially plastic repair products for worn out parts of the body. It is possible to conceive of major replacements of an entire organ such as a kidney or a heart by combining the plastic skills with tissue regeneration efforts that may extend life. This is used to time the heart action. Extensively used are plastic corrugated, fiber (silicone or TP polyester) braided aortas (24).

While it would be difficult to enumerate all of the efforts in the area of implants where plastics are involved, some of the significant ones are: (1) the implanted pacemaker, (2) the surgical prosthesis devices to replace lost limbs, (3) the use of plastic tubing to support damaged blood vessels, and (4) the work with the portable artificial kidney. The kidney application illustrates an area where more than the mechanical characteristics of the plastics are used. The kidney machine consists of large areas of a semi-permeable membrane, a cellulosic material in some machines, where the kidney toxins are removed from the body fluids by dialysis based on the semi-permeable characteristics of the plastic membrane. A number of other plastics are continually under study for use in this area, but the basic unit is a device to circulate the body fluid through the dialysis device to separate toxic substances from the blood. The mechanical aspects of the problem are minor but do involve supports for the large amount of membrane required.

Bioscience

There are two major areas of application for plastics in bioscience. The plastics make interesting materials to be used for mechanical implants into all living systems, including animals and plants where they can serve as repair parts or as modifications of the system. The other applications are based on the membrane qualities of plastics that can control such things as the chemical constituents that pass from one part of a system to another, the electrical surface potential in a system, the surface catalytic effect on a system, and in some cases the reaction to specific influences such as toxins or strong radiation.

Chemically active plastics such as the polyelectrolytes have been used to make artificial muscle materials. This is an unusual type of mechanical power device that creates motion by the lengthening and shortening of fibers made from a chemically active plastic by changing the composition of the surrounding liquid medium, either directly or by the use of electrolytic chemical action. Obviously this form of mechanical power generation is no competitor to thermal energy sources, but it is potentially valuable in detector equipment that would be sensitive to the changing composition of a water stream or other environmental flow situation.

By using direct mechanical action from the artificial muscle, it would be possible to produce reliable sensing and control devices without electrical and electronic equipment. Another interesting application would be to drive prosthetic devices where the action would be similar to the muscle reaction in the body. This unusual type of chemically induced motion should be an interesting one to explore for the solution of unusual problems where conventional approaches do not work.

Surgical Product

The most dramatic use of plastics in medicine can be summarized as being in surgery (Table 4-2). Advances in surgery have enhanced the need for providing plastics that can be utilized in replacing or repairing tissues or organs which have been damaged as a result of trauma or disease. The wide

Table 4-2   Example of plastics used in the medical industry

Acrylics	in bone replacement, corneas, adhesives, hemostatic agents, dentures, contact lens, artificial eyeballs.
Cellulose acetate	in nerve regeneration, packaging material.
Formaldehyde-treated polyvinyl alcohol sponge	in support and growth stimulator for blood vessels out-side and into heart muscle, hemostatic agent in repair of liver and kidney wounds, abdominal aortic grafts, vascular shunts, synthetic skin.
Fluorocarbons	in artificial cornea, blood vessels, heart valve coatings, reconstructive surgery, bone substitution.
Polyamide	in vascular implants, syringes, clamps, blood transfusion sets.
Polycarbonates	in syringes, parts of heart-lung machine, baby bottles, containers.
Polyester fiber	in aortic and peripheral artery transplants.
Polyethylene	in tubing, syringes, oxygen tents, repair of incisional hernias, stomach wall support, repair tissue damage, heart valves, contraceptive implants.
Polypropylene	in syringes, sutures, containers.
Polystyrene	in syringes.
Polyurethanes	in plastic surgery, vascular adhesive, bone adhesive.
Polyvinyl chloride	in surgical tubing, blood collection and administration sets, repair of congenital and traumatic facial defects, surgical drapes, ballon type splints, adhesive bandages.
Polyvinyl pyrrolidone	in artificial membranes for filtration of body fluids.
Silicones	in heart valves, tubing, catheters, defoamers in blood oxygenators, urethral valve, plastic surgery, tendon replacement, lubricants, tissue substitutes.

range of forms such film or fiber and mechanical properties available in plastics continues to make them attractive candidates for such uses. However, even though a plastic may possess the desired physical properties, there is no assurance that it may be successfully utilized in the body. Tissue compatibility is a sine qua non for the long-term utilization of a surgical repair material.

The answers to all questions are not known with certainty, and research toward the solution of such problems will require the combined efforts of the plastic chemist and designer with the physician. It is through the efforts of such multidiscipline groups that surgical repair materials of outstanding long-term utility are produced, studied, evaluated, and made available to the patient.

Plastic implants provide a forum for the cross-fertilization of such prepared minds. There is the utilization in tissues of the postenucleation moveable implant and plastic artificial cornea. Use is made of such plastics as PMMAs, silicones, PTFEs, and hydrophilic polymers. This action is related to the implantation of plastic artificial corneas, retinal detachment surgery, glaucoma drainage tubes, repair of bony defects, replacing the vitreous of the eye, and substitution for the eye's crystalline lens. The concept of the incompletely covered foreign body was evolved in connection with the plastic artificial cornea and postenucleation implant. In each case, the plastic (or metal) is exposed to the exterior environment and epithelial cells do not heal like epithelial cells. The semi-exposed implant has been maintained without extrusion for many years. Each of these applications requires plastics with varying properties.

Complex Environment

It is important to recognize that human bodies have extremely complex environments. They could be identified as having the most horrible environmental situation. Reason for this situation is due to the fact that the many different human bodies have differ-

ent environmental requirements. Thus what can survive in one body usually does not survive in other bodies. This type of reaction requires extensive "prototyping" to ensure that a medical product can survive and meet its requirements in all human bodies.

Dental Product

Most of over six million dentures produced annually in the USA are made of acrylics (PMMAs) that includes full dentures, partial dentures, teeth, denture reliners, fillings and miscellaneous uses. Plastics have been edging into the dental market for over a half century. Even before the introduction of acrylics to the dental profession in 1937, nitrocellulose, phenol-formaldehyde and vinyl plastics were used as denture base materials. Results, however, were not wholly satisfactory because these plastics did not have the proper requisites of dental plastics. Since then, PMMAs have kept their lead as the most useful dental plastics, although many new plastics have appeared and are still being tested. Predominance of PMMAs is not surprising, for they are reasonably strong, have exceptional optical properties, low water absorption and solubility, and excellent dimensional stability. Most denture base materials, therefore, contain PMMA as the main ingredient.

Plastics have not progressed very far as filling materials. About 2 wt% of all fillings are plastics. The low mechanical properties of plastics in comparison with metals limit their application to front teeth where stresses are not so great. It is interesting to note development efforts has taken place in the use of whiskers for reinforcing dental plastic, metal, and ceramic fillings. Some preliminary test results on the addition of randomly distributed chopped, short whiskers to a coating plastic have reversed the previous proportional less of strength with powder additions. Although this is far from theoretical, it is already quite significant in that it allows the addition of pigment for coloring purposes and a restoration of the loss of strength with whisker additions.

Medical Packaging

Plastics are extensively used in medicine to package drugs, ointments, and accessories. Plastics serve to protect medicines, surgical/clinical equipment, medical materials, etc. from contamination and breakage in many ways, from single-service squeeze packs of cough syrup to carrying cases used to ship human eyes between hospital eye banks.

The sophistication of TP processing methods that includes sterilization procedures, has allowed the development of low cost, disposable packages for single-doses of medication, eliminating need for sterilization, etc. Rigid blister packs hold capsules and pills in easily dispensed single servings; semi-rigid and flexible squeeze packs hold single applications of medicines and ointments, etc. Availability of medicines in premeasured single-dose disposable units pleases doctors who previously could only hope that out-patients would dose themselves properly when not supervised.

Biological and Microbial Degradation

Certain plastics such as TP polyesters, polyurethanes, cellulosics, and plasticized PVC can be degraded by microorganisms. It has been observed that enzymes attack noncrystalline regions preferentially. As a result, it has been determined that the resistance of susceptible plastics to microbial degradation is related directly to the degree of crystallinity of these plastics. They remain relatively immune to attack as long as their molecular weight remains high. Most of these plastics are characteristically durable and inert in the presence of microbes.

This stability is important to plastics' long-term performance. However, for some applications only short-term performance is desired before the product is discarded, as in the fast-food and packaging markets. In such cases it is considered advantageous for discarded plastic to degrade when exposed to microbes. There thus exists a requirement to develop or modify plastics possessing the properties required for their service life, but with the capability of degrading in a timely

and safe manner, particularly to handle the worldwide waste situation.

The amount of degradation of plastics under the action of bacteria and fungi is of interest because of land-shortage problems in solid-waste management and litter accumulation and other environmental problems on land and sea. The agricultural use of plastics in mulch, films, seeding pots, and binding twines has increased significantly, making biodegradation a desirable feature in plastics, to minimize disposal and soil-pollution problems. With plastics designed to include degradability, the designer could have a monumental problem in ensuring that degradation will occur only after a product's useful life is over.

Other Form of Degradation

Basically degradation is a deleterious change in characteristics such as the chemical structure, physical and mechanical properties, and/or appearance of plastic. A degraded appearance usually means discoloration. Degradation can occur during heat processing. Factors that determine the rate of degradation are: (1) residence time, (2) stock (melt) temperature and distribution of stock temperature, (3) deformation rate and deformation rate distribution, (4) presence of oxygen or other degradation-promoting additive, and (5) presence of antioxidants and other stabilizers.

The deterioration of plastics by biological agents should be distinguished from other forms of plastic degradation. Many other types of plastic degradation may be classified clearly as chemical in nature. In them a deteriorative agent causes a chemical degradative reaction to occur. Chemical bonds are broken or new ones established. Different molecular species of a molecular size smaller or larger than the original desirable species are formed, and these species no longer have the properties for which the original plastic was chosen. Other forms of degradation include landfill, overheating, photochemistry, photodegradable, photooxidation, radiation induced reaction; ultrasonic degradation, fusion, zymoplastic degradation (substance which, not themselves enzymes, are believed to participate in the formation of enzymes), etc.

This generalization is also true for the degradation caused by heat, electromagnetic radiation, oxygen and ozone, and high-energy nuclear radiation. It is also true for the chemical degradation caused by acids, bases, or other strongly reactive chemical agents. The reaction types include oxidation, ozonization, radical formation, cross-linking, chain scission, and others. The symptoms are described as hardening, embrittlement, softening, cracking, crazing, discoloration, or alteration of specialized properties such as dielectric strength.

The situation with some forms of biological deterioration is somewhat different. Where the agent is macrobiological, as in the case of rodents, insects, and marine borers, the attack is physical in nature, such as by gnawing or boring. The attack is not at the atomic or molecular level. Any breaking of molecular bonds such as in polymer chain shortening is thus accidental. The attack may be said to be at the material's structural level, not the polymer molecule level.

An important item to note is that most commercially used plastics are not single component pure substances. Practically always, the basic polymer itself, rarely if ever a single molecular species, is compounded with other components such as plasticizers, pigments, antioxidants, and other additives. More often than not, then, biological susceptibility is due to the nonpolymer component.

Plastics' deterioration can be classified as either by a microorganism, a macroorganism, or a marine organism (both micro and macro). In the case of microbiological agents, as in fungal and bacterial deterioration, the plastic alterations are caused by chemical attack. This has been demonstrated for the attack on the natural polymer cellulose by fungi through the cellulose enzymes, for many esters, and for many hydrocarbons. It is not yet so clearly proven for the many synthetic polymers, but there is sufficient evidence that may be ascribed to enzyme action as being probably the chief mechanism. Thus, although the medium of attack is biological, the destructive agents are chemical.

Fungal and bacterial deterioration are identified as microbiological and have always caused problems to materials. Fungal attack on plastics has received a great deal of attention beginning with the early days of World War II, when the tropical theaters served to focus attention on the overall problem of materials deterioration.

Microbial deterioration of plastics is intimately involved with the moisture problem, especially with regard to plastics in electronic equipment. For this reason much of the literature treats the two problems together. Furthermore, there is often confusion between the deterioration of the electrical properties of plastics, more often than not a moisture phenomenon, and actual deterioration of the substance of the polymer.

Most investigators agree that in the electronics field moisture accounts for the greater effect. Often, if the moisture problem is solved the fungal aspect is also overcome because of the dependence of organisms on water. Yet not all of this twin problem may be ascribed to moisture, for there are instances where microorganisms are able to destroy the substance of a polymer or attack the nonpolymer constituents of a plastic formulation. Furthermore, as one shifts attention from plastics in electronic equipment to other items where plastics are used, there are clear-cut cases of destruction by fungi. Examples may be found in films, fibers, and coatings.

Dynamic Load Isolator

Thermoplastic elastomer (TPE) components are frequently subjected successfully to dynamic loads where energy and motion controls are required. The products involved range from sporting goods to home appliances to automobiles to buildings to bridges to boats to aircraft to spacecraft. The uses of bonded elastomers for energy and motion control in construction, vehicles, instruments, etc. are extensively used (Chapter 2, **THERMAL EXPANSION AND CONTRACTION, Energy and Motion Control**).

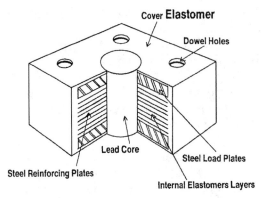

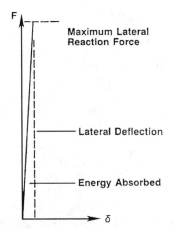

Fig. 4-12(a)   Energy and motion control building isolator.

Fig. 4-12(b)   Typical load deflection that is characteristic of a marine structure.

An example of this type of energy and motion control device is shown in Fig. 4-12(a). It shows a section through a high-load elastomeric building isolator where the elastomer layers are used to ensure horizontal flexibility. Steel reinforcing plates give rigidity for vertical loads. This isolator resists wind loads elastically without perceptible movement, yields under earthquake loads, and deforms plastically dampening side-to-side vibrations (14, 55).

Anyone in a building, on the highway, on a rapid transit vehicle, or on a ship has a bonded elastomer working for them. In a building, it controls vibration and noise from motors and engines. For rapid transit it supports the rail and the vehicle, thus reducing noise and vibration to adjacent buildings. For ships, bonded elastomers absorb their berthing energies, with single units being as large as 3 m (10 ft.) high and weighing 19 tons. For all these applications, elastomers are used either in shear, compression, tension, torsion, buckling, or a combination of two or more uses, depending on the needs of the specific application (Chapters 6 and 7).

Consider, for example, a berthing vessel that is a structure that has to be designed to withstand the energy developed by the vessel. The more rigid the system, the higher the reactive forces must be to absorb the vessel's kinetic energy. The area under the structure's load as against its deflection response curve (that is, of the energy absorbed) is typical to that shown in Fig. 4-12(b).

For economic reasons, designers typically reduce the mass of a structure, but doing so reduces the lateral reaction forces that the structure can withstand. This happens as the structure is allowed to deflect more or an energy-absorbing device is applied. An elastomer is ideal in such an environment because it will not corrode. Metal components can thus be totally encapsulated and protected against corrosion in an elastomer and then bonded to all-metal surfaces. The elastomer can be used in conditions of shear, compression, or buckling. In examining the load deflection characteristics of these three systems, note that the one that results in the lowest reaction force generally also produces the lowest-cost structure. Figures 4-12(c) to 4-12(e) shows six results that could be obtained, compared to an ideal hydraulic system with 100% energy efficiency.

Filter

Water

One of the major problems facing our civilization is the availability of pure water. The largest source of water located near many cities is the ocean, but the ocean is filled with large amounts of dissolved salts. To recover water from the sea by any of the conventional distillation processes is to date extremely wasteful of energy and costly. However in

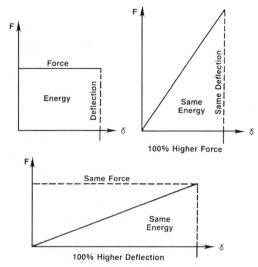

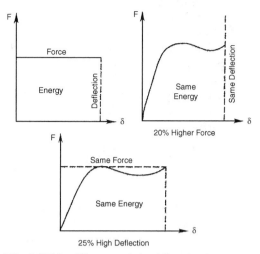

Fig. 4-12(e)   Elastomeric buckling energy capacity as compared to a 100% efficient curve.

Fig. 4-12(c)   Elastomeric shear energy capacity as compared to a 100% efficient curve.

areas such as the Middle East where fresh water is very scarce, ocean salt water has been filtered by different techniques for many decades.

Plastic membranes are being used in systems that could well pave the way to large-scale water recovery from the sea. The process is reverse osmosis. When water is separated from a concentrated solution of a salt by a semi-permeable membrane, there is a pressure that drives the pure water into the solution for dilution. The driving force is the concentration gradient and it is in the form of a pressure that is related to the difference in the vapor pressure of the water and the vapor pressure of the solution at the temperature at which the process takes place. By applying a pressure greater than the osmotic pressure to a solution, the direction of flow of the water is reversed and pure water is removed from the solution.

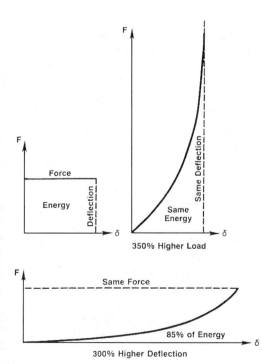

Fig. 4-12(d)   Elastomeric compression energy capacity as compared to a 100% efficient curve.

Plastic membrane   This is done by the use of a water permeable plastic membrane held deep enough under the sea so that the hydrostatic pressure is greater than the osmotic pressure of the seawater. The water distills out of the solution through the membrane and is pumped to the surface. Large areas of the membranes, mechanically supported to withstand the very high pressures are essential to make the process perform rapidly for the most economical production.

Cellulosic plastics are usually used for the membrane, but any water vapor permeable material is a good possibility, provided the film has good mechanical properties.

Designing the membrane structure for a reverse osmosis plant is a difficult project, particularly in view of the fact that in addition to the pressure exposure, the presence of strong concentrations of dissolved minerals is a hostile environment for plastics.

There have been several very ingenious designs for membrane structures using naturally strong shapes such as arches and tubes reduced to a scale where the amount of surface area for diffusion per unit volume is very high. Other innovations in design and fabrication of these large area membrane structures could easily lead to a significant breakthrough in the availability of an unlimited supply of pure water.

Plastics have had a long history of use in water problems using a special type of plastic material. Ion exchange plastics have been used for many years to remove dissolved materials from water and to make high purity water. The quality of ion exchange treated water is probably better than water purified by any other means. The ion exchange plastics are polyelectrolytes. These plastics are polymers that contain either acid or alkaline side groups capable of reacting with dissolved ions in the water. A standard system used is an acid substituent that forms an insoluble salt with ions such as calcium and strontium and removes them from the water and replaces the mineral ion with hydrogen. The ion exchange plastics can be regenerated by passing another solution through the bed such as a mild mineral acid which removes the attached ions and replaces them with hydrogen again so that the resin can be reused.

Gas

Plastics have a unique contribution to make in exploring new environments. Their applications in space vehicles are well known and are generally of a mechanical nature. Plastics are used extensively in equipment for exploring under the sea. One unique application for plastics used in the sea environment and which is important in other areas that range from packaging to space is in filtering gases. This specific application involves the use of membranes made from special plastics such as silicone compounds that permit dissolved oxygen in seawater to permeate the membrane in one direction and to allow carbon dioxide and carbon monoxide to pass through in the opposite direction.

A large membrane pack of this type will act like an artificial gill, permitting a swimmer to breathe like a fish and remain submerged for much longer periods of time than are possible with scuba equipment. Speculative fiction has man returning to live in the seas, and this type of application may make it possible. Their application in spacecraft is obvious as a part of a continuously recycled air support system. The oxygen permeability of silicone materials is just one example of the selective permeability of plastics.

Liner

Plastics provide different performance requirement in providing protective liners in many different applications such as building foundations, pipe and tank liners containing corrosive liquids, etc. As an example Fig. 4-13 shows an RP stack liner being inspected prior to installation in a 682 ft. high reinforced concrete chimney (background) of the 1,500-megawatt Intermountain Power Project near Delta, Utah (1985).

The liner, of PPG glass fiber, protects the concrete shell from the corrosive gases that occur when sulfur dioxide is produced during coal-fired power generation. Fiber glass-TS polyester RPs provided years of service under these operating conditions. Such liners have been used in this type of application since at least the 1970s. They rapidly became a viable construction material against steel and brick liners. The liners in this Utah project are in canlike sections 45 ft. long and 28 ft. in diameter. The sections were filament wound using 46 to 50 wt% TS polyester-impregnated fiber glass rovings. The completed liner contained about 100 thousand miles, or 11 million pounds of fiber glass roving strands.

Fig. 4-13 RP stack liner.

Paper and Plastic

One of the problems that face our civilization is the fact that the pressure on natural resources is reported to be hindering progress. Periodically an energy crisis exists that has led to a so-called materials crisis in plastics and even other materials such as cellulose papers. Petroleum is currently the major source of raw materials for most high volume plastics.

Oil and substitute resources such as coal are supposedly in limited supply (although our government reports we have enough coal for the next 250 years that includes its growth in use during that period), and it may well be that another approach to the problem is required. An example is different raw material sources to produce plastics that involve biotechnology (186), more vegetation, etc. Ingenuity in the applications of materials, the province of the designer, and the use of materials that seem to be uniquely modifiable, such as plastics, are needed. Thus the plastics material suppliers are developing renewable resources for the plastics.

We can take as an example worldwide papermaking that now consumes forests at a rate that is supposedly difficult to replace. Unlike the uses for wood, which are generally long-term use goods, most wood pulp paper is used for newspapers, business world, and periodicals or publications that are read and usually discarded, loading our solid waste disposal system and adding mountains to our trash.

Save the Tree Myth

Plastic papers have been developed as substitutes for these cellulose papers, but the economics are poor since the plastics are more costly. Also plastics weigh tend to be more than the cellulose paper. So it is possible to save the forests (does it really need it since it is easy to replenish as the past century proved). Did you know when America was discovered and up until the end of the 19th century there were literally no trees when compared to those in USA now and any depletion can be replaced and even expanded (as one knows who is learned in this field). Another factor related to this tree myth is that when the world started its Computerized World it was said by many that much less paper would be required. Of course much more is used and required.

Synthetic paper now targets high-priced specialty applications that include beverage labels, restaurant menus, drivers' licenses, recipe books, instruction manuals, maps, and book jackets.

The justification to go to a higher cost paper material would be part of a system where the paper is continuously reused. By using a material which has erasable printing generated by the remote printing terminal, the newspaper or periodical could be printed at its destination.

Plastics can be used to make erasable printing media by a number of different techniques. Photo changing dyes could be incorporated into the structure of the plastics. The printer could change the dye to the colored form to read, and the material can be bleached with another unit that would reverse the photo coloring process. An ionic type plastic can be incorporated into the plastics and used to color the printed area by the use of an indicator type reaction with an organic acid or base. Another method would be to use a thermal printer in conjunction with liquid crystal type materials that would alter the state of the liquid crystals in the printed areas. Applying heat and electrical fields to the printed sheet would erase the printing.

Other schemes involving dichroic dyes with heat and electrical fields are also possible. Each of the possibilities could use the plastic structure of the substrates, its durability, or both. This approach would recycle the material for carrying the printed messages at the point of use, eliminating handling and distribution costs, and would require a fraction of the enormous amount of paper now consumed in delivering news and other literary material. The newspaper or periodical would have the familiar size and appearance and would present little change to the reader. The convenience of real on time home delivery and other built in aspects of the system would make it a useful successor to the present one. (This is just a point to discuss and amuse oneself but it could happen.)

Union Carbide Corp produced the first of the synthetic papers in the late 1960s. Since that time other examples of synthetic papers include DuPont's Tyvek nonwoven paper and Van Leer's Valeron cross-laminated film. This market is now dominated by a few large, worldwide ventures with proprietary processing techniques that extend the use of single and multilayer extruded blown film or cast film.

Developing Idea

Polyelectrolytes such as the ion exchange plastics form an interesting group of materials because of their ability to interact with water solutions. They have been used in medical applications involving the removal of heavy metal ions from the human body. They can be used to interact with external electric fields and change their physical properties drastically as is illustrated by the fact that some electrically active liquid crystals are polyelectrolytes of low molecular weight.

Another application for polyelectrolyte materials is in the forming plastics with unusual physical properties with regard to adhesion. The incorporation of small amounts of organic acid materials into polyolefin structures results in materials that have excellent adhesion to metals, paper, glass, and a variety of other materials. In addition, the materials with electrolytic structures can have a metal ion incorporated into the structure resulting in the formation of the ionomer type of resin that has much better mechanical properties than the basic polymer material.

Several other interesting material developments that may be useful to the designer in the future will be mentioned. Since one key step in any design is the material selection, an important aspect of the designer's responsibility is to be familiar with the range of material possibilities. Plastics interact with high energy radiation by giving bursts of visible light. This can be made selective for particular types of radiation and the effect has been used to make materials for scintillation counters to measure gamma radiation and particle streams such as alpha particles and beta rays.

The ability of some plastic systems to do this may be useful in schemes for handling the radiation output of nuclear devices, including the radiation from the fusion power machines under development. Obviously the application is not for shielding, which the heavy metals do much better, but rather for an energy level reduction system that would convert the high energy radiation to forms which would be more useful in power distribution.

There has been a great deal of interesting work done recently in attaching active enzyme materials to plastics substrates to convert simple organic molecules into the more complex forms used in biological processes. This technique makes available a catalyst bed capable of doing large-scale synthesis of materials such as proteins and carbohydrates that are essential to life processes. With a major food crisis always looming as a result of the rapidly increasing population of the world, it may be necessary to revive the possibility of synthetic food production even though this subject is very controversial. Since farm land is being depleted and recurring drought conditions reduce food supplies, it is likely that synthetic food will be a necessary supplement. The selective action of enzyme membranes may be a way to approach the food synthesis problem.

Joining and Assembling

Joining of a plastic product to another product composed of the same or a different plastic material, as well as other materials such as metal, is often necessary when: (1) the finished assembly is too complex or large to fabricate in one piece or (2) disassembly and re-assembly is necessary.

The success of a specific technique will depend on whether, as a by-product of the technique, sizable stress levels in the plastic product may result. Guarding against potential stresses in the assembly is a very important aspect of complete product design. There are many techniques that provide assembling all kinds of products. Each have technical and/or cost advantages and limitations. Examples of a few are reviewed in this section with more information in Chapter 3, **BASIC FEATURE and FEATURE INFLUENCING PERFORMANCE**.

Molded-In Insert

Plastics perform satisfactorily with metal-molded inserts and expansion-type inserts. To minimize the stresses created at the metal-plastic interface by the differences in thermal expansion rates for molded-in inserts, observe the following safeguards: (1) the design permitting, use plain, smooth inserts; (2) use simple pull-out and torque-retention grooves when high torque and pull-out retention are required; (3) if a knurled insert is used, keep the size of the knurls to a minimum, remove all sharp comers, and round the hidden end of the insert and keep the knurled section away from products' edges (Fig. 4-14), (4) keep the inserts clean, removing chemicals such as oil from them; (5) use high side of mold temperatures to reduce thermal stresses, such as for commodity plastics at 82 to 105°C (180 to 220°F); and (6) provide sufficient material around the insert.

Use the following guidelines for material thicknesses around inserts: with aluminum use 0.8 times the outer radius of the insert, with brass use 0.9 times it, and in steel

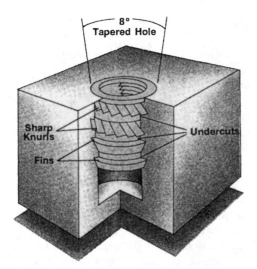

Fig. 4-14 Example of an insert with special features.

use a thickness equal to the outer radius. To ensure a proper interface, prototyping is recommended (Chapter 3, **FEATURE INFLUENCING PERFORMANCES, Injection Molding,** *Molded-in insert*).

When metal inserts require hermetic sealing, consider coating them with a flexible elastomer such as an RTV rubber, polyurethane, or epoxy system. A second method is to design an annular space or reservoir at one end of the insert from which to dispense the flexible elastomers to effectively create a hermetic seal. Flexible sealants are also used to compensate for differences in the thermal coefficient of expansion between metal and plastic.

Holding with Formed Head

A holding head is similar to the head formed during riveting except that in the plastic product there is a protruding stud that fits through a hole in the product to be joined and the head is shaped over it. It is an economical method of joining. Spinning or ultrasonic forming can shape the head.

The spinning operation consists of high-speed rotating and suitably shaped tool that creates frictional heat that will permit the stud to conform to the configuration in the tool. Pressure exerted on the tool and

the time of rotation are accurately controlled. The spinning device can produce joints, at high speed, of good quality.

Ultrasonic head forming and welding is a fast assembly technique. It is a very rapid operation of about 2 seconds or less and lends itself to full automation. In this process high-frequency vibrations and pressure are applied to the products to be joined, heat is generated at the plastic causing it to flow, and, when the vibrations cease, the melt solidifies. The heart of the ultrasonic system is the horn, which is made of a metal that can be carefully tuned to the frequency of the system. The manufacture of the horn and its shape is normally developed by the manufacturer of the equipment. The results from this operation are not only economical, but also most satisfactory from a quality control standpoint.

Snap Fit

Snap fits are widely used for both temporary and permanent assemblies, principally in injection and blow molded products. Besides being simple and inexpensive, snap fits have superior qualities. Snap fits can be applied to any combination of materials, such as plastic and plastic, metal and plastics, glass and plastics, and others. All types of plastics can be used (Chapter 3, **DESIGN CONCEPT, Snap Joint**).

The strength of a snap fit comes from its mechanical interlocking, as well as from friction. Pullout strength in a snap fit can be made hundreds of times larger than its snap in force. In the assembly process, a snap fit undergoes an energy exchange, with a clicking sound. Once assembled, the components in a snap fit are not under load, unlike the press fit, where the component is constantly under the stress resulting from the assembly process. Therefore, stress relaxation and creep over a long period may cause a press fit to fail, but the strength of a snap fit will not decrease with time (84).

When used as demountable assemblies, snap fits can compete very well with screw joints. The loss of friction under vibration can loosen bolts and screws. A snap fit is vibration proof, however, because its assembled products are in a low state of potential energy. There are also fewer parts in a snap fit that means a saving in component and inventory costs.

Successfully designing snap fits depends on observing a set of rules governing the shape, dimensions, materials, and interaction of the mating parts. The interference in a snap fit is the total deflection in the two mating members during the assembly process. Too much interference will create difficulty in assembly, but too little will cause low pullout strength. A snap fit can also fail from permanent deformation or the breakage of its spring components. A drastic change in the amount of friction, created by abrasion or oil contamination, may ruin the snap.

A snap can be characterized by the geometry of its spring component. The most common snaps are the cantilever type, the hollow-cylinder type (as in the lids of pill bottles) and the distortion type (Fig. 4-15). These snaps include those in any shape that is deformed or deflected to pass over interference. The shapes of the mating parts in a hollow cylinder snap is the same, but the shapes of the mating parts in a distortion snap are different, by definition. These classifications are rather nominal, because the cantilever category is used loosely to include any leaf-spring components, and the cylinder type is used also to include noncircular section tubes.

For high-volume production, snap fit designs provide economic, rapid assembly. In many products, such as inexpensive housewares or hand-held appliances, they are designed for one assembly only, with no nondestructive means for disassembling them. Where servicing them is anticipated, provision is made for the release of the assembly with a tool. Other designs, such as those used in the battery compartment covers for calculators and radios, are designed for easy release and reassembly many hundreds of times.

There is always some part of a snap fit that must flex like a spring, usually past a designed-in interference, and quickly return, or at least nearly return to its unflexed position, to create the assembly of two or more parts. The key to successful design is to provide sufficient holding power, without exceeding the elastic limits of the plastic. Fig. 4-16 shows a typical design. Using the beam equations, calculate the maximum stress during assembly. If it stays below the yield point of the plastic, the flexing finger will return to its original position. However, for certain designs there will not be enough holding power, because of the low forces or small deflections.

It has been found that with many plastics the calculated flexing stress can far exceed the yield point stress, if the assembly occurs too rapidly. In other words, the flexing finger will just momentarily pass through its condition of maximum deflection or strain, and the material will not respond as if the yield stress had been greatly exceeded.

A common way to evaluate snap fits is to calculate their strain rather than their stress. Then compare this value with the allowable dynamic strain limit for the particular plastic. In designing the finger it is important to avoid having sharp comers or structural discontinuities that can cause stress risers. A tapered finger provides a more uniform stress distribution, which makes it advisable to use where possible. Snap fits usually require undercuts, so a mold with a side action can be used. Another approach when an opening at the base of the flexing finger is permitted permits no use of a side action (3). There are times when all that has to be done is just pop it off the mold, taking advantage of the plastic's

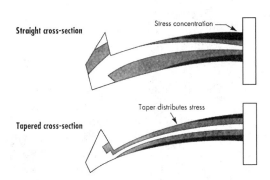

Fig. 4-15 Example of cantilever beam stresses in a snap fit.

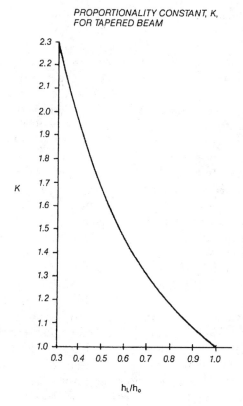

PROPORTIONALITY CONSTANT, K,
FOR TAPERED BEAM

h_L/h_o

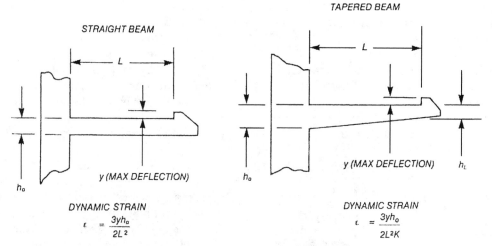

STRAIGHT BEAM

TAPERED BEAM

L

L

y (MAX DEFLECTION)

y (MAX DEFLECTION)

h_o

h_o

h_L

DYNAMIC STRAIN

$$\varepsilon = \frac{3yh_o}{2L^2}$$

DYNAMIC STRAIN

$$\varepsilon = \frac{3yh_o}{2L^2K}$$

Fig. 4-16 Basic snap fit design for a cantilever beam with a rectangular cross-section.

flexibility. Another type of system is the snap on or snap-in kind, used primarily in round products.

A snap fit can be rectangular or of a geometrically more complex cross-section. The design approach for the finger is that either its thickness or width tapers from the root to the hook. Thus, the load-bearing cross-section at any location relates more to the local load. The result is that the maximum strain on the plastic can be reduced and less material will be needed (Table 4-3). With this design approach, the vulnerable cross-section is always at the root (100).

Table 4-3   Examples of different cross section types of snap fit designs

Shape of cross section ▶ / Type of design ▼	A Rectangle	B Trapezoid	C Ring segment	D Irregular cross section
(Permissible) deflection — 1. Cross section constant over the length	$y = 0.67 \cdot \dfrac{\epsilon \cdot l^2}{h}$	$y = \dfrac{a+b_{(1)}}{2a+b} \cdot \dfrac{\epsilon \cdot l^2}{h}$	$y = C_{(2)} \dfrac{\epsilon \cdot l^2}{r_2}$	$y = \dfrac{1}{3} \cdot \dfrac{\epsilon \cdot l^2}{c_{(3)}}$
2. All dimensions in direction y, e.g. h or Δr, decrease to one-half.	$y = 1.09 \cdot \dfrac{\epsilon \cdot l^2}{h}$	$y = 1.64 \dfrac{a+b_{(1)}}{2a+b} \cdot \dfrac{\epsilon \cdot l^2}{h}$	$y = 1.64 \cdot C_{(2)} \dfrac{\epsilon \cdot l^2}{r_2}$	$y = 0.55 \cdot \dfrac{\epsilon \cdot l^2}{c_{(3)}}$
3. All dimensions in direction z, e.g. b and a, decrease to one-quarter.	$y = 0.86 \cdot \dfrac{\epsilon \cdot l^2}{h}$	$y = 1.28 \dfrac{a+b_{(1)}}{2a+b} \cdot \dfrac{\epsilon \cdot l^2}{h}$	$y = 1.28 \cdot C_{(2)} \dfrac{\epsilon \cdot l^2}{r_2}$	$y = 0.43 \cdot \dfrac{\epsilon \cdot l^2}{c_{(3)}}$
Deflection force — 1, 2, 3	$P = \underbrace{\dfrac{bh^2}{6}}_{Z} \cdot \dfrac{E_s \epsilon}{l}$	$P = \underbrace{\dfrac{h^2}{12} \cdot \dfrac{a^2+4ab+b^2}{2a+b}}_{Z} \cdot \dfrac{E_s \epsilon}{l}$	$P = Z_{(4)} \cdot \dfrac{E_s \epsilon}{l}$	$P = Z_{(4)} \cdot \dfrac{E_s \epsilon}{l}$

Welding

It is the joining TP parts by one of several heat-softening processes (2). Not all of them will be equally suited to a shape, size, or joining certain different material or even certain types of the same material. Different type fixtures or jigs are used during welding based on the method used. The different techniques are used to make permanent bonds between materials that can meet different requirements such as shapes, thickness, appearance, bond strength, capability of different being bonded, hermetic seal, or effect of additives or fillers used in the plastics (2).

Once a process is being used, recognize that if the compound additives or fillers are changed or added, bond performance can change or even not exist. As an example an unreinforced plastic can be welded to itself; however with a certain amount of glass fiber fillers (they do not melt) added to the plastic, action in weld strength can be reduced or even eliminated.

An example of welding is frictional spinning that can be applied to two plastic products with circular joints. It is especially suitable for large parts where ultrasonic welding may be impractical or equipment cost prohibitive. In this operation the faces to be joined are pressed together while one part is rotated and the other is held in a fixed position. Frictional heat produces a molten zone that becomes a weld when rotation stops. When alignment is precise and centering means are incorporated in the parts, the result is a good joint in terms of strength and appearance. The approximate parameters are 40 to 50 feet per minute peripheral speed and 300 to 400 psi pressure.

Ultrasonic welding's principle of operation requires that the design of the joining surfaces meet special requirements. The important feature in ultrasonic welding is the energy director that consists of an initial small contact area through which the flow of energy is started. Variations of this design can be adopted readily to larger parts without the need of resorting to proportionately larger welding facilities. The cycle times are fast (less than 2 seconds) and energy consumption is low due to the fact that only a thin layer of material on both components is softened, from which a good welding joint is obtained.

Summary

The joining and assembling of plastics is only limited by the ingenuity and skill of the designer. The only precaution that has to be exerted is that no stresses are generated during the operation. Basically the preferred method is to have the material softened so that it can flow and adjust to the new condition. The softening need be only a few mils thick to have a favorable result. The plastic does not care as long as the temperature is within its melt limits. Not all plastic permit a melting action so they may be difficult to bond. Some of these plastics require special surface treatment to obtain a bond.

Predicting Performance

Avoiding structural failure can depend in part on the ability to predict performance for all types of materials (plastics, metals, glass, and so on). When required designers have developed sophisticated computer methods for calculating stresses in complex structures using different materials. These computational methods have replaced the oversimplified models of materials behavior relied in the past. The result is early comprehensive analysis of the effects of temperature, loading rate, environment, and material defects on structural reliability.

This information is supported by stress-strain behavior data collected in actual materials evaluations. With computers the finite element method (FEA) has greatly enhanced the capability of the structural analyst to calculate displacement, strain, and stress values in complicated plastic structures subjected to arbitrary loading conditions (Chapter 2). FEA techniques have made analyses much more precise, resulting in better and more optimum designs.

Nondestructive testing (NDT) is used to assess a component or structure during its operational lifetime. Radiography, ultrasonics, eddy currents, acoustic emissions, and other methods are used to detect and monitor flaws that develop during operation (Chapter 5).

The selection of the evaluation method(s) depends on various factors such as the specific type of plastic, the type of flaw to be detected, the environment of the evaluation, the effectiveness of the evaluation method, the size of the structure, and the economic consequences of structural failure. Conventional evaluation methods are often adequate for baseline and acceptance inspections. However, there are increasing demands for more accurate characterization of the size and shape of defects that may require advanced techniques and procedures and involve the use of several methods.

Designing a good product requires a knowledge of plastics that includes their advantages and disadvantages (limitations) with some familiarity of the processing methods. Until the designer becomes familiar with processing, a reliable fabricator Or fabricators) must be taken into the designer's confidence early in the development stage and consulted frequently during those early days. The fabricator and the mold or die designer should advise the product designer on plastic materials behavior and how to simplify the design to permit easier processability (Chapter 3, **FEATURES INFLUENCING PERFORMANCE**).

There are material and processing limitations that can influence the fabrication and performances of products. These subjects are reviewed in Chapter 6, **MATERIAL VARIABLE** and Chapter 8, **EQUIPMENT/ PROCESSING VARIABLE**.

Design Verification

DV refers to the series of procedures used by the product development group to ensure that a product design output meets its design input. It focuses primarily on the end of the product development cycle. It is routinely understood to mean a thorough prototype testing of the final product to ensure that it is acceptable for shipment to the customers.

In the context of design control, however, DV starts when a product's specification or standard has been established and is an on-going process. The net result of DV is to conform with a high degree of accuracy that the final product meets performance requirements and is safe and effective. According to standards established by ISO-9000, DV should include at least two of the following measures: (a) holding and recording design reviews, (b) undertaking qualification tests and demonstrations, (c) carrying out alternative calculations, and (d) comparing a new design with a similar, proven design.

Design and Safety

The ultimate requirement of any product is that it performs the function for which it is designed. With many materials the design life of the product is usually not as important as it is with plastics because of the behaviors such as creep of plastics (Chapter 2). In all cases the useful life is an important consideration whether the item is a pan for the kitchen or a bridge to handle traffic in a city.

The people who use the designed product expect it to be properly designed to perform satisfactorily in the intended environment for the indicated life, without endangering any person or becoming functionally useless before the end of the predicted life. This, of course, implies that the user does not abuse the product and maintains it properly. It is the responsibility of the designer to provide the user with sufficient information so that one can intelligently use the product and properly maintain it. No product can be guaranteed to perform properly if it is abused.

The responsibility is always reviewed by different organizations. This has become a matter of legal as well as moral responsibility. Improper design or inadequate safety instructions can lead to litigation in civil and criminal action. It is important for the designer to keep the ultimate user in mind when one designs a product. One must also make adequate allowance for human error and poor judgment to prevent malfunction and possible hazards to the user or to other property (such as a car that can be a real life threatening hazard).

Plastic products are used most everywhere and it would be difficult to describe a typical use situation. One environment familiar to most people is the home and this environment has enough hazardous elements to serve as a useful starting point for this discussion. The kitchen is a particularly difficult environment. The list of hazards includes a surprising range of physical and chemical environments that make the kitchen a torture chamber for plastics (or other materials) used in this area. If the stresses that the plastics encounter in use are included, it is apparent that kitchen items require careful designing to resist the environment. For example, an egg beater has plastics gears that are exposed to chemical attack from food oils, acetic acid (vinegar), fruit juices, cleaning detergents, etc.

Environmental factors for kitchen appliances are many. Examples are as follows: heat, cold, water, water erosion effects, impact, high humidity at elevated temperature, chemicals (soaps and detergents, oils, fats, and greases, fruit juices, phosphates in dishwasher cleaners), fruit acids, caustic bleach, etc.), biological exposures, fungus garbage, microorganisms, enzymes, vermin, mechanical loading (static and/or dynamic), and so on.

This type of caution approach is essential in high risk applications since only certain plastics can be used as the basis for such applications. This chapter has been concerned with the problems of the product in use under the types of environments and potential abuse that may be encountered, and how the designer can prevent premature product failure which is one of their major responsibilities.

Determination of the hazard potential and designing to eliminate the hazard is one element of the solution. The use of carefully designed accelerated and continuous prototype testing is another element. The instruction in use and proper maintenance procedure is the third element. By exercising judgment as to the appropriate combination of these elements consistent with the economic factors involved, the designer can have a product that will perform for its projected design life with a minimum of hazard to the user and with a high degree of satisfaction in use.

The application area is so large that no cookbook advice would be useful. Each type of application must have the potential failure evaluated, the economics calculated, the degree of assurance of performance set, the extent of required testing determined, and the product designed and evaluated to meet the criteria established. There are sources of data available on which to base the approaches to the requirements, but the final combination of factors must be determined by the designer.

Product evaluations must be done more carefully and the testing must be more sophisticated and extensive. A systematic approach is essential for the evaluation of a plastic (or any material) and this must be backed up by a quality control system to insure that the actual products are made to perform properly. A guide to the steps involved follow:

1. Material selection to meet the requirements of the application, followed by materials testing to indicate the performance over the range of operating conditions expected.

2. Construction of test samples that can be subjected to actual exposure to the end-use environments which are tested to destruction to determine the possible modes of failure and the conditions that cause the failure.

3. Design of the product to meet the anticipated environmental stresses as well as the functional requirements. The product must have a designed-in means for reducing the possibility of failure plus either a fail-safe mode or an indication of incipient failure that can be monitored.

4. Extensive testing of the product in simulated service as well as accelerated testing. The testing program should include a method of evaluating the effects of fabrication procedures on the product performance and methods to inspect for compliance with the proper procedures. This may include destructive testing, x-ray inspection, ultrasonic testing, and a wide range of other rather sophisticated testing procedures.

5. Introduction of the product into limited service with constant monitoring of its performance. As the experience factor on the product performance increases its use can be extended. The continuing testing of the products under the procedure in step 4 is essential so that if degradation in the product performance is seen, it will show up in the tests before it becomes a serious problem in the end use.

6. Continuous monitoring in use over a period of several years to insure continued performance and a replacement program that will take products out of service after a conservative use-life to preclude possible failure.

Risk

As reviewed designers in the plastics and other industry have the responsibility to ensure that all products produced will be safe and not contaminate the environment, etc. Recognize that when you encounter a potential problem, you are guilty until proven innocent (or is it supposed to be the reverse). So keep the records you need to survive the legal actions that can develop.

There are many risks people are subjected to in the plant, at home, and elsewhere that can cause harm, health problems, and/or death with plastic products representing very few. Precautions should be taken and enforced based on what is practical, logical, and useful. However, those involved in laws and regulations, as well as the public and, particularly the news media should recognize there is acceptable risk.

Acceptable Risk

This is the concept that has developed decades ago in connection with toxic substances, food additives, air and water pollution, fire and related environmental concerns, and so on. It can be defined as a level of risk at which a seriously adverse result is highly unlikely to occur but it cannot be proven whether or not there is 100% safety. In these cases, it means living with reasonable assurance of safety and acceptable uncertainty.

Examples of this concept exists all around us such as the use of automobiles, aircraft, boats, lawnmowers, foods, medical devices (231), water, air we breathe, news reports, and so on. Practically all elements around us encompass some level of uncertainty and risk. Otherwise as we know it would not exist.

Interesting that about 1995 a young intern at FDA made some interesting calculations. If they permitted the packaging of Coca Cola in acrylic barrier plastic bottles, and if you drank 37,000 gallons of coke per day for a lifetime, you would have a 10% risk of getting cancer. Since normal people have a 25% risk of getting cancer, reducing it by 10% was a real plus for coke and the acrylic barrier plastic bottles. So perhaps a law should be enacted requiring that the public should drink "lots" of coke.

People are exposed to many risks. Some pose a greater threat than others. The following data concerns the probability over a lifetime of premature death per 100,000 people. In USA 290 hit by a car while being a pedestrian, 200 tobacco smoke, 75 diagnostic X-ray, 75 bicycling, 16 passengers in a car, 7 Miami/New Orleans drinking water, 3 lightning, 3 hurricane, and 2 fire.

Perfection

The target is to approach perfection in a zero-risk society. Basically, no product is without risk; failure to recognize this factor may put excessive emphasis on achieving an important goal while drawing precious resources away from product design development and approval. The target or goal should be to attain a proper balance between risk and benefit using realistic factors and not the "public-political panic" approach.

Achievable program plans begin with the recognition that smooth does not mean perfect. Perfection is an unrealistic idea. It is a fact of life that the further someone is removed from a task, the more they are apt to expect so called perfection from those performing it. The expectation of perfection blocks genuine communication between designers, workers, departments, management, customers, vendors, and laws (lawyers). Therefore one can define a smoothly run program as one that designs or creates a product that meets requirements (safety, etc.), is delivered on time, falls within the price guidelines, and stays close to budget.

Perfection is never reached; there is always room for improvements as summarized in the FALLO approach (Fig. 1-3) and throughout history. As it has been stated, to live is to change and to reach perfection is to have changed often (in the right direction).

In addition to the product, the designer, equipment installer, user, and all others involved in production should all consider performing a risk assessment and target in the direction of perfection. The production is reviewed for hazards created by each part of the line when operating as well as when equipment fails to perform or complete its task. This action includes startups and shutdowns, preventative maintenance, QC/inspection, repair, etc.

Plastic/Process Interaction

Material and process interaction and their effects on the performance of plastic products are important factors for the designer to understand. It can result in design limitations such as a selected material meeting performance requirements but not processible by the desired method of fabrication. Most of the limitations that are reviewed in this book can be corrected and do not effect the product performances when qualified people handle the limitations. However they are presented to reduce or eliminate potential problems.

When a product is made from a plastic, in most cases the fabricating process could subject the material to rather severe conditions such as excess elevated temperatures, high pressures, high shear rate flow, and/or chemical changes. These interactions can place limitations on the design approach. Recognizing the limitations or problems that can develop, the successful design will be a compromise between the requirements of function, productibility, and cost. Examples of these limitations or problems are presented. Only the major limitations or problems are presented even though there are usually exceptions. If exceptions are important, they are included.

Even though this review pertains to certain fabricating processes, they can be related to other processes (1, 3, 6, 8, 9, 11, 50, 62, 64, 190).

Molding

Molding can be classified as high or low pressure processing systems. High pressure identifies processes such as injection molding, compression molding, and transfer molding. Low pressure identifies reaction injection molding, rotational molding, and form molding; also identifies processes such as extrusion, thermoforming, blow molding, and casting. Each process has its advantages and limitations as reviewed in Chapter 8. Since more interacting problems exist with the high pressure systems the following review highlights this system. Even though these limitations or problems can be related to low pressure systems, the low pressure systems are usually much easier to eliminate or control.

Injection Molding

In high pressure molding, such as injection molding, the walls and other sections of a TP product represent to the designer the structures required to make the product functional for its intended use. To the mold designer and molder they represent the flow path for the plastics material. This flow takes place at high rates and under the (controllable) complicating conditions of flow in a passage much cooler than the plastic, through a gate (orifice) whose dimensions are severely restricted to reduce the effect on the appearance of the product. With these complications in mind, it is apparent that it may not be possible, or it may be very difficult, to mold certain shapes. Large area products with thin walls represent one class of products that can present difficulties to certain molders, however it is routinely done in qualified operations (Chapter 8, **INJECTION MOLDING**).

Freezing action Because of the heat exchange between the flowing TP melt and the mold walls, the flow may freeze (solidify) before the product is completely filled. Products that have alternate sections with thick and then thin walls can cause problems in flow and cooling that make them difficult to fill. In some cases the plastics that have been selected for the end use requirement are too viscous to flow properly in a mold cavity, and this makes the manufacture difficult.

TS plastic products that are injection, transfer, or compression molded combine thick and thin sections relatively easily since the hardening process is a chemical reaction (Chapter 6). Annular shapes are best made by compression to gain best dimensional control and freedom from distortion. In the compression process, the molding compound is compressed and reduced to the plastic state in the mold. During this process, portions of the material may lie in hard forms in the mold while other portions are flowing rapidly with great force.

Without proper preheating or mechanical plasticizing of the charge, portions of the product may be uncured and low in density. Transfer (compression) molding insures better properties under average conditions. Impact materials that include long fibers can easily be compression or plunger molded. Screw injection will usually break up the long fibers and produce weaker products thus short fibers are usually used.

Thin to large wall Designing around TP problems is the joint responsibility of the product and mold designers. For example, one way to handle the problem of thin to large area walls is by the inclusion of long ribs into the product in the direction of plastic flow. These ribs are not a functional requirement of the product but they act as auxiliary runners attached to the product to facilitate plastic flow in difficult to fill areas. In some instances the ribs may be used as a surface decoration like a corrugation or they may be on the concealed side of the product where they are stiffeners.

Another problem in molding is the existence of contiguous areas of thick and thin sections in the flow direction. In some cases placing the gate in the thinner section can control this problem. However the usual approach is to gate the thicker section to ensure the complete fill in both sections. Where there are several thick sections multiple gates are used. There could be a limitation on this approach because the weld lines produced by the joining of the several plastic flows are a weak point on the molded product.

In some cases the use of a section that spans the thick and thin regions can be used to act as a built-in runner. This may be across the product or it may be a thickened edge or frame around the product. Still another approach would be to redesign the product for a more uniform wall thickness. The additional wall thickness required can be supplied on a mating part.

Melt flow restriction   This preliminary discussion of the flow restriction problem was concerned with the simple and obvious necessity of filling the mold with plastic. There are a number of other consequences of restricted flow in molding which are less obvious and, generally, more significant to the performance of the product. Restricted flow cause high shear rates in the material as it fills the mold. This necessitates the use of higher injection pressures and usually the use of higher melt temperatures and higher melt index materials to fill the mold cavity can result in lower product performance.

Higher melt index materials generally have lower impact and lower strength properties. The use of higher temperatures usually results in degradation of the plastic properties. Monomer or other low molecular weight breakdown products can be produced which drastically reduce the properties of the plastic material. The high shear rates encountered in molding also result in degradation of the molecular weight of the material just from the shearing action. The shear rate is directly dependent on the pressure drop in a channel and the pressure drop is a cubic function of the channel height.

The high shear rate produces two other effects that significantly affect product performance. The plastics molecules become

aligned as a result of the high shear flow so that the material in the walls is highly oriented in the flow direction. This may be a desirable effect. For example, a restrictor bar is used in molding polypropylene products to generate a living hinge effect by orienting the material. In some materials such as polyamides (nylons), the unidirectional orientation results in improved strength in both the flow direction and perpendicular to the flow direction. In most cases the effect is undesirable since the strength in the direction perpendicular to the flow direction is reduced and the product has a tendency to split along a flow line. In addition, the oriented materials have reduced elevated temperature properties in that the orientation tends to be relieved at a fairly low temperature and the product will distort as a result of the deorientation process.

Residual stress   There is a condition that develops, particularly in products with thin walls. This is a frozen-in stress, a condition that results from the filling process. The TP flowing along the walls of the mold is chilled by heat transferring to the cold mold walls and the material is essentially set (approaching solidification). The material between the two chilled skins formed continues to flow and, as a result, it will stretch the chilled skins of plastics and subject them to tensile stresses. When the flow ceases, the skins of the product are in tension and the core material is in compression that results in a frozen-in stress condition. This stress level is added to any externally applied load so that a product with the frozen-in stress condition is subject to failure at reduced load levels.

There are other conditions that result from the frozen-in stresses. In materials such as crystal polystyrene, which have low elongation to fracture and are in the glassy state at room temperature, a frequent result is crazing; it is the appearance of many fine microcracks across the material in a direction perpendicular to the stress direction. This result may not appear immediately and may occur by exposure to either a mildly solvent liquid or vapor. Styrene products dipped in kerosene will craze quickly in stressed areas.

In any event, the crazing effect can lead to premature product failure. An annealing operation may minimize these stresses.

There is another result of frozen-in residual stresses that can be equally damaging to the product function and which affects materials that are not in the glassy state. This may affect an impact grade of material or a crystalline plastic even more drastically than a glassy material. The frozen-in stresses are real loads applied to the material and when even slightly elevated temperatures are applied stresses can cause the product to deform severely.

What has been reviewed are interactions between the molding process and the product design. Poor process control during molding can produce these severe orientation and frozen strains. However, there are designs used particularly with certain plastics where it is impossible to avoid the frozen strain and orientation problems. For transparent products these difficult-to-mold products the condition can be observed by the means of the photo-stress effect. Examination of a transparent molded product by polarized light will show the combined effect of the orientation and the frozen stresses (Fig. 5-2). It is difficult to determine which effect is being observed since both have the same birefringence effect on the polarized light. The use of reflected polarized light from the surface gives a somewhat different reading of the effect. This may be a way to separate the two effects that would be desirable since the result on the performance is different for each condition.

It would be desirable to make sample prototype tooling and analyze the flow effects on a product that is likely to present a flow problem. In addition to the usual physical testing of the product, the use of photo-stress analysis techniques plus the exposure to selected solvents to check for stress crack characteristics would lead to changes in the product to minimize the effects of the molding on the product performance. As an example there have been cases in the past where piano keys with frozen-in stresses have been released from perspiration, leaving open flow lines (Chapter 5, **STRESS ANALYSIS**).

Gate area The gate area on a molded product represents another processing problem in molded products. The gate causes severe restriction to particularly TP melt flow since it is always desirable to have it as small as possible to reduce its visibility on the product. Because of the especially high shear rate on the material as it passes through the gate, the material is heated because a substantial part of the potential energy represented by the pressure on the material is converted to heat by friction.

The effect on the material can be drastic and, in the case of shear sensitive materials, there is substantial degradation in the molecular weight of the material as it passes through the gate. If the material is a filled one, such as a fiber glass material, the severe flow patterns generated at the gate will break up the reinforcing materials and can convert fiber to powder with a substantial loss in the reinforcing properties of the filler. (The same action can occur when a screw plasticizer is used.) These type problems can be reduced or basically eliminated with proper mold design and process control during fabrication.

There is not too much orientation that is permanently added to material as it passes through the gate since the continued flow in the cavity basically tends to produce turbulence that destroys the orientation. The last material to pass through, however, does retain its orientation and the gate area in a molded product is usually highly oriented and could be weak. In the case of jetting, the result is a patch of highly oriented material somewhere on the molded product near where the first material entered the mold.

Jetting Jetting is a condition that results when the mold design has no immediate impediment to flow and the plastics is ejected into a relatively large open volume. This jetted material becomes a weak point on the product and a surface blemish that is difficult to conceal.

It is controlled by changing the gate to direct the material to a nearby wall to slow the initial flow, by changing the size of the gate to reduce flow rate, by changing the shape of the gate, and/or by making adjustments in the fill rate on the product during molding. The design's involvement in these areas is achieved by consultation with the mold designer and molder to indicate which options are available on the product that will not interfere with its function.

Weld line Weld or knit lines where two parts of a melt join while flowing into the mold cavity can result in problems. The quality of the weld depends on the temperature of the material at the weld point and the pressure present in the melt after flowing from the gate. The higher the temperature and pressure, the more complete the weld and the better the product performance and appearance. Bringing the material to the weld point at a higher temperature and pressure requires rapid filling of the mold cavity (Chapter 3, **BASIC FEATURE, Weld Line**).

This action tends to produce flow orientation of the material and the possibility of induced frozen-in stresses that will also detract from product performance. In order to reach a reasonable compromise on these problems, the molder can operate the mold at a higher temperature. However it will increase machine cycle time due to longer cooling time resulting in higher production costs. The product designer can minimize the problem by increasing the wall thickness to permit easier flow, by the use of ribs to act as built in runners to improve and redirect the flow of material in the cavity, and/or by modification of the design to shift and/or eliminate obstructions to flow. In some cases holes may be molded partially through the section to eliminate weld lines so the melt flows through it. Mold makers will suggest mold designs that minimize weld lines.

These are the primary process interactions that the designer must be aware of in order to determine process interference in product performance and design. Specific materials may introduce other problem areas as, for example, air entrapment, differential expansion, and the problem of a level of crystallinity in a crystalline plastic that exceeds the allowed level for stability of a product.

Venting Proper venting of the mold cavity is essential for the successful molding of plastic products. Since venting may influence the product design, it is desirable to consider the different venting techniques used in a mold cavity (1, 3, 9). Included for certain plastics is "breathing" or "bumping" the mold halves to eliminate entrapped air (3).

Molded TSs, whether molded by injection, transfer or compression, also have design restrictions imposed because of the chemical curing action that takes place during the molding and curing. Certain specific problems occur with specific TSs. For example there are phenolic materials and others that evolve gaseous products during the cure. They can have porosity problems caused by insufficient pressure applied to a particular area of the product. This lack of pressure in an injection or transfer molded is caused by filling at too slow a rate so that the pressure is not transferred from the gate to the remote portion of the cavity before the reaction causes the material to set up and block pressure transfer. In some cases this is a result of the product design that has complex paths for the material to fill. It can be overcome by redesigning the product to increase the flow as suggested for TP products. Part of the problem can be corrected by changes in the grade of material. Here, as in the other cases, determining which factors are best changed in cooperation with the mold designer and molder will produce the quality result.

Extrusion

For several basic reasons, the extrusion process does not have the large number of possible process product interactions that the preceding molding methods presented. Due to this situation it can not fabricate the complex shapes and tighter tolerances obtained from molding. The process is a steady-state continuous production operation that can be brought to a condition of control. However it has its share of potential problems (Chapter 8, **EXTRUSION**).

The operating pressures and shear rates in the extrusion process are considerably lower than they are in molding. As it exits the die, but not necessarily when it leaves the process, the material is in an essentially stress-free condition. Depending on the wall thickness of the material and the particular material, there is orientation of the plastic to a greater or lesser controllable degree. Thin walls produce higher orientation in materials such as PP, that is a highly crystalline polyolefin, and which orients much more than materials such as PVC.

Melt flow After the material (extrudate) leaves the die, it is usually drawn in size and passed through a set of auxiliary equipment that basically can form the final desired shape (6). The material is also cooled either by subjecting it to air flows, by immersing it in a water tank, by subjecting the extrudate to a water spray, and/or these combinations. There are other techniques such as having the material drawn through a chilled metal mandrel by the use of vacuum applied into the mandrel. These draw-down, forming, and cooling procedures can and do introduce stresses in the product which can affect the performance of the extruded materials if they are not properly controlled.

Memory One commonly encountered problem with extruded products as a result of processing interaction, particularly with materials such as acrylics and vinyls which have an extensive type of "memory" characteristics, is that the product will shorten in the machine direction and thicken in the cross machine direction with the application of even low heat. This effect is analogous to the molding melt flow orientation reviewed and results from the orientation produced by the draw-down process and frozen-in stresses produced because the draw-down was done at too low a temperature. Generally the die-induced orientation is not a major factor in this effect since it can be corrected usually by changes in the process operating conditions and/or modification of the material. Any orientation can cause this effect at high enough temperature, generally near the glass transition or crystalline melting point.

The designer should be aware of the fact that this is to be considered in designing with extruded products. The designer can exercise little control over this pull back condition except to be guided by the experience of the extrusion processor to indicate which materials are particularly susceptible to this problem and what the recommended wall thicknesses are to minimize the effect. In general, one of the best ways to improve the condition is to slow down the rate of extrusion. As a result, products have a tendency to pull back. They also will be more costly to produce.

Distortion Another problem with extrusions is caused by distortion of the section by the effect of heat and other environmental conditions such as exposure to water or chemical agents that tend to soften the plastic. These distortions are generally reversals of the profile back to the shape that it had exiting in the die. This action indicates that the post die forming operations were done at a lower than desirable temperature which results in a molded-in stress. When the stress is relieved the product distorts. In some instances these stresses cannot be eliminated by process changes so that the product is inherently deficient in performance.

One way the designer can cope with this situation is to indicate to the die designer and extrusion processor what the anticipated operating conditions for the product are, so that the design of the tooling will minimize the potential for distortion. This action provides the processor to: (1) better control of the shape leaving the die achieved by more careful die design and correction so that a minimum of post die shaping is required, (2) operating the line at a higher temperature when shaping jigs are used, (3) careful cooling of the extrudate, and (4) finally by generally operating the process at lower rates to insure better process control.

Dimension One general problem that exists with extruded shapes could be the control of dimensions. Because the production rate can affect the relative dimensions in an extrusion as well as the overall size, dimension control becomes an important economic factor. For economical production of extruded products it is advisable for the product designer to indicate which dimensions, if any, are really critical to the function of the product so that these are controlled, and to indicate the widest acceptable variance on other less critical dimensions. In this way the extrusion operator can adjust the process to the maximum speed consistent with the production of a usable product.

Insistence on all dimensions as critical will result in the process being restricted to one narrow set of operating conditions, usually at a low production rate, with a high scrap rate and high product costs. In cases where the dimensions are critical, it may be that the extrusion process should be discarded in favor of molding or machining of the product. However controlling all dimensions can be accomplished at a cost.

Subject to the limitations indicated, control over extrusion process products is consistent enough to make a uniform repeatable product once the limitations are accommodated. Here, as in other processing, good communication between the processor and the designer will help make for a successful economical product.

Thermoforming

Sheet forming processes, such as vacuum forming, do have effects on the product. The designer should be aware that these will affect the performance of one's product and one should learn how to modify the design to minimize any deleterious effects. Probably the most serious problem encountered in formed film or sheet products results from the fact that the materials are made from film or sheet at temperatures well below the melt softening point of the plastic, usually near the heat distortion temperature for the material. Forming under these condition when the draw down ratio is exceeded for a specific plastic can result in over stretched orientation of the material, the production of frozen-in stresses, poor product reproducibility, and/or immediate or in service failure by the product cracking (Chapter 8, **THERMO-FORMING**).

Stress These conditions are unavoidable by the very nature of the sheet forming processes. The designer must accept the fact that the heat resistance of a sheet formed product and its resistance to other environmental stress factors even though lower than for a molded or extruded products do exist. The objective of the designer should be to minimize the amount of stretching needed to make the product so that the over-all performance will be as close to the molded product as possible. Also by using tighter film or sheet thickness dimensional controls, thermoforming permits more accurate reproduction followed with reduced product costs (1, 6).

There is a variation or degree of stretching that occurs as a film or sheet is drawn down over several different male or female shapes. The variation in stretching can occur in different portions of the film or sheet as the corners become sharper. There is a good correlation between the extent of stretching and the susceptibility of the product to damage; the degree to which it will occur will vary widely from one material to another. A material such as rigid PVC or cellulose acetate propionate will be much less likely to show damage when subjected to thermal or environmental stress than a polystyrene or polyethylene.

Memory The nature of the damage due to "memory" to the TP product varies from one material to another. One temperature effect common to all materials results in products where it tends to revert to the original shape. The extent of this and the temperature at which it will occur will depend on the material, the forming process operating conditions, and the product design. With regard to the design, the most stable TP products are those with generous corner radii and smoothly blended surfaces with a minimum of sharp corners and reentrant curves that will stretch the material excessively.

TP materials are stable when properly molded. The best processing conditions will preheat the material to the highest possible uniform temperature of the material particularly thicknesswise and then form the sheet as rapidly as possible. There are limitations on the process temperature because some materials such as PE have a narrow range of forming temperatures while others such as PVC may be susceptible to thermal degradation.

The speed of closure to the form is a function of machine condition and sheet thickness, with the thicker sheets being more difficult to move rapidly. The designer should indicate this to the sheet former when form stability at elevated temperatures is critical and the process must be tailored to improve this condition. One should also select a material which is intrinsically more stable and easily formed to minimize the possibility of unmolding. These factors must be considered in conjunction with the design of the product to minimize sheet stretching differentials in the part.

Orientation In addition to where the product tends to revert to the original shape, the effects of the stretching of the sheet materials result in two other impairments of the product. Highly stretched sections are usually thin and highly oriented. It has a decided tendency to split in these areas in a direction parallel to the stretching that took place. The designer should consider this potential situation and consider thicker material so that the product will perform adequately in use.

The other effect of having a stretched area is a reduction in resistance to stress cracking. Crazing is a possibility in such areas such as in polystyrenes, and environmental stress cracking caused by solvent substances will occur in the stretched areas. This is a particularly important consideration in vacuum formed products used for packaging food that frequently has some solvent action on the plastics.

Blow Molding

With respect to the BM process, the type of situation where the material is stretched at temperatures below the melt temperature applies in a similar manner to that reviewed above for thermoformed sheet processing, only to a lesser degree. It is convenient to think of extrusion or injection BM as a more generalized form of a plastics reforming process such as sheet forming. The comments on designing to minimize points of sharp stretching and excessive draw mentioned in forming apply to blow molded products as well (Chapter 8, **Blow Molding**).

By extrusion parison control it is possible to minimize the wall thickness variation and the extent of stretching and stretch orientation. These are the province of the processor when the designer is not familiar with BM. Knowledge is required to provide information on what is possible and to select the specific BM process that has the capability to mold the product. The designer should be aware of the possible failure modes and compensate for them in the design. There is little else the designer can do but select the best material and process to make the product.

Complex design BM has started to full fill its rather unlimited capability in producing complex shaped products. The reason is due to designers becoming familiar with the behavior of BM processes and their exceptional design capabilities and few limitations (Chapter 8).

Casting

The process interaction in cast plastic products is mainly involved with the curing processes and with mold filling problems. Voids and porous sections are a frequent problem with castings because the mold filling is done at atmospheric pressure, or low pressure, and if the product has thin sections to fill, the flow may be a problem.

Other than designing to avoid such difficult flow conditions and selecting a material with good flow characteristics that will perform properly, the designer must rely on the skills of the fabricator to make good products. Frequently a casting is selected because of the low tooling requirements and rapidity with which the product can be put into production. After the production level is increased and the requirements for better product quality are imposed, it may be desirable to change to pressure molded products when the higher production level justifies the increased tooling costs. However there are high production casting lines, such as encasing jewelry, emblems, etc., that are more beneficial costwise.

In many cases design modifications can substantially improve the producibility of the products and reduce their cost with improved product quality. As an example if voids exist the problem can usually be corrected by modifying or changing the plastic's composition and/or the use of a vacuum system during the casting. Understanding the effects of the process on the product is essential in making successful products.

Specific information in this area is usually available from the processor who has experience with a wide variety of products and knows the type of problems that have been encountered in the past.

Because of the complexities of the materials and the effects of the processing on the materials, this is an area where predictability based on scientific data is very limited and casting experience is desirable. Successful products will result from close cooperation between designer, tool designer, mold maker, and processor to arrive at design compromises that make the products acceptable and producible at an economical price.

Law and Regulation

The consuming public must assume that the producer of a product has shown reasonable consideration for the safety, correct quantity, proper labeling, and other social aspects of the product. Since the 1960s these types of important concerns have expanded and been reinforced by a recognition of the consumer's right to know as well as by concerns for conservation, ecology, antilittering, and the like. Numerous safety-related and socially responsible laws have been enacted and more are on the way.

A designer's failure to be aware of and comply with existing regulations can lead to legal entanglements, fines, restrictions, and even jail sentences. In addition, there are also the penalties of costly, damaging publicity, lawsuits, and the loss of consumer goodwill. In the meantime, as for all other industries, the goal of reliable companies and associations is to produce products that eliminate potential problems. Unfortunately, nothing is perfect, so problems can develop, which is simply a fact of life. And there is always more to be done, as in the disposal issue (like eliminating wars and having all people like each other).

There are many examples of action to eliminate or reduce problems. As an example there is the Quality System Regulation (**APPENDIX, TERMINOLOGY**). FDA requires details on how products made of different materials (steel, glass, plastic, etc.) such as medical devices be manufactured. The details of the process are documented so that once a product produced in USA is approved, the product can only be produced by following what was in the QSR preparation. No change can be made. The exact plastic composition has to be used, process control settings remain the same, etc. Literally if a waste paper basket had been identified and located in a specific location in the plant, you can not relocate, change its size, etc. It has been reported that to make a change could cost literally over a million dollars. Result of the QSR regulation is too ensure the safety of a person when the medical device is used. The QSR approach should be considered/used by plastic fabricators and/or product customers when rigid requirements exist.

On just the subject of appliance safety the Underwriters Laboratories (UL) have published more than four hundred safety standards to assess the hazards associated with manufacturing appliances. These standards represent basic design requirements for various categories of products covered by the organization. For example, under UL's Component Plastics Program a material is tested under standardized, uniform conditions to provide preliminary information as to a material's strong and potentially weak characteristics.

The UL plastics program is divided into two phases. The first develops information on a material's long- and short-term properties. The second phase uses these data to screen out and indicate a material's strong and weak characteristics. For example, manufacturers and safety engineers can analyze the possible hazardous effects of potentially weak characteristics, using UL standard 746C.

Products manufactured using concepts in UL Standard 746D provide quick verification of material identification, along with the assurance that acceptable blending or simple compounding operations are used that would not increase the risk of fire, electrical shock, or personal injury.

The Standard for Tests for Flammability of Plastic Materials for Parts in Devices and Appliances (UL 94) has methods for determining whether a material will extinguish, or burn and propagate flame. The UL Standard for Polymeric Materials-Short Term Property Evaluations is a series of small-scale tests used as a basis for comparing the mechanical, electrical, thermal, and resistance-to-ignition characteristics of materials.

It is the general consensus within the worldwide "fire community" that the only proper way to evaluate the fire safety of products is to conduct full-scale tests or complete fire-risk assessments. Most of these tests were extracted from procedures developed by the American Society for Testing and Materials (ASTM) and the International Electrotechnical Commission (IEC). Because they are time tested, they are generally accepted methods to evaluate a given property. Where there were no universally accepted methods the UL developed its own.

The advisory committees for developing the test protocol include the following:

American Association of Retired Persons (AARP)

American Furniture Manufacturers Association (AFMA)

American Hotel and Motel Association (AHMA)

American Society for Testing and Materials (ASTM)

Carpet and Rug Institute (CRI)

Consumer Product Safety Commission (CPSC)

Fire Marshals Association of North America (FMANA)

Fire Retardant Chemicals Association (FRCA) General Services Administration (GSA)

International Association of Fire Chiefs (IAFC)

Man-Made Fiber Producers Association (MMFPA)

National Association of Home Builders (NAHB)

National Institute of Standards & Technology (NIST)

National Conference of States on Building Codes and Standards (NCSBCS)

National Electrical Manufacturers Association (NEMA)

Underwriters Laboratories (UL)

U.S. Fire Administration (USFA)

Designing and Legal Matter

In designing a product factors to consider include protecting your design, product liability, and many more. It is important to recognize what laws and legal matters exist or actions can occur unfortunately for even the "good guy." This type of information or action could be considered a competitive situation. It is important to keep up to date on laws and legal matters that can effect your product. The following provides some general information guides.

Accident Report

Fabricators/manufacturers do not plan for their products to fail or to cause harm to people. But if an incident should occur that results in serious injury or death, the problem must be investigated immediately to prevent it from occurring again. U.S. Federal regulations require that a manufacturer report the event to FDA. However, the customer, the patient, his or her family, and the manufacturer all need to know what happened, which makes the investigation of the problem critical. To eliminate any improper investigation, manufacturers should have a trained crisis management committee in place before a complaint is received so that a standard operating procedures is followed defining what actions are to be taken and by whom.

Acknowledgment

The formal document that accepts a customer order; includes a delivery promise, method and time for payment, and identifies any exceptions to the terms and conditions stated on the customer's purchase order.

Chapter 11 Act

US permits legal protection from creditors under Chapter 11 of the US Federal Bankruptcy Act. Interesting, particularly when you study being on both side and what one can get away with legally. Time the law changes for the credible operation even though the change has been discussed for many decades.

Conflict of Interest

They range from personal to legal matters with the usual main conflict between the private interests and the official responsibilities of a person in a position of trust such as the company's top executive officers or a government official or agency.

Consumer Product Safety Act

CPSA is a significant consumer safety law. It is part of U.S. legislative law and augments the common law and case of product liability. Purpose of the law is: (1) to protect the public against unreasonable risks of injury associated with consumer products; (2) to assist consumers in evaluating the comparative safety of consumer products; (3) to develop uniform safety standards for consumer products and to minimize conflicting state and local regulations; and (4) to promote research and investigation into the causes and prevention of product-related deaths, illnesses, and injuries. Overall target is to prevent hazardous material and products or defective designed products from reaching the consumer.

Copyright

It is an intangible property such as the ownership of a design or literary property granted by law.

Defendant

While anyone along the trail of commerce (manufacturer, wholesaler, or retailer) can become a defendant in a law suit, it is usually the manufacturer who is held liable to the injured party. The manufacturer is the one with the "$ deepest pockets" or the one from which the largest award can be obtained.

Employee Assignment Invention

In assigning an invention, usually the employment contract will govern. However, some states have Employee Invention Laws. These laws, in effect, retain personal, non-business related inventions for the employee as long as they are not made on the employer's equipment or time.

Expert Witness

Litigation in the plastic and other industries usually involves patent infringement, theft of trade secret, product liability, or a specific performance. With the usual patent law, the expert is expected to report on the obviousness of an invention. Prior art and knowledge of the requirements for patentability will often be key parts of the expert's testimony. Unfortunately judges who have a weak technical background and little understanding of the patent law hear many of these cases. The job of the expert is to reduce a complex art or science into easy to understand testimony for all in the proceedings.

An expert witness is to be a person with substantial training in a specific field who can look at a set of data and come to a scientific conclusion about the merits of the issues. Target here is that any conflict between objectively and the highly opinionated atmosphere inherent in legal proceedings has always been problematic for technical people with ethics.

Insurance Risk Retention Act

With IRRA companies in the same industry are permitted to form a specialized insurance company to insure themselves. As an example, one was been established in Vermont 1992 that was called the Plastics Industry Risk Retention Group (PIRRG).

Invention

Chief requirement is that (1) it be an unobvious to a person having ordinary skill in the art to which the claim pertains and (2) knowing everything that has gone wrong before is not applicable.

Mold Contractional Obligation

Custom molders have traditionally assumed no responsibility for the legality of the design of the customer's product, the design of the molded product as a component of that product, or products produced to the customer's design and specification. In the event a molded product infringes, or is claimed to infringe, any letters of patents, or copyright, the customer has assumed the responsibility involved. Normally most quotation forms include clauses that explicitly detail the indemnification provisions and mold storage responsibility.

Patent

In USA a patent is awarded to the person first producing an invention, not necessarily who first applied for a patent. The opposite policy prevails in the rest of the world with USA policy probably changing in order to achieve worldwide patent law harmonization. USA utility patents (machines, equipment, etc.) in the past where good for at least 17 years after date the patent was issued. As of 1995, the patent is good for 20 years after the date the patent is filed (prior to the date it is issued) that eliminated those who would file for a patent and let it drag out for many years prior to being issued when it would be needed for infringement, etc.

Patentability

Qualifications for obtaining a patent on an invention or process (USA) are: (1) the invention must not have been published in any country or in public use in USA in either case for more than one year to date of filing application, (2) it must not have been known in USA before that date of invention by the applicant, (3) it must not be obvious to an expert in the art/technology, (4) it must be useful for a purpose not immoral and not injurious to the public welfare, and (5) it must fall within five statutory classes on which only patents may be granted, namely, (a) composition of material, (b) process of manufacture or treatment, (c) machine, (d) design, and (e) plant produces asexually.

Patent Information

Patents tend to be the literature of technology with full disclosure of its invention details. This legal document confers to its owner the right to exclude others from using it.

Patent Infringement

Generally, ignorance of the patent or trademark rights of others is no excuse to an infringing activity. Moreover, it may give rise to costs and risks in withdrawal or recall of products, ads, attorney' fees, etc. These potential costs will probably outweigh the cost of the initial searches or clearances.

Patent Pooling with Competitor

In the past, USA competing companies could not cooperate, such as in R&D, without breaching antitrust laws. Patent pooling, such as collecting and cross-licensing patents, was precluded. Today the antitrust laws are reviewed, interpreted, and enforced less stringently, which permits industrial cooperation in selected and specific areas where poling does exist. This explanation is a simplistic summation to a very complicated situation.

Patent Search

There are three major steps to a patent search. (1) There is the US Patent Classification System that is a sort of subject index to all patents, (2) CASSIS is a computerized software information system provided by the USA patent office, and (3) review the patent that takes time; involves the weekly official worldwide gazettes, magazines, etc. There are many ways available to search the patent database in both US and worldwide, but one web that is particularly useful to the novice or occasional searcher is one offered by IBM locate at: http://www.patents.IBM.com

Patent Term Extension

The PTE complex law of 1984 (USA) offers an opportunity to extend the effective life of patents for new medical inventions up to five years.

Patent Terminology

Preparing a patent and ensuring that proper and protective terms are used (to elim-inate "substitutions") requires time and money to prepare a foolproof patent. Cost per patents has been in the millions of dollars.

Plaintiff

A lawsuit is a civil suit seeking compensation by the plaintiff for damages, usually money, for some type of liability against the responsible party(s). A product liability may arise as a result of a defect in design and/or manufacturer, improper service, breach of warranty, negligence in marketing, etc. Under the doctrine of strict liability the plaintiff must prove factual proof of damage. Before the trial the plaintiff is entitled to certain information by right of discovery. It includes all records that pertain to the alleged damage and depositions of individuals involved. Oral depositions before a court reporter permit both sides of the litigation to discover the important facts of the case.

Processor Collaborative Venture

PCV provides an innovative approach to cost containment (55). One method is to lower the cost of outsourcing parts and components by forming a group-purchasing venture. When legally structured meeting the established guidelines of the Antitrust Div. of the U.S. Dept. of Justice (DOJ), allow members to aggregate purchases in order to obtain competitive volume discounts, reducing their costs, and subsequently the prices charged to their customers. Such collaborations are often viewed by DOJ as not only benign, but procompetitive.

As explained by R. Branand (55) while U.S. manufacturers are well aware of the barriers imposed on U.S. collaborations by antitrust considerations. However they are usually not as familiar with the opportunities that are encouraged. Properly structured the competitor association can help USA companies achieve goals such as expanding into foreign markets, funding expensive innovations efforts, and lowering production and other costs. For over a decade this powerful cost-cutter has been available to U.S. businesses, yet misapplied antitrust concerns have prevented many companies from pursuing its benefits.

Processor Contract

It is usually consider a subgroup to the custom processor. They have little involvement in the business of their customer. They usually just sell machine time.

Product Liability Law

Two types of law are involved; contract and tort. A contract is an agreement between two or more parties that is enforceable in a court of law. A tort is a civil wrong committed by the invasion of any personal or private right that each person enjoys by virtue of federal and state laws. The personal or private right affected must be one that is determined by law rather than by contract. In addition to the tortuous act, there must also be personal injury and/or property damage. Over half the USA states have adopted to varying degrees the doctrine of strict liability tort, that means that the injured person need only prove that a product was unreasonably dangerous to win the case. Various conditions make it easier to win cases. As an example proof that the manufacturer of the product is negligent is no longer required.

Protect Design

Five different methods of protecting your design exists in USA. Each is weighed according to its advantages and disadvantages based on specific needs. They are: (1) contracts/other party agrees not to make, use, etc. without designer's permission; (2) copyrights/protection exists upon creation of design; (3) trade dresses/protection when design is either inherently distinctive or has become distinctive; (4) utility patents/protects the functional and structural features of a product; and (5) design

patents/protect the ornamental appearance of a product without regard to how it functions.

Protection Strategy

For a molder to control secrecy concerning proprietary information, the first approach is to keep it as a personal secret. If people have to be exposed to it, such as present or new employees, visitors, and customers since there exists a need to know, those people should sign a nondisclosure agreement. This agreement could set up problems since a person could already be familiar with the so-called secret.

Quotation

Document quote that states the selling price and other sales conditions of a material, product, etc. Did you know that by law if someone reports that verbally the vender made statements such as "buy this injection molding machine and all you have to do is push a button to make good/acceptable parts"... the vender is in trouble... even if that vendor wins the case (odds are against winning), it will be very expensive to be in court.

Right-to-Know

This law (Fed. Reg. 29 cfr 1910.1200) covers employees' right to know about the chemical hazards to which they are exposed if they exist in a working area.

Shop-Right

It is a term referring to a non-exclusive royalty-fee license given to a employer where an employee uses the employment's time and/or equipment to develop an invention. Shop-rights come into play when there is no assignment agreement.

Software and Patent

The Court of Appeals for the USA Federal Circuit issued (1992) a decision that could strengthen the legal position that so-called pure software could be patented (Arrhythmia Research Technology vs. Corazonix Corp. 22 USPQ2d 103 of CAFC march 12, 1992).

Tariff

It is basically a schedule of duties or cost rates imposed by a government on imported or in some countries exported goods. In certain areas of the world to offset tariff duties, worldwide free-trade agreements exist.

Term

It is important in the workplace and when legal actions occur that terms have their proper definition to ensure accuracy of discussions in the plant and/or in the court room.

Tort Liability

The tort laws have been impeding new biomaterial and medical device developments by the large companies. It is very difficult for them to justify the financial risk incurred from the relatively low level of their sales. Action is being taken to change the laws.

Trademark

TM is a symbol or insignia designating one or more proprietary products or the manufacture of such products, that has been officially registered and approved by the U.S. Patent and Trademark Office (PTO). The acceptable designation is a superior capital R enclosed in a circle, however, quotation marks may be used. There are three levels of TM protection namely: (1) common law covers unregistered TM with limited legal protection; (2) state registration where you register the TM and are protected in that state

only, and (3) federal registration that offers registered TM protection across state lines.

Trade Name

TN is the name or style under which a concern does business. The government concerned alone or with a device such as a surrounding oval may register the TN.

Warranty

There are different items that have warranties such as equipment, products, and materials. Fulfillment of warranties tends to be a two-way situation. As an example when one buys equipment, you are not just buying equipment, you are entering into a relationship. This may sound tripe, but it is demonstrably true in the case of capital equipment. The warranty relationship can be defined in writing by the warranty document. It goes into detail as to what the OEM (original equipment manufacturer) seller promises to do in event of equipment failure due to specific causes. It also details the responsibilities of the equipment owner. Sometimes the expectations of the processor and OEM are seriously mismatched.

The best way to avoid this situation is to clarify understandings before the equipment is delivered. It is usually clear who pays for parts. Make sure you understand, however, the responsibilities vs. the OEM for shipping, travel, and other costs. The details can significantly defer from OEM to OEM.

Design Detractor and Constrain

As reviewed throughout this book, designing acceptable products requires knowledge of the behavior of the different plastics and their processing characteristics (Chapter 6, **MATERIAL VARIABLE** and Chapter 8 **EQUIPMENT/PROCESSING VARIABLE**).

Although there is no limit theoretically to the shapes that can be created, practical considerations must be met such as available and size of processing equipment with cost. These relate not only to the product design, but also the mold or die design, since they must be considered as one entity in the total creation of a usable, economically feasible product.

Troubleshooting Design Problem/Failure

Troubleshooting is the art and science of remedying product defects after the process has demonstrated the ability to produce acceptable production products. Most defects respond to one of a variety of process and/or material changes. The target is knowing when a particular solution will work, and correctly identifying which problem is actually causing the defect. When making adjustments consider: (1) create a mental image of what should be happening, (2) look for obvious differences, (3) make only one change at a time designwise, materialwise, and/or processingwise, and (4) allow the process to stabilize after any change is made.

Studies have determined that probably 60% of defects result from the machine/equipment, 20% mold/die, 10% material, and 10% operator. Available are software programs already installed on the machines processor controller or available as a software package that can provide some troubleshooting help.

Troubleshooting Guide

With all types of equipment, materials, and products, troubleshooting guides are setup (usually required) to take fast, corrective action when products do not meet their performance requirements such as dimensions, shape, surface appearance, and physical and mechanical properties. This problem solving approach fits into the overall fabricating-design interface as summarized in the FALLO approach (Fig. 1-3). Troubleshooting guides are reviewed in throughout this book (1, 3, 6, 7, 20).

Many different guides are provided by equipment and material suppliers. However, the product "problems-to-solutions" are usually developed when setting-up a fabricating line by the processor. A simplified approach to troubleshooting is to develop a checklist that incorporates the rules of a problem-to-solving procedure. (1) Have a plan and keep updating it based on the experienced gained in operating the equipment. (2) Watch the processing conditions. (3) Change only one condition at a time. (4) Allow sufficient time for each change and keep some kind of a log of the action, with results, that are occurring. (5) Check housekeeping, storage areas, dryers, granulators, personnel clothing, and personnel behavior. (6) Narrow the range of areas in which the problem belongs, e.g. material storage and handling, mold/die, specific equipment in the fabrication line (such robot, cooling tank, and puller), specific control, product design, environment (humidity, ventilation location and direction of forced air, dust, etc.), people, and management.

The following provides an example as a guide pertaining to extruded products that starts with common operating problems and possible solutions. When possible start with feeding low bulk density plastic in a starved fed extruder. To avoid aeration and therefore increased potential for volumetric feed limitation, minimize the free fall path from the feeder to the extruder feed throat. If a barrel zone on the barrel constantly overrides or requires too much cooling to maintain a set point, it may be that the melting is being concentrated in that section. This can either exist because of screw design or an improper barrel heat profile. A simple and hopeful solution is to increase the melting prior to the "hot zone" of the screw (6).

Troubleshooting by Remote Control

To aid the manufacturing plants, remote troubleshooting has been available from different equipment manufacturers and service facilities. Users of certain microprocessor equipment need not be concerned about their plant's personnel's ability to service and maintain the equipment. Via a communication link from your computer controller to the services central computer, a specialist and/or automatic device can immediately check out conditions in your controller as well as in the complete production line. This remote diagnostic link can also be used to set-up preventative maintenance programs.

Defining the Trouble

When setting up troubleshooting guides, as well as reviewing any problems or even open discussions on the subject of fabricating, it is important that the terms used to identify a problem be understandable, clear, and properly defined. As an example the word flaw could have different meanings to different people. Flaw could identify blush, burn, discoloration, fill-in, flow marks, glossiness, gouge, haze, inconsistency, misalignment, non-adhesion, nonuniform, pit, porosity, protrusion, runs, scratches, sink mark, smearing, speck, void, and weld line. This is "stretching" or "far-fetching" the term flaw but it should highlight the fact that a proper definition can eliminate problems.

Design Failure Theory

In many cases, a product fails when the material begins to yield "plastically." In a few cases, one may tolerate a small dimensional change and permit a static load that exceeds the yield strength. Actual fracture at the ultimate strength of the material would then constitute failure. The criterion for failure may be based on normal or shear stress in either case. Impact, creep and fatigue failures are the most common mode of failures. Other modes of failure include excessive elastic deflection or buckling. The actual failure mechanism may be quite complicated; each failure theory is only an attempt to explain the failure mechanism for a given class of materials. In each case a safety factor is employed to eliminate failure.

An example of a theory is the Griffith theory. It expresses the strength of a material in terms of crack length and fracture surface energy. Brittle fracture is based on the idea that the presence of cracks determines the brittle

strength and crack propagation occurs. It results in fracture rate of decreased elastically stored energy that at least equals the rate of formation of the fracture surface energy due to the creation of new surfaces.

These failures could be due to the variability of the plastic material and/or fabrication control of equipment. Applying an approach such as the **Troubleshooting Guide** reviewed, can direct you to the solution.

Product Failure

Different techniques or methodologies are used to analyze premature molded product failures in order to meet cost requirements. Various methods of auditing and computer software programs are used or developed by designers and/or fabricators to provide an analysis of potential problems. Interesting is the fact that the actual time and cost to design products may take less than 5% of the total time and cost to fabricate products. Even though this is a relatively small percentage of the overall operation, it has a direct and important influence performancewise and costwise on the success or failure of fabricating molded products.

Avoiding product failures can depend, in part, on the ability to predict the performance of plastic materials and their shapes. With available time, the usual approach of product prototype and/or field-testing provides useful and reliable performance data when conducted properly. As an example designers continue to develop sophisticated computer methods for calculating stresses in complex structures.

The computational methods have replaced the oversimplified models of material behavior formerly relied on. However, for new and very complex product structures that are being designed to significantly reduce the volume of materials used and in turn the product cost, computer analysis is conducted on prototypes already fabricated and undergoing testing. This computer approach can result in early and comprehensive analysis of the effects of conditions such as temperature, loading rate, environment, and material defects on nonstructural and/or structural reliability. The information is supported by stress-strain behavior collected in actual material evaluations.

When required, combined with the use of computers, the finite element analysis (FEA) method can greatly enhanced the capability of the structural analyst to calculate displacement and stress-strain values in complicated structures subjected to arbitrary loading conditions. In its fundamental form, the FEA technique is limited to static, linear elastic analysis. However, there are advanced FEA computer programs that can treat highly nonlinear dynamic problems efficiently.

Important features of these programs include their ability to handle sliding interfaces between contacting bodies and the ability to model elastic-plastic material properties. These program features have made possible the analysis of impact problems that in the past had to be handled with very approximate techniques. FEAs have made these analyses much more precise, providing better direction in locating high stress areas. Final verification of load-carrying capability usually requires actual testing of the fabricated product prototype based on computational analysis.

Managing Failure

Effective management of any product (IM, etc.) is much more than the production of immediate results. As Leonard A. Schlesinger (Harvard Business School) reviews, effective management includes creating the potential for achieving good results over the long run. There is the manager, who as president of a company, can produce spectacular results for a 3- to 10-year period. However that person can hardly be considered effective if, concurrently, people allow plant and equipment to deteriorate, creates an alienated or militant workforce, lets the company develop a bad name in the marketplace, and ignores new product development.

Dealing with current or impending problems is a key reality of people behavior in almost all-modern organizations. Coping with complexities associated with today and the immediate future absorbs the vast majority of time and energy for most managers.

Most managers will readily admit that their ability to predict their company's future is limited. Indeed, with the possible exceptions of death and taxes, the only thing entirely predictable is that things will change. Even for the most bureaucratic company in the most mature and stable environment, change is inevitable.

Over a period of 20 years, it is possible for a company, even one that is not growing, to experience numerous changes in its business, product markets, competition, government regulations, available technologies, business strategy, labor markets, and so on. These changes are the inevitable products of its interaction with a world that is not static.

Business and change Growing organizations tend to experience even more business-related changes over a long period of time. Studies will show that growing businesses not only increase the volume of the products or services they provide, but also tend to increase the complexity of their products or services, their forward or backward integration, their rate of product innovation, the geographic scope of their operations, the number and character of their distribution channels, and the number and diversity of their customer groups. While all of this growth-driven change is occurring, competitive and other external pressures also increase. The more rapid the growth, the more extensive the changes that are experienced.

These types of business changes generally require organizational adjustments. For example, if a company's labor markets change over time, it must alter its selection criteria and make other adjustments to fit the new type of employee. New competitors might emerge with new products, thus requiring renewed product development efforts and a new organizational design to support that effort. In a growing company, business changes tend to require major shifts periodically in all aspects of its organization.

The inability of an organization to anticipate the need for change and to adjust effectively to changes in its business or in its organization causes problems. These problems sometimes take the form of poor collaboration and coordination; they may involve high turnover or low morale. Always, however, such problems affect the organization's performance goals that are not achieved and/or resources are wasted. Because change is inevitable and because it can so easily produce problems for companies, the key characteristic of an effective organization from a long-run viewpoint is its ability to anticipate needed organizational changes and to adapt as business conditions change.

Anticipatory skills can help prevent the resource drain caused by organizational problems, while adaptability helps an organization avoid the problems that change can produce. Over long periods of time, this ability to avoid an important and recurring resource drain can mark the difference between success and failure for an organization.

Bureaucratic dry rot It has been emphasized by a number of social scientists that in the past decades there has been expressed serious concern over what they call bureaucratic dry rot. We all pay a heavy price, they note, for the large, bureaucratic, nonadaptive organizations that are insensitive to employees' needs, ignore consumers' desires, and refuse to accept their social responsibilities.

Existing evidence suggests that although most contemporary organizations cannot be described as adaptive, many managers nevertheless appreciate the benefits of adaptability. When polled managers often respond that "ideally" they would like to have the ideal organization, but they also admit that their current organization does not have all or even some of these characteristics.

Business Failure

There is a formula for business failures based on Dun & Bradstreet, Inc. annually published data. The vast majority of the firms involved are small. Why do failures occur? D&B has offered the following tabular explanation (apparent cause/percent): inadequate sales/49.9, competitive weakness/25.3, heavy operating expenses/13.0, receivables difficulties/8.3, inventory difficulties/7.7, excessive fixed assets/3.2, poor location/2.7, neglect/0.8, disaster/0.8, fraud/0.5, and others/1.1. Numbers do not add up to 100% because some failures are attributed to a combination of apparent causes. One can include that product design directly influences competitive weakness and heavy operating expenses.

5

Testing and Meaning of Test Data

Introduction

This chapter will present examples of testing and their behavior that influence decisions to be made by designers when analyzing data. Testing methods have been prepared and used for over 2,000 years with probably more in the past century that in all the past. Examining and properly applying property test data of plastics is important to designers.

The technology of manufacturing the same basic type or grade of plastics (as with steel and other materials) by different suppliers may not provide the same results. In fact a supplier furnishing their material under an initial batch number could differ when the next batch is delivered and in turn could effect the performance of your product. Taking into account manufacturing tolerances of the plastic, plus variables of equipment and procedure, it becomes apparent that checking several types of materials from the same or from different sources is an important part of material selection and in turn their use.

Based on past performances it has been proven that the so-called interchangeable grades of plastics have to be evaluated carefully by the designer as to their affect on the performance of a product. An important consideration to include as far as equivalent grade of material is concerned is its processing characteristics. There can be large differences in properties of a product and test data if the manufacturing features vary from grade to grade or batch to batch. This situation in most cases does not effect the product performances but could require changing equipment process controls to maximize the product performances and minimize cost.

Overall Responsibility

Should the designer have this type of responsibility. That person gets the credit for a successful product. If the product fails in service regardless of the reason the responsibility should be the designer who did not meet the product's entire requirement. In specifying the specific plastic and/or process to use, their test requirements should have been more complete and/or meet closer requirements. This action would include factors such as the limits on the variabilities of the design configuration (dimensions, etc.), plastic, and process. As an example, quality control (QC) on the plastic and fabricated product is required even if all that is required is limiting ($\pm$) the weigh the material and fabricated product.

The problem of acquiring complete knowledge and control of candidate material grades should be resolved in cooperation with the raw material suppliers. Having the material supplier meet specific performance requirements is important. In turn it may be necessary for testing incoming material even if the supplier provides data you requested. It should be recognized that selection of the favorable material is one of the three basic elements in producing a successful product namely design, material selection, and conversion into a finished product. How to resolve processing problems uses the same approach as reviewed with material control.

So one can say the designer should not carry all these responsibilities. True but if the responsibility is not delegated to one person, you allow for problems to develop. Perhaps a certain qualified manager (the designer's boss) should have the responsibility. For a small operation it is usually its owner who may or may not properly delegate specific responsibilities.

Destructive and Nondestructive Testing

Testing yields basic information about any materials (plastics, steels, etc.), its properties relative to another material, its quality with reference to standards or material inspections, and can be applied to designing with plastics. Examples of static and dynamic tests are reviewed in Chapter 2.

There are destructive and nondestructive tests (NDTs) (2). Most important, they are essential for determining the performance of plastic materials to be processed and of the finished fabricated products. Testing refers to the determination by technical means properties and performances. This action, when possible, should involve application of established scientific principles and procedures. It requires specifying what requirements are to be met. There are many different tests (thousands) that can be conducted that relate to practically any material or product requirement. Usually only a few will be applicable to meet your specific application. Examples of these tests will be presented.

In the familiar form of testing known as destructive testing, the original configuration of a test specimen and/or product is changed, distorted, or usually destroyed. The test provides information such as the amount of force that the material can withstand before it exceeds its elastic limit and permanently distorts (yield strength) or the amount of force needed to break it. These data are quantitative and can be used to design structural products that would withstand a certain load, heavy traffic usage, etc.

NDT examines material without impairing its ultimate usefulness. It does not distort the specimen and provides useful data. NDT allows suppositions about the shape, severity, extent, distribution, and location of such internal and subsurface residual stresses; defects such as voids, shrinkages, and cracks; and others. Test methods include acoustic emission, radiography, IR spectroscopy, x-ray spectroscopy, magnetic resonance spectroscopy, ultrasonic, liquid penetrant, photoelastic stress analysis, vision system, holography, electrical analysis, magnetic flux field, manual tapping, microwave, and birefringence (Fig. 5-1).

There is usually more than one test method to determine a performance because each test has its own behavior and meaning. As an example there are different tests used to determine the abrasion resistance of materials. There is the popular Taber abrasion test. It determines the weight loss of a plastic or other material after it is subjected to abrasion for a prescribed number of the abrader disk rotations (usually 1000). The abrader consists of an idling abrasive speed controlled rotating wheel with the load applied to the wheel. The abrasive action on the circular specimen is subjected to a rotary motion.

Other abrasion tests have other type actions such as back and forth motion, one direction, etc. These different tests provide different results that can have certain relations to the performance of a product that will be subjected to abrasion in service.

A method of evaluating the adhesive bond to a plastic coating substrate is a tape test. Pressure-sensitive adhesive tape is applied to an area of the adhesive coating, which is

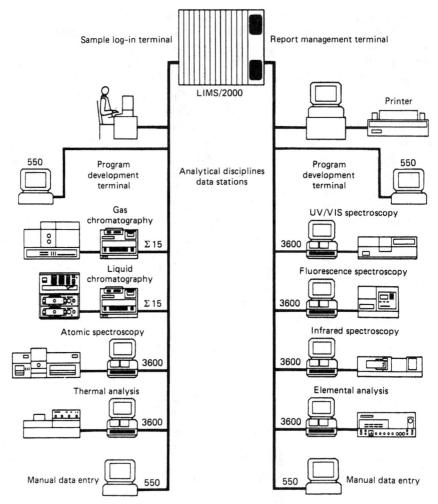

Sample log-in terminal

Report management terminal

LIMS/2000

Printer

550

Program development terminal

Analytical disciplines data stations

Program development terminal

550

Gas chromatography

UV/VIS spectroscopy

Σ 15

3600

Liquid chromatography

Fluorescence spectroscopy

Σ 15

3600

Atomic spectroscopy

Infrared spectroscopy

3600

3600

Thermal analysis

Elemental analysis

3600

3600

Manual data entry   550

550   Manual data entry

Fig. 5-1 Examples of plastics evaluation in a computer-aided chemistry laboratory.

sometimes crosshatched with scratch lines. Adhesion is considered to be adequate if the tape pulls off no coating when it is removed.

A bearing strength test method is used for determining the behavior of materials subjected to edgewise loads such as those applied to mechanical fasteners (plastics, etc.). For plastics, one of the tests uses a flat rectangular specimen with a bearing hole centrally located near one end. It is loaded gradually either in tension or compression. Load and longitudinal deformation of the hole are measured frequently or continuously to rupture with resulting data plotted as a stress-strain curve. For this purpose, strain is calculated by dividing change in the hole diameter in the direction of loading by the original hole diameter. Bearing stress is calculated by divid-

ing the load by the bearing area being equal to the product of the original hole diameter and specimen thickness. Test results are influenced by the edge-distance ratio, that is the ratio between the distance from the center of the hole to the nearest edge of the specimen in the longitudinal direction and hole diameter.

A different type of evaluation is the potential of a material (plastic, etc.) that comes in contact with a medical patient to cause or incite the growth of malignant cells (that is, its carcinogenicity). It is among the issues addressed in the set of biocompatibility standards and tests developed as part 3 of ISO-10993 standard that pertain to genotoxicity, carcinogenicity, and reproductive toxicity. It describes carcinogenicity testing as a means to determine the tumorigenic potential of devices, materials, and/or extracts to either a single or multiple exposures over a period simulating the total life span of the devices. The circumstance under which such an investigation may be required is given in part 1 of ISO-10993.

Interesting that in this highly scientific world, there are what appear to be nonscientific test methods. As an example early during the 1930s, the US navy in Dalgren, VA developed a very successful and useful aircraft canopy "chicken" impact test called the Dalgren test. This test continues to be used providing the required test results to ensure the proper performance of a canopy in service. Basically a 4 lb (2 kg) chicken is fired out of a cannon-like device and is used to evaluate the impact damage on aircraft windows. Your author attempted to develop a replacement, highly scientific test, to replace the Dalgren test without success other than making one more intricate and costly to conduct that provided the same results.

In order to determine the strength and endurance of a material under stress, it is necessary to characterize its mechanical behavior. Moduli, strain, strength, toughness, etc. can be measured microscopically in addition to conventional testing methods. These parameters are useful for material selection and design. They have to be understood as to applying their mechanisms of deformation and fracture because of the viscoelastic behavior of plastics (Chapter 2). The fracture behavior of materials, especially microscopically brittle materials, is governed by the microscopic mechanisms operating in a heterogeneous zone at the crack tip or stress raising flow.

In order to supplement micro-mechanical investigations and advance knowledge of the fracture process, micro-mechanical measurements in the deformation zone are required to determine local stresses and strains. In TPs, craze zones can develop that are important microscopic features around a crack tip governing strength behavior. For certain plastics fracture is preceded by the formation of a craze zone that is a wedge shaped region spanned by oriented micro-fibrils. Methods of craze zone measurements include optical emission spectroscopy, diffraction techniques, scanning electron microscope, and transmission electron microscopy.

Conditioning procedures of test specimens and products are important in order to obtain reliable, comparable, and repeatable data within the same or different testing laboratories. Procedures are described in various specifications or standards such as having a standard laboratory atmosphere [50 ± 2% relative humidity, 73.4 ± 1.8°F (23 ± 1°C)] with adequate air circulation around all specimens. The reason for this type or other conditioning is due to the fact the temperature and moisture content of plastics can affect different properties.

Testing and Classification

Properties of plastics such as physical, mechanical, and chemical are governed by their molecular weight, molecular weight distribution, molecular structure, and other molecular parameters (Chapters 2 and 8); also the additives, fillers, and reinforcements that enhance certain processing and/or performance characteristics. Properties are also effected by their previous history (includes recycled plastics), since the transformation of plastic materials into products is through the application of heat and pressure involving many different fabricating processes. Thus, variations in properties of products can occur even when the same plastic and processing equipment are used. Conducting tests such as those related to molecular characteristics provides a means of classifying them based on test results (2).

Testing and Quality Control

Testing and QC are discussed but often the least understood. Usually it involves the inspection of materials and products as they complete different phases of processing. Products that are within specifications proceed, while those that are out of specification are either repaired or scrapped. Possibly the workers who made the out-of-spec products are notified so "they" can correct "their" mistake.

The approach just outlined is after-the-fact approach to QC; all defects caught in this manner are already present in the product being processed. This type of QC will usually catch defects and is necessary, but it does little to correct the basic problem(s) in production. One of the problems with add-on QC of this type is that it constitutes one of the least cost-effective ways of obtaining quality products. Quality must be built into a product from the beginning of the design that follows the FALLO approach (Fig. 1-3); it cannot be inspected into the process. The target is to control quality before a product becomes defective.

Testing and People

Personnel or operators involved in testing from raw materials to the end of the fabricating line develop capability via proper training and experience. Experience and/or developing the proper knowledge are required to ensure that the correct test procedure is being conducted and test results are accurate and not interrupted incorrectly. At times, with new problems developing on-line, different tests are required that may be available or have to be developed. Unfortunately a great deal of "reinventing the wheel" can easily occur so someone should have the responsibility to be up to date on what is available.

Another unfortunate or fortunate situation exists that a very viable test was at one time developed and used within the industry. In time it was changed many times by different companies and organizations (ASTM, ISO, etc.) to meet new industries needs concerning specific requirements. One studying the potential of using that particular test may not have the access to the basic test that probably is all that is required.

Basic vs. Complex Test

Choosing and testing a plastic when only a few existed that could be used for specific products would prove relatively simple if the selection were limited, but the variety of plastics has proliferated. Today's plastics are also more complex, complicating not only the choice but also the necessary tests. Fillers and additives can drastically change the plastic's basic characteristics, blurring the line between commodity and engineering plastics. Entirely new plastics have been introduced with esoteric molecular structures. Therefore, plastic suppliers now have many more sophisticated tests to determine which plastic best suits a product design or fabricating process.

For the product designer, however, a simple basic test, such as a tensile test, will help determine which plastic is best to meet the performance requirements of a product. At times, a complex test may be required. The test or tests to be used will depend on the product's performance requirements.

To ensure quality control material suppliers and developers routinely measure such complex properties as molecular weight and its distribution, crystallinity and crystalline lattice geometry, and detailed fracture characteristics (Chapter 6). They use complex, specialized tests such as gel permeation chromatography (2, 3), wide- and narrow-angle X-ray diffraction, scanning electron microscopy, and high-temperature pressurized solvent reaction tests to develop new polymers and plastics applications.

Specification and Standard

The industry specifications and standards are regularly updated to aid designers and processors in controlling quality and to meet safety requirements, and thus they will prove useful to anyone who must choose tests and QC procedures. For example, the ASTM, UL, ISO, and DIN (see below) tests are among the most popular and important ones. Organizations involved directly or indirectly in preparing or coordinating specifications, regulations, and standards include the following:

ASTM. American Society for Testing and Materials.

UL. Underwriters Laboratories.

ISO. International Organization for Standardization.

DIN. Deutsches Instut, Normung.

ACS. American Chemical Society.

AMS. Aerospace Material Specification.

ANSI. American National Standards Institute.

ASCE. American Society of Chemical Engineers.

ASM. American Society of Metals.

ASME. American Society of Mechanical Engineers.

AWS. American Welding Society.

BMI. Battele Memorial Institute.

BSI. British Standards Institute.

CPSC. Consumer Product Safety Commission.

CSA. Canadian Standards Association.

DOD. Department of Defense.

DODISS. Department of Defense Index & Specifications & Standards.

DOT. Department of Transportation.

EIA. Electronic Industry Association.

EPA. Environmental Protection Agency.

FMRC. Factory Mutual Research Corporation.

FDA. Food and Drug Administration.

FMVSS. Federal Motor Vehicle Safety Standards.

FTC. Federal Trade Commission.

JAPMO. International Association of Plumbing & Mechanical Officials.

IEC. International Electrotechnical Commission.

IEEE. Institute of Electrical and Electronic Engineers.

IFI. Industrial Fasteners Institute.

IPC. Institute of Printed Circuits.

ISA. Instrument Society of America.

JIS. Japanese Industrial Standards.

MIL-HDBK. Military Handbook.

NADC. Naval Air Development.

NACE. National Association of Corrosion Engineers.

NAHB. National Association of Home Builders.

NEMA. National Electrical Manufacturers' Association.

NFPA. National Fire Protection Association.

NIST. National Institute of Standards & Technology (previously the National Bureau of Standards).

NIOSH. National Institute for Occupational Safety & Health.

OSHA. Occupational Safety & Health Administration.

PLASTEC. Plastics Technical Evaluation Center.

PPI. Plastics Pipe Institute.

QPL. Qualified Products List.

SAE. Society of Automotive Engineers.

SPE. Society of Plastics Engineers.

SPI. Society of the Plastics Industry.

STP. Special Technical Publications of the ASTM.

TAPPI. Technical Association of the Pulp and Paper Industry.

These test procedures and standards are subject to change, so it is essential to keep up to date if one has to comply with them. It may be possible to obtain the latest issue on a specific test (such as a simple tensile test or a molecular weight test) by contacting the organization that issued it. For example, the ASTM issues new annual standards that include all changes. Their Annual Books of ASTM Standards contain more than seven thousand standards published in sixty-six volumes that include different materials and products. There are four volumes specifically on plastics: 08.01-Plastics 1; 08.02-Plastics 11; 08.03-Plastics III, and 08.04-Plastic Pipe and Building Products. Other volumes include information on plastics and RPs. The complete ASTM index are listed under different categories for the different products, types of tests (by environment, chemical resistance, etc.), statistical analyses of different test data, and so on (56, 128, 129).

The ASTM issues other useful information for the designer that are included in its Special Technical Publications (STPs). Some examples of STPs are STP 701, Wear Tests for Plastics: Selection and Use, R. Bayer, ed., 1980, 106 pages; STP 736, Physical Testing of Plastics, R. Evans, ed., 1981, 142 pages; STP 816, Behavior of Polymeric Materials in Fire, E. L. Schaffer, ed., 1983, 121 pages; STP 846, Quality Assurance of Polymer Materials and Products, Green, Miller and Turner, ed., 1985, 142 pages; and STP 936, Instrumental Impact Testing of Plastics and Composite Materials, S. L. Kessler, ed. 376 pages.

Stress Analysis

There are different techniques to evaluate the quantitative stress level in prototype and production products. They can predict potential problems. Included is the use of electrical resistance strain gauges bonded on the surface of the product. This popular method identifies external and internal stresses. Their various configurations are made to identify stresses in different directions. This technique has been extensively used for over a half century on very small to very large products such as toys to airplanes. There is the optical strain measurement system that is based on the principles of optical interference. It uses Moire, laser, or holographic interferometry (2, 3, 20).

Another very popular method is using solvents that actually attack the product. It works only with those plastics that can be attacked by a specific solvent. Immersed products in a temperature controlled solvent for a specific time period identifies external and internal stresses. After longer time period's products could self-destruct. Stress and crack formations can be calibrated using different samples subjected to different loads.

There is the brittle coating system applied on the surface of a product that identifies conditions such as stressed levels, cracks, etc. A lacquer coating is applied, usually sprayed, on the surface of the product. It provides experimental quantitative stress-strain measurement data. As the product is loaded in proportion to loads that would be encountered in service, cracks begin to appear in the coating. The extent of cracks is noted for each increment of load. Prior to this action, the coating is calibrated by spraying it on a simple beam and observing the strain at which cracks appear. This nondestructive test method can be used to aid in placing strain gauges for further measurements.

Photoelastic measurement is a very useful method for identifying stress in transparent plastics. Quantitative stress measurement is possible with a polarimeter equipped with a calibrated compensator. It makes stresses visible (Fig. 5-2). The optical property of the index of refraction will change with the level of stress (or strain). When the photoelastic

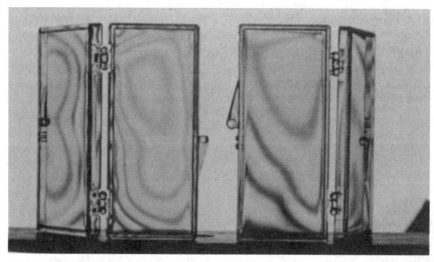

Fig. 5-2   Photoelastic stress patterns for these two molded products during the same production run shows that the processing conditions have changed; right view relates to why the product fails in service.

material is stressed, the plastic becomes birefringent identifying the different levels of stress via color patterns (2, 3, 216).

This photoelastic stress analysis is a technique for the nondestructive determination of stress and strain components at any point in a stressed product by viewing a transparent plastic product. If not transparent, a plastic coating is used such as certain epoxy, polycarbonate, or acrylic plastics. This test method measures residual strains using an automated electro-optical system.

This concept has been known for over a century. Expressed as Brewster's Constant law, it states that the index of refraction in a strained material becomes directional, and the change of the index is proportional to the magnitude of the stress (or strain) present. Therefore, a polarized beam in the clear plastic splits into two wave fronts in the X and Y directions that contain vibrations oriented along the directions of principal stresses. An analyzing filter passes only vibrations parallel to its own transmitting plane (Chapter 4, **TRANSPARENT AND OPTICAL PRODUCT, Polarized Lighting**).

The constructive and destructive interference creates the well known colorful patterns seen when stressed plastic are placed between two polarized filters. Some information about the stress gradients comes from observations of the patterns that provide qualitative analysis. The index of refraction in these directions is different and the difference (or birefringence) is proportional to the stress level.

When light that has experienced such retardation is viewed by a polarizer oriented at 90° to the original plane of light polarization, the two components of the original light beam interfere with one another. This results in a change in color and intensity of the observed light. Observed colors correspond to different levels of retardation at that point, which in turn correspond to stress levels.

To solve the measurement problem and obtain quantitative results (retardation, magnitude of the residual strain, etc.), various techniques are used. An example is using a very simple device known as a wedge compensator (ASTM D 4093). It is placed between the

light coming through the sample and the analyzing filter. The compensator reverses the retarding action of the induced strains in the plastic. Strain is calculated in the compensator by multiplying the birefringence (retardation per unit thickness) by a strain-optic response of the plastic being tested. Equal but opposite retardation is established and when superimposed on the retardation caused by the induced strain that restores a null. The intensity of the transmitted light becomes zero; revealed by a visible black fringe. A scale on the compensator supplies a quantitative reading of retardation.

Flat surfaces that are not readily conducive to stress evaluation by other means can be tested by the nondestructive Moire fringe analysis. Measurements of strains both elastic and plastic as well as evaluation of high temperature effects on the part are possible. A transparent film with a grid of equidistant lines is initially deposed on the product. Deformation in the product due to stresses changes the spacing between the grid lines. When a test grid is superimposed on a nondeformed grid the superposition produces an optical effect known as Moire fringes. If the test product is not strained and the grids are precisely aligned, no fringes will be observed. Visible fringes can be precisely measured to determine the degree of strain in a product.

Flaw Detection

Test methods are used to detect flaws. As an example when flaws or cracks "grow" in plastic, minute amounts of elastic energy are released and propagated in the material as an acoustic wave. A nondestructive acoustic emission test has sensors placed on the surface that can detect these waves providing information about location and rate of flaw growth. These principles form the basis for nondestructive test methods such as sonic testing.

The nondestructive electrical eddy current test is a method in which eddy current flow is induced in the test object. Changes in the flow by variations in the test specimen are reflected into a nearby coil or coils for subsequent analysis by suitable instrumentation and techniques. With the nondestructive electromagnetic test methods, different wavelength regions of electromagnetic energy having frequencies less than those of visible light yields information regarding the quality of materials.

A frequently used test has x-rays or gamma rays passing through a structure that absorbs distinctive flaws or inconsistencies in the material so that cracks, voids, porosity, dimensional changes, and inclusions can be viewed on the resulting radiograph.

The nondestructive temperature differential test by infrared is used. In this method, heat is applied to a product and the surface is scanned to determine the amount of infrared radiation is emitted. Heat may be applied continuously from a controlled source, or the product may be heated prior to inspection. The rate at which radiant energy is diffused or transmitted to the surface reveals defects within the product. Delaminations, unbonds, and voids are detected in this manner. This test is particularly useful with RPs.

With nondestructive ultrasonic test back and forth scanning of a specimen is accomplished with ultrasonics. This NDT can be used to find voids, delaminations, defects in fiber distribution, etc. In ultrasonic testing the sound waves from a high frequency ultrasonic transducer are beamed into a material. Discontinuities in the material interrupt the sound beam and reflect the energy back to the transducer, providing data that can be used to detect and characterize flaws. It can locate internal flaws or structural discontinuities by the use of high frequency reflection or attenuation (ultrasonic beam).

Of historical interest may be the use of a half dollar coin (the lighter weight 25¢ not as efficient). During the early 1940s the coin tap test was used very successfully in evaluating the performances of plastics, particularly RP primary aircraft structures. With a good ear (human hearing ear) there was (and is) a definite different sound between a satisfactory and unsatisfactory RP product. The unsatisfactory product would contain voids, delaminations, defects in fiber distribution, etc. In

the mean time the more elaborate and accurate sonic testing equipment were developed and used.

There is the microtoming optical analysis test. In this procedure thin slices (under 30 μm) of the plastics are cut from the product at any level and microscopically examined under polarized light transmitted through the sample. Rapid quality and failure analysis examination occurs by this technique. This technique has been used for many years in biological studies and by metallurgists to determine flaws, physical and mechanical properties. Examination can be related to stress patterns, mechanical properties, etc.

Limitation of Test

When working with tests it becomes (logically) obvious in most cases that options exist as to how the test is to be conducted. This is true for the different materials or products (plastics, steels, etc.). Different sizes (thicknesses, widths, and/or lengths) and/or shapes of test specimens are usually required with plastics such as those rigid to flexible to brittle. Different speed of testing are used, and so on. The explanation in the test provides a guideline as to what specific test conditions and specimen are used for the different types of materials. Another potential variable relates to specimen shrinkage, which results from the preparation of specimens. After being processed or upon cooling, specimens can develop nonuniform shrinkage (sink marks).

These test methods and the number and complexity of the variables present is related to the level of sophistication of the test. The combination that can influence test data defines the test limitations. Variables are found not only in test methods, but also in other non-test-related areas affecting data generation. Examples include misinterpretation, misuse, or misapplication of the test or any of its integral parts (test setup, test procedure, reporting, etc.) contribute to their limitations (2 to 11, 64, 208).

Test variables are a primary contributor to test limitations. These limitations are determined by the variables within a standard test method (STMs from ASTM) and within statistical analysis. Variables are also associated with materials (Chapter 6), and processability (Chapter 8). Adding to test limitations is a general lack of understanding regarding the language of testing. As usually reported, some people do not know the definition of a test, nor do they know the difference between physical properties (length, temperature, density, etc.) and mechanical properties (tensile, flex, impact, etc.). Also, the term sample and specimen are sometimes used interchangeably that is not correct.

Let it be known that one can say the tests are essential and all provide useful functions. However like design, material and process variables that exist, there are also testing variables. In order to apply them to a design, the designer should understand their meanings and purpose for existing. The result will be the proper use of tests.

Meaning of Data

It is evident that in order to use the data from those you perform to those from material suppliers data sheets, it is imperative to have a thorough understanding of how the data are evolved and what caution is to be exercised when applying the data to product designs or other evaluations. They can easily be interpreted incorrectly to mean something one desire's in their design approach. Interpretations are always made and provide excellent logical approaches to developing a design however they require dedicated concentrations and relationships to the basic meaning of the test.

In reality tests have only certain meanings. The following information provides examples of guides as to the meaning to a test. Tests reviewed are based on ASTM standards. Very limited reviews are provided in each of the following examples regarding size of specimens, speed of testing, and so forth that are provided in the specific standards or specifications. Information in the standards or specifications generally provide much more details on what they mean. When reviewing

a test that is to be used in a purchase order, etc. make sure you have proper identification of the test including where required specimen configuration with test's date of issue. What is important to the designer is if the test relates to the product performance requirements.

Physical Property

Specific Gravity/Density

Specific gravity and density are frequently used interchangeably; however, there is a very slight difference in their meaning. Specific gravity is the ratio of the weight of a given volume of material at 73.4°F (23°C) to that of an equal volume of water at the same temperature. Density is the weight per unit volume of material at 73.4°F (23°C) (Table 5-1). Water is the standard where it has a specific gravity of 1. The density of water is at 62.4 lb/ft^3.

The discrepancy enters because water at 73.4°F (23°C) has a specific gravity slightly less than one. To convert density to specific gravity, the following factor can be used (ASTM D 792):

$$\text{Density, g per cm}^3 = \text{specific gravity} \times 0.99756 \tag{5-1}$$

To the designer, the specific gravity is useful in calculating strength-to-weight and cost-to-weight ratios, and as a means of identifying a material.

Specific volume is a conversion of specific gravity into cubic inches per pound. Since the volume of material in a product is the first bit of information established after its shape is formulated, the specific volume is a convenient conversion factor for weight:

$$\text{Specific volume (in}^3\text{/lb)} = 27.7/\text{specific gravity} \tag{5-2}$$

Many different additives, fillers, and/or reinforcements are used in plastic materials. The weight of the compounds change according to the amount included. Figure 5-3 provides a guide to determining their specific gravities.

Table 5-1 Specific gravity and density comparisons of different materials

Materials	Specific Gravity	Density, lb./cubic in.
Thermoplastics		
ABS	1.06	0.0383
Acetal	1.43	0.0516
Acrylic	1.19	0.0430
Cellulose Acetate	1.27	0.0458
Cellulose Acetate Butyrate	1.19	0.0430
Cellulose Propionate	1.21	0.0437
Ethyl Cellulose	1.10	0.0397
Methyl Methacrylate	1.20	0.0433
Nylon, Glass-Filled	1.40	0.0505
Nylon	1.12	0.0404
Polycarbonate	1.20	0.0433
Polyethylene	0.94	0.0339
Polypropylene	0.90	0.0325
Polybutylene	0.91	0.0329
Polystyrene	1.07	0.0386
Polyimides	1.43	0.0516
PVC—Rigid	1.20	0.0433
Polyester	1.31	0.0473
Thermosets		
Alkyds, Glass-Filled	2.10	0.0758
Phenolic—G.P.	1.40	0.0505
Polyester, Glass-Filled	2.00	0.0722
Rubber	1.25	0.0451
Metals		
Aluminum SAE-309 (360)	2.64	0.0953
Brass—Yellow (#403)	8.50	0.3070
Steel—CR Alloy (Strip & Bar)	7.85	0.2830
Steel—Stainless 304	7.92	0.2860
Magnesium AZ—91B	1.81	0.0653
Iron—Pig, Basic	7.10	0.2560
Zinc—SAE-903	6.60	0.2380

The number of grams per cubic centimeter is the same as the specific gravity. For example, if the specific gravity is 1.47, that substance has a density of 1.47 gms/cm^3.

Water Absorption

The data should indicate the temperature and time of immersion and the percentage of weight gain of a test specimen. The same applies to data at the saturation point of 73.4°F (23°C), and, if the material is usable at 212°F (100°C), also to saturation at this temperature.

Moisture or water absorption is an important design property. It is particularly significant for a product that is used in conjunction with other materials that call for fits and clearances along with other close tolerance dimensions.

The moisture content of a plastic affects such conditions as electrical insulation resistance, dielectric losses, mechanical properties, dimensions, and appearances. The effect on the properties due to moisture content depends largely on the type of exposure (by immersion in water or by exposure to high humidity), the shape of the product, and the inherent behavior properties of the plastic material. The ultimate proof for tolerance of moisture in a product has to be a product test under extreme conditions of usage in which critical dimensions and needed properties are verified. Plastics with very low water-moisture absorption rates tend to have better dimensional stability.

Water Vapor Transmission

There are substantial differences in the rates at which water vapor and other gases can permeate different plastics. For instance, PE is a good barrier for moisture or water vapor, but other gases can permeate it rather readily. Nylon, on the other hand, is a poor barrier to water vapor but a good one to other vapors. The permeability of plastic films is reported in various units, often in grams or cubic centimeters of gas per 100 in.^{2} per mil of thickness (0.001 in.) of film per twenty-four hours. The transmission rates are influenced by such different factors, as pressure and temperature differentials on opposite sides of the film.

The effectiveness of a vapor barrier can be rated in a term such as perms. An effective vapor barrier in buildings should have a rating no greater than, say, 0.2 perm. A rating of one perm means that one ft^2 of the barrier is penetrated by one gram of water vapor per hour under a pressure differential of one in. of mercury. One in. of mercury equals virtually 0.5 psi; one gram is one seven-thousandth of a pound.

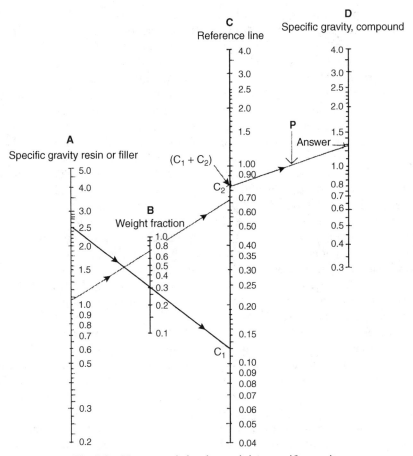

Fig. 5-3   Nomograph for determining specific gravity.

A similar problem is presented by vehicle tires and certain blow molded bottles, which must be virtually impermeable to air and other gases. An example of the use of a very impermeable elastomers is butyl rubber. Because of its impermeability to gases, butyl rubber is used as a roof coating. With plastic bottles, different layers of both coinjected and coextruded plastics (Chapter 8) can be used to fabricate the bottle to make it impermeable to different vapors and gases depending on the barrier plastic included.

Water Vapor Permeability

The material to be tested is fastened over the mouth of a dish that contains either water or a desiccant. This assembly is placed in an environment of constant humidity and temperature. The gain or loss in weight of the assembly is used to calculate the rate of water vapor movement through the specimen under prescribed conditions of humidity inside and outside of the dish. The results are reported in grams per 100 square inches during 24 hours, or equivalent metric units.

It should be recognized that all plastic materials over a time period allow a certain amount of water vapor, organic gas, or liquid to permeate the thickness of the material. It is only a matter of degree of permeation between various materials used as barriers against vapors and gases. It has been found that the permeability coefficient is a function of the solubility coefficient and diffusion coefficient. The process of permeation is explained as the solution of the vapor into the incoming surface of the barrier, followed by diffusion through the barrier thickness, and evaporation on the exit side.

The problem of permeability exists whenever a plastic material is exposed to vapor, moisture, or liquids. Typical cases are electrical batteries, instruments, components installed underground, encapsulated electrical components, food packaging, and various fluid-material containers. In these cases, a plastic material is called upon to form a barrier either to minimize loss of vapor or fluid or to prevent the entrance of vapor or fluid into a product. From the designers' viewpoint, the tolerable amount of permeation established by test under conditions of usage with a prototype product of correct shape and material is the only direct answer.

Different factors influencing permeation. (1) The composition of the barrier, including additives, fillers, colorants, plasticizers, etc. Even when data on permeability are available for a specific industry grade of a plastic, they cannot be used for evaluation because different commercial grades can contain ingredients that can change the values. (2) Crystalline plastics are better vapor barriers than amorphous plastics. Also, TSs have better barrier properties than TPs, especially when the fillers are nonmoisture absorbing. (3) An increase in temperature brings about an increase in permeability. Additionally, an increase in vapor pressure of the permeating agent also causes acceleration of transmission. (4) Product thickness is inversely proportional to permeation, i.e., with double the thickness, there is one-half of the evaporation. (5) Coatings such as epoxy-based finishes will improve resistance to permeation. (6) In the case of organic vapors, the permeation will depend not only on the composition of the barrier, but also on the molecular configuration of both the barrier material and the permeating agent.

Shrinkage

The use of correct shrinkage information is very important, not only for having the desired proportions of a product, but also for functional purposes. The shrinkage data can be shown in a range of two values. The lower figure is intended to apply to thin parts, whereas the higher figure would involve thicker parts. Your interest is in your specific thickness(s). Determining the shrinkage to occur when products (plastics, steels, etc.) are fabricated is not an easy task even when similar products are to be manufactured.

The choice of shrinkage for a selected material and a specific product is the responsibility initially of the designer but also involves the mold or die designers and the fabricators. When the product designer has limited knowledge on how shrinkage is effected by the mold or die during fabrication, these people have to be included in the design. If experience with the selected grade of plastic is limited, the design should be submitted to the material supplier for recommendations or someone how is knowledgeable, and the data coordinated with the interested parties.

Tolerance Where very close tolerances are involved, preparing a prototype of the full size product may be necessary to establish critical dimensions. If this step is not practical, it may be necessary to test a mold or die during various stages of cavity or die opening manufacture with allowances for correction in order to determine the exact shrinkage needed.

Considering the factors that can contribute to variations in shrinkage during the fabrication of the products, it will be fully appreciated how significant it is to select the appropriate numbers. The data on shrinkage have to be approached with much care if one is to avoid dimensional problems with the plastic product. Examples of how shrinkage is influenced by different processes will provide some insights on how critical it is to ensure a degree of repeatability in the materials behavior and the process. Chapters 6 and 7 provide more information.

As an example the shrinkage and in turn its tolerance of injection molded TPs will be affected as follows. (1) Higher cavity pressures will cause lower shrinkages. (2) Thick sections will shrink more than thin ones. (3) A cooler at the time of the product being ejected from the mold cavity will bring about a lower shrinkage. (4) A melt temperature of the material at the lower end of the recommended

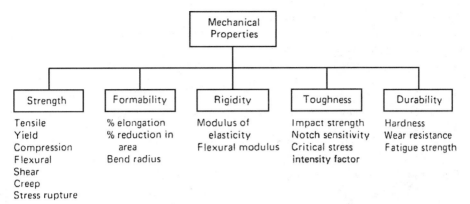

Fig. 5-4   General guides to mechanical property tests.

range will produce a lower shrinkage. (5) A longer cycle time, above the required solidification point, will partially conform the product closer to the mold dimensions, thereby bringing about lower shrinkage. (6) Openings in a product (holes and core shapes) will somewhat interfere with shrinkage in the cavity and thus will bring about lower and varying shrinkages than would be the case in a product without openings. (7) Larger feeding gates to the product will cause lower shrinkage permitting higher pressure buildup in the cavity. (8) Materials that are crystalline have dual shrinkage so they are higher in the direction of material flow and lower perpendicular to it. In a symmetrical part; when center gated, the shrinkage will average out and be reasonably uniform. (9) Most TPs attain their full shrinkage after 24 hours, but there are some which may take weeks to stabilize their dimensions fully (manufacturer of the material usually indicates whether there is a delayed shrinkage effect present). (10) Glass fiber-reinforced or otherwise filled TPs have considerably lower shrinkages than the basic plastic.

The TS plastic compression-molded products will have a higher shrinkage when: (1) cavity pressure is on the low side, (2) when mold temperature is on the high side, (3) when cures are shorter, (4) when products are thicker, (5) when a material is soft flowing (highly plasticized), (6) when a material is preheated at relatively low heat, and/or (7) when a high moisture content is present in the raw material.

Mechanical Property

As summarized in Fig. 5-4 mechanical properties encompass different behaviors.

Tensile Property

It should be recognized that tensile properties would most likely vary with a change of speed of the pulling jaws and with variation in the atmospheric conditions. Figure 2-14 shows the variation in a stress-strain curve when the speed of testing is altered; also shown are the effects of temperature changes on the stress-strain curves. When the speed of pulling force is increased, the material reacts like brittle material; when the temperature is increased, the material reacts like ductile material.

The tensile data show the stress necessary to pull the specimen apart and the elongation prior to breaking. A moderate elongation (about 6%) of a test specimen generally implies that the material is capable of absorbing rapid impact and shock. The area under the stress-strain curve is indicative of overall toughness except for reinforced plastics (Fig. 2-9). A material of very high strength, high rigidity, and little elongation would tend to be brittle in service. For applications where almost rubbery elasticity is desirable, a high ultimate (over 100%) elongation is an asset.

The tensile test and the calculated property data from it provide a most valuable source of information for the designer in determining product dimensions. The important consideration is that use conditions compare reasonably closely to the test conditions as far as speed of load, temperature, and moisture are concerned. Should use conditions differ appreciably, a test should be requested for data that are comparable to service requirements, which would thereby ensure that applicational needs will be based on more exacting data. The test is relatively inexpensive, and, where critical uses are encountered, it will eliminate interpolation and guessing. The tensile data are also useful for comparing various materials in this property. It is to be noted that tensile data should only be applied to short-term stress conditions, such as operating a switch or shifting a clutch gear, etc.

The yield point is the first point on the stress-strain curve at which an increase in strain occurs without an increase in stress. The stress at which a material exhibits a special limiting deviation from the proportionality of stress-to-strain is the yield strength. A material whose stress-strain curve exhibits points of zero slope may be considered to have a yield point such as described in Fig. 2-11. The data sheets usually omit the yield strength when there is a zero slope point on the stress-strain curve in the yield region. In reinforced plastic materials, the values of the yield strength and the tensile strength are very close to each other.

The important tensile modulus (modulus of elasticity) is another property derived from the stress-strain curve. The speed of testing, unless otherwise indicated is 0.2 in./min, with the exception of molded or laminated TS materials in which the speed is 0.05 in./min. The tensile modulus is the ratio of stress to corresponding strain below the proportional limit of a material and is expressed in psi (pounds per square inch) or MPa (megaPascal) (Fig. 2-7).

The proportional limit is the greatest stress that a material is capable of sustaining without any deviation of the proportionality law. It is located on the stress-strain curve below the elastic limit. The elastic limit is the greatest stress that a material is capable of sustaining without any permanent strain remaining upon complete release of the stress.

For materials that deviate from the proportionality law even well below the elastic limit, the slope of the tangent to the stress-strain curve at a low stress level is taken as the tensile modulus. When the stress-strain curve displays no proportionality at any stress level, the secant modulus is employed instead of the tensile modulus (Fig. 2-2). The secant modulus is the ratio of stress to corresponding strain, usually at 1% strain or 85% from the initial tangent modulus.

The tensile modulus is an important property that provides the designer with information for a comparative evaluation of plastic material and also provides a basis for predicting the short-term behavior of a loaded product. Care must be used in applying the tensile modulus data to short-term loads to be sure that the conditions of the test are comparable to those in use. The longer-term modulus is treated under the creep test (Chapter 2).

The tensile data can be applied to the design of short-term (such as 1 or 2 hour duration) or intermittent loads in a product provided the use temperature, the humidity, and the speed of the load are within 10% of the test conditions outlined under the procedure. The intermittent specification merely indicates that there be sufficient time for strain recovery after the load has been removed.

The next step is to determine an allowable working stress. This is done by using a safety factor usually of $1^1/_2$ to $2^1/_2$ on the yield strength or tensile strength. If the type of stress is clearly defined, the $1^1/_2$ factor is adequate; otherwise, it should be higher (Chapter 2, **Safety Factor**).

The final step is to calculate the elongation that the product would experience under the selected allowable working stress to see if such an elongation would permit the proper functioning of the product. The elongation could conceivably become the limiting component, and the working stress can be calculated from:

$$\text{Modulus} = \text{stress/strain} = E \qquad (5\text{-}3)$$

If product use conditions vary appreciably from those of the standard test, a stress-strain curve, derived using the procedure of anticipated requirement, should be requested and appropriate data developed.

Flexural Property

These properties apply to products subjected to bending, and, since many plastic products are involved in uses where bending stresses are generated, this deserves close attention, especially in view of the viscoelastic nature of the materials.

It should be noted that test information would vary with specimen thickness, temperature, atmospheric conditions, and different speed of straining force. This test is made at 73.4°F (23°C) and 50% relative humidity. For brittle materials (those that will break below a 5% strain) the thickness, span, and width of the specimen and the speed of crosshead movement are varied to bring about a rate of strain of 0.01 in./in./min. The appropriate specimen size are provided in the test specification.

The flexural strength is the maximum stress that a material sustains at the moment of break. For materials that do not fail, the stress that corresponds to a strain of 5% is frequently reported as the flexural strength (Fig. 2-15).

As a matter of interest it should be stated that in this test the force of bending and associated amount of deflection is recorded. A formula gives the relationship between deflection and strain:

The flexural yield strength is determined from the calculated data of load-deflection curves that show a point where the load does not increase with an increase in deflection.

The flexural modulus is the ratio, within the elastic limit, of stress to corresponding strain. It is calculated by drawing a tangent to the steepest initial straight-line portion of the load-deflection curve and using an appropriate formula.

In many plastic materials, as is the case with metals, when performing the flexural tests, increasing the speed of deflecting force makes the specimen appear more brittle and increasing the temperature makes it appear more ductile. This is the same relationship as in tensile testing.

When materials are evaluated against each other, the flexural data of those that break in the test cannot be compared unless the conditions of the test and the specimen dimensions are identical. For those materials (most TPs) whose flexural properties are calculated at 5% strain, the test conditions and the specimen are standardized, and the data can be analyzed for relative preference. For design purposes, the flexural properties are used in the same way as the tensile properties. Thus, the allowable working stress, limits of elongation, etc. are treated in the same manner as are the tensile properties.

Compressive Property

The compressive data are of limited design value. They can be used for comparative material evaluation and design purposes if the conditions of the test approximate those of the application. The data are of definite value for materials that fail in the compressive test by a shattering fracture. On the other hand, for those that do not fail in this manner, the compressive information is arbitrary and is determined by selecting a point of compressive deformation at which it is considered that a complete failure of the material has taken place. About 10% of deformation are viewed in most cases as maximum.

The test can provide compressive stress, compressive yield, and modulus. Many plastics do not show a true compressive modulus of elasticity. When loaded in compression, they display a deformation, but show almost no elastic portion on a stress-strain curve; those types of materials should be compressed with light loads. The data are derived in the same manner as in the tensile test. Compression test specimen usually requires careful edge loading of the test specimens otherwise the edges tend to flour/spread out resulting in inacturate test result readings (2–19).

Shear Strength

The specimen is mounted in a punch-type shear fixture, and the punch (1 in. diameter) is pushed down at a rate of 0.05 in./min until the moving portion of the sample clears the stationary portion. Shear strength is calculated as the force per area sheared. Shear strength is particularly important in film and sheet products where failures from this type of load may occur (Fig. 2-21). This property can be used for comparison with other materials and for determination of the forces needed for punching openings (holes, etc.).

Izod Impact

The popular Izod impact tester can use different size specimens depending on the type of plastic and their method of fabrication. The specimen is usually 1/8 in. × $\frac{1}{2}$ in. × 2 in.; other sizes are also used. Specimens can be notched or unnotched. A notch is cut in a specified manner on the narrow face of the specimen. The sample is clamped in the base of a pendulum testing machine so that it is cantilevered upward with the notch facing the direction of impact. The pendulum is released, and the force expended in breaking the sample is calculated from the height the pendulum reaches on the follow-through. The speed of the pendulum at impact is controlled.

The impact test, with its usual notch, indicates the energy required to break notched specimens under standard conditions. It is calculated as foot-pounds per inch (J/m) of notch and is usually calculated on the basis of 1 in. wide specimen, although the specimen may be thinner in the lateral direction.

The Izod value is useful in comparing various types of grades of a plastic within the same material family. In comparing one plastic with another, however, the Izod impact test should not be considered a reliable indicator of overall toughness or impact strength. Some materials are notch-sensitive and develop greater concentrations of stress from the notching operation. It should be noted that the notch serves not only to concentrate the stress, but also to present plastic deformation during impact.

The Izod impact test may indicate the need to avoid inside sharp corners on parts made of such materials. For example, nylon and acetal-type plastics, which in molded products are among the toughest materials, are notch-sensitive and register relatively low values on the notched Izod impact test.

Tensile Impact

Small and long specimens of tensile bar shape specimens have their major change in dimensions in the necked-down section. The specimen is mounted between a pendulum head and crosshead clamp on the pendulum of an impact tester. The pendulum is released and it swings past a fixed anvil that halts the crosshead clamp. The pendulum head continues forward, carrying the forward portion of the ruptured specimen. The energy loss (tensile impact energy) is recorded, as well as whether the failure appeared to be of a brittle or ductile type.

This test has possible advantages over the notched Izod test. The notch sensitivity factor is eliminated, and energy is not used in pushing aside the broken portion of the specimen. The test results are recorded in ft-lb/in.^{2} (kJ/m^2). This allows for minor variations in dimensions of the minimum in cross-section area.

Two specimens are used, S and L (short and long), so that the effect of elongation on the result can be observed. A ductile failure (best observed on the L specimen) results in a higher elongation and, consequently, in a higher total energy absorption than a brittle failure (best observed on the S specimen) in any specific material. The energy for specimen fracture is a function of the force times the distance it travels. Thus two materials showing the same energy values in the tensile impact test (all elements of the test being the same) could consist of two different factors, such as a small force and a large elongation compared to a large force and a small elongation.

If one is to consider the application of these data to a design, the size of the force and its rate of application would have to be obtained and compared with the design requirement. The breakdown of energy into components of force and speed becomes possible by the addition of electronic instrumentation to the testing apparatus, thus enabling the supplier of the material to furnish additional information for material selection.

Impact Strength

The data from the lzod and tensile impact tests can be comparatively evaluated especially when experience has been acquired with any one type of material (Table 5-2). One should keep in mind their individual limitations. Impact resistance is a significant characteristic of a material in many product designs. Associated with impact resistance is the term material toughness. Neither one can be measured in a way that is meaningful to the designer. Per ASTM test procedures test specimens can be of different thicknesses. With certain plastics from different manufacturers the impact strength test result for one thickness can be higher than the other using a different thickness. However in actual service, products from these materials can show the tougher material to have the lower impact strength result; this is a rare situation (Chapter 7, **SELECTING PLASTIC, Property Category,** *Impact*).

The term impact implies a very high speed of the acting force, whereas toughness is not related to any specific speed. Since the two terms are used in conjunction with each other in describing resistance to impact, it appears desirable to correlate those readily obtainable properties that would reflect on speed of impact and toughness.

At a 73.4°F (23°C) test temperature and a (ASTM) speed of acting force listed in each test category, the following results prevail: (1) a high modulus and high lzod points to a very tough material, (2) a high modulus and low Izod points to a brittle material, and (3) a low modulus and high Izod points to a flexible and ductile material.

When use conditions differ from those applied to data sheet tests, certain comparative evaluation can be made. Selecting an established high impact plastic such as polycarbonate as the standard, a tensile test would be made on this material at use speeds of strik-

ing force and end use environmental conditions. This provides a modulus and stress-strain curve. The same kind of test would be made on the materials being evaluated.

The area under a tensile stress-strain curve is a measure of toughness. It thus becomes possible to compare the modulus and areas under the curve and thereby estimate impact strength as a percentage of the standard. The notch sensitivity factor is eliminated and a judgment element is introduced that can prove accurate if the information is diligently analyzed. Where critical design areas involving safety to humans or protecting valuable devices are concerned, the simulation of end use with full size prototype products (including extremes of conditions) is the most desirable way to test selected materials.

There are other types of impact tests for shock loading where energy is required to cause complete failure is reported. Each has their specific behaviors that can be related to specific product performance requirements. Tests include ball burst, ball or falling dart using different weights and heights, bag drop, bullet-type instantaneous impact, Charpy, dart drop, Mullen burst, tear resistance, and tub (2).

Hardness

Hardness basically is the resistance to indentation as measured under specific conditions such as depth of indentation, load applied, and time period. Different tests relate to different hardness behaviors of plastics. They include Barcol, Brinell, durometer, Knoop, Mohs, Rockwell, Shore, and Vicat (2).

Hardness is closely related to strength, stiffness, scratch resistance, wear resistance, and brittleness. The opposite characteristic, softness, is associated with ductility. There are different kinds of hardness that measure a number of different properties (Fig. 5-5). The usual hardness tests are listed in three categories: (a) to measure the resistance of a material to indentation by an indentor; some measure indentation with the load applied, some the residual indentation after it is removed, such as tests using Brinell hardness,

Table 5-2   Room temperature impact resistance data for several plastics

Generic Material Type	Trademark	Grade	Izod Impact Energy for 0.318 cm (0.125 in.) thick, Notched Specimens, J/m	Tensile-impact Short Specimens, kJ/m^2	Energy Long Specimens, kJ/m^2
ABS	Cycolac	DH	235	99–131[a]	
		GSM	374	102–115[a]	
		KJB	214	95	
		L	400	100–120[a]	
Acetal copolymer	Celcon	M25	85		190
		M90	75		150
Acetal homopolymer	Delrin	100	123		350
		500	74.7		200
		900	69.4		150
Acrylic	Plexiglas	V052/045	21[b]		
		MI-7	32[c]		
		DR	64[b]		
Nylon (DAM) (0.2% moisture)	Zytel	101	53	157	504
		ST 801	907		588
		158 L	53	153	611
		211	80		525
Nylon (50% RH) (2.5% moisture)	Zytel	101	112	231	1470
		ST 801	1068		1155
		158 L	75	218	945
Phenolic	Durez	29053[d]	19		
		152[d]	17		
		18441[d]	14		
Polycarbonate	Lexan	141	640–850	473–631	
		940	640	526	
Polyethylene	Dow	08064N	53		88
		10062N	48		105
		04052N	80		140
		08035N	130		81
Phenylene ether copolymer	Prevex	PQA	267	113[a]	
		VKA	293	86[a]	
Polypropylene	Pro-fax	6523	42.7		
		7523	133.5		
		8523	379		
Polystyrene	Fostarene	50	21		
	Hostyren	360	54		
		760	97		
		840	161		
Polysulfone	Udel	P-1700	69	341	
		P-1710	69	421	
		P-1270	69	336	

[a]0.159 cm ($\frac{1}{16}$ in.) thick specimens.
[b]Molded notch.
[c]0.635 cm (0.250 in.) thick specimen with molded notch.
[d]Injection-molded specimens.

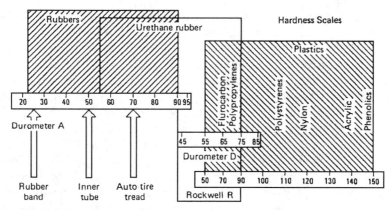

Fig. 5-5   Hardness of different materials using different test methods.

Vickers and Knoop indentors, Barcol hardness, and Shore durometers (2); (b) to measure the resistance of a material to scratching by another material or by a sharp point, such as the Bierbaum hardness or scratch-resistance test and the Moh one for hardness; and (c) to measure rebound efficiency or resilience, such as the various Rockwell hardness tests. The various tests provide different behavior characteristics for plastics, as described by different ASTM standards such as D 785. The ASTM and other sources provide different degrees of comparison for some of these tests.

Some ductile plastics, such as PC and ABS, can be fabricated like metals with punching and cold-forming techniques. These processing techniques are analogous to the hardness tests in that a rigid "indentor" is pressed into a sheet of a less-rigid plastic.

Durometer hardness   An arbitrary numerical value that measures the resistance to intention of a blunt indenter point of the durometer. The higher the number, the greater indention hardness.

Barcol hardness   Also called Barcol impresser. It is a measure of the hardness of a plastic, that includes laminate or reinforced plastic, using a Barber Coleman spring loaded indenter. Gives a direct reading on a 0 to 100 scale; higher number indicates greater hardness. This test is often used to measure the degree of cure for plastics, particularly TS plastics.

Brinell hardness   A common test used to determine the hardness of a material by indentation of a specimen. Pressing a hardened steel ball generally 10 mm diameter down on a specimen carries out the test, and the diameter of the subsequent impression formed provides a basis for calculating hardness.

Knoop hardness   It is a measure of hardness is measured by a calibrated machine that forces a rhomb-shape, pyramidal diamond indenter having specified edge angles under specific small loading conditions into the surface of the test material; the long diagonal in the material is measured after removal of the load.

Mohs hardness   It is a measure of the scratch resistance of a material. The higher the number, the greater scratch resistance with number 10 being termed diamond.

Rockwell hardness   Sheets or plaques at least 0.25 in. thick are used. This thickness may be built up of thinner pieces, if necessary. A steel ball under a minor load is applied to the surface of the specimen. This indents the specimen slightly and assures good contact. The gauge is then set at zero. Basically the major (higher) load is applied for 15 seconds and removed, leaving the minor load still applied. The indentation remaining after 15 seconds is read directly off the testing equipment dial.

The size of the balls used and loadings vary, and values obtained with one set cannot be correlated with values from another set. Rockwell hardness can differentiate the relative hardness of different types of a given plastic; but, since elastic recovery is involved as well as hardness, it is not valid to compare the hardness of various types of plastic entirely on the basis of this test.

Hardness usually implies resistance to abrasion, wear, or indentation (penetration). In plastics it only means resistance to indentation. The scales range from; (1) "R" with a major load of 60 kg; indenter of 0.5 in., (2) "L" with a major load of 60 kg; indenter of 0.25 in., (3) "M" with a major load of 100 kg; indenter of 0.25 in., (4) "E" with a major load of 100 kg; indenter of 0.125 in., and (5) "K" with a major load of 150 kg; indenter of 0.125 in.

The hardness is of limited value to the designer, but can be of some value when comparing these data between materials.

Scleroscope hardness   It is a dynamic indentation hardness test using a calibrated instrument that drops a diamond-tipped hammer from a fixed height onto the surface of the material being tested.

Shore hardness   It is the indentation hardness of a material as determined by the depth of an indentation made with an indenter of the Shore type durometer. The scale reading on this durometer is from zero (corresponding to 0.100 in. depth) to 100 for zero depth. The Shore A indenter has a sharp point, is spring-loaded, and is used for the softer plastics. The Shore B indenter has a blunt point, is spring-loaded at a higher value, and is used for harder plastics.

Vicat hardness   It is a determination of the softening point for TPs that have no definite melting point. The softening point is taken the temperature at which the specimen is usually penetrated to a depth of 1 mm^2 (0.0015 in^2) circular or square cross section, under a 1,000 g load.

Deformation Under Load

The specimen is a small cube, either solid or composite. It is placed between the anvils of the testing machine, and loaded at 1000, 2000, or 4000 psi. The gauge is read 10 s after loading, and again 24 h later. The deflection is recorded in mils. Calculation is made after the specimen is removed from the testing machine. By dividing the change in height by the original height and multiplying by 100, the percentage deformation is calculated. This test may be run at different temperatures.

This test on rigid plastics indicates their ability to withstand continuous short-term compression without yielding and loosening when fastened as in insulators or other assemblies by bolts, rivets, etc. It does not indicate the creep resistance of a particular plastic for long periods of time. It is also a measure of rigidity at service temperatures and can be used as identification for procurement. Data should indicate stress level and the temperature of the test.

Fatigue Strength

The fatigue strength is defined as that stress level at which the test specimen will sustain "N" cycles prior to failure. The data are generated on a machine that runs at 1800 cycles per minute. This test is of value to material manufacturers in determining consistency of their product (Chapter 2).

Long-Term Stress Relaxation/Creep

This review concerns the long-term behavior of plastics when exposed to conditions that include continuous stresses, environment, excessive heat, abrasion, and continuous contact with liquids. This subject has been reviewed in Chapter 2 (**LONG-TERM LOAD BEHAVIOR**) but since it is a very important subject the review is continuing. Tests such as those outlined by ASTM D 2990 that describe in detail the specimen preparations and testing procedure are intended to produce consistency in observations and records by various manufacturers, so that they can be correlated to provide meaningful information to product designers.

The procedure under this heading is intended as a recommendation for uniformity of making setup conditions for the test, as well as recording the resulting data. The reason for this action is the time consuming nature of the test (many years duration), which does not lend itself to routine testing. The test specimen can be round, square, or rectangular and manufactured in any suitable manner meeting certain dimensions. The test is conducted under controlled temperature and atmospheric conditions.

The requirements for consistent results are outlined in detail as far as accuracy of time interval, of readings, etc., in the procedure. Each report of test results should indicate the exact grade of material and its supplier, the specimen's method of manufacture, its original dimensions, type of test (tension, compression, or flexure), temperature of test, stress level, and interval of readings.

When a load is initially applied to a specimen, there is an instantaneous strain or elongation. Subsequent to this, there is the time-dependent part of the strain (creep), which results from the continuation of the constant stress at a constant temperature. In terms of design, creep means changing dimensions and deterioration of product strength when the product is subjected to a steady load over a prolonged period of time.

All the mechanical properties described in tests for the conventional data sheet properties represented values of short-term application of forces, and, in most cases, the data obtained from such tests are used for comparative evaluation or as controlling specifications for quality determination of materials along with short-duration and intermittent-use design requirements.

The visualization of the reaction to a load by the dual component interpretation of a material is valuable to the understanding of the creep process, but meaningless for design purposes. For this reason, the designer is interested in actual deformation or part failure over a specific time span. This means making observations of the amount of strain at certain time intervals which will make it possible to construct curves that could be extrapolated to longer time periods. The initial readings are 1, 2, 3, 5, 7, 10, and 20 h, followed by readings every 24 h up to 500 h and then readings every 48 h up to 1000 h (Chapter 2).

The time segment of the creep test is common to all materials, i.e., strains are recorded until the specimen ruptures or the specimen is no longer useful because of yielding. In either case, a point of failure of the test specimen has been reached.

The stress levels and the temperature of the test for a material is determined by the manufacturer. The guiding determinants are the continuous allowable working stress at room temperature and the continuous allowable working stress at temperatures of potential applications.

The strain readings of a creep test can be more convenient to a designer if they are presented as a creep modulus. In a viscoelastic material, strain continues to increase with time while the stress level remains constant. Since the modulus equals stress divided by strain, we have the appearance of a changing modulus.

The creep modulus, also known as apparent modulus or viscous modulus when graphed on log-log paper, is normally a straight line and lends itself to extrapolation for longer periods of time. The apparent modulus should be differentiated from the modulus given in the data sheets, because the latter is an instantaneous value derived from the testing machine.

The method of obtaining creep data and their presentation have been described; however, their application is limited to the exact same material, temperature use, stress level, atmospheric conditions, and type of test (tensile, compression, flexure) with a tolerance of ±10%. Only rarely do product requirement conditions coincide with those of the test or, for that matter, are creep data available for all grades of material that may be selected by a designer. In those cases a creep test of relatively short duration such as 1000 h can be instigated, and the information can be extrapolated to the long-term needs. It should be noted that reinforced thermoplastics and thermosets display much higher resistance to creep (Chapter 2).

Creep information is not as readily available as short-term property data sheets are. From a designer's viewpoint, it is important to have creep data available for products subjected to a constant load for prolonged periods of time. The cost of performing or obtaining the test in comparison with other expenditures related to product design would be insignificant when considering the element of safety and confidence it would provide. Furthermore, the proving of product performance could be carried out with a higher degree of favorable expectations as far as a plastic material is concerned. Progressive material manufacturers can be expected to supply the needed creep and stress-strain data under specified use conditions when requested by designer; but, if that is not the case, other means should be utilized to obtain required information.

In conclusion regarding creep testing, it can be stated that creep data and a stress-strain diagram indicate whether plain plastic properties can lead to practical product dimensions or whether a RP has to be substituted to keep the design within the desired proportions. For long-term product use under continuous load, plastic materials have to consider creep with much greater care than would be the case with metals.

Summation

Throughout this book all the different types of mechanical properties are presented and reviewed. These mechanical properties include a tremendous range of different types that can usually be characterized by their stiffness, strength, and toughness.

Stiffness The same factors that influence thermal expansion dictate the stiffness of plastics. Thus in a TS the degree of cross-linking and amount of overall flexibility are important. As an example, in a TP its crystallinity and secondary bond's strength control its stiffness.

Strength The subject of strength is much more complex than stiffness, since so many different types exist such as short or long term, static or dynamic, and torsion or impact strengths. Some strength aspects are interrelated with those of toughness. This section reviews certain simplified concepts of strength that are important influences on strength based on long and short term exposure.

The crystallinity of TPs is important for their short term yield strength. Unless the crystallinity is impeded, increased molecular weight generally also increases the yield strength. However, the cross-linking of TSs increases their yield strength substantially but has an adverse effect upon toughness.

Long term rupture strengths in TPs are increased much more readily by increasing the secondary bonds' strength and crystallinity than by increasing the primary bond strength. Fatigue strength is similarly influenced, and all factors that influence thermal dimensional stability also affect fatigue strength. This is a result of the substantial heating that is often encountered with fatigue, particularly in TPs.

Toughness The subject of toughness is usually the most complex factor to define and understand. Tough plastics are usually described as ones having a high elongation to failure or ones in which a lot of energy must be expended to produce failure. For high toughness a plastic needs both the ability to withstand load and the ability to elongate substantially without failing except in the case of reinforced TSs that are tough, which may have high strengths with low elongation.

It may appear that factors contributing to high stiffness are required, but this is not true, because there is an inverse relationship between flaw sensitivity and toughness: the higher the stiffness and the yield strength of a TP, the more flaw sensitive it becomes. However, because some load bearing capacity is required for toughness, high toughness can be achieved by a high trade off of certain factors.

Crystallinity increases both stiffness and yield strength, resulting usually in decreased toughness. This is true below T_g in most amorphous plastics, and below or above the T_g in a substantially crystalline plastic. However, above the T_g in a plastic having only moderate crystallinity increased crystallinity improves its toughness. Furthermore, an increase in molecular weight from low values increases toughness, but with continued increases, the toughness begins to drop.

Cross-linking produces some dimensional stability and improves toughness in a non-crystalline type of plastic above the T_g, but

high levels of cross-linking lead to embrittlement and loss of toughness. This is one of the
problems with TSs for which an increase in T_g
is desired. Increased cross-linking or stiffening of the chain segments increases the T_g, but
it also decreases toughness. A popular way
to increase toughness is to blend, compound,
or copolymerize a brittle plastic with a tough
one. Although some loss in stiffness is usually encountered, the result is a satisfactory
combination of properties.

Thermal Property

Figure 5-6 and Tables 5-3 to 5-5 provide an
introductory guide to the different thermal
properties of plastics. Heat resistance properties of plastics retaining 50% of properties
obtainable at room temperature with plastic exposure and testing at elevated temperatures are shown in Fig. 5-6 for the general
family or group type.

Zone 1: acrylic, cellulose esters, crystal-
lizable block copolymers, LDPE, PS,
vinyl polymers, SAN, SBR, and urea-
formaldehyde.

Zone 2: acetal, ABS, chlorinated polyether, ethyl cellulose, ethylene-vinyl acetate copolymer, furan, ionomer, phe-

noxy, polyamides, PC, RDPE, PET, PP,
PVC, and urethane.

Zone 3: polychlorotrifluoroethylene, and
vinylidene fluoride.

Zone 4: alkyd, fluorinated ethylene-
propylene, melamine-formaldehyde,
phenol-furfural, and polysulfone.

Zone 5: acrylic, diallyl phthalate, epoxy,
phenol-formaldehyde, TP polyester, and
polytetrafluoroethylene.

Zone 6: parylene, polybenzimidazole,
polyphenylene, and silicone.

Zone 7: polyamide-imide, and polyimide;

Zone 8: plastics now being developed using rigid linear macromolecules rather
than crystallization and cross-linking,
etc.

Specific family or group of plastics
(polyethylene, polyvinyl chlorides, etc.) are
compounded or alloyed to provide different
properties and/or processing behaviors. Thus
a plastic listed in Fig. 5-6 could have different
heat resistance properties.

Deflection Temperature Under Load

The DTUL, also called the heat distortion
temperature (HDT) of a plastic is a method
to guide or assess its load-bearing capacity at an elevated temperature. Details on
the method of testing are given in ASTM
D648. Basically a 1.27 cm ($^1/_2$ in.) deep plastic test bar is mounted on supports 10.16 cm
(4 in.) apart and loaded as a beam. A bending
stress of either 66 psi or 264 psi (455 gPa or
1,820 gPa) is applied at the center of the span.

The test is conducted in a bath of oil, with
the temperature increased at a constant rate
of 2°C per minute. The DTUL is the temperature at which the sample attains a deflection of 0.0254 cm (0.010 in.). This test is only
a guide. It represents a method that could
be correlated to product designs, but as with
most other tests conducted on test specimens
and not on a finished product, it is just a guide
(Fig. 5-7).

In this test, if the specimen contains internal stresses the value will be lower than a
specimen with no stresses. In fact, the test can

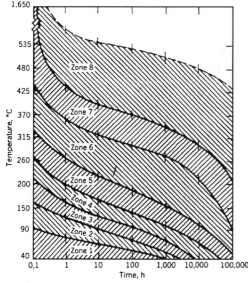

Fig. 5-6 Guide to heat resistance with 50% retention of properties.

Table 5-3 Examples of plastics in elevated temperature applications

Polymer	Comments
Polyphenyls	Decompose at 530°C (986°F); infusible, insoluble polymers.
Polyphenylene oxide	Decomposes close to 500°C (932°F); heat cures above 150°C (302°F) to elastomer; usable heat range −135–185°C (−211–365°F).
Polyphenylene sulfide	Melts at 270–315°C (578–599°F); crosslinked polymer stable to 450°C (842°F) in air: adhesive and laminating applications.
Polybenzyls; polyphenethyls	Fusible, soluble, and stable at 400°C (752°F); low molecular weight.
Parylenes (poly-*p*-xylylene)	Melt above 520°C (968°F); insoluble; capable of forming films; poor thermal stability in air; stable to 400–525°C (752–977°F) in inert atmosphere.
Polyterephthalamides	Melting points up to 455°C (851°F); fibers have good tenacity, elongation, modulus.
Polysulfanyldibenzamides	Melting points up to 330°C (626°F); soluble; good fiber properties.
Polyhydrazides	Dehydrate at 200°C (392°F) to over 400°C (752°F) to form polyoxadiazoles; good fiber properties.
Polyoxamides	Some melting points above 400°C (752°F); give clear, flexible films.
Phenolphthalein polymers	Melting points of 300°C (572°F) to over 400°C (752°F); formable into fiber and film.
Hydroquinone polyesters	Soluble polymers with melting points of 335°C (635°F) to over 400°C (752°F).
Polyhydroxybenzoic acids	Films melt at 380–450°C (716–842°F); stable to oxidation but not to hydrolysis; tough, flexible films; good thermal stability.
Polyimides	Commercial film, coating, and resin stable up to 600°C (1112°F); continuous use up to 300°C (572°F).
Polyarylsiloxanes	Good thermal stability 400–500°C (752–932°F); coatings, adhesives.
Carboranes	Stable in air and nitrogen at 400–450°C (752–842°F); elastomeric properties for silane derivatives up to 538°C (1000°F); adhesives.
Polybenzimidazoles	Developmental laminating resin, fiber, film; stable 24 hours at 300°C (572°F) in air.
Polybenzothiazoles	Stable in air at 600°C (1112°F); cured polymer soluble in concentrated sulfuric acid.
Polyquinoxalines	Stable in air at 500°C (932°F); tough, somewhat flexible resins; make film, adhesive.
Polyphenylenetriazoles	Thermally stable to 400–500°C (752–932°F); make film, fiber, coatings.
Polydithiazoles	Decompose at 525°C (977°F); soluble in concentrated sulfuric acid.
Polyoxadiazoles	Decompose at 450–500°C (842–932°F); can be made into fiber or film.
Polyamidines	Stable to oxidation up to 500°C (932°F); can make flexible elastomer.
Pyrolyzed polyacrylonitrile	Stable above 900°C (1625°F); fiber resists abrasion with low tenacity.
Polyvinyl isocyanate ladder polymer	Soluble polymer that decomposes at 385°C (725°F); prepolymer melts above 405°C (761°F).
Polyamide-imide	Service temperatures up to 288°C (550°F); amenable to fabrication.
Polysulfone	Thermoplastic; use temperature −102°C (−152°F) to greater than 150°C (302°F); acid and base resistant.
Polybenzaylene benzimidazoles (pyrrones)	Thermally stable to 600°C (1112°F); insoluble in common solvents; good mechanical properties.
Polybenzoxazoles	Stable in air to 500°C (932°F); insoluble in common solvents except sulfuric acid; nonflammable; chemical resistant; film.
Ionomer	High melt and tensile strength; tough; resilient; oil and solvent resistant; adhesives, coatings.

(*Continues*)

Table 5-3   (*Continued*)

Polymer	Comments
Diazadiphosphetidine	Thermoplastic up to 350°C (662°F); thermosetting at 357°C (707°F); cured material has good thermal stability to 500°C (932°F); amenable to fabrication.
Phosphorous amide epoxy	Soluble B-staged material; amenable to fabrication; good thermal stability.
Phosphonitrilic	Retention of properties in air up to 399°C (750°F).
Metal polyphosphinates	Polymers stable to better than 400°C (752°F).
Phenylsilesesquioxanes (phenyl-T ladder polymers)	Soluble; high molecular weight; infusible; improved tensile strength; high thermal stability to 525°C (977°F) in air; film forming.

be used to determine the degree of internal stress. Since a stress and the deflection for a certain depth of test bar are specified, this test may be thought of as establishing the temperature at which the flexural modulus decreases to particular values: 35,000 psi (240 MPa) at 66 psi load stress, and 140,800 psi (971 MPa) at 264 psi.

Coefficient of Linear Thermal Expansion

The specimen can be square or round to fit a dilatometer test tube in a free sliding manner. The length is governed by the sensitivity of dial gauge, the expected expansion, and the accuracy desired. The specimen is mounted in the dilatometer and placed in a bath of either −22°F (−30°C) or 87°F (+30°C) until the temperature of the bath is reached. When this takes place, the indicator dial is read showing the expansion or contraction of the specimen. These readings are compared with measurement of specimen length prior to placing it in the dilatometer.

With the application of plastics in combination with other materials, the coefficient of expansion plays an important role in making design allowances for expansions (also contractions) of various materials at different temperatures so that satisfactory functions of products are ensured.

Table 5-4   Examples of ignition and flash temperatures

	Self Ignition		Flash Ignition	
	°F	°C	°F	°C
Polyethylene	662	350	644	340
Polypropylene	1022	550	968	520
Polytetrafluoroethylene	1076	580	1040	560
Polyvinyl chloride	842	450	734	390
Polyvinyl fluoride	896	480	788	420
Polystyrene	914	490	662	350
SBR (Styrene Butadiene Rubber)	842	450	680	360
ABS (Acrylonitrile Butadiene Styrene)	896	480	734	390
Polymethyl methacrylate	806	430	572	300
PAN (Polyacrylonitrile)	1040	560	896	480
Cellulose (paper)	446	230	410	210
Cellulose acetate	878	470	644	340
66 Nylon cast	842	450	788	420
66 Nylon spun and drawn	986	530	914	490
Polyester	896	480	824	440

Table 5-5   Effects of elevated temperature and chemical agents on stability of plastics

Environment

Plastic Material	Acetal Copolymer	Acetal Homopolymer	Nylon 6/6	Thermoplastic Polyester (PBT)	Thermoplastic Polyester (PET)	Polyester Elastomer	Liquid Crystal Polymer*	Polyphenylene Sulfide*	Polyarylate	Polycarbonate	Polysulfone*	Modified Polyphenylene Oxide	Polypropylene	ABS	316 Stainless Steel	Carbon Steel	Aluminum	% Change by Weight
Temperature (°F)	77 (25°C)	200 (93.3°C)	77	200	77	200	77	200	77	200	77	200	77	200	77	200		
Acetals	1–4	2–4	1	2	1–2	4	1–3	2–5	1–5	2–5	5	5	5	5	1	2–3		0.22–0.25
Acrylics	5	5	2	3	5	5	1	3	2	5	4	4–5	5	5	5	5		0.2–0.4
Acrylonitrile-Butadiene-Styrenes (ABS)	4	5	2	3–5	3–5	5	1	2–4	1	2–4	1–4	5	1–5	5	3–5	5		0.1–0.4
Aramids (aromatic polyamide)	1	1	1	1	1	1	2	3	4	5	3	4	2	5	1	2		0.6
Cellulose acetates (CA)	2	3	2	3	3	4	2	3	3	5	3	5	3	5	5	5		2–7
Cellulose acetate butyrates (CAB)	4	5	1	3	3	4	2	4	3	5	3	5	3	5	5	5		0.9–2.0
Cellulose acetate propionates (CAP)	4	5	1	3	3	4	1	2	3	5	3	5	3	5	5	5		1.3–2.8
Diallyl phthalates (DAP, filled)	1–2	2–4	2	3	2	4	2	3	2	4	1–2	2–3	2	4	3–4	4–5		0.2–0.7
Epoxies	1	2	1	2	1–2	3–4	1	1–2	1	2	2–3	3–4	4	4–5	2	3–4		0.01–0.10
Ethylene copolymers (EVA) (ethylene-vinyl acetates)	5	5	5	5	5	5	1	2	1	5	1	5	1	5	2	5		0.05–0.13
Ethylene/tetrafluoroethylent copolymers (ETFE)	1	1	1	1	1	1	1	1	1	1	1	1	1	1	1	1		<0.03
Fluorinated ethylene propylenes (FEP)	1	1	1	1	1	1	1	1	1	1	1	1	1	1	1	1		<0.01
Perfluoroalkoxies (PFA)	1	1	1	1	1	1	1	1	1	1	1	1	1	1	1	1		<0.03
Polychlorotrifluoroethylenes (CTFE)	1	1	1	1	3	4	1	1	1	1	1	1	1	1	1	1		0.01–0.10
Polytetrafluoroethylenes ®(TFE)	1	1	1	1	1	1	1	1	1	1	1	1	1	1	1	1		0
Furans	1	1	1	1	1	1	2	2	2	2	1	1	5	5	1	1		0.01–0.20
Ionomers	2	4	1	4	4	4	1	4	1	4	2	4	1	5	1	4		0.01–1.4
Melamines (filled)	1	1	1	1	1	1	2	3	2	3	2	3	2	3	1	2		0.01–1.30
Nitriles (high barrier alloys of ABS or SAN)	1	4	1	2–4	1–4	2–5	1	2–4	1	2–4	2–5	5	3–5	5	1–5	5		0.2–0.5
Nylons	1	1	1	1	1	2	1	2	2	3	5	5	5	5	1	1		0.2–1.9
Phenolics (filled)	1	1	1	1	1	1	2	3	3	5	1	1	4	5	2	2		0.1–2.0
Polyallomers	2	4	2	4	4	5	1	1	1	1	1	3	1	4	1	3		<0.01

(The ETFE, FEP, PFA, CTFE and TFE rows are grouped under the side-label "Fluorocarbons.")

A rating of 1 indicates greatest stability.

The difference in thermal expansion between the usual commodity plastics and steel is very large. It is to be noted that some plastic material changes in length rather abruptly at some temperatures, beyond the limits of the test condition. In such cases, a special investigation should be instigated, and the coefficient of expansion established under temperatures of usage. However there are plastics that can be compounded to match or even have less thermal expansion than steel, etc.

This test shows the reversible linear thermal expansion. The accuracy of these results may be affected by factors in certain plastics such as loss of plasticizer, solvent, relieving of stresses, etc. When a product demands most precise data, the factors mentioned should be considered for their possible influence on the information.

Brittleness Temperature

The conditioned specimens are cantilevered from the sample holder in the test apparatus, which has been brought to a low temperature (that at which the specimens would be expected to fail). When the

Table 5-5   (*Continued*)

Environment — Temperature (°F). A rating of 1 indicates greatest stability.

Plastic Material	Aromatic Solvents 77 (25°C)	Aromatic Solvents 200 (93.3°C)	Aliphatic Solvents 77	Aliphatic Solvents 200	Chlorinated Solvents 77	Chlorinated Solvents 200	Weak Bases and Salts 77	Weak Bases and Salts 200	Strong Bases 77	Strong Bases 200	Strong Acids 77	Strong Acids 200	Strong Oxidants 77	Strong Oxidants 200	Esters and Ketones 77	Esters and Ketones 200	24-H Water Absorption (% Change by Weight)
Polyamide-imides	1	1	1	1	2	3	1	1	3	4	2	3	2	3	1	1	0.22–0.28
Polyarylsulfones (PAS)	4	5	2	3	4	5	1	2	2	2	1	1	2	4	3	4	1.2–1.8
Polybutylenes (PB)	3	5	1	5	4	5	1	2	1	3	1	3	1	4	1	3	<0.01–0.3
Polycarbonates (PC)	5	5	1	1	5	5	1	5	5	5	1	1	1	1	5	5	0.15–0.35
Polyesters (thermoplastic)	2	5	1	3–5	3	5	1	3–4	2	5	3	4–5	2	3–5	2	3–4	0.06–0.09
Polyesters (thermoset-glass fiber filled)	1–3	3–5	2	3	2	4	2	3	3	5	2	3	2	4	3–4	4–5	0.01–2.50
Polyethylenes (LDPE-HDPE—low density to high density)	4	5	4	5	4	5	1	1	1	1	1–2	1–2	1–3	3–5	2	3	0.00–0.01
Polyethylenes (UHMWPE—ultrahigh molecular weight)	3	4	3	4	3	4	1	1	1	1	1	1	1	1	3	4	<0.01
Polyimides	1	1	1	1	1	1	2	3	4	5	3	4	2	5	1	1	0.3–0.4
Polyphenylene oxides (PPO) (modified)	4	5	2	3	4	5	1	1	1	1	1	2	1	2	2	3	0.06–0.07
Polyphenylene sulfides (PPS)	1	2	1	1	1	2	1	1	1	1	1	1	1	1	2	1	<0.05
Polyphenylsulfones	4	4	1	1	5	5	1	1	1	1	1	1	1	1	3	4	0.5
Polypropylenes (PP)	2	4	2	4	2–3	4–5	1	1	1	1	1	2–3	2–3	4–5	2	4	0.01–0.03
Polystyrenes (PS)	4	5	4	5	5	5	1	5	1	5	1	5	4	5	4	5	0.03–0.60
Polysulfones	4	4	1	1	5	5	1	1	1	1	1	1	1	1	3	4	0.2–0.3
Polyurethanes (PUR)	3	4	2	3	4	5	2–3	3–4	2–3	3–4	2–3	3–4	4	4	4	4	0.02–1.50
Polyvinyl chlorides (PVC)	4	5	1	5	5	5	1	5	1	5	1	5	2	5	4	5	0.04–1.00
Polyvinyl chlorides—chlorinated (CPVC)	4	4	1	2	5	5	1	2	1	2	1	2	2	3	4	5	0.04–0.45
Polyvinylidene fluorides (PVDF)	1	1	1	1	1	1	1	1	1	2	1	2	1	2	3	5	0.04
Silicones	4	4	2	3	4	5	1	2	4	5	3	4	4	5	2	4	0.1–0.2
Styrene acrylonitriles (SAN)	4	5	3	4	3	5	1	3	1	3	1	3	3	4	4	5	0.20–0.35
Ureas (filled)	1	3	1	3	1	3	2	3	2	3	4	5	2	3	1	2	0.4–0.8
Vinyl esters (glass-fiber filled)	1	3	1–2	2–4	1–2	4	1	3	1	3	1	2	2	3	3–4	4–5	0.01–2.50

A rating of 1 indicates greatest stability.

specimens have been in the test medium for 3 minutes, a single impact is administered, and the samples are examined for failure. Failures are total breaks, partial breaks, or any visible cracks. The test is conducted at a range of temperatures producing varying percentages of breaks. From these data, the temperature at which 50% failure would occur is calculated or plotted and reported as the brittleness temperature of the material according to this test.

This test is of some use in judging the relative merits of various materials for low-temperature flexing or impact. However, it is specifically relevant only for materials and conditions specified in the test, and the values cannot be directly applied to other shapes and conditions.

The brittleness temperature does not put any lower limit on service temperature for end use products. The brittleness temperature is sometimes used in specifications.

Thermal Aging

Section UL 746B provides a basis for selecting high-temperature plastics and provides a long-term thermal-aging index, the RTI or relative thermal index. The testing procedure calls for test specimens in selected thicknesses to be oven aged at certain elevated temperatures (usually higher than the expected operating temperature, to accelerate the test), then be removed at various intervals and tested at room temperature.

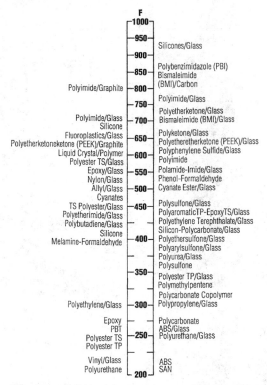

F

1000—
950—
900—   Silicones/Glass
850—   Polybenzimidazole (PBI)
            Bismaleimide
            (BMI)/Carbon
Polyimide/Graphite —800—
            Polyimide/Glass
750—
Polyimide/Glass —700—   Polyetherketone/Glass
Silicone   Bismaleimide (BMI)/Glass
Fluoroplastics/Glass   Polyketone/Glass
Polyetherketoneketone (PEEK)/Graphite —650—   Polyetheretherketone (PEEK)/Glass
Liquid Crystal/Polymer   Polyphenylene Sulfide/Glass
Polyester TS/Glass —600—   Polyimide
Epoxy/Glass —550—   Polamide-Imide/Glass
Nylon/Glass   Phenol-Formaldehyde
Allyl/Glass —500—   Cyanate Ester/Glass
Cyanates
TS Polyester/Glass —450—   Polysulfone/Glass
Polyetherimide/Glass   PolyaromaticTP-EpoxyTS/Glass
Polybutadiene/Glass   Polyethylene Terephthalate/Glass
Silicone   Silicon-Polycarbonate/Glass
Melamine-Formaldehyde —400—   Polyethersulfone/Glass
                               Polyarylsulfone/Glass
                               Polyurea/Glass
                               Polysulfone
350—   Polyester TP/Glass
       Polymethylpentene
       Polycarbonate Copolymer
Polyethylene/Glass —300—   Polypropylene/Glass

Epoxy   Polycarbonate
PBT   ABS/Glass
Polyester TS —250—   Polyurethane/Glass
Polyester TP

Vinyl/Glass   ABS
Polyurethane —200—   SAN

Fig. 5-7 Guide to heat resistance based on the heat-distortion temperature per ASTM D 648 at 264 psi.

are usually 11,000 hours, with a minimum of 5,000 hours. This value is the RTI.

As practiced by the UL, the procedure for selecting an RTI from Arrhenius plots usually involves making comparisons to a control standard material and other such steps to correct for random variations, oven temperature variations, condition of the specimens, and others. The stress-strain and impact and electrical properties frequently do not degrade at the same rate, each having their own separate RTIs. Also, since thicker specimens usually take longer to fail, each thickness will require a separate RTI.

The UL uses RTIs as a guideline to qualify materials for many of the standard appliances and other electrical products it regulates. This testing is done in a conservative manner qualified by judgments based on long experience with such devices; UL does not apply indexes automatically. In general, these RTIs are very conservative and can be used as safe continuous-use temperatures for low-load mechanical products.

Other Heat Test

There are different heat tests, some being specific to a product environment. There are those for temperature and also humidity. With certain materials, humidity combined with elevated temperatures has a significant effect on the material's behavior. This effect would not be evident in the conventional heat distortion test (HDT).

Test specimens can also be used to simulate some degree of warpage. Figure 5-8 compares unreinforced and reinforced glass fiber-TS polyester flexural-type specimens at different temperatures in a droop test (with a center support), sag test (end supports), and an expansion test (bolted at three points). The study for this particular test is conducted at various temperatures.

Thus by analyzing the thermal limits of the various materials available, starting with the maximum and minimum environmental temperatures under which a product must operate and adding any thermal increase

Another reason for using higher temperatures is that for an application requiring long-term exposure a candidate plastic is often required to have an RTI value higher than the maximum application temperature. The properties tested can include mechanical strength, impact resistance, and electrical characteristics. A plastic's position in a test's RTI is based on the temperature at which it still retains 50% of its original properties.

The time required to produce a 50% reduction in properties is selected as an arbitrary failure point. These times can be gathered and used to make a linear Arrhenius plot of log time versus the reciprocal of the absolute exposure temperature. An Arrhenius relationship is a rate equation followed by many chemical reactions. A linear Arrhenius plot is extrapolated from this equation to predict the temperature at which failure is to be expected at an arbitrary time that depends on the plastic's heat-aging behavior, which

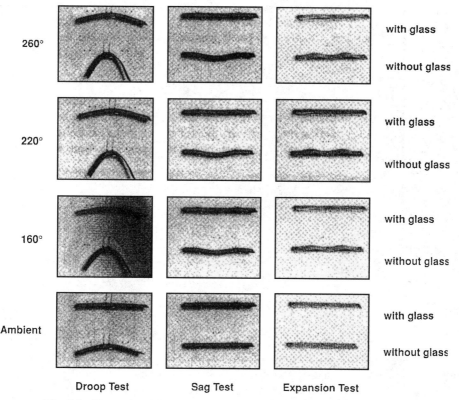

260°			with glass
			without glass
220°			with glass
			without glass
160°			with glass
			without glass
Ambient			with glass
			without glass

Droop Test	Sag Test	Expansion Test

Fig. 5-8 Example of droop, sag, and expansion tests using RP samples.

from hysteresis heat that develops from flex or vibration and so on will at least tell the designer which materials cannot be used.

The ratings given the designer will also provide some idea of the short-term stiffness to be expected of various materials at elevated temperatures, as well as their thermal aging resistance with regard to certain properties. Establishing two parameters, ASTM D 648 and UL 746B, for a variety of materials provides the designer with a reasonable starting point for initially assessing materials for high-temperature applications. Most high-performance plastics are filled compounds, since fillers and reinforcements (Fig. 5-9) generally enhance high temperature strength and stiffness.

A general definition for a high-temperature plastic is one having a thermal value in terms of ASTM D 648 and UL 746B higher than 149°C (300°F). There are numerous plastics that are both processable and have

useful mechanical properties in the 149 to 260°C (300 to 500°F) range (Fig. 5-10). Their costs are usually high, but so is their performance.

High-temperature plastics fall into the usual categories of TSs and TPs. The TSs are used principally by the aircraft and aerospace markets but also for automotive, industrial, medical, and electronic products. Epoxies are principally used, with other plastics being TS polyesters, phenolics, and urethanes. These plastics are usually reinforced with the high strength fibers seen in Fig. 5-9, individually or in combination with S-glass, graphite, aramid, and others. About 85wt% of all RPs used in conventional-temperature environments only require E-glass (14).

High-temperature TPs are available to compete with TSs, metals, ceramics, and other nonplastic materials. The heat-resistant TPs include polyetheretherketone (PEEK) and polyethersulfone (PES), polyamideimide, liquid crystal polymer (LCP) and others.

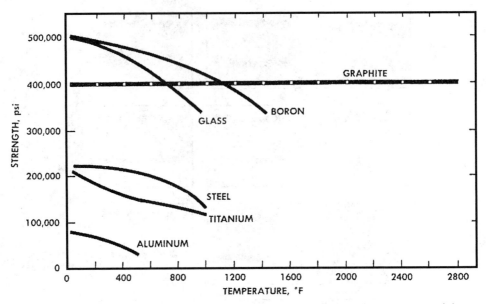

Fig. 5-9 Examples of tensile strengths vs. temperature in high performance materials.

These TPs have high inherent heat resistance and offer such other advantages over TSs as higher toughness and ease of processing. Some of these plastics are amorphous with a high T_g such as PES, and some like PEEK and the liquid crystal polymer (LCP) are highly crystalline. Some high-temperature plastics are commercially available in neat form, with most being available only in the filled or reinforced form for such high-performance products. Many of these plastics can be processed on standard or modified TP processing equipment, which requires melting at higher temperatures than commodity or engineering plastics, but others, like the polyimides, may require machining into shape.

The direction of high-temperature TSs appears to be toward more toughness using new processing techniques for the aviation and aerospace markets. The high-temperature TPs have been receiving considerable attention as possible replacements for TSs in advanced RPs because of their higher toughness, faster processing, and ease of repair. The TPs are being promoted in both unreinforced and reinforced forms as molding and extrusion materials. Their high continuous use temperatures, combined with their good chemical, water, and flame resistance and low smoke generation, are finding new applications, particularly as their price is reduced through higher volume usage. The growth of TP-TS hybrids, offering TPs' ready processing combined with TSs' long-term dimensional stability, are also positive.

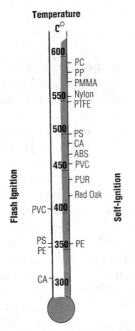

Fig. 5-10 Basic guide to flash ignition and self-ignition temperatures (per ASTM D 1929) for plastics and red wood.

Electrical Property

The resistance of most plastics to the flow of direct current is very high. Both surface and volume electrical resistivities are important properties for applications of plastics insulating materials. The volume resistivity is the electrical resistance of the material measured in ohms as though the material was a conductor. Insulators will not sustain an indefinitely high voltage; as the applied voltage is increased, a point is reached where a drastic decrease in resistance takes place accompanied by a physical breakdown of the insulator. This is known as the dielectric strength, which is the electric potential in volts, which would be necessary to cause the failure of a 1/8-in. thick insulator (Chapter 4, **ELECTRICAL/ELECTRONICS PRODUCT**).

Electrical Resistance

Specimens for these tests may be any practical form, such as flat plates, sheets, or tubes. These tests describe methods for determining the several properties defined below. Two electrodes are placed on or embedded in the surface of a test specimen. Different properties are obtained.

Insulation resistance is the ratio of direct voltage applied to the electrodes to the total current between them; dependent upon both volume and surface resistance of the specimen. In materials used to insulate and support components of an electrical network, it is generally desirable to have insulation resistance as high as possible.

Volume resistivity is the ratio of the potential gradient parallel to the current density.

Surface resistivity is the ratio of the potential gradient parallel to the current along its surface to the current per unit width of the surface. Knowing the volume and surface resistivity of an insulating material makes it possible to design an insulator for a specific application.

Volume resistance is the ratio of direct voltage applied to the electrodes to that portion of current between them that is distributed through the volume of the specimen.

Surface resistance is the ratio of the direct voltage applied to the electrodes to that portion of the current between them that is in a thin layer of moisture or other semiconducting material which may be deposited on the surface. High volume and surface resistance are desirable in order to limit the current leakage of the conductor that is being insulated.

Arc Resistance

This test shows the ability of a material to resist the action of an arc of high voltage and low current close to the surface of the insulation in tending to form a conducting path therein. The arc resistance data are of relative value only for distinguishing materials of nearly identical composition, such as for quality control, development, or identification.

Dielectric Strength

Specimens are thin sheets or plates having parallel plane surfaces and are of a size sufficient to prevent flashing over. Dielectric strength varies with thickness and, therefore, specimen thickness must be reported. The dielectric strength varies inversely with the thickness of the specimen. The dielectric strength of plastics will drop sharply if holes, bubbles, or contaminants are present in the specimen being tested.

Since temperature and humidity affect results, it is necessary to condition each type of material as directed in the specification for that material. The test for dielectric strength must be run in the conditioning chamber or immediately after removing the specimen from the chamber.

The specimen is placed between heavy cylindrical brass electrodes, which carry electrical current during the test. There are two ways of running this test for dielectric strength. In the short-time test the voltage is increased from zero to breakdown at a uniform rate. The precise rate of voltage rise is specified in the governing material specifications. In the step-by-step test the initial voltage applied is 50% of breakdown voltage shown by the short-time test. It is increased at rates specified for each type of material, and the breakdown level is noted. Breakdown by these tests means passage of sudden excessive current through the specimen and can be verified by instruments and by visible damage to the specimen.

This test is an indication of the electrical strength of a material as an insulator. The dielectric strength of an insulating material is the voltage gradient at which electric failure or breakdown occurs as a continuous arc (the electrical property analogous to tensile strength in mechanical properties). The dielectric strength of materials varies greatly with several conditions such as humidity and geometry, and it is not possible to directly apply the standard test values to field use unless all conditions, including specimen dimensions, are the same. Because of this, the dielectric strength test results are of relative rather than absolute value as a specification guide.

Dielectric Constant and Dissipation Factor

The specimen may be a sheet of any size convenient to test, but should have uniform thickness. The test may be run at standard room temperature and humidity, or in special sets of conditions as desired. In any case, the specimens should be preconditioned to the set of conditions used. Electrodes are applied to opposite faces of the test specimen. The capacitance and dielectric loss are then measured by comparison or substitution methods in an electric bridge circuit. From these measurements and the dimensions of the specimen, dielectric constant and loss factor are computed.

The dissipation factor is a ratio of the real power (in-phase power) to the reactive power (power 90° out of phase). It is also defined as: (1) IT is the ratio of conductance of a capacitor in which the material is the dielectric to its susceptance, (2) IT is the ratio of its parallel reactance to its parallel resistance; it is the tangent of the loss angle and the cotangent

of the phase angle, and (3) IT is a measure of the conversion of the reactive power to real power, showing as heat.

The dielectric constant is the ratio of the capacity of a condenser made with a particular dielectric to the capacity of the same condenser with air as the dielectric. For a material used to support and insulate components of an electrical network from each other and ground, it is generally desirable to have a low level of dielectric constant. For a material to function as the dielectric of a capacitor, on the other hand, it is desirable to have a high value of dielectric constant, so that the capacitor may be physically as small as possible.

The loss factor is the product of the dielectric constant and the power factor, and is a measure of total losses in the dielectric material.

Optical Property

Examples of plastics' transparent properties are shown in Table 5-6. A basic behavior of the appearance of a transparent material, that is one that transmits light, is its transmittance: the ratios of the intensities of light passing through and the light incident on the specimen. Similarly, the appearance of an opaque material (one which may reflect light but does not transmit it) is characterized by its reflectance, the ratio of the intensities of the reflected and incident light. A translucent substance is one that transmits part and reflects part of the light incident on it. Gloss is the geometrically selective reflection of a surface responsible for its shiny or lustrous appearance. This property may be measured by the use of various photoelectric instruments or simply by observation.

Haze and Huminous Transmittance

In this test, haze of a specimen is defined as the percentage of transmitted light that, in passing through the specimen, deviates more than 2.5° from the incident beam by forward scattering. Basically it is defined as the ratio of transmitted to incident light.

Table 5-6 Examples of plastics' transparent properties

Properties	ASTM Method	Units	Methyl Methacrylate (Acrylic)	Polystyrene (Styrene)	Polycarbonate	Methyl Methacrylate Styrene Copolymer
Refractive index (n_D)	D 542		1.491	1.590	1.586	1.562
Abbe. value (v)	D 542		57.2	30.9	34.7	35
$dn/dt \times 10^{-5}/°C$			8.5	12.0	14.3	14.0
Haze (%)	D 1003	%	<2	<3	<3	<3
Luminous transmittance (0.125-in. thickness)	D 1003	%	92	88	89	90
Critical angle (i_c)		degree	42.2	39.0	39.1	39.6
Deflection temperature	D 648–56	°F				
3.6 F/min., 264 psi			198	180	280	
3.6 F/min., 66 psi			214	230	270	212
Coefficient of linear thermal expansion	D 696–44	in./in./°F $\times 10^{-6}$	3.6	3.5	3.8	3.6
Recommended max. cont. service temp.		°F	198	180	255	200
Water absorption (immersed 24 hrs. at 73°F)	D 570–63	%	0.3	0.2	0.15	0.15
Specific gravity (density)	D 792		1.19	1.06	1.20	1.09
Hardness (0.25-in. sample)	D 785–62		M 97	M 90	M 70	M 75
Impact strength (Izod Notch)	D 256	ft.-lb./in.	0.3–0.5	0.35	12–17	
Dielectric strength	D 149–64	V/mil	500	500	400	450
Dielectric constant						
60 HZ	D 150		3.7	2.6	2.90	3.40
10^6 Hz			22.2	2.45	2.88	2.90
Power factor						
60 Hz	D 150		0.05	0.0002	0.0007	0.006
10^6 Hz			0.03	0.0002–0.0004	0.0075	0.013
Volume resistivity	D 257	ohm-cm	10^{18}	$>10^{16}$	8×10^{16}	10^{15}

These qualities are considered in most applications for transparent plastics, forming a basis for directly comparing the transparency of various grades and types of plastic. The data are of value when a material is considered for optical purposes. Many transparent plastics do not have water clarity, and, for this reason, the data should indicate whether the material was natural or tinted when tested.

Luminous Reflectance

Opaque specimens should have at least one plane surface. Translucent and transparent specimens must have two surfaces that are plane and parallel. This test is the primary method for obtaining colormetric data. The property determined that is of design interest is luminous transmittance.

Opacity and Transparency

Opacity or transparency is important when the amount of light to be transmitted is a consideration. These properties are usually measured as haze and luminous transmittance. As reviewed haze is defined as the percentage of transmitted light through a test specimen that is scattered more than 2.5° from the incident beam. Luminous transmittance is the ratio of transmitted light to incident light. Table 5-7 provides the optical and various other properties of different transparent plastics.

Some definitions of key terms used in identifying optical conditions are reviewed in Chapter 4, **TRANSPARENT AND OPTICAL PRODUCTS, Properties, Performances, and Products.**

Abrasion and Mar Resistance

In this test for transparent plastics, the loss of optical effects is measured when a specimen is exposed to the action of a special abrading wheel. In one type of test the amount of material lost by a specimen is determined when the specimen is exposed to falling abrasive particles or to the action of an abrasive belt. In another test, the loss of gloss due to the dropping of loose abrasive on the specimen is measured. The results produced by the different tests may be of value for research and development work when it is desired to improve a material with respect to one of the test methods. The variables that enter into tests of this type are

Table 5-7 Notable behaviors of some transparent plastics

Generic Family	Notable Characteristics
Transparent ABS	Good impact properties, good processibility
Acrylic (PMMA)	Excellent resistance to outdoor exposure, crystal clarity
Allyl diglycol carbonate	Good abrasion/chemical resistance, thermoset
Cellulosics	Heat sensitive, limited chemical resistance, good toughness
Nylon, amorphous	Excellent abrasion resistance, moisture sensitive
PET, PETG	Good barrier properties, not weatherable, clarity dependent on processing, orientation greatly increases physical properties
Polyarylate	Excellent UV resistance, high heat distortion
Polycarbonate	Excellent toughness, good thermal/flammability characteristics
Polyetherimide	Good chemical/solvent resistance, good thermal/flammability properties, inherent high color
Polyphthalate carbonate	Good thermal properties, autoclavable
Polyethersulfone	Excellent thermal stability, resists creep
Poly-4–methylpentene–1	UV/moisture sensitive, high crystalline melting point, lowest density of all thermoplastics
Polyphenylsulfone	Excellent thermal stability, resists creep, inherent high color
Polystyrene	Excellent processibility, poor UV resistance, brittle
Polysulfone	Excellent thermal/hydrolytic stability, poor weatherability/impact strength
PVC, rigid	Excellent chemical resistance/electrical properties, weatherable, decomposition evolves HCl gas
Styrene acrylonitrile	Good stress-crack and craze resistance, brittle
Styrene butadiene	Good processibility, no stress whitening
Styrene maleic anhydride	Higher-heat styrenic, brittle
Styrene methyl methacrylate	Good processibility, slightly improved weatherability
Thermoplastic urethane, rigid	Excellent chemical/solvent resistance, good toughness

so numerous that it is questionable how the information from such tests could be used. Some of the factors are type of abrasive, shape of abrading particle, nature of plastic material, speed of action on the plastic, shape of the part and its temperature, the manner in which the abrasive is attached to the backing, and the bonding agent.

Currently, these tests are of no practical value to the designer and the only approach to the problem of scratch, mar, and abrasion resistance is to simulate actual performance needs. For optical purposes, a cast sheet in the allyl family of plastics known as CR39 has been used as a standard of comparison in evaluating scratch and mar resistance of a material. The CR39 is used for eye lenses and other optical products where the advantages of plastics are a consideration. Coatings have been developed for polycarbonate, acrylics, and other plastics that dramatically improve the scratch and mar resistance of these materials.

Weathering

Outdoor Weathering

The specimen has no specified size. Specimens for this test may consist of any standard fabricated test specimen or cut/punch pieces of sheet or machined sample. Specimens are mounted outdoors on racks slanted at 45° and facing south. It is recommended that concurrent exposure be carried out in many varied climates to obtain the broadest, most representative total body of data. Sample specimens are kept indoors as controls and for comparison. Reports of weathering describe all changes noted, areas of exposure, and period of time.

Outdoor testing is the most accurate method of obtaining a true picture of other resistance. The only drawback of this test is the time required for several years' exposure that are usually located in different climatic zones around the world. A large number of specimens are usually required to allow periodic removal and to run representative laboratory tests after exposure.

Accelerated Weathering

The specimen may be any shape. Artificial weathering has been defined by ASTM as "The exposure of plastics to cyclic laboratory conditions involving changes in temperature, relative humidity, and ultraviolet (UV) radiant energy, with or without direct water spray, in an attempt to produce changes in the material similar to those observed after long-term continuous outdoor exposure."

Three types of light sources for artificial weathering are in common use: (1) enclosed UV carbon arc [7.5 UV energy output, approx. (x sunlight)], (2) open-flame sunshine carbon, and (3) water-cooled xenon arc. Selection of the light source involves many conditions and circumstances, such as the type of material being tested, product service conditions, previous testing experience, or the type of information desired.

Since weather varies from day to day, year to year, and place to place, no precise correlation exists between artificial laboratory weathering and natural outdoor weathering. However, standard laboratory test conditions produce results with acceptable reproducibility and in general agreement with data obtained from long-time outdoor exposures.

Fairly rapid indications of weatherability are therefore obtainable on samples of known materials that through testing experience over a period of time have general correlations established. There is no artificial substitute for precisely predicting outdoor weatherability on materials with no previous weathering history. Weatherometers produce conditions to accelerate effects that would be observed in specimens exposed outdoors.

Accelerated Exposure to Sunlight

The Atlas Type FDA-IR Fadeometer is used primarily to check and compare color stability. Besides determining the stability of various pigments needed to provide both standard and custom colors, the Fadeometer is helpful in preliminary studies of various stabilizers, dyes, and pigments compounded in plastics to prolong their useful life.

It is primarily for testing materials to be used in products subject to indoor exposure and to sunlight. Exposure in the Fadeometer cannot be related directly to exposure in direct sunlight, partially because other weather factors are always present outdoors.

Conditioning Procedure

As reviewed it is important that test specimens or products be properly prepared based on available specifications and/or standards that provide controlled conditioning procedures when conducting weathering as well as all other tests. The following is one example. There are other conditions set forth to provide for testing at higher or lower levels of temperature and humidity.

Procedure for conditioning test specimens can call for the following periods in a standard laboratory atmosphere [$50 \pm 2\%$ relative humidity, $73.4 \pm 1.8°F$ ($23 \pm 1°C$): Adequate air circulation around all specimens must be provided. The reason for this test is due to the fact the temperature and moisture content of plastics affects different properties such as the physical and electrical properties. In order to get comparable test results at different times and in different laboratories a standard has been established.

Harmful Component

Review in Chapter 2, **WEATHERING/ ENVIRONMENT, Weather Resistance**.

Environmental Stress Cracking

This test was prepared and is limited to type 1 (low-density) polyethylenes. Specimens are annealed in water or steam at $212°F$ ($100°C$) for 1 h and then equilibrated at room temperature for 5–24 h. After conditioning the specimens are nicked according to directions given. The specimens are bent into a U shape in a brass channel and inserted into a test tube that is then filled with fresh reagent (Igepal). The tube is stoppered with an aluminum-covered cork and placed in a constant temperature bath at $122°F$ ($50°C$).

These specimens are inspected periodically and any visible crack is considered a failure. The duration of the test is reported along with the percentage of failures. The cracking obtained in this test is indicative of what may be expected from a wide variety of stress-cracking agents. The information cannot be translated directly into end-use service prediction, but serves to rank various types and grades of polyethylene categories of resistance to environmental stress cracking. Though restricted to type 1 polyethylene, this test can be used on high and medium density PE materials as well as other plastics, in which case it would be considered a modified test.

Fire

Flammability

Underwriters' Laboratories (UL) Test 94 can be used. The placement of the specimen, the size of the flame, and its position and location with respect to the specimen are described in detail in this important UL specifications. Depending on their nonburning to burning capabilities, results of tests are reported as being materials classed 94V-0, 94V-1, 94V-2, 94-5V, etc. (Chapter 2, **HIGH TEMPERATURE, Flammability**).

Oxygen Index

The test method describes a procedure for measuring the minimum concentration of oxygen in a flowing mixture of oxygen and nitrogen that will support glowing combustion of plastics. The oxygen index is the minimum concentration of oxygen expressed as a volume percent in a mixture of oxygen and nitrogen that will just support glowing combustion of a material initially at room temperature under the conditions of this method. From this description, it is apparent that the lower the oxygen index the more the plastic contributes to the support of combustion.

Many of the basic plastics require additives that will improve their resistance to supporting combustion. These improvements vary in degree, and the designer must be cautioned not to over specify the requirement for flammability. It should be recognized that the highest protection against burning can be costly, can contribute to higher specific gravity, and can adversely affect mechanical properties.

Analyzing Testing and Quality Control

Designers and processors should keep quality under control and demand consistent materials that can be used with minimum of uncertainty. Basically involves inspection and testing of raw materials to the finished products. Plant QC is as important to the end result as selecting the best processing and control conditions with the correct grade of plastic, in terms of both properties and appearance. After the correct plastic has been chosen, any blending, reprocessing, and storage stages of operation need to be frequently or continuously updated. The processor should set up specific measurements of quality to prevent substandard products reaching the customer. QC involve those quality assurance actions which provide a means to control, measure, and establish requirements of the characteristics of plastic materials, processes, and products.

From a practical aspect, when the expression "quality control" is use, we tend to think in terms of a good or excellent product. In industry, it is one that fulfills customer's expectations. These expectations or standards of performance are based on the intended use and selling price of the product. Control is the process of regulating or directing an activity to verify its conformance to a standard/specification and to take corrective action if required. Therefore QC is the regulatory testing process for those activities that measure a product's performance, compare that performance with established standards/specifications, and pursue corrective action regardless of where those activities occur.

There are three phases in the evolution of most QC systems; (1) defect detection where an "army" of inspectors tries to identify defects; (2) defect prevention where the process is monitored, and statistical methods are used to control process variation, enabling adjust-

ments to the process to be made before defects are produced; and (3) total quality control where it is finally recognized that quality must extend throughout all functions and it is management's responsibility to integrate and lead the various functions towards the goals of commitment to quality and customer-first orientation (3).

When using the defect-detection approach to quality control certain problems develop. Inspection does nothing to improve the process and is not very good at sorting good-from-bad. Also, sampling plans developed to support an acceptable quality level (AQL) of 5%, for example, say that a company is content to deliver or reject 5% defects.

There are different methods to apply QC on-line. An example is with infrared measurement. The ability to record IR spectra of plastic melts provides for process monitoring and control in the manufacture process. Precise information on quality can be obtained rapidly. Furthermore, it is also possible to make measurements on unstable intermediates of importance. Although spectroscopy on melts is considerably different from that on solid materials, this does not limit the information content. IR has for many years been an important aid to investigating the chemical and physical properties of molecules. It gives qualitative and quantitative information on chemical constituents, functional groups, impurities, etc. As well as its use in studying low molecular weight compounds, it is used with equal success for characterizing plastics. It is a highly informative method of applying testing.

To ensure that QC and testing procedures are followed a quality control manual should be implemented. It is a document usually setup in a computer's software program that states and provides the details of the plant's quality objectives and how they will be implemented, documented, and followed.

Statistical Process Control and Quality Control

The term statistics basically is a summary value calculated from the observed values in a sample or product. It is a branch of mathematics dealing with the collection, analysis, interpretation, and presentation of masses of numerical data. The word statistic has two generally accepted meanings: (1) a collection of quantitative analysis data (data collection) pertaining to any subject or group, especially when the data are systematically gathered and collated and (2) the science that deals with the collection, tabulation, analysis, interpretation, and presentation of quantitative data.

Statistical process control (SPC) is an important on-line method in real time by which a production process can be monitored and control plans can be initiated to keep quality standards within acceptable limits. Statistical quality control (SQC) provides off-line analysis of the big picture such as what was the impact of previous improvements. It is important to understand how SPC and SQC operate.

There are basically two possible approaches for real-time SPC. The first, done on-line, involves the rapid dimensional measurement of a part or a non-dimensional bulk parameter such as weight that is the more practical method. In the second approach, contrast to weight, other dimensional measurements of the precision needed for SPC are generally done off-line. Obtaining the final dimensional stability needed to measure a part may take time. As an example, amorphous injection molded plastic parts usually require at least a half-hour to stabilize.

The SPC system starts with the premise that the specifications for a product can be defined in terms of the product's (customer's) requirements, or that a product is or has been produced that will satisfy those needs. Generally a computer communicates with a series of process sensors and/or controllers that operate in individual data loops.

The computer sends set points (built on which performance characteristics of the product must have) to the process controller that constantly feeds back to the computer to signal whether or not the set of points are in fact maintained. The systems are programmed to act when key variables affecting product quality deviate beyond set limits (3).

6

Plastic Material Formation and Variation

Introduction

Plastics are the most used materials on a volume basis when compared to steel and aluminum. They are broadly integrated into today's lifestyle and make a major, irreplaceable contribution to virtually all market areas. They are one of a large and varied group of materials totaling over 35,000 worldwide. They usually consist of, or contain as an essential ingredient, an organic substance. Most are produced synthetically; very few occur in nature. There are also plastics that contain inorganic material. Table 6-1 and Fig. 6-1 show their manufacturing stages from raw materials to products.

Practically all plastics at some stage in their manufacture or fabrication can be formed into various simple to extremely complex shapes that can range from being extremely flexible (rubbery/elastomeric) to extremely hard (high performance properties). The use of a virtually endless array of additives, fillers, reinforcements, etc. permits compounding from the raw material suppliers to the fabricators imparting specific qualities to the basic raw materials (polymers) and expanding opportunities for plastics. Compounding relies on the polymerization chemistry to mechanical mixing to combine a base polymer with modifiers, additives, and other plastics to develop new plastics. Clearly these many combinations are endless so that new materials are always on the horizon to meet new industry requirements.

The usefulness of plastics materials is due to the fact that they provide many different environment resistant properties, are light in weight, and have a higher strength per pound than most metals and other material of construction. They can be changed into end products by the relatively simple and relatively inexpensive means of fabricating the material in a liquid or semisolid form and then cooling it to a solid. They usually have good appearance and surface characteristics. The advantages of plastics as compared to competing materials include corrosion resistance, wide range of color and appearance properties, and, in many cases, excellent chemical and weather resistance. Plastics are usually well adapted to mass production methods. Their use often reduces overall manufacturing costs.

Plastics are important for three main reasons.

1. They can be manufactured with a wide variety of properties and may very often be "tailored" for a specific set meeting

Table 6-1 Stages in plastics manufacturing

Process:	Chemical				Physical and Mechanical		
Stages	Basic Chemical	Monomers	Polymerization	Compounding	Processing	Fabricating	Finishing
	Petroleum is converted to *petrochemicals* such as: ethylene, benzene, propylene, acetylene	Petrochemicals plus other chemicals are converted into *monomers* such as: ethylene, vinyl chloride, acrylonitrile, styrene, propylene	One or more monomers are polymerized to form *polymers* or copolymers such as: polyethylene, poly (vinyl chloride), styrene-acrylonitrile, butadiene, copolymer (ABS), polystyrene polypropylene	Plasticizers, stabilizers, color pigments, anti-exidants, inhibitors, and other chemicals are sometimes *added* to the base polymers to form *compounds* suitable for use by processors or as coatings for paper, wood, etc. or in paints and adhesives	The plastics *compounds* are *formed* into a variety of solid shapes such as sheets, tubes, rods, film, and other shapes, by the heat and/or pressure of casting, molding, extrusion, or other means of processing. This step may provide a finished product such as plastic pipe, etc.	These solid shapes may be *fabricated* by thermoforming, machining, etc., to create usable plastics articles such as toys or appliances	In some cases there is a *finishing* step such as the printing of surface designs on vinyl film
State	Gases and liquids	Mostly liquids	Solids and slurries	Solids and slurries	Solids	Solids	Solids
Customers	Monomer manufacturers	Polymerizers	Processors		Fabricators and end users	Finishers and end users	End users
Important trends and improvements	New and larger manufacturing facilities which will lower prices through economies of sale	New manufacturing processes which will lower prices through greater efficiency and use of lower cost raw materials	New techniques of copolymerization and stereospecific polymerization which allow producers to create polymers with specific sets of processing and end use characteristics	New and more effective additives which expand the range of usefulness of plastics	New processing equipment and techniques which can produce very large and/or stress-free parts	Use of butt welding of large plastics parts is extending the range of shapes which may be made out of plastics	New plating methods which increase the environmental resistance and eye appeal of plastics. New graphic finishing techniques such as woodgraining which will allow plastics to compete as decorative items

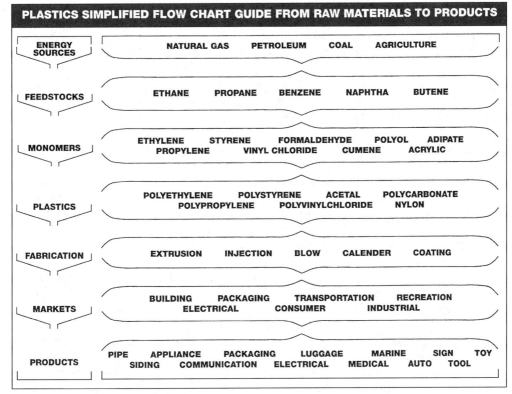

Fig. 6-1 Plastics simplified flow chart guide from raw materials to products.

environmental and end use conditions. This may be done by adjusting the operating conditions during manufacture, by adding fillers, reinforcements, and/or other additives, by copolymerization, by compounding, etc.

2. They may be processed in a wide variety of ways, some of which are adapted to high-speed manufacturing to making very small to very large products.

3. The wide variety of formulations and manufacturing processes available allows the designer to achieve the lowest possible cost for a product, often the lowest cost of any available construction material.

It is unfortunate that plastics do not have all the advantages and none of the disadvantages of other materials but often overlooked is the fact that there are no materials that do not suffer from some disadvantages or limitations. The faults of materials known and utilized for hundreds of years are often overlooked; the faults of the new materials are often overemphasized.

As examples, iron and steel are attacked by the elements of weather but the common practice of coating these with protective plastic paints and then forgetting their susceptibility to attack is all too prevalent. Wood and concrete are useful materials, yet who has not seen a rotted board and cracked concrete. Does this lack of perfection mean that no steel, wood, or concrete should be used? Of course not. The same reasoning should apply to plastics. In many respects, the gains made with plastics in a short span of time far outdistance the advances made in other technologies.

Definition

The term plastic comes from the Greek word "to form." It identifies many different plastic materials. Polymers, the basic ingredients used in practically all plastics, can be defined as high molecular weight organic compounds, synthetic or natural substance consisting of molecules characterized by the repetition (neglecting ends, branch junctions, and other minor irregularities) of one or more types of monomeric units. A repeated small unit, the mer, such as ethylene, rubber, or cellulose, can represent its structure. Practically all of these polymers (base materials) use certain types of ingredients to perform properly during fabrication and meet performance/cost requirements of products in service.

Petroleum is currently the major source of raw materials for most high volume polymers. Also used are gas and substitute resources such as coal that are supposedly in limited supply (although our government reports we have enough coal for the next 250 years that includes its growth expansion in use during that period), and it may well be that another approach to the problem is required. An example is different raw material sources to produce plastics that involve biotechnology (186).

The term's plastic, polymer, resin, elastomer, and reinforced plastic (RP) are somewhat synonymous. However, polymer and resin usually denote the basic material. Whereas plastic pertains to polymers or resins containing additives, fillers, and/or reinforcements. Recognize that practically all materials worldwide contain some type of additive or ingredient. An elastomer is a rubberlike material (natural or synthetic). Reinforced plastics (also called composites although to be more accurate called plastic composites) are plastics with reinforcing additives, such as fibers and whiskers, added principally to increase the product's mechanical properties.

Worldwide the term preferred is plastics. The fact is that: (1) this industry identifies itself as a plastics industry, (2) practically all people worldwide use the term plastics, (3) practically all materials, products, exhibition shows, technical meetings, advertising, etc. use the term plastics, and (4) as it is repeatedly said, this is a World of Plastics. As shown in this book there are terms that overlap and also interfere with each other. A major example is stating that thermoplastics (TPs) are cured during processing; cure occurs only with thermoset plastics (TSs) or when a TP is converted to a TS plastic.

Plastic also refers to a material that has a physical characteristic such as plasticity and toughness. The general term commodity plastic, engineering plastic, advanced plastic, advanced reinforced plastic, or advanced plastic composite is used to indicate different performance materials. These terms and others will be reviewed latter in this chapter. Plastics are made into specialty products that have developed into major markets. An example is plastic foams that can provide flexibility to rigidity as well as other desired properties (heat and electrical insulation, toughness, filtration, etc.).

The term plastic is not a definitive one. Metals, for instance, are also permanently deformable and are therefore plastic. How else could roll aluminum be made into foil for kitchen use, or tungsten wire be drawn into a filament for an incandescent, light bulb, or a 100 ton ingot of steel be forged into a rotor for a generator. Likewise the different glasses, which contain compounds of metals and nonmetals, can be permanently shaped at high temperatures. These cousins to polymers and plastics are not considered plastics within the plastic industry or context of this book.

To better understand the properties of plastics it is important to know about the transitions that occur, such as those that have a glass transition temperature (T_g) (Chapter 7, **THERMAL PROPERTY, Glass Transition Temperature**). Nearly all the mechanical properties of plastics are determined primarily by these transitions and the temperatures at which they occur. With a change in temperature different plastics can have either quick or gradual changes in viscosity and as temperatures increase certain materials can change from basically rigid solids to liquids either quickly or gradually, depending on their chemical structure and composition.

Thermoplastic and Thermoset Plastic

Thermoplastics (TPs) are plastics that repeatedly soften when heated and harden

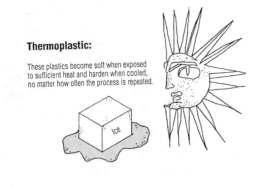

Thermoplastic:

These plastics become soft when exposed to sufficient heat and harden when cooled, no matter how often the process is repeated.

Thermosetting:

The plastics materials belonging to this group are set into permanent shape when heat and pressure are applied to them during forming. Reheating will not soften these materials.

Fig. 6-2 Characteristic of thermoplastics and thermosets.

Table 6-2 Examples of melt-processing temperatures for TPs

Material	Processing Temperature Rate	
	°C	°F
ABS	180–240	356–464
Acetal	185–225	365–437
Acrylic	180–250	356–482
Nylon	260–290	500–554
Polycarbonate	280–310	536–590
LDPE	160–240	320–464
HDPE	200–280	392–536
Polypropylene	200–300	392–572
Polystyrene	180–260	356–500
PVC, rigid	160–180	320–365

Note: Values are typical for injection molding and most extrusion operations. Extrusion coating is done at higher temperatures (i.e., about 600°F for LDPE).

when cooled (Figs. 6-2 and 6-3 and Table 6-2). There are those soluble in specific solvents and burn to some degree. Their softening temperatures vary. Care must be taken to avoid degrading, decomposing, or igniting these materials. Generally, no chemical changes take place during processing. An analogy would be a block of ice that can be softened (turned back to a liquid), poured into any shape mold or die, then cooled to become a solid again. This cycle repeats. TPs generally offer easier processing, and better adaptability to complex designs than do thermosets (TSs).

Most TP molecular chains can be thought of as independent, intertwined strings resembling spaghetti. When heated, the individual chains slip, causing a plastic flow. Upon cooling, the chains of atoms and molecules are once again held firmly. With subsequent heating the slippage again takes place. There are practical limitations to the number of heating and cooling cycles before appearance or mechanical properties are drastically affected. Initially this recycling practically does not effect certain TPs. However there are those that are effected after just one cycle.

Thermosets (TSs) are plastics that undergo chemical change (cross-linking) during processing to become permanently insoluble and infusible (Figs. 6-2 and 6-3). This cross-linking is a curing reaction. Such natural and synthetic elastomers as latex, nitrile, millable polyurethanes, silicone butyl, and neoprene, which attain their properties through the process of vulcanization, are also in this cross-linking behavior pattern. The best analogy of all TSs is that of a hard-boiled egg that has turned from a liquid to a solid and cannot be converted back to a liquid. In general, with their tightly cross-linked structure TSs resist higher temperatures and provide greater

Example of a Thermoplastic Processing Heat-Time Profile Cycle

a. Start of process
b. Plastic melted
c. Plastic hard but can be resoftened

Example of a Thermoset Processing Heat-Time Profile Cycle

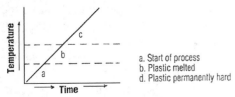

a. Start of process
b. Plastic melted
d. Plastic permanently hard

Fig. 6-3 Melting characteristics of TPs and TSs based on their heat-time processing profiles.

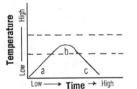

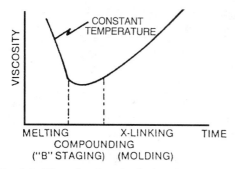

Fig. 6-4 Viscosity changes during the processing of TS. The B-stage represents the start of the heating cycle followed by a chemical reaction (cross-linking) and solidification of the plastic.

dimensional stability and strength than do most TPs.

The structure of TSs, as of TPs, is also chainlike. Prior to fabricating, TSs are similar to TPs. Chemical cross-linking is the principal difference between TSs and TPs. In TSs, during curing or hardening the cross-links are formed between adjacent molecules, resulting in a complex, interconnected network that can be related to its viscosity and performance (Figs. 6-4 and 6-5). These cross-bonds prevent the slippage of individual chains, thus preventing plastic flow under the addition of heat. If excessive heat is added after cross-linking has been completed, degradation rather than melting will occur. TSs generally are not used alone in load bearing products. They must be filled or reinforced with materials such as calcium carbonate, talc, or glass fiber. The most common

reinforcement is glass fiber, but others are also used (16, 25, 227).

Each plastic has its own distinct or special properties and advantages. See Tables 6-3 and 6-4, also Fig. 1-8, for the typical names and properties of plastics. The dividing line between a TP and a TS is not always distinct. For instance, cross-linked TSs are TPs during their initial heat cycle and prior to chemical cross-linking. Plastics, such as a cross-linked polyethylene (XLPE), normally are TPs that have been cross-linked either by high-energy radiation or chemically during processing.

In addition to the broad categories of TPs and TSs, TPs can be further classified in terms of their structure, as either crystalline, amorphous, or liquid crystalline. Other classes (terms) include elastomers, copolymers, compounds, commodity resins, engineering plastics, or neat plastics. Additives, fillers, and reinforcements are other classifications that relate directly to plastics' properties and performance.

Structure and Morphology

In addition to the size of the molecules and their distribution, the shapes or structures of individual polymer molecules also play an important role in determining the properties and processability of plastics. There are those that are formed by aligning themselves into long chains of molecules and others with branches or lateral connections to form complex structures. All these forms exist in either two or three dimensions.

Because of the geometry, or morphology, of these molecules some can come closer together and more orderly than others. These are identified as crystalline and all others that behave like spagetti as amorphous. Morphology influences such properties as mechanical and thermal, swelling and solubility, specific gravity, and other properties (mechanical, physical, chemical, electric, etc.).

This behavior of morphology basically occurs with TP, not TS plastics. When TSs are processed, their individual chain segments are strongly bonded together during a chemical reaction that is irreversible.

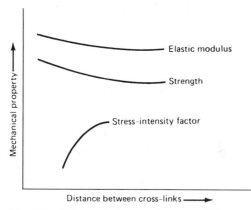

Fig. 6-5 Effect of distance between TS cross-linked sites on compression properties.

Table 6-3   Types of plastics

Acetal (POM)	Polyetherketone (PEK)
Acrylics	Polyetheretherketone (PEEK)
Polyacrylonitrile (PAN)	Polyetherimide (PEI)
Polymethylmethacrylate (PMMA)	Polyimide (PI)
Acrylonitrile butadiene styrene (ABS)	Thermoplastic PI
Alkyd	Thermoset PI
Allyl diglycol carbonate (CR-39)	Polymethylmethacrylate (acrylic) (PMMA)
Allyls	Polymethylpentene
Diallyl isophthalate (DAIP)	Polyolefins (PO)
Diallyl phthalate (DAP)	Chlorinated PE (CPE)
Aminos	Cross-linked PE (XLPE)
Melamine formaldehyde (MF)	High-density PE (HDPE)
Urea formaldehyde (UF)	Ionomer
Cellulosics	Linear LDPE (LLDPE)
Cellulose acetate (CA)	Low-density PE (LDPE)
Cellulose acetate butyrate (CAB)	Polyallomer
Cellulose acetate propionate (CAP)	Polybutylene (PB)
Cellulose nitrate	Polyethylene (PE)
Ethyl cellulose (EC)	Polypropylene (PP)
Chlorinated polyether	Ultra-high-molecular weight PE (UHMWPE)
Epoxy (EP)	Polyoxymethylene (POM)
Ethylene vinyl acetate (EVA)	Polyphenylene ether (PPE)
Ethylene vinyl alcohol (EVOH)	Polyphenylene oxide (PPO)
Fluorocarbons	Polyphenylene sulfide (PPS)
Fluorinated ethylene propylene (FEP)	Polyurethane (PUR)
Polytetrafluoroethylene (PTFE)	Silicone (SI)
Polyvinyl fluoride (PVF)	Styrenes
Polyvinylidene fluoride (PVDF)	Acrylic styrene acrylonitrile (ASA)
Furan	Acrylonitrile butadiene styrene (ABS)
Ionomer	General-purpose PS (GPPS)
Ketone	High-impact PS (HIPS)
Liquid crystal polymer (LCP)	Polystyrene (PS)
Aromatic copolyester (TP polyester)	Styrene acrylonitrile (SAN)
Melamine formaldehyde (MF)	Styrene butadiene (SB)
Nylon (Polyamide) (PA)	Sulfones
Parylene	Polyether sulfone (PES)
Phenolic	Polyphenyl sulfone (PPS)
Phenol formaldehyde (PF)	Polysulfone (PSU)
Phenoxy	Urea formaldehyde (UF)
Polyallomer	Vinyls
Polyamide (nylon) (PA)	Chlorinated PVC (CPVC)
Polyamide-imide (PAI)	Polyvinyl acetate (PVAc)
Polyarylethers	Polyvinyl alcohol (PVA)
Polyaryletherketone (PAEK)	Polyvinyl butyrate (PVB)
Polyaryl sulfone (PAS)	Polyvinyl chloride (PVC)
Polyarylate (PAR)	Polyvinylidene chloride (PVDC)
Polybenzimidazole (PBI)	Polyvinylidene fluoride (PVF)
Polycarbonate (PC)	
Polyesters	
Aromatic polyester (TS polyester)	
Thermoplastic polyesters	
Crystallized PET (CPET)	
Polybutylene terephthalate (PBT)	
Polyethylene terephathalate (PET)	
Unsaturated polyester (TS polyester)	

Table 6-4 Examples of a few properties of a few plastics

Property	Thermoplastics	Thermosets
Low Temperature	TFE	DAP
Low Cost	PP, PE, PVC, PS	Phenolic
Low Gravity	Polypropylene methylpentene	Phenolic/nylon
Thermal Expansion	Phenoxy glass	Epoxy-glass fiber
Volume Resistivity	TFE	DAP
Dielectric Strength	PVC	DAP, polyester
Elasticity	EVA, PVC, TPR	Silicone
Moisture Absorption	Chlorotrifluorethylene	Alkyd-glass fiber
Steam Resistance	Polysulfone	DAP
Flame Resistance	TFE, Pl	Melamine
Water Immersion	Chlorinated polyether	DAP
Stress Craze Resistance	Polypropylene	All
High Temperature	TFE, PPS, Pl, PAS	Silicones
Gasoline Resistance	Acetal	Phenolic
Impact	UHMW PE	Epoxy-glass fiber
Cold Flow	Polysulfone	Melamine-glass fiberglass
Chemical Resistance	TFE, FEP, PE, PP	Epoxy
Scratch Resistance	Acrylic	Allyl diglycol carbonate (C-39)
Abrasive Wear	Polyurethane	Phenolic-canvas
Colors	Acetate, PS	Urea, melamine

Crystalline and Amorphous Plastic

Plastic molecules that can be packed closer together can more easily form crystalline structures in which the molecules align themselves in some orderly pattern. During processing they tend to develop higher strength in the direction of the molecules. Since commercially perfect crystalline polymers are not produced, they are identified technically as semicrystalline TPs (normally up to 85% crystalline and the rest amorphous). In this book and as usually identified by the plastic industry, they are called crystalline.

The amorphous TPs, which have their molecules going in all different directions, are normally transparent. Compared to crystalline types, they undergo only small volumetric changes when melting or solidifying during processing. Tables 6-5 to 6-9 compare the basic performance behaviors of crystalline and amorphous plastics. Exceptions exist, particularly with respect to certain plastic compounds that include additives and reinforcements.

As symmetrical molecules approach within a critical distance, crystals begin to form in the areas where they are the most densely packed. A crystallized area is stiffer and stronger, a noncrystallized (amorphous) area is tougher and more flexible. With increased crystallinity, other effects occur. As an example, with polyethylene (crystalline) there is increased resistance to creep, heat, and stress cracking as well as increased mold shrinkage.

In general, crystalline types of plastics are more difficult (but controllable) to process, requiring more precise control during fabrication, have higher melting temperatures and

Table 6-5 General morphology of TPs

Crystalline		Amorphous
No	Transparent	Yes
Excel	Chemical resistance	Poor
No	Stress-craze	Yes
High	Shrinkage	Low
High	Strength	Low*
Low	Viscosity	High
Yes	Melt temperature	No
Yes	Critical T/T†	No

*Major exception is PC.
† T/T = Temperature/time.

Table 6-6 Distinctive characteristics of polymers

Crystalline	Amorphous
Sharp melting point	Broad softening range
Usually opaque	Usually transparent
High shrinkage	Low shrinkage
Solvent resistant	Solvent sensitive
Fatigue/wear resistant	Poor fatigue/wear

melt viscosities, and tend to shrink and warp more than amorphous types. They have a relatively sharp melting point. That is, they do not soften gradually with increasing temperature but remain hard until a given quantity of heat has been absorbed, then change rapidly into a low-viscosity liquid. If the amount of heat is not applied properly during processing, product performance can be drastically reduced and/or an increase in processing cost occur. This is not necessarily a problem, because the qualified processor will know how to process the plastic.

Amorphous plastics soften gradually as they are heated, but they do not flow as easily during molding as do crystalline materials.

Processing conditions influence the performance of plastics. For example, heating a crystalline material above its melting point, then quenching it can produce a plastic that has a far more amorphous structure. Its properties can be significantly different than if it is cooled properly (slowly) and allowed to recrystallize; during processing it becomes amorphous. The effects of time are similar to those of temperature in the sense that any

Table 6-7 Examples of crystalline and amorphous TPs

Crystalline	Amorphous
Acetal (POM)	Acrylonitrile-butadiene-styrene (ABS)
Polyester (PET, PBT)	Acrylic (PMMA)
Polyamide (nylon) (PA)	Polycarbonate (PC)
Fluorocarbons (PTFE, etc.)	Modified polyphenylene oxide (PPO)
Polyethylene (PE)	Polystyrene (PS)
Polypropylene (PP)	Polyvinyl chloride (PVC)

Table 6-8 Examples of key properties for engineering TPs

Crystalline	Amorphous
Acetal	*Polycarbonate*
Best property balance	Good impact
Stiffest unreinforced	resistance
thermoplastic	Transparent
Low friction	Good electrical
	properties
Nylon	*Modified PPO*
High melting point	Hydrolytic
High elongation	stability
Toughest thermoplastic	Good impact
Absorbs moisture	resistance
Glass reinforced	Good electrical
High strength	properties
Stiffness at elevated	
temperatures	
Mineral reinforced	
Most economical	
Low warpage	
Polyester (glass reinforced)	
High stiffness	
Lowest creep	
Excellent electrical	
properties	

given plastic has a preferred or equilibrium structure in which it would prefer to arrange itself timewise. However, it is prevented from doing so instantaneously or at least on "short notice." If given enough time, the molecules will rearrange themselves into their preferred pattern. Heating causes this action to occur sooner. During this action severe shrinkage and property changes could occur in all directions in the processed plastic products.

This characteristic morphology of plastics can be identified by tests (2, 3). It provides excellent control as soon as material is received in the plant, during processing, and after fabrication.

Liquid Crystalline Polymer

Liquid crystalline polymers (LCPs) are best thought of as being a separate, unique class of TPs. Their molecules are stiff, rodlike

Table 6-9   General properties of TPs during and after processing

Property	Crystalline*	Amorphous†
Melting or softening	Fairly sharp melting point	Softens over a range of temperature
Density (for the same material)	Increases as crystallinity increases	Lower than for crystalline material
Heat content	Greater	Lower
Volume change on heating	Greater	Lower
After-molding shrinkage	Greater	Lower
Effect of orientation	Greater	Lower
Compressibility	Often greater	Sometimes lower

*Typical crystalline plastics are polyethylene, polypropylene, nylon, acetals, and thermoplastic polyesters.
†Typical amorphous plastics are polystyrene, acrylics, PVC, SAN, and ABS.

structures organized in large parallel arrays or domains in both the melted and solid states. These large, ordered domains provide LCPs with characteristics that are unique compared to those of the basic crystalline or amorphous plastics (Table 6-10). They are called self-reinforcing plastics because of their densely packed fibrous polymer chains.

These LCPs provide the designer with unparalleled combinations of properties, such as resisting most solvents and heat. Unlike many high-temperature plastics, LCPs have a low melt viscosity and are thus more easily processed resulting in faster cycle times than those with a high melt viscosity thus reducing processing costs. They have the lowest warpage and shrinkage of all the TPs. When they are injection molded or extruded, their molecules align into long, rigid chains that in turn align in the direction of flow and thus act like reinforcing fibers, giving LCPs both very high strength and stiffness. As the

melt solidifies during cooling, the molecular orientation "freezes" (solidifies) into place. The volume changes are only minute with virtually no frozen-in stresses.

In service, products experience very little shrinkage or warpage. They have high resistance to creep. Their fiberlike molecular chains tend to concentrate near the surface, resulting in products that are anisotropic, meaning that they have greater strength and modulus in the flow direction, typically on the order of three to six times those of the transverse direction. However, adding fillers or reinforcing fibers to LCPs significantly reduces their anisotropy, more evenly distributing strength and modulus and even boosting them. Most fillers and reinforcements also reduce overall cost and place mold shrinkage to zero or near zero. Consequently, products can be molded to tight tolerances. These low-melt-viscosity LCPs thus permit the design of products with long or complex flow paths and thin sections.

Table 6-10   General properties of crystalline, amorphous, and liquid crystalline polymers

Property	Crystalline	Amorphous	Liquid Crystalline
Specific gravity	Higher	Lower	Higher
Tensile strength	Higher	Lower	Highest
Tensile modulus	Higher	Lower	Highest
Ductility, elongation	Lower	Higher	Lowest
Resistance to creep	Higher	Lower	High
Max. usage temperature	Higher	Lower	High
Shrinkage and warpage	Higher	Lower	Lowest
Flow	Higher	Lower	Highest
Chemical resistance	Higher	Lower	Highest

They have outstanding strength at extreme temperatures, excellent mechanical-property retention after exposure to weathering and radiation, good dielectric strength as well as arc resistance and dimensional stability, low coefficient of thermal expansion, excellent flame resistance, and easy processability. Their UL continuous-use rating for electrical properties is as high as 240°C (464°F), and for mechanical properties it is 220°C (428°F). LCPs' high heat deflection value permits LCP molded products to be exposed to intermittent temperatures as high as 315°C (600°F) without affecting their properties. Their resistance to high-temperature flexural creep is excellent, as are their fracture-toughness characteristics. This family of different LCPs resists most chemicals and weathers oxidation and flame, making them excellent replacements for metals, ceramics, and other plastics.

LCPs are exceptionally inert and resist stress cracking in the presence of most chemicals at elevated temperatures, including the aromatic and halogenated hydrocarbons as well as strong acids, bases, ketones, and other aggressive industrial products. Their hydrolytic stability in boiling water is excellent, but high-temperature steam, concentrated sulfuric acid, and boiling caustic materials will deteriorate LCPs. In regard to flammability, LCPs have an oxygen index ranging from 35 to 50%. When exposed to open flame they form an intumescent char that prevents dripping.

Copolymer

Polymer properties can be varied during polymerization. The basic chemical process is carried during their manufacture; the polymer is formed under the influence of heat, pressure, catalyst, or combination inside vessels or tubular systems called reactors. One special form of property variation involves the use of two or more different monomers as comonomers, copolymerizing them to produce copolymers (two comonomers) or terpolymers (three monomers). Their properties are usually intermediate between those of homopolymers, which may be made from the individual monomers, and sometimes superior or inferior to them. (A polymer such as polyethylene is formed from its monomer ethylene, polyvinyl chloride polymer from its vinyl chloride monomer, and so on.)

Compounded/Alloyed Plastic

Since the first plastic cellulosic was produced in 1868, there has been an ever-growing demand for specially compounded plastics. Using a post-reactor technique, plastics can be compounded by alloying or blending polymers in addition to using additives such as colorants, flame retardants, heat or light stabilizers, lubricants, fillers, and/or reinforcements (Fig. 6-6). With reinforcements the resulting reinforced compounds are usually referred to as reinforced plastics (RPs).

Alloy and Blend

Alloys are combinations of polymers that are mechanically blended. They do not depend on chemical bonds, but do often require special compatibilizers. Plastic alloys are usually designed to retain the best characteristics of each constituent. Most often, property improvements are in such areas

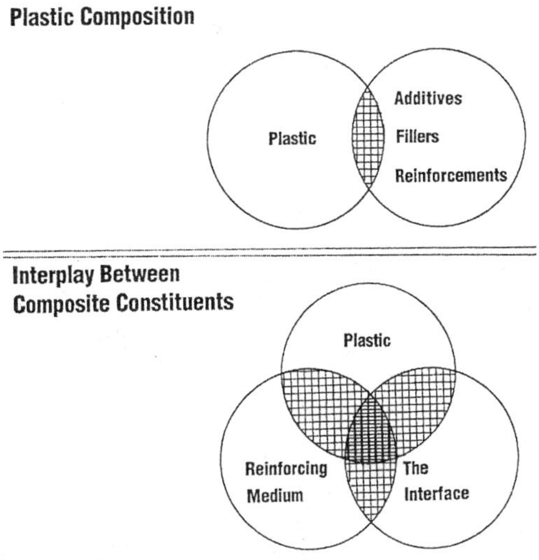

Plastic Composition

Interplay Between Composite Constituents

Fig. 6-6 Composition of plastics.

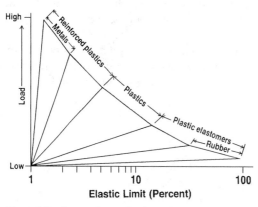

Fig. 6-7 Strength and elasticity of different materials.

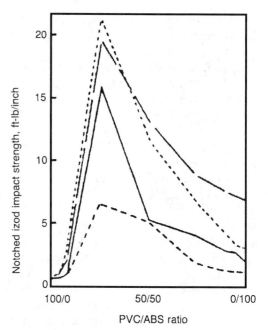

Fig. 6-9 Example of how alloying affects plastic properties; curves reflect four different blends.

as impact strength, weather resistance, improved low-temperature performance, and flame retardation (Figs. 6-7 to 6-10 and Tables 6-11 and 6-13).

The classic objective of alloying and blending is to find two or more polymers whose mixture will have synergistic property improvements (Fig. 6-8). Among the techniques used to combine dissimilar polymers are cross-linking to form what are called interpenetrating networks (IPNs), and grafting, to improve the compatibility of the plastics.

Alloys can be classified as either homogeneous or heterogeneous. The former can be depicted as a solution with a single phase or single glass-transition temperature (T_g). A heterogeneous alloy has both continuous

and dispersed phases, each retaining its own distinctive T_g. Until recently, blending and alloying were either restricted to polymers that had an inherent physical affinity for each other or else a third component, called a compatibilizer, was employed. These constraints severely limited the types of polymers that could be blended without sacrificing their good physical properties. As a rule, incompatible polymers produce a

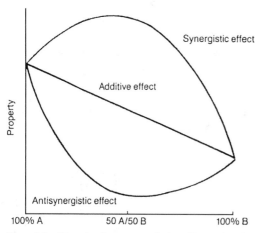

Fig. 6-8 Developing synergistic effects is the most usual objective of compounding plastics to gain significant performance.

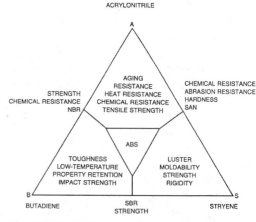

Fig. 6-10 ABS terpolymer properties are shown influencing individual constituent plastic properties.

Table 6-11   Examples of alloying to provided cost-performance improvements

Plastic	Cost index	Yield strength index	Impact strength index
Polypropylene			
Polystyrene			
Impact styrene (alloy)			
ABS			
ABS/PVC (alloy)			
ABS/Polycarbonate (alloy)			
Rigid PVC			
PVC/acrylic (alloy)			
Polyphenyleneoxide (Noryl)			
Polycarbonate			
Polysulfone			
Polysulfone/ABS (alloy)			

100   500   100   200   100   450 1250   3000

heterogeneous alloy with poor physical properties.

The advances in polymer blending and alloying technology have occurred through three routes: (1) similar-rheology polymer pairs, (2) miscible polymers such as polyphenylene oxide and polystyrene, or (3) interpenetrating polymer networks (IPNs). All these systems were limited to specific polymer combinations that have an inherent physical affinity for each other. However with

R&D developments, there is another overall approach to producing blends via reactive polymers.

Interpenetrating Network

IPNs consist of an interwoven matrix of two polymers. A typical method for producing these alloys involves cross-linking one of the monomers in the presence of

Table 6-12   Upgrading PVC by blending

Upgraded Property	Blending Polymer
Impact resistance	ABS, methacyrylate-butadiene-styrene, acrylics, polycaprolactone, polyimide, polyurethanes, PVC-ethyl acrylate
Tensile strength	ABS, methacyrylate-butadiene-styrene, polyurethanes, ethylene-vinyl acetate
Low-temperature toughness	Styrene-acrylonitrile, polyurethanes, polyethylene, chlorinated polyethylene, copolyester
Dimensional stability	Styrene-acrylonitrile, methacrylate-butadiene-styrene
Heat-distortion temperature	ABS, methacyrylate-butadiene-styrene, polyimide, polydimethyl siloxane
Processability	Styrene-acrylonitrile, methacrylate-butadiene-styrene, chlorinated polyethylene, PVC-ethyl acrylate, ethylene-vinyl acetate, chlorinated polyoxymethylenes (acetals)
Moldability	Acrylics, polycaprolactone
Plasticization	Polycaprolactone, polyurethanes, nitrile rubber, ethylene-vinyl acetate, copolyester, chlorinated polyoxymethylenes (acetals)
Transparency	Acrylics, polymide
Chemical/oil resistance	Acrylics
Toughness	Nitrile rubber, ethylene-vinyl acetate
Adhesion	Ethylene-vinyl acetate

Table 6-13 Outstanding properties of some commercial plastic alloys

Alloy	Properties
PVC/acrylic	Flame, impact, and chemical resistance
PVC/ABS	Flame resistance, impact resistance, processability
Polycarbonate/ABS	Notched impact resistance, hardness, heat-distortion temperature
ABS/polysulfone	Lower cost
Polypropylene/ethylene-propylene-diene	Low-temperature impact resistance and flexibility
Polyphenylene oxide/polystyrene	Processability, lower cost
Styrene acrylonitrile/olefin	Weatherability
Nylon/elastomer	Notched Izod impact resistance
Polybutylene terephthalate/polyethylene terephthalate	Lower cost
Polyphenylene sulfide/nylon	Lubricity
Acrylic/polybutylene rubber	Clarity, impact resistance

the other. The need for a chemical similarity between the two types of molecules is thus reduced, because cross-linking physically traps one with the other. The result is a structure composed of two different intertwined plastics, each retaining its own physical characteristics.

Reactive Polymer

A reactive polymer (RP) is simply a device to alloy different materials by changing their molecular structure inside a compounding machine. True reactive alloying induces an interaction between different phases of an incompatible mixture and assures the stability of the mixture's morphology. The concept is not new. This technology is now capable of producing thousands of new compounds to meet specific design requirements.

The relatively low capital investment associated with compounding machinery (usually less than $1 million for a line, compared with many millions for a conventional reactor), coupled with a processing need for small amounts of tailored materials, allows small and mid-sized compounding companies to take advantage of producing reactive polymers.

There are a variety of reactive alloying techniques available to the compounder. They typically involve the use of a reactive agent or compatibilizer to bring about a molecular change in one or more of the blended components, thereby facilitating bonding. They include the grafting process and copolymerization interactions, whereby a functional material is built into the polymer chain of a blend component as a comonomer, with the resultant copolymer then used as a compatibilizer in ternary bonds, such as a PP-acrylic acid copolymer that bonds PP and AA. Another technique is solvent-based interactions, using materials such as polycaprolactone, which is miscible in many materials and exhibits strong polarity, as well as hydrogen bonding, using the simple polarity of alloy components.

Grafting

Grafting two dissimilar plastics often involves a third plastic whose function is to improve the compatibility of the principal components. This "compatibilizer" material is a grafted copolymer that consists of one of the principal components and is similar to the other component. The mechanism is similar to that of having soap improve the solubility of a greasy substance in water. The soap contains components that are compatible with both the grease and the water.

Additive, Filler, and Reinforcement

Compounding to change and improve the physical and mechanical properties of plastics makes use of a wide variety of

Table 6-14   Guide to use of fillers and reinforcements

Filler or Reinforcement	Chemical Resistance	Heat Resistance	Electrical Insulation	Impact Strength	Tensile Strength	Dimensional Stability	Stiffness	Hardness	Lubricity	Electrical Conductivity	Thermal Conductivity	Moisture Resistance	Processability	Recommended For use in*
Alumina, tabular	•	•	•	•		•	•					•	•	S/P
Aluminum powder										•	•			S
Aramid	•	•	•	•	•	•	•	•	•				•	S/P
Bronze							•	•		•	•			S
Calcium carbonate	•	•	•	•		•	•	•					•	S/P
Carbon black		•				•	•			•	•		•	S/P
Carbon fiber										•	•			S
Cellulose				•		•	•	•						S/P
Alpha cellulose			•			•						•		S
Coal, powdered	•											•		S
Cotton				•	•	•	•	•						S
Fibrous glass	•	•	•	•	•	•	•	•				•		S/P
Graphite	•				•	•	•	•	•	•	•	•		S/P
Jute				•		•								S
Kaolin	•	•	•			•	•	•	•			•	•	S/P
Mica	•	•	•			•	•	•	•			•		S/P
Molybdenum disulfide							•	•	•			•	•	P
Nylon	•	•	•	•	•	•	•	•	•				•	S/P
Orlon	•	•	•	•	•	•	•	•					•	S/P
Rayon				•	•	•	•	•						S
Silica, amorphous			•									•	•	S/P
Sisal fibers	•			•		•	•	•				•		S/P
Fluorocarbon							•	•	•	•				S/P
Talc	•	•	•			•	•	•	•			•	•	S/P
Wood flour				•		•								S

*P = thermoplastic, S = thermoset.

ingredients (Fig. 6-6 and Tables 6-14 to 6-17). The major and large market for products such as additives, fillers, colorants, etc. continues to expand as the demand for plastics to function in wider or more extreme markets and under stricter regulatory regimes continue to expand. There are ingredients that are used to improve the processing capabilities of plastics.

In general adding reinforcing fibers significantly increases mechanical properties. Particulate fillers of various types usually increase the modulus, plasticizers generally decrease the modulus but enhance flexibility, and so on. These RPs can also be called composites. However the name composites litterly identifies thousands of different combinations with very few that include the use of plastics (Table 6-18). In using the term composites when plastics are involved the more appropriate term are plastic composites.

Many ingredients, especially those that are conductive, may affect electrical properties. Most plastics, which are poor conductors of current, build up a charge of static electricity. Antistatic agents can be added to attract moisture, reducing the likelihood of a spark or discharge.

In most cases, different additives are used to provide lower cost and different characteristics encompassing specific overall properties. As an example, coupling agents are added to improve the bonding of a plastic to its inorganic reinforcing materials such as glass fibers. A variety of silanes and titanates are used for this purpose. Some extenders (that is fillers) permit a large volume of a given plastic to be produced with

Table 6-15   Trade-off in TPs and RPs

Desired Modification	How Achieved	Sacrifice (from Base Resin)		Comments
		Amorphous	Crystalline	
Increased Tensile Strength	Glass fibers Carbon fibers Fibrous minerals	Ductility, cost Ductility, cost	Ductility, cost Ductility, cost Ductility	Glass fibers are the most cost-effective way of gaining tensile strength. Carbon fibers are more expensive; fibrous minerals are least expensive but only slightly reinforcing. Reinforcement makes brittle resins tougher and embrittles tough resins. Fibrous minerals are not commonly used in amorphous resins.
Increased Flexural Modulus	Glass fibers Carbon fibers Rigid minerals	Ductility, cost Ductility, cost Ductility	Ductility, cost Ductility, cost Ductility	Any additive more rigid than the base resin produces a more rigid composite. Particulate fillers severely degrade impact strength.
Flame Resistance	FR additive	Ductility, tensile strength, cost	Ductility, tensile strength, cost	FR additives interfere with the mechanical integrity of the polymer and often require reinforcement to salvage strength. They also narrow the molding latitude of the base resin. Some can cause mold corrosion.
Increased Heat-Deflection Temperature (HDT)	Glass fibers Carbon fibers Fibrous minerals	Ductility, cost Ductility, cost	Ductility, cost Ductility, cost Ductility	When reinforced, crystalline polymers yield much greater increases in HDT than do amorphous resins. As with tensile strength, fibrous minerals increase HDT only slightly. Fillers do not increase HDT.
Warpage Resistance	5 to 10% glass fibers 5 to 10% carbon fibers Particulate fillers	Ductility, cost, tensile strength	Cost Cost Ductility, cost, tensile strength	Amorphous polymers are inherently nonwarping molding resins. Only occasionally are fillers such as milled glass or glass beads added to amorphous materials, because they reduce shrinkage anisotropically.

(Continues)

Table 6-15 (*Continued*)

Desired Modification	How Achieved	Sacrifice (from Base Resin)		Comments
		Amorphous	Crystalline	
				Addition of fibers tends to balance the difference between inflow and cross-flow shrinkage usually found in crystalline polymers. When a particulate is used to reduce and balance shrinkage, some fiber is needed to offset degradation.
Reduced Mold Shrinkage (Increased mold-to-size capability)	Glass fibers Carbon fibers Fillers	Ductility, cost Ductility, cost Tensile strength, ductility, cost	Ductility, cost Ductility, cost Tensile strength, ductility, cost	Reinforcement reduces shrinkage far more than fillers do. Fillers help balance shrinkage, however, because they replace shrinking polymer. The sharp shrinkage reduction in reinforced crystalline resins can often lead to warpage. The best "mold-to-size" composites are reinforced amorphous composites.
Reduced Coefficient of Friction	PTFE Silicone MoSe Graphite	Cost	Cost	These fillers are soft and do not dramatically affect mechanical properties. PTFE loadings commonly range from 5 to 20%; the others are usually 5% or less. Higher loadings can cause mechanical degradation.
Reduced Wear	Glass fibers Carbon fibers Lubricating additives	— — —	— — —	The subject of plastic wear is extremely complex and should be discussed with a composite supplier.
Electrical Conductivity	Carbon fibers Carbon powders	Ductility, cost Tensile strength, ductility, cost	Ductility, cost Tensile strength, ductility, cost	Resistivities of 1 to 100,000 ohm-cm can be achieved and are proportional to cost. Various carbon fibers and powders are available with wide variations in conductivity yields in composites.

Table 6-16　Influence of fillers and reinforcements on TPs

Resin	Reinforcements	Fillers
Amorphous 　ABS 　SAN 　Amorphous 　Nylon 　Polycarbonate 　Modified PPO 　Polystyrene 　Polysulfones	+Can more than double 　tensile strength +Can increase flexural 　modulus fourfold +Raise HDT slightly ±Toughen brittle resins, embrittle 　tough resins +Can provide 1000 ohm-cm 　resistivity +Reduce shrinkage −Reduce melt flow −Raise cost	−Lower tensile strength +Can more than double flexural 　modulus +Raise HDT slightly −Embrittle resins +Can impact special properties such 　as lubricity, conductivity, flame 　retardance +Reduce and balance shrinkage −Reduce melt flow +Can lower cost
Crystalline 　Aceals 　Nylon 6, 6/6 　6/10, 6/12, 11, 12 　Polypropylene 　Polyphenylene sulfide 　Thermoplastic 　Polyesters 　Polyethylene	+Can more than triple tensile 　strength +Can raise flexural modulus 　sevenfold +Can nearly triple HDT ±Toughen brittle resins, 　embrittle tough resins +Can provide 1 ohm-cm 　resistivity +Reduce shrinkage −Cause distortion −Reduce melt flow −Raise cost	−Lower tensile strength +Can more than triple flexural 　modulus +Raise HDT slightly −Embrittle resins +Can impart special properties such 　as lubricity, conductivity, magnetic 　properties, flame retardance +Reduce shrinkage +Reduce distortion −Reduce melt flow +Can lower cost

relatively little actual plastic. Calcium carbonate, silica, and clay are frequently used as extenders reducing the cost of the plastic.

Many plastics because they are organic are flammable incorporate flame-retardants. Additives that contain chlorine, bromine, phosphorous, metallic salts, and so forth reduce the likelihood that combustion will occur or spread. Lubricants like wax or calcium stearate reduce the viscosity of molten plastic and improve its forming characteristics. Plasticizers are low-molecular-weight materials that alter the properties and forming characteristics of plastics. An important application is the production of flexible grades of PVC.

Colorants must provide colorfastness under the required exposure conditions of light, temperature, humidity, chemical exposure, and so on, but without reducing other desirable properties such as flow during

Table 6-17　Example of carbon black on mechanical properties of an ABS

Filler Content c, %	Tensile Modulus E, N/mm^2 (kips/in.^{2})	Breaking Strength σ_r, N/mm (kips/in.^{2})	Elongation at Break ε_T, %	Impact Strength a_z, kJ/m^2
0	2,280 (331)	30.9 (4.48)	8.2	208
3	2,500 (362)	44.2 (6.41)	3.4	36
5	2,720 (394)	43.2 (6.26)	3.1	43
7.5	2,820 (409)	37.7 (5.47)	2.5	41
10	3,010 (436)	35.1 (5.09)	2.2	31
15	3,540 (513)	27.8 (4.03)	1.9	29
20	4,000 (580)	24.8 (3.60)	1.1	26

Table 6-18 Examples of different composite systems

Matrix Material	Reinforcement Material	Properties Modified
Metal	Metal, ceramic, carbon, glass fibers	Elevated temperature strength Electrical resistance Thermal stability
Ceramic	Metallic and ceramic particles and fibers	Elevated temperature strength Chemical resistance Thermal resistance
Glass	Ceramic fibers and particles	Mechanical strength Temperature resistance Chemical resistance Thermal stability
Organics, Thermosets, Thermoplastics	Carbon, glass, organic fibers, glass beads, flakes, ceramic particles, metal wires	Mechanical strength Elevated temperature strength Chemical resistance Antistatic Electrical resistance EMF shielding Flexibility Wear resistance Energy absorption Thermal stability

processing, resistance to chalking and crazing, and impact strength retention. Colorants are usually classed as either pigments or dyes. Pigments are insoluble particles large enough to scatter light but not to provide the high transparency of dyes that are soluble. But dyes are usually poorer in lightfastness, heat stability, and tendency to bleed and migrate in the plastic system, so that they are much less used than pigments. The various special colorants include rnetallics, fluorescents, phosphorescents, and pearlescent colorings.

Pigments may be organic or inorganic. Organic ones usually provide stronger, more transparent colors, are higher priced (although not necessarily more costly to process), and more soluble in plastic systems. Important organic pigments include monochromes and diazos (in yellow, orange, and red), phthalocyanine (in blues and greens), quinacridone (in gold, maroon, violet, and so on), peryiene, and others.

Inorganics are denser and usually of a larger particle size. Common inorganic pigments include iron oxides in buff colors, titanium dioxide in white, lead and zinc chromates (in yellows, oranges, and reds), and other metal oxides and salts. Carbon blacks are also widely used, both as a colorant and to protect polymers from thermal and UV degradation as well as a reinforcing filler.

Reinforced Plastic

Reinforced plastics (RPs) hold a special place in the design and manufacturing industry because they are unique materials (Figs. 6-11 and 6-12). During the 1940s, RPs (or low-pressure laminates, as they were then commonly known) was easy to identify. The basic definition then, as now, is simply that of a plastic reinforced with either a fibrous or nonfibrous material. TSs such as polyester (Table 6-19) and E-glass fiber dominated and still dominates the field. Also used are epoxies.

What essentially characterizes RPs is their ability to be molded into extremely small but also large shapes well beyond the basic capabilities of other processes, at little or no pressure. Also, there are instances in which

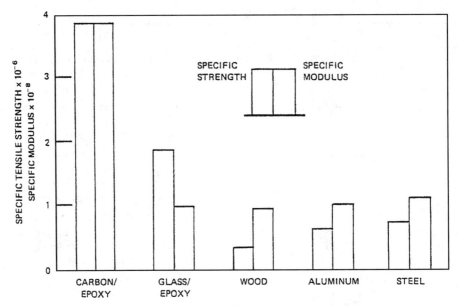

Fig. 6-11   Comparison of RPs with other materials.

less heat is required. Consequently, RPs during the 1940s and 1950s went by the name low-pressure laminates.

In the past, the term high-pressure laminates was reserved for melamine and phenolic impregnated papers or fabrics compressed under high pressures (about 13.8 to 34.5 MPa, or 2,000 to 5,000 psi). They were heated to form either decorative laminates (for example Formica and Micarta) or industrial lami-

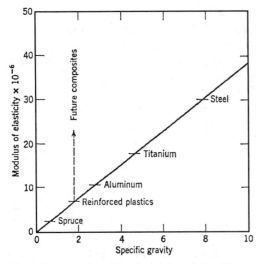

Fig. 6-12   Relationship of modulus of elasticity to specific gravity of materials.

nates for electrical and other industries. The compression molding process was used.

By the early 1960s, the processing of RPs had begun to involve higher pressures, and the name "low-pressure laminates" was dropped in favor of simply reinforced plastics (RPs). But even then, the name referred primarily to reinforced TSs using principally glass fibers and encompassed specialized RP molding processes. By 1970 major changes had occurred. Reinforcements other than glass fiber were in use and TPs as well as TSs were being reinforced. The application of RTS and RTP methods of processing began to increase, using conventional processing techniques like injection molding and rotational molding.

By this time the industry required a more inclusive term to describe RPs, so composite was added. Thus the name in the plastics industry became Reinforced Plastic Composites. More recently they became known only as Composites. However composites identify many other combinations of basic materials (Table 6-18). The fiber reinforcements included higher modulus glasses, carbon, graphite, boron, aramid (strongest fiber in the world, five times as strong as steel on an equal-weight basis), whiskers, and others (Table 6-20 and Figs. 6-13 and 6-14). In

Table 6-19 Characteristics of glass fiber-TS polyester RPs

Polyester Type	Characteristic	Typical Uses
General purpose	Rigid moldings.	Trays, boats, tanks, boxes, luggage, seating.
Flexible resins and semirigid resins	Tough, good impact resistance, high flexural strength, low flexural modulus.	Vibration damping; machine covers and guards, safety helmets, electronic part encapsulation, gel coats, patching compounds, auto bodies, boats.
Light-stable and weather-resistant	Resistant to weather and ultraviolet degradation.	Structural panels, skylighting, glazing.
Chemical-resistant	Highest chemical resistance of polyester group; excellent acid resistance, fair in alkalies.	Corrosion-resistant applications, such as pipe, tanks, ducts, fume stacks.
Flame-resistant	Self-extinguishing, rigid.	Building panels (interior), electrical components, fuel tanks.
High heat distortion	Service up to 500°F, rigid.	Aircraft parts.
Hot strength	Fast rate of cure (hot), moldings easily removed from die.	Containers, trays, housings.
Low exotherm	Void-free thick laminates, low heat generated during cure.	Encapsulating electronic components, electrical premix parts—switch-gear.
Extended pot life	Void-free uniform, long flow time in mold before gel.	Large complex moldings.
Air dry	Cures tack-free at room temperature.	Pools, boats, tanks.
Thixotropic	Resists flow or drainage when applied to vertical surfaces.	Boats, pools, tank linings.

Table 6-20 Examples of different fiber reinforcements

Type of Fiber Reinforcement	Specific Gravity	Density lb./in.^{3} (g/cm^3)	Tensile Strength 10^3 psi (GPa)	Specific Strength 10^6 in.	Tensile Elastic Modulus 10^6 psi (GPa)	Specific Elastic Modulus 10^8 in.
Glass						
E Monofilament	2.54	0.092 (2.5)	500 (3.45)	5.43	10.5 (72.4)	1.14
12-end roving	2.54	0.092 (2.5)	372 (2.56)	4.04	10.5 (72.4)	1.14
S Monofilament	2.48	0.090 (2.5)	665 (4.58)	7.39	12.4 (85.5)	1.38
12-end roving	2.48	0.090 (2.5)	550 (3.79)	6.17	12.4 (85.5)	1.38
Boron (tungsten substrate)						
4 mil or 5.6 mil	2.63	0.095 (2.6)	450 (3.10)	4.74	58 (400)	6.11
Graphite						
High strength	1.80	0.065 (1.8)	400 (2.76)	6.15	38 (262)	5.85
High modulus	1.94	0.070 (1.9)	300 (2.07)	4.29	55[a] (380)	7.86
Intermediate	1.74	0.063 (1.7)	360 (2.48)	5.71	27 (190)	4.29
Organic						
Aramid	1.44	0.052 (1.4)	400 (2.76)	7.69	18 (124)	3.46

[a] Also commercially available up to 100×10^6 psi.

Note: The principal reinforcement, with respect to quantity, is glass fibers, but many other types are used (cotton, rayon, polyester/TP, nylon, aluminum, etc.). Of very limited use because of their cost and processing difficulty are "whishers" (single crystals of alumina, silicon carbide, copper, or others), which have superior mechanical properties.

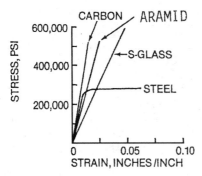

Fig. 6-13   Stress-strain curves of different fiber materials.

Fig. 6-14 specific modulus = modulus/density. Plastics include use of the heat-resistant TPs such as the polimides, polyamide-imide, and others. Table 6-21 provides data on the thermal properties of RPs. To date at least 80 wt% are glass fiber and about 60 wt% of those are polyester (TS) type RPs.

A designer can produce RP products whose mechanical properties in any direction will be both predictable and controllable. This is done by carefully selecting the plastic and the reinforcement in terms of both their composition and their orientation, and following up with the appropriate process (Chapter 8, **REINFORCED PLASTIC**). All types of shapes can be produced: flat and complex, solid and tubular rods or pipes, molded shapes and housings and other complex configurations and structures such as angles, channels, box and I-beams, and so on. The RPs can produce the strongest materials in the world (Fig. 2-6).

The reinforcement type and form chosen (woven, braided, chopped, etc.) will depend on the performance requirements and the method of processing the RP (Fig. 6-15). Fibers can be oriented in many different patterns to provide the directional properties desired. Depending on their packing arrangement, different reinforcement-to-plastic ratios are obtained (Appendix A. **PLASTICS TOOLBOX**).

In its simplest presentation, using glass fiber with plastic, if the fibers were packed as closely as possible (like stacked pipe), the glass would occupy 90.6 vol% (volume) or 95.6 wt%. With a "square" packing (parallel layered fibers so that each layer has fibers 90° to each other) the glass would occupy 78.5 vol% or 88.8 wt%.

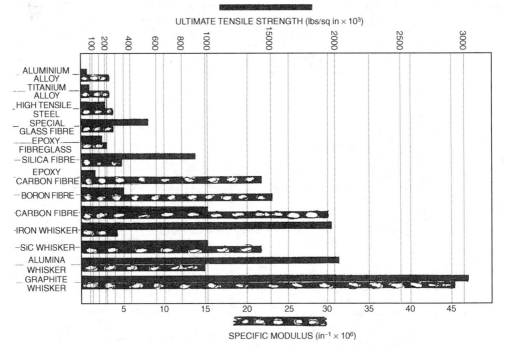

Fig. 6-14   Properties of different RPs that includes whiskers, aluminum, titanium, and steel.

Table 6-21   Thermal properties of reinforcing fibers

Property	E Glass	Carbon	HM Carbon	Aramid
Mean fiber diameter, μ (mils)	10–17 (0.39–0.67)	7 (0.27)	8 (0.31)	12 (0.47)
Therm. Cond.,	7.0	60	97	3.5
BTU-in./hr.-ft.^{2} (W/m·K)	(1.0)	(8.6)	(14)	(0.50)
Specific Heat @ 70°F,	0.192	0.17	0.17	0.34
BTU/lb.°F(J/Kg·K)	(803)	(710)	(710)	(1400)
Coefficient of thermal exp., 10^{-6} in./in./°F (10^{-6} cm/cm °C)				
Longitudinal	1.6 (2.9)	−0.55 (−0.99)	−0.28 (−0.50)	−1.1 (−2.0)
Transverse	4.0 (7.2)	9.32 (16.8)	— (1.8)	33.0 (59.4)
Surface energy, ergs/cm^2	31.0	53.0	—	41.0

Note: One micron = 0.001 cm or = 0.00004 in.
One grain of salt = 100 microns.
One human hair = 70 microns.
The human eye cannot distinguish below 40 microns.
Usually the length of short fibers is less than 3.175 mm (0.125 in.), but generally is 0.76 to 0.52 mm (0.030 to 0.060 in.), and long fibers are longer than 3.175 mm (0.125 in.).

Glass fibers and most other reinforcements require special surface treatment to ensure the bonding and compatibility of the fibers to the plastic in order to maximize performances. Treatments are also used to protect individual filaments during handling and processing (7, 14).

Basic Design Theory

Fiber-reinforced plastics differ from many other materials because they combine two essentially different materials of fibers and a plastic into a single composite. In this way they are somewhat analogous to reinforced concrete, that combines concrete and steel. However, in the RPs the fibers are generally much more evenly distributed throughout the

mass and the ratio of fibers to plastic is much higher (Fig. 6-16).

In designing fibrous-reinforced plastics it is necessary to take into account the combined actions of the fiber and the plastic. At times the combination can be considered homogeneous, but in most cases homogeneity cannot be assumed.

Thus, it is necessary to allow for the fact that two widely dissimilar materials have been combined into a single unit. In the basic design approach certain fundamental assumptions are made. The first, and most important assumption, is that the two materials act together. With a load applied (stretching, compression, twisting, etc.) the fibers and plastic under load is the same; that is, the

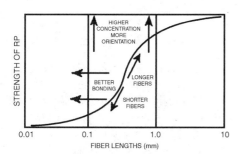

Fig. 6-15   Effect of fiber length on RP strength.

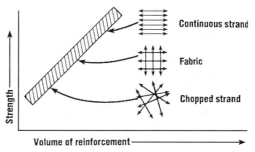

Fig. 6-16   Strength-to-volume relationship for reinforcement used in RPs.

strains in the fiber and plastic are equal. This assumption implies that a good bond exists between the plastic and the fiber to prevent slippage between them and wrinkling of the fiber (7, 10, 14).

The second major assumption is that the material is elastic, meaning that the strains are directly proportional to the stresses applied and when the load is removed the deformation will disappear. In engineering terms the material is assumed to obey Hooke's Law. This assumption is probably a close approximation of the material's actual behavior in direct stress below its proportional limit, particularly in tension, if the fibers are stiff and elastic in the Hookean sense and carry essentially all the stress. This assumption is probably less valid in shear, where the plastic carries a substantial portion of the stress. The plastic may then undergo plastic flow, leading to creep or relaxation of the stresses, especially when the stresses are high.

More or less implicit in the theory of materials of this type is the assumption that all the fibers are straight and unstressed or that the initial stresses in the individual fibers are essentially equal. In practice this is quite unlikely to be true. It is expected, therefore, that as the load is increased some fibers will reach their breaking points first. As they fail, their loads will be transferred to other as yet unbroken fibers, so that the successive breaking of fibers rather than the simultaneous breaking of all of them will cause failure. As reviewed in Chapter 2 (**SHORT TERM LOAD BEHAVIOR, Tensile Stress-Strain,** *Modulus of elasticity*) the result is usually the development of two or three moduli.

The effect is to reduce the material's overall strength and reduce its allowable working stresses accordingly, but the design theory is otherwise largely unaffected, as long as basically elastic behavior occurs. The development of higher working stresses is thus largely a question of devising fabrication techniques like filament winding to make fibers work together to obtain their maximum strength (Chapter 8).

In the following discussion of design theory the values of a number of elastic constants must be known in addition to the strength properties of the plastic, the fibers, and their combination. In the examples used, more-or-less arbitrary values for the elastic constants and strength values have been chosen to illustrate the basic theory, but any other values could have been used just as well.

Theory of combined action Any material when stressed stretches or is otherwise deformed. If the plastic and the fiber in RPs are firmly bonded together, the deformation will be the same in both. For efficient structural behavior high-strength fibers are employed, but these must be more unyielding than the plastic. Therefore for a given deformation or strain a higher stress is developed in the fiber than in the plastic. If the stress-strain relationships of fiber and plastic are known (e.g., from their stress-strain diagrams), the stresses developed in each for a given strain can be computed and their combined action determined.

Figure 6-17 shows stress-strain diagrams for glass fiber and two plastics. Curve A, typical of glass, shows that stress and strain are very nearly directly proportional to each other to the breaking point. Here stiffness, or the modulus of elasticity as measured by the ratio of stress to strain, is high. Curve B represents a hard plastic. Stress here is directly proportional to strain when both are low, but the stress gradually levels off as the strain increases. Its stiffness is much lower than that of glass. It is measured by the tangent to the curve, usually at the origin. Curve C represents a softer plastic intermediate between the hard plastic and the very soft plastics. Stress and strain here are again directly proportional at low levels, but not when the strains become large. The modulus of elasticity, as measured by the tangent to the curve, is lower than for the hard resin.

These stress-strain diagrams may be applied, for example, to the investigation of a rod of which has its total volume is glass fiber and half plastic. If the glass fibers are laid parallel to the axis of the rod, at any cross-section, half the total cross-sectional area is glass and half plastic. If the rod is stretched 0.5%, reference to the stress-strain diagrams

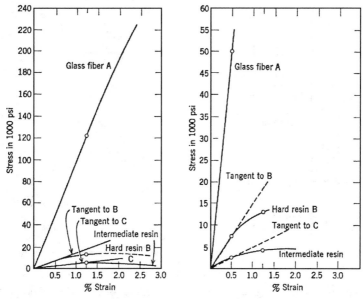

Fig. 6-17 Stress-strain diagrams for RPs.

in Fig. 6-17 will show that the glass is stressed at an intensity of 345 MPa (50,000 psi) and the plastic B at 52 MPa (7,500 psi) or plastic C at 17 MPa (2,500 psi).

If, for example, the rod has a total cross-section of one-half square inch, the glass is one-quarter square inch and the total stress in the glass is one-quarter times 50,000, or 12,500 lb (5700 kg). Similarly, the stress in the plastic B is 1,875 lb (850 kg) and in plastic C is 625 lb (285 kg). The load required to stretch the rod made with plastic B is therefore the sum of the stresses in the glass and plastic, or 14,375 lbs. Similarly, for a rod utilizing plastic C the load is 13,125 lbs. The average stress on the one-half square inch cross-section is therefore 28,750 psi (198 MPa) or 26,250 psi (180 MPa), respectively.

An analogous line of reasoning shows that at a strain of 1.25% the stress intensity in the glass is 125,000 psi (862 MPa) and in plastic B and C at 12,600 and 4,500 psi (87 and 31 MPa), respectively. The corresponding loads on rods made with plastics B and C are 34,400 lb (15,600 kg) and 32,375 lb (14,700 MPa), respectively. Additional detailed information is available concerning this analysis as well as developing data for plain RP plates, composite plates, bending of beams and plates, etc. (10).

Property Range

With RPs different performance capabilities can be obtained. Reason for this capability is because the designer can combine different materials in different proportions. Examples of properties, including other materials of construction, are shown in Figs. 5-9, 6-18a to c and Table 6-22.

Elastomer

An elastomer is a rubberlike material (natural or synthetic) that is generally identified as a material which at room temperature stretches under low stress to at least twice its length and snaps back to approximately its original length on release of the stress (pull) within a specified time period. The term elastomer is often used interchangeably with the term plastic or rubber (2, 14).

Although rubber originally meant a natural thermoset material obtained from a rubber tree, with the development of plastics it identifies a thermoset elastomer (TSE) or thermoplastic elastomer (TPE) material. Different properties identify the elastomers such as strength and stiffness, abrasion resistance, solvent resistance, shock and

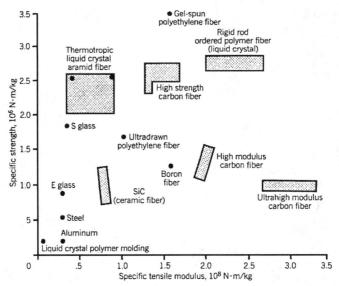

Fig. 6-18a Comparison of specific strength vs. specific modulus of RPs. Specific properties are normalized by plastics density (Pa or N/m³ divided by kg/m³).

vibration control, electrical and thermal insulation, waterproofing, tear resistance, cost-to-performance, etc.

Plastic elastomers are generally lower-modulus flexible materials that can be stretched repeatedly and will return to their approximate original length when the stresses are released. The rubber materials have been around for over a century. They will always be required to meet certain desired properties, but thermoplastic TPEs are replacing traditional TS natural and synthetic rubbers (elastomers). TPEs are also

widely used to modify the properties of rigid TPs, usually by improving their impact strength.

TPEs offer a combination of strength and elasticity as well as exceptional processing versatility. They present creative designers with endless new and unusual product opportunities. More than hundreds of major different groups of TPEs are produced worldwide, with new grades continually being introduced to meet different electrical, chemical, radiation, wear, swell, and other requirements (Tables 6-23 and 6-24).

Quite large elastic strains are possible with minimal stress in TPEs; these are the synthetic rubbers. TPEs have two specific characteristics: their glass transition temperature (T_g) is below that at which they are commonly used, and their molecules are highly kinked as in natural TS rubber (isoprene). When a stress is applied, the molecular chain uncoils and the end-to-end length can be extended several hundred percent, with minimum stresses. Some TPEs have an initial modulus of elasticity of less than 10 MPa (1,500 psi); once the molecules are extended, the modulus increases.

The modulus of metals decreases with an increase in temperature. However, in stretched TPEs the opposite is true, because

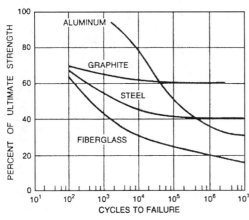

Fig. 6-18b Fatigue property curves of different materials.

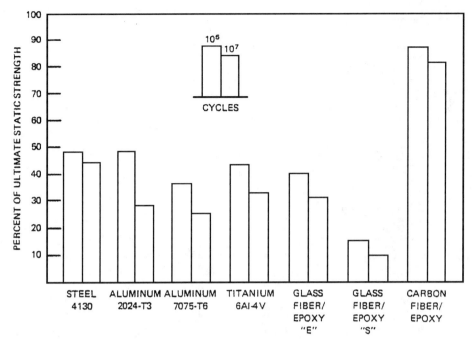

Fig. 6-18c   Fatigue bar graph of different materials.

with them at higher temperatures there is increasingly vigorous thermal agitation in their molecules. Therefore, the molecules resist more strongly the tension forces attempting to uncoil them. To resist requires greater stress per unit of strain, so that the modulus increases with temperature. When stretched into molecular alignment many rubbers can form crystals, an impossibility when they are relaxed and "kinked."

To date, with the exception of vehicle tires, TPEs have been replacing TS rubbers in virtually all applications. Unlike natural TS rubbers, most TPEs can be reground and reused, thereby reducing overall cost. There are types where the need to vulcanize them is eliminated, reducing cycle times, and products can be molded to tighter tolerances. Most TPEs can be colored, whereas natural rubber is available only in black. TPEs also weigh 10 to 40% less than natural rubber (166).

TPEs range in hardness from as low as 25 Shore A up to 82 Shore D (Chapter 5, **MECHANICAL PROPERTY, Hardness**). They span a temperature of −34 to 177°C (−29 to 350°F), dampen vibration, reduce noise, and absorb shock. However,

designing with TPEs requires care, because unlike TS rubber that is isotropic, TPEs tend to be anisotropic during processing as with injection molding. Tensile strengths in TPEs can vary as much as 30 to 40% with direction.

Commodity and Engineering Plastic

About 90 wt% of plastics can be classified as commodity plastics, the others being engineering plastics. The five families of commodities LDPE, HDPE, PP, PVC, and PS account for about two thirds of all the plastics consumed. The engineering plastics such as nylon, PC, acetal, etc. are characterized by improved performance in higher mechanical properties, better heat resistance, higher impact strength, and so forth. Thus, they demand a higher price. About a half century ago the price per pound difference was about 20¢; now it is about $1.00. There are commodity plastics with certain reinforcements and/or alloys with other plastics that put them into the engineering category. Most TSs and RPs are engineering plastics.

Table 6-22   RP trade-offs

Desired Modification	How Achieved	Sacrifice (from Base Resin)		Comments
		Amorphous	Crystalline	
Increased Tensile Strength	Glass fibers Carbon fibers Fibrous minerals	Ductility, cost Ductility, cost NA	Ductility, cost Ductility, cost Ductility	Glass fibers are the most cost effective way of gaining tensile strength. Carbon fibers are more expensive; fibrous minerals are least expensive but only slightly slightly reinforcing. Reinforcement makes brittle resins tougher and embrittles tough resins. Fibrous minerals are not commonly used in amorphous resins.
Increased Flexural Modulus	Glass fibers Carbon fibers Rigid minerals	Ductility, cost Ductility, cost Ductility	Ductility, cost Ductility, cost Ductility	Any additive more rigid than the base resin produces a more rigid composite. Particulate fillers severely degrade impact strength.
Flame Resistance	FR additive	Ductility, tensile strength, cost	Ductility, tensile strength, cost	FR additives interfere with the mechanical integrity of the polymer and often require reinforcement to salvage strength. They also narrow the molding latitude of the base resin. Some can cause mold corrosion.
Increased Heat-Deflection Temperature (HDT)	Glass fibers Carbon fibers Fibrous minerals	Ductility, cost Ductility, cost NA	Ductility, cost Ductility, cost Ductility	When reinforced, crystalline polymers yield much greater increases in HDT than do amorphous resins. As with tensile strength, fibrous minerals increase HDT only slightly. Fillers do not increase HDT.
Warpage Resistance	5 to 10% glass fibers 5 to 10% carbon fibers Particulate fillers	NA NA Ductility, cost, tensile strength	Cost Cost Ductility, cost, tensile strength	Amorphous polymers are inherently nonwarping molding resins. Only occasionally are fillers such as milled glass or glass beads added to amorphous materials because they reduce shrinkage anisotropically. Addition of fibers tends to balance the difference between in-flow and cross-flow shrinkage usually found in crystalline polymers. When a particulate is used to reduce and balance shrinkage, some fiber is needed to offset degradation.

(Continues)

Table 6-22   (*Continued*)

Desired Modification	How Achieved	Sacrifice (from Base Resin)		Comments
		Amorphous	Crystalline	
Reduced Mold Shrinkage (Increased mold-to-size capability)	Glass fibers Carbon fibers Fillers	Ductility, cost Ductility, cost Tensile strength, ductility, cost	Ductility, cost Ductility, cost Tensile strength, ductility, cost	Reinforcement reduces shrinkage far more than fillers do. Fillers help balance shrinkage, however becuase they replace shrinking polymer. The sharp shrinkage reduction in reinforced crystalline resins can often lead to warpage. The best "mold-to-size" composites are reinforced amorphous composites.
Reduced Coefficient of Friction	PTFE Silicone MoS_2 Graphite	Cost	Cost	These fillers are soft and do not dramatically affect mechanical properties. PTFE loadings commonly range from 5 to 20%; the others are usually 5% or less. Higher loadings can cause mechanical degradation.
Reduced Wear	Glass fibers Carbon fibers Lubricating additives	— — —	— — —	The subject of plastic wear is extremely complex and should be discussed with a composite supplier.
Electrical Conductivity	Carbon fibers Carbon powders	Ductility, cost Tensile strength, ductility, cost	Ductility, cost Tensile strength, ductility, cost	Resistivities of 1 to 100,000 ohm-cm can be achieved and are proportional to cost. Various carbon fibers and powders are available with wide variations in conductivity yields in composites.

Neat Plastic

Identifies a plastics with **N**othing **E**lse **A**dded **T**o. It is a true virgin polymer since it does not contain additives, fillers, etc. These are very rarely used.

Structural Foam

The following review concerns structural foams (SFs). Review Chapter 8, **FOAMING** regarding the different TP and TS types of foam available. Most of the foamed plastics are TPs. When compared to solid plastics, a density reduction of up to at least 40% can easily be obtained in SF products. The actual density reduction obtained will depend on the products' thickness, the design's shape, and the melt flow distance during processing such as how much plastic occupies the mold cavity.

Low-pressure SF products can have characteristic surface splay patterns. However, the utilization of increased mold temperatures, increased injection rates, or grained mold surfaces will serve to minimize or hide this surface streaking. Finishing systems like

Table 6-23 Guide to data for elastomers (E = Excellent, G = Good, F = Fair, & P = Poor)

	Natural Rubber	Isoprene Rubber	Styrene-butadiene Rubber	Butadiene Rubber	Butyl Rubber
Abbreviation(s)	NR	IR	SBR	BR	IIR
ASTM D2000 designations	AA, BA	AA, BA	AA, BA	AA, BA	AA, BA CA
Density, Mg/m³	0.92	0.92	0.94	0.93	0.92
Glass transition temp., °C	−70	−70	−60	−105	−65
Temperature serviceability, °C					
lower limit	−55	−55	−45	−70	−50
upper limit, continuous	+70	+70	+70	+70	+100
intermittent	+100	+100	+100	+100	+125
Physical properties					
Hardness range, IRHD	30-100	35-100	40-100	45-90	35-85
Tensile strength, MPa					
gum	24	21	3	3	10
reinforced	28	24	24	17	17
Resilience	E	E	G	E	P
Resistance to:					
tear	E	G-E	G	F	G
abrasion	E	E	E	E	F
compression set	G	G	G	G	F
creep/stress relaxation	E	E	G	G	F
gas permeation	F	F	F	F	E
Electrical resistivity	E	E	E	E	E
Environmental resistance to:					
heat	F	F	F	F	G-E
oxidation	F	F	F	F	G
ozone	P	P	P	P	G
flame	P	P	P	P	P
water	G	G	G	G	E
dilute acids	G	G	G	G	G
concentrated acids	F-G	F-G	F-G	F-G	F-G
alkalis	G	G	G	G	G
aliphatic hydrocarbons	P	P	P	P	P
aromatic hydrocarbons	P	P	P	P	P
halogenated solvents	P	P	P	P	P
oxygenated solvents	F-G	F-G	F-G	F-G	G
animal/vegetable oils	P-G	P-G	P-G	P-G	G

(*Continues*)

sanding, filling, and painting for these structural foam areas are used and have proved to be capable of completely eliminating surface splay.

High-pressure structural foam products have generally been found to require little or no postfinishing. Although high-pressure foam products may exhibit visual splay, their surface smoothness is maintained and no sanding or filling is required.

For SF, mold pressures of approximately 4.1 MPa (600 psi) are required compared to pressures of 34.5 MPa (5,000 psi) and greater in injection molding. As a result, large, complicated parts of 23 kg (50 lb.) and higher can be produced using multi-nozzle equipment, or up to about 16 kg (35 lb.) with single-nozzle equipment and hot runner systems. Product size is, in fact, limited only by the size of existing equipment, tool design, and a material's properties limit product complexity. Product cost can be kept in line through such advantages as product consolidation, function integration, and assembly labor savings.

Table 6-23   (*Continued*)

Ethylene-propylene	Chloroprene Rubber	Nitrile Rubber	Polyurethane	Chlorosulphonated polyethylene	Silicone	Fluorocarbon	Note
EPM, EPDM	CR	NBR	AU, EU	CSM	Q	FKM	1
AA, BA CA, DA	BC, BE	BF, BG	BG	CE	FC, FE, GE	HK	2
0.86	1.23	1.00	1.05	1.18	0.98-1.6	1.85	3
−58	−49	−24	−50	−28	−120	−22	4
−40	−35	−20	−50	−20	−60	−20	5
+125	+100	+100	+70	+125	+200	+200	
+150	+125	+125	+100	+150	+250	+250	
30-90	35-95	40-100	50-100	40-95	40-90	50-95	6
3	17	4	35	24	7	17	7
21	21	21	—	21	10	17	
G	G	P-F	F	F	F	F	8
F	F	P-F	E	F	P	F	
F	E	F	E	G	F	G	9
G	G	G	G	F	G	G	10
F	F	F	F	F	F-G	G	11
G	G	G-E	G	E	P-F	E	
E	F	F	G	G	E	G	
E	G	G	F	E	E	E	12
E	G	G	G	E	E	E	13
E	G	P	E	E	E	E	14
P	F	P	P	F	F	F	15
E	F	F-G	F	G	G	E	16
E	G	G	F	E	F	E	17
G	G	F-G	P	G	F	E	17
G	G	F-G	F	E	F	P-G	
P	G	E	E	F	P	E	
P	P-F	F-G	F	F	P	E	
P	P	P	F	P	F	G	
G	P	P	P	P-F	F-G	P-F	18
G	G	E	G	G	F	E	

When an engineering plastic is used with the structural foam process, the material produced exhibits behavior that is easily predictable over a large range of temperatures. Its stress-strain curve shows a significantly linearly elastic region like other Hookean materials, up to its proportional limit. However, since thermoplastics are viscoelastic in nature, their properties are dependent on time, temperature, and the strain rate. The ratio of stress and strain is linear at low strain levels of 1 to 2%, and standard elastic design principles can thus be applied up to the elastic transition point.

Large, complicated products will usually require more critical structural evaluation to allow better prediction of their load-bearing capabilities under both static and dynamic conditions. Thus, predictions require careful analysis of the structural foam's cross-section.

The composite cross-section of a SF product contains an ideal distribution of material, with a solid skin and a foamed core. The

Table 6-24 Example of elastomers per ASTM D 2000 and SAE J 200

Type ↓ ↓ Class	Typical Rubber
AA	Natural rubber, styrene butadiene, butyl, ethylene propylene, polybutadiene, Polyisoprene
AK	Polysulfide
BA	Ethylene propylene, styrene butadiene (high temperature) Butyl
BC	Chloroprene, chlorinated polyethylene
BE	Chloroprene, chlorinated polyethylene
BF	Nitrile
BG	Nitrile, urethane
BK	Polysulfide, nitrile
CA	Ethylene propylene
CE	Chlorosulfonated polyethylene, chlorinated polyethylene
CH	Nitrile, epichlorohydrin Ethylene/acrylic
DA	Ethylene propylene
DE	Chlorinated polyethylene, chlorosulfonated polyethylene
DF	Polyacrylate (butyl-acrylate type)
DH	Polyacrylate
FC	Silicone (high strength)
FE	Silicone
FK	Fluorinated silicone
GE	Silicone
HK	Fluorinated rubbers

The simply supported beam has a load applied centrally. The upper skin go into compression while the lower one goes into tension, and a uniform bending curve will develop. However, this happens only if the shear rigidity or shear modulus of the cellular core is sufficiently high. If this is not the case, both skins will deflect as independent members, thus eliminating the load-bearing capability of the plastic composite structure.

The fact that the cellular core provides resistance against shear and buckling stresses implies an ideal density for a given foam wall thickness. This optimum thickness is critically important in designing complex stressed products. As a 6.4 mm ($^1/_4$ in.) wall, for example, plastics such as both modified polyphenylene oxide and polycarbonate exhibit the best processing, properties, and cost in the range of a 25% weight reduction. Laboratory tests show that with thinner walls about 4 mm (0.157 in.), this ideal weight reduction decreases to 15%. When the wall thickness reaches approximately 9 mm (0.350 in.), the weight can be reduced by 30%.

However when the SF cross-section is analyzed, its composite nature still results in a twofold increase in rigidity, compared to an equivalent amount of solid plastic, since rigidity is a cubic function of wall thickness. This increased rigidity allows large structural products to be designed with only minimal distortion and deflection when stressed within the recommended values for a particular foamable plastic.

Depending on the required analysis, the moment of inertia can be evaluated three ways. In the first approach, the cross-section is considered to be solid material (Fig. 6-20). The moment of inertia, I_x is:

$$I_x = bh^3/12 \qquad (6-1)$$

where b = the width and h = the height.

manufacturing process can distribute a thick, almost impervious solid skin that is in the range of 25% of the overall wall thickness at the extreme locations from the neutral axis (Fig. 6-19). These are the regions where the maximum compressive and tensile stresses occur in bending.

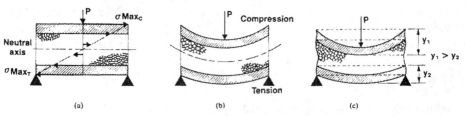

Fig. 6-19 Composite cross-section of a structural foam product.

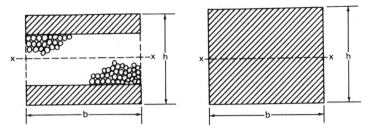

Fig. 6-20 Cross-section of a solid material.

This commonly used approach provides acceptable accuracy when the load-bearing requirements are minimal, for example, in the case of simple stresses and when time and cost constraints prevent more exact analysis.

The second approach ignores the strength contribution of the core and assumes that the two outer skins provide all the required rigidity (Fig. 6-21). The equivalent moment of inertia is then equal to:

$$I_x = b(h^3 - h_1^3)/12 \qquad (6\text{-}2)$$

This formula results in conservative accuracy, since the core does contribute to the stress-absorbing function. It also adds a built-in safety factor to a loaded beam or plate element when safety is a concern.

A third method is to convert the structural foam cross-section to an equivalent I-beam section of solid resin material (Fig. 6-22).

The moment of inertia is then formulated as:

$$I_x = [bh^3 - (b - b_1)(h - 2t_x)]/12 \qquad (6\text{-}3)$$

where $b_1 = b(E_c)/E_s$, E_c = the modulus of the core, E_s = the modulus of the skin, t_s = the thickness of the skin, and h_1 = the height of the equivalent web (core).

This approach may be necessary where operating conditions require stringent load-bearing capabilities without resorting to overdesign and thus unnecessary costs. Such an analysis produces maximum accuracy and would thus be suitable for finite-element analysis on complex products. However, the one difficulty with this method is that the core modulus and the as-molded variations in skin thicknesses cannot be accurately measured.

Plastics with a Memory

Thermoplastics can be bent, pulled, or squeezed into various useful shapes. But eventually when heat is added, they return to their original form. This behavior, known as plastic memory, can be annoying. If property applied, however, plastic memory offers interesting design possibilities for all types of fabricated products.

When most materials are bent, stretched, or compressed, they alter their molecular structure or grain orientation to accommodate the deformation permanently, but this is not so with thermoplastics. They temporarily assume the deformed shape, but they always maintain the internal stresses that want to force the material back to its original shape.

Most TP products can be produced with a built-in memory. That is, their tendency to move into a new shape is included as an

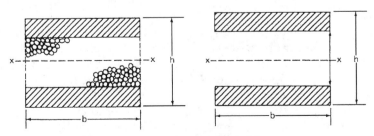

Fig. 6-21 Cross-section of a sandwich structure.

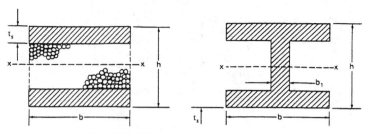

Fig. 6-22 Cross-section of an I-beam.

integral part of the design. So after the products are assembled in place a small amount of heat can coax them to change shape. Plastic products can be deformed during assembly, then allowed to return to their original shape. In this case the products can be stretched around obstacles or made to conform to unavoidable irregularities without their suffering permanent damage.

The time/temperature-dependent change in mechanical properties results from stress relaxation and other viscoelastic phenomena that are typical of these plastics. When the change is an unwanted limitation it is called creep. When the change is skillfully adapted to use in the overall design, it is referred to as plastic memory.

Even though potential memory exists in all TPs, polyolefins, neoprenes, silicones, and other cross-linkable TPs are example of plastics that can be given memory either by radiation or by chemically curing. Fluorocarbons, however, need no such curing. When this phenomenon of memory is applied to fluorocarbons such as TFE, FEP, ETFE, ECTFE, CTFE, and PVF2, interesting high-temperature or wear-resistant applications become possible.

Orientation

A thermoplastic's molecular orientation can be accidental or deliberate. However, excessive frozen-in stress due to orientation can be extremely damaging if products are subject to environmental stress cracking or crazing in the presence of chemicals, heat, and so on. Initially the molecules are relaxed; molecules in amorphous regions are in random coils, those in crystalline regions relatively straight and folded. During processing

the molecules tend to be more oriented than relaxed, particularly when sheared, as during injection molding and extrusion.

After temperature-time-pressure is applied and the melt goes through restrictions (molds, dies, etc.), the molecules tend to be stretched and aligned in a parallel form. The result is developing directional properties and dimensions. The amount of change depends on the type of thermoplastic, the amount of restriction, and, most important, its rate of cooling. The faster the rate, the more retention there is of the frozen orientation. After processing, products could be subject to stress relaxation with changes in performance and dimensions. With certain plastics and processes there is an insignificant change. If changes are significant, one must take action to change processing conditions, particularly controlling the cooling rate (Chapter 8, **PROCESSING AND PROPERTIES, ORIENTATION**).

Material Variable

Even though equipment operations have understandable but controllable variables that influence processing, the usual most uncontrollable variable in the process can be the plastic material. The degree of properly compounding or blending by the plastic manufacturer, converter, or in-house by the fabricator is important. Most additives, fillers, and/or reinforcements when not properly compounded will significantly influence processability and fabricated product performances.

With the passing of time and looking ahead these material and equipment variabilities continually are reduced due to improvement in their manufacturing and process control capabilities. However they still exist. To ensure control of material, as previously reviewed, setting up controls via different tests and setting limits are important. Even set within limits, processing the materials could result in inferior products. As an example the material specification from a supplier will provide an available minimum to maximum value such as molecular weight distribution. It is determined that when material arrives all on the maximum side it produces acceptable products. However when all the material arrives on the minimum value process control has to be changed in order to produce acceptable products.

In order to judge performance capabilities that exist within the controlled variabilities, there must be a reference to measure performance against. As an example, the injection mold cavity pressure profile is a parameter that is easily influenced by variations in the materials. Related to this parameter are four groups of variables that when put together influences the profile: (1) melt viscosity and fill rate, (2) boost time, (3) pack and hold pressures, and (4) recovery of plasticator. Thus material variations may be directly related to the cavity pressure variation. Details on **EQUIPMENT/PROCESSING VARIABLE** are in Chapter 8.

A very important factor that should not be overlooked by a designer, processor, analyst, statistician, etc. is that most conventional and commercial tabulated material data and plots, such as tensile strength, are mean values. They would imply a 50% survival rate when the material value below the mean processes unacceptable products. Target is to obtain some level of reliability that will account for material variations and other variations that can occur during the product design to processing the plastics (Chapter 5, **ANALYZING TESTING AND QUALITY CONTROL**).

Recycling

Also called regrind and reclaiming. Most processing plants have been reclaiming/recycling reprocessable TP materials such as molding flash, rejected products, blown film trim, scrap, and so on. TS plastics (not remeltable) have been granulated and used as filler materials. If possible the goal is to significantly reduce or eliminate any trim, scrap, rejected products, etc. because it has already cost money and time to go through a fabricating process; granulating just adds more money and time.

Also it usually requires resetting the process controls to handle it alone (or even when blending with virgin plastics and/or additives) because it usually does not have uniform particle sizes, shapes, and melt flow characteristics. Keeping the scrap before/after granulating clean is a requirement. These behaviors as well as overheating the plastic during the cutting action of a granulator can significantly influence their processability and performance of fabricated products.

Figure 6-23 shows how regrind levels affect the mechanical properties of certain formulations of plastics "once through" the fabricating process and blended with virgin material. The regrind, or scrap, amount is a percentage by weight. Figure 6-24 is an example of the potential effects of the number of times regrind influences the performance of an injection molded TP mixed with virgin plastics. Figure 6-25 is an injection molding flow diagram example. See in Chapter 7, **THERMAL PROPERTY, Heat History, Residence Time, and Recycling**.

Since scrap can be a mixture ranging from fine dust to large irregular chunks of different shapes, thicknesses, etc., it is important to use a granulator that provides the most uniformity and the least heat and mechanical damage to the scrap. Material having this range of size when granulating will at least influence product performances where critical requirements exist. The material influences the processing characteristics that in turn effect the product's performances.

Overheating during the cutting action of the granulator can cause the most damage. For heat sensitive plastics to eliminate any heat damage cryogenic granulating is used. Recognize that a granulator that handles soft plastics will not work well when granulating hard plastic. One that handles thin plastic is not the proper type to handle thick plastics. Two or more different performing

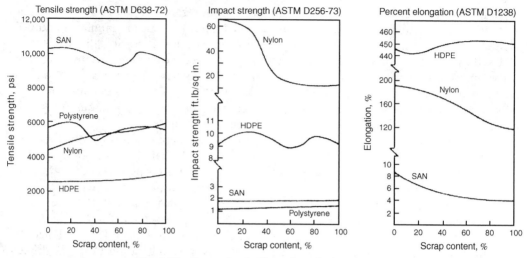

Fig. 6-23   Example of effect of regrind on plastics.

granulators may be required to process thick material to ensure that overheating is minimized. The first granulator is designed to handle the thick material. The next granulator has the capability to process the first granulated plastics. The second granulated plastics may require a finer operating granulator and so on until the desired material is obtained.

In summary recycling will reduce performance properties. The amount of reduction can be very slight to undesirable amounts during the first remelting process. Granulated plastics that have been significantly degraded may be reformulated by the addition of stabilizers, pigments, plasticizers, fillers, reinforcements, and/or other additives. These blends, particularly the general purpose commodity plastics, usually improves their processability and/or product performances.

When RPs are granulated, the lengths of the fibers are reduced. On reprocessing with virgin materials or alone, their processability and product performances are definitely change. So it is important to determine if the change will affect final product performances. If it will, a limit for the amount of regrind mix should be determined or no recycled RP is to be used. Use it in some other product such as simulated wood.

Recycling Energy Consumption

Plastics have many advantages. Included are the facts that they have the lowest energy consumption in the recycling processes of about 2 MJ/kg (2 to 2.5 MJ/I) and when incinerated the highest recovery energy content exists of about 42 MJ/kg. Some comparisons with other materials are provided. (1) Processing waste paper requires 6.7 MJ/kg and as a general rule about twice as much paper is needed compared to plastics for

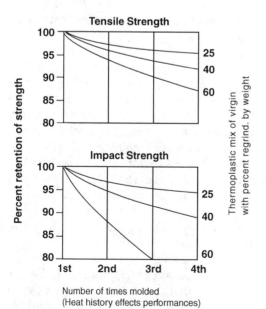

Fig. 6-24   Example of effects of number of times in recycled injection molded plastics.

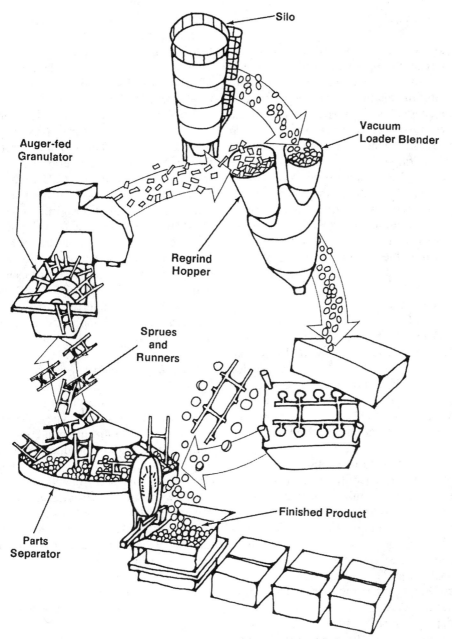

Fig. 6-25 Flow diagram: virgin plastics, molded products with runners, granulating runners, to regrind blend with virgin plastics.

comparable applications. (2) In glass production, if one uses about 10% of recycled glass, this only reduces the energy consumption of the process by about 2%; thus the use of recycled glass requires about 8 MJ/kg, but the comparative figure is higher when considered in relation to each product, as one needs about 10 to 20 times as much material compared with plastics. (3) The energy requirement for processing scrap steel and tin-plate is about 6 MJ/kg. (4) Aluminum recycling requires about 50% of the energy needed to make a product from virgin aluminum; about 50 MJ/kg.

Design Source Reduction

This generally defines the design, manufacture, purchase, or use of materials or products to reduce the amount of material used before they enter the municipal solid waste stream. Because it is intended to reduce pollution and conserve resources, source reduction should not increase the net amount or toxicity of waste generated throughout the life of the product. The EPA has established a hierarchy of guidelines for dealing with the solid waste situation. Their suggestions include (logically) source reduction, recycling, waste-to-energy gains, incineration, and landfill. The target is to reduce the quantity of trash.

Recycling Method

This subject effects designers since many products have the requirement by regulations or otherwise to use recycled plastics. Different methods are used to recycle materials to provide plastics with a continuing life. Method used is influenced by factors such as costs, quantity involved, weight involved, size and shape, complexity of mixed types of plastics, extended of contamination such as metallic particles, continued availability of material, etc. (Recognize that they can also be used as energy sources through incineration that can be combined with production of electricity and/or hot water for example).

In addition to granulating, the processes used include depolymerization to thermal liquefaction and gasification (back to feedstocks or intermediates), chemical pyrolysis, chemical depolymerization such as methanalysis and glycolysis, alcoholysis, catalytic cracking, gasification, hydrogenation, hydrolysis, reactive extrusion, and thermal steam of plastics. Each technique has advantages and drawbacks. Some require careful plastic sorting of mixed materials, and cleaning (2).

The choice of recycling to materials or energy has to be decided by an economic audit. Recycling is preferable to landfill practice, the costs of which are increasing and where the favorable properties of plastics are not used. Municipal authorities have to consider the economics of recycling operations, tak-ing into account the cost of landfill. Factors to consider are: revenue from recycled materials or produced energy, cost of recycling, savings from non-disposal in landfill, and cost of disposal in landfill of the remaining tonnage after recycling.

While recycling can save energy and resources in the manufacturing process, getting recyclables to market and then processing into products also uses energy and generates waste that must be managed. Use of fuels and the environmental impact of preparing, collecting, sorting, and transporting recyclables should be considered when developing an audit. The ideal target is that the recycling results in a profitable venture. An example is when a town takes a positive approach to handling all kinds of waste, such as in Chatham, MA, a profit is made over and above the cost of separating, handling, delivery to customers, etc. the "waste."

Recycling limitation Criteria of logistic technology and properties will determine whether or not it is plausible to reclaim and reuse plastic wastes. These criteria can be assessed economically in a complex way under the aspects of production and economy. Logistic criteria will cover the conditions of accrual according to location and quantity. Technological criteria are the purity of type plastic, cleanliness, and geometry (basic shape and uniformity). Property criteria result from the extent of damage of the material during recycling.

Reactive extrusion recycling This review only concerns one of the recycling method that is also called reactive compounding, or REX (reactive extrusion). It refers to the performance of chemical reactions during plastic processing that includes recycling plastics such as PET (202). The most common reactants are plastic or preplastic melts and gaseous, liquid, or molten low molecular weight compounds. A particular advantage of the extruder as a chemical reactor is the absence of a solvent as the reaction medium.

No solvent-stripping or recovery process is required, and product contamination by solvent or solvent impurities is avoided. The chemical reaction may take place in the melt phase or, less commonly, in the liquid phase, as when bulk polymerization of monomers performed in an extruder, or in the solid phase when the plastic is conveyed through the extruder in a slurry. The types of reactions developed include bulk polymerization, graft reaction, interchain copolymer formation, coupling or branching reaction, controlled molecular weight degradation, and functionalization or functional group modification (2).

Web Site Connect Buyer and Seller of Recycled Plastic

In keeping with its commitment to economically and environmentally responsible recycling, the American Plastics Council (APC) has a Web site feature that connects buyers and sellers of recycled plastics. The site, located at **www.plasticsresource. com/recycling/marketsdb/index.phtml** allows users to select the type of plastic they want and a source location from which to obtain it.

As an example users searching for vinyl sources can browse through hundreds of post-consumer residential, industrial, commercial and institutional sources. This site is in addition to the searchable Directory of North American Companies involved in the Recycling of Vinyl (PVC) Plastics located on the Vinyl Institute's Web page at **www.vinylinfo.org/database/vinyidata2.** These tools provide information to companies that are searching for specific information about recycling to meet their processing needs.

Plastic Future and Biotechnology

We are now in the century of biotechnology. Biological breakthroughs influenced mankind that includes raising life expectancies by 10 years, commercial fuel cells using corn stalks as the energy source to plastics. The advances made by scientists in DNA and RNA have left no industry untouched including plastic (186).

The reliance of fossil fuels has been challenged by lower cost and renewable sources that are more environmentally friendly. The traditional chemical plant has met serious competition from green plants. Many monomers are now made via fermentation, using low-cost sugars as feedstock. Some of the commodity monomers are under siege by chemicals extracted from biomass. Monomer production has been expanded to include many more monomers from nature.

Worldwide suppliers with bioengineering capabilities are displacing established polymers with cost-effective and higher performing plastics. An explosion of novel polymers has been made by enzymatic control. The use of enzymes for polymerization has drastically altered the landscape of polymer chemistry. Processors can request specific properties for each application as opposed to the usual making do with what is available. The supplier can deliver to the processor desired properties requested.

There are methods to manipulate the backbones of polymers in several areas that include control of microstructures such as crystallinity, precise control of molecular weight, copolymerization of additives (flame retardants), antioxidants, stabilizers, etc.), and direct attachment of pigments. A major development with all this type action has been to provide significant reduction in the variability of plastic performances, more processes can run at room temperature and atmospheric pressure, and 80% energy cost reductions.

7

Material Property

Introduction

Designing has never been easy in any material, particularly plastics, because there are so many. Plastics practically provide more types with the many variations that are available than any other material. Of the more than 35,000 different plastics worldwide, only a few hundred are used in large quantities. Unfortunately, some designers view plastics as a single material because they are not aware of all the types available (Appendix A, **PLASTICS DESIGN TOOLBOX**).

Plastics are families of materials each with their own special advantages. The major consideration for a designer is to analyze what is required as regards to performances and develop a logical selection procedure from what is available.

The ranges of properties in plastics encompasses all types of environmental and load conditions, each with its own individual, yet broad, range of properties (Fig. 1-9). These properties can take into consideration wear resistance, integral color, impact resistance, transparency, energy absorption, ductility, thermal and sound insulation, weight, and so forth. There is unfortunately no one plastic that can meet all maximum properties. Therefore, the designer has different options, such as developing a compromise because many product requirements provide options, particularly if cost is of prime importance.

The combination approach permits using plastics that have different properties. They can just be stacked together, but with the available processes they can also be put together so that each material retains its individuality yet has a bond with the adjoining plastics. These processes of coinjection, coextrusion, and so on are reviewed in Chapter 8. Each of the individual plastics can provide such characteristics as wear resistance, being a barrier to water, electrical conductor, and adding strength. Low cost recycled (solid waste) plastics can be "sandwiched" between other expensive, high performance plastics so they only act as a filler, increase strength, etc.

Plastics can also be combined with other materials such as aluminum, steel, and wood to provide specific properties. Examples include PVC/wood window frames and plastic/aluminum-foil packaging material. All combinations require that certain aspects of compatibility such as processing temperature and linear coefficient of thermal expansion or contraction exist.

The designer can use conventional plastics that are available in sheet form, in I-beams or other forms, as is common with many other materials. Although this approach with plastics has its place, the real advantage with plastic lies in the ability to process them to fit the design shape, particularly when it comes to complex shapes. Examples include two or more products with mechanical and electrical connections, living hinges, colors, and snap fits that can be combined into one product.

Like other materials, hot enough fires can destroy all plastics. Some burn readily, others slowly, others only with difficulty; still others do not support combustion after the removal of the flame. There are certain plastics used to withstand the reentry temperature of 2,500°F (1,370°C) that occurs when a spacecraft returns into the earth's atmosphere; the time exposure is parts of a millisecond. Different industry standards can be used to rate plastics at various degrees of combustibility (Chapter 2, **HIGH TEMPERATURE**).

Plastics' behavior in fire depends upon the nature and scale of the fire as well as the surrounding conditions and how the products are designed. For example, the virtually all-plastic 35 mm slide projector uses a very hot electric bulb. When designed with a metal light and heat reflector with an air circulating fan, no fire develops. Therefore, designing in this type product environment requires understanding all the variables so that the proper plastics can be used.

Mechanical Property

Most plastics are used to produce products because they have desirable mechanical properties at an economical cost. For this reason their mechanical properties may be considered the most important of all the physical, chemical, electrical, and other considerations for most applications. Thus, everyone designing with such materials needs at least some elementary knowledge of their mechanical behavior and how they can be modified by the numerous structural factors that can be in plastics (Chapters 2 to 6).

Plastics have the widest variety and range of mechanical properties of all materials (Figs. 1-8 and 7-1 and 7-2). They vary from basically soft to hard, rigid solids. Great many structural factors determine the nature of their mechanical behavior, such as whether a load occurs over the short term or the long

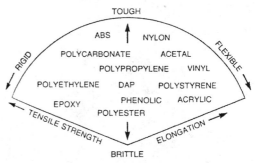

Fig. 7-1 Example of the range of mechanical properties for plastics.

term (Chapter 2). As a rule, design is based on certain minimum strength or minimum deformation criteria.

Short-term testing is important in designs and for quality control of plastics to ensure the required properties of plastics used in production. The short time data provides the designer information that permit comparisons of one material with another. However, a true comparison is possible only if both sets of data were determined in exactly the same way. For example, the speed of loading tensile test specimens influences performance factors such as deformation. Also, comparing the impact resistance of a $\frac{1}{2}$ in. specimen with that of a $1/8$ in. specimen will result in a different analysis of the material's properties. Thus, it is necessary to describe the exact testing conditions along with each set of data sheets. The data from short-term testing give the user an important overall picture of the material (Chapter 5).

The long-term testing of certain plastics allows their strength properties to be identified rapidly. Three of the major control test procedures for long-term testing and predicting product lifetime are creep, fatigue, and impact as reviewed in Chapter 2. Figure 7-3 shows long-term tensile creep curves at 20°C (49°F) for polypropylene and nylon; numbers in parenthesis refer to stress levels in MPa. Figure 7-4 shows long-term tensile fatigue curves for dry nylon 6 that is 4.5 mm (0.18 in.) thick, acrylic (PMMA) that is 6.4 mm (0.25 in.) thick, and polytetrafluoroethylene (PTFE) that is 6.6 mm (0.26 in.) thick. The test frequency is at 1,800 cpm.

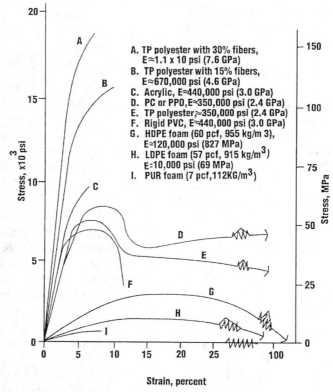

Fig. 7-2 An example of a range in tensile strength, modulus of elasticity, and elongation of some TPs with and without chopped glass fibers by weight and type of reinforcement.

Toughness

Toughness is usually the most complex factor to technically define and understand. Tough TPs are usually described as ones having a high elongation to failure or ones in which a large amount of energy must be expended to produce failure. For high toughness a plastic needs both the ability to withstand load and the ability to elongate

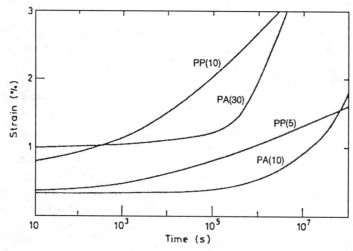

Fig. 7-3 Example of long-term tensile creep curves.

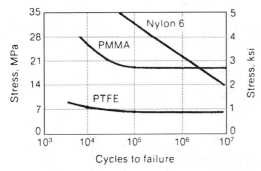

Fig. 7-4 Example of long-term tensile fatigue curves.

substantially without failing except in the case of RPs. The RPs can have high strengths with low elongation.

It may appear that factors contributing to high stiffness are required. This is not true because there is an inverse relationship between flaw sensitivity and toughness; the higher the stiffness and the yield strength of a TP, the more flaw sensitive it becomes. However, because some load bearing capacity is required for toughness, high toughness can be achieved by a high trade-off of certain factors. Crystallinity increases both stiffness and yield strength, resulting usually in decreased toughness. This is true below its glass transition temperature (T_g) in most noncrystalline (amorphous) plastics, and below or above the T_g in a substantially crystalline plastic (Chapter 6). However, above the T_g in a plastic having only moderate crystallinity increased crystallinity improves its toughness. Furthermore, an increase in molecular weight from low values increases toughness, but with continued increases, the toughness begins to drop.

Cross-linking produces some dimensional stability and improves toughness in a noncrystalline type of plastic above the T_g, but high levels of cross-linking lead to embrittlement and loss of toughness. This is one of the problems with TSs for which an increase in T_g is desired. Increased cross-linking or stiffening of the chain segments increases the T_g, but it also decreases toughness. A popular way to increase toughness is to blend, compound, or copolymerize a brittle plastic with a tough plastic. Although some loss in stiffness

is usually encountered, the result is a satisfactory combination of properties.

Deformation and Toughness

Deformation is an important attribute in most plastics, so much so that it is the very factor that has led them to be called plastic. For designs requiring such traits as toughness or elasticity this characteristic has its advantages, but for other designs it is a disadvantage. However, there are plastics, in particular the RPs, that have relatively no deformation or elasticity and yet are extremely tough (Fig. 7-5); (a) toughness related to heat deflection or rigidity and (b) toughness or impact related to temperature for polystyrene (PS) and high impact polystyrene (HIPS).

These types of behavior characterizes the many different plastics available (Table 7-1). Some tough at room temperature, are brittle at low temperatures. Others are tough and flexible at temperatures far below freezing but become soft and limp at moderately high temperatures. Still others are hard and rigid at normal temperatures but may be made flexible by copolymerization or adding plasticizers.

By toughness is meant resistance to fracture. However, there are those materials that are nominally tough but may become embrittled due to processing conditions, chemical attack, prolonged exposure to constant stress, and so on. A high modulus and high strength, with ductility, is the desired combination of attributes. However, the inherent nature of plastics is such that their having a high modulus tends to associate them with low ductility, and the steps taken to improve the one will cause the other to deteriorate.

As previously described (Chapter 2), the area under short-term stress-strain curves provides a guide to a material's toughness and impact performance (Fig. 7-6). The ability of a TP to absorb energy is a function of its strength and its ductility that tends to be inversely related. The total absorbable energy is proportional to the area within the lines drawn to the appropriate point on the curve from the axis. The material in area A is

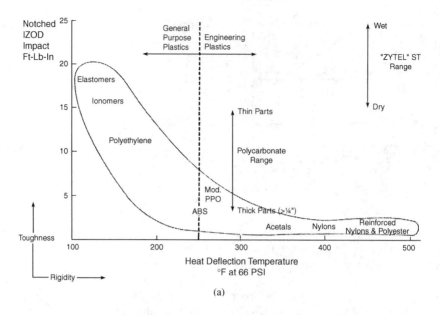

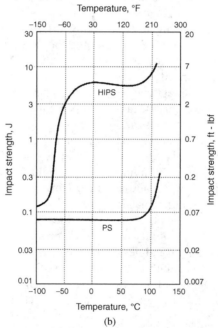

Fig. 7-5   Examples of toughness in plastics.

rubberlike and is just as tough (that is of equal area) as material B, which is metallic. Most plastics, like material B, fall between these extremes, but some fall into both A and C. Toughness can be relate to moisture content in the plastic (Fig. 7-7) based on the area under the tensile stress-strain curves.

Soft, weak materials have a low modulus, low tensile strength, and only moderate elon-gation to break. The elastic modulus or the modulus or elasticity is the slope of the ini-tial straight-line portion of the curve. Hard, brittle materials have high moduli and quite high tensile strengths, but they break at small elongations and have no yield point. Hard, strong plastics have high moduli, high ten-sile strengths, and elongations of about 5% before breaking. Their curves often look as

Table 7-1 Examples of toughness or fracture characteristics for TPs

	Material	Unnotched	Notched
PMMA	Polymethylmethacrylate		
PA	Polystyrene	Brittle	Brittle
SAN	Styrene-acrylonitrile copolymer		
ABS	Acrylonitrile butadiene styrene		
CA	Cellulose acetate		
HDPE	High-density polyethylene		
PA	Polyamide (Nylon)		
PB	Polybutene	Ductile	Brittle
PC	Polycarbonate		
POM	Polyoxymethylene		
PP	Polypropylene		
PTP	Polyethylene terephthalate		
PVC	Polyvinyl chloride		
LDPE	Low-density polyethylene		
PB	Polybutene	Ductile	Ductile
TFE	Polytetrafluoroethylene		

though the material broke about where a yield point might have been expected.

Soft, tough plastics are characterized by low moduli, yield values or plateaus, high elongations of 20 to 1,000%, and moderately high breaking strengths. The hard, tough plastics have high moduli, yield points, high tensile strengths, and large elongations. Most plastics in this category show cold drawing or necking during the stretching operation.

From a practical viewpoint toughness is readily understood, but technically there tends to be no scientific method of measuring it. One definition of toughness is simply the energy required to break the plastic. This energy is equal to the area under the stress-strain curve. The toughest plastics should be those with very great elongations to break, accompanied by high tensile strengths; these materials nearly always have yield points. One major exception to this rule is reinforced plastics that use reinforcing fibers like glass and graphite.

Stress-strain tests may be made in compression as well as tension. A modulus may

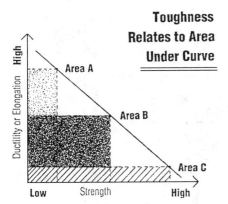

Fig. 7-6 Toughness tends to relate to the area under the stress-strain curve.

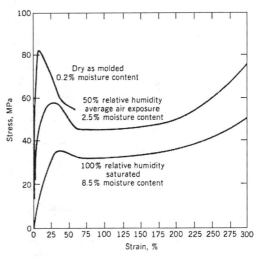

Fig. 7-7 Tensile stress-strain curves of three different moisture contents at 23°C (73°F) and different areas under the curves.

be calculated from the initial slope of its curve. And materials under compression are much less brittle than when under tension. Thus, many plastics that are brittle when tested in tension become ductile and show yield points under compression, as for instance polystyrene. Typical values of ultimate strength in compression for many plastics are about twice that of the tensile strength. Flexural strength tests in which part of a specimen is under tension and part under compression generally give values of ultimate strengths that are between the values for ultimate tension and compression (Chapter 2).

Stiffness

The same factors that influence thermal expansion dictate the stiffness of plastics. Thus, in TS the degree of cross-linking and amount of overall flexibility are important. In a TP its crystallinity and secondary bond's strength control its stiffness.

Strength

The subject of strength is much more complex than stiffness, since so many different types exist: short or long term, static or dynamic, etc. (Chapter 2). Some strength aspects are interrelated with those of toughness. The crystallinity of TPs is important for their short term yield strength. Unless the

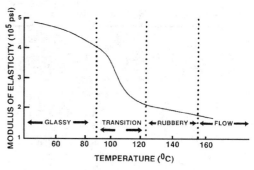

Fig. 7-8 Example of TP modulus during different temperature phases.

crystallinity is impeded, increased molecular weight generally also increases the yield strength. However, the cross-linking of TSs increases their yield strength substantially but has an adverse effect upon toughness.

Factors that influence thermal dimensional stability also affect fatigue strength. This is a result of the substantial heating that is often encountered with fatigue, particularly in TPs.

Temperature Effect

An examination of the effect of temperature on the modulus of elasticity (also viscosity) of a typical TP is shown in Fig. 7-8. As the temperature is increased the plastic changes through different stages from a rigid solid to a liquid through the stages of being glassy, in transition, rubbery, and flow. Figure 7-9 shows the effect of TPs' degree of crystallinity upon modulus vs. temperature.

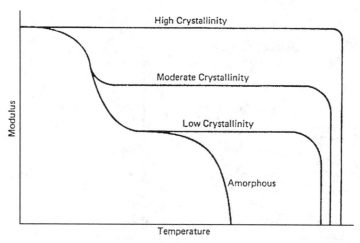

Fig. 7-9 Effect of crystallinity upon modulus vs. temperature.

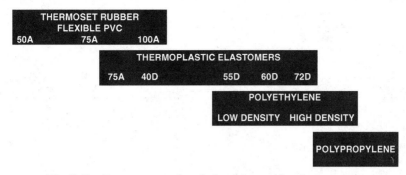

Fig. 7-10 Durometer scale relationships and hardness ranges.

Other

Figure 7-10 shows durometer scale relationships and hardness ranges. The letter designations refer to the Shore hardness test (Chapter 5, **MECHANICAL PROPERTY, Hardness**).

The properties of TP and TS plastics in Tables 7-2 and 7-3 show that there is a wide range of properties exist. Of the over 35,000 plastics available, each have their inherent properties and processabilities.

Electrical Property

Plastics and RPs offer the designer a great degree of freedom in the design and manufacture of products requiring specific electrical properties (Table 7-4). Their combination of mechanical and electrical properties makes them an ideal choice foreverything from micro electroniccomponents to large electrical equipment enclosures. The most notable electrical property of plastics is their ability as good insulators, but there are considerably many other important electrical properties available to the designer working with different plastics (Chapter 4, **ELECTRICAL/ELECTRONIC PRODUCT**).

Tables 7-5 to 7-7 show that there are different orders of magnitude between plastics and metals. Depending on the application, plastics may be formulated and processed to exhibit a single property or a designed combination of electrical, mechanical, chemical, thermal, optical, aging properties, and others. The chemical structure of polymers and the various additives they incorporate provide compounds to meet many different performance requirements.

There exists an extensive amount of all kinds of electrical data worldwide to meet all kinds of plastic products' electrical requirements. Examples of different properties with different plastics are given in Tables 7-8 and 7-9 and Fig. 7-11. The major

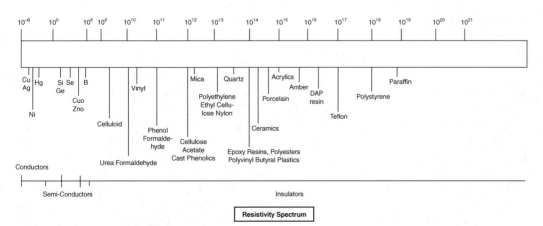

Fig. 7-11 Spectrum of volume resistivity.

Table 7-2　Example of properties for thermoplastics

Resin Material	Impact Strength Notched Izod, ft-lb/in., $\frac{1}{2}$ in. bar	Tensile Strength, psi $\times 10^3$	Tensile Modulus, psi $\times 10^3$	Elongation, (%)	Flexural Strength, psi $\times 10^3$	Compressive Strength, psi $\times 10^3$
1. Acetal	1.2–2.3	8.8–10	400–410	12–75	13–14	18
2. ABS	1.0–9.5	3.5–10.5	200–450	10–100	5–15	5–11
3. Acrylic	0.4	8.7–10.5	380–430	3–6	14–16	14–17
4. Acrylic high impact	0.5–2.3	5.5–8	225–330	23–38	8.5–12	7–12
5. Cellulose acetate	.5–5.6	2.3–8.1	—	10–70	2.2–11.5	2.2–10.9
6. Cellulose acetate butyrate	0.8–6.3	2.6–6.9	—	40–88	1.8–9.3	2.1–9.4
7. Cellulose propionate	0.9–10.2	1.8–7.3	—	30–100	2.8–11	3–9.6
8. Chlorinated polyether	0.4	6	160	60–160	5	—
9. Chlorotrifluoroethylene	3.5	6	150–190	60–190	8–10	6–12
10. Ethyl cellulose	1.7–6	2.3–6.5	—	10–40	3–6.7	—
11. Ethyl vinyl acetate	No break	20–40	3.0–15	500–1500	—	—
12. FEP	No break	2–3.2	60–80	350	—	—
13. Nylon 6	0.9–4	9.5–12.4	200–450	25–300	9–16.6	4–11
14. Nylon 6/6	0.9–2	11.2–13.1	410–480	60–300	14.6	5–13
15. Nylon 6/10	—	7–8.5	160–280	50–300	10.5	4–6
16. Polyallomer	1.5–12	2.9–4.2	100–170	400–650	4–5	—
17. Polycarbonate	2–3	8–9	345	75	11–13.5	12.5
18. Polyethylene low density	No break	1–2.4	14–38	20–800	—	—
19. Polyethylene medium density	No break	1.7–2.8	50–80	80–600	—	—
20. Polyethylene high density	0.5–23	2.8–5	75–160	10–800	1–4	0.8–3.6
21. Polyethylene high molecular weight	>20	5.4	102	525	3.5	2.4
22. Polypropylene	0.5–15	2.3–5	150–250	10–700	4.5–6	6–8
23. Polystyrene	0.25–0.65	5–9	400–500	1–2.5	7–15	11.5–16
24. Polystyrene high impact	0.7–1.5	3.5–8	300–400	10–40	5.5–12.5	8–16
25. Polyurethane	No break	4.5–8	—	100–600	0.7–1	20
26. Poly(vinyl chloride) (flexible)	Varied	1–4	—	100–450	—	—
27. Poly(vinyl chloride) (rigid)	0.4–22	6–9	200–600	5–40	8–15	10–11
28. Poly(vinyl dichloride) (rigid)	0.8–6.3	7.5–8.8	348–450	65	14.2–17	—
29. Styrene acrylonitrile (SAN)	0.3–55	10–12	500	1–3.2	17	15–17.5
30. TFE fluorocarbon	No break	2–5	50–100	75–400	—	4–1.2
31. Ionomer	5.7–14	3.5–5.5	28–40	300–450	—	—
32. Phenoxy	1.5–12	8–9.5	350–410	50–100	12–14.5	—
33. Polyphenylene oxide	—	—	—	50–80	—	—
34. Polysulfone	1.3	10.2	360	50–100	15.4	15.4

(*Continues*)

testing organizations that set the conditions and specifications pertaining to electrical properties are the American Society for Testing and Materials (ASTM), Canadian Standards Association (CSA), Underwriters Laboratories (UL), International Electrotechnical Commission (IEC), International Organization for Standardization (ISO), and American National Standards Institute (ANSI).

Electromagnetic Compatibility

The use of electronics has shown large growth in a variety of kinds of computer equipment, such as for data processing, transportation and industrial controls, automation, and medical devices. As plastic housings become more widespread than traditional metal housings, the issue of electromagnetic compatibility (EMC) developed. EMC is the

Table 7-2   (*Continued*)

Resin Material	Compressive Modulus, psi × 10³	Heat Distortion Temperature, °F, 264 psi	Heat Resistance, Continuous °F	Coefficient Thermal Expansion, in./in.- °C × 10⁻⁵	Thermal Conductivity, cal/cm²-sec- °C-cm × 10⁻⁴	Volume Resistivity, ohm-cm
1. Acetal	410	255	185	8.1–8.5	1.6–5.5	$1–10^{13}$
2. ABS	200–450	185–230	160–235	5.7–10	5–8	$10^{15}–10^{17}$
3. Acrylic	350–430	167–198	130–200	5–8.5	4.4	$>10^{14}$
4. Acrylic high impact	250–360	169–190	140–195	6.5–10.5	4.0	$10^{16}–10^{17}$
5. Cellulose acetate	—	111–195	140–175	8–16	4–8	$10^{10}–10^{12}$
6. Cellulose acetate butyrate	—	113–202	140–175	11–17	4–8	$10^{10}–10^{12}$
7. Cellulose propionate	—	121–228	140–175	11–16	4–8	$10^{12}–10^{16}$
8. Chlorinated polyether	130	185–210	250–275	8	3.13	1.5×10^{16}
9. Chlorotrifluoroethylene	180	160–170	390	5–7	4–6	10^{18}
10. Ethyl cellulose	—	150–200	140–180	10–20	3.8–7	—
11. Ethyl vinyl acetate	—	—	120–170	10–20	8	—
12. FEP	70	124	400	8.3–10.5	5.9	$>10^{18}$
13. Nylon 6	347	150–175	250	8.3	5.9	$3.3^{13}–4.5^{13}$
14. Nylon 6/6	400	200	250	10	5.8	$10^{14}–10^{15}$
15. Nylon 6/10	—	145	220	10	5.5	4.5×10^{12}
16. Polyallomer	—	115–140	250	8–11	2–4	$>10^{16}$
17. Polycarbonate	345	270	250–270	7	4.6	2.1×10^{16}
18. Polyethylene low density	—	—	140–175	10–20	8	$10^{15}–10^{18}$
19. Polyethylene medium density	—	—	150–180	10–20	8	$10^{15}–10^{18}$
20. Polyethylene high density	50–110	110–125	180–250	10–20	8	$6 \times 10^{15}–10^{18}$
21. Polyethylene high molecular weight	110	120	250	13	8	$>10^{16}$
22. Polypropylene	—	140–200	250	6–8.5	2.8–4	6.5×10^{16}
23. Polystyrene	300–560	167–203	150–180	6–8	1.9–3.3	$10^{17}–10^{21}$
24. Polystyrene high impact	—	150–200	120–170	6.5–8.5	1–3	$10^{13}–10^{17}$
25. Polyurethane	—	—	150–180	10–20	5	$2–10^{11}$
26. Poly(vinyl chloride) (flexible)	—	—	—	7–25	3–4	$10^{11}–10^{15}$
27. Poly(vinyl chloride) (rigid)	300–400	140–175	160–165	5–10	3–5	$10^{12}–10^{16}$
28. Poly(vinyl dichloride) (rigid)	—	212–220	185–210	7–8	3–4	10^{15}
29. Styrene acrylonitrile (SAN)	650	200–208	—	7	3	10^{15}
30. TFE fluorocarbon	70–90	132	550	5.5 (25–60°C)	6	$>10^{18}$
31. Ionomer	—	—	140	12–13	5.8	$>10^{16}$
32. Phenoxy	325	175–188	—	3.2–3.8	—	$2.75–5 \times 10^{-3}$
33. Polyphenylene oxide	—	375	—	—	—	10^{17}
34. Polysulfone	370	345	300	$3.1–10^{-5}$ in./in.°F	1.8 BTU/h-sq ft-ft lb	5×10^{16}

(*Continued*)

ability of an electrical device to function normally without interference from or interfering with another electrical device. EMC regulations usually emphasize the containment of electromagnetic interference (EMI) to specific levels across the designed frequency ranges.

The nonconductive characteristics of plastics can become a major drawback in certain applications. Because they are electrical insulators, they do not shield electronic impulses generated by outside sources. Nor do they prevent electromagnetic energy from being emitted from equipment housed in a plastic

Table 7-2   (*Continued*)

Resin Material	Dielectric Constant, 60 Cycles	Dielectric Strength, ST $\frac{1}{8}$-in. Thickness, volts/mil	Power Factor, 60 Cycles	Arc Resistance, sec	Water Absorption, 24 hr, %	Rockwell Hardness
1. Acetal	3.7–3.8	500	.004–.005	129	0.12–0.25	M94, R120
2. ABS	2.6–3.5	300–450	.003–.007	45–90	0.2–0.4	R50–120
3. Acrylic	3.7	450–500	.04–.05	No tracking	0.3	M84–97
4. Acrylic high impact	3.5–3.7	450–480	.04–.05	No tracking	0.2–0.3	M20–67
5. Cellulose acetate	3.5–7.5	290–600	.01–.06	50–130	2.1–4.2	R35–118
6. Cellulose acetate butyrate	3.5–6.4	250–400	.01–.04	—	0.9–2.2	R31–116
7. Cellulose propionate	3.4–4.2	300–450	.01–.04	170–190	1.2–2.8	R15–120
8. Chlorinated polyether	3	400	.01	—	.01	R100
9. Chlorotrifluoroethylene	2.65	450	.015	>360	Nil	R85–112
10. Ethyl cellulose	—	—	—	—	1.3–1.5	R50–110
11. Ethyl vinyl acetate	2.3	—	—	—	<.01	R3–7
12. FEP	2.1	500	.0002	>165	<.01	—
13. Nylon 6	6.1	300–400	0.4–0.6	140	1.5	R107–119
14. Nylon 6/6	3.6–4.0	300–400	.014	140	1.3	R118–123
15. Nylon 6/10	4.0–7.6	300–400	.04–.05	140	0.4	R111
16. Polyallomer	3.2	500–1000	.0001–.0005	—	<.01	R50–85
17. Polycarbonate	3.17	400	.0009	120	15	M80, R118
18. Polyethylene low density	2.28	450–1000	.0001–.0005	Melts	<.02	R10
19. Polyethylene medium density	2.3	450–1000	.0001–.0005	Melts	<.02	R15
20. Polyethylene high density	2.3	450–1000	.0001–.0005	Melts	<.01	R30–60
21. Polyethylene high molecular weight	2.3	710	.0003	—	<.01	R55
22. Polypropylene	2.1–2.27	450–1000	.0001–.0005	13–185	<.01	R30–99
23. Polystyrene	2.5–2.65	500–700	.0001–.0005	60–100	.03–.05	M60–80
24. Polystyrene high impact	2.5–3.5	500	.003–.005	60–90	.05–.10	M25–69
25. Polyurethane	6.7–7.5	450–550	.015–.017	—	0.60–0.80	M28, R60
26. Poly(vinyl chloride) (flexible)	5–9	300–1000	.08–.15	—	0.15–0.75	—
27. Poly(vinyl chloride) (rigid)	3.4	425–1040	.01–.02	—	.07–.40	R100–120
28. Poly(vinyl dichloride) (rigid)	—	1200	—	—	0.11	R118
29. Styrene acrylonitrile (SAN)	2.8–3	400–500	—	—	0.23–0.28	M30–83
30. TFE fluorocarbon	2.1	400	<.0001	No tracking	.01	R58
31. Ionomer	2.4–2.5	1000	0.1	—	0.1–1.4	D60–65
32. Phenoxy	4.1	404–520	.0012–.0009	70	0.13	R113–118
33. Polyphenylene oxide	—	400–500	—	—	—	—
34. Polysulfone	2.82	425	.0008–.0056	122	0.22	M69, R120

(*Continues*)

enclosure. Government regulations have been set up requiring shielding when the operating frequencies are greater than 10 kHz.

Every electronic system has some level of electromagnetic radiation associated with it. If this level is strong enough to cause other equipment to malfunction, the radiating device will be considered a noise source and usually be subjected to shielding regula-tions. This is especially true when EMI occurs within the normal frequencies of communication. When the electronic noise is sufficient to cause malfunctioning in equipment such as medical devices, flight instrumentation, etc. the results could prove life threatening. Reducing the emission of and susceptibility to EMI or radio frequency interference (RFI) to safe levels is thus the prime reason to shield

Table 7-2   (*Continued*)

Resin Material	Flammability, in./min	Specific Gravity	Mold Shrinkage, in./in.	Clarity
1. Acetal	1.1	1.410–1.425	.022	Translucent to opaque
2. ABS	1.0–2	1.01–1.07	.003–.007	Opaque
3. Acrylic	0.6–0.7	1.18–1.18	.002–.006	Transparent
4. Acrylic high impact	1.1–1.2	1.11–1.18	.004–.008	Translucent to opaque
5. Cellulose acetate	0.5–2	1.23–1.34	.001–.007	Transparent
6. Cellulose acetate butyrate	0.5–1.5	1.15–1.22	.003–.006	Transparent
7. Cellulose propionate	0.5–1.5	1.16–1.23	.001–.006	Transparent
8. Chlorinated polyether	Self-ext.	1.4	.004–.006	Semitranslucent to opaque
9. Chlorotrifluoroethylene	Nil	2.90–2.14	.010–.015	Transparent to opaque
10. Ethyl cellulose	—	1.11–1.13	—	Transparent to opaque
11. Ethyl vinyl acetate	Slow burning	0.93–0.95	.01–.02	Transparent
12. FEP	Nonflam.	2.14	.03	Transparent to opaque
13. Nylon 6	Self-ext.	1.13–1.14	.007–.011	Transparent to opaque
14. Nylon 6/6	Self-ext.	1.13–1.15	.007–.015	Transparent to opaque
15. Nylon 6/10	Self-ext.	1.07–1.09	.015	—
16. Polyallomer	Slow burning	0.90–.906	.01–.02	Transparent to opaque
17. Polycarbonate	Self-ext.	1.2	.005–.007	Transparent
18. Polyethylene low density	Slow burning	.910–.925	.01–.03	Transparent to opaque
19. Polyethylene medium density	Slow burning	.926–.940	.01–.035	Transparent to opaque
20. Polyethylene high density	Slow burning	0.94–0.98	.01–.04	Translucent to opaque
21. Polyethylene high molecular weight	Very slow	0.94	.03	Translucent to opaque
22. Polypropylene	Slow burning	0.90–.0908	.008–.025	Translucent to opaque
23. Polystyrene	0.5–2.5	1.05–1.06	.002–.006	Transparent
24. Polystyrene high impact	0.5–2.5	1.04–1.06	.003–.005	Translucent to opaque
25. Polyurethane	Slow burning	1.20–1.26	.008–.012	Translucent to opaque
26. Poly(vinyl chloride) (flexible)	Self-ext.	1.15–1.80	.002–.004	Transparent to opaque
27. Poly(vinyl chloride) (rigid)	Self-ext.	1.33–1.58	—	Transparent to opaque
28. Poly(vinyl dichloride) (rigid)	Self-ext.	1.50–1.54	.007–.008	Translucent to opaque
29. Styrene acrylonitrile (SAN)	0.47–0.7	1.07–1.08	.003–.004	Transparent
30. TFE fluorocarbon	Nonflam.	2.13–2.18	.02–.06	Transparent to opaque
31. Ionomer	9–1.1	0.94–0.96	.001–.005	Transparent
32. Phenoxy	Slow burning, Self-ext.	1.17–1.34	.003–.004	Transparent to opaque
33. Polyphenylene oxide	Self-ext.	1.06	—	—
34. Polysulfone	Self-ext.	1.24–1.25	.0076	Transparent to opaque

medical devices (and other devices) in whatever type of housing exist, including plastic.

The usual plastics alone lack sufficient conductivity to shield EMI and RFI interference. Designers can reduce or eliminate sufficiently electromagnetic emissions from plastic housings like those of medical devices and computers just by shielding the inner emission sources with metal shrouds in the so-called tin can method. They may reach the same effect by designing electronics to keep emissions below standard limits or by incorporating shielding into the plastic housing itself. Designers will often employ all three strategies in a single design. What is most important is to attempt to locate all the shielding in a relatively small volume within the larger housing and then tin can it to provide a simplified solution rather than spreading it out.

Among the shieldings incorporated into housings, the most popular and useful applied technologies are those for conductive coatings, zinc-arc spray, or electroless plating. Other methods include the use of conductive foils or molded conductive plastics, silver reduction, vacuum metalization, and cathode sputtering. Although zinc-arc spraying once accounted for about half the market, conductive coatings surpassed it and now maintains the largest single market share. The properties of various coatings are

Table 7-3   Example of properties for thermoset plastics

Resin Material	Impact Strength, izod ft-lb/in.	Tensile Strength, psi × 10³	Flexural Strength, psi × 10³	Flexural Modulus, psi × 10³	Compressive Strength, psi × 10³	Heat Distortion °F@264 psi
1. Alkyd glass-filled general purpose	1.5–4	5–9	12–18	1700–2000	21–29	>400
2. Alkyd-glass MAI 30	3–4	7–9	14–19	2000	22–24	>400
3. Alkyd-glass MAI 60	8–10	5–9	12–20	2000–2500	24–36	>400
4. Alkyd mineral	.30–.35	3–8	6–15	2200–3000	16–20	350–400
5. DAP glass MIL–M–19833 GDI–30	6–15	7–9.5	17–19	1250–1600	24–45	390–500
6. DAP mineral (MDG) MIL–M–14F MIL–P–4389	.28–.40	5.5–6.5	9.6–10	1200	22–25	300–320
7. DAP orlon MIL–F–14F SD 1–5	.55–4.5	6.8	9–10.5	710	20–30	240–266
8. DAP glass MIL–M–14F SDG	0.4–1.2	6.7–9.2	10–19	1300	25–30	350–500
9. DAP unfilled	0.3	4	7–9	600	22–24	310
10. Epoxy glass	5.5	11.5	22	3000–3200	30–38	500
11. Epoxy mineral	0.35–0.5	6–11	19.5	900–1200	37	260–360
12. Melamine cellulose	.25–.35	5–10	10–16	1300–1600	25–24	350–410
13. Melamine cloth	.55–.90	7–10	12–15	1600	30–35	310
14. Melamine glass MMI 30	4–16	6–10	15–24	2400	20–32	400
15. Phenolic asbestos MFA 30	3.7	8	16	2500	24	400
16. Phenolic cloth CFI–10	1.05–2.2	6.5–7	0.5–10	1000–1200	20–25	300–360
17. Phenolic cloth CFI–20	2.3	6.5–7.5	10–12.5	1000	22–25	300
18. Phenolic flock/flour	.34–.50	6–9	7–12	1000	20–30	270–325
19. Phenolic glass	9–17	7–11	13.5–22	3000	14–35	600
20. Phenolic GP	.26–.38	6–9	7–12	1000	25–35	275–370
21. Phenolic GPI–100	13	16.8	24.9	2865	39	>550
22. Phenolic MFE	0.5	9.2	17.5	2600	32	415
23. Phenolic mineral MFG	0.64–1.5	4–7	8.5–11.5	1400	17.5–20	320–500
24. Phenolic mineral MFH	0.26	5–8	6–9	1500	25–35	300–450
25. Phenolic mineral MFI–20	5	8.7	18.6	2500	34.5	>500
26. Phenolic nylon	0.5	8	12	600	24	290
27. Phenolic rubber/flour	0.5	4	5.5–7	400	16–18	250
28. Urea cellulose	.24–.35	5–10	10–18	1300–1600	25–38	266–380
29. Silicone	8–20	4–8	13–19	2100–2500	10–13	>900
30. Silicone mineral	.25–.39	2.5–4.4	6.8–8	1250–2270	11–18	340–900

(*Continues*)

described in Fig. 7-12 and Table 7-10. Other conductive coatings are also used. Unlike other shielding methods, conductive coatings are usually applied to the interiors of housings and do not require additional design efforts to achieve external aesthetic goals. All offer trade-offs in shielding performance, the physical properties of the plastics, ease in production, and cost.

Often, differences in test measurements and samples' configurations make comparisons difficult. The ASTM has a standard that defines the methods for stabilizing materials measurement, thus allowing relative measurements to be repeated in any laboratory. These procedures permit relative performance ranking, so that comparisons of materials can also be made. Nonetheless, the designer will still have to confirm the suitability of a material's shielding performance for each system through such conventional means as screen-room or open-field testing. Each approach to shielding should also be subjected to simulated environmental conditions, to determine the shield's behavior during storage, shipment, and exposure to humidity, which could accelerate the effects of aging of shielding materials. In this

Table 7-3   (*Continued*)

Resin Material	Heat Resistance, Continuous °F	Thermal Expansion in./°C × 10⁻⁵	Thermal Conductivity, cal/sec-cm²-°C-cm × 10⁻⁴	Volume Resistivity, ohm/cm	Dielectric Constant, 60 Cycles	Dielectric Strength, STI/8V.P.M.
1. Alkyd glass-filled general purpose	300	2–5	10–15	1×10^{15}	6.7	275–375
2. Alkyd-glass MAI 30	350	2–4	10–15	—	6	350
3. Alkyd-glass MAI 60	300	2–5	8–12	10^{12}–10^{14}	5–5.6	375
4. Alkyd mineral	275–400	2–5	15–25	10^{12}–10^{14}	5.7–6.3	375
5. DAP glass MIL–M–19833 GDI–30	365–465	2.7–3.5	6–8	10^{12}	4.5	350–400
6. DAP mineral (MDG) MIL–M–14F MIL–P–4389	350–440	3.5–4.2	13.7	$10^{13}+$	5.2	350–450
7. DAP orlon MIL–F–14F SD 1–5	300–350	4-5.4	6.3–6.9	1×10^{10}	3.8	366–400
8. DAP glass MIL–M–14F SDG	365–465	1.2–3.5	6.7	10^{13}	4.25	420
9. DAP unfilled	350	—	—	2×10^{16}	3.6	450
10. Epoxy glass	300–500	1.08–1.2	8.5	3.8×10^{15}	5.5(@ 100)	360 volts/mil
11. Epoxy mineral	550	3.7	18.1	10^{14}–10^{16}	4.4	400
12. Melamine cellulose	210	2–5.7	7–10.1	0.8–2.0×10^{12}	7.9–9.5	300–400
13. Melamine cloth	250	2.5–3	10.6	1–3×10^{11}	8.1–12.6	250–340
14. Melamine glass MMI 30	300–400	1.2–2	11.5	2×10^{11}	9.7–11.1	170–240
15. Phenolic asbestos MFA 30	450	.5–1	14	—	—	85
16. Phenolic cloth CFI–10	275	2–3	9.3	5–8×10^{11}	6.1-21.2	250–400
17. Phenolic cloth CFI–20	340	1.5–3	7	6×10^{10}	5.2–21	300
18. Phenolic flock/flour	300	2–3.9	7–8	2×10^{11}	6–10	300–375
19. Phenolic glass	325–450	.8–1.6	9.7	1×10^{12}	5.6–7.2	350–400
20. Phenolic GP	300	3–4.1	5–8.1	2×10^{11}	5–10	300–425
21. Phenolic GPI–100	—	—	32	—	5.4(1MC)	382
22. Phenolic MFE	—	1.1	15	—	4.5(1MC)	360
23. Phenolic mineral MFG	325–425	1.5–2.5	15.87	1×10^{12}	40–60	150–250
24. Phenolic mineral MFH	400–450	1.5	10	1×10^{12}	9-15	250–350
25. Phenolic mineral MFI–20	475	1.5	14	1.6×10^{10}	45	85
26. Phenolic nylon	275	7.5	7.50–18	1×10^{12}	4	450
27. Phenolic rubber/flour	275	1.5–4	5	3.4×10^{9}	9	300
28. Urea cellulose	170	2.7	7–10.1	$.5$–5×10^{11}	7–9.5	300–400
29. Silicone	450–700	.8	7.5	3×10^{14}	4.35	250–280
30. Silicone mineral	400–700+	2.5–6	4–10	1×10^{4}	3.6–4.5	280–400

(*Continued*)

Table 7-3   (*Continued*)

Resin Material	Power Factor, 60 cycles	Arc Resistance, sec	Water Absorption %-24 hr	Rockwell Hardness	Specific Gravity	Mold Shrinkage, in./in.
1. Alkyd glass-filled general purpose	.01–.02	180+	.07–.2	—	1.93–2.32	.003–.006
2. Alkyd-glass MAI 30	.005–.01	180+	.05	—	2.3	.001–.005
3. Alkyd-glass MAI 60	.020–.030	180+	.07–.10	—	2.02–2.13	.001–.004
4. Alkyd mineral	.030–.045	180+	.05–.12	E108	2.17–2.24	.004–.007
5. DAP glass MIL–M–19833 GDI–30	.010–.017	130–180	.05–.25	M108	1.57–186	.0009–.004
6. DAP mineral (MDG) MIL–M–14F MIL–P–4389	.06–.40	140	.2–.5	M100	1.58–1.74	.004–.008
7. DAP orlon MIL–F–14F SD 1–5	.025–.035	115	.1–.2	M108	1.31–1.34	.009–.010
8. DAP glass MIL–M–14F SDG	.010–.017	125–180	.05	E80	1.6–1.75	.002–.004
9. DAP unfilled	.008	120	.09	M114–116	1.27	.007
10. Epoxy glass	.015	125–187	.03–.05	M105–115	1.8	.0005
11. Epoxy mineral	.011	180–190	.06–.15	M110	1.85	.002
12. Melamine cellulose	.030–.080	120–140	.1–.6	M118–124	1.47–1.52	.006–.015
13. Melamine cloth	.1–.34	125–150	.3–.6	—	1.4–1.5	.003–.005
14. Melamine glass MMI 30	.14–.23	180–186	.09–.6	—	1.9–2	.001–.003
15. Phenolic asbestos MFA 30	—	190	.35	M112	1.95	.0005–.0015
16. Phenolic cloth CFI–10	.16–.64	—	.8–1	M95–107	1.36–1.40	.002–.005
17. Phenolic cloth CFI–20	.64	4	.9–1	E79	1.37–1.40	.002–.005
18. Phenolic flock/flour	.25–.3	80(15518)	.3–1	M95–116	1.34–1.38	.004–.008
19. Phenolic glass	.02–.05	10–60	.5–1	M90–99	1.70–1.90	.0009
20. Phenolic GP	.05–.5	26–102	.2–1	M95–117	1.34–1.46	.004–.008
21. Phenolic GPI–100	.026(1MC)	125	.03	—	1.69	.0001–.001
22. Phenolic MFE	.013(1MC)	180+	.03	—	1.85	.002
23. Phenolic mineral MFG	.25–.50	100–190	.12–.4	M90–115	1.72–1.86	.001–.002
24. Phenolic mineral MFH	.20	100–180	.2	M100	1.6–2	.002–.005
25. Phenolic mineral MFI–20	.28	165	.04	M105	1.76	.0001–.001
26. Phenolic nylon	.02	80	.20	M108	1.22	.012–.016
27. Phenolic rubber/flour	.14	8–10	1.4–2	M50	1.29–1.32	.007–.010
28. Urea cellulose	.035–.040	80–130	.5–.7	M116–120	1.47–1.52	.006–.14
29. Silicone	.003–.02	175–240	.10–.30	M87	1.88	.0005
30. Silicone mineral	.002–.01	220–240+	.05–.22	M75–90	1.85–2.82	.005–.009

way degradation can be observed, along with other problems that might occur in a product's service life.

The Underwriters Laboratories utilizes a combination of methods for environmental conditioning and adhesion testing to evaluate various approaches to shielding and to determine the plastic types that are suitable for use in electronic devices. Their concern is primarily safety should a metalized plastic delaminate or chip off, creating an electrical short that could cause a fire.

Table 7-4   Electrical property guide

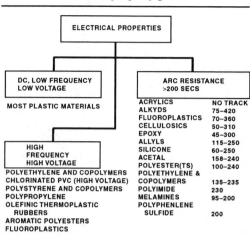

Table 7-6   Examples of conductivities using different additives/fillers

Fillers	Conductivity σ S/cm
Carbon black	0.01 to 0.1
Aluminum platelets	1 to 50
Steel fibers	1 to 50
Carbon fibers	0.1 to 10
Mica coated with nickel	1 to 10

and telephonic installations requiring insulation. Phenolic plastic compounds initially became the basic materials for insulating wires and other related products. Since then many new plastics, predominantly TPs, have been developed and used for widely variant applications.

There are many thousands of outstanding applications where plastics are used in electrical designs. The designers' imaginations have excelled in developing new plastic products. An example is the folded membrane and snap switches in controlling electronic devices.

Membrane technology is the technique of producing flat, thin, lightweight switch arrays by joining two or more membranes. Each membrane features etched or screened conductors placed face to face with a thin material to separate the active elements. A second, surface-printed overlay is attached to the top of the switch assembly to provide a graphic indication of the switch's location. Their advantages are low cost in high volumes, their moderate tooling charges, and their capability of providing attractive, bright, durable frontal graphics.

To maximize results with any product, the designer should reduce the circuit-noise generation and susceptibility of the product to as much as possible. Consider the choice of shielding early on in the design process, before deciding on final packaging to minimize the amount of external shielding required. Doing so will also alleviate last-minute shielding fixes and, of course, a good deal of exposure and delay in marketing the product.

Design Concept

Many ideas for advancing electrical and electronic systems have been adopted since the early 1940s, which saw the start of high electronic frequency radar systems. The earliest major use of plastics for electrical insulation early in last century come with the advent of developments in electrical

Table 7-5   Electrical conductivities of different materials

Material	Conductivity σ S/cm	Density o g/cm^3	σ/o S/(cm^2 g)
Copper	5.9×10^5	8.9	6.6×10^4
Silver	6.3×10^5	10.4	6.0×10^4
Aluminium	3.6×10^5	2.7	1.3×10^5
Polyacetylene with iodine iodine	2.0×10^4	0.8	2.5×10^4
Polypyrrole with phenylsulfonate phenylsulfonate	1.5×10^2	1.3	1.2×10^2
Polystyrene	10^{-16}	1.05	9.5×10^{-17}

Table 7-7   Applications for conductive plastics where different additives and fillers are used

Material	Available Forms	Characteristics	Applications
Thermoplastics			
Acrylics	Powder, solutions	Easily processed. Form strong films. Temperature range 200–250 F. Adaptable for paint or spray.	Coatings.
Fluorocarbons (PTFE, FEP, PVF$_2$)	Powder, emulsions	Excellent high temperature properties. TFE to 500 F. FEP is easier to mold, but maximum use temperature is 400 F. Nearly inert chemically. Nonflammable. Loading with conductive filler improves creep resistance. Low coefficient of friction.	High-temperature cable shielding, gaskets, heat-shrinkable tubing.
Polyimides	Powder, solutions	Excellent high temperature properties, 400 to about 700 F. Difficult to process.	High-temperature cable shielding, conductive film.
Polyolefins (Polyethylene, Polypropylene)	Powder, pellets	Tough and chemical resistant. Weak in creep and thermal resistance. Polyethylene maximum use temperature 210 F, polypropylene 260 F. May be injection and extrusion molded, vacuum formed. Low cost.	Antistatic sheet and tiles, heat-shrinkable tubing, deicer boots.
Vinyls (PVC, PVA, PVAC, Copolymers)	Powder, pellets, organosols	General-purpose material. Many forms available including hard and flexible types. Properties are highly dependent on plasticizer used. May be injection, extrusion, compression molded, vacuum formed. Low cost.	Cable shielding, antistatic sheet and hose, RF gaskets, heat-shrinkable tubing.
Thermosets			
Diallyl Phthalate (DAP, DAIP)	Powder	Excellent humidity resistance. Withstands temperatures over 400 F. Easily processed. Dimensional stability excellent. May be compression, transfer and injection molded. Low cost.	Precision potentiometers, RF connectors, waveguide auxiliaries, attenuators, heating panels, heated battery cases.
Epoxies	Powder, one and two-part liquids and paste	Many types of resins available, providing wide spectrum of properties. Easy to compound. Low shrinkage and excellent dimensional stability. Good to excellent adhesion. May be cast or molded.	Coatings, sealants, adhesives, solderless PC boards.

(Continues)

Table 7-7   (*Continued*)

Material	Available Forms	Characteristics	Applications
Phenolics	Powder, solutions	Excellent thermal stability to over 300 F generally, and over 400 F in special formulations. Broad choice of resins. May be cast or compression, transfer, or injection molded.	Precision potentiometers, RF connectors, heating panels.
Elastomers			
Natural rubber	Solid	Good physical properties and resistance to cutting and abrasion. Low heat and ozone resistance.	Gaskets.
Polyisobutylene	Liquid	Good resistance to ozone and abrasion.	Calking (nonsetting), hose for transfer of flammables, pipe dope, electric fuse.
Silicone	Paste, liquid for foaming	Excellent chemical resistance. High temperature capability to 500 F. New formulations have higher tear strength and lower compression set.	Gaskets, antistatic rollers, RF shielding, heat shrinkable tubing, setting and nonsetting calking, deicer boots, flexible heater tape.
Urethane	Liquid	Exceptional abrasion, cut, and tear resistance. Poor moisture and heat resistance. Variety of formulations leading to different properties including range of durometers without plasticizers.	Antistatic rollers and tires, hose for transfer of flammables, strain gages, pressure transducers.

Thermal Property

In order to select materials that will maintain acceptable mechanical characteristics and dimensional stability designers must be aware of both the normal and extreme thermal operating environments to which a product will be subjected. TS plastics have specific thermal conditions when compared to TPs that have various factors to consider

Table 7-8   Resistivity and dielectric properties

Material	Resistivity		Dielectric Constant/Dissipation Factor					
	Volume	Surface	100 Hz	1 kHz	1 mHz	10 mHz	100 mHz	1.000 mHz
ABS	2×10^{16}	10^{14}	.005/2.9	.006/2.8	.008/2.8	.007/2.8	.005/2.7	.001/2.7
Acrylic	10^{18}	10^{14}	.062/3.6	.058/3.2	.045/3.1	.033/2.9		
Cellulose Ester	3×10^{15}	10^{14}	.006/3.8	.011/3.6	.024/3.3	.022/3.2	.020/3.0	.014/2.1
FEP	10^{18}	10^{16}	.0005/2.1	.0005/2.1	.0005/2.1	.0005/2.1	.0008/2.09	.0007/2.05
Nylon 6	10^{15}	10^{13}	.031/4.2	.024/3.8	.031/3.8	.020/4.0		
Polycarbonate	10^{16}	10^{15}	.001/3.1	.0013/3.1	.007/3.1	.011/3.1	.015/3.1	
Polyethylene	10^{19}	10^{16}	.0001/2.34	.0001/2.34	.0001/2.34	.0001/2.34	.0001/2.34	.0001/2.34
Alkyd	10^{13}	10^{14}	.02/6.0	.02/5.8	.015/5.4			
DAP (SD15)	10^{16}	10^{13}	.026/3.8	.020/3.7	.016/3.6			
Phenolic MFE	10^{14}	10^{9}	.013/5.4	.013/5.3	.033/4.9			
Epoxy	10^{16}	10^{14}	.004/3.22	.004/3.25	.004/3.25			

Table 7-9 Examples of electrical, physical, and mechanical properties

	ASTM	PEI[†]	PES[†]	PPS[†]	PSF[†]	PC[†]
Physical						
Specific gravity, g/c^3		1.27	1.37	1.30	1.55	1.20
Water absorption, % by weight	D 570	0.25	0.43	0.02	0.3	0.26
Electrical						
Dielectric strength, v/mil	D 149	710	400	380	425	425
Arc resistance, sec	D 495	128	120	34	39	115
Volume resistivity, ohm/cm	D 257	10^{18}	10^{16}	10^{15}	10^{17}	10^{17}
Thermal						
HDT at 264 psi. °F	D 648	392	397	275	345	265
Long-term service temp, UL index, °F		338	356	—*	302	255
Oxygen index, %		40	38	47	38	25
Mechanical						
Flexural modulus, psi (MPa)	D 780	480,000 (3,310)	375,000 (2,586)	550,000 (3,792)	390,000 (2,689)	340,000 (2,344)
Impact strength, notched Izod, ft./lb./in.	D 256	1.0	1.6	0.4	1.3	2.2
Tensile strength, psi (MPa)	D 638	15,200 (104.8)	12,200 (84.12)	9,500 (65.5)	10,200 (70.33)	9,500 (65.5)

*Not UL listed.
[†]*Note:* Glass filler can considerably extend the performance of the above polymers. PEI = polyetherimide; PES = polyether sulfone; PPS = polyphenylene sulfide; PSF = polysulfone; PC = polycarbonate.

that influence the product's performances and processing capabilities. TPs' properties and processes are influenced by their thermal characteristics such as melt temperature (T_m), glass-transition temperature (T_g), dimensional stability, thermal conductivity, thermal diffusivity, heat capacity, coefficient of thermal expansion, and decomposition (T_d). Table 7-11 provides some of these data on different plastics.

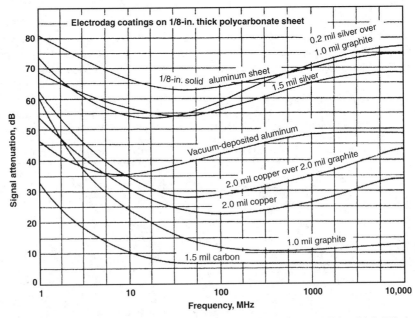

Fig. 7-12 Comparing shielding effect of conductive coatings on a 1/8 in. thick PC sheets.

Table 7-10   Conductive coating systems that provide EMI/RFI shielding on plastics

Shielding System	Advantages	Disadvantages
Conductive Coatings		
Silver	Highly conductive (0.1 ohm per square foot or less); applied by conventional spray equipment; easy application; electrically stable (minimal change in resistance with environmental cycling); easily applied to selected area; field repairable.	High cost
Nickel	Low cost (15–30 cents per square foot); good conductivity (less than 1.0 ohm per square foot); applied by conventional spray equipment; easy application; relatively stable (differs with manufacturer); easily applied to selected area; field repairable.	Lesser quality formulations available; some are stable, some are not.
Copper	Highly conductive (less than 0.5 ohm per square foot); easy application; low cost (15–30 cents per square foot).	Oxidation can reduce conductivity (resistance can change to effectively make copper an insulator); some may be alloys—if layered with silver, cost will rise.
Graphite	Low cost (5–15 cents per square foot); easy application; excellent ESD (electrostatic discharge) performance.	Less conductivity (ranging from 2 ohms to the thousands per square foot, depending upon the amount of graphite); modest shielding capability (up to 30–40 dB).
Arc/Flame Spray	Highly conductive (less than 0.1 ohm per square foot; hard, dense coating.	Requires grit blasting to promote mechanical bonding to plastic; special applications equipment required; requires special applicator safety procedures for dust and fumes; warps thermoplastics; not suitable for thin-walled designs; not field repairable.
Vacuum Metalization/ Ion Plating	Highly conductive (less than 0.1 ohm per square foot); controllable film thickness; not limited to simple housing designs.	Requires primer coat; entire part must be done, forcing exterior painting; not field repairable; specialized application equipment; vacuum chamber size a limiting factor; requires specialized knowledge; subject to corrosion in humid atmosphere unless protected.
Electrolysis Deposition	Highly conductive (both nickel and copper less than 0.1 ohm per square foot).	Requires specialized equipment/ knowledge; entire part must be coated, forcing exterior painting; if copper is used it must be protected by a nickel coating or some other coating.
Conductive Plastics	Good thermal transfer; elimination of secondary operation for shielding.	Requires a secondary operation for grounding.

Table 7-11 Examples of thermal properties of TPs (properties of common materials included for comparison)

Plastics (morphology)	Density g/cm³ (lb/ft³)	Melt Temperature T_m, °C (°F)	Glass Transition Temperature T_g °C (°F)	Thermal Conductivity (10^{-4} cal/s · cm °C) (BTU/lb. °F)	Heat Capacity cal/g °C (BTU/lb. °F)	Thermal Diffusivity 10^{-4} cm²/s (10^{-3} ft.²/hr.)	Thermal Expansion 10^{-6} cm/cm °C (10^{-6} in./in. °F)
PP (C)	0.9 (56)	168 (334)	5 (41)	2.8 (0.068)	0.9 (0.004)	3.5 (1.36)	81 (45)
HDPE (C)	0.96 (60)	134 (273)	−110 (−166)	12 (0.290)	0.9 (0.004)	13.9 (5.4)	59 (33)
PTFE (C)	2.2 (137)	330 (626)	−115 (−175)	6 (0.145)	0.3 (0.001)	9.1 (3.53)	70 (39)
PA (C)	1.13 (71)	260 (500)	50 (122)	5.8 (0.140)	0.075 (0.003)	6.8 (2.64)	80 (44)
PET (C)	1.35 (84)	250 (490)	70 (158)	3.6 (0.087)	0.45 (0.002)	5.9 (2.29)	65 (36)
ABS (A)	1.05 (66)	105 (221)	102 (215)	3 (0.073)	0.5 (0.002)	3.8 (1.47)	60 (33)
PS (A)	1.05 (66)	100 (212)	90 (194)	3 (0.073)	0.5 (0.002)	5.7 (2.2)	50 (28)
PMMA (A)	1.20 (75)	95 (203)	100 (212)	6 (0.145)	0.56 (0.002)	8.9 (3.45)	50 (28)
PC (A)	1.20 (75)	266 (510)	150 (300)	4.7 (0.114)	0.5 (0.002)	7.8 (3.0)	68 (38)
PVC (A)	1.35 (84)	199 (390)	90 (194)	5 (0.121)	0.6 (0.002)	6.2 (2.4)	50 (128)
Aluminum	2.68 (167)	1,000		3000 (72.5)	0.23	4900 (1900)	19 (10.6)
Copper/bronze	8.8 (549)	1,800		4500 (109)	0.09	5700 (2200)	18 (10)
Steel	7.9 (493)	2,750		800 (21.3)	0.11	1000 (338)	11 (6.1)
Maple wood	0.45 (28.1)	400 (burns)		3 (0.073)	0.25	27 (10.5)	60 (33)
Zinc alloy	6.7 (418)	800		2500 (60.4)	0.10	3700 (1430)	27 (15)

* = Crystalline resin. A = Amorphous resin.

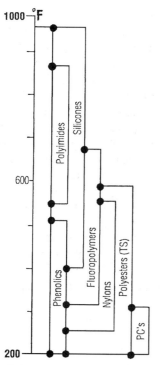

Fig. 7-13 Guide to classifying some of the plastics by range of continuous heat.

All these thermal properties relate to how to determine the best useful processing conditions to meet product performance requirements. There is a maximum temperature or, to be more precise, a maximum time-to-temperature relationship for all materials preceding loss of performance or decomposition. Figure 7-13 provides a temperature guide for continuous heating of plastics.

Residence Time and Recycling

See Chapter 8, **PROCESSING BE-HAVIOR**.

Melt Temperature

The T_m occurs at a relatively sharp point for crystalline materials. Amorphous materials basically do not have a T_m; they simply start melting as soon as the heat cycle begins. In reality there is no single melt point, but rather a range, which is often taken as the peak of a differential scanning calorime-

try (DSC) curve (see Appendix B, **TERMINOLOGY**).

The T_m is dependent on the processing pressure and the time under heat, particularly during a slow temperature change for relatively thick melts. Also, if the T_m is too low, the melt's viscosity will be high and more power will be required for processing. If the viscosity is too high, degradation will occur. There is the right processing window used for the different plastics being melted.

Glass-Transition Temperature

The glass-transition temperature (T_g) is the point below that a TP behaves as glass does; it is very strong and rigid, but brittle. Above this temperature it is neither as strong nor rigid as glass, but neither is it brittle. At T_g the plastic's volume or length starts to increases (Figs. 7-14 to 7-16). The amorphous TPs have a more definite T_g.

A plastic's thermal properties, particularly its T_g, influence its processability in many different ways. The selection of a plastic should take this behavior into account. The operating temperature of a TP is usually limited to below its T_g. A more expensive plastic could cost less to process because of its lower T_g that results in a shorter processing time, requiring less energy for a particular weight, etc.

The glass transition generally occurs over a relatively narrow temperature span and

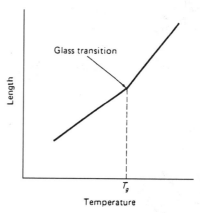

Fig. 7-14 Effect of T_g on the volume or length of TPs.

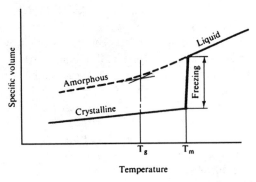

Fig. 7-15 Solidification during processing of glassy amorphous and crystalline TPs.

is similar to the solidification of a liquid to a glassy state; it is not a phased transition. Not only do hardness and brittleness undergo rapid changes in this temperature region, but other properties such as the coefficient of thermal expansion and specific heat change rapidly. This phenomenon has been called second-order transition, rubber transition, or rubbery transition. The word transformation has also been used instead of transition. When more than one amorphous transition occurs in a plastic, the one associated with segmental motions of the plastic backbone chain, or accompanied by the

largest change in properties, is usually considered to be the glass transition.

Designers should know that above T_g, the mechanical properties of TPs are reduced. Most noticeable is a reduction in stiffness by a factor that may be as high as 1,000.

The T_g can be determined readily only by observing the temperature at which a significant change takes place in a specific electric, mechanical, or physical property. Moreover, the observed temperature can vary significantly, depending on the specific property chosen for observation and on details of the experimental technique (for example, the rate of heating, or frequency). Therefore, the observed T_g should be considered to be only an estimate. The most reliable estimates are normally obtained from the loss peak observed in dynamic mechanical tests or from dilatometric data (ASTM D-20).

Mechanical Property and T_g

Figures 7-17 and 7-18 provides examples of modulus vs. T_g for amorphous and crystalline plastics. Temperature can help explain some of the differences observed in plastics. For example at room temperature polystyrene and acrylic are below their respective T_g values, we observe these materials in their glassy stage. In contrast, at room temperature natural rubber is above its T_g [$T_g = -75°C (-103°F); T_m = 30°C (86°F)$], with the result that it is very flexible. When it

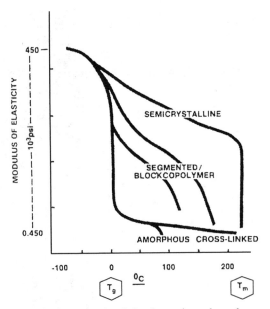

Fig. 7-16 Example of the dynamic and mechanical properties of TPs and TSs in relation to their T_g and T_m.

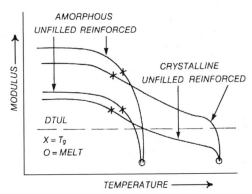

Fig. 7-17 Modulus behavior of amorphous and crystalline plastics showing T_g and melt temperatures.

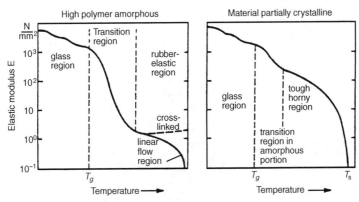

Fig. 7-18 Modulus vs. temperature dependence going through different processing stages.

is cooled below its T_g natural rubber becomes hard and brittle.

Dimensional Stability

Dimensional stability is an important thermal property for the majority of plastics. It is the temperature above which plastics lose their dimensional stability. For most plastics the main determinant of dimensional stability is their T_g. Only with highly crystalline plastics is T_g not a limitation.

Substantially crystalline plastics in the range between T_g and T_m are referred to as leathery, because they are made up of a combination of rubbery noncrystalline regions and stiff crystalline regions. The result is that such plastics as PE and PP are still useful at the higher temperatures.

Thermal Conductivity and Thermal Insulation

Thermal conductivity is the rate at which a material will conduct heat energy along its length or through its thickness. ASTM tests give an indication of how much heat must be added to a unit mass of plastic in order to raise its temperature 1°C. This is an important factor, since plastics are often used as effective heat insulation in heat-generating applications and in structures where heat dissipation is important. The high degree of the molecular order for crystalline TPs makes their values tend to be twice those of the amorphous types.

The conductivity of plastics is dependent on a number of variables and cannot be reported as a single factor. It depends mainly on temperature and molecular orientation. Its dependence can be ascertained. However, the molecular orientation may vary within a product, resulting in a variation in thermal conductivity. It is important for the designer to recognize such a situation.

For certain products, skill is required to estimate a product's performance under steady-state heat-flow conditions, especially those made of RPs (Fig. 7-19). The method and repeatability of the processing technique can have a significant effect. In general, thermal conductivity is low for plastics and the plastic's structure does not alter its value significantly. To increase it the usual approach is to add metallic fillers, glass fibers, or electrically insulating fillers such as alumina. Foaming can be used to decrease thermal conductivity.

Heat Capacity

The heat capacity or specific heat of a unit mass of material is the amount of energy required to raise its temperature 1°C. It can be measured either at constant pressure or constant volume. At constant pressure it can be larger than at constant volume, because additional energy is required to bring about a volume change against external pressure.

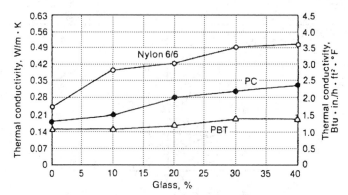

Fig. 7-19   Example of the effect on thermal conductivity by varying the glass fiber content (by weight) in RPs.

The specific heat of amorphous plastics increases with temperature in an approximately linear fashion below and above T_g, but a steplike change occurs near the T_g. No such stepping occurs with crystalline types.

For plastics, heat capacity is usually reported during constant pressure heating. Plastics differ from traditional engineering materials because their specific heat is temperature sensitive.

Thermal Diffusivity

Whereas heat capacity is a measure of energy, thermal diffusivity is a measure of the rate at which energy is transmitted through a given plastic. It relates directly to processability. In contrast, metals have values hundreds of times larger than those of plastics. Thermal diffusivity determines plastics' rate of change with time. Although this function depends on thermal conductivity, specific heat at constant pressure, and density, all of which vary with temperature, thermal diffusivity is relatively constant.

Coefficient of Linear Thermal Expansion

Like metals, plastics generally expand when heated and contract when cooled. Usually, for a given temperature change many TPs have a greater change than metals. The coefficient of linear thermal expansion (CLTE) is the ratio between the change of a

linear dimension to the original dimension of the material per unit change in temperature (per ASTM standards). It is generally given as cm/cm/°C or in./in./°F.

The CLTE is an important consideration if dissimilar materials like one plastic to another or a plastic to metal and so forth that are to be assembled where material expansion or contraction is restricted. The CLTE is influenced by the type of plastic (liquid crystal, for example) and RP (particularly the glass fiber content and its orientation). It is especially important if the temperature range includes a thermal transition such as T_g. Normally, all this activity with dimensional changes is available from material suppliers.

The design of products has to take into account the dimensional changes that can occur during fabrication and during its useful service life. With a mismatched CLTE could be destructive that includes factors such as cracking or buckling.

Expansion and contraction can be controlled in plastic by its orientation, crosslinking, adding fillers or reinforcements, and so on. With certain additives the CLTE value could be zero or near zero. For example, plastic with a graphite filler contracts rather than expands during a temperature rise. RPs with only glass fiber reinforcement can be used to match those of metal and other materials. In fact, TSs can be specifically compounded to have little or no change.

In a TS the ease or difficulty of thermal expansion is dictated for the most part by the degree of cross-linking as well as the overall stiffness of the units between the cross-links. The less flexible units are also more resistant to thermal expansion. Such influences as secondary bonds have much less effect on the thermal expansion of TSs. Any cross-linking of TPs has a substantial effect. With the amorphous type, expansion is reduced. In a crystalline TP, however, the decreased expansion as a result of cross-linking may be partially offset by a loss of crystallinity.

Thermal Stress

If a plastic product is free to expand and contract, its thermal expansion property will usually be of little significance. However, if it is attached to another material, one having a lower CLTE, then the movement of the part will be restricted. A temperature change will then result in developing thermal stresses in the part. The magnitude of these stresses will depend on the temperature change, the method of attachment and relative expansion, and the modulus characteristics of the two materials at the point of the exposed heat.

In its simplest form, thermal stresses can be calculated by using the following equation:

$$\sigma = E_p(\alpha_1 - \alpha_2)(\Delta T) \qquad (7\text{-}1)$$

where: σ = thermal stress, E_p = elastic modulus of elasticity, α_1 = CLTE of material #1, α_2 = CLTE of material #2, and ΔT = temperature change.

The goal is to eliminate or significantly reduce all sources of thermal stress. This can be achieved by keeping the following factors in mind: (1) when adding material for local reinforcement, select a material with the same or a similar CLTE, (2) where plastic is to be attached to a more-rigid material, use mechanical fasteners with slotted or oversized holes to permit expansion and contraction to occur, (3) do not fasten dissimilar materials tightly, and (4) adhesives that remain ductile, such as urethane and silicone, through the product's expected end-use temperature range can be used without causing stress cracking or other problems.

In addition to dimensional changes from changes in temperature, other types of dimensional instability are possible in plastics, as in other materials. Water-absorbing plastics, such as certain nylons, may expand and shrink as they gain or lose water, or even as the relative humidity changes. The migration or leaching of plasticizers, as in certain PVCs, can result in slight dimensional change.

Decomposition Temperature

For applications having only moderate thermal requirements, thermal decomposition may not be an important consideration. However, if the product requires dimensional stability at high temperatures, it is possible that its service temperature or processing temperature may approach its temperature of decomposition (T_d) (Table 7-12). A plastic's decomposition temperature is largely determined by the elements and their bonding within the molecular structures as well as the characteristics of additives, fillers, and reinforcements that may be in them.

Aging at Elevated Temperature

Aging at elevated temperatures typically involves exposing test specimens or products at different temperatures for different extended time periods. Tests are performed at room or the respective testing temperatures for whatever mechanical, physical, or electrical property is of interest. These tests of aging

Table 7-12 Examples of temperature decomposition

Material	°F	(°C)
PP	610–750	(321–399)
PC	645–825	(341–441)
PVC	390–570	(199–299)
PS	570–750	(299–399)
PMMA	355–535	(180–280)
ABS	480–750	(249–399)
PA	570–750	(299–399)
PET	535–610	(280–322)
Fluoropolymer	930–1020	(499–549)

Note: Adding certain fillers and reinforcements can raise decomposition temperatures.

can be used as a measure of thermal stability in design as is done with other materials.

Temperature Index

The Underwriters Laboratories (UL) tests are recognized by various industries to provide continuous temperature ratings, particularly in electrical applications. These ratings include separate listings for electrical properties, mechanical properties including impact, and mechanical properties without impact. The temperature index is important if the final product has to receive UL recognition or approval.

Intumescent Coating

These coatings bubble and foam to form a thermal insulation when subjected to a fire. They have been used for many decades. Such coatings cannot be differentiated from conventional coatings prior to the occurrence of a fire situation. Thereupon, however, they decompose to form a thick, nonflammable, multicellular, insulative barrier over the surface on which they are applied. This insulative foam is a very effective insulation that maintains the temperature of a flammable or heat distortable substrate below its ignition or distortion point. It also restricts the flow of air (oxygen) to fuel the substrate.

These coatings provide the most effective fire-resistant system available but originally were deficient in paint color properties. Since, historically, the intumescence producing chemicals were quite water-soluble, coatings based thereon did not meet the shipping can stability, ease of application, environmental resistance, or aesthetic appeal required of a good protective coating.

More recently there have been developed water-resistant phosphorus-based intumescence catalyst. This commercially available product, as an example Phos-Chek P/30 tradename from Monsanto, can be incorporated (with other water insoluble reagents) into water-resistant intumescent coatings of either the alkyd or latex-emulsion type. These intumescent coatings, formulated ac-

cording to the manufacturer's recommendations, are described as equivalent to conventional products in coating properties and also provide permanent fire resistance to the substrate on which they are applied.

Other

In addition to what has been presented, the commercially available literature provides all kinds of the important behavioral thermal/temperature properties that would be important to designers' for certain specific requirements (181). Examples are given in Figs. 5-9 and 7-20 to 7-23 and Tables 7-13 to 7-15.

Other Behavior

Drying Plastic

Of the various plastics available, such TPs as nylon, PC, PMMA, PUR, PET, and ABS are among those categorized as hygroscopic.

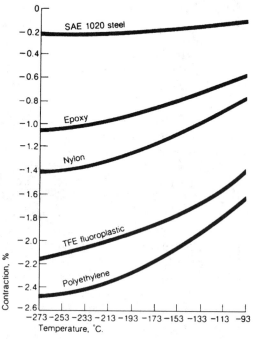

Fig. 7-20   Examples at low temperatures of thermal contraction in unfilled plastics and steel. With RPs using TS plastics change can be significantly reduced or even at zero (using graphite, etc.).

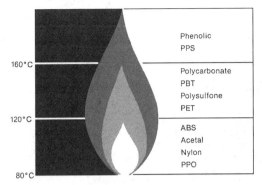

Fig. 7-21 Simplified overview of typical uses temperaturewise.

These absorb moisture, which then has to be carefully removed before the plastics can be fabricated into acceptable products (2, 3). Low concentrations, as specified by the plastic supplier, can be achieved through efficient drying systems and properly handling the dried plastic prior to and during molding, extrusion, etc. (Figs. 7-24 and 7-25). When desired processor can have these hygroscopic plastics properly dried and shipped in sealed containers.

Drying hygroscopic plastics should not be taken casually. The simple tray dryers or mechanical convection hot-air dryers that are

adequate for nonhygroscopic plastics are simply not capable of removing water to the degree necessary for the proper processing of hygroscopic types or their compounds, particularly during periods of high humidity.

For the record let it be known that in the past (half century ago) about 80% of fabricating problems was due to inadequate drying of all types of plastics when a processing problem developed. Now it could be down to 50%.

Moisture Influence

The effect of having excess moisture manifests itself in various ways, depending on the process being employed. The common result is a loss in both mechanical and physical properties for hygroscopic and nonhygroscopic plastics (Fig. 7-26). During injection molding splays, nozzle drool, sinks, and other losses that may occur (3). The effects during extrusion can include gels, trails of gas bubbles in the extrudate, arrowheads, wave forms, surging, lack of size control, and poor appearance (6).

Plastic Memory

Plastic memory is a phenomenon of TPs that has been stretched while hot beyond its heat distortion point to return to its original processed or molded form. Different plastics have varying degrees of this characteristic and degree of return is basically dependent on temperature (Chapter 4, **Thermoforming, Memory**).

Corrosion Resistance

Corrosion is fundamentally a problem associated with metals. Since plastics are electrically insulating they are not subject to this type of damage. Plastics are basically non-corrosive. However, there are those that can be affected when exposed to corrosive environments. It is material deterioration or destruction of materials and properties brought about through electrochemical, chemical,

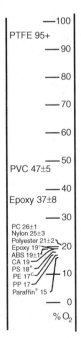

Fig. 7-22 Limiting oxygen index values for a few plastics.

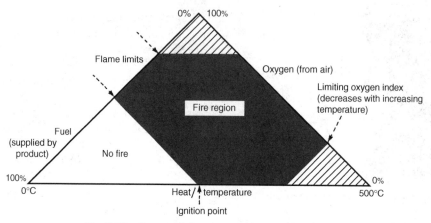

Fig. 7-23 Summarizing the requirements for a fire.

and mechanical actions. Direct attack by an electrochemical, or galvanic action, is the most common. It is the tendency of different metals going into solutions when exposed to natural or man-made electrolytes. The difference in electrical potential can cause damage.

Basically corrosion resistance is the ability of a material to withstand contact with ambient natural factors or those of a particular artificially created atmosphere without degradation or change in properties. Since plastics (not containing metallic additives) are not subjected to electrolytic corrosion, they are widely used where this property is required alone as a product or as coatings and linings for material subjected to corrosion such as in a chemical and water filtration plants, mold/die, etc. Plastics are used as protective coatings on many different products such as steel rod, concrete steel reinforcement, etc.

Table 7-13 Examples of environmental factors that could effect certain plastics

Parameter	Manifestation and Effects
Radiation	Solar, nuclear, thermal
Temperature	Elevated, depressed, cyclic around the 'norm'
Physico chemical factors	Chemical attack, physical changes such as plasticiser bleed
Organic solvents	Vapour absorption, dissolution, stress corrosion cracking
Stress factors	Sustained stress, cyclic stress, compression set (in rubbers) under continuous loading
Biological factors	Microorganisms, fungi, bacteria, animals, insects can destroy materials or change their properties
Wind	Air-borne particulate erosion from dust or water precipitation
Normal air constituents	Oxygen, ozone, carbon dioxide, nitrogen oxides
Air contaminants	Gases: sulphur oxides, halogen compounds
	Mists: aerosols, salt, alkalies
	Particulates: sand, dust, grease
Combined action of wind and water	Surface damage
Water	Solid (snow, ice), liquid (rain, condensation, standing water), vapour (relative humidity). Rain, hail, sleet or snow may have physical effects
Freeze–thaw	Thermal expansion
Use factors	Normal wear and tear, abuse during installation, abuse by user application outside the designed use conditions

Table 7-14 Example of mechanical and physical properties of plastics after sterilization

Polymer Material	Specific Gravity	Sterilization		Visual Clarity	Tensile Strength PSI Yield	Elongation to Break (%)	Stiff or Ductile	Relative Ease Process	Leading Medical Uses
		Steam (@121°C)	Radiation (2.5 Mrd)						
Cellulosics (cellulose-acetate-propionate)	1.19–1.23	Distorts	Yes	Clear	1000–7000	10–50	Ductile	Easy	Burettes/tubes
Fluoroplastics (TFE, FEP)	2.10–2.15	Yes	Marginal	Opaque	3000–5000	175	Ductile	With care	Flexible tubing
Thermoplastic elastomers (TPE)	0.9–1.2	Marginal	Yes	Cloudy	1500–3000	700–1000	Ductile	Easy	Molded parts / Film bags
Natural rubber	1.1–1.4	Yes	Yes	Opaque	3500–4500	350–900	Ductile	Special	Stoppers/parts
Polyurethane (polyether, aliphatic)	1.15	Yes	Yes	Clear	6000–7000	200–1000	Ductile	Easy	Film, tubing, and components
Polyamide (nylon 6–6)	1.04–1.14	Yes	Yes	Clear and cloud	9000–12,000	50–100	Ductile	With care	Packaging film and catheters
Silicone rubber	1.12	Yes	Yes	Clear	1200–2500	300–900	Ductile	Special	Tubing/parts
Butyl rubber	1.1–1.4	Yes	Yes	Opaque	2500–3000	300–500	Ductile	Special	Stoppers/seals
Polyacetal	1.40	Marginal	Damaged	Opaque	8000–9000	40–75	Stiff	Easy	Nonfluid path molded parts
SAN (styrene acrylonitrile)	1.08	Distorts	Yes	High clarity	11,000	3	Very stiff	Easy	Molded parts
Polyethylene (all types)	0.86–0.96	Marginal to poor	Yes	Cloudy	4000	500–1000	Ductile	Easy	Containers Caps
Polypropylene	0.9	Yes	Marginal Stabilized	Cloudy	5000	500–700	Ductile	Easy	Containers Syringes Film bags and tubing
PVC									
Flexible	1.21	Yes	Yes	Clear	2500	350	Ductile	With care	Molded parts
Rigid	1.45	Distorts	Yellows	Clear	6500	0.5–150	Stiff	Can burn	Containers
Polyester PET (polyethylene terephthalate)	1.35	Distorts	Yes	Clear	7800	50–300	Stiff / Also films	With care	Molded parts
Styrene	1.05	Distorts	Yes	High clarity	6000	2–5	Very stiff	Easy	Lab ware
ABS (acrylonitrile butadiene-styrene)	1.06	Distorts	Yes	Opaque and clear	7000	2–30	Stiff	Easy	Molded parts
Polycarbonate	1.20	Yes	Yes	Clear	9000	110	Ductile	With care	Molded parts
Polysulfone	1.25	Yes	Yes	Clear	10,000	20–100	Ductile	With care	Molded parts
Acrylic (polymethyl-methacrylate)	1.19	Distorts	Yellows	High clarity	10,000	2–15	Stiff	Easy	Molded parts
Polymethyl-pentene	0.83	Yes	Marginal	Clear	3400	25–100	Stiff	Easy	Labware/parts

Table 7-15(a)   Guide to relative radiation stabilization of medical plastic devices where dose (kilogray) in ambient air and room temperature at which elongation changes by 25%

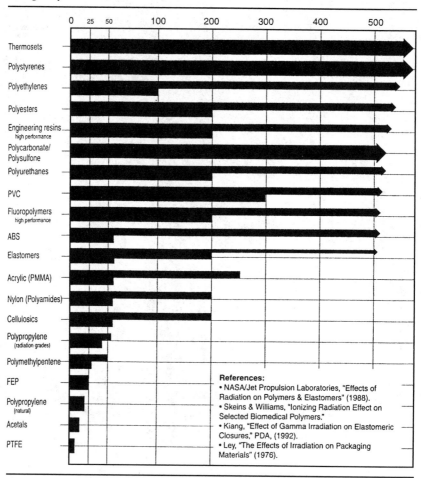

Complex corrosive environments results in at least 30% of total yearly plastics production being required in buildings, chemical plants, transportation, packaging, and communications. Plastics find many ways to save some of the billion dollars lost each year by industry due to the many forms of corrosion.

One example is the use of rigid self-expanding closed cell polyurethane foams as a method to inhibit corrosion of the interior surfaces of metal (steel, etc.) structural cavities exposed to seawater and moisture is one of many example of plastic providing corrosion protection. Unfilled metal cavities are a general feature of various structures or products used in the marine, building, electronics,

automotive, heavy equipment, and aerospace products.

Premature deterioration of the internal surfaces of these cavities is associated with the fact that they are usually poorly protected from corrosion and provide a region for the accumulation and stagnation of salt laden debris and/or moisture during exposure to marine environments. Preferred application of protective paints during products manufacture is often prohibited because welding or other joining operations could damage or destroy preexisting coatings. Rigid polymer foamed in-place are used in regions that are difficult or impossible to access for conventional surface applications of protective plastic paints, etc. (238).

Table 7-15(b)   Guide to radiation stability of plastics

Material	Radiation Stability	Comments
Polystyrene	Excellent	
Polyethylene, various densities	Good/Excellent	High-density grades not as stable as medium- or low-density grades.
Polyamides (nylon)	Good	Nylons 10, 11, 12, 6-6 are more stable than 6. Film and fiber are less resistant.
Polyimides	Excellent	
Polysulfone	Excellent	Natural material is yellow.
Polyphenylene sulfide	Excellent	
Polyvinyl chloride (PVC)	Good	Yellows. Antioxidants and stabilizers prevent yellowing. High-molecular-weight organotin stabilizers improve radiation stability; color-corrected radiation formulations are available.
Polyvinyl chloride/Polyvinyl acetate	Good	Less resistant than PVC.
Polyvinylidene dichloride (Saran)	Good	Less resistant than PVC.
Styrene/acylonitrile (SAN)	Good/Excellent	
Polycarbonate	Good/Excellent	Yellows. Mechanical properties not greatly affected; color-corrected radiation formulations are available.
Polypropylene, natural Polypropylene, stabilized	Poor/Fair	Physical properties greatly reduced when irradiated. Radiation-stabilized grades, utilizing high molecular weights and copolymerized and alloyed with polyethylene, should be used in most radiation applications. High-dose-rate E-beam processing may reduce oxidative degradation.
Fluoropolymers:		When irradiated, PTFE and PFA are significantly damaged. The others show better stability. Some are excellent.
Polytetrafluoroethylene (PTFE)	Poor	
Perfluoro alkoxy (PFA)	Poor	
Polychlorotrifluoroethylene (PCTFE)	Good/Excellent	
Polyinyl fluoride (PVF)	Good/Excellent	
Polyvinylidene fluoride (PVDF)	Good/Excellent	
Ethylene-tetrafluoroethylene (ETFE)	Good	
Fluorinated ethylene propylene (FEP)	Fair	
Cellulosics:		Esters degrade less than cellulose does.
Esters	Fair	
Cellulose acetate propionate	Fair	
Cellulose acetate butyrate	Fair/Good	
Cellulose, paper, cardboard	Fair/Good	
Polyacetals	Poor	Irradiation causes embrittlement. Color changes have been noted (yellow to green).
ABS	Good	High-impact grades are not as radiation resistant as standard-impact grades.
Acrylics (PMMA)	Fair/Good	
Polyurethane	Good/Excellent	Aromatic discolors; polyesters more stable than esters. Retains physical properties.

(Continued)

Table 7-15(b)   (*Continued*)

Material	Radiation Stability	Comments
Liquid crystal polymer (LCP)	Excellent	Commercial LCPs excellent; natural LCPs not stable.
Polyesters	Good/Excellent	PBT not as radiation stable as PET.
Thermosets:		
Phenolics	Excellent	Includes the addition of mineral fillers.
Epoxies	Excellent	All curing systems.
Polyesters	Excellent	Includes the addition of mineral or glass fibers.
Allyl diglycol carbonate (polyester)	Excellent	Maintains excellent optical properties after irradiation.
Polyurethanes:		
Aliphatic	Excellent	
Aromatic	Good/Excellent	Darkening can occur. Possible breakdown products could be derived.
Elastomers:		
Urethane	Excellent	
EPDM	Excellent	
Natural rubber	Good/Excellent	
Nitrile	Good/Excellent	Discolors.
Polychloroprene (neoprene)	Good	Discolors. The addition of aromatic plasticizers renders the material more stable to irradiation.
Silicone	Good	Phenyl-methyl silicones are more stable than are methyl silicones. Platinum cure is superior to peroxide cure; full cure during manufacturing can eliminate most postirradiation effects.
Styrene-butadiene	Good	
Polyacrylic	Poor	
Chlorosulfonated polyethylene	Poor	
Butyl	Poor	Friable, sheds particulates.

Chemical Resistance

Part of the wide acceptance of plastics is from their relative compatibility to chemicals as compared to that of other materials. Because plastics are largely immune to the electrochemical corrosion to which metals are susceptible, they can frequently be used profitably to contain water and corrosive chemicals that would attack metals. Plastics are often used in corrosive environments for chemical tanks, water treatment plants, and piping to handle drainage, sewage, and water supply. Figures 7-27 and 7-28 use glass fiber TS polyester RPs. Structural shapes for use under corrosive conditions often take advantage of the properties of RPs.

However, certain plastics are subject to attack by aggressive fluids and chemicals, although not all plastics are attacked by the same media. It is thus most practical to select a plastic to meet a particular design

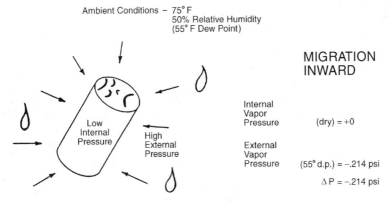

Ambient Conditions – 75° F
50% Relative Humidity
(55° F Dew Point)

MIGRATION
INWARD

Internal
Vapor
Pressure   (dry) = +0

External
Vapor
Pressure   (55° d.p.) = –.214 psi

ΔP = –.214 psi

Fig. 7-24 Mechanics of moisture absorption in plastics.

performance condition. For example, some plastics like HDPE are immune to almost all commonly found solvents. Polytetrafluoroethylene (PTFE) in particular is noted principally for its resistance to practically all-chemical substances. It includes what has been generally identified as the most inert material known worldwide.

It is important to recognize that all materials will have problems in certain environments, whether they are plastics, metals, aluminum, or something else. For example, the chemical effect and/or corrosion of metal surfaces has a damaging effect on both the static and dynamic strength properties of metals because it ultimately creates a reduced cross-section that can lead to eventual failure. The combined effect of corrosion and stress on strength characteristics is called stress corrosion. When the load is variable, the combination of corrosion and the varying stress is called corrosion fatigue.

This problem can be controlled in several ways. One is to select the best material, such as stainless steel, copper alloy, or titanium. Another is to use a nonmetallic protective coating of plastic. Certain systems like plating can reduce fatigue strength. Shot peening rather then plating seems to produce much greater improvement, but shot peening, plating, and then baking can bring the fatigue limit to a point lower even than that of the base metal. The point in this review is that all materials have their limitations and must be critically analyzed if no prior experience exists upon which to draw.

For example, RP underground gasoline storage tanks have this experience. A Chicago service station's May 1963 installation was still leaktight and structurally sound

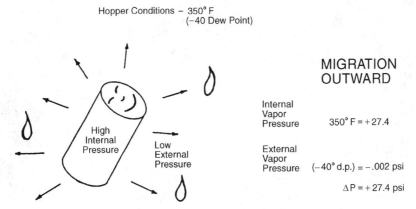

Hopper Conditions – 350° F
(–40 Dew Point)

MIGRATION
OUTWARD

Internal
Vapor
Pressure   350° F = +27.4

External
Vapor
Pressure   (–40° d.p.) = –.002 psi

ΔP = +27.4 psi

Fig. 7-25 Mechanics of moisture migrating out of plastics.

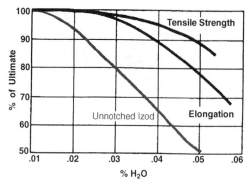

Fig. 7-26 Example of the effects of moisture on the mechanical properties of an IM hygroscopic PET plastic.

when unearthed in May 1988. The tank was one of sixty developed by Amoco Chemical Co. It was fabricated in two semicylindrical sections of fiber glass-woven roving and chopped strand mat impregnated by an unsaturated isophthalic TS polyester plastic selected for its superior resistance to acids, alkalis, aromatics, solvents, and hydrocarbons. The two sections were bonded to each other and to end caps with RP lap joints. Today the cylinder would be a single, unified

Fig. 7-27 Water filtration tank (20 ft D, 32 ft H) could be the largest low-pressure RP molded tank ever built and shipped in one piece.

Fig. 7-28 Gasoline marina 4,000 gallon tanks being installed.

construction as seen in Fig. 7-28. The demand for this type of petroleum storage tank has grown rapidly worldwide as environmental regulations have become more stringent.

Today's underground tanks must last thirty or more years without undue maintenance. To meet these criteria they must be able to maintain their structural integrity and resist the corrosive effects of soil and gasoline, including gasoline that has been contaminated with moisture and soil.

The RP tank just reviewed that was removed in 1988 met these requirements, but two steel tanks unearthed from the same site at that time failed to meet them. There was no record of how long the steel tanks had been in service, but one was dusted with white metal oxide and the other showed signs of corrosion at the weld line. Rust had weakened this joint so much that it could be scraped away with a pocketknife.

Tests and evaluations were conducted on the tank that had been twenty-five years in the ground and also on similarly constructed tanks unearthed at five and a half and seven and a half years that showed the RP tanks could more than meet the service requirements. Table 7-16 provides factual, useful data from these tests.

The chemical and corrosion resistance of plastics is well known. Most materials suppliers have developed long-term data for the commonly used and other chemicals as well. Great care must be taken in selecting them, particularly regarding environmental conditions. For instance, two materials that do not attack a plastic when used separately may be troublesome when used in combination or diluted with water. And additives such as fillers, plasticizers, stabilizers, colorants, and catalysts can decrease or increase the chemical resistance of unfilled or neat plastics. Temperature is also important in all cases; careful tests must be made under the actual conditions of use in making a final selection.

Of especial importance to chemical resistance, particularly in the RPs, is the processing method used. If, for example, a chemical and a mechanical component act simultaneously, cracking or fiber debonding can occur in the plastic, considerably accelerating the diffusion of the aggressive media to the glass fibers. Whereas the diffusion of aggressive media such as acids and alkalis proceeds slowly in plastics, these media advance rapidly along glass fibers. The serviceability of these types of plastics in a corrosive media can be guaranteed only if proper attention is given to processing variables like voids, including the fiber orientation and construction.

Table 7-16 Data on RP underground gasoline storage tanks unearthed after different time periods

Property	Test Results		
	Age at Testing		
	5.5 Years	7.5 Years	25.0 Years
Buried-excavated	1/7/65–8/21/70	4/4/64–10/24/71	5/15/63–5/11/88
Flexural strength: Psi	19,500	24,200	22,400
MPa	134	167	154
Flexural modulus: Psi	725×10^3	795×10^3	635×10^3
MPa	4,992	5,482	4,378
Tensile strength: Psi	10,700	13,600	10,500
MPa	74	94	72
Tensile modulus: Psi	$1,160 \times 10^3$	$1,053 \times 10^3$	$1,107 \times 10^3$
MPa	7,260	8,000	7,630
Tensile elongation: %	1.11	1.25	1.13
Notched Izod			
impact strength: ft.-lb./in.	9.7	11.0	14.1
J/m	518	587	753

Friction, Wear, and Hardness Property

Friction is the resistance against change in the relative positions of two bodies touching one another. If the area of contact is a plane, the relative motion will be a sliding one and the resistance will be called sliding or kinetic friction. If the material in the area of contact is loaded beyond its strength, abrasion or wear will take place. Both phenomena are affected by numerous factors such as the load, relative velocity, temperature, and type material.

Although plastics may not be as hard as metal products, there are those that have excellent resistance to wear and abrasion. Plastic hardware products such as cams, gears, slides, rollers, and pinions frequently provide outstanding wear resistance and quiet operation. Smooth plastic surfaces result in reduced friction, as they do in pipes and valves.

The frictional properties of TPs, specifically the reinforced and filled types, vary in a way that is unique from metals. In contrast to metals, even the highly reinforced plastics have low modulus values and thus do not behave according to the classic laws of friction. Metal-to-thermoplastic friction is characterized by adhesion and deformation resulting in frictional forces that are not proportional to load, because friction decreases as load increases, but are proportional to speed. The wear rate is generally defined as the volumetric loss of material over a given unit of time. Several mechanisms operate simultaneously to remove material from the wear interface. However, the primary mechanism is adhesive wear, which is characterized by having fine particles of plastic removed from the surface.

The presence of this powder is a good indication that the rubbing surfaces are wearing properly. Conversely, the presence of melted plastic or large gouges or grooves at the interface normally indicates that the materials are abrading, not wearing, or the pressure velocity (PV) limits of the materials may be exceeded (Chapter 4, **BEARING, PV Factor**).

The ease and economy of manufacturing gears, cams, bearings, slides, ratchets, and so on with injection-moldable TPs have led to a widespread displacement of metals in these types of applications. In addition to their inherent processing advantages, the products made from these materials are able to dampen shock and vibration, reduce product weight, run with less power, provide corrosion protection, run quietly, and operate with little or no maintenance, while still giving the design engineer tremendous freedom.

These characteristics can be further enhanced and their applications widened by fillers, additives, and reinforcements. Compounding properly will yield an almost limitless combination of an increased load-carrying capacity, a reduced coefficient of friction, improved wear resistance, higher mechanical strengths, improved thermal properties, greater fatigue endurance and creep resistance, excellent dimensional stability and reproducibility, and the like.

Different test results are available to the designer wanting friction and wear data as well as the usual mechanical short and long term data, corrosion resistance, readings, and so on. The data presented include the load and velocity capabilities of a bearing material as expressed by the product of the unit load P based on the projected bearing area and linear shaft velocity V. The symbol PV denotes the important property of the pressure-velocity relationship.

Wear tests are conducted such as using a thrust-washer test apparatus. A sample thrust washer is mounted in an antifriction bearing equipped with a torque arm. The test specimen holder is drilled to accept a thermocouple temperature probe. The raised portion of the thrust washer bears against a dry, cold-rolled, carbon-steel wear ring with a 12- to 16-microinch finish at an 18 to 22 Rockwell C scale hardness at room temperature. Each evaluation is conducted with a new wear ring that has been cleaned and weighed on an analytical balance. The bearing temperature and friction torque is continuously monitored. The test duration is dependent upon the period required to achieve a 360-degree contact between the raised portion of the thrust washer and the wear ring. The average wear factor and duration of this break-in period are then reported. The wear factors reported for each compound is based on its equilibrium wear rate independent of break-in wear.

The coefficient of friction data can be obtained with the same thrust-washer test apparatus. The test specimen is run in against the standard wear ring until a 360-degree contact between the raised portion of the thrust washer and the wear ring is achieved. The temperature of the test specimen is then allowed to stabilize at the test conditions (generally 40 psi, 50 ft./min., room temperature, and dry). After thermal equilibrium occurs, the dynamic frictional torque generated is measured with the torque arm that is mounted on the antifriction bearing. An average of a minimum of five readings is taken.

Although hardness is a somewhat nebulous term, it can be defined in terms of the tensile modulus of elasticity. From a more practical side, it is usually characterized by a combination of three measurable parameters: (1) scratch resistance; (2) abrasion or mar resistance; and (3) indentation under load. To measure scratch resistance or hardness, an approach is where a specimen is moved laterally under a loaded diamond point. The hardness value is expressed as the load divided by the width of the scratch. In other tests, especially in the paint industry, the surface is scratched with lead pencils of different hardnesses. The hardness of the surface is defined by the pencil hardness that first causes a visible scratch. Other tests include a sand-blast spray evaluation.

The material's loss in weight or the change in optical transmission usually measures abrasion resistance and reflectance after a sample has been exposed to an abrasive surface. This is usually done under load, for a predetermined number of cycles or a time period specified by ASTM methods.

Tests for indention under load are performed basically like the ASTM measure the hardness of other materials, such as metals and ceramics. There are at least four popular hardness scales in use. Shore A and Shore D is for soft to relatively hard plastics and elastomers. Barcol is used from the mid-range of Shore D to above it as well as RPs. Rockwell M is used for very hard plastics (Chapter 5, **MECHANICAL PROPERTY, Hardness**),

Plastic-to-Metal Wear

Most studies on the wear and friction characteristics of plastics have concentrated on plastic versus plastic or plastic versus steel wear rings in the same finish and hardness. However, the increased use of aluminum in structural and bearing components has resulted in available, reliable wear-property data involving plastics run against aluminum surfaces. In addition, cost-reduction programs in the business-machine and appliance industries, which have led to the elimination of some parts-finishing operations, have resulted in characterizing the action between rough metal surfaces and plastics.

Plastic-to-Plastic Wear

The wear characteristics of one plastic as opposed to another vary widely, even among materials that have good natural lubricity. When an application calls for plastic-to-plastic bearings, shafts, gears, or other wear members, the combination of materials must be chosen carefully. Because plastics are not rigid, they do not behave according to the classic laws of friction. It is these deviations that cause some of the unexpected results when plastics are run against metals.

Frictional forces are not proportional to load-friction increases with increasing speed, and the static coefficient of friction is lower than its dynamic one. When two viscoelastic low-modulus materials are run against each other, additional inconsistencies result.

Despite these differences, one trend remains clear. It is the wear factor generated when TP is run against itself is extremely high, unless it is operating temperature and pressure are quite low. In applications requiring all-plastic components, the wear rate can be reduced, if crystalline plastics are being used, by running dissimilar plastics against each other. If amorphous plastics are involved, or if environmental or manufacturing procedures require that only a single compound be used, that compound should contain an internal lubricant, like PTFE at loadings of 15 to 20 wt%.

Wear is often greater on a moving surface when dissimilar NEAT plastics are paired. Similar behavior occurs with pairs consisting of lubricated, unreinforced plastics running against themselves or against dissimilar lubricated plastics. The addition of a reinforcing fiber generally produces increased wear in a mating, unreinforced plastic. The addition of reinforcing fibers to both surfaces may result in decreased wear, compared to that in unreinforced plastics.

The wear factors of glass-fiber RPs are lower than those with carbon-fiber materials when run against a carbon-reinforced material, because glass fibers are much harder than carbon ones. Lubricating RPs with PTFE dramatically reduces the wear factors in both similar or dissimilar mating plastics. During the initial break-in period a film of PTFE is transferred to the mating surface, thus creating a PTFE-to-PTFE bearing condition that lowers the wear factors for both the moving and stationary surfaces. The addition of a PTFE lubricant to the mating material reduces the detrimental effects of glass fibers, with respect to wear, on an opposing surface.

Selecting Plastic

Much of the market success or failure of a plastic product can be attributed to the initial choice of material. Even though the range of plastics has become large and the levels of their properties so varied that in any proposed application only a few of the many plastics will be suitable.

A compromise among properties, cost, and manufacturing process generally determines the material of construction. Selecting a plastic is very similar to selecting a metal. Even within one class, plastics differ because of varying formulations, just as steel compositions vary (tool steel, stainless steel, etc.). There are, of course, products for which no plastics is satisfactory, and the interests of the producer and consumer alike are best served by using some other material.

For many applications, however, plastics have superseded metal, wood, glass, natural fibers, etc. Many developments in the electronics and transportation industries and in packaging and domestic goods, have been made possible by the availability of suitable plastics. Thus comes the question of whether to use a plastics and if so, which one.

As an initial step, the product designer must anticipate the conditions of use and the performance requirements of the product, considering such factors as life expectancy, size, condition of use, shape, color, strength, and stiffness. These end use requirements can be ascertained through market analysis, surveys, examinations of similar products, testing, and general experience. A clear definition of product requirements will often lead directly to choice of the material of construction. At times incomplete or improper product requirement analysis is the cause for a product to fail.

Whether the product is a new model of an established commodity or a completely new development, a list can be made of the properties the material or group of materials to be employed must possess, and of those that are also most desirable. By reference to the relevant material properties and prices, an analysis can be made to determine the plastic most likely to be suitable from all requirements.

As reviewed within each one of the major classes of plastics (PE, PVC, PC, etc.) there are usually a very wide variety of specific formulations, each of which has slightly different properties and/or processing capabilities at various costs. Prices, too, will tend to vary depending upon the supplier, the current state of the market, and the volume of plastic that the processor is prepared to purchase.

Computerized Database

The use of computers in design and related fields is widespread and will continue to expand. It is increasingly important for designers to keep up to date continually with the nature and prospects of new computer hardware and software technologies. For example, plastic databases, accessible through computers, provide product designers with property data and information on materials and processes. To keep material selection accessible via the computer terminal, there are design database that maintain such information as graphic data on thermal expansion, specific heat, tensile stress and strain, creep, fatigue, programs for doing fast approximations of the stiffening effects of rib geometry, educational information and design assistance, and more.

With the over 35,000 plastics reviewing what is available as well as keeping up with the constantly proliferating new and replacement types for a specific set of design requirements can seem daunting. Nevertheless, with a logical approach to design this can be done in a practical manner. However, it would probably be impossible to keep up to date manually even for the veteran. Manual searching capable of doing the job at an affordable cost has become difficult to arrange. On-line computerized databases can cut through this information overload by organizing a material's properties into a manageable format (Appendix A: **PLASTICS DESIGN TOOLBOX**). An example of a simplified readout is shown in Table 7-17. Such programs not only significantly reduce time but also present a host of new options.

Besides doing a relatively fast, efficient materials search on what is available today, some databases also offer integration with CAD/CAE/CAM systems to support designing, finite element analysis, processing, testing, and other programs. To make the databases more practical and useful, major international agreements are being arrived at to set uniform methods for sample preparation and test methods. Basically, numerous test standards exist that in many cases are either not in accord with the different data available or are only regionally.

In order to meet the rising demand for information thousands of databases are available worldwide. Nearly all supply technical literature, economic information, patent references, and manufacturers' addresses. Materials databases with numerical values are a relatively small part of these programs. Because the majority of these databases are from individual manufacturers of plastics, there is only limited comprehensive, neutral information on most materials in these software programs.

The German federalist ministries of the Economy and of Research and Technology recognized this situation and during the 1980s launched programs to assist in the development of comprehensive factual databases. Within this framework, the Deutsches Kunststoffinstitut (DKI being the German

Table 7-17 Example of a simplified readout for TPs' toughness or fracture behavior Izod impact test results

	Material	Unnotched	Notched
PMMA	Polymethylmethacrylate		
PA	Polystyrene	Brittle	Brittle
SAN	Styrene-acrylonitrile copolymer		
ABS	Acrylonitrile butadienne styrene		
CA	Cellulose acetate		
HDPE	High-density polyethylene		
PA	Polyamide (Nylon)		
PB	Polybutene	Ductile	Brittle
PC	Polycarbonate		
POM	Polyoxymethylene		
PP	Polypropylene		
PTP	Polyethylene terephthalate		
PVC	Polyvinyl chloride		
LDPE	Low-density polyethylene		
PB	Polybutene	Ductile	Ductile
TFE	Polytetrafluoroethylene		

Plastics Institute) established the materials database called Polymat. This program brings greater availability into a plastics market in which a general perspective is becoming increasingly difficult to obtain. This database contains information on plastics and elastomers, supplying about thirty to fifty properties for each material. Initially some six thousand plastics, from about seventy manufacturers, were stored.

The concept of the Polymat database was based on the following criteria: (1) the data-base is neutral, independent of raw-material manufacturers; (2) anyone can use the database; (3) all the products on the European market should, if possible, be included; (4) since testing is carried out in accordance with a variety of different international standards, the relevant standard, as well as the testing conditions, is registered; (5) during the search, all properties should be capable of being linked with one another as desired; and (6) the sources used for the database are the technical data sheets and additional information supplied by raw-material manufacturers, and various lectures, publications, and measured data from different institutes.

In order for such an extensive project to remain manageable, certain requirements were necessary. Initially the data were confined to TPs, TSs, TPEs, and casting plastics. To be included in this group were the TSEs, reinforced plastics, foams, semifinished products, and others. Polymat completed its initial work in 1989. New plastics products on the market and updated additional information on existing products are continually added. Data no longer available are still accessible to the user in a memory file.

Each plastic in this database is first characterized by descriptive data such as its trade name, manufacturer, product group, form of supply, or additives. Then follows complex technical information on each material, with details of fields of application, recommended processing techniques, and special features. The central element of this material database is the numerical values it gives on a wide range of mechanical, thermal, electrical, optical, and other properties. All these items can be searched for individually or in the combination of properties that was the subject of the enquiry.

A number of material suppliers offer information on their products on electronic devices (floppy discs, CDs, etc.) for use on personal computers. An important one, called Campus, is a database concept started by four German material manufacturers who use a uniform software. This database, initially developed jointly by BASF, Bayer, Hoechst, and Hulls, provided for other manufacturers to join. The present consortium has more than 50 materials suppliers worldwide. It is given in the form of diskettes in German, English, French, Italian, or Spanish. Each diskette contains the uniform test and evaluation program and the range of the respective material producers. It runs on IBM-compatible personal computers under the MS-DOS operating system.

In order to understand the possibilities of these two databases, a comparison can be made of a central database like Polymat and Campus. The Polymat central database provides the following: (1) all the products of the various firms represented are included; (2) the search is independent of the manufacturers and can be performed for all the products of all the manufacturers; (3) available are not only the values contained in the list of basic values but other data specified in such standards as DIN, ASTM, and BS (although a search can nevertheless be confined to products whose data conform to the list of basic values); (4) the information is presented only once and is then maintained centrally; and (5) a selection can be made between a greater number of materials and manufacturers.

The manufacturers' database, Campus, provides use that is free of charge and no charges for data transmission. The actual value of the table of basic values described in Campus lies in its effect on the standardization and streamlining of testing. In the long term, the nonparticipants in the material market will not be able to remain outside this development.

This comparison shows how these two electronic information systems are not comparable, because they pursue completely different objectives. The material manufacturers' databases provide information on products whose manufacturer and product classes are already known. Polymat also gives information if the manufacturer and product classes are not known, but only if the requirements with regard to the material can be described. Polymat can also be used if a replacement material is being sought for a product that can no longer be supplied. Furthermore, Polymat is also capable, for example, of answering the question of possible manufacturers of nylon 6, or of how many different nylon 6 grades an individual manufacturer can supply. If only the trade name is known, its manufacturer or distributor can be traced. This is especially important in the case of foreign products sold by a trading company under the same trade name.

When Polymat was being established, it was necessary to decide on whether to charge the material manufacturer but not the user, or make no charge to the manufacturer but charge the user. DKI chose not to charge the user.

Neither database generally offers the possibility of integrating into it the greater number of values and test data that may already be established by users or processors. These organizations have data for their own internal use, and their goal has been to integrate all these types of data sources. Such in-house databases are at present available under operating system BS 2000 and in conjunction with the database software known as Adabas.

With regard to the common European market, the European Economic Community (EEC) has undertaken numerous activities concerned with materials and material information systems. In one demonstration program for material databases eleven such databases from various countries in the EEC are being cooperatively developed with joint standards for terminology, data presentation, database access, and the user interface of search commands, aids, and menus. For the materials class of plastics, Polymat was selected to participate in this cooperative work. Interesting developments occur from which the users of central material databases in the entire EEC area can benefit.

Electronic marketplace/E-commerce In addition to the many databases available and person-to-person contacts, E-commerce in plastics has been conducted through suppliers' web sites or the dot-commerce independent web sites that link material buyers with sellers in transactions or auction formats. During the year 2000 five plastic producers/suppliers and various elastomer producers/suppliers created a new and important business model of a joint-venture web site. It provides multiple companies to join forces to do business. This is a strategy some observers call competition and others regard as just another form of selling in an electronic format. Regardless of how it is perceived, the model will help propel e-commerce into the mainstream of processor procurement due to the size and wealth of the companies involved. The plastic model example is the largest online business-to-business site todate.

The five major TP producers agreed during April 2000 to form a joint-venture web site offering their materials and related goods to injection molders. BASF, Bayer, Dow, DuPont, and Ticana/Celanese signed a letter of intent to form a neutral business-to-business market-place focused on delivering products and related services (including plastics from other suppliers) to injection molders around the globe. The injection molding market worldwide is a \$50 billion/yr. business. This site can also serve other processes like extrusion.

A few days after the above announcement GE Plastics, a pioneer in the use of the Web, said that it would enhance the e-commerce offerings available through its Polymerland distribution unit. It is targeted to expand the site beyond transactional e-commerce.

Also in existence is the joint-venture web site for leading elastomers suppliers that include Bayer, CK Witco Corp., DSM Elastomers, DuPont-Dow Elastomers L.L.C., Flexsys, M.A. Hanna Rubber Compounding, and Zeon Chemicals L.P. They targeted to create **www.ElastomerSolutions.com**, a global marketplace and customer-focused, web-based community devoted solely to the sale of elastomers and associated products.

During the year 1999 TradeXchange was formed as a joint-venture web site through which General Motors, Ford, and Daimler-Chrysler source their purchases. Their total purchases are at $240 billion annually.

All this marketplace action of the new electronic is targeted to enable customers all over the world to purchase high-quality TPs, other plastic-related materials, molding equipment, tooling, maintenance supplies, packaging materials, and related services.

RAPRA free internet search engine The number of plastic-related web sites is increasing exponentially, yet searching for relevant information is often laborious and costly. During 1999 RAPRA Technology Ltd., the UK-based plastics and rubber consultancy, launched what is believed to be the first free Internet search engine focused exclusively in the plastics industry. It is called Polymer Search on the Internet (PSI). It is accessible at **www.polymersearch.com**. Companies involved in any plastic-related activity are invited to submit their web-site address for free inclusion on PSI. RAPRA Technology's

USA office is in Charlotte, NC (tel. 704-571-4005).

Selection Worksheet

Selecting an optimal material for a given product must obviously be based on analysis of the requirements to be met. A simplified approach involves comparing the specific service requirements to the potential properties of a plastic. What follows is a simplified but practical material-selection approach. This "longhand" system is a basis of the fastest computerized databases.

It follows these steps: (1) select the design criteria from a worksheet [Table 7-18(a)] and check off only the major criteria across the worksheet, keeping it simple but realistic; (2) refer to the selection chart [Table 7-18(b)] and transfer the bold-faced numerical rating in each selected criteria column to the worksheet, for example, if toughness is one criterion, list 6, 4, 1, 2, 4, and 2 on the worksheet from top to bottom in the toughness column; (3) add up the numbers across the worksheet

Table 7-18(a)   Selection worksheet

Material Characteristics	Strength and Stiffness	Toughness	Short-Term Heat Resistance	Long-Term Heat Resistance	Environmental Resistance	Dimensional Accuracy in Molding	Dimensional Stability	Wear and Frictional Properties	Point Subtotal	Cost	Point Total
G/R Design Criteria Resin Groups											
Styrenics ABS SAN Polystyrene											
Olefins Polyethylene Polypropylene											
Other Crystalline Resins Nylons 6 6/6 6/10, 6/12 Polyester Polyacetal											
Arylates Modified PPO Polycarbonate Polysulfone Polyethersulfone											
High Temp. Resins PPS Polyamide-imide											
Fluorocarbons FEP ETFE											

Ratings: 1—most desirable; 6—least desirable. Large numbers indicate group classification, small numbers the specific resins within that group.

Table 7-18(b)   Glass-reinforced TP compound selection sheet

G/R Resin Groups (Design Criteria)	Strength and Stiffness	Toughness	Short-Term Heat Resistance	Long-Term Heat Resistance	Environmental Resistance	Dimensional Accuracy in Molding	Dimensional Stability	Wear and Frictional Properties	Cost
Styrenics (group)	**3**	**6**	**6**	**6**	**6**	**1**	**5**	**6**	**2**
ABS	2	1	1	1	1	3	2	3	3
SAN	1	2	2	2	2	1	1	1	2
Polystyrene	3	3	3	3	3	2	3	2	1
Olefins (group)	**5**	**4**	**4**	**5**	**3**	**5**	**5**	**3**	**1**
Polyethylene	2	2	2	2	2	1	1	2	1
Polypropylene	1	1	1	1	1	1	1	1	2
Other Crystalline Resins — Nylons (group)	**1**	**1**	**2**	**4**	**4**	**4**	**4**	**2**	**3**
6	2	2	2	2	5	1	4	3	1
6/6	1	3	1	1	4	2	3	2	2
6/10, 6/12	3	1	3	3	3	2	2	3	4
Polyester	4	4	2	1	2	2	1	4	1
Polyacetal	5	5	5	2	1	3	2	1	1
Arylates (group)	**3**	**2**	**3**	**3**	**5**	**1**	**2**	**4**	**4**
Modified PPO	4	3	4	4	3	4	4	4	1
Polycarbonate	2	1	3	3	4	1	3	3	2
Polysulfone	2	2	2	2	2	2	2	1	3
Polyethersulfone	1	3	1	1	1	3	1	2	4
High Temp. Resins (group)	**2**	**4**	**1**	**1**	**2**	**4**	**1**	**4**	**5**
PPS	1	2	2	2	1	1	2	2	1
Polyamide-imide	2	1	1	1	2	2	1	1	2
Fluorocarbons (group)	**6**	**2**	**2**	**1**	**1**	**6**	**6**	**1**	**6**
FEP	2	1	2	1	1	2	2	1	2
ETFE	1	2	1	2	2	1	1	2	1

Ratings: 1—most desirable; 6—least desirable. Large numbers indicate group classification, small numbers the specific resins within that group.

in Table 7-18(a) to the "point subtotal" column to find the plastic group with the lowest-point subtotal, that will be the best for a given application on a performance basis; (4) add in the cost factor and total it, to find again the plastic group with the lowest number, again the best for the application on a cost-performance basis.

Finally, repeat the first four steps, but this time use the small numbers on the selector chart and only for the plastic group that was found to be the best. The plastic with the lowest final total will be the best for the application on a cost-performance basis.

Such a selector worksheet can include specifically what the designer requires, with appropriate numerical ratings. Tables 7-18(c) and 7-18(d) provide examples of how to use these simplified worksheets in evaluating different products.

Other Guide

The material information and data presented in this chapter and other sections have provided a variety of useful selection guides; they are assembled in the INDEX under **Selection guide**. Additional guides are in Tables 7-19 to 7-27. It should be remembered that the values given here and elsewhere in this book are representative rather than precise. These values vary depending on the specific type of material, the manufacturing process, and the condition and method of testing. Thus, for example, the tensile strength of a PC given in one table could be quite different from that in another table. The procedure to follow is to properly identify a plastic, usually by its manufacturer's name, its trade name, the manufacturer's grade or identification listing, and by what its data sheet says about its properties.

Preliminary Consideration

The remainder of this chapter provides a summary overview on plastic materials. This section reviews different types of plastics. The descriptions are brief listing a few of their main characteristics. Details cannot be included since so many different formulations

Table 7-18(c)   Gasoline powered chain-saw housing resulting in Nylon 6 ot 6/6

Material Characteristics	Strength and Stiffness	Toughness	Short-Term Heat Resistance	Long-Term Heat Resistance	Environmental Resistance	Dimensional Accuracy in Molding	Dimensional Stability	Wear and Frictional Properties	Point Subtotal	Cost	Point Total
G/R Resin Groups — Design Criteria	X	X	X		X						
Styrenics: ABS, SAN, Polystyrene	3	6	6		6				21	2	23
Olefins: Polyethylene, Polypropylene	5	4	4		3				16	1	17
Other Crystalline Resins — Nylons 6	2 (**1**)	2 (**1**)	2 (**2**)		5 (**4**)				11 (**8**)	1 (**3**)	12 (**11**)
Nylons 6/6	1	3	1		4				9	2	11
Nylons 6/10, 6/12	3	1	3		3				10	4	14
Polyester	4	4	2		2				12	1	13
Polyacetal	5	5	5		1				16	1	17
Arylates: Modified PPO, Polycarbonate, Polysulfone, Polyethersulfone	3	2	3		5				13	4	17
High Temp. Resins: PPS, Polyamide-imide	2	4	1		2				9	5	14
Fluorocarbons: FEP, ETFE	6	2	2		1				11	6	17

Table 7-18(d)   Impeller for chemical handling pump resulting in PPS

Material Characteristics	Strength and Stiffness	Toughness	Short-Term Heat Resistance	Long-Term Heat Resistance	Environmental Resistance	Dimensional Accuracy in Molding	Dimensional Stability	Wear and Frictional Properties	Point Subtotal	Cost	Point Total
G/R Resin Groups — Design Criteria	X		X	X	X						
Styrenics: ABS, SAN, Polystyrene	3		6	6	6				21	2	23
Olefins: Polyethylene, Polypropylene	5		4	5	3				17	1	18
Other Crystalline Resins — Nylons 6, 6/6, 6/10, 6/12, Polyester, Polyacetal	1		2	4	4				11	3	14
Arylates: Modified PPO, Polycarbonate, Polysulfone, Polyethersulfone	3		3	3	5				14	4	18
High Temp. Resins: PPS	1 (**2**)		2 (**1**)	2 (**1**)	1 (**2**)				6 (**6**)	1 (**5**)	7 (**11**)
Polyamide-imide	2		•1	1	2				6	2	8
Fluorocarbons: FEP, ETFE	6		2	1	1				10	6	16

Table 7-19 Example of the range of mechanical properties for plastics

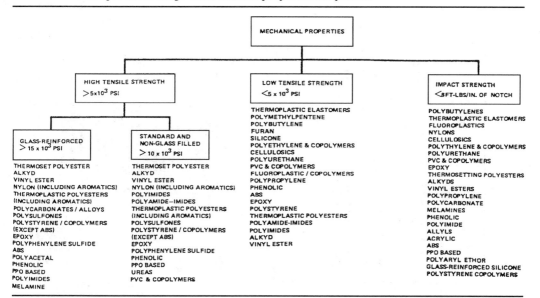

are available that in turn provide many different characteristics. They are available via hard copy or software and kept up to date by material suppliers and other organizations. Should a need arise for data at conditions different from those available for the design, it would not be too difficult or costly to obtain the needed information.

As a general rule, it is considered desirable to examine the properties of three or more materials before making a final choice. Material suppliers should be asked to participate in type and grade selection so that their experience is part of the input. The technology of manufacturing plastic materials, as with other materials (steel, wood, etc.) results in that the same plastic compounds supplied from various sources will generally not deliver the same results in a product. As a matter of record, even each individual supplier furnishes their product under a batch number, so that any variation can be tied down to the exact condition of the raw-material production. Taking into account manufacturing tolerances of the plastics, plus variables of equipment and procedure, it becomes apparent that checking several types of materials from the same and/or from different sources is an important part of material selection.

Experience has proven that the so-called interchangeable grades of materials have to be evaluated carefully by the designer as to their affect on the quality of a product. An important consideration as far as equivalent grade of material is concerned is its processing characteristics. There can be large differences in properties of a product and test data if the processability features vary from grade to grade. It must always be remembered that test data have been obtained from simple and easy to process shapes and do not necessarily reflect results in complex product configurations.

The problem of acquiring complete knowledge of candidate material grades should be resolved in cooperation with the raw material suppliers. It should be recognized that selection of the favorable materials is one of the basic elements in a successful product-configuration design, material selection, and conversion into a finished product (Appendix A: **PLASTICS DESIGN TOOLBOX**).

Individual families of plastics such as polyolefins, polystyrenes, nylons, and polyvinyl chlorides are compounded to produce many different individual plastics. The polyolefin is actually made up of its families of polyethylenes, polypropylenes, etc. In turn the

Table 7-20   Plastic selection guide based on different properties

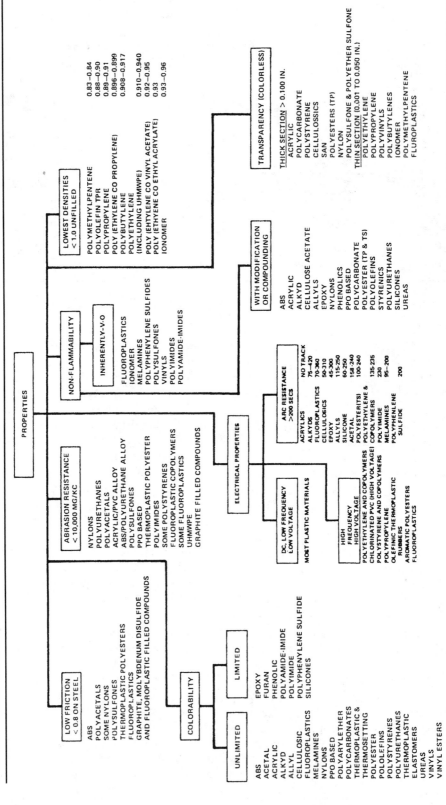

Table 7-21 Examples of plastics' continuous operating temperatures

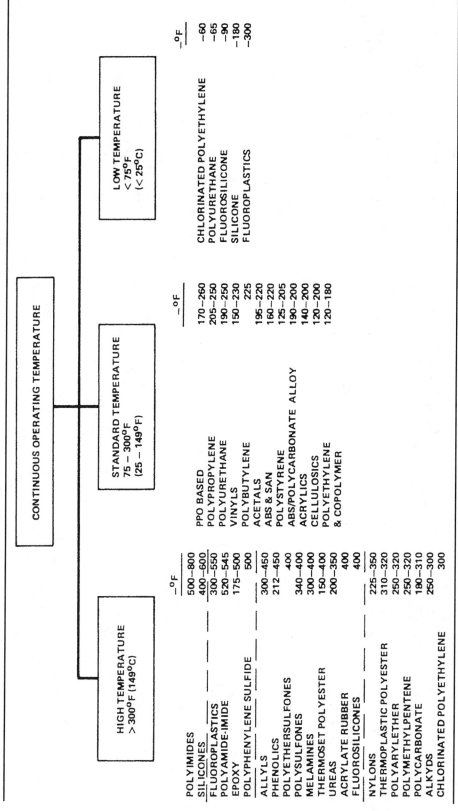

CONTINUOUS OPERATING TEMPERATURE

HIGH TEMPERATURE
> 300°F (149°C)

	°F
POLYIMIDES	500–800
SILICONES	400–600
FLUOROPLASTICS	300–550
POLYAMIDE-IMIDE	520–545
EPOXY	175–500
POLYPHENYLENE SULFIDE	500
ALLYLS	300–450
PHENOLICS	212–450
POLYETHERSULFONES	400
POLYSULFONES	340–400
MELAMINES	300–400
THERMOSET POLYESTER	150–400
UREAS	200–350
ACRYLATE RUBBER	400
FLUOROSILICONES	400
NYLONS	225–350
THERMOPLASTIC POLYESTER	310–320
POLYARYLETHER	250–320
POLYMETHYLPENTENE	250–320
POLYCARBONATE	180–310
ALKYDS	250–300
CHLORINATED POLYETHYLENE	300

STANDARD TEMPERATURE
75 – 300°F
(25 – 149°F)

	°F
PPO BASED	170–260
POLYPROPYLENE	205–250
POLYURETHANE	190–250
VINYLS	150–230
POLYBUTYLENE	225
ACETALS	195–220
ABS & SAN	160–220
POLYSTYRENE	125–205
ABS/POLYCARBONATE ALLOY	190–200
ACRYLICS	140–200
CELLULOSICS	120–200
POLYETHYLENE & COPOLYMER	120–180

LOW TEMPERATURE
< 75°F
(< 25°C)

	°F
CHLORINATED POLYETHYLENE	–60
POLYURETHANE	–65
FLUOROSILICONE	–90
SILICONE	–180
FLUOROPLASTICS	–300

Table 7-22   Examples of plastics' extreme temperature applications

Polymer	Comments
Polyphenyls	Decompose at 530°C (986°F); infusible, insoluble polymers.
Polyphenylene oxide	Decomposes close to 500°C (932°F); heat cures above 150°C (302°F) to elastomer; usable heat range −135 to 185°C (−211 to 365°F)
Polyphenylene sulfide	Melts at 270 to 315°C (578 to 599°F); cross-linked polymer stable to 450°C (842°F) in air; adhesive and laminating applications.
Polybenzyls; polyphenethyls	Fusible, soluble, and stable at 400°C (752°F); low molecular weight.
Parylenes (poly-*p*-xylylene)	Melt above 520°C (968°F); insoluble; capable of forming films; poor thermal stability in air; stable to 400 to 525°C (752 to 977°F) in inert atmosphere.
Polyterephthalamides	Melting points up to 455°C (851°F); fibers have good tenacity, elongation, modulus.
Polysulfanyldibenzamides	Melting points up to 330°C (626°F); soluble; good fiber properties.
Polyhydrazides	Dehydrate at 200°C (392°F) to over 400°C (752°F) to form polyoxadiazoles; good fiber properties.
Polyoxamides	Some melting points above 400°C (752°F); give clear, flexible films.
Phenolphthalein polymers	Melting points of 300°C (572°F) to over 400°C (752°F); formable into fiber and film.
Hydroquinone polyesters	Soluble polymers with melting points of 335°C (635°F) to over 400°C (752°F).
Polyhydroxybenzoic acids	Films melt at 380 to 450°C (716 to 842°F); stable to oxidation but not to hydrolysis; tough, flexible films; good thermal stability.
Polyimides	Commerical film, coating, and resin stable up to 600°C (1,112°F); continuous use up to 300°C (572°F).
Polyarylsiloxanes	Good thermal stability 400 to 500°C (752 to 932°F); coatings, adhesives.
Carboranes	Stable in air and nitrogen at 400 to 450°C (752 to 842°F); elastomeric properties for silane derivatives up to 538°C (1,000°F); adhesives.
Polybenzimidazoles	Developmental laminating resin, fiber, film; stable 24 hours at 300°C (572°F) in air.
Polybenzothiazoles	Stable in air at 600°C (1,112°F); cured polymer soluble in concentrated sulfuric acid.
Polyquinoxalines	Stable in air at 500°C (932°F); tough, somewhat flexible resins; make film, adhesive.
Polyphenylenetriazoles	Thermally stable to 400 to 500°C (752 to 932°F); make film, fiber, coatings.
Polydithiazoles	Decompose at 525°C (977°F); soluble in concentrated sulfuric acid.
Polyoxadiazoles	Decompose at 450 to 500°C (842 to 932°F); can be made into fiber or film.
Polyamidines	Stable to oxidation up to 500°C (932°F); can make flexible elastomer.
Pyrolyzed polyacrylonitrile	Stable above 900°C (1,625°F); fiber resists abrasion with low tenacity.
Polyvinyl isocyanate ladder polymer	Soluble polymer that decomposes at 385°C (725°F); prepolymer melts above 405°C (761°F).
Polyamide-imide	Service temperatures up to 288°C (550°F); amenable to fabrication.
Polysulfone	Thermoplastic; use temperature −102°C (−152°F) to greater than 150°C (302°F); acid and base resistant.
Polybenzaylene benzimidazoles (pyrrones)	Thermally stable to 600°C (1,112°F); insoluble in common solvents; good mechanical properties.

(Continues)

Table 7-22 *(Continued)*

Polymer	Comments
Polybenzoxazoles	Stable in air to 500°C (932°F); insoluble in common solvents except sulfuric acid; nonflammable; chemical resistant; film.
Ionomer	High melt and tensile strength; tough; resilient; oil and solvent resistant; adhesives, coatings.
Diazadiphosphetidine	Thermoplastic up to 350°C (662°F); thermosetting at 357°C (707°F); cured material has good thermal stability to 500°C (932°F); amenable to fabrication.
Phosphorous amide epoxy	Soluble B-staged material; amenable to fabrication; good thermal stability.
Phosphonitrilic	Retention of properties in air up to 399°C (750°F).
Metal polyphosphinates	Polymers stable to better than 400°C (752°F).
Phenylsilesesquioxanes (phenyl-T ladder polymers)	Soluble; high molecular weight; infusible; improved tensile strength; high thermal stability to 525°C (977°F) in air; film forming.

Table 7-23 Examples of tensile stress relaxation of TP RPs at elevated temperatures

Base Resin	Glass (wt%)	Decrease in Applied Stress (%) with Time* at Temperature						
		73°F (23°C)	200°F (93.3°C)	300°F (149)	350°F (176)	400°F (204)	450°F (232)	500°F (260)
PES	30	7/8/9	20/21/25	33/35/39	35/40/57	61/74/90	X/X/X	X/X/X
PEI	30	7/9/11	13/16/25	32/34/38	34/39/55	58/69/86	X/X/X	X/X/X
PPS	40	3/5/9	20/21/22	26/27/28	26/28/32	26/33/34	X/X/X	X/X/X
PEEK	30	13/14/16	17/21/23	25/28/30	28/32/35	30/33/40	32/38/40	32/38/41
PEEK	40†	19/21/25	21/23/27	29/32/37	33/35/40	35/37/42	36/38/43	38/39/44
HTA	30	7/7/8	14/16/22	23/27/35	30/35/50	39/47/59	45/63/55	X/X/X
PEK	30	12/13/15	15/18/20	18/21/24	23/25/29	26/27/30	27/28/31	28/29/32

*Three values indicate percent stress relaxation for 1 h, 5 h, and 15 h. Example: 7/8/9 indicates 7% at 1 h, 8% at 5 h, and 9% at 15 h.
†Long-carbon (Verton) composite (ICI-LNP). X indicates sample would not sustain the test load. Initial stress for all tests was 2,500 psi.

How They Rank in Stress Relaxation

Base Resin	Glass (wt%)	Temperature (°F)						
		73 (23°C)	200 (93.3)	300 (149)	350 (176)	400 (204)	450 (232)	500 (260)
PES	30	3	6	6	7	X	X	X
PEI	30	4	5	5	6	X	X	X
PPS	40	1	3	2	2	2	X	X
PEEK	30	6	4	3	3	3	2	2
PEEK	40*	7	7	7	4	4	3	3
HTA	30	2	2	4	5	5	4	X
PEK	30	5	1	1	1	1	1	1

The lower the number, the higher the retained stress at the indicated temperature.
*Long-carbon (Verton) composite (ICI-LNP).

Table 7-24   Overview of plastics' chemical resistance

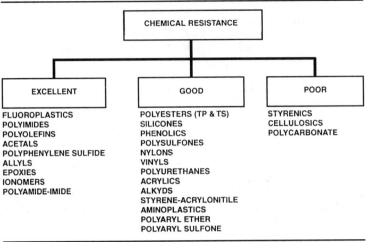

Table 7-25   Examples of qualitative plastics environmental ratings

Material Family	Abrasion Resistance	Weather Ability (Natural)	Paint Ability*	Transparent	Translucent
ABS	F	F–P	No	Yes	Yes
Acetal	G	F	No	Yes	
Acrylic	P	G	No	Yes	
Allyl	G	F	No		
ASA	F	P	No		Yes
Cellulosic	F–P	F–G	No	Yes	
Epoxy	G	F	Yes	Yes	
Fluoroplastic	G	E	No		Yes
Melamine-formaldehyde	G	F–G	Yes		Yes
Nylon	G	F–P	Yes		Yes
Phenol-formaldehyde	G	G	Yes		Yes
Poly (amide-imide)	VG	F	No		Yes
Polyarylether	G	F	No		Yes
Polybutadiene	G	F–G	No		
Polycarbonate	F	F	No	Yes	
Polyester (TP)	G	F	No	Yes	Yes
Polyester-fiberglass (TS)	G	G	Yes	Yes	Yes
Polyethylene	G	P	No		Yes
Polyimide	VG	F–P	No		
Polyphenylene oxide	G	F–G	Yes		
Polyphenylene sulfide	G	G	No		
Polypropylene	G	F–P	No		Yes
Polystyrene	P	F–P	No	Yes	
Polysulfone	G	F–P	No	Yes	
Polyurethane (TS) (TP)	VG	E–G	No		Yes
SAN	F	F	No	Yes	Yes
Silicone	F	VG	No	Yes	Yes
Styrene butadiene	G	G	No	Yes	
Urea formaldehyde	G	F	Yes		
Vinyl	G	G	No	Yes	

*Those with "No" require special paint, primer, or prepainting surface preparation.
Code: E = Excellent; VG = Very Good; G = Good; F = Fair; P = Poor.

Table 7-26   Examples showing permeability of plastics

Type of Polymer	Specific Gravity (ASTM D 792)	Water Vapor Barrier	Gas Barrier	Resistance to Grease and Oils
ABS (acrylonitrile butadiene styrene)	101–1.10	Fair	Good	Fair to good
Acetal—homopolymer and copolymer	1.41	Fair	Good	Good
Acrylic and modified acrylic	1.1–1.2	Fair	—	Good
Cellulosics acetate	1.26–1.31	Fair	Fair	Good
Butyrate	1.15–1.22	Fair	Fair	Good
Propionate	1.16–1.23	Fair	Fair	Good
Ethylene vinyl alcohol copolymer	1.14–1.21	Fair	Very good	Very good
Ionomers	0.93–0.96	Good	Fair	Good
Nitrile polymers	1.12–1.17	Good	Very good	Good
Nylon	1.13–1.16	Varies	Varies	Good
Polybutylene	0.91–0.93	Good	Fair	Good
Polycarbonate	1.2	Fair	Fair	Good
Polyester (PET)	1.38–1.41	Good	Good	Good
Polyethylene				
Low density	0.910–0.925	Good	Fair	Good
Linear low density	0.900–0.940	Good	Fair	Good
Medium density	0.926–0.940	Good	Fair	Good
High density	0.941–0.965	Good	Fair	Good
Polypropylene	0.900–0.915	Very good	Fair	Good
Polystyrene				
General purpose	1.04–1.08	Fair	Fair	Fair to good
Impact	1.03–1.10	Fair	Fair	Fair to good
SAN (styrene acrylonitrile)	1.07–1.08	Fair	Good	Fair to good
Polyvinyl chloride				
Plasticized	1.16–1.35	Varies	Good	Good
Unplasticized	1.35–1.45	Varies	Good	Good
Polyvinylidene chloride	1.60–1.70	Very good	Very good	Good
Styrene copolymer				
(SMA) Crystal	1.08–1.10	Fair	Good	Fair
Impact	1.05–1.08	Fair	Good	Fair

materials in a family could have extremely different properties; some having relatively opposite properties. This action results in having one family, such as polyethylene with its relatively many thousands of formulations, having all kinds of properties. This is an excellent situation since it can be said; "there is a plastic for your design."

In the following list of materials brief explanations of performances and limited applications are given for a few of the different plastics. What is presented will provide some degree of familiarity with the variations of properties existing in the different plastics. Throughout this book many different properties are reviewed. The order in which the following descriptions of plastics are arranged are alphabetically with TPs in the first list followed by the TS plastics.

Thermoplastic

TPs come in greater variety than TSs. They also tend to be more readily to specialty compounding as copolymers, multipolymers, alloys and blends, often customized for cost-effective adaptation to specific application requirements. Unlike TSs, they are in most cases reprocessable without serious losses of properties.

When compared to TSs, they can have limitations of heat-distortion temperatures, cold flow and creep, and are more likely to be

Table 7-27   Guide to elastomers vs. performances

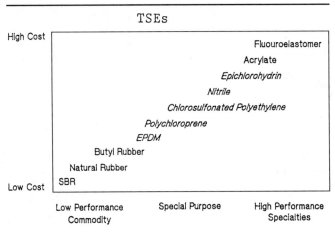

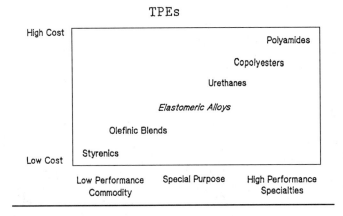

damaged by chemical solvent attack from paints, adhesives, and cleaners. When injection molded, dimensional integrity and ultimate strength are more dependent on the proper process control molding parameters than is generally the case with TSs (where cross-linking offset such problems).

Acetal   This crystalline plastic is strong, stiff, and has exceptional resistance to abrasion, heat, chemicals, creep and fatigue. With a low coefficient of surface friction, it is especially useful for mechanical products such as gears, pawls, latches, cams, cranks, plumbing parts, etc. It is chrome platable.

Acrylic   These polymethyl methacrylate (PMMA) plastics have high optical clarity,

excellent weatherability, very broad color range, and hardest surface of any untreated thermoplastic. Chemical, thermal and impact properties are good to fair. Normally an exterior material used as optical lenses, automotive taillights, decorative nameplates, aircraft glazing, illuminated signs, medical devices, etc. Used as an opaque colored sheeting thermoformed to produce an outer coating behind which glass-fiber-reinforced TS polyester plastics are sprayed to produce camper tops, swimming-pool steps, plumbing fixtures with weatherability and repairability reported superior to polyester gel coats. Like plywood, there are outdoor weather resistant grades and indoor nonweather resistant grades. (Plastic used in plywood determines their outdoor grade.)

Acrylonitrile-butadiene-styrene ABS is a terpolymer that provides a tough, hard, rigid plastic with adequate chemical, electrical and weathering characteristics, low water absorption, and resistance to hot-and-cold water cycles. Used for electronic instrument housings, telephones, sports gear, automotive grilles, and furniture. It is electroplatable, good as a structural foam, and available as a tinted transparent.

Cellulosic They are tough, transparent, hard or flexible natural polymers made from plant cellulose feedstock. With exposure to light, heat, weather and aging, they tend to dry out, deform, embrittle and lose gloss. Molding applications include tool handles, control knobs, eyeglass frames. Extrusion uses include blister packaging, toys, holiday decorations, etc. Cellulosic types, each with their specialty properties, include cellulose acetates (CAs), cellulose acetate butyrates (CABs), cellulose nitrates (CNs), cellulose propionate (CAPs), and ethyl celluloses (EC).

Chlorinated polyether They are corrosion and chemical resistant whose prime use has been to fabricate products and equipment for the chemical and its processing industries. Uses also include pumps, water meters, bearing surfaces, etc.

Chlorinated polyethylene CPEs provide a very wide range of properties from soft/elastomeric to hard. They have inherent oxygen and ozone resistance, have improved resistance (compared to PEs) to chemical extraction, resist plasticizers, volatility, and weathering. Products do not fog at high temperatures as do PVCs and can be made flame retardant.

Ethylene-vinyl acetate EVAs (in the polyolefin family) have exceptional barrier properties, good clarity and gloss, stress-crack resistance, low temperature toughness/retains flexibility, adhesion, resistance to UV radiation, etc. They have low resistance to heat and solvents.

Expandable polystyrene EPS is a modified PS prepared as small beads containing pentane gas which, when steamed, expand to form lightweight, cohesive masses for forms used to mold cups and trays, package fragile products for shipment, etc. Similar dimensionally stable forms molded from EPS are used as cores for such products as automobile sun visors with surface overlays, etc.

Fluoroplastic FPs have superior heat and chemical resistance, excellent electrical properties, but only moderate strength. Variations include PTFE, FEP, PFA, CTFE, ECTFE, ETFE, and PVDF. Used for bearings, valves, pumps handling concentrated corrosive chemicals, skillet linings, and as a film over textile webs for inflatables such as pneumatic sheds. Excellent human-tissue compatibility allows its use for medical implants.

Ionomer They are in the polyolefin family. Their inter chain ionic bonding distinguished them from the other plastics. Performances include being extremely tough, very high tensile strength, and excellent abrasion resistance. Clarity, strength, and good adhesion of ionomer films to metal surfaces are the important properties that have led them into its widespread use in food packaging. Often as a heat-seal layer in TP structures (coextruded films, etc.).

Nylon (Polyamide) PA is a crystalline plastic and the first and largest consumption of the engineering thermoplastic. This family of TPs are tough, slippery, with good electrical properties, but hygroscopic and with dimensional stability lower than most other engineering types. Also offered in reinforced and filled grades as a moderately priced metal replacement.

Parylene The melting point of these film and coating plastics ranges from 290 to 400°C (554 to 752°F), and T_g from 60 to 100°C (14 to 212°F). Their cryogenic performances are excellent. Physical properties are unaffected by thermal cycles from 2°K to room temperature. Good thermal endurance in air, absence of air, and inert atmosphere. They are generally insoluble up to 150°C (302°F). Weather resistance is poor.

Phenylene oxide Based (PPO) plastic that is a choice for electrical applications, housings for computers and appliances, both neat and in structural foam form. It has superior dimensional stability, moisture resistance due to styrene components, which, however, cause some sacrifice of weather and chemical resistance. Use includes automobile wheel covers, pool plumbing, consumer electronic external and internal components.

Polyarylate It is a form of aromatic polyester (amorphous) exhibiting an excellent balance of properties such as stiffness, UV resistance, combustion resistance, high heat-distortion temperature, low notch sensitivity, and good electrical insulating values. It is used for solar glazing, safety equipment, electrical hardware, transportation components and in the construction industry.

Polybutylene It is a polyolefin used for cold and hot water piping. As a blown film it is used for food packaging.

Polycarbonate It is a tough, transparent plastic that offers resistance to bullets and thrown projectiles in glazing for vehicles, buildings, and security installations. It withstands boiling water, but is less resistant to weather and scratching than acrylics. It is notch-sensitive and has poor solvent resistance in stressed molded products. Use includes coffee makers, food blenders, automobile lenses, safety helmets, lenses, and many nonburning electrical applications.

Polyester, thermoplastic TP polyesters have different grades. **Polybutylene terephthalate** (PBT) a crystalline polymer and an excellent engineering material. It has marginal chemical resistance but resists moisture, creep, fire, fats, and oils. Molded items are hard, bright colored, and retain their impact strength at temperatures as low as −40°F (−40°C). Uses include auto louvers, under-the-hood electricals, and mechanical parts.

Polyethylene terephthalate (PET) an amorphous polymer is available in an engineering grade. It is extensively used in beverage bottles and films.

Polyetheretherketone PEEK is a high-temperature, crystalline engineering TP used for high performance applications such as wire and cable for aerospace applications, military hardware, oil wells and nuclear plants. It holds up well under continuous 450°F (323°C) temperatures with up to 600°F (316°C) limited use. Fire resistance rating is UL 94 V-0; it resists abrasion and long-term mechanical loads.

Polyetherimide This is an engineering amorphous thermoplastic. It has superior strength, heat resistance, flame resistance, UV resistance and is transparent, although of amber brown color. Solvent resistance is especially good against aircraft grade fuels and lubricants, but methylene chloride and trichloroethane attack it. Resistance to creep at lower stress loadings and good retention of strength at sustained high levels of heat are claimed to exceed those of other high performance engineering thermoplastics. Applications include printed circuit boards, heater housings, electrical components, steam sterilizable disposable and reusable parts.

Polyethylene PEs are the leading plastics family in total volume sold worldwide. These **polyolefin** materials are relatively inexpensive, easy to process and so versatile that they dominate the packaging and disposable fields. Crystalline in structure, they are varied by chain length, or molecular weight into *low density* (LDPE), *linear low density* (LLDPE), *medium density* (MDPE), *high density* (HDPE) *ultra high density* (UHMWPE), etc. There are also *cross-linked polyethylene* (XLPE) that by chemical or irradiation treatment becomes a TS with outstanding heat resistance and strength. As a family PEs are strong and flexible, very highly chemical resistant, and require special treatments to cement or paint. All kinds of markets use them with packaging being its major outlet. They are blow molded into containers and bottles and molded into boxes, buckets, etc. They are extruded for films, trash bags, and laminated coatings.

Polyimide It is a high-cost heat and fire resistant plastic, capable of withstanding 500°F (260°C) for long periods and up to 900°F (482°C) for limited periods without oxidation. It is highly creep resistant with good low friction properties. It has a low coefficient of expansion and is difficult to process by conventional means. It is used for critical engineering parts in aerospace, automotive and electronics components subject to high heat, and in corrosive environments.

Polyphenylene sulfide PPS is able to resist 450°F (232°C), and has good low temperature strength as well. It has low warpage, good dimensional stability, low mold shrinkage. Use includes hair dryers, cooking appliances, and critical under-the-hood automotive and military parts.

Polypropylene One of the high volume plastics has superior resistance to flexural fatigue stress cracking, with excellent electrical and chemical, properties. This versatile polyolefin overcomes poor low temperature performance and other shortcomings through copolymer, filler, and fiber additions. It is widely used in packaging (film and rigid), and in automobile interiors, under-the-hood and underbody applications, dishwashers, pumps, agitators, tubs, filters for laundry appliances and sterilizable medical components, etc.

Polystyrene One of the high volume plastics, is relatively low in cost, easy to process, has sparkling clarity, and low water absorption. But basic form (crystal PS) is brittle, with low heat and chemical resistance, poor weather resistance. High impact polystyrene is made with butadiene modifiers: provides significant improvements in impact strength and elongation over crystal polystyrene, accompanied by a loss of transparency and little other property improvement. PS is used in many different formulations.

Polysulfone It is a high performance amorphous plastic that is tough, highly heat resistant, strong and stiff. Products are transparent and slightly clouded amber in color. Material exhibits notch sensitivity and is attacked by ketones, esters, and aromatic hydrocarbons. Other similar types in this group include polyethersulfone, polyphenylsulfone, and polyarylsulfone. Use includes medical equipment, solar-heating applications and other performance applications where flame retardance, autoclavability and transparency are needed.

Polyurethane, thermoplastic TPU has excellent properties except for heat resistance (usually only up to 250°F 121°C). It is used in alloys with ABS or PVC for property enhancement. Typical uses are in automobile fascias and exterior body parts, tubing, cord, shoe soles, ski boots and other oil and wear resistant products.

Polyvinyl chloride PVCs are high-volume plastics extensively used that are low in cost with moderate heat resistance and good chemical, weather and flame resistance. They qualify for packaging, pipe and outdoor construction products (siding, window profiles, etc.), and a host of low-cost disposable products (including FDA-grade medical uses in blood transfusion, storage, etc.). PVCs come in a variety of grades, flexible to rigid. They are tough, can be transparent (as in blow molded bottles and jugs), and are also a good alloying plastic to improve properties and reduce costs (ABS/PVC, etc.). There are **polyvinyl acetates** (PVAs), **polyvinyl alcohols** (PVAs), **polyvinyl butrals** (PVBs), **chlorinated PVCs**, etc.

Styrene-acrylonitrile Related to ABS, SAN is hard, rigid, and transparent. It has no butadiene. Excellent chemical and heat resistance, good dimensional stability, and ease of processing characterize it. Special grades are available that have improved UV stability, vapor-barrier characteristics, and weatherability. SAN is used for tinted drinking glasses, low-cost blender jars and water pitchers, and other consumer goods with longer life expectancies than ordinary PS.

Styrene maleic anhydride SMA is a copolymer made with or without rubber modifiers. They are sometimes alloyed with ABS and offer good heat resistance, high impact strength and gloss but with little appreciable improvement in weatherability or chemical resistance over other styrene based plastics.

Thermoset Plastic

Alkyd They are easy to mold, have high heat resistance, and excellent electrical performance, and may be light-colored.

Allyl They have high heat and moisture resistance, good electrical performance in automotive and aerospace uses, good chemical resistance, dimensional stability, low creep (see Diallyl phthlate).

Amino (melamine and urea) Melamine formaldehyde (MF) have excellent electrical properties, heat and moisture resistance, abrasion resistance (good for dinnerware and buttons); in high-pressure laminates it is resistant to alkalies and detergents. They are used as the plastic for counter tops. Urea (**urea formaldehyde**) has properties similar to melamine and is used for wall switch plates, light-colored appliance hardware, buttons, toilet seats, and cosmetics containers. Unlike MFs they are translucent, giving them a brightness and depth of color somewhat similar to opal glass.

Diallyl phthlate DAPs' major use is in electrical connectors since they perform well in electrical circuits. Used also in RP laminates and molding compounds competing with TS polyester types. They offer longer shelf life in the B-stage, less shrinkage during curing, higher heat resistance, etc.

Epoxy They have overall the highest performance of all the thermosets. Properties include very high strength in tension, compression, flexural loadings, very low shrinkages, hard, superior adhesion to other materials, etc. Used with glass cloth to make RP circuit boards, tooling surfaces for metals, and RP castings. Can be cured chemically with or without heat.

Phenolic phenol formaldehydes (PFs) are the low-cost workhorse of the electrical industry (particularly in the past); low creep, excellent dimensional stability, good chemical resistance, good weatherability. Molded black or brown opaque handles for cookware are familiar applications. Also used as a caramel colored impregnating plastics for wood or cloth laminates, and (with reinforcement) for brake linings and many under-the-hood automotive electricals. There are different grades of phenolics that range from very low cost (with low performances) to high cost (with superior performances). The first of the thermosets to be injection-molded (1909).

Polyester, thermoset TS polyesters have an excellent balance of properties, a room-temperature cure and is a major plastic used to make glass-fiber reinforced parts for automobiles, boats, and aircraft parts; used also with other type reinforcing agents. Commodity types have moderate weatherability, high molding shrinkages with wavy surfaces and warpage. Low-profile (polystyrene, etc.) additives reduce shrinkage and surface waviness to almost nil. This has led to major growth in such applications as automobile exterior body panels, instrument housings and microwave dinnerware in the form of **bulk molding compounds** (BMC) and **sheet molding compounds** (SMC).

Polyurethane, thermoset TSUs have durometers range from soft cushion to glass hard with superior wear resistance. Use includes skateboard wheels, solid tires, floor coatings, marine finishes, etc. A major use for soft-foam is automotive bumpers; another is upholstery. Property improvements are made with different added fibers and fillers in reaction molded products to improve cut strength resistance, stiffer moduli, reduce waviness caused by heat and weathering, etc.

Silicone They have excellent heat resistance up to 260°C (500°F), chemical resistance, good electricals, compatible with human body tissues, etc. and a high cost. There are the ***room temperature vulcanizing*** (RTV) types that cure and cross-link at ambient temperatures, catalyzed by moisture in the air. It is a good sealant and excellent for making flexible molds for casting. It is widely used for human implants.

Property Category

A guide to selecting materials is provided (not in any order of priority) by using examples of categories with a few properties. Many more categories as well as many more properties exist since many different plastics exist with their many modifications (via additives, alloying, etc.). They can be used as a guide to meet specific requirements for the product being designed.

Elasticity If the product requires flexibility, examples of the choices includes polyethylene, vinyl, polypropylene, EVA, ionomer, urethane-polyester, fluorocarbon, silicone, polyurethane, plastisols, acetal, nylon, or some of the rigid plastics that have limited flexibility in thin sections.

Odor and taste Polystyrene, styrene-acrylonitrile, polyethylene, acrylic, ABS, polysulfone, EVA, polyphenylene oxide, and many other TPs are examples of satisfactorily odor-free. FDA approvals are available for many of these plastics. Food packaging and refrigerating conditions will also eliminate certain plastics. There are TPs and melamine as well as urea compounds that are suitable for this service.

Temperature Thermal considerations will quickly eliminate many materials. For products operating above 450°F (232°C), ex-

amples of plastics used include the silicones, polyirnides, hydrocarbon plastics, methylpentenes, or glass-bonded mica plastics may be required. A few of the organic plastic-bonded inorganic fibers such as bonded ceramic wool perform well in this field.- Epoxy, diallyl phthalate, and phenolic-bonded glass fibers may be satisfactory in the 450 to 550°F (232 to 288°C) range. A limited group of ablation materials are made for atmospheric reentry of space vehicles (Chapter 2, **HIGH TEMPERATURE**).

Between 250 and 450°F (121 and 232°C), plastics used include glass or mineral-filled phenolics, melamines, alkyds, silicones, nylons, polyphenylene oxides, polysulfones, polycarbonates, methylpentenes, fluorocarbons, polypropylenes, and diallyl phthalates. The addition of glass fillers to the thermoplastics can raise the useful temperature range as much as 100°F and at the same time shortens the molding cycle.

In the 0 to 212°F (−18 to 100°C) range, a broad selection of materials is available. Low temperature considerations may eliminate many of the TPs. Polyphenylene oxide can be used at temperatures as low as −275°F (−165°C). TS materials exhibit minimum embrittlement at low temperatures.

Flame resistance The underwriters ruling on the use of self-extinguishing plastics for contact-carrying members and many other components introduces critical material selection problems. All TSs are basically self-extinguishing. Nylon, polyphenylene oxide, polysulfone, polycarbonate, vinyl, chlorinated polyether, chlorotrifluoroethylene, vinylidene fluoride, and fluorocarbon are examples of TPs that may be suitable for applications requiring self-extinguishing properties. Cellulose acetate and ABS are also available with these properties. Glass reinforcement improves these materials considerably.

Impact As reviewed although impact strength of plastics is widely reported, the properties have no particular design values and can be used only to compare relative response of materials. Even this comparison is not completely valid because it does not solely reflect the capacity of the material to withstand shock loading, but can pick up discriminatory response to notch sensitivity (Chapter 5, **MECHANICAL PROPERTY, Izod Impact**).

A better value is impact tensile, but unfortunately this property is not generally reported. The impact value, with this limitation, can broadly separate those that can withstand shock loading versus those that are poor in this response. Therefore, only broad generalizations can be obtained on these values. Comparative tests on sections of similar size which are molded in accordance with the proposed product must be tested to determine the impact performance of a plastics material. Polycarbonate and ultrahigh molecular weight PE are outstanding in impact strength. The laminated plastics such as glass-filled epoxy, melamine, and phenolic are outstanding in impact strength.

Electric arc resistance   Electrical devices often require arc resistance, as a high-current, high-temperature are will ruin many plastics. Some special arc resisting plastics are available. The most serious cases may require materials such as glass-bonded mica or mineral-filled fluorocarbon products. Lesser arcing problems may be solved by the use of polysulfone, TS polyester-glass, DAP-glass, alkyd, melamine, urea, or phenolic. With low-current arcs, general-purpose phenolic, glass-filled nylon or polycarbonate, acetal, and urea are examples of what may be used very satisfactorily. A coating of fluorocarbon film will improve are resistance in some cases. All circuit breaker problems must be scrutinized with respect to product performance under short-circuit conditions and mechanical shock (Chapter 5, **ELECTRICAL PROPERTY**).

Radiation   In general, plastics are superior to elastomers in radiation resistance but are inferior to metals and ceramics. The materials that will respond satisfactorily in the range of 10^{10} and 10^{11} erg per gram include polyurethane, polystyrene, mineral-filled TS polyester, silicone, glass or asbestos-filled phenolic, certain epoxies, and furane. The next group of plastics in order of radiation resistance includes polyethylene, melamine, urea formaldehyde, unfilled phenolic, and silicone plastics. Those materials that have poor radiation resistances include methyl methacrylate, unfilled TS polyester, cellulose, polyamide, and fluorocarbon.

Transparency   Examples of maximum transparency is available in acrylic, polycarbonate, polyethylene, ionomer, and styrene compounds. Many other TPs may have adequate transparency.

Applied stress   There are TPs that will craze or crack under certain environmental condition. Products that are highly stressed mechanically must be checked very carefully. Polypropylene, ionomer, chlorinated polyether, phenoxy, EVA, and linear polyethylene are examples that offer greater freedom from stress crazing than some other TPs. Solvents may crack products held under stress. TSs is generally preferable for products under continuous loads.

Color   Urea, melamine, polycarbonate, polyphenylene oxide, polysulfone, polypropylene, diallyl phthalate, and phenolic are examples of what is needed in the temperature range above 200°F (94°C) for good color stability. Most TPs will be suitable below this range.

Moisture   Deteriorating effects of moisture are well known as reviewed early in this chapter (**OTHER BEHAVIOR, Drying Plastic**). Examples for high moisture applications include polyphenylene oxide, polysulfone, acrylic, butyrate, diallyl phthalate, glass-bonded mica, mineral-filled phenolic, chlorotrifluoroethylene, vinylidene, chlorinated polyether chloride, vinylidene fluoride, and fluorocarbon. Diallyl phthalate, polysulfone, and polyphenylene oxide have performed well with moisture/steam on one side and air on the other (a troublesome combination), and they also will withstand repeated steam autoclaving. Long-term studies of the effect of water have disclosed that chlorinated polyether gives outstanding performance. Impact styrene plus 25% graphite and high density polyethylene with 15% graphite give long-term performance in water.

Chemical The chemical resistance of plastics is well known as reviewed early in this chapter (**OTHER BEHAVIOR, Chemical Resistance**). The data serves as an excellent initial guide. Most material makers have developed long-term data for commonly used chemicals. Great care must be exercised in this selection because environmental conditions are very impertinent to include in the selection. Note when two materials do not attack a plastic when used separately may be troublesome when used in combination or diluted with water.

Chlorinated polyether is formulated particularly for products requiring, good chemical resistance. Other materials exhibiting good chemical resistance include all of the fluorocarbon plastics, ethylpentenes, polyolefins, certain phenolics, and diallyl phthalate compounds. Additives such as fillers, plasticizers, stabilizers, colorants, and type catalysts can decrease the chemical resistance of unfilled plastics. Certain chemicals in cosmetics will affect plastics, and tests are necessary in most cases with new formulations. Temperature condition is also very important to include in the evaluation. Careful tests must be made under actual use conditions in final selection studies.

Surface wear Hardness is not necessarily the proper index for scratch resistance. In general, the TSs have the best abrasion resistance. Acrylic, ABS, and SAN are examples of plastics that have good fingernail scratch resistance. Tests simulating actual conditions are necessary to obtain the best results. When abrasive wear is the problem, ultrahigh molecular weight polyethylene, urethane, high density polyethylene, Nylon 11, and polyester film are examples of good performers (Chapter 7, **OTHER BEHAVIOR, Friction, Wear, and Hardness**).

Permeability Different plastics provide different permeability properties. As an example polyethylene will pass wintergreen, hydrocarbons, and many other chemicals. It is used in certain cases for the separation of gases since it will pass one and block another. Chlorotrifluoroethylene and vinylidene fluoride, vinylidene chloride, polypropylene, EVA, and phenoxy merit evaluation (Chapter 4, **PACKAGING, Permeability**).

Electrical Electrical considerations will limit the use of certain plastics, and published data are reasonably comparable on similar sections. Final tests must be made on the actual section under field environmental conditions to make sure that the design is adequate. High frequency, high temperature applications are the most difficult to solve. High altitude, high voltage applications with the included ozone problems are often solved by use of glass-bonded mica, which matches the thermal expansion rate of steel and prevents corona gap formation. Vacuum impregnation with epoxy plastic may be suitable for some products. All organic plastics except polyimide can give off vapors that may cause contact failure when used in a vacuum (Chapter 4, **ELECTRICAL/ELECTRONIC PRODUCT**).

Dimensional stability There is plastics with very good dimensional stability, and they are suitable where some age and environmental dimensional changes are permissible. These materials include polyphenylene oxide, polysulfone, phenoxy, mineral-filled phenolic, diallyl phthalate, epoxy, rigid vinyl, styrene, and various RPs. Such products will gain from an after-bake for dimensional stabilization. Glass fillers will improve the dimensional stability of all plastics.

Materials using plasticizers could be a problem. Materials that exhibit substantial moisture absorption are not stable dimensionally. Many organic plastics show a high thermal expansion differential in comparison with mating metal products, and this can cause serious trouble if tight dimensional relations and tightly bonded inserts must be maintained.

Phenolic glass and a diallyl phthalate glass material are available with very low shrinkage. Glass and other mineral fillers minimize the thermal expansion differential problem. Phenoxy and polyphenylene oxides are examples of being low in shrinkage and thermal expansion.

The TPs change dimensions rapidly as they approach the cold flow point. Great care must be taken in the selection of critical dimensional control products for applications such as machines and instruments. Critical dimensions may be held best by included or assembled inserts in materials that have questionable stability (Chapter 4, **JOINING AND ASSEMBLY**).

Weathering   Many plastics has short lives when exposed to outdoor conditions. The better materials include acrylic, chlorotrifluorethylene, vinylidene fluoride, chlorinated polyether, polyester, alkyd, and black linear poly-ethylene. Black materials are best for outdoor service. Some of the styrene copolymers are suitable for certain outdoor uses (Chapter 2, **WEATHERING/ ENVIRONMENT**).

8

Plastic Processing

Overview

The successful design and fabrication of good plastic products requires a combination of sound judgment and experience. Designing acceptable products requires behavioral knowledge of plastics that includes their advantages and disadvantages (limitations) and some familiarity with processing methods. Until the designer becomes familiar with processing, a qualified fabricator must be taken into the designer's confidence early in development and consulted frequently during those early days. The fabricator and mold or die designer should advise the product designer on materials behavior and how to simplify processing. Understanding only one process and in particular just a certain narrow aspect of it should not restrict the designer.

Worldwide extrusion consumes approximately 36wt% of all plastics. IM follows by consuming 32wt%. However there are just in USA about 80,000 injection molding machines (IMMs) and about 18,000 extruders operating to process the many different types of plastics. Consumption by other processes is estimated at 10wt% blow molding, 8% calendering, 5% coating, 3% compression molding 3%, and others 3%. Thermoforming, which is the fourth major process used, consumes principally at least 30% of the extruded sheet and film that principally goes into packaging.

Growth of the processing equipment can be related to the prediction made in mid-1999 by the Freedonia Group Inc. (Cleveland, OH, tel. 440-646-0484). Their *Plastics Processing Machinery Report* reviews that USA machinery sales demand will rise at 5.8% per year to $1.5 billion by year 2003. IMM is the largest category that accounts for 51% of all the machinery sales. By 2003 blow molding machines will grow the fastest reaching $505 million, extrusion will reach $440 million, and thermoforming will reach $455 million. They also reported that there are now over 350 USA machinery builders with five having over 50% of sales.

Influence on Performance

It is important to understand what happens to plastics during the manufacture of products and how it fits into the overall manufacturing operation (Table 8-1). With proper fabricating controls, products are reproduced meeting performance requirements at the minimum cost.

A design evaluation of the product will often hinge on what is the best process of fabrication for the product in question. Sometimes the necessity for certain elements in the design such as thin sections, long delicate inserts, requirements of exact concentricity, or

Table 8-1   Manufacturing analysis diagram

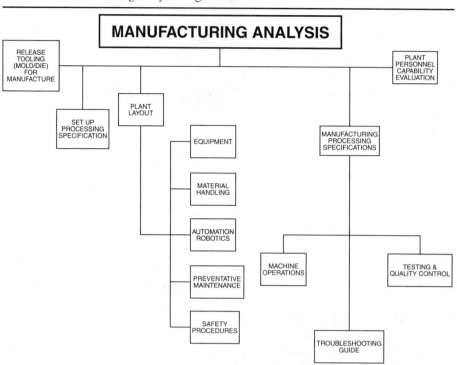

extremely high accuracy of dimensions, make it desirable to use one technique of processing rather than another. There is also the possibility that the process desired for a selected material could not be used because the plastics melt flow and/or other characteristics are not suitable. Result could be determining if the selected material can be modified for the process desired with no loss in meeting performance requirements, finding another plastic, or developing a compromise.

The different fabricating methods and processing behaviors are reviewed explaining effects on the properties of the different plastic that relate to design. Examples include processing parameters such as melt temperature with flow rate or fill rate, die or mold temperature, and packing pressure in injection molding, draw-down ratio in extrusion, and draw ratio in thermoforming (Fig. 8-1). Details on how fabricating equipment is to be setup and operated are reviewed in many sources (1–3, 6–10, 20, 37).

Excessively high processing temperatures can increase the rate of thermal decompo-

sition of plastics or develop deleterious effects. Too low a processing temperature can cause melt viscosity of these materials to be too high, thereby requiring excessive pressure (energy) to fill the mold or extrude the sheet or profile. High melt viscosity and high processing pressure can lead to high shearing stresses in the material and may impart increased molecular orientation to the final product. Molecular orientation makes the product anisotropic; its properties are not the same when measured in different directions. Thermoforming at sheet temperatures that are too low can lead to excessive tensile stresses. Likewise, low fill rate, high packing pressure, and low mold temperature in injection molding, high draw-down ratio in extrusion, or high draw ratio in thermoforming may cause increased bulk orientation in the resulting products.

Basically, with the higher pressures it is possible to develop tighter dimensional tolerances with higher mechanical performance, but there is also a tendency to develop undesirable stresses (orientations) if the processes

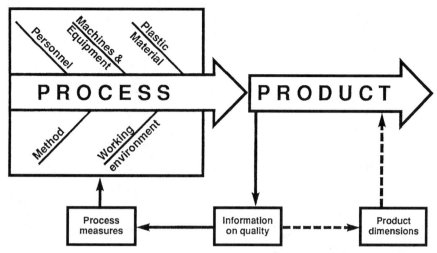

Fig. 8-1   Simplified diagram applicable to fabricating.

are not properly understood or controlled. A major exception is reinforced plastics processing at low or contact pressures. Regardless of the process used, its proper control will maximize performance and minimize undesirable process characteristics.

Practically all processing machines can provide useful products with relative ease, and certain machines have the capability of manufacturing products to very tight dimensions and performances. The proper coordination of plastic and machine facilitates these type of performances. This interfacing of product and process requires continual updating because of continuing new developments in manufacturing operations that provide advantages such as meeting tighter tolerances, reduce cost, etc.. The information presented throughout this book will make past, present, and future developments understandable in a wide range of applications.

Most products are designed to fit processes of proven reliability and consistent production. Various options may exist for processing different shapes, sizes, and weights (Table 8-2). Parameters that will help one to select the right options are (1) setting up specific performance requirements; (2) evaluating materials' requirements and their processing capabilities; (3) designing products on the basis of material and processing characteristics, considering product complexity and size (Fig. 8-2) as well as a product and process

cost comparison (Tables 8-3 and 8-4); (4) designing and manufacturing tools (molds, dies, etc.) to permit ease of processing; (5) setting up the complete line, including auxiliary equipment, testing, and providing quality control, from delivery of the plastics to the equipment through production to the product; and (7) interfacing all these parameters by using logic and experience or obtaining a required update on technology.

Plastics usually are obtained in the form of granules, powder, pellets, and liquids. Processing mostly involves their physical change (thermoplastics), though in some cases chemical reactions occur (thermosets) (Chapter 6). The following reviews primarily pertain to processing TPs since over 90wt% of all plastics processed are TPs. Processing TSs will also be included in appropriate processes. The common features of these processes are

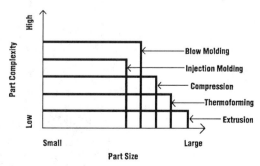

Fig. 8-2   Overview guides for processing characteristics in regard to size and complexity.

Table 8-2 Examples of competitive processes vs. different products

	Injection Molding	Extrusion	Blow Molding	Thermoforming	Reaction Injection Molding	Rotational Molding	Compression and Transfer Molding	Matched Mold Spray-up
Bottles, necked containers, etc.	2, A	—	1	2, A	—	2	—	2
Cups, trays, open containers, etc.	1	—	—	1	1	—	1	2
Tanks, drums, large hollow shapes, etc.	—	—	1	2, A	—	1	—	2
Caps, covers, closures, etc.	1	—	—	2	2	—	1	—
Hoods, housings, auto parts, etc.	1	—	2	2	2	—	1	1
Complex shapes, thickness changes, etc.	1	—	—	—	—	—	1	2
Linear shapes, pipe, profiles, etc.	2, B	1	—	—	—	—	2, B	—
Sheets, panels, laminates, etc.	—	1, C	—	—	—	—	2	2

1. Prime process.
2. Secondary process.
A. Combine two or more parts with ultrasonics, adhesives, etc.
B. Short sections can be molded.
C. Also calendering process.

Table 8-3 Detailed guide to product size

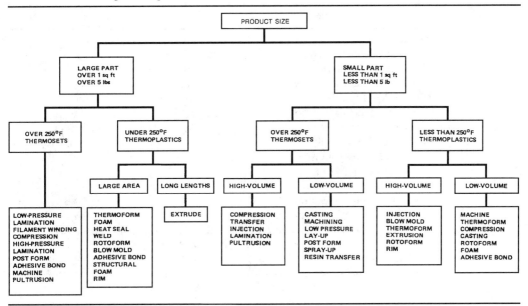

Table 8-4 Cost comparison of products vs. processes (cost factor × material cost = purchased cost of product)

Process	Cost Factor	
	Overall	Average
Blow molding	$1\frac{1}{16}$ to 4	$1\frac{1}{8}$ to 2
Calendering	$1\frac{1}{2}$ to 5	$2\frac{1}{2}$ to $3\frac{1}{2}$
Casting	$1\frac{1}{2}$ to 3	2 to 3
Centrifugal casting	$1\frac{1}{2}$ to 4	2 to 4
Coating	$1\frac{1}{2}$ to 5	2 to 4
Cold pressure molding	$1\frac{1}{2}$ to 5	2 to 4
Compression molding	$1\frac{3}{8}$ to 10	$1\frac{1}{2}$ to 4
Encapsulation	2 to 8	3 to 4
Extrusion forming	$1\frac{1}{16}$ to 5	$1\frac{1}{8}$ to 2
Filament winding	5 to 10	6 to 8
Injection molding	$1\frac{1}{8}$ to 3	$1\frac{3}{16}$ to 2
Laminating	2 to 5	3 to 4
Match-die molding	2 to 5	3 to 4
Pultrusion	2 to 4	2 to $3\frac{1}{2}$
Rotational molding	$1\frac{1}{4}$ to 5	$1\frac{1}{2}$ to 3
Slush molding	$1\frac{1}{2}$ to 4	2 to 3
Thermoforming	2 to 10	3 to 5
Transfer molding	$1\frac{1}{2}$ to 5	$1\frac{3}{4}$ to 3
Wet lay-up	$1\frac{1}{2}$ to 6	2 to 4

(1) mixing, melting, and plasticizing; (2) melt transporting and shaping; and (3) finishing. Mixing, melting, and plasticizing produce a plasticized melt, usually made in a screw (extruder or injection). Melt transporting and shaping involves applying pressure to the hot melt in order to move it through a die or into a mold. The final feature of processing, finishing, is the usual solidification of the melt. As summarized in Table 8-5, certain processes are unique to certain plastics.

Many product designs are inherently limited by the economics of the process that must be used to make them. For example, to date TSs are not blow molded, and they have limited extrusion possibilities. Many hollow products, particularly very large ones, may be produced more economically by the rotational process than by blow molding. The need for a low quantity of products may eliminate certain molding processes and indicate the use of casting or others.

The extrusion process has fewer process control problems with TPs than does injection molding but has greater problems in dimensional control and shape. During

Table 8-5 Example of properties and processes for the major TS plastics in RPs

Thermosets	Properties	Processes
Polyesters	Simplest, most versatile, economical, and most widely used family of resins; good electrical properties, good chemical resistance, especially to acids	Compression molding, filament winding, hand lay-up, mat molding, pressure bag molding, continuous pultrusion, injection molding, spray-up, centrifugal casting, cold molding, encapsulation
Epoxies	Excellent mechanical properties, dimensional stability, chemical resistance (especially to alkalis), low water absorption, self-extinguishing (when halogenated), low shrinkage, good abrasion resistance, excellent adhesion properties	Compression molding, filament winding, hand lay-up, continuous pultrusion, encapsulation, centrifugal casting
Phenolic resins	Good acid resistance, good electrical properties (except arc resistance), high heat resistance	Compression molding, continuous lamination
Silicones	Highest heat resistance, low water absorption, excellent dielectric properties, high arc resistance	Compression molding, injection molding, encapsulation
Melamines	Good heat resistance, high impact strength	Compression molding
Diallyl *o*-phthalate	Good electrical insulation, low water absorption	Compression molding

extrusion when the plastic leaves the die, it can be relatively stress-free. It is drawn in size and passed through downstream equipment from the extruder to form its shape as it is cooled, usually by air and/or water. The dimensional control and the die shape required to achieve the desired product shape are usually solved by trial-and-error settings. With the more experience one has in the specific plastic and equipment to be used, less or no trial time is needed.

Other analyses can be made. Compression and injection molds, which are expensive and relatively limited in size, are employed when the production volume required is great enough to justify the molds' costs and the sizes are sufficient to fit available equipment's limitations. Extrusion produces relatively uniform profiles of unlimited lengths. Casting is not limited by pressure requirements and large products can be produced. Calendered sheets are limited in their width by the width of the material's rolls, but are unlimited in length. Vacuum forming is not greatly limited by pressure, although even a small vacuum distributed over a large area

can build up an appreciable load. Blow molding is limited by equipment that is feasible for the mold sizes. Rotational molding can produce relatively large parts.

The tendency of injection molding and extrusion to align long chain molecules in the direction of flow results in their having markedly greater strength in that direction than at right angles. With an extruded pressure pipe, for example, its major strength could be in the axial/machine direction. With changes in processing controls, different directional properties can be obtained such as providing higher strength in the circumferential direction. If in an injection mold the plastic flows in from several gates, the melts must unite or weld where they meet causing points of potential weakness and undesirable markings on the product's surface. Careful gating with proper processing control can eliminate or reduce weld lines.

The nature of the process may have profound influence on such properties as impact strength. Figure 8-3 compares the impact strengths of three PP formulations that were processed either by injection or compression

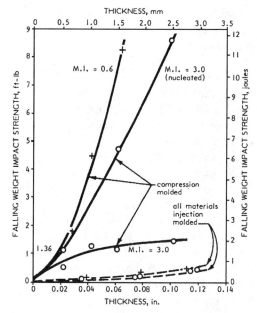

Fig. 8-3 Example of the effect of processing conditions on impact for PP with different melt index flow behaviors.

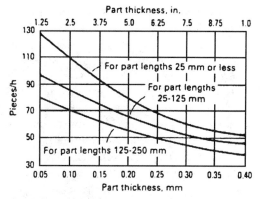

Fig. 8-4 Example of cycle time during IM of TPs as a function of product thickness.

molding. In this example the IM process resulted in a drastic reduction of impact strength over that offered by the CM. Varying the processing conditions can reverse this situation.

Sometimes certain products are most economically produced by fabricating them with conventional machining out of compression molded blocks, laminates or extruded sheets, rods or tubes. Also at times it may be advantageous to design a product for the postmolding assembly of inserts to reduce fabricating time and gain cost benefits. However specialized equipment is available for very efficient automatic insert molding during IM that provides gains over secondary operations.

If the processing is to be subcontracted, the choice of fabricator should place no limits on a design. There is a way to make a product if the projected values justify the price; any job can be done "at a price." The real limiting factors are factors such as tool design considerations, material shrinkages, subsequent assembly or finishing operations, dimensional tolerances, undercuts, insert inclusions, parting lines, fragile sections, production rate or cycle time (Fig. 8-4), and cost to fabricate.

Applying the following principles, applicable to virtually all manufacturing processes,

will aid the designer in specifying products that can be produced at minimum cost: (1) maintain design simplicity; (2) use standard materials and components; (3) specify liberal tolerances; (4) employ the most processable plastics; (5) collaborate with manufacturing people; (6) avoid secondary operations; (7) designing what is appropriate to the expected level of production; (8) utilize special process characteristics; and (9) avoide processing restrictions.

In light of the many types of behavior plastics that can manifest and the considerable effect this behavior can have on the performance of the finished product, it behooves designers to become familiar with specific behavior characteristics of each plastic considered for an application. Recognize potential problems. A major cause for problems is not of poor product design but instead that the processes operated outside of their required operating window. This subject will be reviewed latter in this Chapter under **PROCESS CONTROL**.

As an example to plastic behavior, in-mold decorating with coatings and/or decorative material is routine and very successfully accomplished. However there are several coating treatments that can embrittle the surface of a plastic product, including contamination with certain chemical agents, oxidative degradation or cross-linking of the plastic molecules by heat aging or exposure to ultraviolet light. Under certain circumstances cracks starting in this brittle surface layer can propagate into the underlying ductile plastic at sufficiently high velocity to induce a brittle failure of the entire product. Surface embrittlement can have a significant effect on the observed stress/strain, fatigue, and impact behavior of the plastic.

Plastic products are made by a variety of basic manufacturing processes. As an example a major method such as extrusion has subdivisions that include profile, pipe, tube, film, sheet, coating, post forming, etc. equipment In injection molding there are subdivisions such as coinjection, gas assist, foam, in-mold decorating, etc. equipment. There are literally hundreds of processes used with only about the dozen, as reviewed in this chapter, that are principally used (2).

Processing and Material Behavior

The flow patterns resulting from the conditions of a particular fabricating process are very important in influencing product performances. The melting of plastics follows different phases that effect performances. An example is its modulus of elasticity as shown in Fig.7-8. As the temperature increases, the plastic goes through the phases of glassy, transition, rubbery, to melt flow.

Reinforcing fibers, specifically the glass fibers, are brittle. Thus, when they are used in conjunction with a brittle matrix, as are certain TSs, it might be expected that the composite would have low fracture energy. In fact, this is not true, and the impact strength of most glass-TS RPs is many times greater than the impact strengths of either the fibers or the matrix. Impact strength is higher if the bond between the glass fibers and the matrix is relatively weak, because if it is so strong that it cannot be broken, cracks will propagate across the matrix and fibers, and very little energy will be absorbed. Thus, there is a conflict between the requirements for maximum tensile or flexural modulus or strength (long glass fibers and strong interface bonds) and maximum impact strength.

Tolerance and Dimensional Control

To predict product dimensions and the fluctuation of dimensions (tolerances) during a production run, many variables have to be considered. These include the plastic materials with their variabilities, geometry of the product that includes thicknesses, toolmaking quality applied in producing the die or mold, and very important the fabricating conditions and processes fluctuations inherent during processing. Computer programs developed have made it possible to provide model guides that tend to understand the complex (controllable) interactions of these many factors. This allows molders to more accurately predict product dimensions and to model the relationship between the control of the molding process and the product tolerances.

This interplay of the many variables is extremely complex and involves a matrix of the many variables. As an example in the molding simulation TMconcept system programmed Molding & Cost Optimization (MCO) of Plastics & Computer Inc., Dallas, TX, there are well over 300 variables. It is not reasonable to expect a person using manual methods to calculate these complex interactions even if molding only a modest shaped product without omissions or errors. Computerized process simulation is a practical tool to monitor the influence of design alternatives on the processability of the product and to select molding conditions that ensure the required product quality (3).

Process used provides different control capabilities. As an example closed molding (injection, compression, etc.) provides fine detail on all surfaces. Open molding (blow molding, thermoforming, spray-up, etc.) provides detail only on the one side in contact with the mold, leaving the second side free-formed. Continuous production (extrusion and pultrusion) yields products of continuous length. Hollow (rotational or blow) produces hollow products. These processes can be used creatively to make different types of products. For example, two molded or thermoformed components can be bonded together to form a hollow product, or they can be blow molded.

Shrinkage After preliminary study, the designer has to define the geometry. This process usually passes through several stages, beginning with preliminary drawings and sketches that indicate the basic design and functions. More detailed sketches will show the appropriate wall thickness, ribs, radii, and other structures, including the tolerances that are required to be met. Tables 8-6 to 8-8 provide guides to shrinkages and tolerances (Chapter 2, **INSTABILITY BEHAVIOR, Shrinkage/Tolerance**).

Using the calculated shrinkage theory can dictate how much oversize to cut the tool (mold or die) if a product has a relatively simple shape. For other shapes some critical key dimensions of the product will, more often than not, not be as predicted from the shrink allowance, particularly if the item is long, complex, or require tight tolerance. The important factors that influence the shrinkage of a specific plastic in using a specific machine, such as injection molding, by causing it to vary and not follow the values like those in Table 8-6 to 8-8, are flow direction, wall thickness, flow distance, and the presence of reinforcing fibers.

Determining shrinkage involves more than just applying the appropriate correction factor from a material's data sheet. Shrinkage is caused by a packing pressure and volumetric change in a plastic as it cools from a molten to a solid form. Shrinkage is not a single event but occurs over a period of time. Most of it happens in the mold or die, but it can continue for up to twenty-four to forty-eight hours after being molded. This so-called postmold shrinkage may require a constraining cooling fixture. Additional shrinkage can occur when frozen-in stresses are relieved by annealing or exposure to high service temperatures.

The main considerations in mold or die design affecting shrinkage are to provide adequate cooling (required temperature control), and structural rigidity. Cooling conditions is the most critical especially for crystalline plastics. The cooling system must be adequate for the heat load. Slow cooling increases shrinkage by giving plastic molecules more time to reach a relaxed state. In crystalline types, having longer cooling time leads to a higher level of crystallinity, which in turn accentuates shrinkage. Proper cooling, along with having an overall melt-flow analysis of

how the material will react in the mold, by the mold designer, will eliminate or at least be capable of controlling the potential problems of shrinkage and warpage. This analysis can also include the best gate locations in molds.

A number of the computer-aided flow-simulation programs now offer modules designed (targeted) to forecast shrinkage and, to a limited degree, warpage from the interplay of plastic and mold temperatures, cavity pressures, stress, and other variables in mold-fill analysis. The predicted shrinkage values in various areas of the product should be used as the basis for sizing the mold cavity, either by manual input or feed-through to the mold dimensioning program.

Computer-aided flow-simulation programs are also available for dies. All the programs can successfully predict a certain amount of shrinkage under specific conditions that can be applied to experience. The actual shrinkage is finally determined after molding or extruding the products. When not in spec process control changes can meet the requirements unless some drastic error had been included in the analysis.

Inspection and tolerance Inspection variations are often the most critical and most overlooked aspect of the tolerance of a fabricated product. Designers and processors base their development decisions on inspection readings, but they rarely determine the tolerances associated with these readings. The inspection variations may themselves be greater than the tolerances for the characteristics being measured, but without having a study of the inspection method capability this can go unnoticed.

Inspection tolerance can be divided into two major components: the accuracy variability of the instruction and the repeatability of the measuring method. The calibration and accuracy of the instrument are documented and certified by its manufacturer, and it is periodically checked. Understanding the overall inspection process is extremely useful in selecting the proper method for measuring a specific dimension. When all the inspection methods available provide an acceptable level of accuracy, the most economical method should be used.

Table 8-6   Guidelines for nominal TP mold shrinkage rates per ASTM $\frac{1}{4}$ & $\frac{1}{2}$ in. thick test specimens

Material	Avg. Rate* per ASTM D 955	
	0.125 in. (3.18 mm)	0.250 in. (6.35 mm)
ABS		
Unreinforced	0.004	0.007
30% glass fiber	0.001	0.0015
Acetal, copolymer		
Unreinforced	0.017	0.021
30% glass fiber	0.003	NA
HDPE, homo		
Unreinforced	0.015	0.030
30% glass fiber	0.003	0.004
Nylon 6		
Unreinforced	0.013	0.016
30% glass fiber	0.0035	0.0045
Nylon 6/6		
Unreinforced	0.016	0.022
15% glass fiber + 25% mineral	0.006	0.008
15% glass fiber + 25% beads	0.006	0.008
30% glass fiber	0.005	0.0055
PBT polyester		
Unreinforced	0.012	0.018
30% glass fiber	0.003	0.0045
Polycarbonate		
Unreinforced	0.005	0.007
10% glass fiber	0.003	0.004
30% glass fiber	0.001	0.002
Polyether sulfone		
Unreinforced	0.006	0.007
30% glass fiber	0.002	0.003
Polyether-etherketone		
Unreinforced	0.011	0.013
30% glass fiber	0.002	0.003
Polyetherimide		
Unreinforced	0.005	0.007
30% glass fiber	0.002	0.004
Polyphenylene oxide/PS alloy		
Unreinforced	0.005	0.008
30% glass fiber	0.001	0.002
Polyphenylene sulfide		
Unreinforced	0.011	0.004
40% glass fiber	0.002	NA
Polypropylene, homo		
Unreinforced	0.015	0.025
30% glass fiber	0.0035	0.004
Polystyrene		
Unreinforced	0.004	0.006
30% glass fiber	0.0005	0.001

*Rates in in./in. (*Courtesy* ICI-LNP)

Table 8-7 Example of wall thickness ranges and tolerances for RPs

| Molding Method | Thickness Range* | | Maximum Practicable Buildup Within Individual Part | Normal Thickness Tolerance, mm (in.) |
	Min., mm (in.)	Max., mm (in.)		
Hand lay-up	1.5 (0.060)	30 (1.2)	No limit; use cores	±0.5 (0.020)
Spray-up	1.5 (0.060)	13 (0.5)	No limit; use many cores	±0.5 (0.020)
Vacuum-bag molding	1.5 (0.060)	6.3 (0.25)	No limit; over three cores possible	±0.25 (0.010)
Cold-press molding	1.5 (0.060)	6.3 (0.25)	3–13 mm ($^1/_8$–$^1/_2$ in.)	±0.5 (0.020)
Casting, electrical	3 (.125)	115 (4.5)	3–115 mm ($^1/_8$–4 $^1/_2$ in.)	±0.4 (0.015)
Casting, marble	10 (.375)	25 (1)	10–13 mm; 19–25 mm ($^3/_8$–$^1/_2$ in; $^3/_4$–1 in.)	±0.8 (.031)
EMC molding	1.5 (0.060)	25 (1)	Min. to max. possible	±0.13 (0.005)
Matched-die molding: SMC	1.5 (0.060)	25 (1)	Min. to max. possible	±0.13 (0.005)
Pressure-bag molding	3 (.125)	6.3 (.25)	2:1 variation possible	±0.25 (0.010)
Centrifugal casting	2.5 (0.100)	4.5% of diameter	5% of diameter	±0.4 mm for 150-mm diameter (0.015 in. for 6-in. diameter); ±0.8 mm for 750-mm diameter (0.030 in. for 30-in. diameter)
Filament winding	1.5 (0.060)	25 (1)	Pipe, none; tanks, 3:1 around ports	Pipe, ±5%; tanks, ±1.5 mm (0.060 in.)
Pultrusion	1.5 (0.060)	40 (1.6)	None	1.5 mm, ±0.025 mm ($^1/_{16}$ in. ±0.001 in.); 40 mm, ±0.5 mm (1 $^1/_2$ in. ±0.020 in.)
Continuous laminating	0.5 (0.020)	6.3 (1/4)	None	±10% by weight
Injection molding	0.9 (0.035)	13 (0.5)	Min. to max. possible	±0.13 (0.005)
Rotational molding	1.3 (0.050)	13 (0.5)	2:1 variation possible	±5%
Cold stamping	1.5 (0.060)	6.3–13 (0.25–0.50)	3:1 possible as required	±6.5% by weight; 6.0% for flat parts

*Thickness may be varied within parts, but prolonged cure times, slower production rates, and the possibility of warpage may result. If possible, the thickness should be held uniform throughout a part.

As the overall fabricating tolerance is analyzed into the sources of its variation components, the potential advantage of analytical programs comes into play with their ability to efficiently process all these factors. All the empirical tolerance ranges for each tooling method and inspection method are stored in data files for easy retrieval. For each critical dimension the program sums all the component tolerances and computes a ± overall tolerance for each critical dimension. The program then provides a tabulated estimate of the achievable processing tolerances and pinpoints the areas that contribute most of the overall required tolerance. This information is useful in identifying the needed tolerances that in practice can be expected to exceed the initial design tolerance.

Table 8-8 Recommended dimensional tolerances for RPs

Dimension, mm (in.)	Class A* (Fine Tolerance), mm (in.)	Class B† (Normal Tolerance), mm (in.)	Class C** (Coarse Tolerance), mm (in.)
0–25 (0–1)	±0.12 (±0.005)	±0.25 (±0.010)	±0.4 (±0.016)
25–100 (1–4)	±0.2 (±0.008)	±0.4 (±0.016)	±0.5 (±0.020)
100–200 (4–8)	±0.25 (±0.010)	±0.5 (±0.020)	±0.8 (±0.030)
200–400 (8–16)	±0.4 (±0.016)	±0.8 (±0.030)	±1.3 (±0.050)
400–800 (16–32)	±0.8 (±0.030)	±1.3 (±0.050)	±2.0 (±0.080)
800–1,600 (32–64)	±1.3 (±0.050)	±2.5 (±0.100)	±3.8 (±0.150)
1,600–3,200 (64–128)	±2.5 (±0.100)	±5.0 (±0.200)	±7.0 (±0.280)

*Class A tolerances apply to parts compression molded with precision matched-metal molds. BMC, SMC, and preform are included.
†Class B tolerances apply to parts press molded with somewhat less precise metal molds. Cold-press molding, casting, centrifugal casting, rotational molding, and cold stamping can apply to this classification when molding is done with a high degree of care. BMC, SMC, and preform compression molding can apply to this classification if extra care is not used.
**Class C applies to hand lay-up, vacuum bag, and other methods using molds made of RP/C material. It applies to parts that would be covered by Class B when they are not molded with a high degree of care.

Establishing initial design tolerances is often done on an arbitrary, uninformed basis. If the initial estimated tolerance proves too great, a lower shrink plastic could be used to reduce the shrinkage range. However, if the key dimension was across a main parting line, the tooling could be redesigned to eliminate the condition and consequently reduce variation from tool construction. Even with all these data processed by a computer, estimating tolerances is difficult if they are not properly interrelated with the highly dependent factors of the material behavior and tooling designs.

Viscoelasticity

The flow of plastics is compared to that of water in Fig. 8-5 to show their different behaviors. The volume of a so-called Newtonian fluid, such as water, when pushed through an opening is directly proportional to the pressure applied (the straight dotted line), the flow rate of a non-Newtonian fluid such as plastics when pushed through an opening increases more rapidly than the applied pressure (the solid curved line). Different plastics generally have their own flow and rheological rates so that their non-Newtonian curves are different.

With plastics there are two types of deformation or flow; viscous, in which the energy causing the deformation is dissipated, and elastic, in which that energy is stored. The combination produces viscoelastic plastics. See Chapter 2, **MATERIAL BEHAVIOR, Rheology and Viscoelasticity**, regarding their effects on fabricated products.

Not only are there two classes of deformation, there are also two modes in which deformation can be produced: simple shear and simple tension. The actual action during melting, as in the usual screw plasticator is extremely complex, with all types of shear-tension combinations. Together with engineering design, deformation determines the pumping efficiency of a screw plasticator and

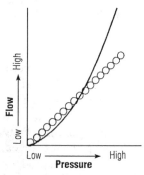

Fig. 8-5 Rheology and flow properties of plastics (solid curve) and water (dotted curve).

controls the relationship between output rate and pressure drop through a die system or into a mold.

Shear rate   When a melt moves in a direction parallel to a fixed surface, such as with a screw barrel, mold runner and cavity, or die wall, it is subject to a shearing force. As the screw speed increases, so does the shear rate, with potential advantages and disadvantages. The advantages of an increased shear rate are a less viscous melt and easier flow. This shear-thinning action is required to "move" the melt.

A disadvantage observed with higher shear rates is that too high a heat increase may occur, potentially causing problems in cooling, as well as degradation and discoloration. A high shear rate can lead to a rough product surface from melt fracture and other causes. For each plastic and every processing condition there is a maximum shear rate beyond which such problems can develop.

When water (a Newtonian liquid) is in an open-ended pipe, pressure can be applied to move it. Doubling the water pressure doubles the flow rate of the water. Water does not have a shear-thinning action. However, in a similar situation but using a plastic melt (a non-Newtonian liquid), if the pressure is doubled the melt flow may increase from 2 to 15 times, depending on the plastic used. As an example, linear low-density polyethylene (LLDPE), with a low shear-thinning action, experiences a low rate increase, which explains why it can cause more processing problems than other PEs. The higher-flow melts include polyvinyl chloride (PVC) and polystyrene (PS).

Model/Prototype Building

Model building or prototyping is a key step in product design. This stage, which is usually expensive and time consuming, often account for more than half the time taken for design. Almost every type of new product design involves creating one or more prototypes prior to production. Different modeling systems that have recently become available now per-

mit making them in a matter of a few hours or overnight where previously modeling was only available that took from days to months (Chapter 3, **PROTOTYPE** and Chapter 4, **BOOK SHELVE**).

For designers using CAD, even seeing a product on a high-resolution graphics screen is sometimes not enough. The physical design can bring to life a high-tech design, along with formerly unnoticed flaws. By quickly forming 3-D conceptual models from design ideas, designers can evaluate a design concept, demonstrate its feasibility, and then sell the new idea.

Prototyping basically provides a 3-D model suitable for use in the preliminary evaluation of form, design, performance, and material processing of products, molds, dies, etc. When properly used this automatic/fast system can accelerate product development, improve product quality, and time to the market for a product.

Processing Behavior

Understanding and measuring melt flow during processing is important for two reasons. First, it provides a means for determining whether a plastic can be formed into a useful product such as a usable extruded extrudate, completely fill a mold cavity, provide mixing action in a screw, meet product thickness requirements, etc. Second, the flow is an indication of whether its final properties will be consistent with those required by the product. The target is to provide the necessary homogeneous-uniformly heated melt during processing to have the melt operate completely stable and working in equilibrium.

In practice, even though with the developments that has occurred in the past and continues this perfect stable situation is never achieved and there are variables that affects the output. If the process is analyzed one can decide that two types of variables affect the quality and output rate. They can be identified as (1) the variables of the machine's design and manufacture and (2) the operating or dynamic variables that control how the machine is run.

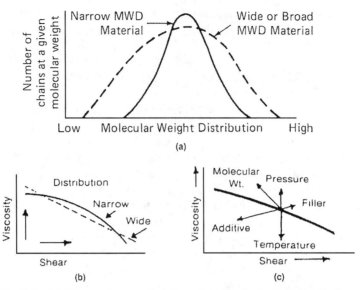

Fig. 8-6   Melt flow rates as a function of molecular weight distribution.

Rheology and Melt Flow

Rheology is the science of the deformation and flow of matter under force. It is concerned with the response of plastics to mechanical force. An understanding of rheology and the ability to measure rheological properties such as molecular weight and melt flow is necessary before flow behavior can be controlled during processing. Such control is essential for the fabrication of plastic materials to meet product performance requirements (Chapter 2, **MATERIAL BEHAVIOR, Rheology and Viscoelasticity**).

Molecular Weight Distribution and Melt Flow

One method of defining plastics is to use their molecular weight (MW), a reference to the plastic molecules' weight and size. Here MW refers to the average weight of plastics that is always composed of different weight molecules. These differences are important to the processor, who uses the molecular weight distribution (MWD) to evaluate materials. A narrow MWD enhances the performance of plastic products. Wide MWD permits easier processing. Melt flow rates is dependent on

the MWD. Figure 8-6 relates melt flow to MWD starting with (a) MWD curves and followed with (b) viscosities vs. shear rates as related to MWD and (c) factors influencing viscosities. With MWD differences of incoming material the fabricated performances can be altered; the more the difference, the more dramatic change occurs in the products.

MW is the sum of the atomic weights of all the atoms in a molecule. It represents a measure of the chain length for the molecules that make up the polymer and in turn the plastic that influences processing performances to meet product performance (2). The MWD is basically the amount of component polymers that go to make up a polymer. Component polymers, in contrast, are a convenient term that recognizes the fact that all plastic materials comprise a mixture of different polymers of differing molecular weights.

The average molecular weight is the sum of the atomic masses of the elements forming the molecule, indicating the relative typical chain length size of the polymer molecule. Many techniques are available for the determination of these molecular weight characteristics that can be related to processing performance. For example, two plastics may have exactly the same or similar average MWs but very different MWDs with the result that processing performances and performance properties of the products vary. Result could be OK but also not OK.

Melt Flow and Viscosity

When reviewing the subject of plastic melt flow, the subject of viscosity is involved. Basically viscosity is the property of the resistance of flow exhibited within a body of material. Ordinary viscosity is the internal friction or resistance of a plastic to flow. It is the constant ratio of shearing stress to the rate of shear. Shearing is the motion of a fluid, layer by layer, like a deck of cards. When plastics flow through straight tubes or channels they are sheared and the viscosity expresses their resistance.

The melt index (MI) or melt flow index (MFI) is an inverse measure of viscosity. High MI implies low viscosity and low MI means high viscosity. Plastics are shear thinning, which means that their resistance to flow decreases as the shear rate increases. This is due to molecular alignments in the direction of flow and disentanglements.

Viscosity is usually understood to mean Newtonian viscosity in which case the ratio of shearing stress to the shearing strain is constant. In non-Newtonian behavior, which is the usual case for plastics, the ratio varies with the shearing stress (Fig. 8-5). Such ratios are often called the apparent viscosities at the corresponding shearing stresses. Viscosity is measured in terms of flow in Pa's, with water as the base standard (value of 1.0). The higher the number, the less flow.

Newtonian flow   It is a flow characteristic where a material (liquid, etc.) flows immediately on application of force and for which the rate of flow is directly proportional to the force applied. It is a flow characteristic evidenced by viscosity that is independent of shear rate. Water and thin mineral oils are examples of fluids that posses Newtonian flow.

Non-Newtonian flow   Plastic melts are non-Newtonian. They have basically abnormal flow response when force is applied. That is, their viscosity is dependent on the rate of shear. They do not have a straight proportional behavior with application of force and rate of flow. When proportional, the behavior has a Newtonian flow. Deviations from this ideal behavior may be of several different types. One type called apparent viscosity may not be independent of the rate of shear; it may increase with shear rate (shear thickening or shear dilatancy or decrease with rate of shear (shear thinning or pseudoplasticity). The latter behavior is usually found with plastic melts and solutions. In general such a dependency of shear stress on shear rate can be expressed as a power law. Another type is where the viscosity may be time dependent, as for material exhibiting thixotropy or rheopexy (2, 3, 6).

Melt Flow Rate

MFR tests are used to detect degradation in fabricated products where comparisons, as an example, are made of the MFR of pellets to the MFR of product. MFR has a reciprocal relationship to melt viscosity. This relationship of MW to MFR is an inverse one; as the MFR increases, the MW drops. MW and melt viscosity is also related; as one increases the other increases.

Melt index test   The melt indexer (extrusion plastometer) is the most widely used rheological device for examining and studying plastics in many different fabricating processes. It is not a true viscometer in the sense that a reliable value of viscosity cannot be calculated from the flow index that is normally measured. However, it does measure isothermal resistance to flow, using an apparatus and test method that are standard throughout the world. The standards used include ASTM D 1238 (U.S.A.), BS 2782-105C (U.K.), DIN 53735 (Germany), JIS K72 IO (Japan), ISO RI 133/R292 (international), and others.

In this MI instrument the plastic is contained in a barrel equipped with a thermometer and surrounded by an electrical heater and an insulating jacket. A weight drives a plunger that forces the melt through the die

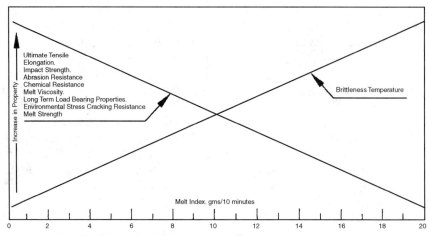

Fig. 8-7   Effect of MI on the properties of polyethylene.

opening, using a standard opening of 2.095 mm (0.0824 in.) and a length of 8 mm (0.315 in.). The standard procedure involves the determination of the amount of plastic extruded in 10 min. The flow rate (expressed in g/10 min.) is reported. As the flow rate increases, viscosity decreases. Depending on the flow behavior, changes are made to standard conditions (die opening size, temperature, etc.) to obtain certain repeatable and meaningful data applicable to a specific processing operation.

The Ml test equipment is easy to operate, provides repeatable results, and low cost to operate. It is widely used for quality control and for distinguishing between members of a single family of plastics. Specifically, this MI makes a single-point test that provides information on resistance to flow at only a single shear rate. Because variations in branching or MWD can alter the shape of the viscosity curve, the MI may give a false ranking of plastics in terms of their shear rate resistance to flow. To overcome this problem, extrusion rates are sometimes measured for two loads, or other modifications are made.

In summary, the MI is an indicator of the average molecular weight (MW) of a plastic and is also a rough indicator of processability. Low MW materials have high Mls and are easy to process. High MW materials have low Mls and are more difficult to process, as they have more resistance to flow, but they are processable. End-use physical properties

improve as the Ml decreases (Figs. 8-7 and 8-8). Because processability simultaneously decreases, Ml selection for a given application is a compromise between properties and processability. Table 8-9 lists typical MI ranges for the more common plastic processes and materials. Materials with other Mls are still processable, but they usually require more sophisticated start-up procedures and operating process controls. Table 8-10 shows how MI and density interrelate.

Melt Flow and Elasticity

As a melt is subjected to a fixed stress or strain, the deformation versus time curve will show an initial rapid deformation followed by a continuous flow. Elasticity and strain are compared in Fig. 8-9 that provides (a) basic deformation vs. time curve, (b) stress-strain deformation vs. time with the creep effect, (c) stress-strain deformation vs. time with the stress-relaxation effect, (d) material exhibiting elasticity, and (e) material exhibiting

Table 8-9   Typical melt index ranges for common process

Process	MI Range
Injection molding	5–100
Rotational molding	5–20
Film extrusion	0.5–6
Blow molding	0.1–1
Profile extrusion	0.1–1

Table 8-10 Performance influenced by MI and density of plastics

	With Increasing Melt Index	With Increasing Density
Rigidity	—	Increases
Heat resistance	Decreases	Increases
Stress crack resistance	Decreases	Decreases
Permeation resistance	—	Increases
Abrasion resistance	—	Increases
Clarity	—	Decreases
Flex life	Decreases	Decreases
Impact strength	Decreases	Decreases
Gloss	Increases	Increases
Vertical crush resistance	—	Increases
Cycle	Decreases	Decreases
Flow	Increases	Decreases
Shrinkage	Decreases	Increases
Parison roughness	Decreases	Increases
Parison sag	Increases	Decreases
Pinch quality	Increases	—
Parting line difference	—	Increases

plasticity. The relative importance of elasticity (deformation) and viscosity (flow) depends on the time scale of the deformation. For a short time elasticity dominates, but over a long time the flow becomes purely viscous. This behavior influences processes.

Deformation contributes significantly to process-flow defects. Melts with only small deformation have proportional stress-strain behavior. As the stress on a melt is increased, the recoverable strain tends to reach a limiting value. It is in the high stress range, near the elastic limit, that processes operate.

Molecular weight, temperature, and pressure have little effect on elasticity; the main controlling factor is MWD. Practical elasticity phenomena often exhibit little concern for the actual values of the modulus and viscosity. Although MW and temperature influence the modulus only slightly, these parameters have a great effect on viscosity and thus can alter the balance of a process.

Flow Performance

In any practical deformation there are local stress concentrations. Should the viscosity

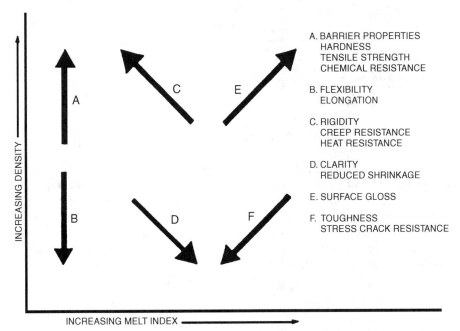

Fig. 8-8 Effect of density and MI changes on the properties of polyethylene, with properties increasing in the direction of the arrows.

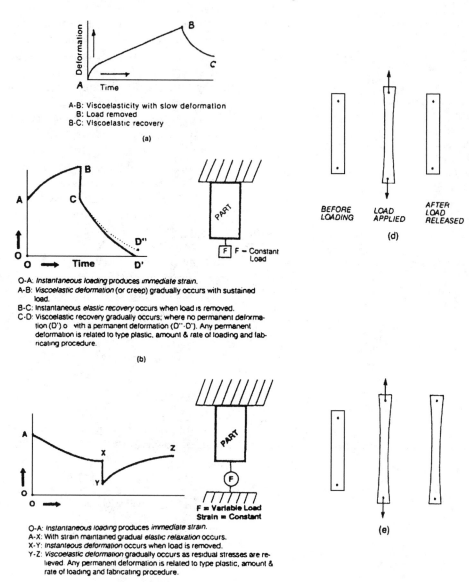

Fig. 8-9 Elasticity and strain curves.

increase with stress, the deformation at the stress concentration will be less rapid than in the surrounding material. The stress concentration will be smooth and the deformation stable. However, when the viscosity decreases with increased stress, any stress concentration will cause catastrophic failure.

Flow Defect

Flow defects, especially as they affect the appearance of a product, play an important role in many processes. Flow defects are not always undesirable, as for example in producing a matt finish. Six important types of defects can be identified, which are applied here to extrusion because of its relative simplicity. These flow analyses can be related to other processes and even to the complex flow of injection molding.

Nonlaminar flow Ideally, it is a melt flow in a steady, streamlined pattern in and/or out of a tool (die, mold, etc.). Actually, the melt is distorted, causing defects called melt fracture or elastic turbulence. To reduce or eliminate this problem, the entrance to the die or mold is tapered or streamlined.

Sharkskin During flow the melt next to the metal tends not to move, whereas that in the center flows rapidly. When the melt flow pressure is relieved, its flow profile is abruptly changed to a uniform velocity. This change requires rapid acceleration of the surface layer, resulting in high local stress. If this stress exceeds some critical value the surface breaks, giving the rough appearance called sharkskin. With the rapid acceleration, the deformation is primarily elastic. Thus the highest surface stress, and worst sharkskin, will occur in plastics with a high modulus and high viscosity, or in high molecular weight plastics of narrow MWD at low temperatures and high processing rates. The addition of controlled heating, locally reducing the viscosity, is effective in reducing sharkskin.

Nonplastication This condition produces uneven stress distribution, with consequent lumpiness. The product could appear ugly or have a fine matt finish. With a wide MWD their could be a lack of gloss.

Volatile There are plastics that contain small quantities of material that boil at processing temperatures, or they may be contaminated by water absorbed from the atmosphere. These volatiles may cause bubbles, a scarred surface, and other defects. Processing methods of removing volatiles are used such as drying materials to be processed, vented plasticator barrels, etc. (Chapter 7, **OTHER BEHAVIOR, Drying Plastic**).

Shrinkage The transition from room temperature to a high processing temperature may decrease a plastic's density by up to 25%. Cooling causes possible shrinkage (up to 3%) and may cause surface distortions or voiding with internal frozen strains. As discussed in other chapters, this situation can be reduced or eliminated by special techniques, such as controlled cooling under pressure.

Melt structure High shear at a temperature not far above the melting point may cause a melt to take on too much molecular order. In turn, distortion could result.

Thermodynamic

With the heat exchange that occurs during processing, thermodynamics becomes important. It is the high heat content of melts (about 100 cal/g) combined with the low rate of thermal diffusion (10^{-3} cm^2/s) that limits the cycle time of many processes. Also important are density changes, which for crystalline plastics may exceed 25% as melts cool. Melts are highly compressible; a 10% volume change for a force of 700 kg/cm^2 (10,000 psi) is typical. A surface tension of about 20 g/cm may be typical for film and fiber processing when there is a large surface-to-volume ratio (Appendix B, **TERMINOLOGY, Thermodynamic**).

Residence Time

It is the amount of time a plastic is subjected to heat during the fabrication of plastics such as in a plasticator during extrusion, injection molding, or blow molding. Too long a residence time (and/or to high a heat) can result in over heating the melt. Depending on the plastic being melted a relatively minor to a definite major problem could result lowering the fabricated product's performance. With recycled plastics the residence time includes the previous melting action time plus the reprocessing melt action time. If there is a third recycling of the same plastic, add that time period . . . . and so on.

Chemical Change

The chemical changes that can occur during processing and effect product performances include (1) continued polymerization and cross-linking, which increases viscosity; (2) depolymerization or damaging of molecules, which reduces viscosity; and (3) complete changes in the chemical structure, which may cause color changes. Already degraded plastics may catalyze further degradation.

Trend

Because melts have different properties and there are many ways to control processes, detailed factual predictions of final output are difficult to arrive. Research and hands-on operation have been directed mainly at explaining the behavior of melts or plastics like with other materials (steel, glass, and so on). Modem equipment and controls are overcoming some of this unpredictability. Ideally, processes and equipment should be designed to take advantage of the novel properties of plastics rather than to overcome them.

Processing And Property

Problem/Solution

In order to understand potential problems and their solutions, it is helpful to consider the relationships of machine capabilities, plastics processing variables, and part performance. Chapter 3, **FEATURE INFLUENCING PERFORMANCE** provides a preliminary analysis to this subject.

A distinction should be made between machine conditions and processing variables. Machine conditions are basically temperature, pressure, and processing time (such as screw rotation/rpm, and so on) in the case of a screw plasticator, die and mold temperature and pressure, machine output rate (lb./hr), and the like. Processing variables are more specific such as the melt temperature in the die or mold, melt flow rate, and pressure used.

The distinction between machine conditions and fabricating variables is a necessary one to avoid mistakes in using problem-and-solution or cause-and-effect relationships to advantage. If the processing variables are properly defined and measured, not necessarily the machine settings, they can be directly

correlated with the products' properties. For example, if one increases cylinder temperature, melt temperatures do not necessarily also increase. Melt temperature is also influenced by screw design, screw rotation rate, back pressure, and dwell times. It is much more accurate to measure melt temperature and correlate it with properties than to correlate cylinder settings with properties.

The problem-solving approach that ties the processing variables to products' properties includes considering melt orientation, polymer degradation, free volume/molecular packing and relaxation, cooling stresses, and other such factors. The most influential of these four conditions is melt orientation, which can be related to molded-in stress or strain.

Plastic degradation can occur from excessive melt temperatures or abnormally long times at temperature, called the residence time or heat history from plasticator to cooling of the product. Excessive shear can result from poor screw design, too much screw flight-to-barrel clearance, cracked or wornout flights, and such. Orientation in plastics refers simply to the alignment of the melt-processing variables that definitely affect the intensity and performance of orientation.

Plastic with a Memory

Thermoplastics can be bent, pulled, or squeezed into various useful shapes. But eventually, especially if heat is added, they return to their original form. This behavior, known as plastic memory, can be annoying. If properly applied, however, plastic memory offers interesting design possibilities for all types of fabricated parts (Chapter 6, **PLASTICS WITH A MEMORY**).

Orientation

A plastic's molecular orientation can be developed during processing. Initially the melted molecules are relaxed. During processing the molecules tend to be more

Table 8-11   Effects of orientation of polypropylene films

Properties	Stretch (%)				
	None	200	400	600	900
Tensile strength, psi	5,600	8,400	14,000	22,000	23,000
(MPa)	(38.6)	(58.0)	(96.6)	(152.0)	(159.0)
Elongation at break, %	500	250	115	40	40

Properties	As Cast	Uniaxial Orientation	Balanced Orientation
Tensile strength, psi (MPa)			
MD*	5,700 (39.3)	8,000 (55.2)	26,000 (180)
TD†	3,200 (22.1)	40,000 (276)	22,000 (152)
Modulus of elasticity. psi			
MD	96,000 (660)	150,000 (1,030)	340,000 (2,350)
TD	98,000 (680)	4000,000 (2,760)	330,000 (2,280)
Elongation at break, %			
MD	425	300	80
TD	300	40	65

*MD = Machine direction.
†TD = Transverse direction and direction of universal orientation.

oriented than relaxed, particularly when sheared, as during flow in an injection mold or through an extrusion die. After temperature-time-pressure is applied and the melt goes through restrictions (molds, dies, etc.), the molecules tend to be stretched and aligned in a parallel form. The result is a change in directional properties and dimensions.

The amount of change depends on the type of thermoplastic, the amount of restriction, and, most important, its rate of cooling. The faster the rate, the more retention there is of the frozen orientation. After processing, products could be subject to stress relaxation, with changes in performance and dimensions. With certain plastics and processes there is an insignificant change. If changes are significant and undesirable, one must take action to change the processing conditions, particularly increasing the cooling rate.

By deliberate stretching, the molecular chains of a plastic are drawn in the direction of the stretching, and inherent strengths of the chains are more nearly realized than they are in their naturally relaxed configurations. Stretching can take place with heat during or after processing by blow molding, extruding

(review in this Chapter **EXTRUSION, Orientation**), thermoforming, etc. Products can be drawn in one direction (uniaxially) or in two opposite directions [biaxially, also called bioriented (BO)], in which case many properties significantly increase uniaxially or biaxially (Table 8-11 and Fig. 8-10).

Molecular orientation results in increased stiffness, strength, and toughness (Table 8-12); as well as resistance to liquid and gas permeation, crazing, microcracks, and others in the direction or plane of the orientation. The orientation of fibers in reinforced plastics causes similar positive influences. Orientation in effect provides a means of tailoring and improving the properties of plastics.

Considering a fiber or thread of nylon-66, which is an unoriented glassy polymer, its modulus of elasticity is about 2,000 MPa (300,000 psi). Above the T_g its elastic modulus drops even lower, because small stresses will readily straighten the kinked molecular chains. However, once it is extended and has its molecules oriented in the direction of the stress, larger stresses are required to produce added strain. The elastic modulus increases.

Table 8-12   Effect of molecular orientation on the impact properties of polypropylene films

ASTM Tensile Impact, ft.-lb./in.^{2}		
	Temperature	
Material	Room	−20°F (−29°C)
Unoriented PP	40	0
Oriented PP	avove test limit	500

High-Energy Fatigue Impact (55 lb. weight @ 50-in. height) (24.9 kg weight at 127-cm height)	
Material	Number of drops to failure
Steel	12
Unoriented PP	
41 × 10^3 psi tensile	1
Oriented PP	
28 × 10^3 psi, 32%	
elongation	130

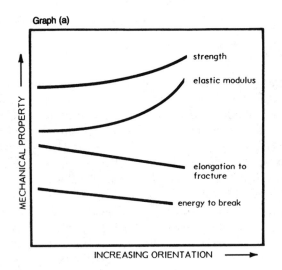

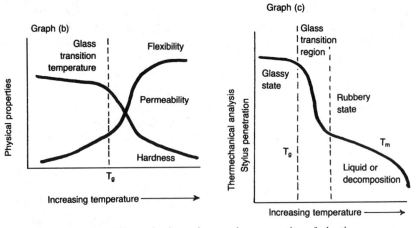

Fig. 8-10   Effect of orientation on the properties of plastics.

The next stop is to cool the nylon below its T_g without removing the stress, retaining its molecular orientation. The nylon becomes rigid with a much higher elastic modulus in the tension direction [15,000 to 20,000 MPa (2 to 3 × 10^6 psi)]. This is nearly ten times the elastic modulus of the unoriented nylon-66 plastic. The stress for any elastic extension must work against the rigid backbone of the nylon molecule and not simply unkink molecules. This procedure has been commonly used in the commercial production of man-made fibers since the 1930s via DuPont.

Another example of the many oriented products is the heat-shrinkable material found in flat or tubular films or sheets. The orientation in this case is terminated downstream of an in-line extrusion-stretching operation when a cool enough temperature is achieved. Orientation can also be performed as a secondary operation. Reversing the operation, shrinkage occurs. The reheating and subsequent shrinking of these oriented plastics can result in a useful property. It is used, for example, in heat-shrinkable PE wrapping of packages, flame-retardant PP tubes, flat communication cable wraps, furniture webbings, pipe fittings, medical devices, and many other products (Chapter 4, **TRANSPARENT AND OPTICAL PRODUCT, Polarized Lighting**).

Directional Property

Another important orienting fabricating procedure concerns applying directional properties to reinforced plastics. This subject is reviewed in Chapter 3, **DESIGN CONCEPT, Reinforced Plastic Directional Property**.

FALLO Approach

All processes fit into a scheme that requires the interaction and proper control of different operations. An example is the FALLO (Follow **ALL O**pportunities) approach that makes one aware that many steps are involved in processing and all must be properly understood and coordinated (Fig. 1-3). Basically the FALLO approach diagram consists of: (1) designing a product to meet performance and manufacturing requirements at the lowest cost; (2) specifying the proper plastic material that meet product performance requirements after being processed; (3) specifying the complete equipment line by (a) designing the tool (mold, die) "around" the product, (b) putting the "proper performing" fabricating process "around" the tool, (c) setting up auxiliary equipment (up-stream to down-stream) to "match" the operation of the complete line, and (d) setting up the required "complete controls" (such as testing, quality control, troubleshooting, maintenance, data recording, etc.) to produce "zero defects"; and (4) purchasing and properly warehousing plastic materials. Using this type of approach leads to maximizing product's profitability.

Tooling

Tools include dies, molds, mandrel, jigs, fixtures, punch dies, etc. for shaping and fabricating parts (Tables 8-13 and 8-14).

Mold

Molds are used in many plastic processes with many of the molds having common assembly parts (Fig. 8-11). Many molds, particularly for injection molding, have been preengineered as standardized products that can be used to include cavities, different runner systems, cooling lines, unscrewing mechanisms, etc. (Table 3-17).

There are different types of injection molds that permit processing plastics by different techniques (Fig. 8-12). To meet different product shape requirements mold design operations range from simple to very complex (Figs. 8-13 to 8-15) designs. Target is to design products that will require the easiest mold to design. A mold can be a highly sophisticated/ expensive piece of equipment. It can

Table 8-13   Guide to tooling selection

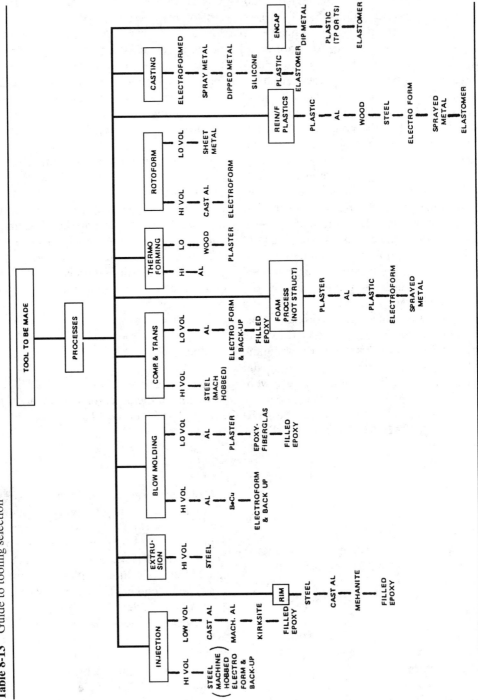

Table 8-14 Average properties and relative cost of certain tool alloys

Alloy Designation	Tensile Strength 0.2% Yield	Working Hardness	Coeff. of Thermal Expansion	Thermal Conductivity	Cost Index
	KSI	Rockwell C	$10^6 /°F$	$BTU/FT^2/HR/°F$	(AISI 4140 = 1)
AISI 4140	100	27–30	12.7	19	1.0
AISI P 20	120	28–35	7.1	17	1.3
AISI H 13	180	40–45	7.1	17	3.5
UNS S42000	200	28–30	6.5	15	2.5
PH 15.5	175	38–40	6.2	12	4.5
MAR 18(300)	290	48–56	5.6	17	7.5

comprise of many parts requiring high quality metals and precision machining. To capitalize on its advantages, the mold may incorporate many cavities, adding further to its complexity.

Production molds are usually made from steel for pressure molding that requires heating or cooling channels, strength to resist the forming forces, and/or wear resistance to withstand the wear due to plastic melts, particularly that which has glass and other abrasive fillers. However most blow molds are cast or machined from aluminum, beryllium copper, zinc, or Kirksite due to their fast heat transfer characteristics. But where they require extra performances steel is used.

The design and construction of a mold plays a significant role in the dimensional integrity of its product. The cavity that forms the final product can be shaped using a variety of steel removal methods from jig grinding to wire-feed electrical discharge machining. Each removal method has a corresponding range of variation. Although the tolerances of these processes are geometry dependent, some processes are more accurate than others. Studies can be performed to determine an average range of variation for each method. Among other important tool design and construction considerations for determining tool performance for a given dimension are the mold's construction details, such as its main parting lines. All these factors

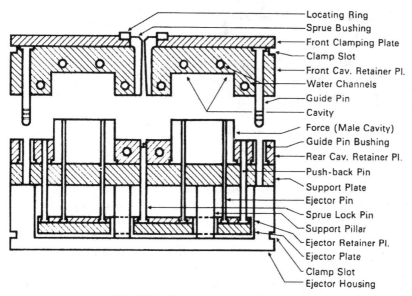

- Locating Ring
- Sprue Bushing
- Front Clamping Plate
- Clamp Slot
- Front Cav. Retainer Pl.
- Water Channels
- Guide Pin
- Cavity
- Force (Male Cavity)
- Guide Pin Bushing
- Rear Cav. Retainer Pl.
- Push-back Pin
- Support Plate
- Ejector Pin
- Sprue Lock Pin
- Support Pillar
- Ejector Retainer Pl.
- Ejector Plate
- Clamp Slot
- Ejector Housing

Fig. 8-11 Two-part standard mold.

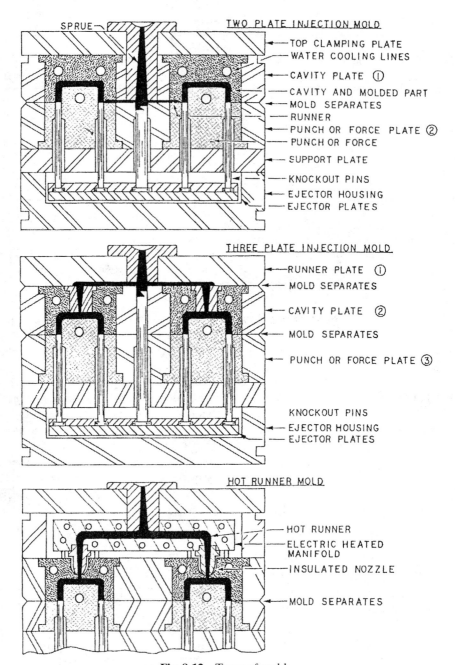

SPRUE

TWO PLATE INJECTION MOLD

— TOP CLAMPING PLATE
— WATER COOLING LINES
— CAVITY PLATE ①
— CAVITY AND MOLDED PART
— MOLD SEPARATES
— RUNNER
— PUNCH OR FORCE PLATE ②
— PUNCH OR FORCE
— SUPPORT PLATE
— KNOCKOUT PINS
— EJECTOR HOUSING
— EJECTOR PLATES

THREE PLATE INJECTION MOLD

— RUNNER PLATE ①
— MOLD SEPARATES
— CAVITY PLATE ②
— MOLD SEPARATES
— PUNCH OR FORCE PLATE ③
KNOCKOUT PINS
— EJECTOR HOUSING
— EJECTOR PLATES

HOT RUNNER MOLD

— HOT RUNNER
— ELECTRIC HEATED MANIFOLD
— INSULATED NOZZLE
— MOLD SEPARATES

Fig. 8-12   Types of molds.

add to the overall variations in related dimensions of the mold's products (3, 7, 9, 10, 20).

Die

The function of a die is to accept the available melt from an extruder and deliver it to takeoff equipment as a shaped extrudate (profile, film, sheet, pipe, filament, etc.) with minimum deviation in cross-sectional dimensions and a uniform output by weight, at the fastest possible rate (Figs. 3-34 to 3-36). A well-designed die should permit color and compatible plastic changes quickly with little off-grade material. It will distribute the

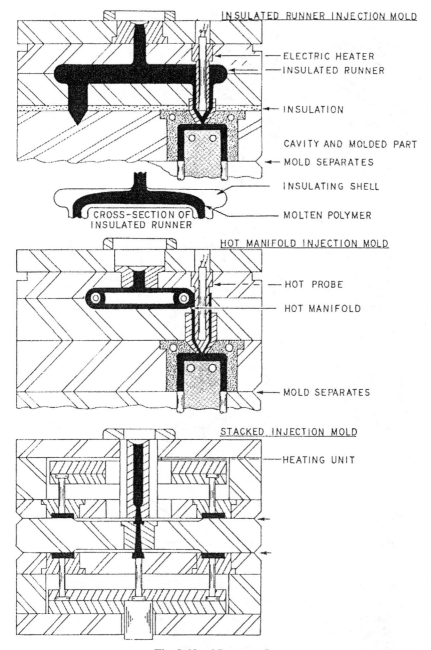

Fig. 8-12   (*Continued*)

melt in the die flow channels so that it exists with a uniform density and velocity. Figure 8-16 is an example of one type of die; different designs are use to extrude different products.

The flow rate is influenced by all the variables that can exist in preparing the melt during extrusion such as die heat and pressure with time in the die. Unfortunately, in spite of all the sophisticated plastic flow analysis and the rather mechanical computer-aided design capabilities, it is very difficult to design a die. Experience (most important) with an empirical approach must be used, as it is quite difficult to determine the optimum flow channel geometry from engineering calculations. It is important to employ rheological flow

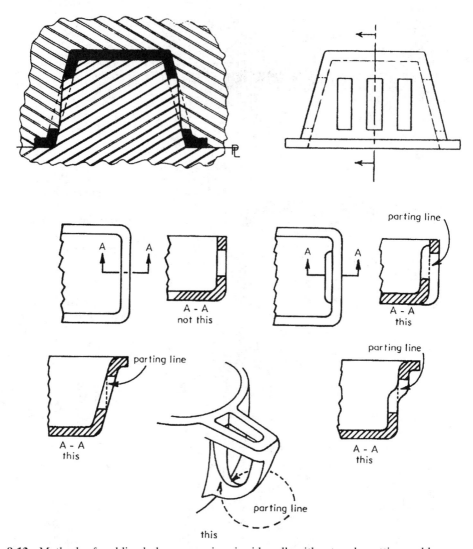

Fig. 8-13   Methods of molding holes or openings in sidewalls without undercutting mold movement.

properties and other melt behavior via the applicable CAD programs for the type of die required. The most important ingredient is experience, which, for the novice, is hopefully properly recorded in a computer program. Nevertheless, die design has remained more

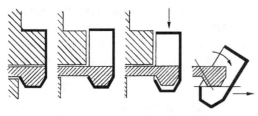

Fig. 8-14   Slide within a slide mold.

of an art than any other aspect of process design. Dies can work only if the operator of the processing line has developed the important ability to debug through proper process controls.

A well-built die with adjustments (temperature changes, restrictor/choker bars, valves and other devices) may be used with a particular group of materials. Usually a die is designed for a specific plastic meeting its particular rheological behaviors. To simplify the processing operation, the die design should consider certain factors. If possible the goals are to have the extrudate (product) of a uniform wall thickness (otherwise the heat

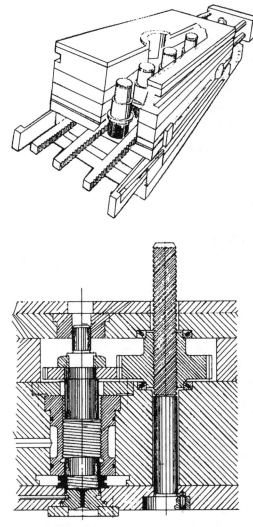

that when a melt is extruded from the die, there is some swelling (as reviewed latter in the **Extrusion** section). After exiting the die, it is usually stretched or drawn down to a size equal to or smaller than the die opening. The dimensions are then reduced proportionally so that in an ideal plastic the drawn-down section is the same as the original section.

Because of the melt-elasticity effects of the material, it does not draw down in a simple proportional manner; thus, the draw-down process is a source of errors in the profile. Errors are significantly reduced in a balanced situation such as circular extrudate. These errors must be corrected by modifying the die and takeoff equipment.

There are substantial influences on a material created by the flow orientation of the molecules, so there are different properties in the flow direction and perpendicular to the flow. These differences have a significant effect on the performance of the part (Chapter 2).

The pumping pressure required on the melts entering the different designed die heads differs to meet their melt flow patterns within the die cavities. The pressure usually varies as follows: (1) blown and lay-flat films at 13.8–41.3 MPa (2000–6000 psi); (2) cast film, sheet, and pipe at 3.5–27.6 MPa (500–4000 psi); (3) wire coating at 10.3–55.1 MPa (1500–8000 psi), and (4) monofilament at 6.9–20.7 MPa (1000–3000 psi).

Fig. 8-15 Method of molding threaded caps.

transfer problem is magnified); to minimize the use of hollow sections; to minimize narrow or small channels; and to use generous radii on all comers, such as a minimum of 0.5 mm (0.02 in.). An "impossible" or difficult process can still be designed, but it requires extensive experience (both practical and theoretical), with trial-and-error runs, to make it practical.

Basics of Flow

The non-Newtonian behavior of a plastic melt makes its flow through a die somewhat complicated. One characteristic of plastic is

Injection Molding

Continued development in the IM process is due to its worldwide large sales of IM equipment that are required to meet new processing demands. Equipment involved is approximately $4.5 billion/yr. (USA $1.35 billion/yr) with estimates at 30% in machines, 60% in molds, 6% in robots, and 4% in hot runners. Marketwise 55% are technical products (electronic, mechanical, medical, etc.), 20% automotive, 10% packaging, and 15% others. Worldwide approximately $180 billion/yr. sales exist for IM products (2, 3).

The IM process is greatly preferred by designers because the manufacture of products

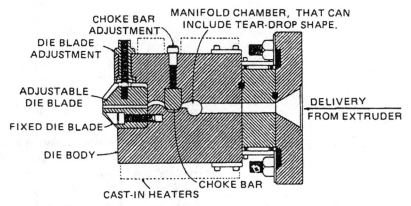

Fig. 8-16   Simplified schematic of an extruder sheet die.

in complex 3-D shapes can be more accurately controlled dimensionwise and tolerancewise with IM than with other processes. As its method of operation is much more complex than others, IM requires a thorough understanding of the process. There are basically two IMM systems that are the reciprocating and two-stage systems (3).

Figure 8-17 shows a reciprocating schematic of the load profile that highlights the way in which the melt is plasticized (softened) and forced into the mold, the clamping system for opening and closing the mold under pressure; the type of mold used, and the machine controls. There are many different types or designs of IMMs (IM machines) that permit molding many different type products based on factors such as quantities, sizes (from micro to large sizes), shapes, product performances, type plastic and/or economics (3).

Plastic moves from the hopper onto the feeding portion of the reciprocating extruder screw. The flights of the rotating screw cause the material to move through a heated extruder barrel where it softens (is made fluid) so that it can be fed into the shot chamber (front of screw). This motion generates pressure [usually 50–300 psi (0.35–2.07 MPa)], which causes the screw to retract. When the preset limit switch (or a position transducer) is reached the shot size is met and the screw stops rotating. Basically at a preset time the screw acts as a ram to push the melt into the mold. Injection takes place at high pressure [usually 2,000–5000 and also up to 30,000 psi (14–35 up to 210 MPa)]. There is also limited use of lower pressure operations [usually down to 50–300 psi (0.34–2.07 MPa)]. The low pressure systems have been used in molding foams that are small to very large

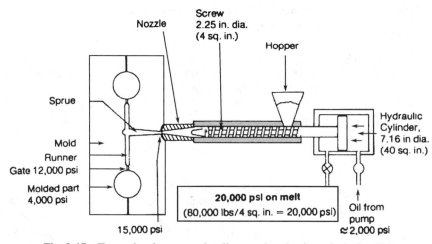

Fig. 8-17   Example of pressure loading on the plastic melt during IM.

products, decorative products such as over- or in-molding, etc. Adequate clamping pressure must be used to eliminate mold opening that would cause flashing. The melt pressure within the mold cavity can range from 1 to 15 tons/sq. in. The pressure used is dependent on the plastic's rheology/flow behavior (Chapter 2).

Time, pressure, and temperature controls indicate whether the performance requirements of a molded product are being met. The time factors include the rate of injection, duration of ram pressure, time of cooling, time of plastication, and screw RPM. Pressure requirement factors relate to injection high and low pressure cycles, back pressure on the extruder screw, and pressure loss before the plastic enters the cavity which can be caused by a variety of restrictions in the mold. The temperature control factors are in the mold (cavity and core), barrel, and nozzle, as well as the melt temperature from back pressure, screw speed, frictional heat, and so on in the plasticator.

Even though most of the literature on processing specifically identifies or refers to thermoplastics (TPs) as in this book, some thermosets (TSs) are used (TS polyesters, phenolics, epoxy, etc.). The TPs reach maximum heat prior to entering "cool" mold cavities, whereas the TSs reach their maximum temperature in "hot" molds (Fig. 6-3).

IM is a repetitive process in which melted (plasticized) plastic is injected or forced into a mold cavity(s) where it is held under pressure until removed in a solid state, basically duplicating the cavity of the mold. The mold may consist of a single cavity or a number of similar or dissimilar cavities, each connected to flow channels or runners that direct the flow of the melt to the individual cavities. Three basic steps occur: (1) raising the plastic temperature in the injection or plasticizing unit so that it will flow under pressure, (2) allowing the plastic melt to solidify in the mold via the mold's cooling action, and (3) opening the mold to eject the molded product(s).

There can be a single gate or multiple gates in a mold cavity that depends on melt flow pattern desired in the cavity. Other considerations exist such as determining whether a high flow rate or low melt flow rate is required in the cavity. Target is to eliminate or reduce any potential weld lines that can effect strength and/or appearance (Fig 8-18), and whether a smooth process sequence is used or an abrupt change is introduced during the cycle. These are only a few of the considerations that have to be made in order to produce a suitable product (3, 223).

Depending on how the melt flow enters and is distributed in the cavity, its strength (and other properties) can be orientated to meet performance requirements. Note that in Fig. 8-19 melt spreads in a branching pattern. Fig. 8-20 is an example where the highest stress is parallel in the orientation direction. An example of locating a gate to obtain required performance of a product that is being subjected to flexing in service is shown in

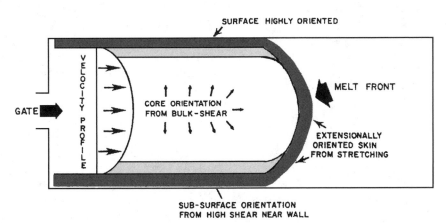

Fig. 8-18 Flow paths are determined by product shape and gate locations.

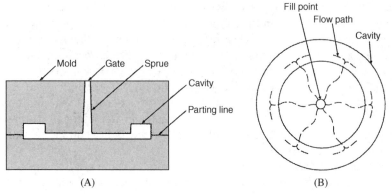

Fig. 8-19   Flow pattern in a center gated circular product.

Fig, 8-21: (a) edge gated, (b) center gated, and (c) edge gated on left side failed whereas edge gated (between fingers) did not fail.

Melt flow paths are determined by product shape and gate location. As shown in Fig. 8-22 flow fronts that meet head on will weld together, forming a weld line. Parallel fronts tend to blend, however they can produce a less distinct meld line that usually results in a stronger bond. If the flow fronts do not have the ideal melt conditions (low temperature, insufficient melt packing, etc.) bond could be very weak; in fact there could be no bond. Adjusting melt conditions will result in maximizing the bond strength (Figs. 8-23 and 8-24).

The primary inherent features of molded products remain unchanged despite some very major hardware developments and widely varying techniques for controlling the

process technology that keeps simplifying the IM process. The final specific properties and property distributions will dependent on the selected process parameters and conditions. The evolution from traditional techniques to very sophisticated, adaptive methods have been accelerated since about the 1970s. Distinct product performance and cost advantages are gained when special sequencing is implemented with advanced process control.

The pace of development has increased with the commercialization of more engineering plastics and high performance plastics that were developed for load-bearing applications, functional products, and products with tailored property distributions. Polycarbonate compact discs, for example, are molded into a very simple shape, but upon characterization reveal a distribution of highly complex optical properties requiring extremely tight dimension and tolerance controls (3, 223).

Many of the new plastics, blends, and material systems require special, enhanced processing features or techniques to be successfully injection molded. The associated materials evolution has resulted in new plastics or grades, many of which are more viscoelastic. That is, they exhibit greater melt elasticity. The advanced molding technology has started to address the coupling of viscoelastic material responses with the process parameters. This requires an understanding of plastics as viscoelastic fluids, rather than as purely viscous liquids, as is commonly held

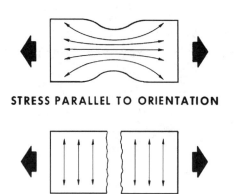

STRESS PARALLEL TO ORIENTATION

STRESS PERPENDICULAR TO ORIENTATION

Fig. 8-20   Effect of orientation during melt flowing through a cavity.

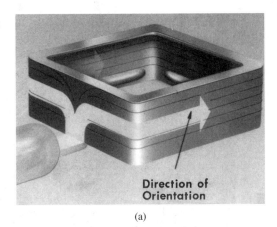

(a)

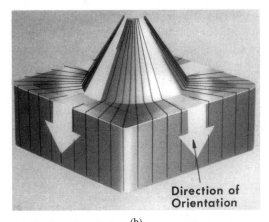

(b)

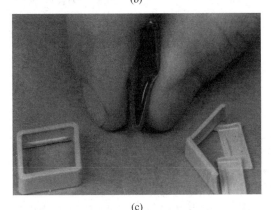

(c)

Fig. 8-21   Effect of edge (a) or center gating (b) a mold showing (c) edge performed but center failed (view on right side).

in the past. The importance of this new insight is apparent when a root-cause analysis is performed on some of today's molded defects, such as surface blemishes and variations in gloss. Exceptions in the materials development arena are high flow plastics for

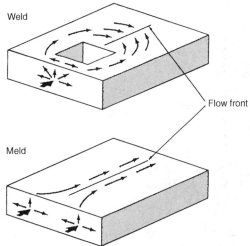

Fig. 8-22   Examples of weld and meld lines.

thin wall products such as cell phones, mobile modems, and personal organizers. These high-flow plastics typically have lower average molecular weights, resulting in reduced viscoelastic behavior.

The basic steps of the IM process produce unique structures in all molded products, whether they are miniature (micro) electronic components, compact discs, or large automotive bumpers. These structures have frequently been compared to plywood with several distinct layers, each with a different set of properties. In all IM products, a macroscopic skin-core structure results from the flow of melt into an empty cavity. Identifiable zones or regions within the skin are directly

Fig. 8-23   Example of weld lines where the gate was located at the top center of the telephone handset.

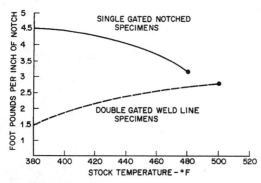

Fig. 8-24   Izod impact strength of ABS plastic vs. stock (melt) temperature.

related to the flow conditions and the temperature difference between the cold cavity walls and the hot plasticized melt.

The basics observed in molded products are always the same; only the extent of the features varies depending on the process variables, material properties, and cavity contour. That is the inherent hydrodynamic skincore structure characteristic of all IM products. However, the ratio of skin thickness to core thickness will vary basically with process conditions and material characteristics, flow rate, and melt-mold temperature difference. These inherent features have given rise to an increase in novel commercial products and applications via coinjection, gas-assisted, low pressure, fusible-core, in-mold decorating, etc.

The IM process is a manufacturing technology that has been modified, extended, and refined for over a century. Many different methods and techniques have been introduced to improve the process and make it more economical in the manufacturing environment. Although these advances have been significant, a technical analysis reveals that relatively few conceptual difference exist between the early process patents and today's methodology (223, 224).

Thickness of Section and Rib

Plastic products should be designed with the minimum wall thickness that will provide the specified structural requirements. This results in saving material and higher molding output due to the rapid transfer of heat from the molten plastic to the cooler mold surfaces. Wall thicknesses should be made as uniform as possible to eliminate distortion, internal stresses, and cracking. Ribs can be used to increase product strength without increasing wall thickness. This approach not only provides strength, but improves material flow and helps prevent distortion during cooling (Appendix A: **PLASTICS DESIGN TOOLBOX**).

Holes are often required in molded products. They should be designed and located so as to introduce a minimum of weakness and to avoid complication in production. This means, for example, that several holes should not be located close together unless a thicker wall section is provided. Where many design problems posed problems due to holes, it is often less expensive to drill after molding, particularly when holes must be deep in proportion to their diameter. However incorporating holes in a mold can be routine.

Designing bucket   By way of another example consider a common household bucket. The analysis that a typical molder might make follows. The mechanical properties required are moderate namely tensile strength of 6,000–8,000 psi and impact strength of about 1 ft-lb/in. (Izod). It must be sufficiently rigid to hold 2 gal of water but some flexibility is desirable in order to absorb bumps and knocks during its use. Electrical properties are clearly unimportant, but water resistance and dimensional stability at moderately elevated temperatures must be good. Heat resistance must be adequate to deal with very hot water (80°C) but not necessarily with any heat over 100°C.

It must he produced in large quantities, over 100,000 per year and, being a domestic utility item, it should be priced low such as $3.00 at retail. Market studies indicate that it should be available in a wide range of colors. The size, quantity needed, and price suggest an IM in a TP material. A study of the cost and properties of the various plastics will show that in general the TPs will have the required properties at a lower cost than will the TSs. In addition they are more adaptable to high speed mass production techniques, such as IM. Problems will arise when a specific manufacturing process must be used due to its availability at the manufacturing company. This situation will dictate a new analysis based on the limitations of the available process.

As an example the cellulosics are ruled out due to poor dimensional stability at elevated temperatures. They will begin to distort at around 50°C (122°F). The polystyrenes, even the rubber-modified types, are also subject to heat distortion above 80°C (176°F). Looking farther down the list of mechanical properties we note that high density polyethylene (HDPE) is available in grades which do not distort until they reach temperatures above 110°C (230°F). Moreover, the past experience of other processors has shown that HDPE is often a logical and effective choice for this type of product. Further analysis shows that the tensile and impact strength is sufficiently great for our purpose and that the material is available in a wide range of colors. Finally, the current published prices of the material place it well within the range of the economics of the decision.

Much of this design is usually based on the molder's experience. It does not use very scientific selection principles but uses a practical approach (Fig. 1-4). The most important source of product design ideas is usually, when available, from the competitor's product lines. The list of materials quoted cannot be considered exhausted. Very often special grades of various plastics are made to meet a demand for some particular modification of one or other property. New advances in copolymers have greatly increased the range of properties available to the plastic designer. In general, the more rigid the specification covering end use, the easier the selection becomes.

Productivity

In order to fabricate a cost-performance effective molded product and understand potential problems vs. solutions, it is helpful to consider the relationships of machine-to-

mold capabilities, plastics processing variables (Fig. 8-25), and product performances. As reviewed a distinction has to be made between machine conditions and processing variables.

The nomograph shown in Fig. 8-26 is a simplified approach in determining shrinkage of a molded plastic. Each plastic has its own mold shrinkage behavior that is related to specific molding conditions. As an example by over pressure packing the cavity one could literally have zero shrinkage immediately after molding. However it could have excess dimensional changes there after. In the nomograph a straight line connects a thickness, as an example, of 3 mm (1.2 in.) on line (1) to a gate area of 4.5 mm² (0.0072 in²) on line (3). [A gate could be 1.5 mm (0.06 in.) thick times 3 mm (0.12 in.) wide.] Line from (1) to (3) intersects line (2) at 0.020 mm/mm (0.020 in./in.) that is the estimated plastic shrinkage at a mold temperature of 93°C (200°F).

This simplified approach is only a guide to show what is happening when only considering thicknesses and gate sizes. Since many other variables exist (change in thicknesses, melt flow behavior, melt temperature/pressure/speed, heat transfer through the mold, etc.) experience provides the guide lines that involve the product design and processing behavior. Software programs have been developed to analyze these type variables that provides some type of relation to shrinkage.

Productivity is directly related to cycle time. There usually is considerable common knowledge about a geometry and process conditions that will provide a minimum cycle time. Practices such as using thinner wall sections, cold or hot runners for TPs or hot or cold ones for TSs narrow sprues and runners, the optimal size and location of coolant (or heat) channels, and lower melt or mold heat, will decrease the solidification time reducing the cycle time.

Modified IM Technique

A few of these techniques will be reviewed that provide designers different capabilities

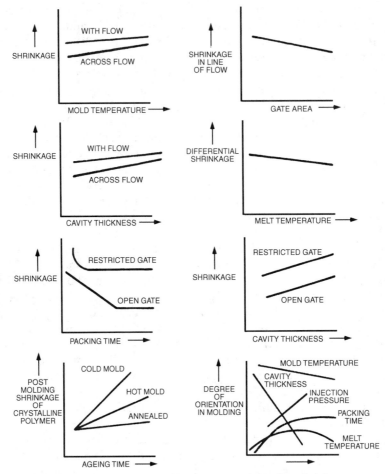

Fig. 8-25   Examples of how IM behaves, including shrinkage.

to fabricate products. An example is the profile molding. During traditional molding, the cavity walls are stationary. In some cases, it is advantageous to move the walls during the filling step or the cavity packing step. Different methods are used in the mold that includes movement of the cavity walls perpendicular to the parting line, and rotation or sliding of the cavity walls. Rotating a core during the filling step adds a biaxial orientation to the plastic especially on the surface. Flexural properties, as well as other mechanical properties, are greatly improved. Polystyrene drinking cups and polypropylene syringes are two examples that readily show large improvements.

Coinjection molding Coinjection molding produces products that can help one visualize the unique structures created in IM products (3). As an example plastic "A" is injected first from one plasticator and fills only a portion of the cavity. Next, Plastic "B" is injected sequentially behind "A" from another plasticator and maintains the basic pressure-driven flow field. When "A" and "B" are metered in the correct proportions for the relative size of the skin region and core region, the result is a molded product that exhibits a core "B" completely encased by a skin ("A"), when the cross section is viewed. For cosmetic applications, a second small portion of skin material "A" is injected after "B" to complete the skin formation at the gate. Coinjected products with two different colored plastics yield an easy identification of the skin and core regions. They have been producing different type shaped structures since the 1950s

Nomograph in English units

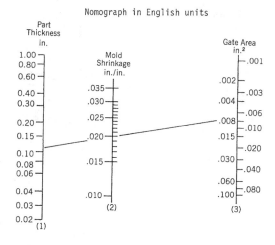

Nomograph in SI units

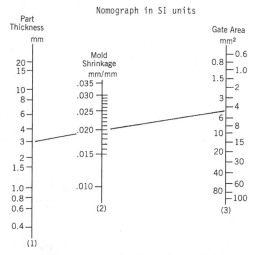

Fig. 8-26   Example in determining shrinkage.

Gas-assist injection molding   The GAIM process provides the designer unique molding capabilities such as exceptional structural strength at low weight. It uses a gas, usually nitrogen with pressures up to 3,000 psi (21 MPa), with the melt in the mold so that channels are formed within the melt. Different systems are used but basically they are all similar (3). Gas can be injected through the center of the IMM nozzle as the melt travels to the cavity or it can be injected separately into the mold cavity. In a properly designed tool run under the proper process conditions, the gas with its much lower viscosity than the melt remains isolated in the gas channels of the product without bleeding out into any thin-walled areas in the mold producing a balloon-like pressure on the melt.

The gas channels are those areas that have been thickened to achieve functional utility in the product or to promote better melt flow during cavity filling. This action provides a high degree of packing the melt against the cavity walls. Gas pressure is held until the melt solidifies. This coring action results in reducing cycle time and quantity of plastic used while developing a more structurally sound product (increases section stiffness), ability to improve surface flatness, reduce warps and sinks over thick sections, etc. Thick parts can easily be made without voids, sink marks, etc.

In this process the plastic is injected and only partially fills the cavity. Gas is then injected. In all cases the gas pressure in the core advances the melt front in the cavity until filling is completed, and prevents the plastic skin from collapsing away from the cavity walls during solidification. An integral skin is in contact with the cavity walls and the gas remains in the core region of the molded product. Because the gas is at a pressure greater than atmospheric pressure, the gas pressure must be reduced before the product is ejected to avoid distorting the product as the restraining cavity walls are removed. Control of the size and location of the gas core is more difficult with highly compressible gas, but techniques and sequences are being refined to yield products that exhibit an acceptable level of reproducibility.

such as sandwich, corrugated, tubular, T, U, etc. IM machines and their special molds are available that can produce two or more coinjection plastics.

It is important to understand that a similar skin and core exist in all injection molded products, although it is generally difficult to distinguish the skin-core interface without an enhancing characterization technique. The two different materials must have a certain degree of compatibility (Table 8-15). They can be highly filled, fiber reinforced, impact modified, UV stabilized, foamed, or using 100% recycle core with a skin that protects the core and provides desirable performances. If the two plastics are not compatible, a third can be used as an interlayer providing the proper bonding.

Table 8-15   Guide to compatibility of plastics for coinjection

Materials	ABS	Acrylic ester acrylonitrile	Cellulose acetate	Ethyl vinyl acetate	Nylon 6	Nylon 6/6	Polycarbonate	HDPE	LDPE	Polymethyl-methacrylate	Polyoxymethylene	PP	PPO	General-purpose PS	High-impact PS	Polytetramethylene terephthalate	Rigid PVC	Soft PVC	Styrene acrylonitrile
ABS	+	+	+				+	−	−	+		−		−	−	+	+	0	+
Acrylic ester acrylonitrile	+	+		+						−					0				+
Cellulose acetate	+		+	−															
Ethyl vinyl acetate		+	−	+				+	+			+			+		+	0	
Nylon 6					+	+	−	−				−			−				
Nylon 6/6					+	+	−	−	−			−			−				
Polycarbonate	+						−	+				−	0						+
HDPE	−		+	−	−		+	+	−	−	0								
LDPE	−		+	−	−		+	+	−	−		+		−					
Polymethylmethacrylate	+							−	−	+		−		−	0		+	+	+
Polyoxymethylene								−	−		+	−							
PP	−	−	+		−	−	0	+	−	−	+	−	−	−					
PPO												−	+	+	+				−
General-purpose PS	−		+				−	−	−	−		−	+	+	+		−	−	
High-impact PS	−	0		−	−		0			0		−	+	+	+		−	−	
Polytetramethylene terephthalate	+															+			+
Rigid PVC	+		+							+		−					+	+	
Soft PVC	0	0						−		+		−		−	−		+	+	+
Styrene acrylonitrile	+	+					+			+		−	−	−			+	+	+

ᵃ + = good adhesion, − = poor adhesion, 0 = no adhesion, blank indicates no recommendation (combination not yet tested). The addition of fillers or reinforcements leads to a deterioration of adhesion between raw materials for skin and core.

Injection-compression molding Also called ICM, coining, and injection stamping. ICM is a variant of injection molding. The essential difference lies in the manner in which the thermal contraction in the mold cavity that occurs during cooling (shrinkage) is compensated. With conventional injection molding, the reduction in material volume in the cavity due to thermal contraction is compensated basically by forcing in more melt during the pressure-holding phase.

By contrast with ICM, a compression mold design is used where male plug fits into a female cavity rather than the usual flat surface parting line mold halves for IM (Fig. 8-27). The melt is injected into the cavity as a short shot thereby not filling the cavity. The melt in the cavity is literally stress-free; it is literally poured into the cavity. Prior to receiving melt, the mold is slightly opened so that a closed cavity exists; the male and female parts are engaged so the cavity is closed. After the melt is injected, the mold automatically closes producing a relatively even melt flow. Upon controlled closing, a very uniform pressure is applied to the melt. Sufficient pressure is applied to provide a molded product without stresses.

Soluble core molding The soluble core technology (SCT) is called by different names such as soluble fusible metal core technology (FMCT), fusible core, lost-core, and lost-wax techniques (3). In this process, a core [usually molded of a low melting alloy (eutectic mixture) but can also use water soluble TPs, wax formulations, etc.] is inserted into a mold such as an injection molding mold. This core can be of thin wall or solid construction.

If the product design permits, it can be supported by the mold halves or spider type pin supports that are used to have it "floating"

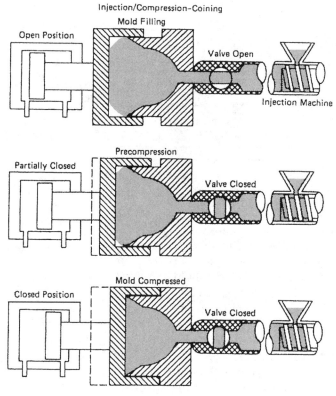

Fig. 8-27 Schematic of coining that combines injection molding and compression molding.

within the cavity; during plastic molding, the pins will melt. After the plastic solidifies, the core is removed by applying a temperature below the melting point of the plastic. Core material is poured through an existing opening or will require drilling a hole in the plastic. This technique is a take off and similar to the lost wax molding process started during the ancient Egyptian times fabricating jewelry. Also the 1944 all plastic airplane used the lost wax process to bag mold its RP sandwich construction (review latter in this chapter **REINFORCED PLASTIC, Process,** *Lost-wax*).

Over-molding Over-molding is also called in-mold assembly, two-color rotary, or two-color shuttle. Two materials are molded so that the first molded shot is over-molded by the second molded shot; first molded part is positioned so the second material can be molded around, over, sections, or through it. The two materials can be the same or different and they can be molded to bond together or not bond together. If materials are not compatible, the materials will not bond so that a product such as a universal or ball-and-socket joint can be molded in one operation. If they are compatible, controlling the processing temperature can eliminate bonding. A temperature drop at the contact surfaces can occur in relation to the second hot melt shot to prevent the bond.

In addition to universal or ball-and-socket joints, other examples of this products using this technique include inner-door panels for automobiles where woven or nonwoven textiles are placed in a mold and the melt is injected. In-mold labeling is another application that goes beyond just a printed message. Individual labels or continuous film can be indexed in the mold at the beginning of each cycle. Besides printing on the film, the film can serve many other functions (increasing impact, toughen plastic, etc.), or the film can contain additives and stabilizers to protect the surface of the molded product.

Certain applications of over-molding are not restricted to a low pressure. Two-shot

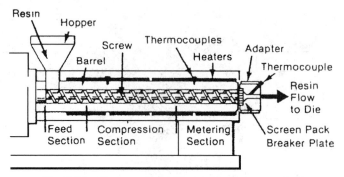

Fig. 8-28   Cross-section of a single-screw extruder.

molding is an example where a plastic is molded into a shape, then placed in another cavity before a second plastic is injected. The first plastic injected serves as a solid mold wall for the second cavity. Keys for a computer keyboard, knobs, and other items are often molded this way to provide information that does not fade or wear away with use. Many variations exist, including molding an elastomer over a rigid plastic, and molding a frame around a lens or optical window. Each step in over-molding is essentially the standard molding process even though the integral structure might resemble a product molded by coinjection molding.

Extrusion

The extruder, that offers the advantages of a completely versatile processing technique, is unsurpassed in economic importance by any other process. This continuously operating process, with its relatively low cost of operation, is predominant in the manufacture of shapes such as profiles, films, sheets, tapes, filaments, pipes, rods, in-line postforming, and others. The basic processing concept is similar to that of injection molding (IM) in that material passes from a hopper into a plasticating cylinder in which it is melted and dragged forward by the movement of a screw. The screw compresses, melts, and homogenizes the material. When the melt reaches the end of the cylinder, it may be forced through a screen pack prior to entering a die that gives the desired shape with no break in continuity

(Fig. 8-28). The screen pack is a filter that restricts unmelted plastic and/or contaminants from entering the die (6).

Practically only thermoplastics go through extruders; no major markets have been developed to date for extruded thermosets.

A major difference between extrusion and IM is that the extruder processes plastics at a lower pressure and operates continuously. Its pressure usually ranges from 1.4 to 10.4 MPa (200 to 1,500 psi) and could go to 34.5 or 69 MPa (5,000 or possibly 10,000 psi). In IM, pressures go from 14 to 210 MPa (2,000 to 30,000 psi). However, the most important difference is that the IM melt is not continuous; it experiences repeatable abrupt changes when the melt is forced into a mold cavity. With these significant differences, it is actually easier to theorize about the extrusion melt behavior as many more controls are required in IM.

Good-quality plastic extrusions require homogeneity in terms of the melt-heat profile and mix, accurate and sustained flow rates, a good die design, and accurately controlled downstream equipment for cooling and handling the product. Four principal factors determine a good die design: internal flow length, streamlining, construction materials, and uniformity of heat control. Heat profiles are preset via tight controls that incorporate cooling systems in addition to electric heater bands. Barrels external surfaces can include the use of forced air ad/or water jackets to aid in controlling the melt temperature. In some machines a water bubbler channel is located within the screw.

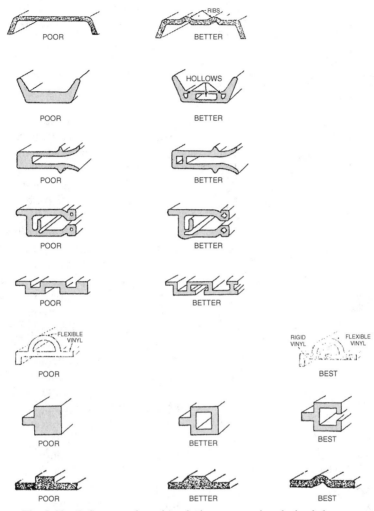

Fig. 8-29   Influence of product design on meeting desired shapes.

The design of the product should consider ease of processing (Figs. 8-29 to 33 and Table 8-16).

Excellent guides for determining the initial orifice die openings for different profiles are shown in Fig. 8-34 (6).

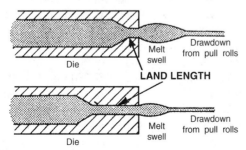

Fig. 8-30   Effect of die land length on melt swell.

On leaving the extruder, the melt (extrudate) is drawn by a pulling device, at which stage it is subject to cooling, usually by water and/or blown air. This is an important aspect of downstream control if tight dimensional requirements are to exist or conservation of plastics is desired. The processor's target is to determine the tolerance required for the pull rate and to see that the device meets the requirements. Even if tight dimensional requirements are not required, the probability is that better control of the pull speed will permit tighter tolerances so that a reduction in the material's output will occur.

As the molecules of the melt flow are aligned in the direction of the output from the die, the strength of the plastic is

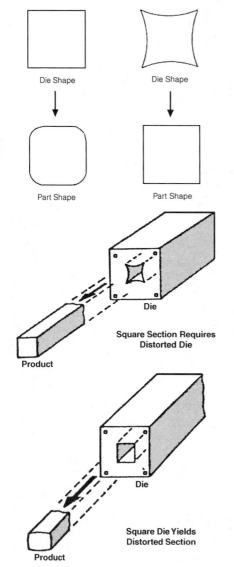

Fig. 8-31 Effect of die orifice shape on the extrudate.

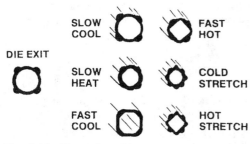

Fig. 8-32 Examples of how temperature, pressure, and takeoff speed (time) variations can potentially influence the shape of the extrudate.

characteristically greater in that direction than at right angles. Depending on the product's use, this may or may not be favorable. Using appropriate devices and controls the degree of orientations can be varied. If desired a balanced strength in both directions can be produced (6).

The success of any continuous extrusion process depends not only upon uniform quality and conditioning of the raw materials but also upon the speed and continuity of the feed of additives or regrind along with virgin plastic upstream of the extruders hopper. Variations in the bulk density of materials can exist in the hopper, requiring controllers such as weight feeders, etc.

In extrusion an extensive theoretical analysis has been applied to facilitate understanding and maximize the manufacturing operation. However, the real world must be understood and appreciated as well. The operatorhas to work within the limitations of the materials and equipment (the basic extruder and all auxiliary upstream and downstream equipment). The interplay and interchange of process controls can help to eliminate problems and aid the operation in controlling the variables that exist. The greatest degree of instability is due to improper screw design, or using the wrong screw for the plastic being processed. Screws are designed to meet the melt requirements for a specific type of plastic.

For uniform and stable extrusion it is important to check periodically the drive system, the take-up device, and other equipment, and compare it to its original performance. If variations are excessive, all kinds of problems will develop in the extruded product. An elaborate process-control system can help, but it is best to improve stability in all facets of the extrusion line. Some examples of instabilities and problem areas include (1) non-uniform plastics flow in the hopper; (2) troublesome bridging, with excessive barrel heat that melts the solidified plastic in the hopper and feed section and reduces or stops the plastic flow; (3) variations in barrel heat, screw heat, screw speed, screw power drive, die heat, die head pressure, and the take-up device; (4) insufficient melting or mixing

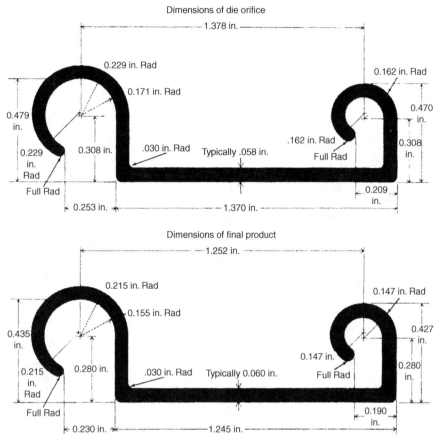

Fig. 8-33 Example of changes in a PVC profile shape from the die orifice to the product where no dimensions remained the same.

capacity; (5) insufficient pressure-generating capacity; (6) wear or damage of the screw or barrel; and (7) melt fracture/sharkskin, and so on. Finally, one must check the proper alignment of the extruder and the downstream equipment. Proper alignment is a must for high-quality, high-speed output.

Regardless of their particular designs, all extruders have the function of conveying plastic and converting it into a melt. For this purpose, both single- and multiple-screw extruders are suitable, but they all have individual characteristic features. Practical and theoretical data show that each type has its place (6). The single-screw machine dominates. However, other types are available, such as twin-screw extruders, which are often used to achieve improved dispersing and mixing, as in the compounding of additives.

Modified Extrusion Technique

A few of these techniques will be reviewed.

Coextrusion Coextrusion provides multiple molten layers usually using two or more extruders with two or more melts going through one die where they are bonded together. This technique permits using melt heat to bond the various plastics (Tables 8-17 to 19), or using the center layer as an adhesive. Coextrusion is an economical competitor to conventional laminating processes by virtue of its reduced materials handling costs, raw materials costs, and machine-time cost. Pinholing is also reduced with coextrusion, even when it uses one extruder and divides the melt into at least a two-layer structure. Other gains include elimination or reduction of delamination and air entrapment.

Table 8-16 Guide for dimensional tolerances for profiles

Dimension	Rigid Vinyl (PVC)	Polystyrene	ABS	Polypropylene	Flexible Vinyl (PVC)	Polyethylene
Wall thickness	±8%	±8%	±8%	±8%	±10%	±10%
Angles	±2°	±2°	±3°	±3°	±5°	±5°
Profile dimensions, ±mm (in.)						
0–3 (0–$\frac{1}{8}$)	0.18 mm (0.007 in.)	0.18 mm (0.007 in.)	0.25 mm (0.010 in.)	0.25 mm (0.010 in.)	0.25 mm (0.010 in.)	0.30 mm (0.012 in.)
3–13 ($\frac{1}{8}$–$\frac{1}{2}$)	0.25 mm (0.010 in.)	0.30 mm (0.012 in.)	0.50 mm (0.020 in.)	0.38 mm (0.015 in.)	0.38 mm (0.015 in.)	0.63 mm (0.025 in.)
13–25 ($\frac{1}{2}$–1)	0.38 mm (0.015 in.)	0.43 mm (0.017 in.)	0.63 mm (0.025 in.)	0.50 mm (0.020 in.)	0.50 mm (0.020 in.)	0.75 mm (0.030 in.)
25–38 (1–$1\frac{1}{2}$)	0.50 mm (0.020 in.)	0.63 mm (0.025 in.)	0.68 mm (0.027 in.)	0.68 mm (0.027 in.)	0.75 mm (0.030 in.)	0.90 mm (0.035 in.)
38–50 ($1\frac{1}{2}$–2)	0.63 mm (0.025 in.)	0.75 mm (0.030 in.)	0.90 mm (0.035 in.)	0.90 mm (0.035 in.)	0.90 mm (0.035 in.)	1.0 mm (0.040 in.)
50–75 (2–3)	0.75 mm (0.030 in.)	0.90 mm (0.035 in.)	0.94 mm (0.037 in.)	0.94 mm (0.037 in.)	1.0 mm (0.040 in.)	1.1 mm (0.045 in.)
75–100 (3–4)	1.1 mm (0.045 in.)	1.3 mm (0.050 in.)	1.3 mm (0.050 in.)	1.3 mm (0.050 in.)	1.7 mm (0.065 in.)	1.7 mm (0.065 in.)
100–125 (4–5)	1.5 mm (0.060 in.)	1.7 mm (0.065 in.)	1.7 mm (0.065 in.)	1.7 mm (0.065 in.)	2.4 mm (0.093 in.)	2.4 mm (0.093 in.)
125–180 (5–7)	1.9 mm (0.075 in.)	2.4 mm (0.093 in.)	2.4 mm (0.093 in.)	2.4 mm (0.093 in.)	3.0 mm (0.125 in.)	3.0 mm (0.125 in.)
180–250 (7–10)	2.4 mm (0.093 in.)	3.0 mm (0.125 in.)	3.0 mm (0.125 in.)	3.0 mm (0.125 in.)	3.8 mm (0.150 in.)	3.8 mm (0.150 in.)

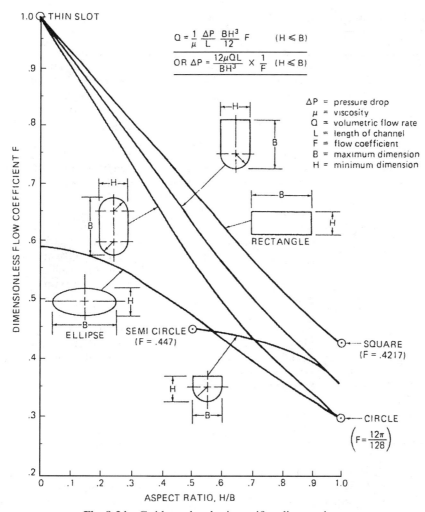

Fig. 8-34 Guide to developing orifice die openings.

In the past, a processor desiring to enter the field had little choice of equipment, but the increased interest in coextrusion has produced a proliferation of equipment. With rapidly changing market conditions and the endless introduction of useful materials, the design of machines has become much more involved. It is important that the processor have flexibility in making selections, but not at the expense of performance, dependability, or ease

Table 8-17 Examples of compatibility between plastics for coextrusion

	LDPE	HDPE	PP	Ionomer	Nylon	EVA
LDPE	3	3	2	3	1	3
HDPE	3	3	2	3	1	3
PP	2	2	3	2	1	3
Ionomer	3	3	2	3	3	3
Nylon	1	1	1	3	3	1
EVA	3	3	3	3	1	3

Code: 1. Layers easy to separate.
2. Layers can be separated with moderate effort.
3. Layers difficult to separate.

Table 8-18   Comparison of feedblock and multimanifold coextrusion dies

Characteristic	Feedblock	Multimanifold
Basic difference	Melt streams brought together outside die body (between extruder and die) and flow through the die as a composite	Each melt stream has a separate manifold: each polymer spreads independently of others: they meet at die pre-land to die exit
Cost	Lower	Higher
Operation	Simplest	—
Number of layers	Not restricted: seven- and eight-layer systems are commercial	Generally restricted to three or four layers
Complexity	Simpler construction: no adjustments basically	More complex
Control flow	Contains adjustable matching inserts, no restrictor bar	Has restrictor bar or flow dividers in each polymer channel; but with blown film dies control is by individual extruder speed or gearboxes
Layer uniformity	Individual layer thickness correction of ±10 percent	Restrictors and manifold can meet ±5 percent
Thin skins	Better on dies >40 in.	Better on dies <40 in.
Viscosity range	Usually limited to 2/1 or 3/1 viscosity range of materials	Range usually much greater than 3/1
Degrudable core material	Usually better	—
Heat sensitivity	More	Less
Bonding	Potentially better: layers are in contact longer in die	—

of operation. One should provide for the material or layer thickness necessary in product changeover without allowing high scrap rates. The goal should be to incorporate scrap regrind within the layered construction.

It is important to be able to control the individual layer distribution across the width of the die. As the viscosity ratio or thickness ratio of the plastics being combined increases, the individual layer distributions of the

Table 8-19   General comparison of metalized coextruded polyethylene and aluminum foil

	Metalized Coextruded Polyolefin	Foil Laminate*
Tensile strength MD	18–19	18–19
CD	12–13	11
Mullen strength	19–20	17
Gurley stiffness MD	70–75	117–112
CD	42–47	72–77
WVTR (gm/csl/24 hrs.)	Approx. .05	.0006
Oxygen transmission (cc/csl/24 hrs.)	Approx. 10	Less than .004
Light transmission	Slightly less than 1%	Approx. 0%
Seal type	Fin only[†]	Fin or Lap
Seal range (40 psi, 15 sec.)	350–500°F	160–350°F
Deadfold (subjective) (1–10 Scale)	4	7
Flex crack resistance (subjective) (1–10 Scale)	8	5

*.0003″ gauge foil, wax laminated to $12^1/_2$# paper, wax laminated to $8^1/_2$# tissue.
[†]While a lap seal is technically possible, the bond is too weak to be considered commercial.

plastic composite film can become displaced. To offset this situation proper die design with control of the line is required.

A number of techniques are available for coextrusion, some of them patented and available only under license. Basically, three types exist: feedblock, multiple manifolds, and a combination of these two (Table 9-18). Productions of coextruded products are able to meet product requirements that range from flat to complex profiles. Figure 8-35 (a) shows a typical 3-layer coextrusion die and (b) examples of rather complex profiles that are routinely extruded.

Special shape Some special dies, shown in Figs. 8-36 and 8-37, produce interesting flow patterns and products, such as tubular to flat netting shapes. Figure 8-37 has a mesh produced by extrusion from a counter rotating die design, originally patented in 1956. A postextrusion stretching process follows the exiting mesh. For circular output a counter rotating mandrel and orifice have semicircular shaped slits through which the melt flow emerges. If one part is held stationary, a rhomboid or elongated pattern is formed; if both parts rotate, a true rhombic mesh is formed. When the slits overlap, a crossing point is formed where the emerging threads are "welded." For flat netting, the slide action is in opposite directions.

In-line postforming In-line postforming, or postextrusion processing, refers to the processing that may be done to the extrudate, usually just after it emerges from the die but before the material has a chance to cool. When the material is worked in such a state it is known as in-line processing, as opposed to cutting, forming, or other processing, which is done on the cold extrusion in a secondary operation.

In-line processing is done automatically, with little or no extra labor on the part of the machine operator. Heat required can be retained from the extrusion operation. The extra processing, which may involve shaping, cutting, re-forming, or surface modification of the extrudate, can considerably increase the value of the extrusion without materially

increasing its processing cost. It may also be done to enable the use of a lower-cost die, as for example flattening a tubular extrusion into an oval so that a much lower cost circular die can be used.

This is a popular forming technique that has provided both performance and cost advantages, principally for long production runs. It is applied as the plastic sheet, film, or profile exits an extruder. Upon leaving the die, and retaining heat, the plastic is continuously postformed. With this type of in-line system the hot plastic is reduced only to the desired heat of forming. All it may require is a fixed distance from the die opening. Cooling can be accelerated with blown air, a water spray, and/or a water bath. This equipment, like others, requires precision tooling with perfect registration.

Examples of some postforrning techniques are shown in Fig. 8-38 (a) in-line postforming embossing, (b) in-line vacuum postforming embossing with water cooled temperature control, (c) in-line vacuum-pressure postforming using water-cooled dies/molds, (d) in-line coil postforming, and (e) in-line fixed-rotating ring postforming.

Coating

Extrusion coating is used extensively in a varity of different applications that include wire coating, steel coating (protect steel), wood coating (decorative), etc. Figure 8-39 shows sections of a profile (could be any shape in any length) in line with an extruder where a crosshead diehead is used to apply the plastic coating (6).

Orientation

The following information is a continuation of what has been reviewed earlier in this chapter (**PROCESSING AND PROPERTY, Orientation**). Orientation consists of a controlled system of stretching heated plastic material (molecules) to improve their strength, stiffness, optical, electrical, and other properties. This process, which has been

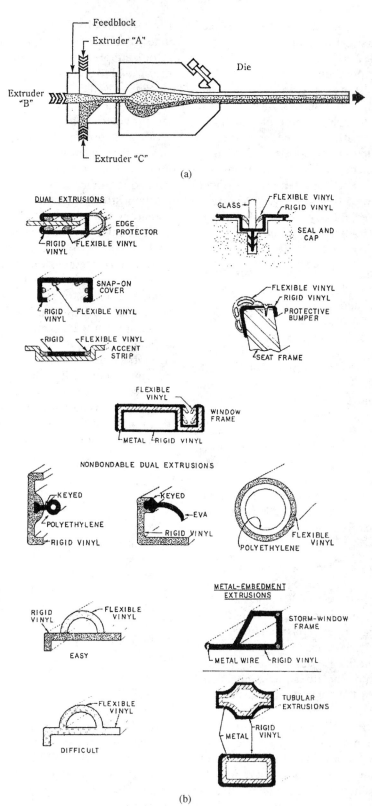

Fig. 8-35   (a) Three layer die and (b) coextruded profile shapes.

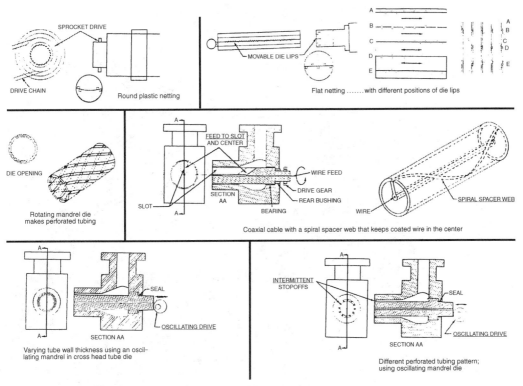

Fig. 8-36   Examples of special action dies that produce round and flat products.

used for almost a century, became prominent during the 1930s for stretching fibers up to ten times. Later it was adapted to stretching extruded film and sheet and, more recently

Fig. 8-37   Example of netting as it exits a die that is available in almost every conceivable form.

other processes such as blow-molded products. Many other products take advantage of its benefits for producing tape, pipe, profile, and thermoformed products, etc. (Table 8-20). Practically all plastics can undergo orientation, although certain types find it particularly advantageous (PET, PP, PVC, PE, PS, PVDC, PVA, and PC). Of the 15 million tons of plastic film sales annually worldwide, about 16% are sales of oriented material.

In extrusion the most important orienting processes are used with flat film and sheet, blown film, and blow molding. During these processes the stretch or blow-up ratios determines the degree of flat or circumferential orientation and the pull rate of the hot plastic determines degree of orientation (6). The optimum stretching heat for amorphous plastics (PVC, etc.) is just above the glass transition temperature; for the crystalline types (PET, PE, and so on) it is just below the melting point. During the stretching process the molecular structure changes, thus usually necessitating an increase in heat

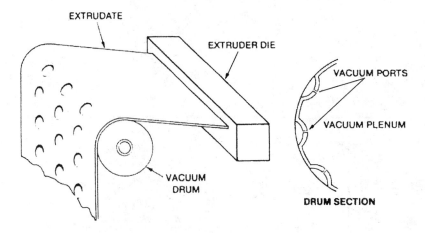

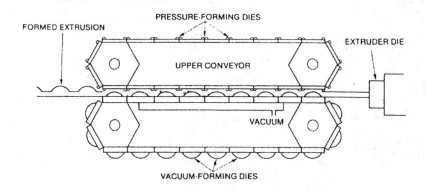

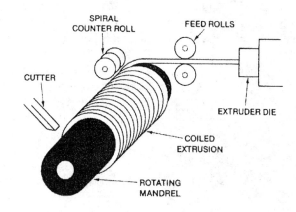

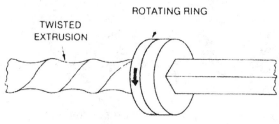

Fig. 8-38   Different products that are postformed in line during extrusion.

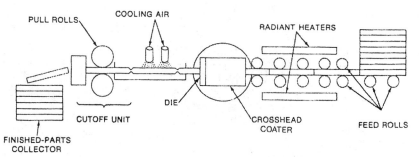

Fig. 8-39 Continuous in line extrusion system coating short length parts.

if further deformation is planned. Afterward, the orientation is "frozen in" by lowering the heat or, with crystalline types, set by increasing the crystalline portion.

With orientation, the thickness is reduced and the surface enlarged. If film is longitudinally stretched in the elastic state, its thickness and width are reduced in the same ratio. A conventional unorienting extrusion film or sheet line is shown in Fig. 8-40. In orienting film or sheet the processor uses a tentering frame (typically used in textile weaving), which is enclosed in a heat-controlled oven, with a very accurate and gentle air flow used to hold the oven air temperature at the required orienting heat (Fig. 8-41). The frame has continuous speed control and diverging tracks with holding clamps. As the clamps move apart at prescribed diverging angles the hot plastic is stretched in the transverse direction, resulting in single orientation (0). To obtain bidirectional orientation (BO) an in-line series of heat-controlled rolls are located between the extruder and tenter frame. The rotation of each succeeding roll is increased, based on the longitudinal stretched properties desired.

In this Fig. 8-41 view (a) the feeder-roll speed to puller-roll speed ratio can be set, such as 1:4, and simultaneously the ratio of width can be set as 1:4. The machined direction ratio is usually accomplished prior to the plastic's entering the temperature controlled oven that contains the tenter frame, by having it move around heat-controlled rolls where the rotational speed of the rolls increases from one roll to the next. View (b) is a schematic of the drawdown phenomenon with swell to produce orientation in the machined (longitudinal) direction.

Blow Molding

BM can be divided into three major processing categories: (1) extruded BM (EBM) with continuous or intermittent melt (called a parison) from an extruder and which principally uses an unsupported parison (Fig. 8-42), (2) injection BM (IBM) with noncontinuous

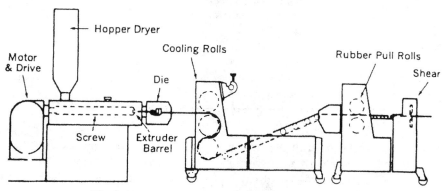

Fig. 8-40 Conventional extruded film or sheet line.

Table 8-20   Examples of different oriented TP products

Ranges of Application	Demands Made	Rate of Stretching	Thermoplastic
Carpet basic weave	Low shrinkage	1:7	PP
	High strength	1:5	PETP
	Temperature stability		
	Specific splicing tendency		
	Matt surface		
Tarpaulins	High strength	1:7	PP
			PE
Sacks	High strength	1:7	PE
	High friction value		PE
	Specific elongation		
	Weather resistant		
Ropes	High tensile strength	1:9 to 1:11 (15)	PP
	Specific elongation		
	Good tendency to splicing		
Twine	High tensile strength	1:9 to 1:11	PP
	High knotting strength		PP/PE
Separating weave	High strength	1:7	PP
Filter weave	Low shrinkage	1:7	PP
	Abrasion resistant	1:5	PETP
Reinforcing weave	Low shrinkage	1:7	PP
	Specific elongation	1:5	PETP
	Temperature resistance		
Tapestry and home textiles	UV-resistance	1:7	PE
	Low static charge		
	Uniform coloration		
	Textile-type handle		
Outdoor carpets	Low shrinkage	1:7	PP
	Wear resistance		
	Weather resistance		
	Elastic recovery		
	Uniform coloration	1:5	PETP
	Defined splicing		
Decorative tapes	Effective surface	1:6	PP with blowing agent
	Low specific gravity		
Knitted tapes, sacks, and other packagings, seed and harvest protective nets	High knotting strength	1:6.5	PP
	Low splicing tendency		PE
	Supplencess		
	UV-resistance		
Packaging tapes	High strength	1:9	PP
	Low splicing tendency	1:7	PETP
Fleeces	Fiber properties	1:7	PP and blends

melt (called a preform) and which principally uses a preform supported by a metal core pin (Fig. 8-43), and (3) stretched/oriented EBM and IBM to obtain bioriented products pro-viding significantly improved performance-to-cost advantages (Fig. 8-44). These BM processes offer different advantages in pro-ducing different types of products based on

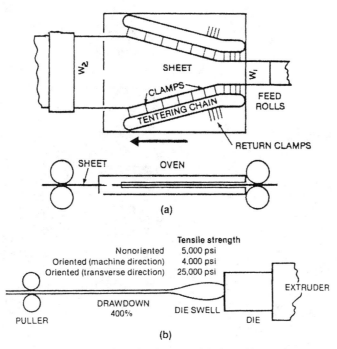

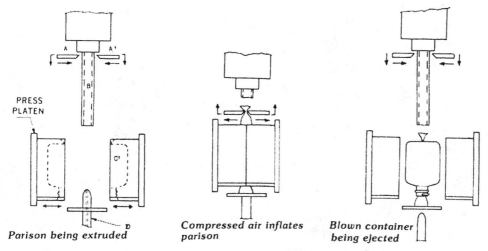

Fig. 8-41 Use of tender frame to biorient film or sheet.

the plastics to be used, performance requirements, production quantity, and costs (20).

Basically the BM lines have an extruder
with a die or an injection machine with a mold
to form the parison or preform, respectively.
In turn the hot parison or preform is located
in a mold. Air pressure through a tubular pin-
type device located usually at the parting line
of the mold will expand the parison or preform to fit snugly inside their respective mold

cavities. Blow molded products are cooled via
the water cooling systems within mold channels. After cooling, the blown products are
removed from their respective molds.

The nature of these processes requires the
supply of clean compressed air to "blow" the
hot melt located within the blow mold. Other
gases can be used, such as carbon dioxide,
to speed up cooling of the blown melt in
the mold. The gas usually requires at least a

Fig. 8-42 Basic extrusion blow molding process.

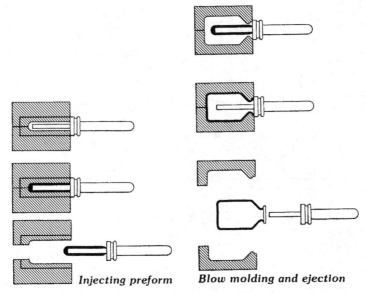

Injecting preform   **Blow molding and ejection**

Fig. 8-43   Basic injection blow molding process.

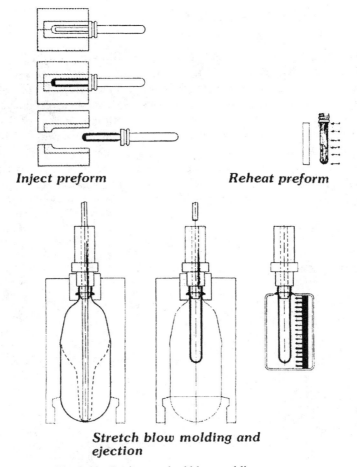

Inject preform   *Reheat preform*

*Stretch blow molding and
ejection*

Fig. 8-44   Basic stretched blow molding process.

Table 8-21 Hollow and structural BM shapes

Industry	Application	Required Properties
Automotive	Spoilers	Low temperature, impact, cost
	Seat backs	Heat distortion, strength/weight
	Bumpers	Low temperature, impact dimensional stability
	Underhood tubing	Chemical resistance, heat
Furniture	Work stations	Flame retardance, appearance
	Hospital furniture	Flame retardance, cleanability
	Office furniture	Flame retardance, cost
	Outdoor furniture	Weatherability, cost
Appliance	Air-handling equipment	Flame retardance, hollow
	Air-conditioning housings	Heat distortion, cost
Business machine	Housings	Flame retardance, cost
	Ductwork	Cost
Construction	Exterior panels	Weatherability, cost
Leisure	Flotation devices	Low temperature, impact strength cost, weatherability
	Marine buoys	Low temperature, impact strength cost, weatherability
	Sailboards	Low temperature, impact strength cost, weatherability
	Toys	Low temperature, impact strength cost, weatherability
	Canoes/kayaks	Low temperature, impact strength cost, weatherability
Industrial	Tool boxes, ice chests	Low temperature, impact strength, cost
	Trash containers, drums	Low temperature, impact strength, cost
	Hot-water tanks	Low temperature, impact strength, cost

pressure of 30 to 90 psi (0.21 to 0.62 MPa) for EBM and 80 to 145 psi (0.55 to 1 MPa) for IBM. Some of the melts may go as high as 300 psi (2.1 MPa). However, stretch EBM or IBM often requires a pressure up to 580 psi (4 MPa). The lower pressures generally create lower internal stresses in the solidified plastics and a more proportional stress distribution; the higher pressures provide faster molding cycles and ensuring conforming to complex shapes. With any lower pressures or lower melt stresses is improved resistance to all types of strain (tensile, impact, bending, environment, etc.).

Production can increase usually by at least 20 to 40% by using aggressive, turbulent chilled air at about −35°C (−30°F) that is allowed to escape following the blowing action. This action can provide several changes of air through the blow pin during a single blowing cycle. Blow molded bottles are predominantly fabricated using the extrusion and injection molding process with or without stretching/orientation.

Plastic shrinkage is dependent on many factors, such as plastic density, stock temperature, mold temperature, product thickness, and blowing air pressure. Once the operating conditions are established, tolerances of ±5% may be expected. Typical polyethylene blow molding shrinkage is as follows:

Low-Density Polyethylene:

Thickness up to 0.075 in.: 0.010–0.015 in./in.
Thickness over 0.075 in.: 0.015–0.030 in./in,

High-Density Polyethylene:

Thickness up to 0.075 in.: 0.020–0.035 in./in.
Thickness over 0.075 in.: 0.035–0.055 in./in.

Different products are blow molded (Table 8-21). Table 8-22 provides cost data comparing different BM processes with different plastics.

Complex Irregular Shape

Extruded blow molded 3-D products are produced. This approach provides the designer with a relatively very important approach in the art of hollow formed products. It is an ideal approach that has many cost-to-performance advantages. Complex, irregular

Table 8-22   Fabricating cost comparison of 16 oz. BM bottles

	Standard Extrusion Blow-molding 2-Parison Head 4-Fold	Stretch Blow Molding PVC (2) Single-parison Heads 4-Fold	Stretch Blow Molding PET
1.0—Machine cost incl. head, molds, ancillaries (lic. fee, stretch PVC and PET)	$270,000	$450,000	$850,000
2.0—Hourly machine costs Depre'n, 5 yr. 30 K hr, $/hr.	$9.00	$14.85	$28.33
Financing cost, 5 yr. 12.5%	2.80	4.65	10.20
Labor, 1 man	13.00	13.00	13.00
Energy at $.06 per kWh	2.50	5.35	11.00
Floor space	1.50	2.00	4.00
Maintenance and consumable mtl.	2.25	3.75	4.50
Total hourly mc costs	$31.05/hr.	$43.60/hr.	$71.03/hr.
3.0—Bottle specs. hourly/annual prod.			
3.1—16 oz. finish wt. (454 g) —regular 37 g (1.3 oz) —stretch PVC 20 g (0.7 oz) —stretch PET 20 g (0.7 oz)			
Cycle time/bottles per hour	8.4 sec./1,714	7.5 sec./1,920	4,000
bottles per yr., millions	10,286	11,520	24,000
4.0—Annual costs			
4.1—16 oz. (454.g) Resin:—37 g $.70/lb. ($70/0.45 kg or $ 1.54 kg)	$585,200		
—20 g $.66/lb ($1.46 kg)		$334,950	
—20 g $.60/lb ($1.32 kg)			$634,360
Machine costs	186,300	261,600	426,180
total p.a.	$771,500	$596,550	$1,060,540
Royalty (PET) Du Pont-per year			30,000
Cost per thousand	$75.00	$51.78	$45.44

Notes: 1. Figures are not to be considered as absolute costs, but rather reflect comparisons between various machine options.

2. All calculations are based upon 100 percent efficiency.

3. All bottle weights are finish weights (flash being considered 100 percent reusable).

shapes can be blown to meet different practical structures, from small to large products. For example, double-walled components are easily produced. Unfortunately very few designers are familiar with this technique.

Examples of different blow molded shapes are shown in Figs. 8-45 to 8-48.

This technique is also called nonaxisymmetric blow molding. In conventional EBM the parison enters the mold rather in a straight tube. In 3-D BM the parison is laid or oriented in the mold prior to closing. It is manipulated in the tool cavity providing complex geometric products that can have

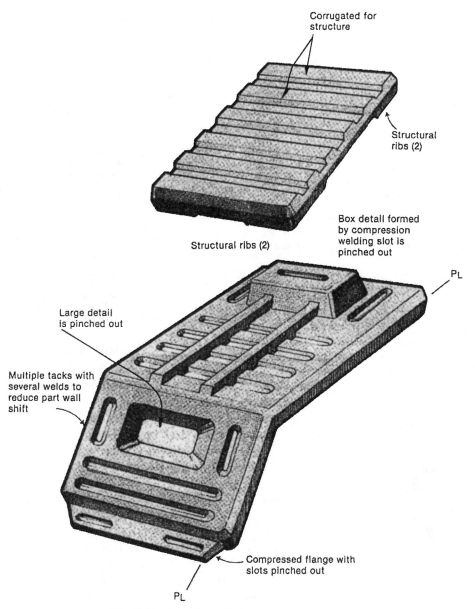

Corrugated for
structure

Structural
ribs (2)

Box detail formed
by compression
welding slot is
pinched out

Structural ribs (2)

PL

Large detail
is pinched out

Multiple tacks with
several welds to
reduce part wall
shift

Compressed flange with
slots pinched out

PL

Fig. 8-45   Double-walled structural BM panels.

uniform or nonuniform wall thicknesses, corrugated or noncorrugated sections, and so on. There are different techniques used for placing the parison into position such as; (1) articulate the extruder nozzle, (2) articulate the mold platen, and (3) robotically orient the parison before the mold closes. Thus, different shaped BM products are fabricated. Sequential BM can be used to integrate hard and soft regions of different plastics on a single tubular structure (parison).

Coextrusion or Coinjection

The use of BM multi-layer plastics is a technology that provides the advantages of taking advantage of using differing materials, including plastic foams that are systematically combined to meet cost to performance requirements (Fig. 8-49). Techniques are basically similar to what has been reviewed for **INJECTION MOLDING** and **EXTRUSION**.

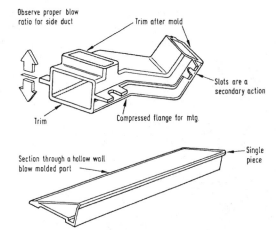

Observe proper blow ratio for side duct

Trim after mold

Slots are a secondary action

Trim

Compressed flange for mtg.

Section through a hollow wall blow molded part

Single piece

Fig. 8-46　Example of BM auto spoiler air duct.

Collapsibility Container

An interesting and practical design involves a bellows-collapsible bottle produced in conventional EBM equipment. These "foldable" in contrast to "passive" bottles provide advantages and conveniences such as (1) reducing storage, transportation, and disposal space; (2) prolonging product freshness by reducing oxidation and loss of carbon dioxide because as the contents is removed its contents is also reduced; and (3) providing continuous surface access to foods such as mayonnaise and jams. Best of all, they provide the consumer with a different, futuristic, fascinating package (Figs. 3-16 and 3-17).

EBM and IBM Comparison

With EBM, compared to IBM, the advantages include lower tooling costs and incor-

poration of blown handle-ware, etc. Disadvantages could be controlling parison swell, producing scrap, limited wall thickness control and plastic distribution, etc. If desired, solid handles can be molded during the blow molding process. Trimming can be accomplished in the mold for certain designed molds, or secondary trimming operations are included in the production lines.

With IBM, the main advantages are that no flash or scrap occurs during processing, it gives the best of all thicknesses and plastic distribution control, critical bottle neck finishes are easily molded to a high accuracy, it provides the best surface finish, etc. Disadvantages could include its high tooling costs, only solid handle-ware, and it "was" reported in the past that they were restricted or usually limited to very small products (however large and complex shaped products were fabricated once the market developed). Similar comparisons exist with biaxial orienting EBM or IBM. With respect to coextrusion, the two methods also have similar advantages and disadvantages, but mainly more advantages for both.

Blow Molding-Compression-Stretched

Processes basically go through the following stages: (1) starts with an extruded sheet, (2) circular blanks are cut from the sheet, (3) compression molded into the desired preliminary shape; during the compression action, the blank can be simultaneously stretched, (4) blow stretching can take place after compression molding, and (5) any

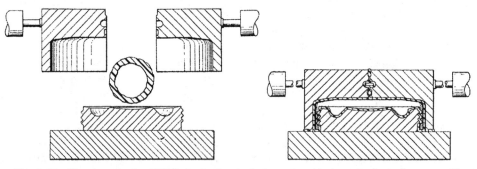

Fig. 8-47　Simple complex EBM shape that includes a threaded core using a 3-part mold.

Fig. 8-48 Corrugated/bellow shape.

trimming that may be required is the final step. There are the CSBM (compression stretch-blow molding) patents that include: (1) those held by Valyi Institute for Plastic Forming (VIPF) located at the University of Massachusetts-Lowell; (2) Dynaplast S.A. has the Co-Blow system; (3) American Can's OMNI container; (4) Petainer's cold forming process; (5) Dow Chemical's solid phase forming; (6) Dow Chemical's coforming (COFO); and (7) others.

Thermoforming

Thermoformed (3-D shape) plastics provide a great variety and quantities of marketable products, in a wide size range from millions of drinking cups or containers (each in ounces) to millions of pick-up truck storage wells (each about 100 lb.) and so on to complex shapes. The process of thermoforming is considered one of the four major fabricating processes following extrusion, injection molding, and blow molding. Since the plastic sheets and films used in thermoforming are products from extruders, the name extrusion-thermoforming is sometimes used. The use of the terms thermoforming or forming in the plastics industry do not include such operations as molding, casting, extrusion, etc. in which shapes or products are "formed" (1).

At least 30wt% percent of all extruded products are thermoformed. They have many advantages over other manufacturing methods. For the mass production of products (packaging, picnic dishes, cups, etc.) sheets and films can be produced in-line with

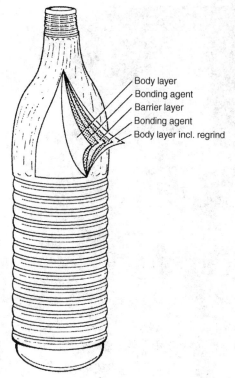

Body layer
Bonding agent
Barrier layer
Bonding agent
Body layer incl. regrind

Fig. 8-49 Coextruded blow molded bottle.

thermoforming equipment. The other major procedure is a secondary operation where rolls or flat sheets or films of materials are feed into the thermoforming equipment. Extruding sheet or film in-line requires dedica-

tion and control to ensuring that the extruder is operating efficiently as well as the thermoformer. For those with this type production, advantages exist including cost savings.

The process basically forms the sheet after it has been heated to the point at which it is soft and flowable, and then applying differential pressure (atmospheric pressure, air pressure, vacuum, or their combinations) to make the sheet or film conform to the shape of the male or female mold producing many different products (Table 8-23). The more precise and controlled pressure applied, the more efficient in reproducing products at the lowest cost occurs (Fig. 8-50).

The following different thermoforming techniques are used to form/shape plastics: air-assist, air-slip, billow, blister package, blowing, bubble, clamshell, cold forming, comoform cold forming, drape, drape with bubble stretching, draw, form and spray, form, fill, and seal, form, fill, and seal with zipper on-line, forging, plug, plug and ring, prebillow, preprinting, pressure, sag, scrapless, shrink wrapping, slip, snap-back, solid-phase pressure, stretch, vacuum, etc. (1, 6, 9). The different techniques influence the capability to provide different depth-to-width forming ratios (Chapter 3, **Thermoforming**). Unless a scrapless forming process is used, thermoformed products require trimming. Different

Table 8-23   Examples of thermoformed products

Polystyrene	Various packaging applications, including transparent meat trays, trays for cookie and candy boxes, blister packages
Polystyrene foam	Meat trays, egg cartons, take-out food containers
Acrylic	Signs and other outdoor applications like motorcycle windshields, snowmobile hoods, and recreational-vehicle bubble tops
Rigid vinyls	Lighting panels, signs, relief maps, bus-interior panels, dishes and trays for chemicals, blister packages, automobile dashboards
Acrylonitrile butadiene styrene (ABS)	Recreational-vehicle components, luggage, refrigerator liners, business-machine housings
Cellulose acetate	Blister packages, rigid containers, machine guards
Cellulose propionate	Machine covers, safety goggles, signs, shipping trays, displays
Cellulose acetate butyrate	Skylights, outdoor signs, pleasure-boat tops, toys
High-density polyethylene	Camper tops, canoes, sleds
Nylon	Reusable trays, outdoor signs, surgical equipment, meat trays
Polycarbonate	Outdoor lighting, face shields, machine guards, aircraft panels and ducts, signs
Polypropylene	Truck-fender liners, drinking cups, juice and dairy-product containers and lids, test-tube racks

Fig. 8-50 ABS 76 × 230 in. sheets are conveyed to an IR heating oven in back of the console and in turn formed into 15 ft outboard-powered runabouts in less than 10 minutes.

type cutters are used depending on shape of product and plastic used.

Also popular is the use of thermoforming coextruded sheets or films (Figs. 8-51 and 52).

Temperature Control

Forming requires thorough, fast, and uniform radiant heat from the surface to the core to the surface of the sheet or film. As a general guide, to achieve these conditions, sheet plastics over 0.040 in. (1.02 mm) should use sandwich-type (under and over the sheet) heater banks. To ensure that sufficient heat

is used, heaters should have capacities of at least 4 to 6 kW/sq. ft. Various type heating elements are used.

The cycle time is controlled by the heating and cooling rates, which in turn depend on the following factors: the temperature of the heaters and the cooling medium, the initial temperature of the sheet, the effective heat transfer coefficient, the sheet thickness, and thermal properties of the sheet.

Different plastics absorb radiant heat more efficiently at various wavelengths, which in turn are effected by the temperature of the emitting heater. Thus it requires that the proper wavelength be used for what the

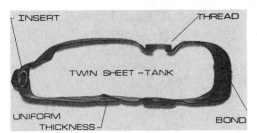

Fig. 8-51 Bonding two coextruded thermoformed parts to produce a gasoline fuel tank.

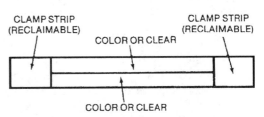

Fig. 8-52 The addition of an extruded single plastic clamping strip at each side of a coextruded sheet permits scrap reclaim of the trim waste.

material requires to perform most efficiently performancewise and energy costwise. With an in-line operation from an extruder delivering heated sheet to the thermoformer, energy savings from 30 to 40% can occur with reduced floor space.

Thermoforming Thermoset Plastic

Practically all thermoforming material used are TPs, however TS plastics can be thermoformed. As an example commercial roll-fed thermoforming machines are forming TS polyimide film (for electrical parts) at temperatures as high as 540°C (1000°F). Prior to being roll-fed, they are sheet-fed. Other TS plastics formed include CPET and B-staged reinforced TS polyester plastics. The TPs are formed above their glass transition temperatures. Specialty products such as TS polyester-glass fiber reinforced plastics have been made into boat hulls, etc.

Thermoforming vs. Injection Molding

Thermoforming (TF), wherever applicable, can offer certain advantages over injection molding (IM). TF has lower capital equipment investments, particularly molds and mold delivery times and maintenance. Very large products can be formed. Products with different thickness can usually be formed on the same mold only requiring minor heating and cooling cycle changes. Fewer stresses can occur with no weld lines.

Many products that are IM cannot be TF; the most popular competition is producing products such as drinking cups where both sides have advantages and disadvantages. The major advantage with IM is providing products that meet tighter tolerances and quality control as well as a multitude of rigid shapes. However, TF cups can meet the customer's performance requirements at a lower cost particularly when they are stretched-oriented during their forming operation. Other applications for TF products exist such as automobile parts (Chapter 4, **World's First All-Plastic Car Body**). To date over ten times more plastic is IM than TF.

Foaming

All markets practical use foamed plastic materials. Guide to consumption by weight is insulation 24%, packaging 18%, cushioning 15%, transportation 12%, consumer 4%, furniture 6%, flooring 6%, bedding 4%, appliance 2%, and other 9%. Many different foamed plastic products are produced, and practically all the processes can be used to make them, particularly extrusion, injection molding, calendering, casting, and RIM. Almost all plastics can be used to make these cellular-core structures, which range from flexible to very rigid objects (Table 8-24). Basically, the plastic is mixed with a blowing agent classified as either physical or chemical.

Foamed plastics, whether TPs or TSs, are a special category in the plastics family. They are available with open-celled construction, closed or interconnecting construction, or in combination. Their densities range from 1.6 to over 960 kg/m^3 (0. 1 to over 60 lb/ft^3). They can be rigid, semirigid, or flexible, and colored or plain. The range of properties they offer in terms of their insulating value, rigidity, compressive strength, cushioning and loading, structural characteristics, and others can be very extensive (Tables 8-25 and 8-26). Their performance depends to a great extent on the type of base plastic used, the type of blowing system, and the method of processing. Each plastic can include fillers or reinforcements to provide certain improved desirable properties.

There are many ways in which foams can be processed and used: as slabs, blocks, boards, sheets, molded shapes, sprayed coatings, extruded profiles, "foamed in place" in existing cavities, in which the liquid material is poured and allowed to foam, and as structural foams (Chapter 6, **STRUCTURAL FOAM**). Conventional equipment such as extruders, injection, or compression machines is used. However specially designed machines are available to just produce foamed products.

The foaming methods vary widely. One approach is to whip air into suspension or a solution of the plastic, which is then hardened by heat curing. A second is to dissolve a gas

Table 8-24 Properties of a few rigid plastics foams

Property	ASTM Test	Phenolics Foamed in Place	Phenolics Syntactic Castable	Polyvinyl Chloride Rigid Closed Cell	Phenylene Oxide Foamable Resin	Polycarbonate	Polystyrene Medium-Density Foam	Polystyrene Molded	Polystyrene Extruded	Polyurethane Rigid Closed Cell
Density, lb./ft.³ (kg/m³)		2–5 (32–80)	50–60 (800–960)	2–4 (32–64)	50 (800)	50 (800)	5.5–7.0 (88–112)	2.0 (32)	2–5 (32–80)	4–8 (64–130)
Tensile strength, psi (MPa)	D 1623	20–54	1000 (6.89)	1,000 (6.89)	3,300 (22.7)	5,500 (37.9)	110–210 (0.76–1.45)	42–68 (0.29–0.47)	180–200 (1.24–1.38)	90–290 (0.62–2.00)
Compression strength at 10% deflection, psi (MPa)	D 1621	22–85 (0.15–0.59)	8,000–13,000 (55.1–89.6)	—	5,500 (37.9)	7,500 (51.7)	2–18 (0.014–0.12)	25–40 (0.17–0.28)	100–180 at 5% (0.69–1.24)	70–275 (0.48–1.90)
Impact strength, ft.-lb./in.		—	19–21	—	18	45	—	0.21	—	—
Maximum service temperature dry, °F (°C)		Continuous at 300 225 (149)	275 (135)	—	200 (93.3)	270 (132)	180–200 (82–93)	165–175 (74–79)	165–175 (74–79)	200–250 (93–121)
Thermal conductivity BTU/in./hr.-ft.²-°F (W/mK)	D 2326	0.20–0.22 (0.29–0.032)	1.0 (0.14)	2.0 (0.29)	—	—	0.32–0.34 (0.046–0.049)	0.23 (0.033)	0.17–0.21 (0.024–0.030)	0.15–0.21 (0.022–0.030)
Coefficient of linear expansion, 10⁻⁶ in./in.-°F	D 696	5	100	40–60	38	25	—	30–40	30–40	40

Table 8-25   Examples of properties for flexible foams

Material	Specific Gravity	Elongation at Break (%)	Heat Deflection Temperature °C (°F)	Thermal Conductivity W/mK (BTU-in./hr. ft.^{2}· °F)
Polyphenylene oxide	0.8	15.0	96 (205)	0.124 (0.860)
Polycarbonate	0.8	4.0	128 (262)	0.151 (1.05)
Epoxy resin	0.78	—	—	0.7 (4.8)
Isocyanurate	0.032	—	—	0.1 (0.69)
Polyether	0.08	—	—	—
Polystyrene	0.17	—	—	0.65 (4.5)
Polystyrene	1.04	3.5	101 (214)	—
Polyurethane	0.11	—	—	0.3 (2.1)
ABS	0.86–1.1	—	72 (162)	—
Acetal	1.130	—	153 (307)	—
Nylon 6/6	0.97	4.1	255 (491)	—
Polybutylene terephthalate	1.1	1.3	207 (405)	—
Polyimide		—	277 (531)	—
Polysulfone	0.87	3.5	177 (351)	—
Polyvinylchloride	0.6	370.0	—	—

in a mix, then expand it when the pressure is reduced. Another is to let a liquid component of a mix be volatilized by heat. Similarly, water produced in an exothermic chemical reaction can be volatilized within the mass by the heat of reaction. A different technique lets carbon-dioxide gas be produced within the mass by chemical reaction. A related way is for a gas such as nitrogen to be liberated within the mass by the thermal decomposition of a chemical blowing agent. Also tiny beads of plastics (EPS, etc.) or even glass microballoons can be put into a plastic mix or syntactic foam or the like.

Table 8-26   Examples of thermal properties of foams compared to wood

Property	Polystyrene	Urethane	Polyethylene	Wood (red oak)
Density, lb./cu. ft.	1.0–3.0	1.5–2.5	2.0	0.7
(kg/m^3)	(16–48)	(24–40)	(32)	
Insulation Value (K factor,				—
BTU-in./hr. °F ft.^{2})	0.24–0.30	0.14–0.16	0.35	
(W/m . K)	(0.030–0.043)	(0.020–0.023)	(0.050)	
Linear coefficient of thermal	4×10^{-5}	5×10^{-5}	8×10^{-5}	—
expansion, in./in. °F				
Maximum temperature for	170–180	250	160	—
continuous use, °F (°C)	(77–82)	(121)	(71)	
Heat of Combustion				
BTU/lb. (MJ/kg)	16,000 (37.18)	11,000 (25.56)	16,000 (37.18)	8,000 (18.59)
BTU/cu. ft.	32,000	22,000	32,000	320,000
BTU/board ft.	2,660	1,840	2,660	26,600
Ignition temperature				
(ASTM D 1929-62T)				
Flash ignition temp., °F (°C)	650–700 (343–371)	600 (316)	650 (343)	500 (260)
Self-ignition temp., °F (°C)	735–915 (391–490)	975 (524)	660 (349)	500 (260)
Surface flame spread				
(ASTM E 84–61	10–25	40–80	Non FR	100
"Tunnel Test")				

Blowing Agent

Also called foaming agents. Depending on the basic plastic and process, different blowing agents are used to produce gas and thus to generate cells or gas pockets in the plastics. They are divided into the two broad groups of physical blowing agents (PBAs) and chemical blowing agents (CBAs).

With PBAs the compressed gases often used are nitrogen or carbon dioxide. These gases are injected into a plastic melt in the screw barrel under pressure (higher than the melt pressure) and form a cellular structure when the melt is released to atmospheric pressure or low pressure. The volatile liquids are usually aliphatic hydrocarbons, which may be halogenated, and include materials such as carbon dioxide, pentane, hexane, methyl chloride, etc. Polychlorofluorocarbons were formerly used but they have now been phased out due to environment problems.

CBAs, generally solid materials, are of two types: inorganic and organic. Inorganics include sodium bicarbonate, by far the most popular, and carbonates such as zinc or sodium. These materials have low gas yields and the cell structure they create is not uniform. Organics are mainly solid materials designed to evolve gas within a defined temperature range, usually called the decomposition temperature range. This is their most important characteristic and allows control over gas developments through both pressure and temperature. This increased control of the CBAs produces a finer and more uniform cell structure as well as better surface quality on the foamed plastic. There are over dozens of different types available that decompose at temperatures from at least 220 to 700°F (105 to 370°C) and possible higher. Many of these CBAs can be made to decompose below their decomposition temperature through the use of activators.

Recognize that only certain CBAs can be used with certain plastics. They have to be compatible chemically and start gassing at the required temperature. If they are not compatible different problems develop such as discoloration, property losses, etc. A CBA with a temperature over the melting temperature of the plastic will not gas to form gas, etc.

Formation and Curing of Rigid Polyurethane Foam

PUR are a broad class of highly cross-linked plastics prepared by multiple additions of poly-functional hydroxyl or amino compounds. Typical reactants are polyisocyanates [toluene diisocyanate (TDI)] and polyhydroxyl molecules such as polyols, glycols, polyesters, and polyethers. The cyanate group can also combine with water; this reaction is the basis for hardening of the one-part foam formulations.

They are foamed by the expansion and capture of gaseous blowing agents that are added to the reactants or formed during the polymerization reaction. Inert low temperature boiling hydrocarbons or fluorocabons are often incorporated into foam precursors that are kept in pressurized containers. When the pressure is released the gas expands within the plastic producing the frothing action. Reaction of the cyanate group with water produces carbon dioxide gas as a blowing agent. Also produced in this reaction are ammines that further combine with isocyanate to form substituted ureas.

Most rigid polyurethane foams have a closed cell structure. Closed cells form when the plastic cell walls remain intact during the expansion process and are not ruptured by the increasing cell pressure. Depending on the blowing process a small fraction (5–10%) of the cells remain open. Closed cell structures provide rigidity and obstruct gaseous or fluid diffusional processes.

One component formulation consists of prepolymers that are intermediate between monomers and the final polymer product. When released from a pressurized container the foaming gas expands and the prepolymer (containing unreacted cyanate groups) reacts with the moisture (water) in air to complete the polymerization reaction and cure. Because curing depends on the presence of moisture, when foam forming reactants are applied to occluded areas, such as cavities, the rate of solidification will be dependent on the venting conditions available. Poor venting will inhibit expansion of the blowing agent and the chemical reactions necessary for a cure. Cure times depend on the ability of moisture to reach the expanded prepolymer material.

Processing of rigid foams from two part formulations involves combining measured quantities of the polyisocyanate with a polyhydroxyl such that there are no or limited reactive isocyanate functional groups. Moisture is not required to complete the cure. Once the reactants are combined the mixture is poured into a form where expansion and polymerization take place simultaneously. Cure times are usually very fast, on the order of minutes.

Expandable Polystyrene

EPS molding illustrates the use of blowing agents. Plastic beads containing a blowing agent are supplied to the molder in solid form. Each about 0. 1 to 0. 3 mm in diameter, these beads or spheres contains a small amount of a hydrocarbon liquid, usually pentane that is used as the blowing agent.

The process involves two major steps. The first consists of a preexpansion of the virgin beads by heat (steam, hot air, radiant heat, or hot water). Steam is the most used medium as it is the most practical and most economical.

The next step conveys these beads, usually through a plastic transport tube by air, to the mold cavity. The final expansion occurs in the mold usually with steam heat, either by having live steam go through perforations in the mold itself or by means of steam probes in the cavity that are withdrawn as the beads are expanding. During expansion the beads melt together, adhering to each other and form a relatively smooth skin, filling the cavity or cavities. With some products multiple cavities can be used. After the heat cycles the cooling cycle starts. Because the EPS is an excellent thermal insulator, it takes a relatively long time to remove its heat prior to demolding compared to a solid plastic. With insufficient time the product will distort. Directing a water spray on the mold usually

does the cooling. To facilitate removal, particularly for complex shapes, mold-release agents are used.

An outstanding property of EPS is its extremely low density (when compared to other processes), that by alteration of the preforming treatment can be varied according to the end use. Other types of plastics are employed to produce expandable plastic foam (EPF), including PE, PP, PMMA, and ethylene-styrene copolymers. They can use the same equipment, with only slight modifications. These plastics have different properties from those of EPS and open up different markets. They provide improved sound insulation, resistances to additional heat deformation, better recovery of shapes in moldings, and so on.

Syntactic Foam

In syntactic foams, instead of employing a blowing agent to form bubbles in the mass, preformed bubbles of glass, ceramics, and/or plastic are embedded in a matrix of an unblown plastic. The preformed bubbles are combined with a foamed plastic to provide cells. Reducing weight is one obvious objective, but this change may be accompanied by other properties. A mixture of microspheres and the plastic can be formulated into a moldable mass that can then be shaped or pressed into cavities and molds much as molding sand and clay. The properties of the finished hardened or cured mass can then be tailored by a suitable plastic formulation. Synthetic wood, for example, can be created by a mixture of TS polyester plastic and small hollow glass spheres (Table 8-27).

Syntactic foam contains an orderly arrangement of hollow sphere fillers. They are usually glass microspheres approximately 100 microns (4 mils) in diameter, provide strong, impervious supports for otherwise weak, irregular voids. As a result, syntactic foam has attracted considerable attention both as a convenient and relatively lightweight buoyancy material and as a porous solid with excellent shock attenuating characteristics. The latter characteristic is achieved

Table 8-27   Example of syntatic foams used in deepwater floatation material

Foam Density lb./ft.^{3}	Glass Microballoons	Epoxy Macroballoons	Uniaxial Compressive Yield Strength, psi*	Hydrostatic Compressive Strength, psi*	Method of Preparation	Resin System
24	Yes	Yes	1,700	1,800	Pack-in-place	Polyester
27	Yes	Yes	2,800	2,300	Pack-in-place	Epoxy
30	Yes	Yes	2,500	3,000	Infiltration	Polyester
32	Yes	No	7,300	7,000	Pack-in-place	Polyester
34	Binary	No	—	16,000	Infiltration	Epoxy
35	Yes	No	10,500	14,000	Vacuum cast	Epoxy
38	Yes	No	10,000	14,000	Cast	Polyester
42	Yes	No	10,000	15,000	Cast	Polyester

*To convert psi to pascals (Pa), multiply by 6.895×10^3.

through crushing of the spheres and filling in the voids with plastic.

Static and dynamic property   The uses of these foams or porous solids are used in a variety of applications such as energy absorbers in addition to buoyant products. Properties of these materials such as a compressive constitutive law or equation of state is needed in the calculation of the dynamic response of the material to suddenly applied loads. Static testing to provide such data is appealing because of its simplicity, however, the importance of rate effects cannot be determined by this one method alone. Therefore, additional but numerically limited elevated strain-rate tests must be run for this purpose.

Interest in the use of syntactic foam as a shock attenuator led to studies of its static and dynamic mechanical properties. Particularly important is the influence of loading rate on stiffness and crushing strength, since oversensitivity of either of these parameters can complicate the prediction of the effectiveness of a foam system as an energy absorber.

Results of uniaxial strain static and gas gun compression tests on syntactic foam have been conducted. The foam was buoyant and composed of hollow glass microspheres (average diameter 100 microns) embedded in an epoxy plastic. Static testing consists of compressing a 0.25 cm × 2.5 cm dia. wafer between carefully aligned 2.5 cm dia. steel pistons. Lateral expansion of the wafer is suppressed by mounting it in a thick-walled (10 cm OD, 2.5 cm ID) cylinder, The degree of expansion is monitored by a circumferential strain gage mounted on the outside of the cylinder. Dynamic testing is conducted using a wave generated gas gun.

Accordingly an experimental and analytical program was undertaken to establish the magnitude of the rate effect over the range of interest. From a materials testing standpoint, it is clear that the rate range is bounded from below zero, which is well approximated by the so-called static testing machines, and from above by rates achieved under instantaneous impact, which are approached in gas gun tests. Assuming monotonicity of the material behavior over this extended rate range, one may argue that data from the bounding experiments will exhibit the maximum discrepancy and hence provide a gross measure of the material sensitivity to rate effects. The experiments conducted in this program were designed to follow this philosophy.

Test results provides the hypothesis that syntactic foam is rate insensitive and that the static uniaxial strain stress-strain curve actually represents the general constitutive relation. Disagreement between the experimental data and the predicted behavior is greatest at low stresses (1 kbar) where experimental stresses are about double those predicted analytically. The discrepancy decreases at the higher stress levels and virtually disappears at and beyond 7 kbar. This range

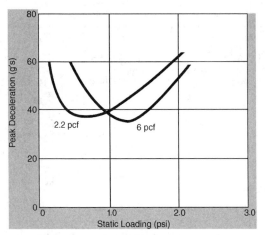

Fig. 8-53 Example of how a closed-cell PE foam density affects both its cushioning and its loading.

of disagreement would extend somewhat further (to about 9 kbar) were the transient current readings rather than the plateau values used in the intermediate stress range.

Cushioning Design

When plastic foams are used to cushion products, as they are in packaging, there are specific design approaches to use. It is widely held that the lower density closed-cell foams that are usually priced lower provide superior cushioning performance, but this assumption is usually incorrect. Figure 8-53 illustrates how closed-cell PE foams with differing densities but the same type and thickness behave under the same dynamic cushioning conditions. These curves represent the amount of mechanical shock experienced during an impact. The lower the curve goes, the greater the cushioning efficiency. For densities above 2 pcf (lb/ft^3) the maximum cushioning efficiency of each material is not significantly different, but the loading at which this maximum efficiency will occur will vary dramatically.

Density effect If a 40 g package were to be designed according to Fig. 8-53 using a 6.0 pcf foam, the foam would measure 3 in. thick at a loading of about 1.35 psi. If an identical package were then produced using a 2.2 pcf foam, its shock performance would not go as

low as 40 g's but would instead produce about 60 g's, or 50% more shock. In order to return to 40 g's, the 2.2 pcf package would need to be redesigned. One approach would be to greatly increase the thickness of the pads constructed from the lower density foam, to provide adequate protection. This approach would, however, increase the package size, impair handling and shipping efficiency, and possibly result in higher costs. The 6.0 pcf foam could, however, be reliably used at 1.2 psi in the thickness shown in Fig. 8-53.

Another approach is to keep the 2.2 pcf foam thickness the same, but decrease the loading from 1.35 to 0.87 psi, to get back to the 40 g level. Although this approach keeps the package size the same, nearly twice as much foam must be used to meet the lower loading. The lower density foam must therefore cost less than half as much as the high density type on a cost-per-unit volume basis if using the lower density one is to result in a cost savings. Below a density of about 2.2 pcf the cushioning efficiency can begin to change with the density. This situation is shown in Fig. 8-54 where the test results for PE foams in several densities below 3 pcf are compared.

Creep resistance Thus, lowering the density produces a considerably higher deceleration and reduces cushioning performance. Also significant is the narrower range of usable static loadings at the bottom of the

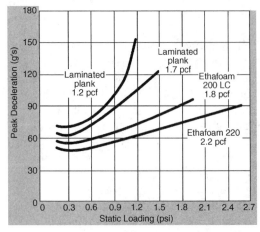

Fig. 8-54 A closed-cell PE foam at different densities compared to its cushioning efficiency.

curves that resulted when the density was reduced. Important consideration in comparing foams of different densities is their compressive creep resistance, and their ability to resist undergoing a permanent thickness loss during their time under load. As the density decreases, so does the creep resistance.

Although it may seem logical for a lower-density foam to cost less to produce because it contains less plastic, this is not necessarily true. The manufacturing rates, the amount and cost of the blowing agent, and the amount and cost of the base plastic all influence the final cost. As a result, very low density foams can actually be more costly to make than others. Thus, it should not be assumed that the cost of a foam will be proportional to its density. Incidentally, this cost situation can also occur with solid plastics in certain shapes, thicknesses, and types, but in solids the problem is rare. Cushioning performance is therefore not improved merely by increasing a foam's density. In order to be certain that a material selected is appropriate and efficient, the designer should carefully compare documented performance data.

Foam Reservoir Molding

This low pressure process, also known as elastic reservoir molding, consists of making basically a sandwich of plastic-impregnated open-celled flexible polyurethane foam between the face layers of fibrous reinforcements. When this plastic composite is placed in a mold and squeezed, the foam is compressed, forcing the plastic outward and into the reinforcement. The elastic foam exerts sufficient pressure to force the plastic-impregnated reinforcement into contact with the heated mold surface. Other plastics are used.

Reinforced Plastic

The term reinforced plastic (RP) refers to combinations of plastic (matrix) and reinforcing materials that predominantly come in fiber forms such as chopped, continuous, woven and nonwoven fabrics, etc. and also in other forms such as powder, flake, etc. They provide significant property and/or cost improvements than the individual components; primary benefits include high modulus, high strength, oriented strength, lightweight, high strength-to-weight ratio, high dielectric strength and corrosion resistance, and long term durability (7, 10, 62).

Also called composite. The term composite denotes the thousands of different combinations of two or more materials that include, in comparison, a few RPs. If referring to composites that incorporate plastics, consider calling them plastic composites. However the more descriptive and popularly used worldwide term is reinforced plastic. In USA annual consumption of all forms of RPs is over $3\frac{1}{2}$ billion lb (1.6 billion kg). Both thermoset (TS) and thermoplastic are used. At least 90wt% use glass fiber and about 45wt% of them use TS polyester plastic. This RP market started during the early 1940s using contact or low pressure TS polyester plastics-glass fiber fabricating systems that were practically all hand lay up with bag molding. In the mean time many different plastics with different reinforcements have been used with rather many different RP processes. All these combinations meet different requirements. In the mean time their products have gone worldwide into the deep ocean waters, on land, and into the air including landing on the moon and in spacecraft.

The RP industry is a mature industry. Improved understanding and control of processes continue to increase performance and reduce variability. Fiber strengths have risen to the degree that 2-D and 3-D RPs can be used producing very high strength and stiff RP products having long service lives of over a half century. Thermoplastic RPs (RTPs), even with their relatively lower properties when compared to thermoset RPs (RTSs) are used in about 55wt% of all RP parts. The RTPs are practically all injection molded with very fast cycles using short glass fiber producing highly automated and high performance parts (Fig. 6-15). Included in these RTPs are stampable reinforced thermoplastics.

RP Characterization

RPs can be characterized many different ways such as type and construction of reinforcement used to those with high impact and fatigue strength properties. Testing for tensile stress-strain (S-S) properties over a range of high-test rates with areas under the S-S curves is a potential method for estimating relative toughness. Comparing fatigue strength for notched and unnotched conditions at various ratios of minimum to maximum stress is useful in structural design (Chapter 2).

Depending on construction and orientation of stress relative to reinforcement, it may not be necessary to provide extensive data on time-dependent stiffness properties since their effects may be small and can frequently be considered by rule of thumb using established practical design approaches. When time dependent strength properties are required, creep and other data are used most effectively. There are many RP products that have had super life spans of many decades. Included are products that have been subjected to different dynamic loads in many different environments from very low temperatures to very high corrosive conditions, etc. An example is aircraft primary structures (10, 14, 62).

RP Directional Property

With RPs an opportunity exists to optimize design by focusing on a material's composition, product geometry, and orientation. However what is involved basically is in "tailor-making" the RP material. The arrangement and the interaction of the usual stiff, strong fibers dominate the behavior of RPs with the less stiff, weaker plastic matrix (TS or TP). A major advantage is that directional properties can be maximized. Basic design theories of combining actions of plastic and reinforcement have been developed and used successful since the 1940s (7, 10, 37).

When compared to unreinforced plastics, the analysis and design of RPs is simpler in some respects and perhaps more complicated in others. Simplifications are possible since the stress-strain behavior of RPs is frequently

fairly linear to failure and they are less time-dependent. For high performance applications, they have their first damage occurring at stresses just below ultimate strength. They are also much less temperature-dependent, particularly RTSs (reinforced TSs). The potential complications that arise relate principally to the directional effects resulting from the fiber construction (Chapter 6, **REINFORCED PLASTIC**).

When constructed from any number and arrangement of RP plies, the stiffness and strength property variations may become much more complex for the novice. Like other materials, there are similarities in that the first damage that occurs at stresses just below ultimate strength. Any review that these type complications cause unsolvable problems is incorrect. Reason being that an RP can be properly designed, fabricated and evaluated to take into account any possible variations; just as with other materials. The variations may be insignificant or significant. In either case, the designer will use the required values and apply them to a safety factor; similar approach is used with other materials (Appendix A: **PLASTIC DESIGN TOOLBOX**).

The fabricator has a variety of alternatives to choose from regarding the kind, form, amount of reinforcement to use, and the process vs. requirements (Table 8-28). With the many different types and forms (organics, inorganics, fibers, flakes, and more) available, practically any performance requirement can be met and molded into any shape. Possible shapes range from very small to extremely large, and from simple to extremely complex.

Orientation of reinforcement The behavior of RPs is dominated by the arrangement and the interaction of the stiff, strong fibers with the less stiff, weaker plastic matrix. The features of the structure and the construction determine the behavior of RPs that is important to the designer. A major advantage is the fact that directional properties can be maximized in the plane of the sheet. As shown in Fig. 8-55 they can be isotropic, orthotropic, etc. Basic design theories of combining actions of plastics and reinforcements

Table 8-28　Overview of RP properties and processes

Process	Reinforcement wt%	Tensile Strength		Tensile Modulus		Flexural Strength		Compressive Strength		Impact Strength		Thermal Conductivity		Heat Distortion at 1.8 MPa		Dielectric Strength	
		MPa	ksi	GPa	10⁶ psi	MPa	ksi	MPa	ksi	J/m	ft·lbf/ft	W/m·K	Btu·in./h·ft²·°F	°C	°F	kV/cm	kV/in.
Spray(a)	30–50 glass-polyester	60–120	9–18	5.5–12	0.8–1.8	110–190	16–28	100–170	15–25	210–640	48–144	0.17–0.23	1.2–1.6	175–205	350–400	80–160	200–400
Compression(a)	15–30 glass-SMC	55–140	8–20	11–17	1.6–2.5	120–210	18–30	100–210	15–30	430–1,150	96–264	0.19–0.25	1.3–1.7	205–260	400–500	120–180	300–450
Compression(a)	25–50 glass mat-polyester	170–210	25–30	6.2–14	0.9–2.0	70–280	10–40	100–210	15–30	530–1,050	120–240	0.19–0.26	1.3–1.8	175–205	350–400	120–240	300–600
Filament winding(a)	30–80 glass-epoxy	550–1,700	80–250	28–62	4.0–9.0	690–1,850	100–270	310–480	45–70	2,150–3,200	480–720	0.27–0.33	1.9–2.3	175–205	350–400	120–160	300–400
Pultrusion(b)	40–80 glass mat-polyester	410–1,050	60–150	28–41	4.0–6.0	690–1,050	100–150	210–480	30–70	2,400–3,200	540–720	0.27–0.33	1.9–2.3	205–260	400–500	80–160	200–400
Pultrusion(b)	30–50 glass mat-polyester	80–210	12–30	6.9–17	1.0–2.5	170–210	25–30	210–340	30–50	530–1,350	120–300	0.22–0.27	1.5–1.85	95–150	200–300	80–120	200–300
Pultrusion(c)	30–55 glass mat and roving-vinyl ester resin	70–280	10–40	6.9–21	1.0–3.0	100–280	15–40	140–340	20–50	270–1,600	60–360	0.22–0.33	1.5–2.3	175–230	350–450	80–130	200–325
Pultrusion(c)	30–55 glass mat and roving-polyester resin	50–240	7–35	5.5–17	0.8–2.5	70–210	10–30	100–280	15–40	210–1,350	48–300	0.22–0.33	1.5–2.3	175–205	350–400	80–120	200–300

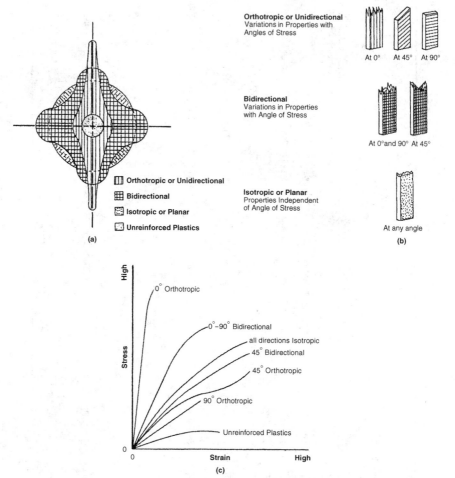

Fig. 8-55   Overview of RPs directional properties: (a) polar directional, (b) different fiber orientations and tensile fracture characteristics, and (c) stress vs. strain diagrams of RPs.

have been developed and used successfully. As an example, woven fabrics that are generally directional in the 0° and 90° angles contribute to the mechanical strength at those angles. The rotation of alternate layers of fabric to a layup of 0°, +45°, 90°, and −45° alignment reduces maximum properties in the primary directions, but increases in the +45° and −45° directions. Different fabric and/or individual fiber patterns are used to develop different property performances (Figs. 8-56 and 57).

A microscopic view of an RP reveals groups of fibers surrounded by the matrix. For example, glass fibers at about 0.01 mm (4×10^{-4} in.) in diameter may comprise from 10 to 90wt% of the area of a given cross-

section. Theories are available to predict overall behavior based on the properties of fiber and plastic constituents' (10). In a practical design approach, the behavior can use the original approach analogous to that used in wood, where individual fiber properties are neglected; only the gross properties, measured at various directions relative to the grain, are considered. This was one of the initial evaluation approaches used during the 1940s (10).

Terminology regarding directional properties used with RPs include the following:

1. *Anisotropic construction*   It is one in which the properties are different in different directions along the laminate flat plane; a

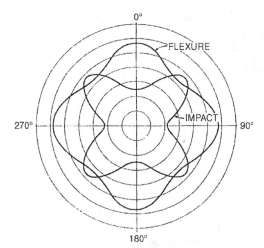

Fig. 8-56 Properties of style 181 glass fabric (bidirectional type); parallel lay-up with 60wt% glass content.

material that exhibits different properties in response to stresses applied along the axes in different directions.

2. *Balanced construction* In woven RPs, equal parts of warp and fill fibers exist. Its construction is one in which reactions to tension and compression loads result in extension or compression deformations only, and

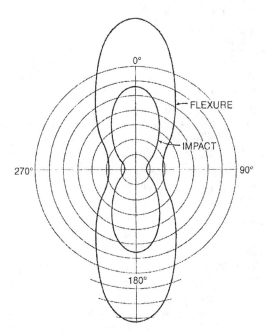

Fig. 8-57 Properties of style 143 glass fabric (unidirectional type); parallel lay-up with 60wt% glass content.

which in flexural loads produce pure bending of equal magnitude in axial and lateral directions. It is an RP in which all laminae at angles other than O° and 90° occur in ± pairs (not necessarily adjacent) and are symmetrical around the central line.

3. *Biaxial load* It is a loading condition in which a specimen/product is stressed in two different directions in its plane, i.e., as an example it is a loading condition of a pressure vessel under internal pressure and with unrestrained ends.

4. *Bidirectional construction* It is an RP with the fibers oriented in various directions in the plane of the laminate usually identifies a cross laminate with the direction 90° apart.

5. *Isotropic construction* Identifies RPs having uniform properties in all directions. The measured properties of an isotropic material are independent on the axis of testing. The material will react consistently even if stress is applied in different directions; stress-strain ratio is uniform throughout the flat plane of the material.

6. *Isotropic transverse construction* Refers to a material that exhibits a special case of orthotropy in which properties are identical in two orthotropic dimensions but not the third. Having identical properties in both transverse but not in the longitudinal direction.

7. *Nonisotropic construction* A material or product that is not isotropic; it does not have uniform properties in all directions.

8. *Orthotropic construction* Having three mutually perpendicular planes of elastic symmetry.

9. *Quasi-isotropic construction* It approximates isotropy by orientation of plies in several or more directions.

10. *Unidirectional construction* Refers to fibers that are oriented in the same direction, such as unidirectional fabric, tape, or laminate, often called UD. Such parallel alignment is included in pultrusion and filament winding applications.

11. *Z-axis construction* In RP, it is the reference axis normal (perpendicular) to the X-Y plane (so-called flat plane) of the RP.

Hetergeneous/homogeneous/anisotropic
RP materials are heterogeneous; which means varied. The material's composition varies from point to point in a heterogeneous mass. However, for design purposes, many heterogeneous materials are treated as homogeneous. This is because a "reasonably" small sample of material cut from anywhere in the body has the same properties as the body. The term homogeneous means uniform. As one moves from point to point in a homogeneous material, the materials' composition remains the same. An unfilled (unreinforced) TP is an example of this type of material.

In an anisotropic material, the properties depend on the direction in which they are tested. For example, rolled metals, which are anisotropic, tend to develop a crystal orientation in the rolling direction. Thus rolled and sheet-metal products have different mechanical properties in the two major directions. Also, extruded plastic film can have different properties in the machine and transverse' directions. These materials are oriented biaxially and are anisotropic. (As reviewed above under **EXTRUSION, Orientation**).

The designer must be aware that as the degree of anisotropy increases, the number of constants or moduli required to describe the material increases; with isotropic construction one could use the usual independent constants to describe the mechanical response of materials, namely, Young's modulus and Poisson's ratio (Chapter 2). With no prior experience or available data for a particular product design, uncertainty of material properties along with questionable applicability of the simple analysis techniques generally used require end use testing of molded products before final approval of its performance is determined.

RPs are either constructed from a single layer or built up from multiple layers. The properties of each layer are usually orthotropic, which is a special case of anisotropy. Fibers that remain straight in the single layer are desired. However, with many fabrics, they are woven into configurations that kink the fiber bundles severely. Such fabric constructions may be very practical since

they drape better over double-warped molds than do fabrics that contain predominantly straight fibers. To reduce the number of kinks in a fabric and develop different formabilities and properties, satin weave fabrics are used. They have different constructions such as weaving a fiber over 9 fibers rather than using the more conventional square weave. The square weave has one fiber over another.

Fiber bundles in lower cost woven roving are convoluted or kinked as the bulky rovings conform to the square weave pattern. Kinks produce repetitive variations in the direction of reinforcement with some sacrifice in properties. Kinks can also induce high local stresses and early failure as the fibers try to straighten within the matrix under a tensile load. Kinks also encourage local buckling of fiber bundles in compression and reduce compressive strength. These effects are particularly noticeable in tests with woven roving, in which the weave results in large scale renforcement. Fiber content can be measured in percent by weight of the fiber portion (wt%). However, it is also reported in percent by volume (vol%) to reflect better the structural role of the fibers, which is related to volume (or area) rather than to weight. When content is only in percent, it usually refers to wt%.

The fiber content in mat or random fiber RPs is usually somewhat lower than for an isotropic laminate which is comprised of a number of unidirectional plies. Both laminates may, for example, be planar-isotropic. The random criss-cross nature of chopped fibers in a mat does not permit close packing of the bundles, and thus the fiber content is usually lower. With a lay-up of unidirectional plies, the packing of fibers within a ply may be very close, and the fiber content can be very high. The higher fiber content made from individual plies tends to make it stiffer and stronger than the mat construction.

There is a relationship between the way the glass is arranged and the amount of glass that can be packed in a given product. By placing continuous strands, such as round glass fibers in a filament winding pattern, next to each other in a parallel arrangement, more glass can be placed in a given volume. Glass content can range from 65 to 95.6 wt% or up to 90.8 vol%. When one-half of the strands are placed at right angles to each half, glass loadings range from 55 to 88.8 wt% or up to 78.5 vol%.

Advanced Reinforced Plastic

An advanced RP (ARP) typically refers to a plastic matrix reinforced with very high strength, high modulus fibers and/or other properties. Examples of these type fibers include carbon, graphite, aramid, boron, S-glass and ZenTron-glass. ARPs can provide the designer with specific properties or characteristics such as strength, stiffness and lower density used in different environments. They can be at least 50-times stronger and 25 to 150-times stiffer than the matrix. As an example ARPs can possess the desirable properties of low density (1.4–2.7 g/cm^3), high strength (3–5 GPa) and high modulus (60–550 GPa). With proper processing these ARPs provide certain properties equal or exceeding those of most other materials.

Micromechanic

It is analyzing the mechanical behavior of RPs by considering the properties, concentration, geometry, and packing of the individual components. This contrasts with macromechanics by recognizing the inhomogeneous nature of RP. By making various approximations of the packing geometry and stress fields within an element of the matrix, the average properties of the element may be calculated (62).

Material

Plastics offer the opportunity to optimize RP design by focusing on material composition in conjunction with reinforcement orientation, as well as product structural geometry. This interrelation affects processing methods, product performances, and costs. This action also gives the designer great flexibility and provides freedom not possible with most other materials. However, it requires a greater understanding of the interrelations to take full advantage of RPs.

Certain plastics provide higher strength and stiffness; a broad range of properties exit. Even though there are literally over 35,000 plastics available worldwide (for all plastic fabricating processes) only a few hundred are used in RPs. In turn only a few of those are predominantly used in most of the RPs. The thermoplastics (TPs) include principally nylons and polypropylenes, as well as polycarbonates, acetals and polyesters. Thermosets (TSs) include predominantly polyesters as well as epoxies, phenolics and urethanes.

TSs and RTSs generally are more suitable to meet the tighter tolerances. The crystalline RTPs, particularly unreinforced TPs, can be more complicated if the designer does not understand their behavior. Crystalline plastics generally have significant different rates of melt flow shrinkage, particularly during injection molding, in the longitudinal melt flow and transverse directions; less transversewise. With reinforcement and/or certain fillers these differences can be reduced or eliminated. With amorphous TPs basically no difference occurs. Compensation for any potential undesirable differences can be made during product and mold designs, reinforcement/filler selection and/or during processing (Chapter 6).

Reinforcements are discrete (usually) inert inclusions used to significantly improve the structural characteristics of a TP or TS plastic. They can be in continuous forms (fibers, filaments, woven or non-woven fabrics, tapes, etc.), chopped forms having different lengths (Fig. 8-58), or discontinuous in form (whiskers, flakes, spheres, etc.). The reinforcements can allow the RP materials to be tailored to the design, or the design tailored to the material. The required approach often results in better designs for certain products and can be less complicated to fabricate. However when permitted, it is best to use prepreg or molding compounds where the plastic and reinforcement are prepared and ready for fabricating into a product. These materials include sheet molding compounds (SMC), bulk molding compounds (BMC)

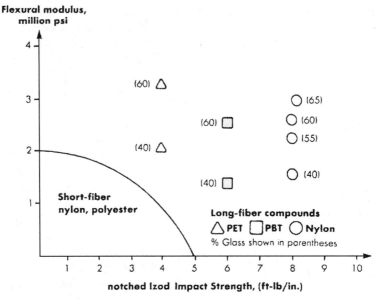

Fig. 8-58   Long glass fiber RP molding compounds have higher and more metal-like than conventional shorter fiber compounds.

and stampable sheets. They can be prepared and processed to meet different directional properties or product performances.

Flexible RP   These materials are used with elastomeric materials providing special engineered products such as conveyor belts, mechanical belts, high temperature or chemical resistant suits, wire and cable insulation, and architectural designed shapes.

Preimpregnation   Also called prepreg. It is the practice of mixing usually TS plastic (hot melt or solvent system; also wet system without solvent) and reinforcement and stored for use at a latter time or for shipping to a molder. The reinforcement can be of any style such as glass fiber mat or fabric. It is partially cured (B-stage) ready-to-mold material in web form that may have a substrate of glass fiber mat, fabric, roving, etc.; paper; cotton cloth; etc. With proper storage condition of temperature, their shelf life can be controlled lasting at least 6 months.

Bulk molding compound   BMC is also called dough molding compound (DMC). It is a molding compound (extruded log form) that is not produced in sheet form. It consists

of the mixture used in sheet molding compound (SMC), except that it contains only short fibers. IMMs are used with special screw designs and some type of ramming system to feed the screw (3).

Sheet molding compound   SMC is a ready-to-mold glass fiber reinforced TS polyester material primarily used in compression molding. SMC is usually TS polyester plastic (generally cross-linked with styrene) with glass fibers and additives such as pigments, fillers, etc. that have been compounded and processed into sheet form to facilitate handling in the molding operations. This B-stage material has a good shelf life particularly when stored in a cool place.

Different methods are used for manufacturing SMC. A few will be reviewed. Figure 8-59 shows continuous glass-filament rovings going through a chopper (where the length

Fig. 8-59   SMC production line with chopped glass fibers.

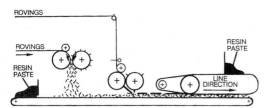

Fig. 8-60 SMC production line incorporating long fibers.

of the chopped fibers can be changed by the location of the cutting blades). The belt moves at a controlled rate and the plastic paste compound is controlled by a doctor blade that provides an opening for the paste to move over the speed-controlled revolving conveyor belt. Plastic carrier film (usually PE that is not shown) placed on the SMC eliminates the sticking problem of B-stage TS compounds, permit ease of handling for shipment, cool room storage, and lay-up for fabrication. The films are removed prior to fabrication lay-up.

Figure 8-60 is a schematic that shows the production of SMCs incorporating long, high-performance fiber reinforcements oriented in either the machine direction or positioned in any direction desired, using single or multiple fibers and rovings to obtain the desired orientation and directional properties. Figure 8-61 shows a schematic of the off-line production process used when required to cut directional-type SMC to conform to a specific mold contour to significantly reduce or even eliminate unwanted wrinkles during lay-up.

Surfacing reinforced mat It is a very thin mat, usually 180 to 510 μm (7 to 20 mil) thick of highly filamentized glass fiber. It is used to produce a smooth surface on glass fiber reinforced plastics.

Gel coat In RP processing a gel coat on the outer surface can be used to ensure a smooth surface appearance and a tough surface. It could contain a thin synthetic fiber veil to improve performance of the gel coat and/or a surfacing mat. It is a quick setting plastic and gelled prior to reinforcement lay-up. The gel coat becomes an integral part of the finished RP product.

RP cost Important to recognize that a major cost in the production of RPs, going from the design concept to the finished product, is materials of construction. They can range from 40 to 90% of the total cost. Thus, it is important to understand how best to use the materials based on the design and processing requirements. It calls for the ability to recognize situations in which certain approaches may be used and to develop problem-solving methods to fit specific design requirements.

An important criteria is to understand and properly apply the interrelations of design requirements with materials of construction and fabricating methods. RPs has some mechanical, formability, and other characteristics that differ from other materials (steel, aluminum, wood, etc.). So what is new; all materials have certain characteristics that

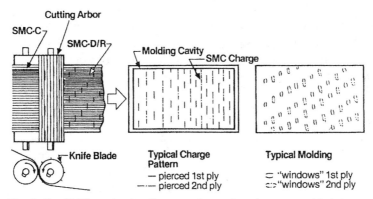

Fig. 8-61 SMC production line permits conforming to a molded shape.

differ from others. It is a fact that RPs have not come near to realizing their great potential in a multitude of applications usually due to cost limitations. An example of its growth is its expanding use in automobile constructions. Cost to performance advantages definitely exist.

Process

Different fabricating processes are employed to produce RP products that represent about 5wt% of all plastic products produced worldwide. They range in fabricating pressures from zero (contact), through moderate, to relatively high, at temperatures ranging from room to well over 100°C (212°F). Equipment may be simple/low cost to rather expensive specialized computer control of the basic machine with auxiliary equipment. Labor costs range from very high to very low. Each process provide capabilities such as meeting production quantity (small to large), performance requirements, proper ratio of reinforcement to matrix, fiber orientation, reliability/quality control, surface finish(s), and so forth versus cost (equipment, labor, utilities, etc.) (Tables 8-29 and 8-30).

Common processes are injection molding (IM), pultrusion, compression molding (CM), contact molding methods (hand lay up, spray, etc.), matched mold methods (modified IM or CM, resin transfer, pressure bag, etc.), spray up, and filament winding. Other processes include autoclave molding; rotational molding, reaction IM reinforced, continuous laminating, and centrifugal casting.

The usual process for processing TP-short glass fiber RPs is IM consuming about 55wt% of all RPs. Specially designed IMMs process TS-RPs with materials such as BMCs (bulk molding compounds). The other processes primarily use TS plastic matrices.

Selecting the optimum process encompasses a broad spectrum of possibilities (shape, size, material used, quantity, tolerance, time schedule, cost, etc.). There are designs when only one process can be used but there can be applications where different processes can be used. Each process, like each material of construction have their capabilities (or limits). Material or product performances are frequently strongly influenced by the process used (2, 10, 14, 62, 92).

Autoclave molding Very high pressures can be obtained for processing RPs enclosed within a bag that initially contains a vacuum. Process may or may not employ an initial vacuum. When required vacuum is used to improve RP performance such as reducing or eliminating entrapped air/gas thus increasing mechanical properties, etc. Air or steam pressures of 100–200 psi (690–1380 kPa) are commonly achieved. If still higher pressures are required, a hydroclave may be used, employing water pressures as high as 10,000 psi (68.9 MPa). The bag must be well sealed to prevent infiltration of high pressure air, steam, or water into the molded product.

Bag molding Process applies an impermeable tailored flexible bag (parting film, elastomer, etc.) over an uncured thermoset RP product located in a mold cavity (male or female), sealing the edges (bagging), and introducing a vacuum and/or compressed air pressure (or water) and heat around the bag. It provides a means of evacuating air and other gases as pressure is applied. Hand operated serrated rollers are usually used to squeeze out voids, air, etc. This high labor technique can produce compact structures that meet tight thickness tolerance simulating injection molded products. This technique is also applied with other RP fabricating processes.

Bag molding Hinterspritzen This patented process allows virgin or recycled TPs such as PP, PC/ABS, etc. to thermally bond with the backing of multilayer PP based fabrics providing good elasticity. This one step molding technique provides a low cost approach for in-mold fabric lamination that range from simple to complex shapes.

Contact molding Also called open molding or contact pressure molding. It is a process for molding RPs in which the reinforcement and plastic are placed in a mold cavity. Cure is either at room temperature using a catalyst-promoter system or by heating

Table 8-29   Process comparison of various RP manufacturing techniques

	Resin Transfer Molding	Open Molding		Compression Molding	
		Spray-up	Hand Lay-up	Mat-Preform	Sheet Molding Compound
Mold construction	FRP† spray metal, cast aluminum: gusket seal, air vents, self-sealing injection port	FRP	FRP, spray metal, cast aluminum, pinch (land)	Metal, shear edge	High grade steel; shear edge
Pressure	Pressure feed pumping equipment req'd: mold halves clamped (methods range from clamp frame to pressure pod)	None	Lows pressure press, capable of 50 psi (hydraulic or pneumatic mechanical); resin dispensing equipment not req'd but recommended	Hydraulic press, normal range of 100-500 psi (0.69-3.05 MPa)	Hydraulic: as high as 2,000 psi (138 MPa)
Cure system		Room temperature		Heated: normal range of 225-325°F (107-163°C)	Heated: normal range of 275-350°F (135-177°C)
Resin compounding equipment	High shear type	Not needed		High shear type	
Reinforcement	Continuous strand mat, preform, woven roving	Continuous roving	Chopped strand mat, woven roving, cloth	Continuous strand mat, preform, woven roving	Continuous roving (specific orientations for higher strength)
Part trim equipment		Yes		With optimum shear edges, minor trimming only	
Generally expected mold life (parts)	3,000	1,000	3,000	150,000+	150,000+

*Courtesy Owners-Corning Corp.
†FRP = Fiberglass reinforced plastics.

Table 8-30   Guide to the directional RP properties vs. processes

Unidirectional Fiber Orientation		REINFORCEMENT TYPES: Filaments, Rovings PROCESSES: Pultrusion, Filament Winding, RTM, RIM
Bidirectional Fiber Orientation	Percentage of Fiberglas reinforcement increases strength in direction of fiber orientation	REINFORCEMENT TYPES: Filaments, Rovings, Woven Fabrics, Braiding PROCESSES: Filament Winding, Hand Layup, Compression Molding, Injection Molding, Vacuum Bag, Stamping, Coining, Pultrusion
Multidirectional Fiber Orientation		REINFORCEMENT TYPES: Chopped Strands, Milled Fibers, Mats PROCESSES: Hand Layup, Compression, Injection, Spray-Up, Vacuum Bag, Autoclave, RTM, RIM, Rotational, Stamping, Autoclave, Coining

in an oven without pressure or using very little (contact) pressure.

Hand lay-up   This is the oldest and in many ways the simplest and most versatile process for producing RP products. However, it is slow and very labored intense. It consists of hand tailoring and placing of layers of (usually glass fiber) mat, fabric, or both on a one-piece mold cavity and simultaneously saturating the layers with a liquid plastic (usually TS polyester). Depending on the plastic additives, the material in the mold can be cured with or without heat, and commonly without pressure. An alternative is to use preimpregnated, B-stage TS polyester such as sheet molding compound (SMC), but in this case heat is applied with low pressure via a impermeable sheet over the material (Fig. 8-62).

Boat.   By far the most important application of RP in marine structures, particularly with respect to volume consumed, has been in boat construction. This has occurred in both civilian and military markets. Growth continues where it already dominates the small boats with the larger boat market growing.

The U.S. Navy pioneered in glass-TS polyester RP (hand lay-up) large boat construction with the production of an 8.5 m

(28 ft) hull in 1947. RP Navy boats (in the U.S. and other countries) range from 3.7 m (12 ft) to over 30.5 m (100 ft).

In 1972, the British launched the world's largest (at that time) RP ship. The H.M.S. Wilton was 46.7 m (153 ft). It became the forerunner of a new class of minehunters that involved other countries (Netherlands, Belgium, Germany, France, Italy, USA, and others). The British program followed with 45.8 to 61 m (100–200 ft) RP minesweepers.

Untraditional hull design.   The U.S. Navy upgraded its minehunter fleet (1991) with a successful ship design based on using glass fiber-TS polyester. This "Osprey" class minehunter was designed and built by Interimarine S.P.A. of Sarzana, Italy (Fig. 8-63). Unlike traditional ships, the new minehunter class does not have longitudinal or transverse framing inside the hull. The design and material combine to provide enough strength and resiliency to withstand underwater explosions. The unstiffened hull is engineered to deform elastically as it absorbs the shock waves of a detonated mine. Judicious design simplifies inspection and maintenance from within the structure.

Basically, the hand lay-up molding process was used, with 98wt% of the structure via

Fig. 8-62 Example of hand lay-up using glass matt-TS polyester RP.

a semiautomated lay-up process. Each mat layer was unrolled and sent through an impregnation liquid plastic bath. Up to six layers were laid-up, wet-on-wet, as a package. A crane laid the wet lay-up along a path in the ship's huge female stainless steel mold. Decks, similarly fabricated, were form-fitted to the hull and bolted in place. Not all the RPs were hand lay-ups. Storage tanks for fuel and water used the filament winding process, etc.

The ship's RP hull was up to 17.8 cm (7 in) thick in the thickest sections. No core materials were used. The glass-to-plastic ratio was 1:1. Final outfitting with gear and equipment resulted in a 55 m (I 88 ft) long warship that holds a crew of 44 people.

In addition to their use in boat hull construction, RPs has been used in a variety of shipboard structures (internal and external). RPs was used generally to save weight and/or to eliminate corrosion problems inherent in the use of aluminum and steel or other metallic constructions. Applications included masts, booms, spinnaker poles, deckhouses, bridge housings, radio rooms, storage tanks (potable water, fuel, etc.), ventilation ducts, piping systems, reefer boxes, hatch covers, sonar domes, radomes, floats, buoys, small safety boats, and more. Much more history exists on RP boats/ships in the literature.

Filament winding FW basically produces high strength and light weight products that consist of two reinforced plastic ingredients that are the reinforcement and a plastic matrix. The process uses a continuous reinforcement (glass, carbon, graphite, PP, wire, and other materials in filament, yarn, tape, etc. forms) either previously impregnated (prepreg) or impregnated at the machine

Fig. 8-63   After removal from the mold, work continued on the ship to which the super structure was added.

with a plastic matrix that is placed on a revolving (removable) mandrel followed with curing. Reinforcements have set pattern lay-ups to meet performance requirements; target is to have them uniformly stressed (Chapter 4, **RP PIPES, Filament Wound Structure**) (7, 10, 14, 37, 62).

Examples of different winding patterns are shown in Table 8-31. The different patterns meet different shape and performance requirements (37). Figure 8-64 shows the use of an isotensoid pattern with only glass fibers and also a view of the cured TS polyester-glass fiber RP structure that provides the strongest product for this shape when compared to any other material (steel, etc.). Figure 8-65 is a racetrack fabricating technique that was used to fabricate a very large tank (rocket motor case) for NASA. This 150,000 gallon tank was $12\frac{1}{2}$ ft diameter by 21 ft long; used a 100 T steel mandrel; contained about 156 million miles of glass fibers; and the textile creel containing 60 spools traveled up to $4\frac{1}{2}$ mph.

Figure 8-66 RP (glass fiber-TS polyester plastic) tank trailers are used for hauling corrosive and hazardous materials. In Canada they meet code 312 to operate.

Injection molding The RTPs (reinforced thermoplastics) are practically all injection molded with very fast cycles using short glass fiber producing highly automated and high performance products. The TPs used include nylons, acetals, polyethylenes, and polypropylene. Of all the RP materials used, about 55wt% represent these RTPs.

Injection-compression molding See in this chapter **INJECTION MOLDING, Modified IM Technique, Injection-compression molding**.

Lost-wax Also called RP molding, fusible-core. A bar (or any shape) of wax is wrapped with RP. After the RP is cured (bag molding, etc.) in a simplified restrictor mold to keep the RP-wax shape, the wax is removed by drilling a hole or removing the end caps by applying a low temperature so that the RP is not effected (review in this chapter **INJECTION MOLDING, Modified IM Technique, *Soluble core molding***. This process can produce high performance products. It was used in fabricating the first all-plastic airplane (Fig. 4-11).

Marco process This process was popular during the 1940s–1950s. Like resin transfer molding (RTM) and bag molding, the reinforcements are laid up in any desired pattern. Low cost matched molds (wood, etc.) confine the reinforcement. A pool of liquid catalyzed TS polyester surrounds the bottom of the mold above its partially opened parting line. From a central opening (hole) in one of the mold halves a vacuum is applied so that the plastic flows through the reinforcements. With proper melt flow, wet-out of fibers occurs and voids are eliminated. This method when first used was the reverse of RTM. The Marco method eventually incorporated pressure plugs at the parting line and also had a push-pull action where pressure was applied in the center hole similar to RTM and use could also include intermittently the vacuum pool action. Eventually only pressure was applied through the center hole; latter became known as RTM.

Pressure bag molding This is a take-of to vacuum bag molding where the bag and mold is placed in a closed system and is subjected to pressure during the curing cycle.

Pultrusion A continuous process for fabricating RPs that usually have a constant cross sectional shape. The reinforcing fibers are pulled through a plastic (usually TS) liquid impregnation bath through rollers, etc. and then through a shaping die followed with a curing action. There are also systems where no plastic bath is used and the plastic is impregnated in the die that is a take-off in extruding wire and cable providing controlled impregnation.

Resin transfer molding With vacuum assisted RTM, this process can be called infusion molding. RTM usually uses liquid TS plastics that is transferred or injected into an enclosed mold usually at low pressures of about 60 psi (410 kPa) in which reinforcement

Table 8-31 Examples of different winding patterns

Type of winding	Considerations	Machinery Required
Hoop or circumferential	High winding angle. Complete coverage of mandrel each pass of carriage. Reversal of carriage can be made at any time without affecting pattern.	Simple equipment. Even a lathe will suffice.
Helix with wide ribbon	Complete coverage of mandrel each pass of carriage. Reversal of carriage can be made at any time without affecting pattern.	Simple equipment with provision for wide selection of accurate ratios of carriage-to-mandrel speeds. Powerful machine and many spools of fiber required for large mandrel.
Helix with narrow ribbon and medium or high angle	Multiple passes of carriage necessary to cover mandrel. Programmed relationship between carriage motion and mandrel rotation necessary. Reversal of carriage must be timed precisely with mandrel rotation. Dwell at each end of carriage stroke may be necessary to correctly position fibers and prevent slippage.	Precise helical winding machine required. Ratio of carriage motion to mandrel rotation must be adjustable in very small increments. Relationship of carriage to mandrel positions must be held in selected program without error through carriage reversals and dwells. Relationship between carriage position and mandrel rotation must be progressive so that pattern will progress.
Helix with low winding angle	Fibers positioned around end of mandrel close to support shaft. Characteristics of "helix with narrow ribbon" apply. Fibers tend to go slack and loop on reversal of carriage. Fibers tend to group from ribbon into rope during carriage reversal. Mandrel turns so slowly that extremely long delay occurs at each end of carriage stroke and speed-up of mandrel at each end of carriage stroke is highly desirable to shorten winding time.	Similar machinery required as for "helix with narrow ribbon." Take-up device for slack fibers is necessary if cross-feed on carriage is not used. Cross-feed on carriage is required for very low winding angles. Programmed rotating eye can be used to keep ribbon in flat band at carriage reversal. Mandrel speed-up device must be programmed exactly with carriage motion or pattern will be lost. Polar wrap machine can be used for narrow ribbons with winding angle below about 15° without take-up device or mandrel speed-up being required.

Type	Description	Machine requirements
Zero or longitudinal	Mandrel must remain motionless during pass of carriage and then rotate a precise amount near 180° while carriage dwells. Fibers must be held close to support shaft during mandrel motion or fibers will slip.	Precise mandrel indexing required. Simple two-position cross feed on carriage sufficient. Vertical mandrel machine and pressure follower for ribbon sometimes required to preserve ribbon integrity.
Polar wrap	Low angle wrap. Fibers may be placed at different distances from centers at each end of wrap. For when geodesic (non-slipping) path does not have to be followed.	Polar wrap machine with swinging fiber delivery arm desirable for high-speed winding. Helical machine with programmed cross-feed will wind polar wraps more slowly.
Cone	General considerations same as for helical winding except that carriage motion is not uniform.	Programmed non-linear carriage motion required. Other machine requirements same as for helical winding.
Simple spherical	Planar windings at a particular angle result in a heavy build-up of fibers at ends of wrap. For more uniform strength, successive windings at higher angles are required.	Sine wave motion of carriage is required for carriage with no cross-feed. At low angles of wind, cross-feed is necessary because carriage travel becomes excessive. Polar wrap machine may be used if range of axis inclination is large enough.
Simple ovaloid	Similar to simple spherical winding but with different carriage or cross-feed motion.	Helical machine with programmed carriage or cross-fee. Polar wrap machine can be used where geodesic (non-slipping) path is in a plane.
True spherical	Path of fibers programmed to give uniform wall thickness and strength to all areas on sphere.	Special machine best approach. Otherwise complex programming of all motions of helical machine required.
Miscellaneous	For successful filament winding, it must be possible to hand-wind with no sideways slipping of fibers on mandels surface.	Machine to reproduce motions of hand winding. Programmed motions in several axes may be required.

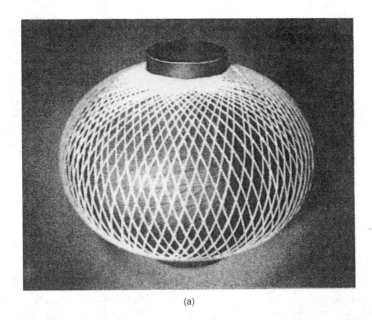

(a)

(b)

Fig. 8-64   (a) FW layup shows isotensoid pattern of the reinforcing fibers and (b) plastic molded isotensoid case/container.

has been placed. The reinforcement is usually glass fiber woven, nonwoven, and/or knitted fabric. Plastic flows through the reinforcement targeting to remove air through release ports and/or openings where its parting line exists. Cure can be with a heated mold or catalyzed so that it develops its own heat based on a prescribed time schedule (Tables 8-32 to 8-34). See also above the Marco process.

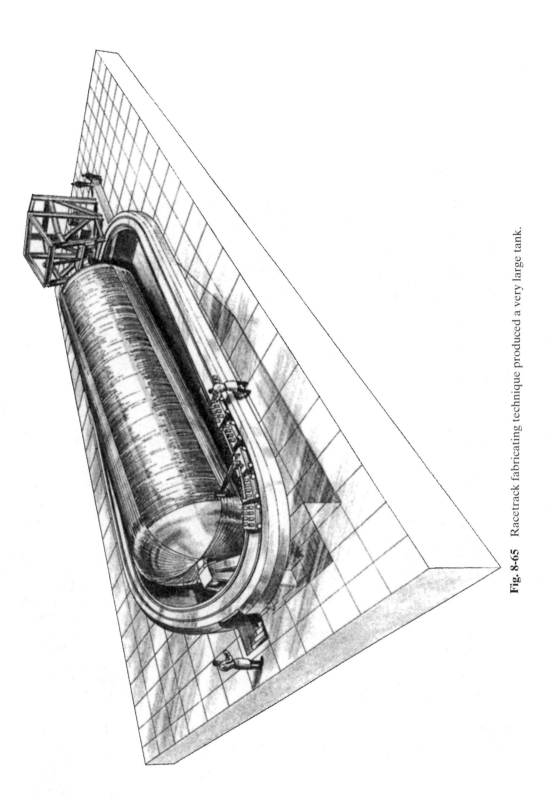

Fig. 8-65   Racetrack fabricating technique produced a very large tank.

Fig. 8-66   RP tank trailer.

SCRIMP process   This Seeman Composites Resin Infusion Process (SCRIMP) is described as a gas-assist resin transfer molding process. As an example glass fiber fabrics/ thermoset vinyl ester polyester plastic and polyurethane foam panels (for insulation) are placed in a segmented tool. A vacuum is pulled with a bag so that a huge amount of plastic can be drawn into the mold (**Marco process** approach). Its curved roof is made separately and bonded to the box with mechanical and adhesive fastening. It is similar to various reinforced plastics molding processes.

Spray-up   Popular system with reinforced plastic production. An air spray gun includes a roller cutter that chops usually glass fiber rovings to a controlled short length before being blown in a random pattern (manually or automatically) onto a surface of the mold; simultaneously the gun sprays catalyzed TS polyester plastic. The chopped fibers are plastic coated as they exit the gun's nozzle. The resulting, rather fluffy, RP mass is consolidated with serrated rollers to squeeze out air and reduce or eliminate voids. A closed mold

with appropriate temperature and pressure produce products.

Stamping   In the stamping process, usually a reinforced TP sheet material is precut to the required sizes. The precut sheet is preheated in an oven, the heat depending on the TP used (such as PP or nylon, where the heat can range upward from 520 to 600°F). Dielectric heat is usually used to ensure that the heat is quick and, most important, to provide uniform heating through the thickness and across the sheet. After heating, the sheet is quickly formed into the desired shape in cooler matched-metal dies, that can use conventional stamping presses or SMC-type compression presses.

Stamping is a highly productive process capable of forming complex shapes with the retention of the fiber orientation in particular locations as required. The process can be adapted to a wide variety of configurations, from small components to large box-shaped housings and from flat panels to thick, heavily ribbed products. Reinforced TS plastic B-stage sheet material can be used with its required heating cycle. However the most popular is to use TP sheets.

Table 8-32 Comparison of resin transfer, compression, and injection molding RP processes

		Process	
	RTM	SMC Compression	Injection
Process operation:			
Production requirement, annual units per press	5,000-10,000	50,000	50,000
Capital investment	Moderate	High	High
Labor cost	High	Moderate	Moderate
Skill requirements	Considerable	Very low	Lowest
Finishing	Trim flash, etc.	Very little	Very little
Product:			
Complexity	Very complex	Moderate	Greatest
Size	Very large parts	Big flat parts	Moderate
Tolerance	Good	Very good	Very good
Surface appearance	Gel coated	Very good	Very good
Voids/wrinkles	Occasional	Rarely	Least
Reproducibility	Skill dependent	Very good	Excellent
Cores/inserts	Possible	Very difficult	Possible
Material usage:			
Raw material, cost	Lowest	Highest	High
Handling/applying	Skill dependent	Easy	Automatic
Waste	Up to 3 percent	Very low	Sprues, runners
Scrap	Skill dependent	Cuts reusable	Low
Reinforcement flexibility	Yes	No	No
Mold:			
Initial cost	Moderate	Very high	Very high
Cycle life	3,000–4,000 parts	Years	Years
Preparation	In factory	Special mold-making shops	
Maintenance	In factory	Special machine shops	

Vacuum bag molding Also called just bag molding; see previous review on **Bag molding**. RP can be prepared for TS plastic curing in an open mold with a flexible membrane or bag over the RP. A vacuum is drawn inside the enclosure [commonly resulting in atmospheric pressures of 10–14 psi (69–97 kPa)] with or without heat (depending on how the plastics was prepared). The result is a molded product with a very smooth surface against the mold surface. Figure 8-67 shows a completely automated vacuum bag molding process.

RP Future

There is always a growing need in many areas to find alternatives to heavy structures made from iron and steel. The modern lightweight RPs are being used in applications where they provide savings in raw materials, energy, and/or installation costs.

Calendering

Basically the calendering process is used in the production of plastic films and sheets. It converts plastic into a melt and then passes the pastelike mass through roll nips of a series of heated and rotating speed-controlled rolls into webs of specific thickness and width. The web may be polished or embossed, either rigid or flexible (9). One of its sheets major worldwide markets is in credit cards. At the low cost side these lines can start a $ million. A line, probably the largest in the world

Table 8-33   Cost comparison of RTM vs. injection and SMC moldings

	Process		
	RTM	SMC Compression	Injection
Process operation:			
Production requirement, annual units per press	5,000–10,000	50,000	50,000
Capital investment	Moderate	High	High
Labor cost	High	Moderate	Moderate
Skill requirements	Considerable	Very low	Lowest
Finishing	Trim flash, etc.	Very little	Very little
Product:			
Complexity	Very complex	Moderate	Greatest
Size	Very large parts	Big flat parts	Moderate
Tolerance	Good	Very good	Very good
Surface appearance	Gel coated	Very good	Very good
Voids/wrinkles	Occasional	Rarely	Least
Reproducibility	Skill-dependent	Very good	Excellent
Cores/inserts	Possible	Very difficult	Possible
Material usage:			
Raw material, cost	Lowest	Highest	High
Handling/applying	Skill dependent	Easy	Automatic
Waste	Up to 3 percent	Very low	Sprues, runners
Scrap	Skill dependent	Cuts reusable	Low
Reinforcement flexibility	Yes	No	No
Mold:			
Initial cost	Moderate	Very high	Very high
Cycle life	3,000–4,000 parts	Years	Years
Preparation	In factory	Special mold-making shops	
Maintenance	In factory	Special machine shops	

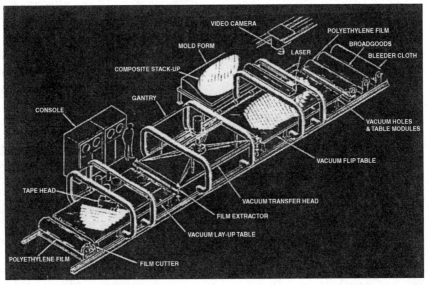

Fig. 8-67   Automated-integrated RP vacuum lay-up process that use prepreg sheets that are in the B-stage.

Table 8-34 Property comparison and design guidelines for RTM vs. other RP processes

Design Parameter	Resin-Transfer Molding	Spray-up	Hand Lay-up	Mat/Preform	Sheet Molding Compound
Minimum inside radius, in. (mm)	$^1/_4$ (6.35)	$^1/_4$ (6.35)	$^1/_4$ (6.35)	$^1/_4$ (6.35)	$^1/_{16}$ (1.59)
Molded-in holes	No	Large	Large	Yes	Yes
In-mold trimming	No	No	No	Yes	Yes
Core pull and slides	Difficult	Difficult	Difficult	No	Yes
Undercuts	Difficult	Difficult	Difficult	No	Yes
Minimum recommended draft (deg.)	2 to 3	0	0	$^1/_4$ to 6-in. depth 1 to 3; above 6-in. depth 3, or as required	
Minimum practical thickness, in. (mm)	0.080 (2.0)	0.060 (1.5)	0.060 (1.5)	0.030 (0.76)	0.050 (1.3)
Maximum practical thickness, in. (mm)	0.500 (12.7)	No limit	No limit	0.500 (12.7)	1 (25.4)
Normal thickness variation, in. (mm)	±0.010 (±0.25)	±0.020 (±0.50)	±0.020 (±0.50)	±0.008 (±0.2)	±0.005 (±0.1)
Maximum thickness buildup, heavy buildup (ratio)	2:1	Any	Any	2:1	Any
Corrugated sections	Yes	Yes	Yes	Yes	Yes
Metal inserts	Yes	Yes	Yes	Yes	Yes
Bosses	Difficult	Yes	Yes	Difficult	Yes
Ribs	Difficult	No	No	Yes	Yes
Hat section	Yes	Yes	Yes	Difficult	No
Raised numbers	Yes	Yes	Yes	Yes	Yes
Finished surfaces	2	1	1	2	2

processing PVC sheet, build by Kleinewefers Kunststoffanlagen GmbH, Munich, Germany, cost $33 million (1999). It is a 5-roll using L-type configuration. They have 3500 mm roll-face widths and 770 mm diameters with an output rate at 4000 kg/h.

The calender was developed over a century ago to produce natural rubber products. With the developments of TPs, these multimillion dollar extremely heavy calender lines started using TPs and more recently process principally much more TP materials. The calender consists essentially of a system of large diameter heated precision rolls whose function is to convert high viscosity plastic melt into film, sheet, or coating substrates. The equipment can be arranged in a number of ways with different combinations available to provide different specific advantages to meet different product requirements. Automatic web-thickness profile process control is used via computer, microprocessor control.

The calendering configuration of rolls may consist of two to at least seven rolls. The number of rolls and their arrangement characterizes them. Examples of the layout of the rolls are the true "L", conventional inverted "L", reverse fed inverted "L", "I", "Z", and so on. The most popular are the four-roll inverted "L" and "Z" rolls. The "Z" calenders have the advantage of lower heat loss in the film or sheet because of the melts shorter travel and the machines simpler construction. They are simpler to construct because they need less compensation for roll bending. This compensation occurs because there are no more than two rolls in any vertical direction as opposed to three rolls in a four roll inverted "L" calender and so on.

The nip is the radial distance or "V" formed between rolls on a line of centers. In-going safety devices in the nip areas are built into these machines. They protect the hands of operators. An emergency stop device is placed in an accessible location on the upstream side.

If a problem develops, the machines immediately stop.

Calendering in the manufacture and surface finishing of plastic products, such as non-wovens and woven fabrics, requires roll systems to meet stringent control of their nip pressure requirements. In this respect, products of uniform quality and thickness, with defined properties, call for an adjustable nip and/or controllability of the nip pressure. Control across the full roll width is achieved by various methods such as: suitable compensation of the deflection of a pair of rolls, mechanical-geometrical compensation such as roll bending, axis crossing and crowning of the rolls, and hydraulic compensation systems. Bowl deflection can occur. It is the distortion suffered by calender rolls resulting from the pressure of the plastic running between them. If nor corrected the deflection produces a sheet or film thicker in the middle than the edges.

Variations in these multimillion dollar calender lines are dictated by the very high forces exerted on the rolls to squeeze the plastic melt into thin film or sheet web constructions. High forces at least up to 6000 psi (41 MPa) could (if rolls were not properly designed and installed) bend or deflect the rolls, producing gauge variations such as a web thicker in the middle than at the edges. During calendering, particularly film, roll-separating forces in the final nip may be as high as 6000 psi. This potential problem is counteracted by different methods that include the following: (1) crowned rolls, which have a greater diameter in the middle than the edged; (2) crossing the rolls slightly (rather than having them truly parallel), thus increasing the nip opening at both ends of the roll; and (3) roll bending, where a bending moment is applied to the end of each roll by having a second bearing on each roll neck, which is then loaded by a hydraulic cylinder. Controls are used to perform any roll bending and crossing of the rolls.

Compounding Material

Important to their success includes the preparation of the material or compound.

It is usually done by computer controlled electronic weighing scales that supply precise amounts of each ingredient to a high intensity mixer. The still-dry, free-flowing blend is then charged to a feed hopper where it is screw fed into a continuous mixer such as an extruder and/or kneader. Under the action of a mixer's reciprocating screw in the confined volume of the mixer chamber, the blend begins to flux or masticate into the required plastic state.

Usually the next step is to force it out of the barrel of the mixing chamber through a die producing strands. The strands can exit as a continuous rope or be chopped into small baseball size buns. This hot plastic material may be passed through a two-roll mill and/or be directly conveyed to the top of the calender rolls. The (usual) parallel rolls have extremely flat surfaces and rotate at possibly the same speed but usually at slightly different speeds depending on the plastic being processed. Although plastic forming occurs in the calender itself, down-stream precision cooling rolls operating equipment are needed to produce the TP film or sheet.

Coating

One special application of calendering is the coating of paper, textile, and/or plastic. For one-sided coating a calender with three rolls is usually sufficient, although four rolls are frequently used for extremely thin coatings. Double-sided coating can either be done simultaneously on both sides using a four-roll or sequentially by two three-roll calenders. Frictional calendering is the process whereby an elastomeric compound is forced into the interces of woven or cord fabrics while passing through calender rolls.

Calendering or Extrusion

Film and sheet can, in principal, be made by calendering or by extrusion. Factors that govern the advantages and disadvantages (limitations) of each process can interact in a complex way. Factors to be considered include: (1) type of material to be processed, (2) quantity of product to be produced, (3) thickness and uniformity required on film or sheet, and (4) costs. The capital equipment and replacement parts in calendering lines are more expensive. The very small-unsophisticated lines start about the million-dollar range compared to the much lower cost extrusion lines. In general, plastic materials such as PE, PP, and PS film and sheet are usually produced through the rather conventional extrusion lines. To produce PVC film and sheet in large quantities, calendering is almost always used since the process is less likely to cause degradation than is extrusion as well as having dimensional and cost advantages.

A web thickness between 0.002 to 0.020 in (0.05 to 0.50 mm) is generally the kind of plasticized film and sheeting produced by calender lines. For extremely light gauges, those under 0.001 in (0.02 mm), calendering could become impractical or damaging to the equipment. The reasons include factors such as for certain materials their exists poor strength of the thin webs and also very high forces developed on the matting heavy duty rolls.

For very heavy/thick gauges such as sheeting over 0.020 in (0.50 mm), calendering may not be the optimum method of production. Reason is that there may not be enough shearing action that can be put into the rolling banks to keep the compound at uniform temperature. In addition, the separating forces on the rolls get so low that gauge variations could become prohibitive.

It can be said that basically the up-stream and down-stream procedures are similar in production lines whether calenders or extruders are used. For a given quantity of output, it is usually necessary to have more extruders than calenders. This situation makes the extrusion lines more flexible and more able to handle relatively short production runs. The extrusion flexibility, when compared to calendering, includes ease of changing product thicknesses, widths, and materials.

Calenders are capable of higher production speeds. Thus, there are situations where they provide a favorable situation for long runs. For these long runs, cost advantages exist. Tolerancewise the calender is easier to produce products that can meet tighter minimum-to-maximum thicknesses on sheets and films. Calendering also provides product uniformity. Constant in-process monitoring and continuous profile adjustments are usually a significant advantage of calendering over other methods.

Compression and Transfer Molding

CM and TM are two methods used to produce molded products from generally thermoset (TS) plastics. CM was the major method of processing plastics during the first half of the last century because of the development and extensive use of phenolic plastics (TSs) in 1909. By the 1940s this situation began to change with the development and use of thermoplastics (TPs) in extrusion and injection molding (IM) processes. CM originally processed about 70wt% of all plastics, but by the 1950s its share of total production was below 25%, and now that figure is about 3% of all plastic products produced worldwide. This change does not mean that CM is not a viable process; it just does not provide the much lower cost-to-performance benefit of TPs, particularly at high production rates. In the early 1900s plastics were almost entirely TS (95wt%), but that proportion had fallen to about 40% by the mid-1940s and now is about 10%.

TSs has experienced an extremely low total growth rate, whereas TPs have expanded at an unbelievably high rate. Regardless of the present situation, CM and TM are still important, particularly in the production of certain low-cost products as well as heat-resistant and dimensionally precise products. CM and TM are classified as high-pressure processes, requiring 13.8–69 MPa (2,000 to 10,000 psi) molding pressures. Some TSs, however, require only lower pressures of down to 345 kPa (50 psi) or even just contact (zero pressure).

CM is the most common method of molding TSs. In this process, material is compressed into the desired shape using a press containing usually a two-part closed mold and is cured with heat and pressure. This process is not generally used with TPs. TM, also called compression-transfer molding is a method principally used with TS plastics. The plastic is first softened by heat and pressure in a transfer chamber (pot) and then forced by the chamber ram at high pressure through suitable sprues, runners, and/or gates into a closed mold to produce the molded product or products using one or more cavities. Usually dielectrically preheated circular preforms are fed into the TM pot and also CM cavity (2, 7, 9).

Reaction Injection Molding

The RIM process involves the high-pressure impingement mixing of two or more reactive liquid components and injection of the mixture into a closed mold at low pressures. Large and thick products can be molded using fast cycles with relatively low-cost materials. Its low energy requirements with relatively low investment costs make RIM attractive (9).

Different materials can be used such as nylon, polyester (TS), and epoxy, but TS polyurethane (PUR) is predominantly used. Almost no other plastic has the range of properties of PUR. Modulus of elasticity range in bending is 200 to 1,400 MPa (29,000–203,000 psi) and heat resistance from 90 to over 200°C (122–392°F). The higher values are for chopped glass-fiber-reinforced RIM (RRIM).

RIM is very similar to RTM (see above **REINFORCED PLASTICS, Processes**). In the reinforced RIM (RRIM) process a dry reinforcement preform is placed in a closed mold. Next a reactive plastic system is mixed under high pressure in a specially designed mixing head. Upon mixing, the reacting liquid flows at low pressure through a runner system to fill the mold cavity, impregnating the reinforcement in the process. Once the mold cavity is filled, the plastic quickly completes its reaction. The complete cycle time required to produce a molded thick product can be as little as one minute.

The advantages of RRIM are similar to those listed for RTM. However, RRIM uses preforms that are less complex in construction and lower in reinforcement content than

those used in RTM. The RRIM plastic systems currently available will build up viscosity rapidly, resulting in a higher average viscosity during mold filling. This action follows the initial filling with a low-viscosity plastic.

Liquid Injection Molding

LIM has been in use longer than RIM; the two processes are practically similar. The advantages it offers in the automated low-pressure processing of (usually) TS plastics is faster molding cycles, low labor cost, low capital investment, energy saving, and space. LIM is very competitive to potting, encapsulating, compression transfer, and injection molding, particularly when insert molding is required.

Rotational Molding

RM is a simple, basic, four-step process that uses a thin-walled mold with good heat-transfer characteristics. Its closed mold requires an entrance for insertion of plastic and, most important, the capability to be "opened" so that solidified products can be removed. These requirements are no problem. Liquid or dry-powder plastic equal to the weight of the final product is put into the mold cavity(s), which rotates simultaneously about two axes located perpendicular to each other (Fig. 8-68). These two rotation speeds can be varied to permit more

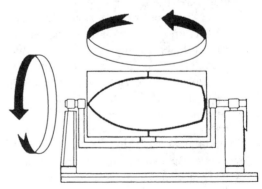

Fig. 8-68   Mold and its rotating mechanism is an example of the RM process.

evenly flow in a complex mold cavity. With slow rotation about each axis, the material inside the mold tumbles to the bottom, creating a continuous path that covers all mold surfaces equally (9).

The next step involves heating the mold while it is rotating. Molds can be heated by a heated oven, a direct flame, a heat-transfer liquid (either in a jacket around the mold or sprayed over the mold), or electric-resistance heaters placed around the mold. With uniform heat transfer through the mold, the plastic melts to build up a layer of molten plastic on the molds inside surface.

After the required heat-time cycle is completed, the mold is ready for cooling, which is accomplished with the mold rotating continually. Cooling is usually done by air from a high-velocity fan and/or by a fine water spray over the mold. After cooling, the final step is to remove the solid hallow product and reload the mold with plastic.

This process is capable of molding small to large hollow items with relatively uniform wall thicknesses, using certain plastics. Its production rates, compared to those of other processes, can be low. The total cost of equipment and the production time for moderate-sized and, especially, large products are also low. Large products range up to at least 85,200 L (22,000 gal.) in size, with a wall thickness of 3.8 cm (1.5 in.). One tank used 2.4 t (5,300 lb.) of XHDPE; the first charge was about 1.5 t (3,300 lb.), followed by 0.45 t (1,000 lb.), and finally another 0.45 t (1,000 lb.).

Molds can be of any shape and can include corrugated or rib constructions to increase their stability and stiffness (large, flat walls can be difficult if not impossible). The thickness of their walls is limited to allow heat penetration.

Encapsulation

Also called conformal coating. It encloses a product in a closed envelope of plastic by immersing the product (solenoids, ornament, sensors, motor components, integrated circuits, and other articles.) in an unheated or heated plastic. Different processes can be used that range from casting to injection molding.

Casting permits applying different techniques. As an example half or part of the casting can gel. The product such as an ornament is placed on the gelled plastic followed with the final pouring of the plastic.

The typical TP encapsulation process is an insert injection molding or liquid injection molding operation. The insert, a coil, or an integrated circuit, for example, is placed in a mold equipped with either fixed spider type supports or retractable pins or other features to support it when molten TP is injected.

This technique with insert molding is a clean, repeatable process that lends itself to automation and cellular manufacturing, and fits well with total quality management (TQM). With off-the-shelf process controls and systematic production methods, manufacturers can deliver repeatable, high-quality products that come out of the tool ready for assembly. The products generally do not require costly trimming or deflashing, as do many TS encapsulated products when using processes such as compression molding (136).

Although horizontal clamp injection molding equipment can be used for encapsulation, vertical-clamp machines allow easier insert placement and greater insert stability during mold clarnping movement. For high-volume production, a vertical machine with a shuttle or rotary table is highly efficient. For example, a two-station table fitted with two lower mold halves allows molding at one station while an operator or robot unloads finished products and loads inserts at the other shuttle station.

Casting

Some TPs and TSs begin as liquids that can be cast and polymerized into solids. In the process various ornamental or utilitarian objects can be embedded in the plastic. By definition, casting applies to the formation of an object by pouring a fluid plastic solution into an open mold where it completes its solidification. Casting can also lead to the formation of film or sheet, made by pouring the liquid plastic onto a moving belt or by precipitation in a chemical bath. Casting differs from many of the other techniques described in this book in that it generally does not involve pressure or vacuum casting, although certain materials and complex products may require one or the other.

Powder Coating

Powder coating is a solventless coating system that is not dependent upon a sacrificial medium such as a solvent, but is based on the performance constituents of solid TP or TS plastics. It can be a homogeneous blend of the plastic with fillers and additives in the form of a dry, fine-particle-size compound similar to flour. The three basic methods are the fluidized bed, electrostatic spray, and electrostatic fluidized bed processes (9).

The advantages of the process include its minimizing air pollution as well as water contamination, and increased product performance when coated, resulting in cost savings. This is basically a chemical coating, so it has many of the same problems as solution painting. If not properly formulated, the coating may sag at high thicknesses, show poor performance when not completely cured, reveal imperfections such as craters and pinholes, and have poor hiding, with low film thickness. It is extensively used to coat wire trays used in dishwashing machines, protecting steel products subjected to salt water, etc.

Textiles, paper, and other flexible substrates such as fusible interlining, interlinking drapery and upholstered fabric, and carpets are examples of large volume applications. Another important market is with metals and other rigid materials. Included are plastics (pipes, tanks, screens, etc.) that can provide protective coatings using its variety of processing techniques.

There are different coating techniques. The woven and nonwoven fabrics normally involves three steps as it passes from the unwind roll to the rewind roll. Powder is metered onto the fabric, heated in an oven (gas or electric) that are usually divided into several heating zones, and cooled by a chill roll.

Coating plastics, metals, etc. steps generally involve surface preparation, preheating substrate, powder applications, and post-heating.

Vinyl Dispersion

Vinyl dispersions are fluid suspensions of special fine-particle-size polyvinyl chloride (PVC) plastics in plasticizing liquids. When the PVC is heated to about 148 to 180°C (300 to 355°F), fusion or mutual solubilization of the plastic and the plasticizer takes place. The dispersion then turns into a homogeneous hot melt. When the melt is cooled below 50 to 60°C (122 to 140°F), it becomes a tough vinyl product. With vinyl dispersions the processor can use convenient liquid-handling techniques such as spraying, pouring, spread coating, and dipping. This system permits products to be made that would otherwise require costly and heavy melt-processing equipment.

The term plastisol is used to describe a vinyl dispersion that contains no volatile thinners or diluents. Plastisols often contain stabilizers, fillers, and pigments, along with the essential dispersion plastics and the liquid plasticizer. All ingredients exhibit very low volatility under processing and use conditions. Plastisols can be made into thick fused sections with no concern for solvent or water blistering, as with solution or latex systems, so they are described as being 100 percent solids.

It is convenient in some instances to extend the liquid phase of a dispersion with organic volatiles, which are removed during fusion. The term organosol applies to these dispersions. Organosol are a suspension of a finely divided vinyl plastic in a plasticizer (diluent) together with a volatile organic liquid. The plastic does not dissolve appreciably in the liquid at room temperature but does at elevated temperatures at which time the liquid evaporates. Upon cooling a homogeneous plastic mass is produced.

Process Control

Fabricating controls involve many facets of the machine operation and the behavior

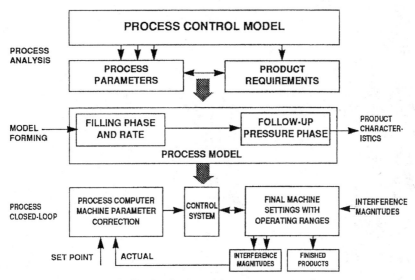

Fig. 8-69   Designing with a process control.

of the plastic. Most important is the interaction between the machine operation and plastic behavior. Basically the processing pressure and temperature vs. time determine the quality of the molded product. The design of the control system has to take into consideration the logical sequence of all these basic functions and their ramifications. Basically developing a process control (PC) flow diagram requires a combination of experience (at least familiarity) of the process and a logical approach to meet the objective that has specific target requirements. PCs range from very simple/standard types to advance complex types.

Control of machines continually enters new eras that dramatically improve ease of machine setup, allow uninterrupted operation, simplify remote handling, reduce fabricating times, cut energy costs, boast part quality, and so on. The process of making a product has many dynamic fragments that must come together properly for successful results. Lack of sufficient PC over each of these fragments will result in a less than desirable product. For success, there are three key ingredients: sufficient dynamic performance, sufficient repeatability, and very important is the selection of proper control parameters. A lack of these ingredients can result in unacceptable products, higher scrap rate, longer cycles, higher part cost, etc. The control unit

is composed of input, signal processing, and power stages (Fig. 8-69).

PCs all have one thing in common. They monitor the process variables, compare them to values known to be acceptable, and make appropriate corrections without operator intervention. The acceptable range of values can be determined by using melt flow analysis software and/or trial and error when the machine is first starting its production. Using the software approach, the acceptable process values are known before the mold or die is ever built. It should be noted that none of the PC solutions address the problem of the lack of skilled setup people. Most of the PC systems available today are rather complex and require skilled people to use them efficiently or at least start up the line.

Adequate PC and its associated instrumentation are essential for product quality control. The goal in some cases is precise adherence to a single control point. In other cases, maintaining the temperature within a comparatively small range is all that is necessary. For effortless controller tuning and the lowest initial cost, the processor should select the simplest controller (of temperature, time, pressure, melt-flow, rate, etc.) that will produce the desired results.

Examples of the more sophisticated controls used with injection molding are seen in Figs. 8-70 and 8-71. Based on the process

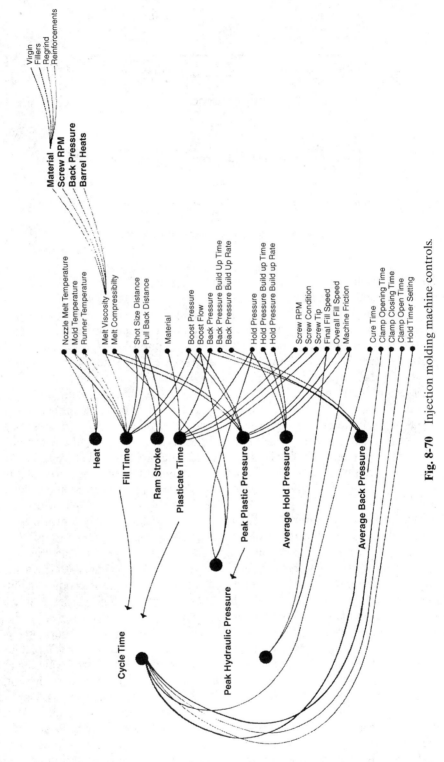

Fig. 8-70 Injection molding machine controls.

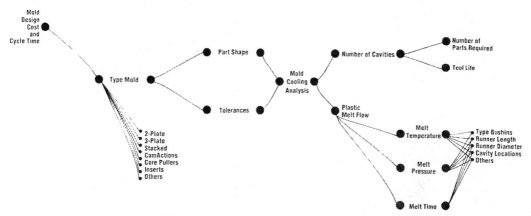

Fig. 8-71   Controls important to mold operation.

control settings, different behaviors of the plastics will occur. Some examples of these behaviors are shown in Figs. 8-25 for injection molding and 8-72 for extrusion.

Processing Window

Regardless of the type of controls available, the processor setting up a machine uses a systematic approach based on experience or that should be outlined in the machine or control manuals. Once the machine is operating, the processor methodically makes one change at a time, to determine the result for each change.

Two basic examples are presented in Figs. 8-73 and 8-74 to show a logical approach to evaluating the changes made with any processing machine that results in an operating window. Within the area (Fig. 8-73) or volume (Fig. 8-74) basically all products meet the performance requirements. However because of machine and plastic variations, rejects can develop at their edges. As the injection-molding machine is very complex with all the controls required to set it up, these examples refer to the injection-molding process. Note that a major cause for problems with any process is not of poor product design but instead that the processes operated outside of their required operating window.

The term process control is often used when machine control is actually performed. As the knowledge base of the fundamentals of the molding process continues to grow,

the control approach is moving away from press control and closer to real process control where material response is monitored and then moderated or even managed. The designer should note that changes in process parameters, such as injection rate, can have dramatic effects on moldings, especially mechanical properties, meeting tolerances, and surface properties (Appendix A. **PLASTICS DESIGN TOOLBOX**).

Auxiliary Equipment

Even though modern primary processing machines with all their ingenious molds/dies and microprocessor control technology is in principle suited to perform flexible tasks, it nevertheless takes a whole series of peripheral auxiliary equipment to guarantee the necessary degree of flexibility. Examples of this action includes: (1) up stream material supply systems; (2) mold or die transport facilities; (3) mold or die preheating banks; (4) mold or die changing devices that includes rapid clamping and coupling equipment; (5) plasticizer cylinder changing devices (screws, barriers, etc.); (6) molded or extruded product handling equipment, particularly robots with interchangeable arms allowing adaptation to various types of production; and (7) transport systems for finished products and handling equipment to pass products on to subsequent production stages (Figs. 8-75 and 8-76) (1–3, 6–9, 10, 20, 37).

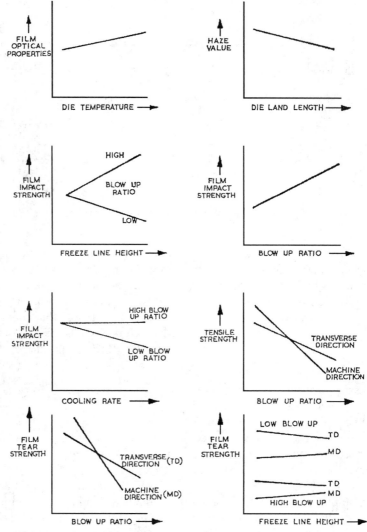

Fig. 8-72   Examples of how extrusion settings affect certain properties of plastics.

Secondary Equipment

Ideally, fabricating TP or TS products will be finished as processed. For example many types of texture or surface finish can be on extruded products or molded into the product, as can almost any geometric shape, hole, or projection. There are situations, however, where it is not possible, practical, or economical to have every feature in the finish product. Typical examples where machining might be required are certain undercuts, complicated side coring, flashing, or places where parting line irregularity is unacceptable. Another common machining/finishing operation with plastics is the removal of the molded remnant of the sprue or gate if it is in an appearance area or critical tolerance region of the part.

Many plastic products are decorated to make them multi-colored, add distinctive logos, or allow them to imitate wood, metal and other materials. Some plastic products are painted since their as-molded appearance is not satisfactory, as may be the case with reinforced, filled or foamed plastics. Painting or coating is also for product protection. The following section will discuss some of the secondary operations frequently used with plastics. Since plastics vary widely in their ability to be machined and to accept finishes,

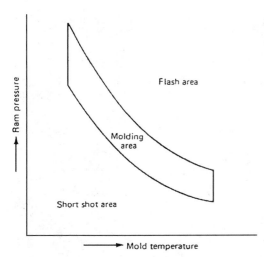

Fig. 8-73 A 2-D molding area diagram (MAD) that plots injection pressure (ram pressure) vs. mold temperature.

this discussion will be general in nature, with details left to other literature dealing with specific plastics.

Machining and Prototyping

All plastics can be shaped and finished with common equipment used for machining

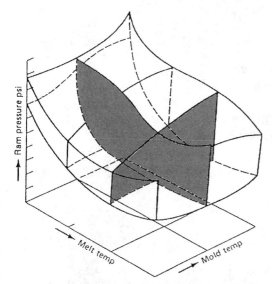

Fig. 8-74 After a 3-D molding volume diagram (MVD) is constructed, it can be analyzed to find the optimum combination of melt temperature, mold temperature, and injection or ram pressure.

metals. In addition, many tools specifically used for woodworking, such as routers, shapers and sanders, are well suited for plastic materials. Since many materials are available in the form of sheets, blocks, slabs, rods, tubes and other cast and extruded shapes, initial prototypes (Chapter 3) are frequently made entirely by machining. Also their may be a requirement for only a few products making it more economical to machine.

The main problems encountered when machining plastics, particularly TPs, are due to the heat built up by friction. As the plastic and cutting tools begin to heat up, the plastic can distort or melt. This can produce reduction in performances, poor surface finish, tearing, localized melting, welding together of stacked products, and jamming of cutters. It is important to prevent the product and cutting tool from heating up to the point where significant softening or melting takes place. There are cutting tools specifically design to cut plastic that eliminate or reduce the heating problem. Some plastic materials machine much easier and faster than others due to their physical and mechanical properties. Generally, a high melting point, inherent lubricity, and good hardness and rigidity are factors that improve machinability.

Drilling and Reaming

In addition to the building of prototype products, drilling and reaming are often required to enlarge, deepen, or remove the draft from, as an example, a molded hole. In some cases, secondary drilling is a more economical and precise solution than side cores in a mold. Although specific requirements will vary with material, the following guidelines apply to almost all TPs.

1. Standard drill presses, as well as other drilling equipment used for metals and wood, are appropriate for drilling and reaming thermoplastics. Speeds and feeds must be controlled to avoid heat build-up.

2. Wood and metal drill bits can usually be used, but best results are obtained with commercially available bits designed for plastics.

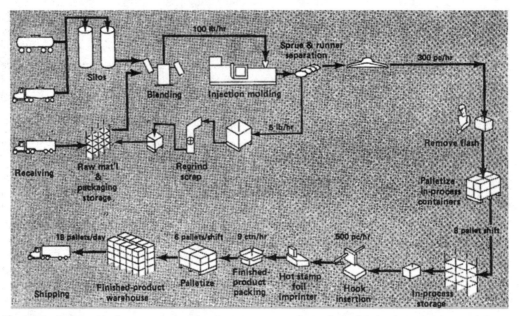

Fig. 8-75 Production line sequence going from upstream, through the fabricating machine, and downstream.

These special drills usually have one or two highly polished or chrome plated flutes, narrow lands, and large helix angles to quickly expel chips and minimize frictional heating.

For holes in thin sections, circle cutters, or drills which only cut the circumference and eject a round thin plug of material, will often be preferred for production.

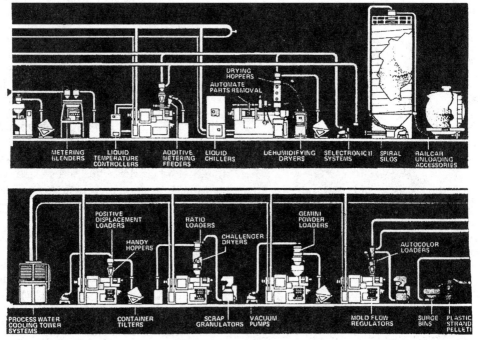

Fig. 8-76 Examples of auxiliary equipment in a production line that starts at the top right end through to the bottom left end.

3. In practice the drill speed and feed rate can be increased for maximum production provided that there is no melting, burning, discoloration or poor surface finish. For deep drilling, frequent withdrawal of the drill may be necessary for chip ejection.

4. Drill bits and reamers must be kept sharp and cool for good results. For high production, carbide tools are sometimes preferred, especially with glass reinforced materials. The first choice for cooling is clean, compressed air, since there is no contamination of the product, and chip removal is improved. If a liquid coolant/lubricant is required for deep drilling, water or some aqueous solution can be used. Metal cutting fluids and oils should be avoided since they may degrade or attack the plastic and create a cleaning problem.

5. Plastic products must be firmly held, fixed or clamped during drilling and reaming operations to prevent dangerous grabbing and spinning of the work.

Thread Tapping

Many plastic products use self-tapping screws, threaded metal inserts, molded-in threads or other fastener systems. When a machine thread must be added after molding, standard metal cutting taps and dies may be used provided that the same precautions regarding heat, chip removal, tool maintenance, and lubrication discussed for drilling are observed. For high production or with filled plastics, carbide taps are recommended. Drilled or molded holes should generally be larger than those specified for steel, and threads finer than 28 threads per inch should be avoided.

Sawing, Milling, Turning, Grinding, and Routing

These cutting operations are usually used only for machined prototypes, or very low volume production of simple shapes. High speed routing is sometimes used for slotting or gate removal on injection molded products. Standard end mills (two-flute), circular cutters, tool bits, wood saw blades, router bits, files, rasps, and sandpaper can usually be used. As with drilling, tools must be kept sharp and cool, and speeds and feeds may be increased until overheating, gumming, or poor finish becomes a problem. All machining operations should provide for dust control, adequate ventilation, safety guards, and eye protection. Inquire about machining information for a specific plastic.

Finishing and Decorating

Since most fabricated products are attractive as well as inherently corrosion and rust resistant when fabricated they usually do not require any finishing or decoration. For others there are paints, coatings, and other surface treatments that usually are used mainly to enhance eye appeal. Tables 8-35 to 8-37 provide some guidelines.

Many reinforced and filled plastic products, as well as structural foam molded or extruded products, emerge with an uneven appearance, and paint may be necessary in critical appearance applications. Common decorative finishes applied to plastic are spray painting, vacuum metallizing, hot stamping, silk screening, metal plating, printing and the application of self-adhesive labels, decals, and border stripping. In some cases, the finish will give the product added protection from heat, ultraviolet radiation, chemicals, scratching, or abrasion.

Some conductive coatings are applied to the inside of the product for dissipation of static electricity and/or provide electromagnetic shielding. These coatings are common in computer and other products such as electronic and medical equipment housings. With all coatings and finishes, a clean surface is essential for a good bond. Care must be used to avoid contamination. Common sources of contamination include oils, release agents (particularly silicone), environment, and handling. In addition to cleaning with solvents and detergents, some plastics require primers, etching, sanding, or flame treatment to enhance adhesion. The following is a brief description of several widely used processes.

Table 8-35   Guide to decorating selection

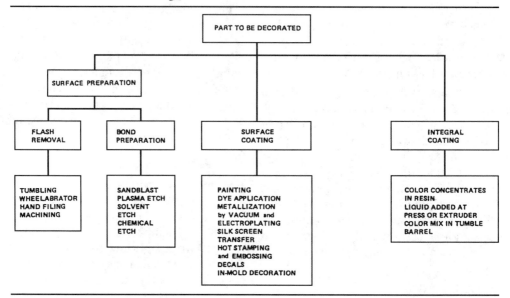

In-Mold Decoration

This term is used both to describe designs that are etched or engraved in the mold surface and the process of inserting a printed film into the mold, to be produced as an integral component of the finished product. Etched surfaces can be drawn both parallel and perpendicular to a parting line of molds or postforming in an extrusion line.. However, be alert with molds to the fact that parallel to the parting line additional draft is required. A wide selection of patterns is available and new ones can be readily created.

Engraved designs and lettering normally have greater depth and fine detail. Parallel to the parting line, a side action (to clear the engraving) will be required in most cases. Therefore, they should be used only when absolutely necessary. Recessed letters and designs are to be avoided whenever possible. They collect dirt and they are costly to put into the tooling. Raised letters are less expensive to make and to maintain. In either case, sharp points, such as those found in the letters N, M, and W, are prone to breaking out when subjected to molding pressures over a period of time.

Hence, it is wise to place designs and lettering as an insert in the mold. This will create an outline around the lettering (which can hopefully be incorporated into the design) however it will make repairs and revisions far less costly. Generally, one should avoid the use of serif typefaces unless the letters are very large indeed. Artwork is prepared for engraving in the same manner as for printing. Regardless of which finishing method is used, it is important to consider its design requirements from the beginning. Many a designer, preoccupied with the mechanical requirements of the product, postpones consideration of this aspect until the very end of the project, only to discover that major design revisions are necessary in order to meet appearance requirements.

Inserting printed film in a mold provides the advantages gained in appearance, hiding defects, and so on. In addition it can provide increasing strength and other properties in the plastic thus reducing the amount of plastics required. Required is compatibility with the fabricated plastic material. The inserts can range from flat to rather complex shapes.

Painting

Most plastics can be painted, though some are a lot more difficult than others.

Table 8-36 Printing and decorating systems

The Process	How It Works	Equipment	Applications	Effect
Painting: Conventional spray	Paint's sprayed by air or airless gun(s) for functional or decorative coatings. Especially good for large areas, uneven surfaces, or relief designs. Masking used to achieve special effects.	Spray guns, spray booths, mask washers often required; conveying and drying apparatus needed for high production.	Can be used on all materials (some require surface treatment).	Solids, multicolor, overall or partial decoration, special effects such as woodgraining possible.
Electrostatic spray	Charged particles are sprayed on electronically conductive parts; process gives high paint utilization; more expensive than conventional spray.	Spray gun, high-voltage power supply; pumps; dryers. Pretreating station for parts (coated or preheated to make conductive).	All plastics can be decorated. Some work, not much, being done on powder coating of plastics.	Generally for one-color, overall coating.
Wiping	Paint is applied conventionally, then paint is wiped off. Paint is either totally removed, remaining only in recessed areas, or is partially removed for special effects such as woodgraining.	Standard spray-paint setup with a wipe station following. For low production, wipe can be manual. Very high-speed, automated equipment available.	Can be used for most materials. Products range from medical containers to furniture.	One color per pass; multicolors achieved in multistation units.
Roller coating	Raised surfaces can be painted without masking. Special effects like stripes.	Roller applicator, either manual or automatic. Special paint feed system required for automatic work. Dryers.	Can be used for most materials.	Generally one-color painting, though multicolor possible with side-by-side rollers.
Screen Printing	Ink is applied to part through a finely woven screen. Screen is masked in areas that won't be painted. Economical means for decorating flat or curved surfaces, especially in relatively short runs.	Screens, fixture, squeegee, conveyorized press setup (for any kind of volume). Dryers. Manual screen printing possible, for very low-volume items.	Most materials. Widely used for bottles; also finds big applications in areas like TV and computer dials.	Single or multiple colors (one station per color).

(*Continued*)

Table 8-36 (Continued)

The Process	How It Works	Equipment	Applications	Effect
Hot Stamping	Involves transferring coating from a flexible foil to the part by pressure and heat. Impression is made by metal or silicone die. Process is dry.	Rotary or reciprocating hot stamp press. Dies. High-speed equipment handles up to 6,000 parts/hr.	Most thermoplastics can be printed; some thermosets. Handles flat, concave, or convex surfaces, including round or tubular shapes.	Metallics, wood grains or multicolor, depending on foil. Foil can be specially formulated (e.g., chemical resistance).
Heat Transfers	Similar to hot stamp but preprinted coating (with a release paper backing) is applied to part by heat and pressure.	Ranges from relatively simple to highly automated with multiple stations for, say, front and back decoration.	Can handle most thermoplastics. A big application area is bottles. Flat, concave or cylindrical surfaces.	Multicolor or single color; metallics (not as good as hot stamp).
Electroplating	Gives a functional metallic finish (matte or shiny) via electrodeposition process.	Preplate etch and rinse tanks; Koroseal-lined tanks for plating steps; preplating and plating chemicals; automated systems available.	Can handle special plating grades of ABS, PP, polysulfone, filled Noryl, filled polyesters, some nylons.	Very durable metallic finishes.
Metallizing: Vacuum	Depositing, in a vacuum, a thin layer of vaporized metal (generally aluminum) on a surface prepared by a base coat.	Metallizer, base, and topcoating equipment (spray, dip or flow), metallizing racks.	Most plastics, especially PS, acrylic, phenolics, PC unplasticized PVC. Decorative finishes (e.g., on toys), or functional (e.g., as a conductive coating).	Metallic finish, generally silver but can be others (e.g., gold, copper).
Cathode sputtering	Uniform metallic coatings using electrodes.	Discharge systems—to by provide close control of metal buildup.	High-temperature materials. Uniform, precise coatings for applications like microminiature circuits.	Metallic finish. Silver and copper generally used. Also gold, platinum, palladinum.
Spray	Deposition of a metallic finish by chemical reaction of water-based solutions.	Activator, water-clean and applicator guns; spray booths, top- and base-coating equipment if required.	Most plastics. For decorative items.	Metallic (silver and bronze).

Method	Description	Equipment	Application	Color
Tamp Printing	Special process using a soft transfer pad to pick up image from etched plate and tamping it onto a part.	Metal plate, squeegee to remove excess ink, conical-shaped transfer pad, indexing device to move parts into printing area, dryers, depending on type of operation.	All plastics. Specially recommended for odd-shaped or delicate parts (e.g., drinking cups, dolls' eyes).	Single- or multicolor—one printing station per color.
In-the-Mold Decorating	Film or foil inserted in mold is transferred to molten plastics as it enters the mold. Decoration becomes integral part of product.	Automatic or manual feed system for the transfers. Static charge may be required to hold foil in mold.	Most plastics, especially polyolefins and melamines. For parts where decoration must withstand extremely high wear.	Single- or multicolor decoration.
Flexography	Printing of a surface directly from a rubber or other synthetic plate.	Manual. Semi- or automatic press, dryers.	Most plastics. Used on such areas as coding pipe and extruded profiles.	Single- or multicolor.
Offset Printing	Roll-transfer method of decorating. In most cases less expensive than other multicolor printing methods.	Ranges from low-cost hand presses to very expensive automated units. Drying, destaticizers, feeding devices.	Most plastics. Used in applications like coding pipe.	Multicolor print or decoration.
Valley Printing	Uses embossing rollers to print in depressed areas of a product.	Embosser with inking attachment or special package system.	Used largely with PVC, PE for such areas as floor tiles, upholstery.	Generally two-color maximum.
Labeling	From simple paper labels to multicolor decads and new preprinted plastic sleeve labels.	Equipment runs the gaunt from hand dispensers to relatively high-speed machines.	Can be used on all plastics. Used mostly for containers and for price marking.	All sorts of colors and types.

Table 8-37   Guide to plastic-decorating methods

	Economics	Aesthetics	Product Design	Chemistry	Manufacturing	Comments
			Done in the Mold			
Engraved mold	*Unit cost:* low *Labor cost:* low *Investment:* moderate	Limited	Unrestricted	Not critical Good durability	No extra operations	Best for simple lettering and texture.
In-mold label	*Unit cost:* high *Labor cost:* high *Investment:* none to moderate	Unlimited	Somewhat restricted	Critical Good durability	Longer molding cycles	Good for thermoplastics and thermosets. Automatic loading equipment becoming available.
Inserted nameplates	*Unit cost:* high *Labor cost:* high *Investment:* moderate	Partially limited	Restricted	Not critical Good durability	Longer molding cycles	Allows three-dimensional as well as special effects.
Two-shot molding	*Unit cost:* high *Labor cost:* high *Investment:* moderate to high	Limited	Somewhat restricted	Not critical Good durability	Two molding operations	Good where maximum abrasion resistance necessary.
			Done after Molding			
Applique	*Unit cost:* high *Labor cost:* high *Investment:* moderate to high	Somewhat limited	Unrestricted	Not critical Good durability	Hand operation	Allows unusual effects.
Electrostate	*Unit cost:* low to moderate *Labor cost:* low *Investment:* moderate to high	Limited	Somewhat restricted	Critical Moderate to good durability		Dry process, no tool contact with product.

Method	Cost					Comments
Flexographic	*Unit cost:* low *Labor cost:* low *Investment:* moderate to high	Somewhat limited	Restricted	Critical — Moderate durability	Automates well	Wet process, tool contacts product. Sometimes requires top coat.
Hand painting	*Unit cost:* high *Labor cost:* high *Investment:* low	Somewhat limited	Unrestricted	Critical — Good durability	Hand operation	Wet process, tool contacts product.
Heat transfer	*Unit cost:* low to moderate *Labor cost:* low to moderate *Investment:* low to moderate	Unlimited	Somewhat restricted	Critical — Good durability	Requires little floor space	Dry process, tool contacts product. Multicolor graphics.
Offset intaglio	*Unit cost:* low *Labor cost:* moderate *Investment:* moderate	Limited	Unrestricted	Critical — Moderate to good durability	Requires little floor space	Wet process, tool contacts product. New process.
Silk screening	*Unit cost:* moderate *Labor cost:* moderate *Investment:* moderate	Somewhat limited	Somewhat restricted	Critical — Good durability	Flexible operation	Wet process, tool contacts product.
Spray	*Unit cost:* moderate *Labor cost:* moderate *Investment:* moderate to high	Limited	Unrestricted	Critical — Good durability	Requires much floor space	Wet process, no tool contact with product.

(Continued)

Table 8-37 (*Continued*)

	Economics	Aesthetics	Product Design	Chemistry	Manufacturing	Comments
Woodgraining	*Unit cost:* high; *Labor cost:* high; *Investment:* moderate to high	Specialized	Specialized	Critical / Good durability	Mostly hand operated	Wet process, tool contacts products.
Hot stamping	*Unit cost:* low; *Labor cost:* low to moderate; *Investment:* low to moderate	Limited	Somewhat restricted	Critical / Good durability	Requires little floor space	Dry process, tool contacts product. Produces bright metalics.
Labeling	*Unit cost:* low to moderate; *Labor cost:* low to moderate; *Investment:* low to moderate	Unlimited	Somewhat restricted	Less critical / Moderate to good durability	Adaptable to many situations	Dry process, no tool contact with product at times. Multicolor graphics.
Metallizing	*Unit cost:* moderate to high; *Labor cost:* moderate to high; *Investment:* high	Limited	Somewhat restricted	Critical / Good durability	Requires special technological know-how	Wet and dry process, no tool contact with product. Produces bright metallics.
Nameplates	*Unit cost:* high; *Labor cost:* moderate to high; *Investment:* low to moderate	Unlimited	Somewhat restricted	Less critical / Good durability	Adaptable to many situations	Dry process, tool contacts product. Multicolor graphics.
Offset	*Unit cost:* low; *Labor cost:* moderate; *Investment:* high	Unlimited	Restricted	Critical / Moderate to good durability	Automates well	Wet process, tool contacts product. Multicolor graphics.

Materials such as polyethylene, polypropylene and acetal, which have waxy surfaces, or other crystalline plastics that are very solvent resistant, can be difficult to paint and require special primers or pre-treatments (flame, etc.) for satisfactory adhesion. Many amorphous plastics easily accept a wide variety of paint coatings.

Although rolling and dipping are sometimes used, power spray painting is the usual method of paint application. Among the coatings used are polyurethane, epoxy, acrylic, alkyd and vinyl based paints. With paints that are oven cured, products must have sufficient heat resistance to survive without distortion, etc.

Vacuum Metallizing and Sputter Plating

In these processes, a special base coat is applied to the surface of the plastic product to be metallized. The product is then placed in a vacuum chamber in which a metallic vapor is created and deposited on the product. A protective clear top coat is then applied over the thin metal layer for abrasion and environmental resistance. The simplest vacuum metallizing processes use resistance heating to melt and vaporize the metal.

These processes are generally limited to pure metals; typically aluminum but also silver, copper and gold. There are vacuum metallizing process that uses an electron beam to vaporize the metal. The sputter plating process uses a plasma to produce the metallic vapor. Both the electron beam and plasma heating methods can be used satisfactorily with alloys such as brass. These are economical metallizing processes that can produce an attractive high gloss finish. However, the adhesion is generally low.

Electroplating

After special pretreatments, specific grades of plastics can be put through electroplating processes similar to those used in the plating of metals. Electroplated plastic products are very durable and provide lightweight replacements for die castings and sheet metal in demanding applications such as automotive grilles and wheel covers.

Flame Spray/Arc Spray

In these processes, specialized equipment actually deposits a fine spray of molten metal on the plastic surface. The relatively thick, rough surface is generally used in non-appearance internal surfaces for electromagnetic and radio frequency shielding, as well as static electricity dissipation.

Hot Stamping

This is a one step economical process for selectively transferring a high quality image to a plastic product. A heated die transfers the pattern from selected transfer tape to a flat plastic surface. Lettering or decorative designs can be transferred in pigmented, wood grain, or metallic finishes.

Sublimation Printing

Sublimation (diffusion) printing is a textile process in which color patterns in dry die crystals are transferred from a release film to the fabric under high heat and pressure. The process has been adapted to plastics. The equipment used is very similar to that used for hot stamping. Under heat and pressure, the dye crystals sublime (go directly to the vapor phase from the solid phase without melting) and the vapor penetrates the plastic product. As a result, the decoration is very durable and wear resistant. It is also cost competitive against other processes such as two-step injection molding or silk screening.

The process is generally limited to certain plastics such as TP polyester and polyester based alloys due to the availability of dye technology from the textile industry. However, new dyes are under development and the application of the process to more plastics is anticipated.

Printing

Lettering and decoration can be applied to most materials using various printing methods. Offset printing, silk screening, and pad printing are among the methods adapted to plastics.

Decal and Label

These are usually self-adhesive, precut, printed patterns on a substrate that are simply adhered to the surface of a product. Decals generally use a transparent plastic film while labels are usually on an opaque plastic, metallic and multilayer sandwich base. Labels of sufficient thickness are useful for hiding unavoidable appearance problems such as gate and sprue removal areas, sink marks, blushes, splays, and weld lines.

Surface Treatment

Different treatments are used to provide a surface that is more receptive to inks, coatings, adhesives, etc. They include chemical solvents and corona treatments. As an example the corona treatment is an effective and efficient process that is commonly used to increase the surface tension of a wide variety of products. Highly consistent and controllable, the process is continually being adapted for new applications using both standard and innovative techniques. When considering the purchase of a corona treating system, one should be certain to investigate the flexibility of the proposed unit as well as its overall efficiency. Selecting the proper system will ensure that the equipment does not become the limiting factor in the event of future manufacturing changes such as an alteration in materials or increased production rates.

Joining And Assembly

Different methods are used for joining or fastening and assembling plastic products and plastic to other materials. It is im-portant to both designer and end-user that the techniques, advantages, and limitations of these methods are understood so that intelligent choices can be made. As an example, different materials that include plastic-to-plastic and plastic-to-metal could have different thermal expansions and could cause failure of the assembly when the individual materials are not free to expand or contract.

Parts to be assembled for TP for high volume production include solvent bonding, adhesive bonding, ultrasonic welding, hot tool welding, electromagnetic and induction bonding, and dielectric heat welding. For low volume TP assembly include gas welding, adhesive bonding, ultrasonic tool welding, hot tool welding, and spin welding. TS plastics include for high volume production molded-in inserts, mechanical fasteners, adhesive bonds, and electromagnetic or induction heating of adhesives; TS plastics include for low volume production adhesive bonding and mechanical fastening. Tables 8-38 to 8-43 provide some guides.

Troubleshooting

With all types of plastic processes, troubleshooting guides are set up to take fast, corrective action when products do not meet their performance requirements. This problem-solving approach fits into the overall fabricating-design interface. One brief example of troubleshooting an RP/composite is in Table 8-44.

A simplified approach to troubleshooting is to develop a checklist that incorporates the basic rules of problem solving such as (1) have a plan, and keep updating it, based on the experience gained; (2) watch the processing conditions; (3) change only one condition or control at a time; (4) allow sufficient time for each change, keeping an accurate log of each; (5) check housekeeping, storage areas, granulators, etc.; and (6) narrow the range of areas in which the problem belongs-that is, primary or secondary machines, molds or dies, operating controls, materials, part designs, and management (1–3, 6–9, 10, 20, 37).

Table 8-38 Methods for part assembly

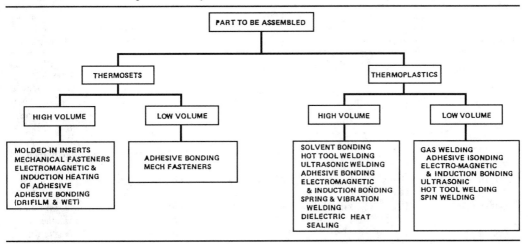

Safety And Processing

All process equipment has procedures to operate and meet safety requirements. They include a checklist that includes preparation (moving material, etc.), startup and shutdown procedures, tooling changes, and to cleanup of all equipment. Most equipment generates high heats and pressures. They are built to run safely, but they must be treated with "respect".

With plastics that decompose, there may be hazards such as personal burns or wounds and air contamination. Faulty controllers and/or freeze-off can cause the overheating situation from a burned out heater. Safety devices should be used that alert the plant when problems develop; people have to be aware of these possible situations. Recognize that personnel injury in plants due to machinery represents 10% of all accidents (Fig. 8-77).

Table 8-39 Assembly methods guide

Thermoplastics	Adhesive Bonding	Dielectric Welding	Induction Bonding	Mechanical Fastening	Solvent Welding	Spin Welding	Staking	Thermal Welding	Ultrasonic Welding
Polyimide	X			X					
Polypropylenes	X		X	X		X	X	X	X
Propylene copolymers	X		X	X		X		X	
Polystyrencs	X		X	X	X	X	X	X	X
Polysulphone	X			X			X		X
Polyvinyl chloride	X	X	X	X				X	
Polyvinyl chloride copolymers	X	X	X	X				X	
PVC—acrylic compounds	X		X	X					X
PVC—ABS compounds	X	X						X	
Styrene acrylo nitrile	X		X	X	X	X	X	X	X
Thermoplastic polyesters	X			X	X	X	X		X

Table 8-40   Reference chart in selecting method of fastening TPs

Thermoplastics	Mechanical Fasteners	Adhesives	Spin and Vibration Welding	Thermal Welding	Ultrasonic Welding	Induction Welding	Remarks
ABS	G	G	G	G	G	G	Body type adhesive Recommended
Acetal	E	P	G	G	G	G	Surface treatment for adhesives
Acrylic	G	G	F–G	G	G	G	Body type adhesive Recommended
Nylon	G	P	G	G	G	G	
Polycarbonate	G	G	G	G	G	G	
Polyester TP	G	F	G	G	G	G	
Polythylene	P	NR	G	G	G–P	G	Surface treatment for adhesives
Polypropylene	P	P	E	G	G–P	G	Surface treatment for adhesives
Polystyrene	F	G	E	G	E–P	G	Impact grades difficult to bond
Polysulfone	G	G	G	E	E	G	
Polyurethane TP	NR	G	NR	NR	NR	G	
PPO modified	G	G	E	G	G	G	
PVC rigid	F	G	F	G	F	G	

E = Excellent, G = Good, F = Fair, P = Poor, NR = Not recommended.

Processing equipment has standard procedures to operate and meet safety requirements. Safety information and standards are available from various sources that include the equipment suppliers, Society of Plastics Industry (SPI), and American National Standards Institute (ANSI). For the past century we have observed increasing activity on the part of equipment manufacturers to upgrade safety; also fabricating plants.

Examples of safety features are many and differ for the different equipment in the lines. Safety interlocks ensure that equipment will not operate until certain precautions have been taken. Safety machine lockout procedures are set up for action to be taken in

Table 8-41   Reference chart in selecting method of fastening TSs

Thermosets	Mechanical Fasteners	Adhesives	Spin and Vibration Welding	Thermal Welding	Ultrasonic Welding	Induction Welding	Remarks
Alkyds	G	G	NR	NR	NR	NR	
DAP	G	G	NR	NR	NR	NR	
Epoxies	G	E	NR	NR	NR	NR	
Melamine	F	G	NR	NR	NR	NR	Material notch sensitive
Phenolics	G	E	NR	NR	NR	NR	
Polyester	G	E	NR	NR	NR	NR	
Polyurethane	G	E	NR	NR	NR	NR	
Silicones	F	G	NR	NR	NR	NR	
Ureas	F	G	NR	NR	NR	NR	Material notch sensitive

E = Excellent, G = Good, F = Fair, P = Poor, NR = Not recommended.

Table 8-42 Percent tensile strength retention using welding techniques

	Original Tensile Strength (psi)*	Hot-Air Welding	Friction Welding	Hot-Plate Welding	Dielectric Welding	Solvent Welding	Adhesive Bonding	Polymerization Welding
Thermosetting plastics								
Epoxy	7,000–13,000	—	10–15	10–15	—	—	50–80	60–100
Melamine	7,000–13,000	—	—	—	—	—	50–80	60–100
Phenolic	6,000–9,000	—	—	—	—	—	50–80	60–100
Polyester	6,000–13,000	—	—	—	—	—	50–80	60–100
Thermoplastics								
Acrylonitrile butadiene styrene	2,400–9,000	50–70	50–70	50–70	50–80	30–60	40–60	—
Acetal	8,000–10,000	20–30	50–70	20–30	—	—	—	—
Cellulose acetate	2,400–8,500	60–75	65–80	65–80	—	90–100	50–60	—
Cellulose acetate butyrate	3,000–7,000	60–75	65–80	65–80	—	90–100	50–80	—
Ethyl cellulose	2,000–8,000	50–70	50–70	50–70	—	80–90	50–80	—
Methyl methacrylate	8,000–11,000	30–70	30–50	20–50	—	40–60	40–60	60–90
Nylon	7,000–12,000	50–70	50–70	50–70	—	—	20–40	—
Polycarbonate	8,000–9,500	35–50	40–50	40–50	—	40–60	5–15	—
Polyethylene	800–6,000	60–80	70–90	60–80	—	—	10–30	—
Polypropylene	3,000–6,000	60–80	70–90	60–80	—	—	20–40	—
Polystyrene	3,500–8,000	20–50	30–60	20–50	—	25–50	20–50	—
Polystyrene acrylonitrile	8,000–11,000	20–60	20–50	20–50	30–50	25–60	20–50	—
Polyvinyl chloride	5,000–9,000	60–70	50–70	60–70	60–70	50–70	50–70	—
Saran	3,000–5,000	60–70	50–70	60–70	60–70	50–70	50–70	—

*To convert psi to Pascals, multiply by 6,895.

Table 8-43 Plastics' behavior with ultrasonic welding

Material	Percent of Weld Strength*	Spot Weld	Staking and Inserting	Swaging	Welding Near Field[†]	Welding Far Field[†]
General-purpose plastics						
ABS	95–100+	E	E	G	E	G
Polystyrene unfilled	95–100+	E	E	F	E	E
Structural foam (styrene)	90–100[1]	E	E	F	G	P
Rubber modified	95–100	E	E	G	E	G–P
Glass filled (up to 30%)	95–100+	E	E	F	E	E
SAN	95–100+	E	E	F	E	E
Engineering plastics						
ABS	95–100+	E	E	G	E	G
ABS/polycarbonate alloy (Cycoloy 800)	95–100+[2]	E	E	G	E	G
ABS/PVC alloy (Cycovin)	95–100+	E	E	G	G	F
Acetal	65–70[3]	G	E	P	G	G
Acrylics	95–100+[4]	G	E	P	E	G
Acrylic multipolymer (XT-polymer)	95–100	E	E	G	E	G
Acrylic/PVC alloy (Kydex)	95–100+	E	E	G	G	F
ASA	95–100+	E	E	G	E	G
Methylpentene	90–100+	E	E	G	G	F
Modified phenylene oxide (Noryl)	95–100+	E	E	F–P	G	E–G
Nylon	90–100+[2]	E	E	F–P	G	F
Polyesters (thermoplastic)	90–100+	G	G	F	G	F
Phenoxy	90–100	G	E	G	G	G–F
Polyarylsulfone	95–100+	G	E	G	E	G
Polycarbonate	95–100+[2]	E	E	G–F	E	E
Polyimide	80–90	F	G	P	G	F
Polyphenylene oxide	95–100+	E	G	F–P	G	G–F
Polysulfone	95–100+[2]	E	E	F	G	G–F
High-volume, low-cost applications						
Butyrates	90–100	G	G–F	G	P	P
Cellulosics	90–100	G	G–F	G	P	P
Polyethylene	90–100	E	E	G	G–P	F–P
Polypropylene	90–100	E	E	G	G–P	F–P
Structural foam (polyolefin)	85–100	E	E	F	G	F–P
Vinyls	40–100	G	G–F	G	F–P	F–P

Code: E = Excellent, G = Good, F = Fair, P = Poor.

*Weld strengths are based on destructive testing. 100 + % results indicate that parent material of plastic part gave way while weld remained intact.

[†]Near field welding refers to joint $1/4$ in. or less from area of horn contact: far field welding to joint more than $1/4$ in. from contact area.

[1]High-density foams weld best.

[2]Moisture will inhibit welds.

[3]Requires high energy and long ultrasonic exposure because of low coefficient of friction.

[4]Cast grades are more difficult to weld due to high molecular weight.

Table 8-44 Troubleshooting RP processes

Problem	Possible Cause	Solution
Nonfills	Air entrapment	Additional air vents and/or vacuum required
	Gel and/or resin timet too short	Adjust resin mix to lengthen time cycle
Excessive	Improper	Check weight and lay-up and/or check
thickness	clamping	clamping mechanisms such as
variation	and/or lay-up	alignment of platens
Blistering	Demolded too soon	Extend molding cycle
	Improper catalytic action	Check resin mix for accurate catalyst content and dispersion
Extended	Improper catalytic	Check equipment, if used, for proper catalyst
curing	action	metering
cycle		Remix resin and contents: agitate mix to provide even dispersion

proper lockout of the machine's operation such as electrical and mechanical circuits. There are preloaded pressure bolts around dies, pressure rupture disks on barrels, turret winder emergency stops, coextrusion line alarm if one extruder stops, drop bar between platens (IMM, CMM, etc.), and so on. The operating environment is continuously upgraded with reduced sound and noise in the operating areas.

To protect operating personnel from recognized hazards, American National Standards Institute (ANSI) voluntary standards have been prepared to assign responsibilities to machine manufacturers, re-manufacturers, modifiers, and employers to ensure safety measures are taken. They are updated periodically. An SPI group (mid-1970s) that included D. V. Rosato as a working member started these ANSI standards on different primary and auxiliary equipment. As of 1992 OSHA adopted these standards making them mandatory. Revisions and up-dating continually occurs.

Equipment/Processing Variable

In addition to material variables, as reviewed in Chapter 6, there are a number of factors in equipment hardware and controls that cause processing variabilities. They include factors such as accuracy of machining component products, method and degree of accuracy during the assembly of component products, temperature and pressure control capability particularly when interrelated with time and heat transfer uniformity in metal components such as those used in molds and dies.

These variables are controllable within limits to produce useful products. What is important to appreciate is that during the past many decades' improvements in equipment have made exceptional strides in significantly reducing operating variabilities or limitations (Fig. 8-78). This action will continue

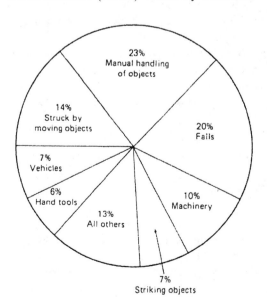

Fig. 8-77 National Safety Council, Chicago, IL provides updates on where accidents occur in all types of manufacturing plants.

Fig. 8-78 You can not expect a 20 year old machine to compete with a machine built to today's standards.

into the future since there is a rather endless improvement in performance of steels and other materials and methods of controlling such as fuzzy controls. Growth is occurring in applying fuzzy logic that in 1981 was based on the idea to mimic the control actions of the human operator (2, 3).

Unfortunately these variables and problems exist in all industries. As an example a major situation occurred that was catastrophic in the aerospace industry. (Nothing is perfect on earth.) The Challenger shuttle spacecraft exploded 28 January 1986 above Cape Canaveral, FL. (Fig. 8-79)

Fig. 8-79 Challenger shuttle spacecraft exploded 28 January 1986 above Cape Canaveral, FL.; photo taken by D. V. R. from Route 95 Florida.

Combining Variable

There are many different processing factors that could influence the repeatability of meeting performance requirements such as tolerances. Some products may require only the compliance of one or two processing factors, but others will require many. Computer programs have been developed to provide the capability of integrating all the applicable factors, thus replacing traditional trial-and-error methods. These programs improve with time since providing improved controls is an endless effort.

Most computer-integrated systems have been developed for injection molding, since a much bigger market exists with IM. Other computer systems are available for the other processes.

Selecting Process

For any given product, the most important processing requirements should be determined based on the plastic to be processed, the quantity, and the dimensions of size and the tolerances. Process selection is a critical step in product design. Failure to select a viable process (and material) during the initial design stages can dramatically increase development costs and timing. It is important to recognize that the process can have a significant effect on the performance of the finished product. The following examples of the considerations in choosing a process are based on what has been reviewed throughout this book.

1. The nature of the process may have a profound influence on a product's mechanical strength.

2. Excessive heat during processing can consume sacrificial heat stabilizers for certain plastics, rendering stabilization levels insufficient to ensure long life at elevated temperatures and/or outdoor weathering. Thermal degradation usually results in embrittlement (tests can be conducted to determine the remaining levels, Chapter 5).

3. The slow cooling of crystalline polymers, such as HDPE and PP, can allow large crystal formations to develop. Such crystals

embrittle the plastic and make it prone to stress cracking.

4. The rapid cooling of certain plastic products can result in "frozen in" stresses and strains (particularly with injection molding). The stresses may decay with time in a viscoelastic manner. However, they will act like any other sustained stress to aggravate cracking or crazing in the presence of aggressive media and hostile environments like UV radiation.

5. Annealing at temperatures below the T_g (glass transition temperature, Chapter 7) where material becomes leathery is not necessarily beneficial. For example, annealing a PC greatly accelerates both its crazing and rupture under sustained loading. In general, the annealing of plastics results in lowering its properties; however, its dimensional stability may be improved. Heating a material to above its T_g, however, results in the relief of internal stresses.

6. Knit or weld lines form where the melt flow during processing meets after flowing through separate gates in an injection mold or after being parted by either "spiders" in an extruder die or bosses in an injection mold. Because the material is not well mixed in the zone of the knit or weld line, the seam thus formed can be weak or brittle under long-term or impact loads. This problem can easily arise with fiber reinforced plastics, where under ideal molding conditions 40 to 60% of their strength can be lost, since fibers fail to knit together at their seams.

7. In RPs, insufficient compaction and consolidation before plastic solidification or cure will result in air pockets, incomplete wet-out and encapsulation of the fibers, and/or insufficient fiber or uniform fiber content. These deficiencies lead to loss of strength and stiffness and susceptibility to deterioration by water and aggressive agents.

These examples show the kinds of alterations that the processing of plastics can have on the performance of the product. As discussed throughout this book, the many different plastics tend all to behave in different patterns, so where a particular problem could develop during processing with one material, it might have little or no effect on another, even if the base plastics are the same but contain different additives or reinforcements. Regardless, the problems that might arise should be eliminated at the outset.

In some cases the designer will not have the ability to choose freely from all the design, material, and process alternatives. For example, a design is often heavily constrained by the need to fit an existing assembly, and the material and process may be determined largely by the need to use existing processing facilities. However, to optimize results the designer should establish the extent of any design freedom early in the design process and explore the design, material, and process alternatives within these bounds. Before final selection of the process, the entire process of production should be considered, including such secondary operations as painting and decorating (Appendix A, **PLASTICS DESIGN TOOLBOX**).

Shape

Both shape and design details are heavily process related. The ability to mold ribs, for example, may depend on material flow during a process or on the flowability of a plastic reinforced with glass. The ability to produce hollow shapes depends on the ability to use removable cores, including air, fusible or soluble solids, and even sand. Hollow shapes can also be produced using cores that remain in the product, such as foam inserts in RTM or metal inserts in IM.

The geometric symmetry of a product influences process selection. For example, an axis of symmetry in a long, narrow product may suggest selecting an extrusion or pultrusion process. Similarly, the need for hollow sections in the product could suggest blow molding or rotational molding. In order to handle materials that melt, flow, and solidify quickly, it may be necessary to use a mechanical process such as injection molding. It is a process that could still be limited by the time available with the particular inadequate machine in question to fill the mold cavity before the melt solidifies; thus, higher pressures are required to increase the speed of mold filling, etc.

Table 8-45 Basic processing methods as a function of product design

Part Design	Process										
	Blow Molding	Casting	Compression	Extrusion	Filament Winding	Injection	Matched Die Molding	Rotational	Thermo-forming	Transfer Compression	Wet lay-up (Contact Molding)
Major shape characteristics	Hollow bodies	Simple configurations	Moldable in one plane	Constant cross section profile	Structure with surfaces of revolution	Few limitations	Moldable in one plane	Hollow bodies	Moldable in one plane	Simple configurations	Moldable in one plane
Limiting size factor	Material	Material	Equipment	Material	Equipment	Equipment	Equipment	Material	Material	Equipment	Mold Size
Maximum thickness. in. (mm)	>0.25 (6.4)	None	0.5 (12.7)	6 (150)	3 (76)	6 (150)	2 (51)	0.5 (12.7)	3 (76)	6 (150)	0.5 (12.7)
Minimum inside radius, in. (mm)	0.125 (3.18)	0.01–0.125 (0.25–3.18)	0.125 (3.18)	0.01–0.125 (0.25–3.18)	0.125 (3.18)	0.01–0.125 (0.25–3.18)	0.06 (1.5)	0.01–0.125 (0.25–3.18)	0.125 (3.18)	0.01–0.125 (0.25–3.18)	0.25 (6.4)
Minimum draft (deg.)	0	0–1	>1	NR[2]	2–3	<1	1	1	1	1	0
Minimum thickness. in. (mm)	0.01 (0.25)	0.01–0.125 (0.25–3.18)	0.01–0.125 (0.25–3.18)	0.001 (0.02)	0.015 (0.38)	0.005 (0.1)	0.03 (0.8)	0.02 (0.5)	0.002 (0.05)	0.01–0.125 (0.25–3.18)	0.06 (1.5)
Threads	Yes	Yes	Yes	No	No	Yes	No	Yes	No	Yes	No
Undercuts	Yes[1]	Yes[1]	NR[2]	Yes	NR[2]	Yes[1]	NR[2]	Yes[3]	Yes[1]	NR[2]	Yes
Inserts	Yes	Yes	Yes	Yes	Yes	Yes	Yes	Yes	NR[2]	Yes	Yes
Built-in cores	Yes	Yes	No	Yes[4]	Yes	Yes	Yes	Yes	No	Yes	Yes
Molded-in holes	Yes	Yes	Yes	Yes	No	Yes	Yes	Yes	Yes	Yes	Yes
Bosses	Yes	Yes	Yes	Yes	No[5]	Yes	No[6]	Yes	Yes	Yes	Yes
Fins or ribs	Yes	Yes	Yes	Yes	No	Yes	Yes	Yes	Yes	Yes	Yes
Molded in designs and nos.	Yes	Yes	Yes	No	No	Yes	Yes	Yes	Yes	Yes	Yes
Surface finish[7]	1–2	2	1–2	1–2	5	1	4–5	2–3	1–3	1–2	4–5
Overall dimensional tolerance (in./in., plus or minus)	0.01	0.001	0.001	0.005	0.005	0.001	0.005	0.01	0.01	0.001	0.02

[1] Special mold required.
[2] Not recommended.
[3] Only flexible material.
[4] Only direction of extrusion.
[5] Possible with special techniques.
[6] Fusing premix/yes.
[7] Rated 1 to 5 (1 = very smooth. 5 = rough).

Table 8-46   Guide to compatibility of processes and RP materials

	Thermosets					Thermoplastics										
	Polyester	Polyester SMC	Polyester BMC	Epoxy	Polyurethane	Acetal	Nylon 6	Nylon 6/6	Polycarbonate	Polypropylene	Polyphenylene sulfide	ABS	Polyphenylene oxide	Polystyrene	Polyester PBT	Polyester PET
Injection molding	•		•	•	•	•	•	•	•	•	•	•	•	•	•	•
Hand lay-up	•			•												
Spray-up	•			•												
Compression molding	•	•	•	•							•					
Preform molding	•			•												
Filament winding	•			•												•
Pultrusion	•			•												
Resin transfer molding	•													•	•	•
Reinforced reaction injection molding	•			•	•	•										

Each process has certain characteristics that can be summarized by determining whether (1) its ribs and bosses are feasible, depending on whether one or both sides of the product reproduce the tool (mold or die) surface; (2) the sequence of material injection or some other process and tool closure allows of having deep vertical sections in the surface wall; (3) the material's viscosity is high enough to allow the use of slides and cores in the tool without their being gummed up with material flowing into the slide mechanism; (4) hollow sections or containers are feasible; and (5) whether hollow or foam-filled box sections can be produced to increase section stiffness (Appendix A: **PLASTICS DESIGN TOOLBOX**).

Size

Product size is limited by the size of available equipment and a process's available pressure such as melt and clamping pressure. The ability to achieve specific shape and design detail is dependent on the way the process operates. Generally, the lower the processing pressure, the larger the product that can be produced. Other restrictions are the length of flow s it relates to residence time and the material's reaction time. With most labor-intensive methods, such as hand lay-up, slow-reacting TSs can be used and there is virtually no limit on size.

The functions and property characteristics of a product will be largely determined by the performance requirements and material selected for fabrication. The basic requirement of the process is its capability of handling a suitable material. For example, if a major function requirement is for resistance to creep under high loads, it is probable that a long-fiber RP will be necessary. Thus it would immediately eliminate such processes as blow molding and conventional injection molding.

Thickness tolerance   With some processes, thickness is limited only by the size of the equipment that is either available or can be produced. A general guide to practical processing thickness limitations is (in inches): injection molding, 0.02 to 0.5; extrusion, 0.001 to 1.0; blow molding, 0.003 to 0.2; thermoforming, 0.002 to 1.0; compression molding, 0.05 to 4.0; and foam injection molding of 0.1 to 5.0.

Table 8-47 Molding processes guide to plastic materials

Material Family	Injection	Compression	Transfer	Casting	Cold Molding	Coating	Structural Foam	Extrusion	Laminating	Sheet Forming	RP Molding — FRP	RP Molding — Filament	Dip and Slush	Blow	Rotational
ABS	X						X	X		X				X	X
Acetal	X						X	X		X				X	X
Acrylic	X	X		X		X		X		X				X	
Allyl			X		X	X			X			X			
ASA	X			X		X	X	X	X	X				X	X
Cellulosic	X					X	X	X	X	X					
Epoxy	X	X	X	X	X	X			X						
Fluoroplastic	X	X	X					X		X		X			
Melamine-formaldehyde	X	X	X	X		X			X						
Nylon	X			X		X	X	X						X	X
Phenol-formaldehyde	X	X	X	X	X	X			X						
Poly (amide-imide)	X	X	X												
Polyarylether	X	X						X							
Polybutadiene	X		X					X						X	X
Polycarbonate	X	X					X	X		X				X	X
Polyester (TP)	X		X			X		X						X	X
Polyester-fiberglass (TS)		X						X	X		X	X			
Polyethylene	X	X				X	X	X(TP)	X	X				X	X
Polyimide	X	X				X		X	X						
Polyphenylene oxide	X						X	X						X	X
Polyphenylene sulfide	X	X				X		X							
Polypropylene	X						X	X		X				X	X
Polystyrene	X						X	X		X				X	X
Polysulfone	X	X						X		X					
Polyurethane (TS) (TP)	X	X	X	X		X	X	X(TP)	X	X				X	X
SAN	X						X	X	X	X					X
Silicone	X	X		X		X			X						
Styrene butadiene		X						X	X						
Urea formaldehyde	X	X							X						
Vinyl	X	X			X	X	X	X					X	X	X

Table 8-48   General information relating processes to properties of plastics

Thermosets	Properties	Processes

Polyesters
  Properties shown
  also apply to
  some polyesters
  formulated for
  thermoplastic
  processing by
  injection molding
	Simplest, most versatile. economical and most widely used family of resins, having good electrical properties, good chemical resistance, especially to acids	Compression molding
		Filament winding
		Hand lay-up
		Mat molding
		Pressure bag molding
		Continuous pultrusion
		Injection molding
		Spray-up
		Centrifugal casting
		Cold molding
		Comoform[1]
		Encapsulation

Epoxies	Excellent mechanical properties, dimensional stability, chemical resistance (especially alkalis), low water absorption, self-extinguishing (when halogenated), low shrinkage, good abrasion resistance, very good adhesion properties	Compression molding Filament winding Hand lay-up Continuous pultrusion Encapsulation Centrifugal casting
Phenolics	Good acid resistance, good electrical properties (except arc resistance), high heat resistance	Compression molding Continuous laminating
Silicones	Highest heat resistance, low water absorption, excellent dielectric properties, high arc resistance	Compression molding Injection molding Encapsulation
Melamines	Good heat resistance, high impact strength	Compression molding
Diallyl phthalate	Good electrical insulation, low water absorption	Compression molding
Thermoplastics		
Polystyrene	Low cost, moderate heat distortion, good dimensional stability, good stiffness, impact strength	Injection molding Continuous laminating
Nylon	High heat distortion, low water absorption, low elongation, good impact strength, good tensile and flexural strength	Injection molding Blow molding, Rotational molding
Polycarbonate	Self-extinguishing, high dielectric strength, high mechanical properties	Injection molding
Styrene-acrylo-nitrile	Good solvent resistance, good long-term strength, good appearance	Injection molding
Acrylics	Good gloss, weather resistance, optical clarity, and color; excellent electrical properties	Injection molding Vacuum forming Compression molding Continuous laminating
Vinyls	Excellent weatherability, superior electrical properties, excellent moisture and chemical resistance, self-extinguishing	Injection molding Continuous laminating Rotational molding
Acetals	Very high tensile strength and stiffness, exceptional dimensional stability, high chemical and abrasion resistance, no known room temperature solvent	Injection molding
Polyethylene	Good toughness, light weight, low cost, good flexibility, good chemical resistance; can be "welded"	Injection molding Rotational molding Blow molding

(Continues)

Table 8-48   (*Continued*)

Thermosets	Properties	Processes
Fluorocarbons	Very high heat and chemical resistance, nonburning, lowest coefficient of friction, high dimensional stability	Injection molding Encapsulation Continuous pultrusion
Polyphenylene oxide modified	Very tough engineering plastic, superior dimensional stability, low moisture absorption, excellent chemical resistance	Injection molding
Polypropylene	Excellent resistance to stress or flex cracking, very light weight, hard, scrach-resistant surface, can be electroplated: good chemical and heat resistance: exceptional impact strength, good optical qualities	Injection molding Continuous laminating Rotational molding
Polysulfone	Good transparency, high mechanical properties, heat resistance, electrical properties at high temperature, can be electroplated	Injection molding

Surface Finish

Another consideration is the ability of a material to provide a surface that is compatible with the requirements of the application: a smooth finish for extruded profiles, molded-in colors, textured surfaces, etc. The compatibility of the major processes with in-mold coating and other insert-surfacing materials, and their compatibility with surface decoration secondary processes, could also be important.

It should be recognized that surface finish can be more than just a cosmetic standard. It also affects product quality, mold or die cost, and delivery time of tools and/or products. The surface can be used not only to enhance clarity for the sake of appearance but to hide surface defects such as sink and parting marks. The Society of Plastics Engineers/Society of Plastics Industries standards range from a No. 1 mirror finish to a No. 6 grit blast finish. A mold finish comparison kit consisting of six hardened tool steel pieces and

Table 8-49   Guide to process selection

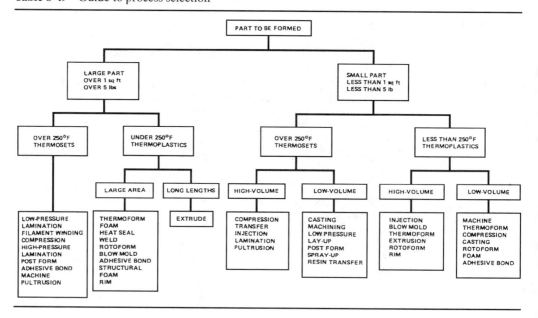

Table 8-50   Economic comparison of three different processes

Production Considerations	Structural Foam	Injection Molding	Sheet Molding Compound
Typical minimum number of parts a vendor is likely to quote on for a single setup	250 (using multiple nozzle equip. with tools from other sources designed for the same polymer and ganged on the platen)	1,000 to 1,500	500
Relative tooling cost, single cavity	Lowest. Machined aluminum may be viable, depending on quantity required	20 percent more. Hardened-steel tooling	20 percent to 25 percent more. Compression-molding steel tools
Average cycle times for consistent part reproduction	2 to 3 minutes ($1/4$ in. nominal wall thickness)	40 to 50 seconds	$1^1/_2$ to 3 minutes
Is a multiple-cavity tooling approach possible to reduce piece costs?	Yes	Yes. Depends on size and configuration, although rapid cycle time may eliminate the need.	Not necessarily. Secondary operations may be too costly and material flow too difficult
Are secondary operations required except to remove sprue?	No	No	Yes, e.g., removing material where a "window" is required (often done within the molding cycle)
Range of materials that can be molded	Similar to thermoplastic injection molding	Unlimited; cost depends on performance requirements	Limited; higher cost
Finishing costs for good cosmetic appearance	40 to 60 cents per sq. ft. of surface (depending on surface-swirl conditions)	None, if integrally colored; 10–20 cents per sq. ft. if painted	None, if secondary operations such as trimming are not required. Otherwise 20 to 30 cents per sq. ft. of surface

associated molded pieces is available through SPE/SPI.

Various types of surface finishes are available for plastics and RP products, such as smooth, textured, molded-in color, and in-mold coating. A textured product surface can be obtained through either a textured mold cavity or a postmold paint process. The former method is the most commonly used. A wide variety of texture designs are available. The surface smoothness, and to some degree the texture, of a plastic or RP is as dependent on the materials used in it as on the surface of the cavity. For example, certain chemically etched textured surfaces can be obtained only if the proper steel or metal is used in the mold cavity. Also, with any mold or die the proper cavity steel based on the plastic processed will

Table 8-51　Design recommendations for choosing an RP process

	Contact Molding. Spray-up	Pressure Bag	Filament Winding	Continuous Pultrusion	Premix/ Molding Compound	Matched Die Molding with Preform or Mat
Minimum inside radius, in.	$1/4$	$1/2$	$1/8$	N.A.*	$1/32$	$1/8$
Molded-in holes	Large	Large	N.R.*	N.A.	Yes	Yes
Trimmed-in mold	No	No	Yes	Yes	Yes	Yes
Built-in cores	Yes	Yes	Yes	N.A.	Yes	Yes
Undercuts	Yes	Yes	No	No	Yes	No
Minimum practical thickness, in. (mm)	0.060 (1.5)	0.060 (1.5)	0.010 (0.25)	0.037 (0.94)	0.060 (1.5)	0.030 (0.76)
Maximum practical thickness, in. (mm)	0.50 (13)	1 (25.4)	3 (76.2)	1 (25.4)	1 (25.4)	0.25 (6.4)
Normal thickness variation, in. (mm)	±0.020 (±0.51)	±0.020 (±0.51)	±0.010 (±0.25)	±0.005 (±0.1)	±0.002 (±0.05)	±0.008 (±0.2)
Maximum buildup of thickness	As desired	As desired	As desired	N.A.	As desired	2 to 1 maximum
Corrugated sections	Yes	Yes	Circumferential only	In longitudinal direction	Yes	Yes
Metal inserts	Yes	Yes	Yes	No	Yes	Yes
Surfacing mat	Yes	Yes	Yes	Yes	No	Yes
Limiting size factor	Mold size	Bag size	Lathe bed length and swing	Pull capacity	Press capacity	Press dimensions
Metal edge stiffeners	Yes	N.R.	Yes	No	Yes	Yes
Bosses	Yes	N.R.	No	No	Yes	Yes
Fins	Yes	Yes	No	N.R.	Yes	N.R.
Molded-in labels	Yes	Yes	Yes	Yes	No	Yes
Raised numbers	Yes	Yes	No	No	Yes	Yes
Gel coat surface	Yes	Yes	Yes	No	No	Yes
Shape limitations	None	Flexibility of the bag	Surface of revolution	Constant cross-section	Moldable	Moldable
Translucency	Yes	Yes	Yes	Yes	No	Yes
Finished surfaces	One	One	One	Two	Two	Two
Strength orientation	Random	Orientation of ply	Depends on wind	Directional	Random	Random
Typical glass percent by weight	30–45	45–60	50–75	30–60	25	30

*Note: N.A.: Not applicable.
　　　N.R.: Not recommended.

significantly reduce its wear and tear and extend its useful life.

Cost

The production flexibility of the fabrication process is often the single most important economic factor in a plastic product. The component's size, shape, complexity, and required production rate can be primary determinants. As an example, small numbers of large objects tend economically to favor casting, as well as the RP's hand lay-up or spray-up process, with a minimal tooling cost and maximum freedom for

Table 8-52   Guide to reinforced TS plastics' process selection

Table 8-53   Comparison of structural foam with five other processes

	Foam vs. Sheet Metal	Foam vs. Die Casting	Foam vs. Sheet Molding Compound	Foam vs. Hand Lay-up Fiberglass	Foam vs. Injection Molding
Advantages	1. Fabrication economy: less assembly time; tighter dimensional tolerances; increased product integrity; less final product-inspection time 2. Fewer parts required for assembly 3. Dent resistance 4. Elimination of oil canning 5. Greater design freedom 6. Better sound damping 7. Reduced damage from shipping 8. Reduced tooling costs for complex configurations	1. Much lower tooling costs 2. Longer tool life, lower maintenance 3. No trim dies required 4. Lighter weight 5. Higher impact resistance 6. Better sound damping 7. Better strength-to-weight 8. Better impact resistance	1. Uniform physical properties throughout the part 2. Warping and sink marks reduced or eliminated 3. No resin-rich areas to cause configuration problems 4. Higher impact resistance 5. Greater inherent structural capabilities 6. Lower shipping costs 7. Large parts more economical 8. Lower tooling costs 9. Better sound damping	1. More consistent part reproduction 2. Lower labor 3. Simplified assembly 4. Better dimensional stability 5. More design freedom 6. More uniform physical properties 7. Better sound damping	(Many process similarities exist) 1. Flexibility for functional engineering 2. Better low- to medium-volume economics 3. Lower tooling costs 4. Better large-part capability 5. Better sound damping 6. Lower internal stresses 7. Sink marks reduced or eliminated 8. Inherent structural strength
Limitations	1. Smaller variety of finishes available, such as chrome or baked enamel 2. No R.F.I. and grounding capabilities 3. Harder to retrofit to frame or skins 4. Thicker wall 5. Higher tool costs than with brakeforming	1. No heat sink capabilities 2. No R.F.I. and grounding capabilities 3. Fewer available finishes for cosmetic appearance 4. Higher finishing costs 5. Thicker walls 6. Possible internal voids	1. Increased finishing costs (surface swirl) 2. Heat distortion 3. Thicker wall 4. Lower physical properties 5. Possible internal voids	1. More prone to heat distortion 2. Poorer economies of part size vs. quantity 3. Thicker walls 4. Higher tooling costs	1. Poorer surface finish 2. Application of cosmetic detail for appearance parts 3. Longer cycle time 4. Thicker walls 5. Poorer high-volume economics 6. Less equipment available for various shot sizes
Potential Savings from Structural Foam (in %)	50 + *	15 to 30	Up to 30	50 +	15 to 20†

*Even with limited quantities.
†Depending on unit volume and part size.

design changes. Many products favor injection and compression molding or long runs in extruders, with their automation capabilities to minimize labor costs. Often the shape dictates the process, such as centrifugal casting or filament winding being used for cylindrical products, rotational molding for complex hollow shapes or extremely large products, and pultrusion for constant cross-sections requiring extremely high strength and stiffness. These general examples should be considered broadly, since individual processes can all be designed for a specific product capability to meet performance requirements at the lowest cost.

A further important point is that major costs can be incurred in the operations required after fabricating: trimming, finishing, joining, attaching hardware, and so on. Observing the following design practices will help reduce costs and improve processing and performance, whatever fabrication method is selected: (1) strive for the simplest shape and form; (2) use the shape of the product to provide stiffness, reducing its required number of stiffening ribs; (3) combine the parts into single moldings or extrusions as much as possible, to minimize assembly time and eliminate designing fasteners and so on; (4) use a uniform wall thickness wherever possible, and make changes in thickness gradually, to reduce stress concentrations; (5) use shape to satisfy functional needs like slots for hoisting, hand grips, and pouring; (6) provide the maximum radii that are consistent with the functional requirements; and (7) keep tolerances as liberal as possible, but once in production aim for tighter tolerances, to save plastic material and probably reduce production cost.

Minimizing cost is generally an overriding goal in any application, whether a process is being selected for a new product application or opportunities are being evaluated for replacing existing materials and equipment. The major elements of cost include capital equipment, tooling, labor, and inefficiencies such as scrap, repairs, waste, and machine downtime. Each element must be evaluated before determining the most cost-effective process from among the available alternatives. To a great extent, the selection of a process for a new plastic or plastic composite product will be dictated, or at least restricted, by the limitations imposed by designs and cost factors. These factors frequently coincide to indicate the preferred processes. Some plastics will have a limited scope of processes available, but others will be more flexible. And products for limited production or small volumes may require processes with low capital and tooling costs to make products economical.

In summary, when considering alternative processes for producing plastic and RP products, the major concerns usually involve (1) limitations that may be imposed by the material, because not all materials can be processed by all methods; (2) limitations imposed by the design, such as the size, single-piece versus multiple-piece construction, a closed or open shape, and the level of dimensional and tolerance accuracy required; (3) the number of products required; and (4) the available capital equipment in-house or subcontracted. Certain equipment may be available, although it may not necessarily be the one needed for the lowest production cost. There have always been designing companies that have in-house fabricating capability that discard their existing equipment in favor of purchasing new equipment in order to reduce costs. Their approach is to design the product, put a mold or die around it, and in turn put a machine around the mold or die so that maximum cost-efficiency exist.

Summary: Matching Process and Plastic

As reviewed throughout this book it can be said that there is a basic thought of designing. It is that the plastic and the process selected profoundly affect the quality and appearance of the product. For this reason, it is usually unwise to create a design first, and then decide on material and process; at least consider all three factors from the start. This seems obvious and logical, but it is frequently ignored in practice, especially when converting a metal design to a plastic material.

As reviewed there is much to consider. Examples include cooling as the product sets up results in different shrinkage rates for thicker versus thinner sections in the different processes. This results in either external waviness or sink marks, or warpage and internal voids, as the product contracts. Flat surfaces are difficult to maintain but not impossible to attain using certain processes. High speed of flow to fill the cavity of the mold is impeded going around square corners, so provision for radii and fillets are important.

Attempting to flow past thin sections to fill wider sections is difficult or may be impossible, because the flow thickens enroute acts like plaque forming along walls of a human artery. Even if both of these ills are avoided, the final result may still contain areas of high shear stresses invisible to the eye but waiting in ambush to cause failure later under extreme conditions previously thought to be well within the material's specifications.

However, when beginning to design a product, it is often wiser to permit the creative mind to freewheel, especially in the initial concept phases. Undue concern for technical aspects may inhibit creativity. After this initial conceptual stage, though, when ideas have been set down graphically and as the design begins to crystallize, it is wise to look carefully and explore the total aspects determining the potential for the most well established bond of plastic and process.

Preliminary consideration of candidate materials, processes and tooling factors, configuration, thicknesses in section, ribs, bosses, holes, surface characteristics, color, graphics, decoration, and assembly methods will begin to impose some discipline on the product design as it evolves. In the middle and latter phases of the design cycle, two or three concepts should make their validity apparent to all involved. With luck (logic), one will

Table 8-54　Guide to tool to be made

Table 8-55   Material guide to tool material selection

	Machinability	Toughness	Corrosion Resistance	Abrasion Resistance	Heat Transfer	Thermal Fatigue	Polishability	Weldability (Repair)	Comments
Aluminum	5	1	3	1	5	1	1	5	Prototype, short run and structural foam molding.
P-20	2	2	2	2	3	3	4	4	Large cavities, cores - eliminate heat treat process and associated warpage and cracking.
H-13	3	2	4	2	3	5	4	4	Thermal fatigue resistance, polishability. Mostly chosen for zinc, aluminum die casting.
420 S.S.	3	2	5	2	1	2	2	4	Corrosion resistance (poor thermal conductivity).
P-6	4	4	2	2	3	3	4	4	Easily machinable. Welds, repairs well. Low carbon steel - not dimensionally stable in heat treatment.
O-1	3	2	2	2	3	2	3	3	Oil hardening - pins, small inserts, etc.
S-7	3	5	3	3	3	3	2	4	Shock resisting steel - long cores - where subjected to mechanical loads (slides, lifters)
A-2	3	4	4	4	3	2	4	1	Good abrasion resistance, polishability. Air hardening - heat treat stable.
D-2	1	4	4	5	3	1	4	1	Extreme abrasion resistance. Used as gate inserts, etc. for filled resin applications.

Range: 5 (best) to 1 (poorest)

Table 8-56   Interrelating to the plastics machinery and equipment sector

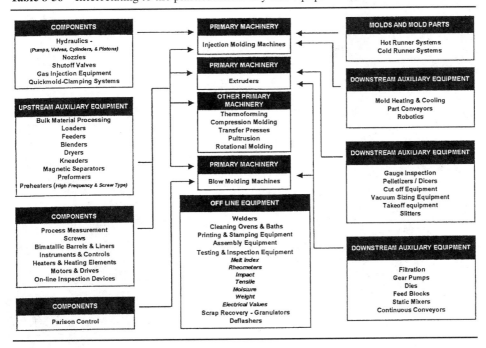

then be so obviously right that it will stand out without question. When seeking the best choice among candidate materials and processes the possibilities can become numerous, and as time passes they will become even more apparent.

In narrowing the options and your experience has not developed, it is best to seek advice from the appropriate people. When choosing a fabricator, seek out those with experience close to the materials, end-use categories and sizes of your project. When pro-totype products are made, test them under the worst possible conditions, and if possible, simulating a longer time frame than the product is designed to withstand in normal service.

The processing information (as well as material information) presented in this and other chapters has provided a variety of useful selection guides (see **INDEX, Selection guide** for guides throughout this book). Additional process selection guides are in the following Tables 8-45 to 8-55.

9

Cost Estimating

Introduction

The cost to produce products involves many different categories such as materials and equipment, method of purchasing materials and equipment, additives used, and fabricating costs (Fig. 9-1). All these factors have a direct effect on a designed product even though the designer is not involved in most of them. However it could very easily influence the expected initial cost that the designer had to include in the design evaluation. As experience tells one, cost of material can be very variable.

Interesting is the fact that the actual time and cost to design products may take less than 5% of the total time and cost to fabricate products. Even though this is a relatively small percentage of the overall operation, it has a direct and important influence performancewise and costwise on the success or failure of fabricating products.

A major cost advantage for fabricating products is their usual low processing cost. The most expensive part of the product is the cost of plastic materials. Since the material value in a plastic product is roughly up to one-half (possibly up to 80%) of its overall cost, it becomes important to select a candidate material with extraordinary care particularly on long production runs. Cost to fabricate using most processes and particularly

in long runs usually represents about 5% of total cost. Thus with the product to be fabricated in-house, it can be economically beneficial to replace existing equipment to meet a lower product cost, as many do. The expensive equipment cost would be justifiable.

In purchasing equipment consider what John Ruskin (1819–1900) stated that it is unwise to pay too much, but it is unwise to pay too little unless you know that the machine is capable of meeting the requirements you set.

It is a popular misconception that plastics are cheap materials. There are low cost types but there are also the more expensive types. Important that one recognizes that it is possible to process a more expensive plastic because it provides for a lower processing cost (1–10, 20, 37, 64).

To put plastics in its proper cost perspective, it is usually best to compare materials based on volume rather than the weight used. On a weight basis most plastics can be more expensive than steel, and only slightly less than aluminum (Fig. 9-2). Examples of other cost factors to consider are reviewed in Figs. 9-3 to 9-5.

Cost-effective production of high-quality plastic products is the prime target of the plastic processor. As an example, the continually growing pressure from the competition demands the choice of IMMs with the smallest possible injection and clamping units in order

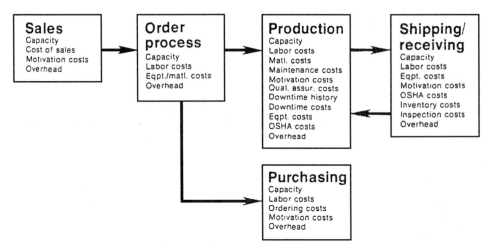

Fig. 9-1   Examples of cost factors.

to minimize the product costs by reducing investment and (important) operation costs. At the same time, however, these savings must not be made at the expense of product quality. This example is applicable to other processes such as extrusion, blow molding, and thermoforming.

To obtain the equipment needed a simple basic approach can be used. Design the product determining the plastic material to be used. Next design and plan building the mold or die to meet the product requirements. Now you are ready to determine the fabricating equipment required. It will be based on two factors: (1) the mold or die size and movements it requires and (2) the plastic material processing requirements. Thus the equipment purchased will only meet the requirements it has to meet. Probably one of the most difficult aspects of purchasing

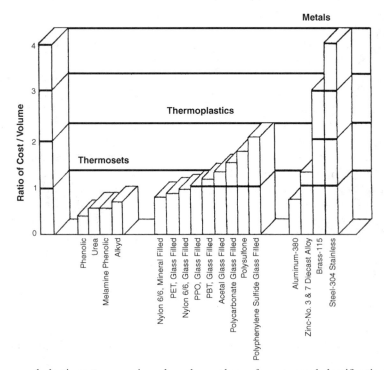

Fig. 9-2   A general plastic cost comparison, based on volume, for a general classification of materials.

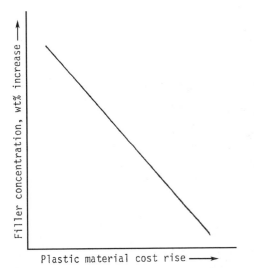

Fig. 9-3 Example why fillers become more attractive as plastic prices go up.

When designing a plastic product, it is frequently desirable to compute an approximate cost in order to determine whether the considered design is within desired economical limits. Based on the preliminary design configuration and type of material to be used, a determination has to be made regarding the fabricating process to be used. Considerations used include the number of products to be made and probably a time schedule that has to be met. There may be options where different plastics are being considered and also different processes. Ideally if history exists on similar products, available data can be used as a guide to develop a cost analysis.

If the designer is not familiar with fabricating plastic products, the usual practice is to contact a reliable fabricator(s), obtain a rough estimate(s), and proceed on the basis of this information to either redesign or finish the design on hand. This is time consuming, and on many occasions even this step is neglected because of time not available. To be efficient in designing products a qualified person should be included who is familiar with the different processing techniques.

equipment is ensuring that the quotes solicit from different machine manufacturers are comparable. With the preparation of a complete detailed specification, particularly when unusual requirements exist on what is required, the quotes will be more compatible.

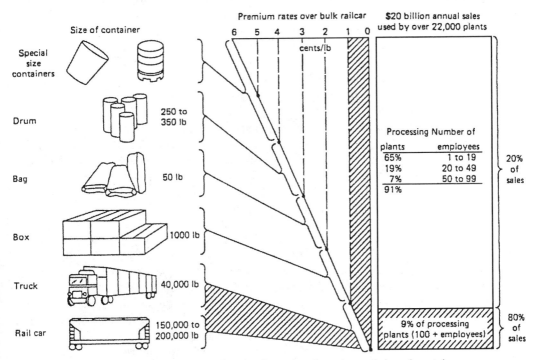

Fig. 9-4 Estimated plastic dollar purchases by plant size and size of containers.

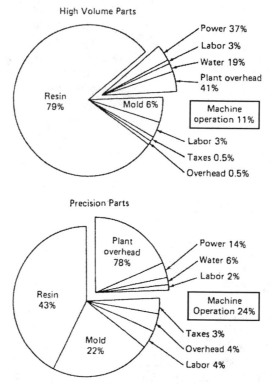

Fig. 9-5   Cost example of injection molding high-volume and precision products.

What may be an inefficient approach to costing a process is for the design group to have a specific manufacturing operation in-house. The companies target is to use their equipment that includes certain sizes. Perhaps the only jobs they take can only fit in their equipment which is OK. However, if they imprevize there is the possibility that a competitor will provide lower cost products using just the appropriate equipment. In the past many decades for companies with the big production products, they design the products, the molds or dies are designed to produce the products, and literally they design the manufacturing equipment around the mold or die. In turn they order new equipment to meet their specifications usually replacing in-house existing equipment. Many of these companies at that time, as at present have better qualified people in designing equipment than certain equipment producers.

The following information concerns injection molding a product. It is a practical and very simplified guide of costing. It covers the main and significant factors that include 85–90% of product direct value and should suffice for the requirements of a product designer when various designs are comparatively evaluated. The two basic ingredients that determine the value of a plastic product are the cost of the plastic material and the cost of fabrication.

Calculating the volume (in.^{3} or cm^3) of the product and multiplying it by cost per the volume unit derives the cost of the plastic material content. Once the volumes of the product are established, one will utilize these data as the basis for obtaining the remainder of the information.

Cost of processing can start by taking the thickest portion of the product that will determine the processing time required. With a contemplated yearly activity and an ordering frequency of say every 45 days, one can tentatively decide on the number of cavities needed when molds are involved. The number of cavities should be such that it would take the 45-day requirement to produce in two hundred hours. Thus the number of cavities times the volume per cavity leads to the machine size.

There are many cases in which the volume of material for all the cavities is small and the projected areas of the cavities are large. In this example, the choice of machine size is governed by the tonnage of the clamp required to keep the mold from opening during the fabricating process (3).

As an example for most plastics the usual is 2–5 tons/in.^{2}. For certain plastics pressure could go up to 15 tons/in.^{2}. Both the material volume and tonnage have to be satisfied while deciding on a machine size and, therefore, machine cost per hour. The larger press of the two requirements should prevail. When machine size in terms of volume and tonnage is decided upon, the molding cost can now be established. Molding cost per hour divided by pieces per hour gives molding cost per piece.

During the checking of volume against the press tonnage, only 70% of the machine volumetric capacity should be considered as the useful volume. The reason for this downgrading is that the heating capacity is based on polystyrene and most other materials require more energy for plasticating than does polystyrene. If the calculated volume for a product is more than 70% of rating, the selection calls for a higher tonnage press.

When using other processes, such as extrusion, blow molding, rotational molding, etc., cost of fabrication is usually related to the size of the product and output rate. Those familiar with any of these processes can easily and quickly provide cost information. What exists and required is experience in the different processes. For the designer not familiar with processes relying on the best-cost analysis can be a problem particularly if time does not exist to contact different fabricators. Recognize that those providing you a cost includes in their estimate factors such as their interest in obtaining the job, the odds of obtaining the job, how busy is their operation, are they helping a competitor, etc. Final quoted costs may differ from approximate cost estimates because quotations will include such features as close tolerances and special performance requirements not previously listed.

We now have the means of obtaining the material cost as well as the manufacturing cost per piece. Consider the sum of these two costs multiplied by 1.2 gives the complete cost. The 20% addition is to cover other elements of expense (probably accidentally not included) as well as a safety factor. The description of the method of cost computing may give the impression of a lengthy procedure; however, the actual performance of the estimating will prove brief and will be worthwhile in doing. The described cost estimates are intended for comparative design evaluation.

Effective Control

Properly training employees will help (and even eliminate) variable costs. In addition to other forms of training, such as shop floor training, seminars, video, and reading, implementing an interactive in-house training program becomes an important cost-effective form of educating the workforce. Effective training is essential to the survival and growth in today's world of plastics. Interactive training has proved to be the best way to provide employees with skills and knowledge that ultimately creates a more confident and productive workforce. An example of this service is available through different schools and organizations such as the University of Massachusetts-Lowell or Penn State at Behrend, PA.

Technical Cost Modeling

A wide range of processes, materials, and economic consequences characterizes the adoption of any technology for producing manufactured products. Although considerable talent can be brought to bear on the processing and designing aspects, economic questions can remain. Cost problems are particularly acute when the technology that will be employed is not fully understood, as much of cost analysis is based on historical data, past experience, and individual accounting practices.

Historically, technologies have been introduced on the shop floor incrementally, with their economic consequences measured directly. Although incorporating technical changes in the plant to test their viability may have been appropriate in the past, it is now economically infeasible to explore today's wide range of alternatives in this fashion. Technical cost modeling (TCM) has thus been developed as a method for analyzing the economics of alternative manufacturing processes without the prohibitive economic burden of trial-and-error innovation and process optimization (3, 20).

TCM is an extension of conventional process modeling that particularly emphasizes capturing the cost implications of process variables and economic parameters. By coordinating cost estimates with processing knowledge, critical assumptions (processing rates, energy used, materials consumed, scrap, etc.) can be made to interact in a consistent, logical, and accurate framework of economic analysis, producing cost estimates under a wide range of conditions.

For example, TCM can be used to determine the plastic process that is best for production without extensive expenditures of capital and time. Not only can TCM be used to establish direct comparisons between processes, but it can also determine the ultimate performance of a particular process, as well as identifying the limiting process steps and parameters.

TCM uses an approach to cost estimating in which each of the elements that contribute to the total cost is estimated individually. These individual estimates are derived from basic principles and the manufacturing process. This reduces the complex problem of cost analysis to a series of simpler estimating problems and brings processing expertise rather than intuition to bear on solving these problems.

In dividing cost into its contributing elements the first distinction to be made is that some cost elements depend upon the number of products produced annually, whereas others do not. For example, the cost contribution of the plastic is the same regardless of the number of items produced, unless the material price is discounted because of high volume. On the other hand, the per-piece cost of tooling will vary with changes in production volume that is influenced by maintenance, wear, etc. These two types of cost elements, which are called the variable and fixed costs, respectively, create a natural division of the elements of manufacturing product cost.

Basically the variable cost elements are those elements of piece cost whose values are dependent on the number of pieces produced. For most plastics fabrication processes the principal variable cost elements are the material, direct labor, and energy costs.

Fixed costs are those elements of piece cost that are a function of the annual production volume. Fixed costs are called fixed because they typically represent one-time capital investments (buildings, silos, processing machines, etc.) or annual expenses unaffected by the number of products produced (building rent, engineering support, administrative personnel, etc.). Typically, these costs are distributed over the total number of products produced in a given period. For plastics processes the principal elements are main machine cost, auxiliary equipment cost, tooling cost, building cost, overhead labor cost, maintenance cost, and the cost of capital.

To demonstrate the use of such a comparative cost analysis, the production of a panel was analyzed according to different processes (Fig. 9-6). In these case studies the following conditions existed: (1) the panels measured 61 × 91 cm (24 × 36 in.) with the wall thickness dictated by the process and part requirements so that the weights of the panels differed; (2) production was at a level of 40,000/yr.; (3) the plastics for all panels were of the same type, except that different grades had to be used, based on the process requirements, so that costs changed; (4) each panel received one coat of paint, except that the structural foam also had a primer coating; and (5) costs were allocated as needed to those processes that required trimming and other secondary operations.

TCM can keep cost data current, based on cost changes from day to day, region to region, and so on. Of course, the means of keeping these data updated require that those

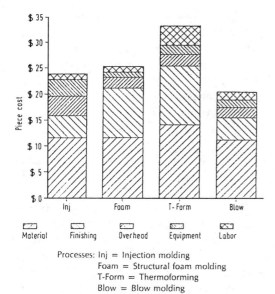

Processes: Inj = Injection molding
Foam = Structural foam molding
T-Form = Thermoforming
Blow = Blow molding

Fig. 9-6 Cost comparison of panel production using TCM program that shows blow molding with the lowest product cost.

costs be obtained on a regular basis and incorporated into the TCM.

Mold/Die Cost

Cost breakdown of a high production mold or die is approximately as follows: Material cost used to about 12 to 20%, design about 5 to 10%, mold building hours about 40 to 60%, and profit at about 5 to 10%. In general they are very expensive with the major cost principally in machine building labor. The proper choice of materials of construction for their different parts is paramount to quality, performance, and longevity of a mold/die. Add good machinability of component metal parts, material which will accept the desired finish (polished, textured, etc.), ability to transfer heat rapidly and evenly, capability of sustained production without constant maintenance, etc.

Using low cost material to meet high performance requirements will compromise their integrity. As an example, the cost of a mold cavity and core materials, for more than 90% of the molds, is less than 5% of the total mold cost. Thus it does not make sense to compromise mold integrity to save a few

dollars; use the best material for the application. There are molds that run a few hundreds to many millions. Design and construction that relates to cost of a mold or die depends on the lifetime required.

An unwritten rule says that a mold or die should cost almost half the cost of the basic machine (injection, extrusion, blow molding, etc.). If it does not then something is wrong such as you probably have an oversized machine for the job for the lower cost mold or die.

Cost Analysis Method

Cost-Benefit Analysis

CBA is the economic analysis, such as with design developments and research programs, in which both the inputs to produce the intervention (or costs) and its consequences or benefits are expressed in monetary terms of net savings or a benefit-cost ratio. A positive net saving or a benefit-cost ration greater than one indicates the intervention saves money.

Direct and Indirect Cost

They are the operating quality costs of prevention and appraisal that are considered to be controllable quality costs. Add that in year 2000 the IRS decided to let companies deduct ISO 9000 costs as a business expense. Also there are the internal and external failure costs. As the controllable cost of prevention and appraisal increases, the uncontrollable cost of internal and external failure decreases. At some point the cost of prevention and appraising defective product exceeds the cost of correcting for the product failure. This point is the optimum operating quality cost.

In addition to the direct operating quality costs, the indirect quality costs and their effect on the total cost curve must be considered. Indirect quality costs can be divided into three categories: customer-incurred quality costs, customer-dissatisfaction quality costs, and loss-of-reputation costs. These intangible, indirect quality costs are difficult to measure; however, they do effect the total quality cost curve. This influence is apparent when the indirect quality costs are added to the direct cost curve. When the optimum point increases, it indicates the need for a lower product defect level. A lower product defect level can be obtained by increasing the prevention and appraisal costs, which subsequently lowers the external failure costs. A lower external failure has a desirable influence on the direct costs. The measurement of the actual indirect costs may be impossible. However, a knowledge that these costs exist and their relationship to the direct costs can aid in their control (3).

Cost Effectiveness

Minimizing costs is generally an overriding goal in any application, whether designing a product, selecting a material, or a process is being selected for a new product or opportunities are being evaluated for replacing existing materials. The major elements of cost, are equipment and material with those that could be called inefficiencies such as scrap, repairs, waste, and machine downtime. Even though the scrap is recycled, it cost to granulate, handle, and possible slow down the line. Each of these elements must be evaluated before determining the most cost-effective approach.

Cost-Effectiveness Analysis

CEA is the economic analysis in which the consequences or effects of intervention are expressed in improvements such as fabricating successful products, years of service, etc.

Cost Estimating

Estimating is a critical aspect that ranges from designing to fabrication. Particularly in fabrication it is often practiced with very little logic. It is shrouded in mystery and rarely discussed by processors. Indeed, it is considered among the dullest of topics. It is extraordinary if one estimate in ten produces a successful

bid. In other words, a 90% failure rate is terrific. No wonder estimating seems like some bizarre sacrificial rite. That does not include those estimates you just go through the motions of preparing because another company is going through the motions of getting three bids, and you have no chance at all of landing the job. Or that company's supplier is overloaded or had an accident, so the customer needs "temporary" help and you know it; in which case provide the high end of the bid (to be safe).

But what more directly represents the heart of a fabricator's business than estimating? You are pulling together every facet of your operation, distilling it, putting numbers on it, and putting your company on the "line." You are stating that this is what we can do, and this is what we must charge to make a normal profit. There are probably as many estimating techniques as there are estimators.

Cost Estimating Factor

Much contemporary estimating follows very vague procedures. The number of factors assembled to reach the appropriate figures is sometimes alarmingly small. Some may not consider scrap, coloring, setup time with trial and error, and so on to mention some of the more obvious omissions. Some estimates, that can work, are simple creations such as determining the part volume or weight, cost of plastic, and processing time; scribbling down some numbers; and adding a fudge factor (possible a little prayer). Some companies do not even have their own standard forms whereby they could develop some useful history. Of course what influences how one estimates generally relates to the fact that plastic processing is a highly competitive industry, so logically spend the time to prepare quotes where a payoff has a possibility to occur.

Cost Reduction

When it is possible, observing the following usual practices for extrusion will help to reduce costs (relate them to your process):

(1) strive for the simplest shape and form; (2) combine parts into single extrusions or use more than one die to extrude products/use multiple die heads and openings; (3) make gradual changes in thickness to reduce frozen stress. (4) where bends occur, use maximum permissible radii. (5) purchase plastic material as economical as possible. (6) keep customers tolerance as liberal as possible, but once in production aim for tighter tolerances to save material costs and also probably reduce production costs.

Cost Target

The production flexibility of the plastics fabrication processes is often the single most important economic factor in producing a product. The products size, shape, complexity, strength, orientation, etc. can be primary determinants but not impossible to produce. Thus, processing takes on the task of doing the "impossible" at the lowest cost (Fig. 9-7). Economics can be improved by targeting various factors. (1) reduction in the use of material by minimizing tolerances. (2) improvement in product quality in terms of strength and/or other mechanical-physical characteristics, (3) reduction in setting-up times of start-up aids and automation systems, and (4) savings in electricity consumption by the optimization of the plasticizing and the use of efficient heating and cooling systems.

Variable Cost

Processing cost variations may be due to one or more of the following factors: (1)

improper or unattainable performance requirements; (2) improper plastic selection; (3) improper in-line and off-line hardware and control selections; (4) improper selection of the complete line; (5) improper collection and/or handling at the end of the line; and (6) improper setup for testing, quality control, and troubleshooting.

Energy Cost

Cost savings via energy conservation can be considered from the viewpoints of machine operation, the plastic material, and the finished product. Fabricating machines are usually energy intensive. Thus it becomes obvious to reduce the energy requirements where it is possible starting with the purchase of any equipment in the line that provides reducing energy consumption.

Product Cost

In a production line that has a relatively long run, the cost for equipment in relationship to producing the product including its financial amortization, usually is about 5% with probably maximum of 10% Plastic material cost could be at about 50% with as high as 80% for high volume production. The other costs include power, water, labor, overhead and taxes. With precision, short runs, costs could be equipmentwise at 20 to 30%, material 45 to 50%. Thus, as it is usually stated, do not buy equipment just because it cost less since more profit could occur with the more expensive equipment; study what is to be purchased. Of course the reverse is possibly true. So, you the buyer, have to know what you want and are ordering to a specification properly determined based on the designed product requirements.

Designing Product

In product design there has always been the desire to use less of any material, because the result is usually a lower-cost product. On the other side of the issue is the use of more

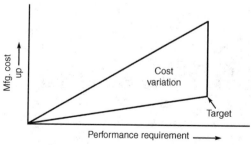

Fig. 9-7 Target to meet performance requirements at the lowest cost.

material to provide for a higher design safety factor beyond what is required. Thus, unfortunately, there are designs using more material than needed. It is inexperience in designing with plastics that causes this problem.

Many designers lack the knowledge of at least relating a material's performance to the processing variables that directly influence safety factors and the amount of a plastic to be used. With the flexibility that exists in designing with plastics, there are different approaches that may be used to reduce product volume or weight, such as applying different shapes like internal ribbing, corrugations, sandwich structures, and orienting or prestretching.

All this activity is aimed at producing products that use less in the way of materials and in turn let less material enter the solid-waste stream. Some designers have habitually listed in product design specifications that specific environmental requirements should be met. A designer sometimes has an opportunity to use a material that provides no problem in the solid-waste stream or to use a design that lets lower-cost recycled plastics be used. In fact, blends of virgin (not previously processed) plastics with recycled plastics could permit the meeting of required product performance and environmental requirements (Chapter 6, **RECYCLING**).

This approach has been used for the past century, but now there will be more use of it, as more and more recycled plastics become available. However, the designer must take into account the potential lower performances that could occur with recycled plastics. Interesting that at times recycled materials is more expensive than virgin plastics. They are used in many cases since there is a regulation (government or industry) that states the product has to contain some amount, such as 50 wt%, of recycled plastics.

The recycled plastic will also have a degree of different contaminants that would eliminate its use in certain devices or products, such as in medicine, electronics, and food packaging. However, within these market applications there are acceptable designs with three-layer coextruded, coinjected, or laminated structures having the contaminated plastic as the center layer, isolated by "clean" plastics around it and no migration occurring.

Another method of reducing the quantity of plastics that has been used in certain products is to use engineered plastics with higher performance than the lower-cost commodity plastics. When applicable, this approach permits using less material to compensate for its higher cost. With a thinner-walled construction there could also be additional cost savings, since less processing heat, pressure, and time cycle is required.

Energy

To produce and process plastics requires less energy than practically any other material (Fig. 9-8). In contrast, glass requires much more energy than any of the materials listed.

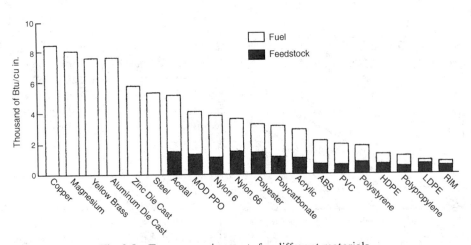

Fig. 9-8   Energy requirements for different materials.

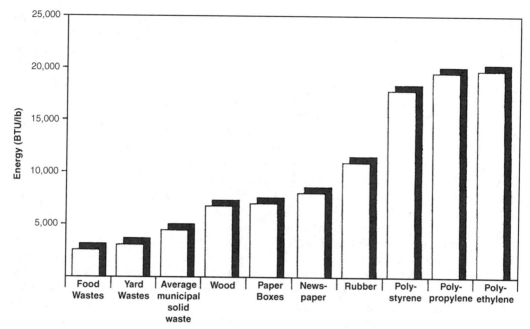

Fig. 9-9   Incineration and energy output.

Solid Waste

Only about 7 wt% of the solid waste produced are plastics. Incinerating, recycling, landfill, and other methods are used to handle the worldwide plastics (and other materials) waste problems. Incineration of plastics produces a high energy content. For example, polystyrene has nearly twice the energy content of coal, without its ensuing problems of ash, acid rain pollution, or harmful emissions. Plastics are just one of many materials that produce solid waste, and, as with other materials, there are good and bad disposal solutions. As shown in Fig. 9-9, plastic scrap and waste does provide cost savings when compared to other materials.

Competition

For over a century plastics have successfully competed with other materials in old and new applications providing cost-performance advantages, etc. In fact within the plastic industry there is extensive competition where one plastic competes with another plastic. Examples include many such as thermoplastic elastomers vs. thermoset

rubbers, clarity film LLDPE vs. LDPE and PVC.

This action will continue and expand as is evident by new plastics being developed. DuPont's iron and cobalt single-site catalyst system that makes HDPE with higher melting points and performances such as adhesion, barrier performances, etc. Other examples include the expanding process techniques, and applications that are always on the next horizon. Thus to help plastics expand, there has been and will continue to exist plastic-to-plastic competition to meet all kinds of requirements, including high and safe performance in all kinds of environments. Recognize that in the future different basic raw materials will be used in addition to those already being used.

In general, companies in the plastics industry can obtain patents upon the processes they use to manufacture new materials (Chapter 4, **DESIGNING AND LEGAL MATTER**). However, since a processed patent discloses a great deal of information that may be useful to competitors even though they are not using or do not wish to use the exact process as that described in the patent. Some firms in the industry do not patent each new process (or material) in order to maintain strict secrecy. To prepare a practical fool proof patent requires lots of money.

Plastic materials' manufacturing is primarily a large-volume, low-cost, low-unit profit margin business with great overall economies. The plastics generally compete with each other on a "money value" basis in which an economic analysis takes into account the differing densities of the various plastics in order to judge them on a cost per pound or volume basis.

Conventionally, we think of competition as being between essentially similar products on a price, quality, and service basis. In plastics, competition is much broader and often more intense.

Competition, at each stage in the plastics industry, is in their raw materials. Many monomers can be made from alternate raw materials such as polyvinyl chloride that may begin with either ethylene or acetylene. Most plastic products may be made from a variety of plastics such as pipe that may be extruded from PVC, polyethylene, ABS, and so on.

Competition is also within processes. Both plastics and finished products may be made by entirely different routes, requiring different procedures and different equipment. For example, polyvinyl chloride may be made by three different processes and thermoplastic sheets may be cast, injection molded, or extruded. Plastic competition is within the industry both locally and particularly international as well (210, 211). Plastics can compete with the products of other industries for the same application; for example, against wood, metal, and concrete in construction application, against natural fibers for textiles, and against animal glue in the adhesives field. In addition, plastics is an international business with growing industries in other countries vying with USA firms for markets both abroad and at home.

An example of the way in which process competition works in the manufacture of plastics is the story of acrylonitrile. The first process for the production of this plastic was based upon the reaction between hydrogen cyanide and acetylene, both hard to handle, poisonous, and explosive chemicals. The raw material costs were relatively low as compared to materials for other monomers, but the plant investment and manufacturing costs were too high. As a result, originally acrylonitrile monomer (1950s) sold for about 30 cents per pound and the future of the material looked dim as other plastics such as polyethylene became available at much lower prices due to their lower production costs.

During the late 1950's, chemists at the Standard Oil Company of Ohio worked

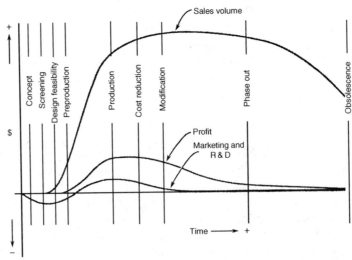

Fig. 9-10   Example of factors to consider in marketing a product.

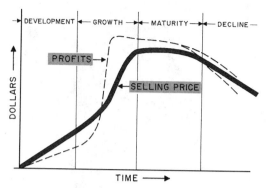

Fig. 9-11 Example for the life of a product.

Fig. 9-13 Any other operation not showing a profit.

on the development of a new process for the manufacture of acrylonitrile. In 1961, SOHIO built the first plant using their new process to produce acrylonitrile from propylene and ammonia, both materials being readily available and easy to handle. This new process used more expensive raw materials than the old one, but required only one production operation to produce the acrylonitrile monomer. This meant lower plant investment and lower manufacturing costs, and the price of the monomer went down to around 14 cents per pound.

This new lower price changed the comparative economic advantages of some of the newer plastics and led to a search for new uses of acrylonitrile and its polymers and copolymers. A new route to Dacron was developed by du Pont using this lower priced acrylonitrile and the use of acrylic fabrics grew rapidly. There was also an increase in uses of ABS and acrylonitrile production capacity.

On the other side of costs, the higher priced plastics continue to be marketable since they meet required performances for certain products. As an example polyaryletherketone

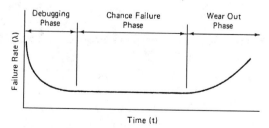

Fig. 9-12 Life-history curve.

(PAEK) plastic that cost $40/lb was being market in dental implants, bone replacement joints, and components for the hip, elbow, finger, knee, spine, and other body products. And so all these type of actions continue in the plastic industry worldwide.

Quotation

Document quote that states the selling price and other sales conditions of a material, product, etc. has different meanings. Did you know that by law if someone reports that verbally the vender made statements such as "buy this injection molding machine and all you have to do is push a button to make good/acceptable products" the vender is legally in trouble. Even if that person wins the case (rare), it will be very expensive to pay for the court case.

Market

Throughout this book reviews have been made on products that literally are used in many different markets. This action fits the usual statement that "this is the World of Plastics" Important with all the cost analysis is that profits have to be included. Influencing factors that involve profits are summarized in Figs. 9-10 to 9-13. The life-history curve, Fig. 9-11, shows the basic format of a typical product cycle for an infinite number of products. It is also called a "bathtub" curve.

10

Summary

Overview

Design is essentially an exercise in predicting performance. The designer of plastic products must therefore be knowledgeable in such behavioral responses of plastics as those to mechanical and environmental stresses. This book presented important basic concepts of plastics that define their range of design behaviors. It provides the background needed to understand performance analysis and the design methods available to the performing designer, as well as for those less familiar with conventional design and engineering practices. Products made of plastics can then be designed using the logical approach that also applies to such other materials as steel, wood, glass, and concrete, which have their own specific techniques of behaviors and analysis. The design approach includes factors such as the proper product evaluation (Table 10-1) to the release of the product (Table 10-2).

Many plastic products seen in everyday life are not required to undergo sophisticated design analysis because they are not required to withstand high static and dynamic loads (Chapter 2). Examples include containers, cups, toys (Fig. 10-1), boxes, housings for computers, radios, televisions and the like, electric iron (Fig. 10-2), recreational products (Figs. 10-3 and 10-4) and nonstructural or secondary structural products of various kinds like the interiors in buildings, automobiles, and aircraft. In fact many of these only require a practical approach (Fig. 1-4).

Designing is, to a high degree, intuitive and creative, but at the same time empirical and technically influenced. An inspired idea alone will not result in a successful design; experience plays an important part, that can easily be developed. An understanding of one's materials and a ready acquaintance with the relevant processing technologies are essential for converting an idea to an actual product. In addition, certain basic tools are needed, such as those for computation and measurement to testing of prototypes and/or fabricated products to ensure that product performance requirements are met.

For these reasons design is spoken as having to be appropriate to the materials of its construction, its methods of manufacture, and the product performances involved. Where all these aspects can be closely interwoven, plastics are able to solve design problems efficiently in ways that are economically advantageous.

Design Success

Plastics provide the designer with many different materials and processes useful

Table 10-1 Product evaluation

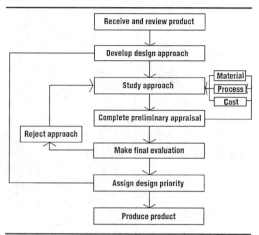

Fig. 10-1 Toys all around us.

toward meeting many of the varying types of product requirements. They are also capable of producing from simple to complex shapes and are economically beneficial. They can be made to have a long life, they resist corrosive environments, and are recyclable, degradable, and can meet practically any performance requirements. They also permit the fabrication of products whose manufacturing would be difficult if not impossible in other materials. Different design approaches are used such as those described in Figs. 10-5 to 10-8.

However, designers must routinely keep up to date on developments with the more useful plastics and acquire additional infor-

Table 10-2 Product release evaluation

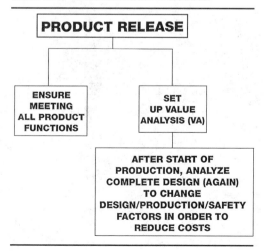

mation on how the plastics behave during processing. The emphasis throughout this book has been that it is not difficult to design with plastics and to produce many different sizes and shapes of thermoplastic and thermoset commodities and engineering plastics, whether unreinforced or reinforced.

Some plastics can be worked by many different processes, but others require a specific process (Fig. 10-9). Process selection can take place before material selection, when a range of materials may be available, or made first to meet performance requirements and only then have the applicable process or processes chosen. (Chapter 7, **SELECTING PLASTIC** and Chapter 8, **SELECTING PROCESS**) Usually, in the latter situation only one special process can be used to provide the best performance-to-cost advantages. A particular design group may have its own processing capabilities. Unfortunately, some operations use just whatever equipment is available. This situation could either be very unprofitable, limit profitability, or restrict product

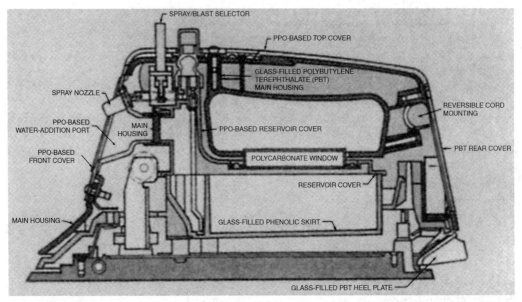

Fig. 10-2   View of an electric iron showing their use of different plastics that use different fabricating processes.

performance. It is important to recognize that the fabrication process can markedly influence all aspects of product performance, including cost.

Compared to other material-based industries, plastics have enjoyed an impressive growth rate over a century since their inception, but particularly since about 1940. The product-design community was quick to recognize the design freedom and great versatility that plastics' materials and processing techniques afforded. Recognizing a growing marketing opportunity worldwide, international plastics material suppliers started an endless cycle of developing new and improved materials to meet continually new design needs. Processing machinery builders worldwide respond with improved equipment and even totally new processes, as conventional tool shops everywhere expanded their capabilities to include mold and die manufacturing for the plastics industry.

Fig. 10-3   Practical all glass fiber-TS polyester RP (hand layup) recreational vehicle.

Fig. 10-4 Examples of recreational plastic products (paddles, surface boards, inflatable boat, etc.).

Important for the reader to recognize is that new plastic material data as well as plastic fabricated product data is endless because of what can be done in the production of plastic materials as well as fabricating equipment. Thus the importance of keeping up to date on this information tends to be an endless project. However when designing with plastics, the basics will remain rather constant as reviewed in this book. Just keep up to date on plastic behavior (Appendix A, **PLASTICS DESIGN TOOLBOX**).

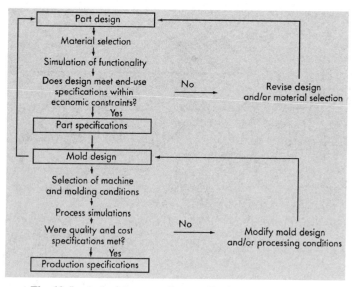

Fig. 10-5 A decision tree diagram for integral designing.

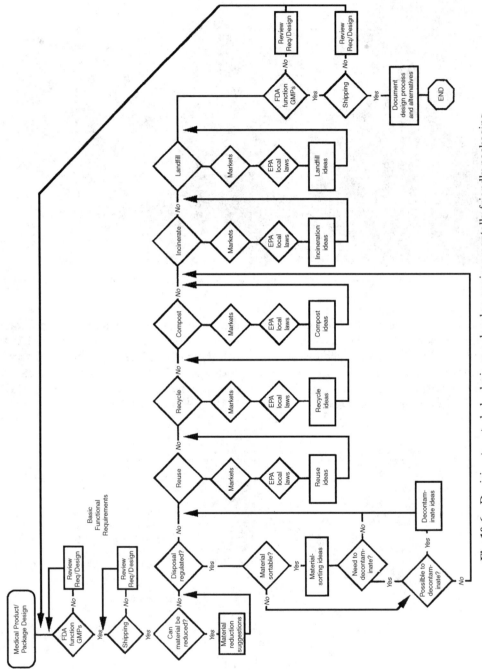

Fig. 10-6   Decision tree to help designers develop environmentally friendly packaging.

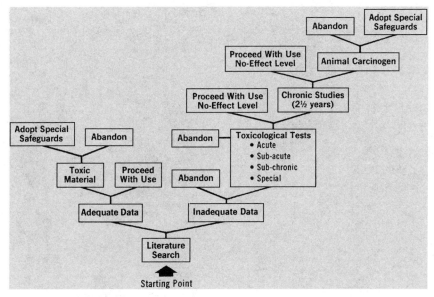

Fig. 10-7   Example of a typical toxicology study for a plastic for industrial us, non-food contact.

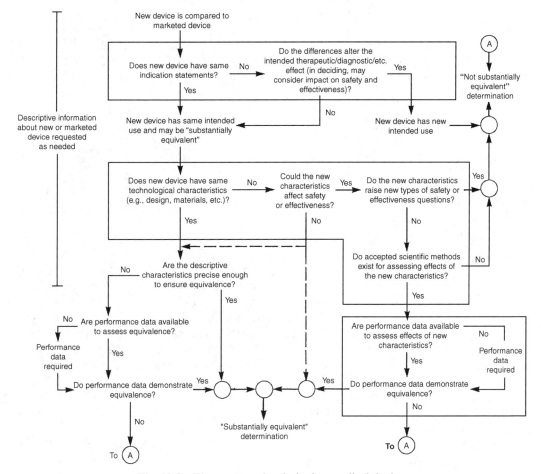

Fig. 10-8   Flow pattern for designing medical device.

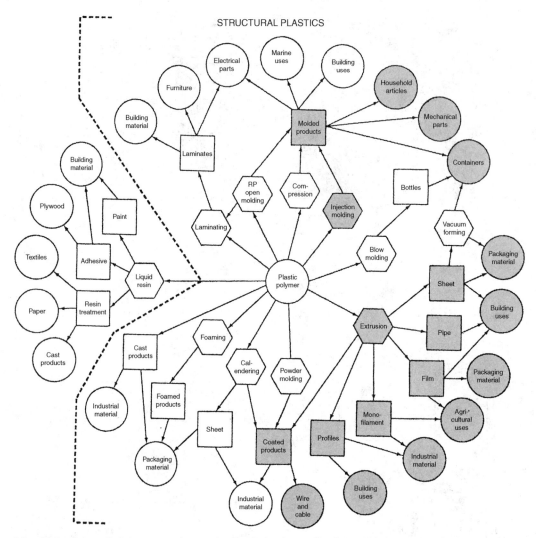

Fig. 10-9 Interrelation among the methods of plastic applications and processing for the family of plastic materials.

Challenge

There is always a challenge to utilize advanced techniques without overlooking the understanding of its basic operation. They include: (1) the different plastic melt flow behaviors, (2) operational monitoring and process control systems, (3) importance of the different molds or dies and auxiliary equipment, (4) fundamentals of product designs, (5) design features that influence mold or die designs and product performances, (6) testing and quality controls, (7) statistical analysis, (8) setting up troubleshooting and main-

tenance guides, (9) detail cost factors that influence products final costs, and (10) analyze competition (both those in the plastic business and those using other materials). Knowledge of this type information ensures the elimination or significant reduction of potential problems. This type of understanding is required in order to be successful in the design-through-prototype-through-manufacture-to-profitable sales of the many different, marketable, plastic products worldwide.

The reputation of plastics periodically has been harmed a great deal by the fact that in certain cases designers and engineers have, after deciding tentatively to try to introduce plastics, lavishly copied the material (metal, aluminum, etc.) used in a product it was suppose to replace. Too much emphasis can not be laid down upon the general principle that if plastics are to be used with maximum advantage and with minimum risk of failure, it is essential for the unfamiliar or limited knowledge designer in working with plastics to do some homework and become familiar and keep up to date with the plastic processes and materials.

Challenge Requires Creativity

In order to find unique, creative solutions to difficult challenges that were not resolved by past tried and true techniques, one must get away from the conventional state of mind that is often unimaginative, frustrating, repetitive, and negative. Examples of this type challenge exist all around us (Fig. 10-10). The nature of some problems tends to invite unimaginative suggestions and attempts only to use past approaches. Problem solving in designing and producing products, as in business and personal problems, generally requires taking a systematic approach. If practical, make rather small changes and allot time to monitor the reaction of result. With whatever time is available, patience and persistence are required.

However, when a problem is particularly difficult or only limited time exists, consider a new and imaginative approach with techniques that previously generated creative ideas. First generate as many ideas as possible that may be even remotely related to the problem. During the idea-generating phase it is of critical importance to be totally positive: no ideas are bad. Evaluation comes later, so do not attempt to provide creativity and evaluation at the same time; it could be damaging to your creativity. Look for quantity of ideas, not quality, at this point. Now all ideas are good; the best will become obvious later.

If possible, relate the problem to another situation and look for a similar solution. This approach can stimulate creative thinking toward other ideas. Try humor; do not be afraid to joke about a problem. The next step is to evaluate all the ideas. Consider categorizing the list, then add new thoughts, select the best, and try them.

After all this action, if nothing satisfactory occurs, rather than give up look for that really creative solution because it is out there. You may be too close to the problem. Get away from the trees and look at the forest. Climb up one of the trees and look at things from a different perspective.

Use your creative talents but be positive. You have now creatively worked through the frustration and negativism that problems seem to generate. Your increasingly creative input will generate future opportunities.

Now let us take the thoughts above and improve on them. In doing so let us avoid saying in effect "My mind is made up-do not give me the facts." Rather, let us use the approach that there is always room for improvement and resolving the problem.

Value Added/Analysis

Well designed products have the value added approach that results in customer satisfaction and profit gains. Value is an amount regarded as a fair equivalent for something, that which is desirable or worthy of esteem, or product of quality having intrinsic worth.

Aside from technology developments, there is always a major emphasis on value added services. It includes the design concept to fabricator that continually tries to find ways to augment or reduce steps during manufacture with the target of reducing costs.

While there are many definitions of value analysis, the most basic is the following formula where $VA = $ (function of product)/(cost of the product). Immediately after the product goes into production, the next step that should be considered is to use the value design approach and the FALLO approach (Fig. 1-3). These approaches are to produce products to meet the same performance requirements but produced at a lower cost.

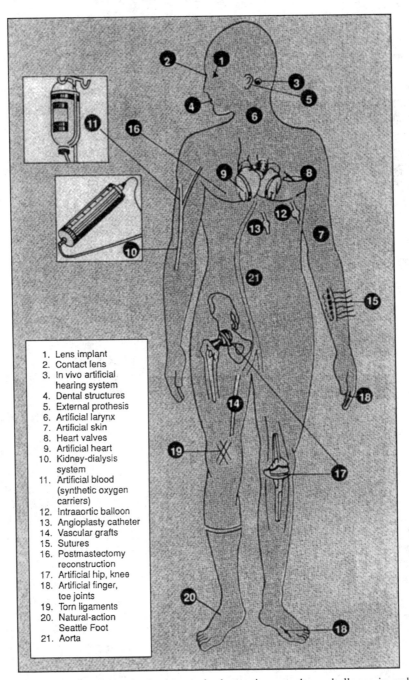

1. Lens implant
2. Contact lens
3. In vivo artificial hearing system
4. Dental structures
5. External prothesis
6. Artificial larynx
7. Artificial skin
8. Heart valves
9. Artificial heart
10. Kidney-dialysis system
11. Artificial blood (synthetic oxygen carriers)
12. Intraaortic balloon
13. Angioplasty catheter
14. Vascular grafts
15. Sutures
16. Postmastectomy reconstruction
17. Artificial hip, knee
18. Artificial finger, toe joints
19. Torn ligaments
20. Natural-action Seattle Foot
21. Aorta

Fig. 10-10 Application of plastics in the human body continues to be a challenge in order to meet performance and biocompatibility requirements.

If you do not take this approach, then your competitor will take the cost reduction approach. VA is not exclusively a cost-cutting discipline. With VA you literally can do "it all" that includes reduce cost, enhance quality, and/or boost productivity.

Plastic Industry Size

Plastic product industry is ranked as the 4th largest USA manufacturing industry and growing 3 to 4 times that of the total national products (Fig. 1-5). Motor vehicles are in 1st place, petroleum refining in 2^{ed} place, and automotive parts in 3^{ed} place. Plastic is followed by computers and their peripherals, meat products, drugs, aircraft and parts, industrial organic chemicals, blast furnace and basic steel products, beverages, communications equipment, commercial printing, fabricated structural metal products, grain mill products, and dairy products (in 15th place). At the end of the industry listings are plastic materials and synthetics in 24th place and ending in the 25th ranking are the paper mills.

Worldwide total sales for the category of plastic products and plastic materials is now well over $275 billion/year. Machinery sales (primary, auxiliary, secondary, etc.) in the plastic industry are estimated to be above $7.5 billion/year (1995).

Recognize that the USA economy has been changing from a manufacturing society to an information and service society (as first reported during 1939 by a college professor who stated it actually started during the start of the 20th century). In 1998 the US Department of Labor reported that about 93 million people are no longer in manufacturing but are in an information and services. Considering this situation the USA plastic fabricating industry continues to grow.

Fabricating Employment

In USA the yearly man-hours employment producing all plastic products by all processes is estimated at 650 million, second to motor vehicles at 845 million. Following plastic products (in millions) are aircraft at 570, commercial printing at 560, newspapers at 475, meat at 460, metal structural products at 350, and computers at 325. The USA plastics Industry is growing and creating jobs faster than the other manufacturing sectors.

Future

In addition to many of the major present markets expanding, the design opportunities for plastics materials in the future will be in such areas where the special properties of the material with their processing capabilities can be made to accomplish a unique result. Ingenuity in the application of materials has been the thrust of the plastics industry, and it will present new opportunities in the future.

The versatility of the plastic materials permits them to be varied to perform special functions. By applying ingenuity to these amazingly adaptable materials, we can produce products that add to the worldwide survival capability of life under severe environments and improve the quality of life under normal environments. The designer's role in fitting the possibilities to the needs is one that is increasingly important.

Research and Development

The extent to which plastics are used in any industry in the future will depend in part upon the continued total R&D activity carried on by materials producers, processors, fabricators, and users in their desire to broaden the scope of plastic applications. An R&D example is the rail car hopper, called the Grasshopper (10, 14). It is literally all plastic that provides improved load and operating performances over metals (Fig. 10-11).

The bulk of such research expenditure is done by the materials producers themselves and the rest by the additive and equipment industries who do more than the processors and fabricators whose share is very small. Important to plastic growth have been government projects in basic research and new applications, particularly the military. Their work in turn expands into the industrial industry.

The project of communicating new technology to processors and users is the subject of much discussion in the industry. The lag time from laboratory discovery to end user benefits is generally three years or more.

Theoretical vs. Actual Value

Through the laws of physics, chemistry, and mechanics, in 1944 theoretical data was determined for different materials (42). These are compared to the present actual values

Fig. 10-11 Plastic railcar hopper.

Table 10-3 Comparison of theoretically possible and actual experimental values for modulus of elasticity and tensile strength of various materials

Type of Material	Modulus of Elasticity			Tensile Strength		
		Experimental			Experimental	
	Theoretical, N/mm² (kpsi)	Fiber, N/mm² (kpsi)	Normal Polymer, N/mm² (kpsi)	Theoretical, N/mm² (kpsi)	Fiber, N/mm² (kpsi)	Normal Polymer N/mm² (kpsi)
Polyethylene	300,0000 (43,500)	100,000 (33%) (14,500)	1,000 (0.33%) (145)	27,000 (3,900)	1,500 (5.5%) (218)	30 (0.1%) (4.4)
Polypropylene	50,000 (7,250)	20,000 (40%) (2,900)	1,600 (3.2%) (232)	16,000 (2,300)	1,300 (8.1%) (189)	38 (0.24%) (5.5)
Polyamide 66	160,000 (23,200)	5,000 (3%) (725)	2,000 (1.3%) (290)	27,000 (3,900)	1,700 (6.3%) (246)	50 (0.18%) (7.2)
Glass	80,000 (11,600)	80,000 (100%) (11,600)	70,000 (87.5%) (10,100)	11,000 (1,600)	4,000 (36%) (580)	55 (0.5%) (8.0)
Steel	210,000 (30,400)	210,000 (100%) (30,400)	210,000 (100%) (30,400)	21,000 (3,050)	4,000 (19%) (580)	1,400 (6.67%) (203)
Aluminum	76,000 (11,000)	76,000 (100%) (11,000)	76,000 (100%) (11,000)	7,600 (1,100)	800 (10.5%) (116)	600 (7.89%) (87)

For the experimental values the percentage of the theoretically calculated values is given in parenthesis, as (47).

in Table 10-3. With steel, aluminum, and glass the theoretical and actual experimental values are practically the same, whereas for polyethylene, polypropylene, nylon, and other plastics they are far apart, and have the important potential of reaching values that are far superior to the present values.

When polyethylene was first produced in the early 1940s, physicists in England, USA, and Germany predicted a tremendous potential for it. At that time the properties of PEs were much lower than those presently available. Out of that original general-purpose PE, have been developed specific PEs in this polyolefin family of plastics such as LDPE, HDPE, UHMWPE, and so on. In turn their different properties, as well as other plastics, continually increase and their variables continue to be reduced and/or easier to process to tighter tolerances.

Design Demand

It can be said that the challenge of design is to make existing products obsolete or at least offer significant improvements. Despite this level of activity there are always new fields of industry to explore. Plastics will continue to change the shape of business rapidly. Today's plastics tend to do more and cost less, which is why in many cases they came into the picture in the first place. Tomorrow's requirements will be still more demanding, but with sound design plastics will satisfy those demands, resulting not only in new processes and materials but improvements in existing processing and materials.

Research will no doubt become even more adept at manipulating molecules to the extent that the range of materials offered to industry will continue to present new

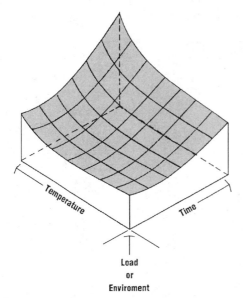

Temperature

Time

Load
or
Enviroment

Fig. 10-12   Product performances in its simplest form relates to this 3-D plot.

opportunities and allow existing businesses to enjoy profitable growth. Also ahead are the different raw material sources to produce plastics that involve biotechnology (186). A reading of the literature and patents being issued indicates that there is a great deal of commercially oriented research being aimed at further improvement and modification into the plastic family. However recognize that the basic analysis for designing plastic products continues to be related to temperature-time-load or environment (Fig. 10-12).

Unfortunately sometimes a new design concept is not accepted or may simply be ahead of its time. In 1483 Leonardo da Vinci designed what he called a spiral screw flying machine. In 1942 Igor Sikorsky developed the R4B helicopter processes (included plastics parts). One could say in a joking manner taking 459 years to bring a designed product to market seams a failure in materials or perhaps the interoffice communication.

Alexander Graham Bell believed the photophone, not the telephone, was his greatest invention. His photophone carried the spoken voice by reflected sunbeams instead of wire, but did not find any practical application a century ago. Because light has 20,000 times shorter frequency than microwaves, it can carry 20,000 time more information. Only since the onset of ("plastic") computers has this ability been needed. It would seem that Alex Bell was ahead of his time.

Fortunately designers did not have to design the human body. The human body is the most complex structure ever "designed" with its so-called 2,000 parts (with certain parts being replaced with plastics). As an example the heart recirculates all the blood in the body every 20 minutes, pumping it through 60,000 miles of blood vessels. Can one image designing the human body. Even with our extensive technology, it would be a total disaster.

The past events in designing plastic products have been nothing short of major worldly achievements. Designers' innovations and visionary provides the required high level of sophistication that is applied to problems that exist with solutions that follow. Ahead is a continuation of meeting new challenges with these innovations and idealism that continues to make plastics a dynamic and visionary industry. The statement that we are in the World of Plastics is definitely true. In fact one can say that plastic products has made life easier for all worldwide.

Appendix A

Plastics Design Toolbox

Here are examples in the selection of the many resources available to the plastics designer and other plastics users; also review references 3, 6, 10, 14, 20, 29, 31, 36, 37, 39, 43 to 125.

Contents

1. Plastics Databases, Electronic

The optimum selection of materials is becoming increasingly important for cost controls and innovation in engineering design. The following databases are tools of choice to help the designer and others meet this need.

1.1 IDES INC.

Tel: 800-788-4668/307-742-9227
Fax: 307-745-9339
(http://www.idesinc.com/Products_1.htm)

1.1.1 Prospector web Prospector Web is an interactive database used to find and compare plastic materials. You can specify your application requirements to search a catalog of nearly all North American plastics and increasing amounts of plastics data from European and Pacific Rim material suppliers. Materials can be searched by any of 200+ properties, reviewed, sorted, and compared to determine the material best fitted to your needs. Test data is available in both English or Metric units and ASTM or ISO format. Both Prospector Web and its sister product Prospector Desktop come with IDES's exclusive Plastics Materials Hotline to help with questions that may arise.

1.1.2 Prospector desktop Prospector Desktop is a disk-based version of the popular Prospector Web. Prospector Desktop also contains multi point data graphs. (Available on CD-ROM or diskette for Windows and Macintosh).

1.1.3 Electronic product catalog Electronic Product Catalog is a sales tool available to material producers and distributors. Potential customers looking for materials can be directed to a website through direct links from datasheets accessed by customers on Prospector Web.

1.1.4 FreeMDS

http://www.freemds.com
FreeMDS from IDES (http://www.idesinc.com/Products_1.htm) is a no-charge service that provides Plastic Material Data Sheets. This growing database contains over 35,000 Material Data Sheets from North America, Europe, and the Pacific Rim.

1.2 PLASPEC Materials Selection Database

http://www.plaspec.com (Tel: 212-592-6570) There are now 12775 grades of plastic materials in the PLASPEC Materials Selection Database. Searches may be conducted for materials using:

- Supplier Name
- Generic Family
- General information
- Processing/Physical Characteristics
- Mechanical Properties
- Thermal Properties
- Electrical Properties
- Optical Properties
- Pricing Information
- Features/Characteristics

1.3 CenBase Materials on WWW

http://www.centor.com/cbmat/CenBASE/ Materials on WWW is a searchable document base on over 35,000 thermoplastics, thermosets, elastomers, and rubbers, composites and fibers, ceramics and metals from over 300 manufacturers product catalogs worldwide. In addition to complete property data, it includes application data, chemical resistance, MSDS and advanced engineering graphs. The database is also available on CD-ROM, and contains the equivalent of over 150,000 pages of data. Engineers, scientists and purchasing professionals use it for competitive analysis, materials selection, materials research, vendor selection and materials engineering education.

1.4 CAMPUS®, The Plastics Database®

registered trademark of CWFG GmbH, Frankfurt/Main, 1991
http://www. CAMPUSplastics.com

1.4.1 CAMPUS, the plastics database

CAMPUS is an internationally known database software for plastic materials, developed by close cooperation with leading plastics producing companies. It is available worldwide from leading material suppliers. More than 50 plastics producers are participants of CAMPUS. Information about the latest list of participants and distribution addresses can be found at the CAMPUS homepage: http://www.CAMPUSplastics.com/. This web site also includes extensive information about the data content of CAMPUS and links to the participants' web sites. It is important to emphasize that only CAMPUS participants distribute CAMPUS diskettes. Each plastic producer distributes his own diskette to his customers without charge.

The plastics properties catalogue includes single-point data, multi-point data, processing data, product description texts and customer service information. You can select plastic products for your specific application by using the query options. The main feature of the CAMPUS philosophy is comparable data. The properties are based on the international standards ISO 10350 for Single-Point data and ISO 11403-1,-2 for Multi-Point data. CAMPUS is available in English, German, Spanish, French and Japanese.

CAMPUS uses a uniform database structure and uniform interface for all participating suppliers, with frequent updates of the property data. It allows preselection or screening of materials, suitable for specific applications, from a worldwide range of commercial plastics, while continuously being developed further with respect to its properties base. CAMPUS is based on two international standards for comparable data, that use meaningful properties based on unambiguous selection of specimen types and conditions for processing and testing, (ISO 10350 [Single-point data] and ISO 11403 [Multi-point data]). Interfaces between CAMPUS and other systems, especially CAE systems are possible via MCBase, the CAMPUS merge program, which is available from M-Base Engineering + Software GmbH, Aachen.

1.4.2 MCBase, the CAMPUS merge database distributed by: M-Base Engineering + Software GmbH,
Dennewartstr. 27 D-52068 Aachen
Germany
Tel: +49 241 963 1450
Fax: +49 241 963 1469
http://www.m-base.de
MCBase is distributed by:

USA—The Madison Group,
505 S. Rosa Rd.
Madison, WI, USA 53719-1257
Tel: 608 231-1907
Fax: 608 231-2694
E-mail: info@madisongroup.com

Germany—KI, Kunststoff Information Verlagsgesellschaft GmbH, Bad
Homburg, Germany

France-SYSTIA Plasturgie,
Centre Hermès, 48, Rue des Grives, F-38920
CROLLES
MCBase offers the possibility to load the original CAMPUS data of different suppliers from version 3.0 and higher into one database, which allows direct comparison. It has been developed in close cooperation with the CAMPUS consortium. For more information see: http://www.m-base.de/. MCBase is user friendly and offers extremely efficient handling of material data. All CAMPUS options are available: define search profiles; define and sort tables; print tables and data sheets; curve overlay; scatter plots. In addition MCBase 4.1 offers search in curves; search for comparable grades; text search; update via Internet; calculation of simulation parameters. A French version of MCBase is available from the distribution agent in France.

1.5 Plastics and Rubbers Data Collection

Plastics Design Library (PDL), William Andrew Inc.
http://www.williamandrew.com
The PDL Electronic Databooks (also available in hardcopy) provide properties of thermoplastics, elastomers, and rubbers. The world's largest collection of phenomenological data, information is provided as concise textual discussions, tables, graphs and images on chemical resistance, creep, stress strain, fatigue, tribology, the effects of UV light and weather, sterilization methods, permeability, film properties, thermal aging, effects of temperature. The Databooks are available on a single CD-ROM as a complete set (the Plastics and Rubbers Data Collection) or as individual topics. They are updated annually.

Features of the complete set include:

- information and data for 180+ material families-thermoplastics, elastomers, alloys and rubbers, 5,000 chemical reagents and exposure media, 175,000 material and reagent combinations
- difficult-to-find information
- search, sort and compare across the entire database
- customize charts, tables and curves to compare performance characteristics
- print out or export to your favorite word processor, spreadsheet or database
- 6,000 curves that enable you to display the coordinates of a chosen point, fit curves to data points, do a trend analysis, export data points and images
- 100,000 tables-search by key-words, numerical ranges and indices completely source referenced

Collections and individual Databooks are available for the following topics:

- Polymer Degradation Collection
- Chemical Resistance of Plastics and Elastomers, Volumes I and II
- Effects of Sterilization Methods on Plastics and Elastomers
- Effect of UV Light and Weather on Plastics and Elastomers

- Performance Properties of Plastics Collection
- The Effect of Creep and Other Time Related Factors on Plastics
- Permeability and Other Film Properties of Plastics and Elastomers
- The Effect of Temperature and Other Factors on Plastics
- Fatigue and Tribological Properties of Plastics and Elastomers
- Dynamical Mechanical Analysis for Practical Engineering

1.5.1 PDLCOM. Available through NACE.   http://www.nace.org/naceframes/ Store/pdlindex.htm

Published by the Plastics Design Library, PDLCOM is an exhaustive reference source of how exposure environments influence the physical characteristics of plastics. Data include resistance to thousands of chemicals, weathering and UV exposure (i.e. color change after accelerated weathering or outdoor exposure); sterilization (radiation, ethylene oxide, steam, etc.); thermal air and water aging; environmental stress cracking and much more.

Description of samples tested, specific test methods used, exposure medium notes, solubility parameters, and other important details are provided. Emphasis is on providing all relevant information so the most informed conclusions and decisions can be made by the user. Over 60,000 individual entries (specific tests) are covered in the database. Classes of materials covered include thermosets, thermosetting elastomers, thermoplastics, and thermoplastic elastomers. Approximately 700 different trade name and grade combinations representing over 130 families of materials are included. Over 3300 exposure environments are represented.

Records can quickly be grouped by generic family, exposure medium or trade name and grade. In addition, records can be searched, sorted and displayed by exposure temperature, exposure time, exposure medium concentration, and supplier or using the PDL resistance rating. Complete information can then be viewed on any individual record.

1.6   Plascams Computer-Aided Materials Selector

(Access is regulated by user ID and password).
RAPRA Technology Ltd. Shawbury, Shrewsbury, Shropshire SY4 4NR, U.K.
Tel: +44-1939-250-383
Fax: +44-1939-251-118
http://www.rapra.net

The system works interactively with the user to select the best material for the specified application, educating the novice and informing the expert. Users can access definitions of materials, their advantages and disadvantages, compare graphs of flexural modulus vs. temperature, review data sheets and explore materials selection examples. The system is also hyper-linked to complete material supplier information and online help.

The first interactive electronic encyclopedia for users of plastics, materials selection is carried out using 3 search routines. The "Chemical Resistance Search" eliminates materials that cannot meet user specified chemical resistance requirements. The other search routines ("Elimination" and "Combined Weighting") eliminate candidate materials based on 72 properties, falling within one of the following groups: General and Electrical, Mechanical, Cost Factors, Production Methods and Post Processing. All data is evaluated and based on independent tests conducted in RAPRA's laboratories.

1.7   POLYMAT

FIZ CHEMIE BERLIN
Postfach 12 03 37
D-10593 Berlin
Tel: +49 (0)30 / 3 99 77-0
Fax: +49 (0)30 / 3 99 77-134
E-mail: Info@FIZ-CHEMIE.DE
http://www.fiz-chemie.de/en/katalog/

1.7.1 POLYMAT materials data for plastics POLYMAT Materials Data for Plastics contains property values, e.g. mechanical, thermal, electrical, optical, rheological properties and text fields, e.g. special characteristics, preferred applications, preferred processing techniques, additives. In all, 109 numeric and 19 text fields are available. The properties are retrieved in tabular form. The long term behavior of plastics is represented in diagrams, which may also be used in searches. An editor is also available for customizing the database with the user's own data.

- Type: in-house numerical database with editor for the construction of a customized database
- Field: thermoplastics, thermoplastics elastomers, thermosetting resins
- Product form: in-house database with editor function
- Language: German or English
- Content: approximately 12,000 materials from 140 manufacturers, approximately 40 measured properties for each material, approximately 15 product class values per material in the absence of experimental values
- Updates: semi-annually
- Source materials: manufacturers' technical bulletins, handbooks, other specialized literature
- Producers/suppliers: FIZ CHEMIE BERLIN and TDS Herrlich GmbH
- Host: at the beginning of 2000, online available at TDS Herrlich GmbH (www.polybase.com)
- Operating systems: MS/DOS 3.1 or higher, WINDOWS 3.1 or higher
- Remarks: also available-POLYMAT light as a version with less data for less costs with direct access to the full version via internet
- Editor: 50 numeric properties and 15 text fields are available

1.7.2 POLYMAT light

POLYMAT light Materials Data for Plastics is a manufacturer independent, materials database for plastics and contains properties of thermoplastics, thermoplastic elastomers and blends. In total, data from approximately 13,000 commercial products of 170 manufacturers are available; products and data can be retrieved via searching in 35 different numerical properties and 15 text fields.

POLYMAT light can be used for: time saving and comprehensive selection of materials according to a customer's application profile; employment of reasonably priced alternative materials with comparable properties in plastics manufacture; searching for alternative manufacturers in case of delivery problems; comparison of different plastics materials for a single production task; market analyses, e.g. a search for manufacturers producing PA 6 with a content of 30% carbon fibers.

- Type: in-house numerical database
- Field: thermoplastics, thermoplastic elastomers, blends
- Product form: in-house database, CD-ROM
- Language: German
- Content: approx. 13,000 materials of 170 manufacturers
- Updates: semi-annually
- Source materials: manufacturers' technical bulletins, handbooks, other specialized literature
- Producers/Suppliers: FIZ CHEMIE BERLIN and TDS Herrlich GmbH
- Operating system: WINDOWS 3.1 or higher
- User aids: handbook, help functions

1.8 SOFINE: Grade Specific Selection of Plastics Materials

TIMPAS OY Kauppakatu 34
Fin-80100 Joensuu Finland
Tel: 358-(0)500-780 011
Fax: 358-(0)50-8532 7850
E-mail: timo.ture@wanadoo.fr
http://www.plasticsselection.com/
Distributed:

UK, worldwide by RAPRA Technology Ltd., Shawbury, Shrewsbury, Shropshire, SY4 4NR, UK
Tel: +44 (0) 1939 250 383
Fax: +44 (0) 1939 251 118
E-mail: techproducts@rapra.net
http://www.rapra.net

Italy, by EUROCAD.
Via Bottola 3,33070 POLCENIGO (PN), Italy

Tel: +39 0434 749 609
Fax: +39 0434 749 921
E-mail: eurocad@tin.it

France, by CERAP S.A.
27, bd du 11 Novembre 1918, BP2132, 69603
Villeurbanne Cedex, France
Tel: +33 (0) 4.72.69.58.30
Fax: +33 (0) 4.78.93.15.56
E-mail: cerap.ingenierie@wanadoo.fr
http://www.cerap-ingenierie.com

Spain, by Diego Ramon Larios S.L.
Plastics and Consulting.
Avda Montevideo, 68, 08340 Vilassar de Mar,
    Barcelona, Spain.
Tel: +34 93 750 21 90
Fax: +34 93 750 23 70
E-mail: drl-plastics@cambrescat.es
http://www.cambrabcn.es/drl-plastics

Sweden, by Plamako Ab
gatan 13, 33421 ANDERSTORP, Sweden
Tel: +46 (0)371 58 82 80
Fax: +46 (0)371 185 85
E-mail: info@plamako.se

The Netherlands, Germany, by Schouenberg
    & Partners V.O.F.
Burg. Stolklaan 16, 4002 WJ Tiel, The
    Netherlands
Tel: +31 (0) 344 616 161
Fax: +31 (0) 344 631 014
E-mail: info@schouenberg.demon.nl
SOFINE includes detailed technical infor-
mation of more than 11,000 plastics (ther-
moplastics and elastomers); plastic materials
from over 100 producers; continuous updat-
ing of the program; operates in 7 languages.
SOFINE versions available include:

1.8.1 Standard version Search of plastics by name, with technical limitations, price
level, producer, processing method etc., and
search of/comparison of equivalent materials:

- Compares different plastic materials on the
  screen.
- Language: Gives the possibility of choosing the working language for the software (English, French, German, Spanish,
  Italian, Finnish or Swedish).

- Country: Gives the possibility of choosing the country for which the program
  gives the contact information of the pro-
  ducer/distributor.
- Currency: Gives the possibility of choosing
  the currency.
- Gives the possibility of transferring files
  from SOFINE to ACCESS and EXCEL.
- Search: by name, technical limitations, or
  equivalent material.

1.8.2 Interactive version Includes the
possibility of adding and modifying all information:

- Properties: Gives the possibility of modifying, adding or deleting plastic data.
- Agency: Gives the possibility of modifying the list of producers and/or distributors with their addresses, telephone and fax
  numbers.
- Personal properties: Gives the possibility
  of adding own properties.

1.8.3 Tailor-made version Made according to the needs and demands of the customer:

- Versions are produced for customers with
  customized possibilities and specific data,
  which can also be updated at agreed
  intervals.
- Links to customer's own database on
  request.

1.9 RUBSCAMS

RUBSCAMS Computer-aided Materials
    Selector for Elastomers
RAPRA Technology Ltd. Shawbury,
    Shrewsbury, Shropshire SY4 4NR, U.K.
Tel: +44-1939 250 383
Fax: +44-1939 251 118
http://www.rapra.net
Rubacams is a computer aided materials
selection routine for elastomeric materials.
Covering 99 generic types of rubber, each
material is cross referenced with over 190
chemical agents and materials property data
including physical, chemical mechanical and
process related properties. Search results

identify the most suitable materials, a detailed description of the elastomer and supplier details.

2. Hard-Copy Data Sources

2.1 Polymers and Elastomers

Polymer and elastomer data are typically producer-specific. Sources therefore tend to be scattered and incomplete. While no single hard-copy source is all encompassing, the following are worth consulting.

2.1.1 "Modern plastics world encyclopedia", A Chemical Week Associates Publication (800-525-5003) updated annually. Includes primer type descriptions and a global listing of key plastic resin and compound properties for a wide range of material grades based on filler and additive content plus primary processing method and supplier; auxiliary equipment and components, fabricating and finishing also covered. Extensive Buyer's Guide included.

2.1.2 "Handbook of Plastics Materials and Technology", Irvin I. Rubin; John Wiley & Sons; (1990); ISBN: 0471096342. Essential information from acetal to XT polymer. This single source comprises 119 chapters of in-depth basic information about plastic materials, properties, processing, assembly, decorating and industry practices–all presented in a readily accessible and consistent format. Also features a wealth of useful auxiliary information and tables.

2.1.3 "Plastics technology manufacturing handbook and buyers guide", Bill Communications (212-592-6570). Updated annually. The Handbook and Buyers' Guide is a comprehensive tool for locating suppliers of primary machinery, materials (thermoplastics and thermosets), auxiliary and secondary equipment and controls, chemicals and additives, and a variety of specialized services. Contains extensive equipment and materials specifications.

2.1.4 "Performance of plastics", W. Brostow; Hanser Gardner Publs; (1999); ISBN: 1569902771. Comprehensively covers the behavior of the most important polymer materials. Subject areas range from Computer Simulations of Mechanical Behavior to Reliability and Durability of aircraft structures made of fiber-reinforced hydrocarbons.

2.1.5 "Saechtling international plastics handbook: for the technologist, engineer & user", 3rd edition, Dr. Hansjurgen Saechtling; Hanser Gardner Publs; (1995); ISBN: 1569901821. Very comprehensive hard-copy data-sources for polymers. Covers key facts about the plastics industry, from basic materials and theoretical concepts to manufacturing, with detailed descriptions of individual plastics, their properties and applications. Contains more than 100 tables of plastics properties and plastics data in ASTM, ISO and DIN standards. Also includes a buyer's guide.

2.1.6 "Plastics for engineers: materials, properties, applications", Hans Domininghaus; Hanser Gardner Publs; (1993); ISBN: 1569900116. Provides a comprehensive overview in text, tables and graphs, of properties and applications for all plastics of current technical and commercial interest.

2.1.7 "The plastics compendium–volumes 1 and 2", ISBN: 18599570585

- **"Volume 1: Key Properties and Sources"**, M.C. Hough, & R. Dolbey; Rapra Technology Ltd.; (1995). Volume 1 contains data on 351 generic and modified material types. Information provided includes property and commercial data sheets covering: advantages and disadvantages; typical applications; materials data listing values of 24 key properties; and source data listing suppliers and their trade names in the USA and Europe.

- **"Volume 2: Comparative Materials Selection Data"**, M.C. Hough, S.J. Allan, & R. Dolbey; (1999). Volume 2 provides comparative materials selection data for the 351 thermoplastic and thermosetting materials covered in volume 1. Each material has been assigned one of six ranking values for each of 62 properties ranging from excellent to not applicable. The information is based upon the numerical rankings contained within Rapra's PLASCAMS (see 1.6).

2.1.8 "Handbook of plastics, elastomers and composites", 3rd edition, Charles A. Harper; McGraw-Hill; (1996); ISBN: 007026693X. This comprehensive source of at-a-glance plastics design data includes property and performance data; application guidelines; costs; joining techniques; fabrication method trade-offs; processing procedures for laminates and reinforced plastic materials; protective and decorative coatings; advanced composite materials; liquid and low-pressure resin systems; thermoplastic elastomers. Treatment of the chemical, mechanical, and electrical properties of plastics, elastomers, and composites; gives complete coverage of plastic compositions and optimizations of plastic product design; advances in thermoplastic elastomers; new developments in applying and processing advanced composite materials; plastics and elastomers for high-volume, high-performance automotive and packaging applications; important factors in the recycling of plastics.

2.1.9 "ASM engineered materials handbook", Vol 2. engineering plastics, ASM; (1988); ISBN: 0871702800. ASM International, 9639 Kinsman Rd. Materials Park, Ohio 44073-0002 (www.asm-intl.org). Engineering Plastics is designed and written for working engineers. The book opens with general design considerations. The volume's Guide to Engineering Plastics Families describes 40 major engineering plastics families, 29 thermoplastics and 11 thermosets. Content includes typical costs, major applications, competitive materials, significant characteristics, performance properties, design and processing considerations, and major suppliers. Other sections include manufacturing process considerations, such as

function and properties requirements, size, shape and design detail considerations, and surface requirements; properties modification by polymer/polymer mixtures and use of additives; and properties of thermoplastic structural foam. Structural Analysis and Design covers the use of engineering formulas.

2.1.10 "International plastics selector", 9th edition, Int. Plastics Selector, San Diego, CA; (1987). Thermoplastics, thermosets, elastomers, and key property areas critical to plastics are extensively specification defined.

2.1.11 "Pocket specs for injection molding", 4th edition, (IDES) (http://www. idesinc.com/Products_1.htm). Quick reference processing guide. Released in January 1999, this newest edition has the most current processing information available in a book. Pocket Specs covers 13,000 injection moldable materials and 15 key processing properties and provide a compact guide for the injection molding of thermoplastic and thermoset materials. Data is provided for individual grades of molding materials from more than 130 manufacturers. The data, provided in tabular from, gives basic information for determining regrind levels, material drying temperatures and times, and initial machine settings for injection pressure, barrel heats, and mold temperature. Additional physical property data includes specific gravity, shrink data, melt flow, and processing temperature ranges.

2.1.12 "Pocket performance specs for thermoplastics", 1st edition, (IDES) (http:// www.idesinc.com/Products_1.htm). Over 13,000 thermoplastic materials from more than 100 manufacturers with 15 different engineering properties make this book the ideal "take anywhere" partner for the plastics industry. This book contains the information needed for quick, accurate, and convenient design and materials information, providing a basic guide to selecting thermoplastic materials. The information in the tables is intended to give you the basic information for determining   the   general   performance characteristics of a plastic material in order to screen for candidate plastic materials.

2.1.13 "Handbook of elastomers", A.K. Bhowmick and H.L. Stephens; Marcel Dekker; (1988); Series: Plastics Engineering, Volume 19; ISBN: 0824778006. This handbook systematically addresses the manufacturing techniques, properties, processing, and applications of rubbers and rubber-like materials. The Handbook of Elastomers provides authoritative information on natural rubbers, synthetic rubbers, liquid rubbers, powdered rubbers, rubber blends, thermoplastic elastomers, and rubber-based composites—offering solutions to many practical problems encountered with rubber materials.

2.1.14 "Technical data sheets", Malaysian Rubber Producers Research Association, Tun Abdul Razak Laboratory, Brickendonbury, Herts. SG13 8NL (1995). Data sheets for various blends of natural rubber.

2.2 All Materials

The five hard-copy data-sources listed below attempt in different ways to span the full spectrum of materials and properties.

2.2.1 "ASM engineered materials reference book", 2nd *edition*, Michael L. Bauccio., ASM International; (1994); ISBN: 0871705028; (www. asm-intl.org). Compact compilation of numeric data for metals, polymers, ceramics and composites. This is an excellent reference for persons involved in nonmetallic materials selection, design, and manufacturing. Sections include:

- Composites (fibers, fillers, and reinforcements, design, tooling and manufacturing)
- Ceramics (single and mixed oxides, carbides, nitrides, borides, glasses, and traditional ceramics)
- Plastics (thermoplastics, thermosets, and production and machining)
- Electronic Materials (properties devices and manufacturing methods)

2.2.2 The "CRC-Elsevier materials selector", 2nd *edition*, N.A. Waterman, and M.F. Ashby; CRC Press; (1996); ISBN: 0412615509. (Now, also available on CD-ROM). Basic reference work. Three-volume compilation of data for all materials; includes selection and design guide. The Materials Selector is the most comprehensive and up-to-date comparative information system on engineering materials and related methods of component manufacture. It contains information on the properties, performance and processability of metals, plastics, ceramics, composites, surface treatments and the characteristics and comparative economics of the manufacturing routes which convert these materials into engineering components and products.

- **Volume 1** addresses the initial stages in solving a materials selection problem, provides the background to all aspects of materials behavior, and discusses manufacturing processes.
- **Volume 2** details the performance of metals and ceramics.
- **Volume 3** covers the performance of polymers, thermosets, elastomers, and composites.

2.2.3 "Handbook of industrial materials", 2nd edition, I. Purvis, Elsevier; (1992); ISBN: 0946395837. A very broad compilation of data for metals, ceramics, polymers, composites, fibers, sandwich structures, and leather. Contents include:

- *Ferrous Metals*: Cast iron, carbon steel, BS970, replacing en steel, alloy steel, spring steel, and casting steel.
- *Non-ferrous Metals and Alloys*: Diaphragm material, metal composite, refractory metal.
- *Non-metallic Materials*: Carbides, carbon, ceramic fiber, ceramic, cermet, composite, cork, elastomer, felt, fiber, glass, glycerin, non-metallic bearing material, rubber (natural), rubber (synthetic), silicone, wood, leather.
- *Thermoplastics*: ABS, acetal and polyacetal, acrylic (methyl methacrylate), cellulose plastic, EVA, fluorocarbon, PTFE,

Ionomer, methylpentene, (TPX), PBA, PETB, polyisobutylene (PIB), nylon (polyamides), polyethylene, polyether-sulphone, polypropylene oxide (PBO), polystyrene, PVC, polyvinylcarbazole, SAN, PBT (thermoplastic polyester), polycarbonate, polymers, polypropylene, POM.

- *Thermoset Plastics*: Alkyd, amino resin, thermosetting acrylic resin, casein, epoxy, phenolic, polyester, polyamide, silicone.
- *Other*: Laminated plastic (industrial laminate), sandwich molding, 'filled' plastic, cellular plastic, glass reinforced plastic (GRP), carbon fiber reinforced plastic (CFRP).

2.2.4 "Materials Handbook", 14th edition, George S. Brady, Henry R. Clauser and John Vaccari; McGraw Hill; (1996); ISBN: 0070070849. Covers metals, ceramics, polymers, composites, fibers, sandwich structures, leather. This one-volume encyclopedia of materials, known simply as "Brady's" and published since 1929, is now in its 14th edition. This unique tool provides a one-stop source of comprehensive information on virtually every material and substance used in industry and engineering.
The Fourteenth Edition gives you:

- A-to-Z organization for easy access;
- Coverage of more than 13,000 materials;
- Details on chemicals, metals, minerals, fuels, plastics, textiles, finishes, woods, elastomers, ceramics, coatings, composites, industrial substances, and natural plant & animal substances;
- Entries on new materials, including recyclate plastics, fullerenes, hard-surfaced polymers, dendrimers, transflective materials, rapid prototyping materials, silicone nitride, supercritical fluids, bulk molding compounds, conversion coatings, folic acid, replacements for chloro-fluorocarbons;
- Properties and characteristics of materials, including composition, production methods, uses, and commercial designation or trade names

2.2.5 Materials selector", Materials Engineering, (now Advanced Materials and

Processes), (ASM), Special Issue; Penton Publishing; (1994). Basic reference work-up dated annually. Tables of data for a broad range of metals, ceramics, polymers and composites.

3. Process Simulation Software

3.1   Moldflow (and C-Mold, a Division of Moldflow)

Moldflow Corporation, 91 Hartwell Avenue, Lexington, MA 02421 USA
Phone: +1-781-674-0085
Fax: +1-781-674-0267
http://www.moldflow.com
http://www.cmold.com/

3.1.1   Dr. C-Mold   Molding intelligence for plastic professionals. Easy to use for quick and dynamic evaluation of critical design and manufacturing variables. Evaluate part and mold design, processing conditions, and material options. No CAD model, meshing, or special training is necessary.

3.1.2   C-MOLD advanced solutions   C-MOLD Advanced Solutions are designed for dedicated simulation users who need in-depth predictions for all phases of design, manufacturing, and resulting part quality. Advanced simulation products cover a wide range of thermoplastic and reactive molding processes, including injection molding, gas-assisted injection molding, co-injection, injection-compression molding, rubber injection molding, reaction injection molding, structural reaction injection molding (SRIM), resin transfer molding (RTM), and microchip encapsulation. Advanced Solutions address all aspects of product and mold design, molding process conditions, and part quality. Evaluate part thickness, part size, gate placement, cooling system placement and efficiency, optimize process conditions, evaluate resulting part size, shape, and structural integrity.

3.1.3   Desktop   products   C-MOLD Desktop Products are designed to meet the needs of design, tooling, manufacturing, and process engineers who may not have access to advanced simulation capabilities or who want to get simulation feedback with high accuracy, but without the time requirements to run an advanced simulation.

- **Desktop tools** address design and manufacturing concerns such as gate placement, injection time/rate, injection pressure, melt and mold temperatures, packing time and pressure, cooling time requirements, and machine size requirements.
- **Project Engineer** uses numerical input values of maximum flow length, nominal wall thickness, and projected area to describe the part geometry.
- **3D QuickFill** uses STL-format, solid-model geometry to show geometry-specific simulation results on the solid part model.

3.1.4 Moldflow plastics advisers Moldflow Plastics Advisers are used in the early stages of part and mold design when the cost of change is minimal and allows the designer to take control of early part and mold design optimization to eliminate potential manufacturing problems downstream. Injection molded plastics parts can be designed for "manufacturability" at the same time as form, fit and function. **Part Advisor** provides automated tools to help the designer optimize a part before the mold is cut. **Mold Adviser** allows mold designers to easily layout and optimize the gate and runner systems for single cavity, multi-cavity or family molds as well as predict clamp tonnage, shot size and cycle time requirements—all during preliminary design and before the part geometry is finalized. Companies that identify and eliminate problems at early or conceptual stages of design, achieve significant benefits, via time and cost savings and the capture of timely market opportunities.

3.1.5 Plastics insight (In-depth analysis of plastics part and mold designs) To undertake in-depth validation of part and mold designs prior to manufacture. Plastics Insight is an integrated suite of CAE analysis software that makes it possible for plastics part design, mold design, and machine processing conditions to be optimized during

the design stage, saving time and money. Plastics Insight solves complicated injection molding problems for all geometry types, thin or thick, simple or complex, that are encountered during filling, packing and cooling, as well as warpage problems. Moldflow Plastics Insight works with all CAD model geometry types including wire frame and surface models, thin-walled solids and thick or difficult-to-midplane solids. Moldflow Plastics Insight products can simulate plastics flow, mold cooling, part warpage, stiffness and shrinkage, and the behavior of fiber-reinforced materials in plastics. In addition, MPI includes products that simulate the gas and thermoset injection molding processes.

3.1.6 Moldflow plastics xpert Moldflow Plastics Xpert injection molding simulation enables process engineers and molders to quickly optimize machine set-up, reduce cycle times, and monitor and correct molding processes during production by providing shop-floor solutions to the problems associated with injection molding machine setup, process optimization and production part quality. While no combination of software and hardware can transform a bad design into good parts, Plastics Xpert quickly differentiates between process-related problems and inherent design problems. Integrated with the molding machine's controller, MPX provides real-time process optimization and feedback that help remedy the production problems. Plastics Xpert also has a remote control capability that allows process optimization and monitoring to be done away from the actual molding machine.

Plastics Xpert is comprised of three Xpert Systems:

- **The Setup Xpert:** Provides an intuitive, systematic, and documentable method for establishing the combination of process parameters that produce good molded parts.
- **The Optimization Xpert:** Automated design of experiments that builds on the foundation established in Setup Xpert and allows users to further optimize the combination of processing parameters to determine a robust "good parts" processing window.

- **The Production Xpert:** A comprehensive production monitoring and control system that will maintain the optimized processing conditions determined with MPX's automated design of experiments.

3.2 MSC.Mvision

900 Chelmsford Street, Lowell, MA 01851-8103
Tel: 800 642-7437/978 453-5310 × 2551
Fax: 978 454-9555
http://www.mechsolutions.com/products/mvision/index.html
MSC.Mvision provides materials information for predictive engineering, ensuring consistent data for engineers evaluating new designs and reducing cycle time by integrating materials data directly into CAD/CAE. The software enables MSC customers to integrate internal materials test information and published materials data directly with their processes. With MSC.Mvision, companies automate the flow of materials data from test through design and analysis, maintaining clear audit trails, thereby ensuring:

- Increased efficiency in the design process
- Reduction in product development and support costs
- Faster and more innovative designs
- More representative analyses
- Consistent usage of materials data
- Reduced materials testing requirements
- Increased confidence in use of materials data
- Fewer inappropriate materials selections, and thus
- Fewer redesigns and warranty recalls

MSC.Mvision provides integrated access to materials information from within the MSC.Patran and Pro/ENGINEER environments, and generates formatted input data for MSC.Nastran and other analysis programs. Customers can also integrate MSC.Mvision readily within their proprietary computer aided engineering environments. MSC provides "off the shelf" materials databanks developed and maintained with authoritative Partners including:

- Battelle Memorial Institute
- University of Dayton Research Institute
- Plastics Design Library, from William Andrew, Inc.
- Materials Sciences Corporation
- GE Plastics
- Penton Publishing
- Information Handling Services, Inc.
- ASM International

3.3 The Madison Group

505 S. Rosa Rd., Suite 124; Madison, WI 53719-1257
Tel: 608-231-1907
Fax: 608-231-2694
E-mail: tmg@prowler.madisongroup.com
http://www.madisongroup.com/Products/products.html
The Madison Group: Polymer Processing Research Corporation was incorporated in 1993 by University of Wisconsin-Madison researchers to permit technology transfer from academia to industry. Several simulation packages for the polymer processing industry have been developed to help various industry design plastic parts and solve processing problems. The Madison Group offers a wide range of software solutions from commercially available packages to custom software development.

3.3.1 Cadpress-SMC (Thermoset compression molding simulation) Cadpress-SMC is a general purpose finite element compression molding simulation program which calculates the mold filling, pressure and velocity distributions, fiber orientation, anisotropic material properties, curing behavior, and the shrinkage and warpage of the final part. Cadpress, developed over the past two decades, is a finite element based simulation package that has become the standard for the compression molding industry. The software simulates the entire molding process for fiber reinforced thermoset compression molding, from mold filling to prediction of residual stresses and warpage of the final part. Through accounting for fiber orientation, it also predicts

the anisotropic material properties of the final part and can be interfaced to other simulation packages, e.g. ANSYS. Cadpress allows engineers to design parts for optimal strength with minimal shrinkage and warpage.

3.3.2 Cadpress-BMC (injection/compression molding simulation) Cadpress-BMC, purchased as an add-on module to the basic Cadpress molding software, was developed to simulate the multiphases of the injection/compression molding process of thermoset molding. With this module, the injection/compression molding process of Bulk Molding Compound (BMC) or vinyl esters can be simulated. The software can calculate the mold filling during the injection phase, the initial fiber orientation after injection, the filling during the subsequent compression phase, and the final fiber orientation as well as shrinkage and warpage of the finished part. The add-on BMC module allows the injection phase of flow to be calculated, and also predicts the initial fiber orientation in the part before compression. Differences between the standard Cadpress and the add-on BMC module are mainly contained in the calculation routines, with minimum user-perceivable changes to the front end of Cadpress. Except for the additional injection inputs, the user interface remains identical to Cadpress.

3.3.3 Cadpress-GMT (Glass-mat thermoplastic compression molding simulation) Cadpress-GMT (Glass-Mat Thermoplastic Simulation Package) developed at the IKV (EXPRESS) simulates the complete compression molding process for fiber reinforced thermoplastic compounds. Using Cadpress-GMT during the design process helps to reduce costs, shorten development times, minimize changes on the final mold and to incorporate design improvements at a very early stage. Express is based on the Finite-Element Method. It is embedded in the CAE design process and offers interfaces to a variety of familiar CAE pre- and postprocessors, including I-DEAS, Patran, and COSMOS.

3.3.4 FiberScan (In situ fiber detection system) Fiber orientation caused during processing of composites has a significant influence on final mechanical properties of a reinforced part. Though simulation programs can predict fiber orientation in a molded part, it is much more difficult to determine the actual fiber orientation experimentally. Traditional burn-out tests can show low fiber content regions or knitlines but the actual fiber orientation is nearly impossible to determine. FiberScan provides a quick, efficient method to experimentally determine the fiber orientation field in a molded part. With any of the following inputs, FiberScan can determine the resulting fiber orientation field from an as-molded part: X-ray of the specimen; photograph of transparent samples; photographs of burnt samples. FiberScan uses image processing technology to analyze the digital input and determine the distribution of fibers in a sample. Using algorithms to increase contrast, x-rays may be analyzed without the need to use lead impregnated fibers. It is also possible to use FiberScan for in situ measurement during production for quality assurance purposes.

3.3.5 DSCfit (Differential scanning calorimetry curve fitting software) For simulation programs to correctly describe the behavior of thermoset polymers during processing, the curing process must be well understood and described in a manner consistent with numerical methods. Programs such as Cadpress and Cure3D (see 3.3.9) characterize curing behavior using empirically based models that relate heat release to degree of cure. To use such models, experimentally obtained DSC data must be properly fit with a rigorous model. DSCfit is a utility program developed to provide the capability of fitting experimental data with the appropriate numerical model.

Because most thermoset composites cure by a thermally activated reaction, a complicated heat transfer process occurs during solidification, the result of an exothermic cross linking reaction in the resin. The complications of thermoset resin curing are compounded by the competing mechanisms of chemical kinetics and molecular diffusion. A further complication is the fact that numerous reactive processes occur as the SMC cure process involves free radical chain growth with the 3 stages of initiation, propagation, and termination. While this chemical process has been well studied and characterized, taking into account the functions of initiators, inhibitors and monomers in SMC, the resulting models are too complex for practical applications. A more convenient method relies upon empirical kinetic models based on the work of Kamal and Sourour who developed an autocatalytic model to represent the exothermic reaction of various molding compounds. The six constant model can directly fit to DSC data for a particular resin formulation.

3.3.6 Application database (custom database applications; Controlling data for design decisions) In the plastics industry, material selection is usually made by very experienced engineers. These specialists use their expertise to make material selections in a heuristic and unstructured manner. This knowledge is fixed to the individual person and is lost when that person leaves the company. A more reliable method is to make the relevant data available in a unified, structured application that is useable by many decision makers. Such an application databank encompasses the critical design information from previous applications along with pertinent material properties. Information describing the functionality of the application, abstractions that relate application properties and useful graphics can be combined in an integrated and searchable way to be useful to the designer or engineer.

The material data base CAMPUS has led to reliable and comparable quantitative values for material selection and design, making the design process more systematic and objective. However, it is still necessary to use application experience in design decisions, particularly when considering material properties that cannot be easily quantified. Among these, is the impact strength of a component relative to the geometry, load conditions, and material properties. While impact strength of a material is measurable and readily available, using this property in design of a new component is not readily accomplished. In addition, end-use properties such as surface quality and optical properties, not linked to a specific material property, cause further problems for systematic design.

The Madison Group and its European partner, M-Base, have developed, based on years of research and experience with material data systems, the concepts and software for the management of such application databases. This application database is searchable by part and application. Capability for general component information, multiple classifications, images and text and links to material properties is included.

Many successful projects have shown that application information can be divided into the following four categories: Terminology, Special Characteristics, Abstract Functions, and Graphics. Although these categories are the basis for the searchable application data base, each project requires a conceptual phase to define how to focus.

- *Terminology:* Components and applications must be handled with consistent and correct names in order to allocate and search through them in a repeatable manner.
- *Special Characteristics:* Information that describes the functionality of the application needs to be handled in a specific manner. This information can include such properties as size, weight, geometric form, special functional elements, processing and assembly methods, and other product specific information.
- *Abstract Functions:* An important aspect of an application database is to support the user in finding analogies between old, well known cases and new design ideas. To make these analogies independent of the imagination and experience of a single user, an abstract description of all functions is recommended. In this case, elemental functions, as they are known from design theory, should be used. Based on such abstract descriptions, the software can help find relations between applications.

- *Graphics:* In most cases a graphical presentation of each application is necessary. Here, it is possible to store photographs, sketches, CAD drawings, computer simulation results, and experimental results for use in the database.

The function of the database can be further enhanced by the inclusion of links to actual material data for the component. Using such existing material database tools as MCBase, the user can quickly look-up the material properties of a component with the simple click of a button.

3.3.7 BEMflow (boundary element simulation—extrusion mixing)
BEMflow is a boundary element simulation package for designing, optimizing, and analyzing polymer processes and equipment. The boundary-only approach of the boundary element method (BEM) makes it attractive for modeling flows in extruders, mixing heads, extrusion dies, internal batch mixers, among others. The boundary element method is founded on rigorous mathematical theory that reduces the dimensionality of the problem. Now optimization can be done on complex 3-dimensional geometries in a realistic time frame on desktop computers.

3.3.8 MiniFlow (the essential tool for injection molding)
MiniFlow is a tool used to simulate the mold filling stage of injection molding. Using material models to fully describe the flow and heat transfer during mold filling, MiniFlow is capable of predicting the flow length and cooling for thermoplastics. This aids the designer/engineer during the critical first steps of material and machine selection in the design stage. This software provides for the selection of polymers, machines, or molding conditions. Generic polymers, specified by manufacturer, or user defined, can be selected. The material parameters for the selected polymer are automatically used for any calculation. Pressure and flow parameters are automatically used for calculations based on injection molding machine selection from a manufacturer data bank or other injection molding machines added into a user defined data bank.

3.3.9 3-D curing (3D thermoset molding simulation)
When molding with thermoset resins, there often arise problems with cycle times or excessive residual stress in the final part. These parts can have non-uniform temperature distributions due to the heat generated during curing, which lead to non-uniform curing and unacceptable residual stress fields. Traditional methods of experimentally modifying molding conditions to yield an acceptable part, are time consuming and expensive. Additionally, the acceptable molding conditions may lead to unacceptably long cycle times.

Cure3D is used to alleviate many of the problems with molding thick sections out of thermoset materials. Through the use of the finite element method, a new design can be evaluated for problems during processing and then modified as appropriate to give both good material properties and improved processability. As three dimensional geometry of the part is considered, accurate solutions for the temperature, cure, and residual stress field are obtained from the simulation model.

Typical thermoset polymeric reactions are exothermic in nature and can release a great deal of heat during cure. The relatively low thermal conductivity of the polymer causes heat to be stored in the polymer and for the bulk temperature to rise. Since most systems are also temperature dependent, the added heat causes an increase in the rate of reaction and a compounding effect on the temperature. The process is further exacerbated in thick sections, where the heat rise can actually exceed the degradation temperature of the polymer system and cause dramatic decreases in the mechanical performance of the material. The problem is further compounded because the evolution of mechanical properties is directly dependent upon the degree of cure of the material. As the material cures, residual stress begins to be formed and built into the structure. If the cure does not progress in a uniform manner, gradients in the stress can result and lead to gross deformation of the finished part. Cure3D solves the coupled equations of energy and stress to accurately determine the temperatures, degree of cure, residual stress, and deformed shape of the component during the molding process.

3.4   SDRC Solutions

SDRC World Headquarters, 2000 Eastman
   Drive, Milford, OH 45150-2400
Tel: 513-576-2400
http://www.sdrc.com/

3.4.1   I-DEAS® master series   SDRC's mechanical design automation (MDA) software, used by manufacturers for the design, analysis, testing, and manufacturing of mechanical products.

3.4.2   Metaphase®   Metaphase offers a Web-centric information infrastructure that harnesses its customers' intellectual capital to drive product innovation and manage the complete product life cycle.

3.4.3   Imageware   Imageware is SDRC's advanced 3D surface modeling and verification technology for the automotive, aerospace, and consumer products industries.

3.4.4   FEMAP and VisQ   With FEMAP, users can define analysis models, integrate the appropriate solver technologies and review, interpret and document their results quickly and efficiently. VisQ speeds and automates the batch solution process by integrating remote computer servers across networks and the Internet with FEMAP on the engineer's desktop.

3.4.5   Product catalog   Organized to help you find the mechanical engineering and data management software tools you need to boost the productivity of individual designers, drafters, analysts, test engineers, and manufacturing engineers, as well as the entire product development team.

3.4.6   Experteam(SM) services   SDRC's ExperTeam services—a world-class engineering organization—provides integration, customization and implementation services, training, and support to help facilitate engineering processes.

3.4.7   Aerospace center of excellence   The SDRC Aerospace Center of Excellence, based in SDRC's San Diego, California facility includes over 30 engineers with expertise in aerospace product development methods. The Center offers customers a process for enabling significant gains in efficiency, productivity, reliability, and overall quality in their product development operations.

3.5   Finite Element Analysis Software

3.5.1   ANSYS, Inc.
Southpointe
275 Technology Drive
Canonsburg, PA 15317
E-mail: ansysinfo@ansys.com
http://www.ansys.com
Tel: 724. 746.3304/800 937-3321
Fax: 724.514.9494
Toll Free Mexico: 95.800.9373321
ANSYS, Inc. has developed two product lines that allow you to make the most of your investment. DesignSpace® gives you access to CAD models through an intuitive, consistent interface while ANSYS® provides the functionality you need to create state-of-the-art, high quality products and the flexibility to work with other CAD software. These two product lines allow users to choose which product works best in their environment. Combining the power of ANSYS analysis with DesignSpace CAD integration establishes a truly collaborative engineering environment for companies to optimize their product designs and internal processes.

3.5.2   Algor, Inc.
150 Beta Drive
Pittsburgh, PA 15238-2932
Phone: +1 (412) 967-2700
Fax: +1 (412) 967-2781
Europe (UK): +44 (1784) 442 246
Algor Publishing Division phone number: (1-800-482-5467)
Information: to learn more about Algor's complete line of CAD/CAE interoperability,

finite element modeling, FEA and Mechanical Event Simulation products, E-mail: info@algor.com for further information on Algor or any of Algor's products.

Tel: +1 (412) 967-2700

Fax: +1 (412) 967-2781

Europe (UK): +44 (1784) 442 246

Algor Publishing Division phone number: (1-800-482-5467)

Algor Finite Element Analysis and other training through Live Webcasts. For webcast schedules see website page http://www.algor.com/webcast/training.htm

3.5.3   Noran Engineering, Inc.

NE/NASTRAN

5182 Katella Ave., Suite 201

Los Alamitos, CA 90720.

Tel: (714) 895-5857

E-mail: info@noraneng.com

Web address: www.nenastran.com

3.5.4   FEAMAP

Free FEMAP 300-Node Demo software from Enterprise Software Products, Inc.

By downloading the basic FEAMAP with Parasolid and ACIS add-ons, the demo license FEAMAP is enabled as a full FEMAP Professional.

http://sai-mtab.com/software/download.htm

3.5.5   STARDYNE

Research Engineers, Inc.

22700 Savi Ranch Pkwy

Yorba Linda, CA 92887

General Inquiries

Email: info@ca.reiusa.com

Ftp Site: ftp.reiusa.com

Phone: (714) 974-2500

Fax: (714) 974-4771

http://www.reiworld.com/

http://www.reiusa.com/sdyn/sdyn0.htm

As the world's first commercially available Finite Element Analysis software, STARDYNE has been at the forefront of technology since 1967. Its comprehensive array of Finite Element capabilities allows the engineer to perform in a wide variety of fields—from space vehicles to missiles to nuclear power plants to sophisticated machinery. Extremely reliable and easy-to-use, STARDYNE offers Linear/Nonlinear Static, Dynamic, Seismic, Buckling, Heat transfer, Fatigue, Fracture analysis; efficient graphical modeling and result verification.

3.5.6   Structural research & analysis corporation

Developers of COSMOS/^{TM} Applications

U.S. Headquarters 12121 Wilshire Blvd. 7th Floor

Los Angeles, CA 90025

Phone: 310.207-2800

Fax: 310.207-2774

E-mail: info@srac.com

http://www.cosmosm.com

Eastern Regional Office

Developers of COSMOS/^{TM} Applications

5000 McKnight Road, Suite 402

Pittsburgh, PA 15237

Phone: 412.635-5100

Fax: 412.635-5115

E-mail: info@srac.com

European Headquaters

Developers of COSMOS/^{TM} Applications

PO Box 98 Ashford, Kent TN24 9WZ

England

Phone: +44-0-1233 642104

Fax: +44-0-1233 642106

E-mail: bob@srac.com

3.5.7   FEMur: Finite element method universal resource   Introduction to FEMur

Learning Modules for the Finite Element Method (FEMur-LAM)

Interactive Learning Tools for Finite Element Method (FEMur-CAL)

Finite Element Resources on WWW

FEMur Developers

Financial, Hardware, and Software Supporters

http://femur.wpi.edu/main_menu.html

3.5.8   Internet finite element resources

Lists public domain and shareware programs, a selection of pointers to commercial packages, and other finite element resources.

http://www.engr.usask.ca/~macphed/finite/fe_resources/fe_resources.html

3.6   Other Plastics Design Computer Software

3.6.1 "Computer aided analysis of stress/strain behavior of high polymers", 2ⁿᵈ edition, Technomic Publs. The Book provides a review of the fundamentals of polymer viscoelasticity, the measurement of viscoelastic properties, and the use of these measurements in a constitutive model. The model is incorporated into the Computer Simulator for predicting the behavior and performance of polymeric materials during processing. Examples of the use of the Computer Simulator in specific applications are included. The software provides for the creation of rheology models from dynamic modulus data to simulate real polymer processing situations. The book and software provide the link between fundamentals and practical applications in the design of polymer manufacturing processes as well as for insight into the enduse performance of polymer products. See http://www.4spe.org for availability.

3.6.2 "Computer program for formed material cost comparisons", developed by Rohm & Haas Co., and distribution rights given to the Thermoforming Div. SPE. This program provides a direct cost comparison of up to 5 resins (ABS, HDPE, HIPS, PETG and PP). Three programs are included; the 1ˢᵗ computes costs for thermoformed parts, the 2ⁿᵈ adds extrusion to the cost comparison. The 3ʳᵈ program is a metric version of the thermoforming/extrusion program. See http://www.4spe.org for availability.

3.6.3 Design of foam-filled structures (with PC Disk)", 2ⁿᵈ edition, John A. Hartsock; Technomic Publs; (1991); ISBN: 0877627452. A comprehensive guide and reference for the structural design of foam filled panel systems. Combines the theory and the calculations to enable an engineer to design the foam filled building panels used for their thermal insulation, load carrying ability, and ease of erection. Also briefly considers other foam filled products. The second edition has been expanded to include information on multiple spans and design values. The PC disk

is provided to save reader time in solving equations developed in the text.

3.7   Internet Collaboration Tools

(Software designed specifically for collaborating online)

3.7.1 CollabWare GS-Design is a solids modeling 3D CAD system specifically designed to work over the Internet. Collabware™ is web based software for the Collaborative Engineering and Design community. Software provides design teams around the world with the ability to design and manage product development using custom tools accessed through the familiar internet browser interface. http://www.collabware.com

3.7.12 Viewcad.com A free CAD sharing site allows users to share CAD drawings securely with anyone in the world. The site provides a simple method for publishing and sharing CAD drawings on the Internet. The site is made available by Arnona Internet Software, provider of Internet solutions for the CAD market. A standard account includes space for 5 drawings and the CADViewer Light viewer is free. Account access is restricted by password protection. http://www.viewcad.com

3.7.3 Parametric technology corporation-windchill Windchill provides a collaborative environment for the sharing and visualization of product and process knowledge. Information can be accessed through a Web browser and used to identify, visualize, and markup models, providing engineers fast and accurate responses to inquiries. http://www.ptc.com/products/windchill/index.htm

4.   Plastics Design Books

4.1   Plastics Design Reference Books

4.1.1 "Designing with plastics: based on material & process behaviors", Donald V. Rosato, Marlene G. Rosato, and Dominick V. Rosato, Kluwer Academic Publishers (2000).

This book provides a simplified and practical approach to designing with plastics that fundamentally relates to the load, temperature, time, and environment subjected to a product. It will provide the basic behaviors in what to consider when designing plastic products to meet performance and cost requirements. Important aspects are presented such as understanding the advantages of different shapes and how they influence designs.

Important are behaviors associated and interrelated with plastic materials (thermoplastics, thermosets, elastomers, reinforced plastics, etc.) and fabricating processes (extrusion, injection molding, blow molding, forming, foaming, reaction injection molding, etc.). They are presented so that the technical or non-technical reader can readily understand the interrelationships.

The data included provides examples of what are available. As an example static properties (tensile, flexural, etc.) and dynamic properties (creep, fatigue, impact, etc.) can range from near zero too extremely high values. They can be applied in different environments from below the surface of the earth, to over the earth, and into space.

This comprehensive resource recognizes that effective design is an interdisciplinary process involving the ability to match application situations with techniques and to develop problem-solving methods to fit the specific design requirements. Detailed chapters' cover processing methods available for manufacture and effective techniques for evaluating plastic properties and applying quality control. Coverage of the complete design cycle also explores:

- Unique plastic performance capabilities and adaptabilities
- Selecting the right plastic or composite for the end use product
- Optimizing material performance during processing
- Maximizing product cost performance with flexible production procedures
- Effective problem analysis to minimize production difficulties

4.1.2 "Injection molding handbook, third edition", Donald V. Rosato, Marlene G.

Rosato, and Dominick V. Rosato, Kluwer Academic Publishers (2000). This third edition has been written to thoroughly update the subject of the Complete Injection Molding Operation in the World of Plastics. By updating the book, there have been changes with extensive additions to over 50% of the 2nd Edition content. Many examples are provided of processing different plastics and relating them to critical factors, that range from product designs-to-meeting performance requirements-to-reducing costs-to-zero defect targets. Changes have not been made that concern what is basic to injection molding (IM). However, more basic information has been added concerning present and future developments, resulting in the book being more useful for a long time to come. Detailed explanations and interpretation of individual subject matters (1500 plus) are provided using a total of 914 figures and 209 tables. Throughout the book there is extensive information on problems and solutions as well as extensive cross-referencing on its many different subjects.

This book represents the ENCYCLOPEDIA on IM, as is evident from its extensive and detailed text that follows from its lengthy Table of CONTENTS and INDEX with over 5200 entries. Even though the worldwide industry literally encompasses many hundreds of beneficial computer software, plastic related programs, this book explains with a brief list these numerous beneficial programs (ranging from operational training to product design to molding to marketing); no one or series of software programs can provide the details obtained and the extent of information contained in this single source-book.

4.1.3 "Designing with reinforced composites: Technology, performance, economics", Dominick V. Rosato; Hanser Gardner Publs; (1997); ISBN: 1569902119. This book presents essential information on how to succeed in meeting product performance requirements while simultaneously producing at the lowest cost with zero defects. The information presented ranges from basic design principles to designs of different sized molded parts produced by different RP processes.

4.1.4   "Design data for reinforced plastics: A guide for engineers and designers",   Neil L. Hancox, Rayner M. Mayer; Chapman & Hall (Kluwer Academic Publishers); (1994); ISBN: 0412493209. In this book, the authors have assembled a systematic set of design parameters describing short and long term mechanical, thermal, electrical, fire and environmental performance, etc. for composites based primarily on continuous glass, aramid and carbon fibers in thermosetting and thermoplastic matrices.

4.1.5   "Design data for plastics engineers", Natti S. Rao, Keith T. O'Brien; Hanser Gardner Publs; (1998); ISBN: 156990264X. Whether working on product design or process optimization, engineers need a multitude of polymer property values. This book provides a quick reference on basic design data for resins, machines, parts, and processes, and shows how to apply these data to solve practical problems.

4.1.6   "Design formulas for plastics engineers",   Natti S. Rao; Hanser Gardner Publs; (1991); ISBN: 1569900841. The formulas in this book are classified for specific areas, including rheology, thermodynamic properties, heat transfer, plastic and part type.

4.1.7   "Flow analysis of injection molds", Peter Kennedy; Hanser Gardner Publs; (1995); ISBN: 1569901813. For mechanical engineers, polymer engineers, and applied mathematicians who want to increase their understanding of flow analysis technology, this book is a thorough introduction to computer simulation of the injection molding process, including MOLDFLOW. Provides mechanical and polymer engineers with the theoretical background and hundreds of equations for using the many software packages now available that apply flow analysis to the design of plastic parts to be manufactured by injection molding. Among the topics are material properties, governing equations of fluid flow, mathematical models, finite element formulations, and numerical solutions.

4.1.8   "Designing with plastics",   G.W. Ehrenstein, G. Erhard; Hanser Publs; (1984); ISBN: 0029487706. Key book sections include (1) design influencing factors, (2) environmental effects on plastics, (3) lifecycle assessment and prediction, (4) cost estimation and (5) design guidelines.

4.2   Fabricated Plastic Product Design Books

4.2.1   "Designing with plastics: based on material & process behaviors",   Donald V. Rosato, Marlene G. Rosato, and Dominick V. Rosato, Kluwer Academic Publishers (2000). This book provides a simplified and practical approach to designing plastic products that fundamentally relates to the load, temperature, time, and environment subjected to a product. It will provide the basic behaviors in what to consider when designing plastic products to meet performance and cost requirements. Important aspects are presented such as understanding the advantages of different shapes and how they influence designs.

Important are behaviors associated and interrelated with plastic materials (thermoplastics, thermosets, elastomers, reinforced plastics, etc.) and fabricating processes (extrusion, injection molding, blow molding, forming, foaming, reaction injection molding, etc.). They are presented so that the technical or non-technical reader can readily understand the interrelationships.

The data included provides examples of what are available. As an example static properties (tensile, flexural, etc.) and dynamic properties (creep, fatigue, impact, etc.) can range from near zero too extremely high values. They can be applied in different environments from below the surface of the earth, to over the earth, and into space.

4.2.2   "Injection molding handbook, third edition",   Donald V. Rosato, Marlene G. Rosato, and Dominick V. Rosato, Kluwer Academic Publishers (2000). This third edition has been written to thoroughly update the subject of the Complete Injection

Molding Operation in the World of Plastics. By updating the book, there have been changes with extensive additions to over 50% of the 2nd Edition content. Many examples are provided of processing different plastics and relating them to critical factors, that range from product designs-to-meeting performance requirements-to-reducing costs-to-zero defect targets. Changes have not been made that concern what is basic to injection molding (IM). However, more basic information has been added concerning present and future developments, resulting in the book being more useful for a long time to come. Detailed explanations and interpretation of individual subject matters (1500 plus) are provided using a total of 914 figures and 209 tables. Throughout the book there is extensive information on problems and solutions as well as extensive cross-referencing on its many different subjects.

This book represents the ENCYCLOPEDIA on IM, as is evident from its extensive and detailed text that follows from its lengthy Table of CONTENTS and INDEX with over 5200 entries. Even though the worldwide industry literally encompasses many hundreds of beneficial computer software, plastic related programs, this book explains with a brief list these numerous beneficial programs (ranging from operational training to product design to molding to marketing); no one or series of software programs can provide the details obtained and the extent of information contained in this single source-book.

4.2.3 "Blow molding handbook", Donald V. Rosato and Dominick V. Rosato Hanser Gardner Publs; (1989).

4.2.4 "Plastic part design for injection molding: An introduction", Robert A. Malloy; Hanser Gardner Publs; (1994); ISBN: 1569901295. This reference reflects the common problems an engineer faces while designing a plastic part and assists the designer in the development of parts that are functional, reliable, manufacturable, and aesthetically pleasing. With wide use of injection molding in the manufacture of plastic parts, understanding the integrated design process

is essential to achieving economical and functional design.

4.2.5 "Injection molding alternatives: A guide for designers and product engineers", Jack Avery; Hanser Gardner Publs; (1998); ISBN: 1569902518. This guide covers a wide range of processes, variations of injection molding techniques, and low volume production techniques used for prototyping and pre-production. The fit, advantages, disadvantages, materials used, and design, application and tooling considerations are reviewed for each process covered. Innovations such as deep-draw blow molding, multi-live feed molding, gas-assisted injection molding and in-mold decoration are discussed. Includes process comparison charts.

4.2.6 "Blow molding design guide", Norman C. Lee; Hanser Gardner Publs; (1998); ISBN: 1569902275. This book provides an understanding of plastic blow molded parts, materials, and processes. It also compares the benefits and limitations of various processes, mold engineering, decoration, assembly techniques, and other topics. Issues relating to manufacturability and cost are emphasized.

4.2.7 "Handbook of package engineering", 3rd edition, J.F. Hanlon, R.J. Kelsey, H.E. Forcinio; Technomic Publs; (1998); ISBN: 1566763061. The standard industry reference on packaging materials and engineering, the 3rd edition includes development of environmentally-sensitive packaging. This reference work presents the basic engineering aspects of packaging: materials, package designs, function and performance, production and graphics, machinery and equipment and standards and regulation. Text also introduces the increasing web of laws and regulations controlling virtually all packaged products in efforts to reduce the impact of packaging disposal on landfill.

4.2.8 "Rotational molding: Design, materials & processing", Glenn Beall; Hanser Gardner Publs; (1998); ISBN: 1569902607. A highly versatile process, rotational molding allows for incredible design flexibility with the added benefit of low production costs. This guide to the rotational molding process explains how to make full use of the capabilities of this manufacturing technique. Emphasis is on when to specify rotational molding and how to design and develop hollow plastic products that can be efficiently produced. The book also reviews the origins of the process, its present status, and future prospects, and discusses design considerations, materials, and molds.

4.3   Industrial Design Reference Books

4.3.1 "Joining of plastics: Handbook for designers and engineers", Jordan I. Rotheiser; Hanser Gardner Publs; (1999); ISBN US: 1569902534; German: 3446174184. This book takes the joining of plastics to a new level by dealing with the special considerations necessary to apply the principal assembly methods to parts manufactured by the 22 major processing methods and made of the 34 most commonly used plastics. This handbook emphasizes the relationship between the assembly methods, materials, and the manufacturing process. In addition, the subjects of design for disassembly, recycling, cost reduction, and the complete elimination of joining operations are addressed. The book provides a chapter for a description of each of the 14 principle fastening and joining methods used to assemble plastics today. The advantages and disadvantages of each method are listed and rapid guidelines for joining of plastics are also provided. The author has gone to considerable lengths to make information retrieval quick and effective. The book is extensively indexed and contains a detailed table of contents.

4.3.2 "American plastic: A cultural history", Jeffrey L. Meikle; Rutgers Univ Press; (1997); ISBN: 0813522358. Meikle traces the course of plastics from 19th-century celluloid and the first wholly synthetic bakelite, in 1907, through twentieth-century science, technology, manufacturing, marketing, design, architecture, consumer culture and the proliferation of compounds (vinyls, acrylics, nylon, etc.) to recent ecological concerns. Winner of the 1996 Dexter Prize from the Society for the History of Technology. 70 illustrations.

4.3.3 "1950s plastics design: Everyday elegance", 2nd edition, Holly Wahlberg; Schiffer Publishing, Ltd.; (1999); ISBN: 0764307835. This book presents a factual discussion of the wide variety of colorful and popular plastics housewares made between 1945 and 1960. Advertisements that announced to the world what new designs were possible with this experimental material are shown. Many color photographs of today's highly collectible plastics objects demonstrate the variety of colors and useful forms that were manufactured. Vinyl, Lucite, Melamine and Formica, to name but a few, have become common household names since their introduction in this era. Here are chairs, tables, dishes, cups, radios, lampshades, draperies, cooking containers, car interiors, floors and more-all made of plastics. A very useful Guide, providing information about all the major manufacturers and trade names, is organized by product types for easy reference.

4.3.4 "Designing with plastics", P.R. Lewis; RAPRA Review Report, No 64; (1993); ISBN: 0902348752. Dr. Lewis surveys the current state of the art in designing with plastics, in terms of materials properties and processing technologies. He also considers the legal implications of intellectual property and product liability, as well as ergonomic and aesthetic design, parts consolidation and recyclability.

4.3.5 "Product design with plastics, A practical manual", Joseph B. Dyn; Industrial Press Inc.; (1983); ISBN: 0831111410. A classic, applied, practical plastic design book. Topics covered include: (1) introduction to the application of plastics, (2) description and derivation of short term and long term properties, (3) polymer formation, variation, and characteristics, (4) product design features, (5) designing the plastic product, (6) joining or assembly techniques, (7) description of processing plastics, and (8) cost estimating of plastic parts for product designers.

4.3.6 *"Plastic product design", 2ⁿᵈ Edition,*

Ronald D. Beck; Van Nostrand Reinhold, (Kluwer Academic Publishers); (1980); ASIN: 0442206321. This book serves very well as a basic guide to the study and application of plastic product design. The main topics discussed are: mold design for part requirements; molded holes and undercuts; threads; inserts; fastening and joining plastics; decorating plastics; extrusion design and processing; reinforced plastics; and tests and identification of plastics.

4.3.7 *"Plastics product design engineering handbook",*

Sidney Levy and T. Harry DuBois; Van Nostrand Reinhold, (Kluwer Academic Publishers); (1977); ASIN 0412005115. A classic design course converted to book format that provides a very good introduction to a multitude of basic design features and environments focused on specific examples and end use application areas.

4.4 *Plastic Design Reference Books: Special Topics*

4.4.1 *Plastics mold design*

- **"Injection Molds 108 Proven Designs"**, Hans Gastrow, E. Lindner (Editor), P. Unger (Editor); Hanser Gardner Publs; (1993); ISBN: 1569900280. This classic belongs on the desk of everyone involved in designing or building injection molds. Invaluable for the working engineer, this book demonstrates problem solving in toolmaking for injection molding and contains a wealth of information, practical tips and proven shortcuts.
- **"Injection Molds and Molding: A Practical Manual"**, **2ⁿᵈ Edition**, Joseph B Dym; Kluwer Academic Publishers; (1987); ISBN: 0442217854. Highlights include a description of CAD/CAM potential and process control capabilities, and a

method of mold maintenance that prolongs the period of operation. Also features a guide for cooling time that can be used for comparing mold cooling performance.

- **"Mold-Making Handbook for the Plastics Engineer, 2ⁿᵈ Edition**, Klaus Stoeckhert, Gunter Mennig (Editor); Hanser Gardner Publs; (1998); ISBN: 1569902615. Stoeckhert's classic provides all of the fundamental and engineering aspects of mold construction and manufacturing. Designed to permit a direct comparison of different molds used in plastics processing, this comprehensive review covers molds for various processing methods (injection, compression and transfer molds, etc.); mold materials (steel, bronzes, aluminum and zinc alloys, materials for prototype molds); and manufacturing and machining processes (including computer-integrated manufacturing and electroforming). Other topics include mold maintenance and the latest developments in CAD and rapid prototyping technology.
- **"How to Make Injection Molds"**, **2ⁿᵈ Edition**, G. Menges, P. Mchren; Hanser Gardner Publs.; (1993); ISBN: 1569900820. This is a comprehensive, classic handbook for the design and manufacture of injection molds. It covers all practical aspects involved such as material selection, fabricating cavities and cores, general mold design, hot runner systems, venting, mechanical/dimensional/thermal design, demolding techniques and devices, maintenance of injection molds, standard elements, hardware, and design/construction procedures. Geared to the applied industrial technologist and academic. Practical problem solutions illustrated throughout the text.

4.4.2 *Plastics failure analysis*

- **"Plastics Failure Guide: Cause and Prevention"**, Myer Ezrin; Hanser Gardner Publs; (1996); ISBN: 1569901848. The focus of this book is on actual field and product failures. This comprehensive volume emphasizes cause and prevention and illustrates how and why a variety of plastic products fail due to fracture, appearance change, loss of adhesion, and many other problems. Topics include the nature, causes, and consequences of plastics failure; fundamental materials variables affecting processing and product performance or failure; failure related to design and material selection; processing-related factors in failure; failure related to service conditions; failure analysis and test procedures; quality control; legal aspects of failure.

- **"Failure Of Plastics"**, Witold Brostow, Roger D. Corneliussen; Hanser Gardner Publs; (1986); ISBN: 1569900086. Complete reference on the mechanical failure of plastics. Covering theory and practice, this book describes an expert knowledge base and provides directions for future work toward elimination of mechanical failure of plastics under varied conditions.

4.4.3 "Handbook of plastics testing technology", 2nd edition, Vishu Shah; John Wiley & Sons; (1998); ISBN: 0471182028. This handbook is the most complete compilation of the tests currently used in the plastics industry. It provides descriptions and diagrams of testing procedures, and explains the significance and advantages and limitations of the tests. Properties that can be tested by the methods described include mechanical, thermal, electrical, weathering, optical, chemical, and flammability. In addition, chapters also discuss conditioning procedures, identification of plastics, characterization and analysis, testing of foam plastics, quality control, professional and testing organization, product liabilities, failure analysis, and uniform global testing standards.

4.4.4 "Designing plastic parts for assembly, 3rd Edition, Paul A Tres; Hanser Gardner Publs; (1998); ISBN: 1569902437. This practical design book facilitates cost-effective design decisions and helps to ensure that the plastic parts and products designed stand up under use. The book describes good joint design and joint purpose, the geometry and nature of the component parts, the types of loads involved and other basic information important in plastic part assembly.

4.4.5 "Guide to short fiber reinforced plastics", Roger F. Jones with Mitchell R. Jones and Donald V. Rosato; Hanser Gardner Publs; (1998); ISBN: 1569902445. Written from the perspective of the product design engineer, the emphasis is on practical aspects of basic design considerations in the selection, use, and automated fabrication of short fiber reinforced thermoplastics and thermoset materials. This book examines the principles characteristics of these materials and their strengths and weaknesses in practical terms for design engineers. It examines the strengths and limitations of these growth industry materials valued at over 1.5 billion dollars (US, 1997). Includes illustrations, suppliers, applications, and references with a more theoretical bent.

4.4.6 "Molded thermosets: A handbook for plastics engineers, molders, and designers", Ralph E. Wright; Hanser Gardner Publs; (1991); ISBN: 1569901120. This handbook provides in-depth coverage of every important family of thermoset polymer systems and their molding—from the raw material through the finished molded part.

4.4.7 "Composites—design manual", Jim Quinn; Jim Quinn Associates Ltd; (1996); available from SPE. A concise handbook of composites related information that engineers, designers and specifiers will find valuable. Contents include specifications for a wide range of reinforcements, and overview of initiators, and other product information, properties, processes, construction analysis, property prediction, and design.

5.   Design Education

5.1   Design Education Books

5.1.1 "Concise encyclopedia of plastics: Fabrication & industry", Donald V. Rosato, Marlene G. Rosato, and Dominick V. Rosato, Kluwer Academic Publishers (2000). This practical and comprehensive book reviews virtually the "A-to-Z" of the plastics industry by using over 20,000 entries. Each of the major subjects (entries) could represent a separate book. Where common information exists, they are cross-referenced. There is extensive cross-referencing where one subject is defined and related to many other subjects, thus significantly reducing the size of the book. Its brief and concise format goes from understanding basic factors such as a plastic's melt flow behavior during processing to designing and fabricating products targeted to meet performance and cost requirements with zero defects. This type of understanding is required in order to initially design, prototype fabricate, and volume manufacture the many different marketable products reviewed in this book and that exist worldwide.

More importantly, this extensive cross-referencing provides information on how the many subjects interrelate. All pertinent information for a subject is included in the definition and/or its cross- referenced component. They are searchable under their own headings based on the reader's needs. In order to cover the needs of different individual interests, many of the subjects have very extensive cross-referencing. Thus, the readers can cross-reference those subjects that meet their needs. This approach simplifies understanding any single subject and, most importantly, shows very vividly the many common similarities and interactions that exist between the subjects in the World of Plastics.

5.1.2 "Plastic injection molding: Material selection and product design fundamentals", Douglas Bryce; Society of Manufacturing Engineers; (1997); ISBN: 0872634884. Shows how to identify the optimum material for a particular product based on the product's design, manufacturing, and end-use parameters. Available from Injection Molding Magazine Bookclub (See 6.2), or Society of Manufacturing Engineers.

5.1.3 "Plastic injection molding: Mold design & construction fundamentals", Douglas Bryce; Society of Manufacturing Engineers; (1998); ISBN: 0872634957. Shows how to design and build injection molds; specifically, how to design gate location,

shape, and size, as well as how to use venting properly. A mold design checklist is also included. Available from Injection Molding Magazine Bookclub (See 6.2), or Society of Manufacturing Engineers (see 7.5).

5.1.4 "Plastics: Product design and process engineering", Harold Belofsky; Hanser Gardner Publs; (1995); ISBN: 1569901791. This textbook, designed for undergraduate mechanical engineering courses, integrates product design with a study of mechanical and physical properties, processing machinery and tooling, and materials and process selection. The focus is on applications rather than training for academic research. Many illustrative examples and quantitative homework problems are included.

5.1.5 "Understanding Product design injection molding", Herbert Rees; Hanser Gardner Publs, (Hanser Understanding Books); (1996); ISBN: 1569902100. This book highlights many of the questions and decisions engineers will face while designing products. The designer using this book will have a better understanding of process and material selection, and how to design an injected mold product that will work as expected and be produced efficiently.

5.1.6 "Computer modeling for polymer processing: Fundamentals", Charles L. Tucker III (Series Editor: Ernest C. Bernhardt); Hanser Gardner Publs; (1989); ISBN: 1569901015. Computer simulations have become an important tool for the engineering of polymer processing operations. This book looks inside this important technology, showing how to use computers and numerical methods to simulate flow, heat transfer and structure development in polymer processing operations.

5.1.7 "Plastics engineering handbook of the society of the plastics industry, inc.", *5th edition*, Michael L. Berins (Editor); Chapman & Hall (Kluwer Academic Publishers); (1991); ISBN: 0412991810. Since 1947, the most comprehensive reference available on plastics processing methods, equipment, and materials. Sponsored by the Society of the Plastics Industry, Inc., the revised and updated fifth edition incorporates all major advances in the plastics industry. It covers the state of the art in both materials—high-temperature thermoplastics, liquid crystal polymers, and thermoplastic composites—and processing—resin transfer molding, structural reaction injection molding, gas-assisted injection molding, stretch blow molding, automation, and process control.

5.2   Plastics Design Training (Seminars and Interactive CD-Roms)

5.2.1   Glenn Beall plastics technology seminars
Glen Beall Plastics, LTD.
32981 North River Road,
Libertyville, Illinois 60048
Tel: 413 733-8588
Fax: 413 733-9325
http://www.pcn.org/Beall.htm
Glenn L. Beall, a Kunststoffe, and IMM contributing editor, has been intimately involved with both plastics product design and injection molding for 40 years. He has been doing preproduction engineering of plastics components since 1957 and today is a recognized authority in this area. Glenn Beall Plastics Ltd., concentrates on product design, consulting and plastics technology seminars. Mr. Beall holds more that 35 patents in the plastics area and works extensively as a designer, consultant and expert witness on projects involving plastics technology.

5.2.2   Paulson training programs, Inc.,
15 No. Main St., P.O. Box 366,
Chester, CT 06412 USA
Phone: 860-526-3099
Fax: 860-526-3454
E-mail: sales@paulson-training.com
http://www.paulson-training.com

5.2.3   A. Routsis associates, Inc.
275 Donohue Road, Suite 14
Dracut, MA 01826 USA
Phone: 978-957-0700
Fax: 978-957-1860

E-mail @netway.com
http://www.plastics-training.com

5.2.4   Society of plastics engineers (SPE)
PO Box 403
Brookfield, CT 06804-0403
Series of SPE sponsored seminars and workshops are held in various locations throughout the year. Subjects range from "Die Design Principles to Plastic Part Design for Economical Injection Molding. A full range of seminars are also conducted in conjunction with SPE's annual technical conference (ANTEC).
http://www.4spe.org

5.2.5   RAPRA training courses
Raspra Technology Ltd.
Shawbury, Shrewsbury, Shropshire SY4 4NR, U.K.
Tel: +44-1939-250383,
Fax: +44-1939-251118)
http://www.rapra.net
Rapra Training Courses are held at Rapra or on-site and may also be developed to meet clients' specific requirements. Popular courses include: Plastics Materials and Products; Designing and Engineering with Rubber; Testing and Specification of Polymer Products; Plastics in Packaging.

5.2.6   Hanser gardner CD-ROM training for plastics
Hanser Gardner Publications
6915 Valley Avenue
Cincinnati, OH 45244-3029
http://www.hansergardner.com
Plastic Part Design Series: Based on Dr. Robert Malloy's book, *Plastic Part Design for Injection Molding*, this new design series consists of six CDs: Product Development and Prototype Process; Mechanical Behavior of Polymers; Mold Filling, Gating, and Weld Lines; Shrinkage, Warpage, and Ejection; Mechanical Fasteners, Press and Snap Fits; Welding and Adhesive Bonding Technology.

5.2.7   IMM Book Club—Training CD-ROMs
Paulson Training Programs
A. Routsis Associates Training Programs

Desktop Dimension International Training Programs
http://www.immbookclub.com/store/
    training.html

5.2.8   GE plastics e-seminars

GE Plastics virtual conference center. e-Seminars offers "live" online conferences.
e-Seminar examples include:
Material Selection
Materials for Single-Use Microwave Food Packaging
http://www.geplastics.com/resins/
    designsolution/seminar/

5.2.9   Nypro online (NYPRO Inc.)

101 Union Street, Clinton, Massachusetts 01510
Tel: 978-365-9721
Fax: 978-368-0236
E-mail: information@nypro.com
http://www.nypro.com
Nypro Online is a strategic training initiative of Nypro Inc., the largest multinational custom injection molder in the world. Nypro Online is the first global plastics education provider offering college education and focused plastics training over the internet jointly with its academic partners, the University of Massachusetts at Lowell, and Paulson Training Programs.

6.   Trade Publications

6.1   Plastics Technology

355 Park Avenue, South; New York, NY 10010-1789
Tel: 212 592-6570
http://www.plasticstechnology.com

6.2   Injection Molding Magazine

59 Madison Ave., Suite 770; Denver, CO 80206
Tel: 303 321-2322
http://www.immnet.com
Injection Molding Bookclub
http://www.immbookclub.com

6.3   Kunststoffe

Carl Hanser Verlag
Kolbergerstraße 22
81679 München
Tel: 089-99830 621
Fax: 089-99830 625
E-mail: kunststoffe@hanser.de
http://www.hanser.de/zeitschriften/KU/
    index.htm

6.4   Design News

275 Washington Street,
Newton, Massachusetts, 02458-1630
Tel: 617-558-4660
Fax: 617-558-4402.
http://www.manufacturing.net/magazine/dn/
    maginfo/aboutdn.html
Design News, in print for over 50 years, has a circulation of over 180,000. It covers the latest tools, components, and materials used in mechanical and electromechanical design of a broad range of products. Articles feature successful engineering projects and new technologies that will spark ideas and assist readers in the design of new products. All articles are written by the DN staff of editors, many of whom are engineers themselves. Design News online provides readers with an ask the Expert feature. Questions are normally answered within 48 hours during the workweek.

6.5   Global Design News

Global Design News, sister publication of Design News magazine in the U.S., was launched in response to the demands of European design engineers for technology from around the world.
http://www.manufacturing.net/magazine/gdn/

6.6   Design & Materials

(Biweekly Newsletter, plastics focus)
Market Search Inc.
2727 Holland Road, Suite A
Toledo, OH 43615
Tel: 415 535-7899

6.7   Product Design and Development

Product Solutions for Design Engineers
© 2000 Cahners Business Information, a division of Reed Elsevier Inc.,
8773 S. Ridgeline Boulevard,
Highlands Ranch, CO 80126.
Fax: 1-303-470-4546
E-mail: webmaster@denver.cahners.com
http://www.pddnet.com/
Product Design & Development is a monthly magazine with 170,000 design engineers and engineering management readers in the original equipment market. Design engineers read the magazine to keep up to date on the components, materials and systems they need to design high-quality end products.

6.8   Desktop Engineering Magazine

Helmers Publishing, Inc
174 Concord Street, PO Box 874
Peterborough NH 03458 USA
Tel: 603-924-9631
Fax: 603-924-6746

6.9   Machine Design

Penton Media, Inc.
1100 Superior Avenue,
Cleveland, Ohio 44114 USA
http://www.machinedesign.com/

6.10   Computer-Aided ENGINEERING

1100 Superior Avenue
Cleveland, OH 44114-2543, USA
Tel: 216.696.7000
Fax: 216.696.1267
http://www.caenet.com

6.11   ANSYS Solutions

ANSYS, Inc.
Southpointe
275 Technology Drive
Canonsburg, PA 15317
http://www.ansys.com
ANSYS Solutions is a quarterly publication focused on software applications for mechanical simulation and engineering processes. Its goal is to provide objective and authoritative information to help readers understand and apply finite element analysis and other technology developed and supported by ANSYS, Inc. Its readers include designers and engineering managers, in addition to a wide variety of industry analysts and partners. The editorial director for ANSYS Solutions is John Krouse, a well-known and respected writer in the CAD/CAM field. John previously served as CAD/CAM editor for Machine Design magazine, Editor-In-Chief and Publisher of Computer-Aided Engineering magazine, and has written several books in the field including, "What Every Engineer Should Know About CAD/CAM."

6.12   CAD User Magazine

BTC Ltd. CAD User
24 High Street, Beckenham,
Kent, BR3 1AY
Tel: +44 (0) 181 663 3818
Fax: +44 (0) 181 663 6776
cad.user@btc.co.uk
The Premium CAD Magazine for the UK and Ireland. CAD User magazine covers the technical aspects of integrating CAD in a multi-discipline, multi-product environment. With indepth reviews of new products, case studies and technical tips, CAD User magazine and its online version, CADUser.Com, are useful for CAD information.

6.13   Materialprüfung

Verschaffen Sie sich den Einblick: Das aktuelle Heft
Ihr Ansprechpartner in der Redaktion von Materialprüfung:
Carl Hanser Verlag
Frauke Zbikowski
Kolbergerstraße 22
81679 München
Tel: 089-99830 614
Fax: 089-99830 623
zbikowski@hanser.de
http://www.hanser.de/zeitschriften/MP/index.htm

6.14 Medical Device & Diagnostic Industry

3340 Ocean Park Blvd., Suite 1000
Santa Monica, CA 90405
Tel: (310) 392-5509

6.15 Reinforced Plastics

Garrard House, 2-6 Homesdale Rd.
Bromley, BR2 9WL, UK
Fax: +44 (0) 208 402 8383

6.16 World Plastics & Rubber Technology

Essex House, Regent St.
Cambridge CB2 3AB, England.

7. Trade Associations

7.1 Industrial Designers Society of America

1142 Walker Rd.
Great Falls, VA 22066
Tel: 703.759.0100
Fax: 703.759.7679
E-mail: idsa@erols.com
http://www.isda.org

- IDSA is dedicated to communicating the value of industrial design to society, business and government
- Publishes Innovation, the professional journal of industrial design practice and education in America
- Organizes a national conference each year, the largest gathering of industrial designers, educators and business executives in the US
- Conducts the annual Industrial Design Excellence Awards (IDEA) under the sponsorship of Business Week magazine
- Organizes five annual District Conferences in concert with the education community
- Publishes the annual Directory of Industrial Designers
- Supports a network of 25 active chapters located in cities across the US
- Distributes Design Perspectives, the monthly newsletter to members
- Performs as a vital member of the International Council of Societies of Industrial Design (ICSID)

7.2 International Council of Societies of Industrial Design

ICSID Secretariat oversees the daily activities of the council; it has been in Helsinki, Finland since 1985
Kaarina Pohto, Secretary General
ICSID Secretariat
Erottajankatu 11 A-18
00130 Helsinki, Finland
Tel: +358 9 696 22 90
Fax: +358 9 696 22 910
E-mail: icsidsec@icsid.org
http:www.icsid.org
ICSID members are professional organizations, promotional societies, educational institutions, government bodies, companies and institutions which aim to contribute to the development of the profession of industrial design. Today ICSID consists of 149 Member Societies, representing 52 countries from all continents (except Antarctica!). These Societies collaborate to establish an international platform through which design institutions worldwide can stay in touch, share common interests and new experiences, and be heard as a powerful voice.

7.3 Color Marketing Group (CMG)

5904 Richmond Highway,
Suite 408, Alexandria, VA 22303 USA
Tel: 703 329-8500
FAX: 703 329-0155
E-mail: cmg@colormarketing.org
http://www.colormarketing.org/
Color Marketing Group, is the premier International Association that forecasts colors for manufactured products. Founded in 1962, this not-for-profit, international Association of 1,600 Color Designers is involved in the use of color as it applies to the profitable marketing of goods and services. CMG provides a forum for the exchange of non-competitive information on all phases of color marketing: color trends and combinations; design influences; merchandising and sales; and education and industry contacts. CMG members are highly qualified Color Designers who interpret, create, forecast and select colors in order to enhance the function, saleability

and/or quality of a product. Two International Conferences are held each year during which CMG members forecast Color Directions® one to three years in advance for all industries, manufactured products and services.

7.4   Society of Plastics Engineers (SPE)

PO Box 403 Brookfield, CT 06804-0403 USA
Tel: 203 775-0471
Fax: 203 775-8490
E-mail: info@4spe.org
http://www.4spe.org
SPE is the recognized medium of communication amongst scientists and engineers engaged in the development, conversion and applications of plastics. An international Society, with many of its 35,000 members residing outside the United States, The SPE mission is to provide and promote the knowledge and education of plastics and polymers worldwide.

7.5   Society of Manufacturing Engineers (SME)

One SME Drive, P.O. Box 930
Dearborn MI, 48121-0930
Tel: 313 271-1500/800 733-4763
http://www.sme.org
SME, headquartered in Michigan is the world's leading professional society serving the manufacturing industry. Founded in 1932, SME has some 60,000 members in 70 countries and supports a network of chapters worldwide. Through publications, expositions, professional development resources and member programs, SME influences more than 500,000 manufacturing executives, managers and engineers.

7.6   ASM International

9639 Kinsman Rd.
Materials Park, Ohio 44073-0002 USA
(440) 338-5151
1-800-336-5152/1-888-336-5152
http://www.asm-intl.org

ASM International is a society whose mission is to gather, process and disseminate technical information. ASM fosters the understanding and application of engineered materials and their research, design, reliable manufacture, use and economic and social benefits. This is accomplished via a unique global information-sharing network of interaction among members in forums and meetings, education programs, and through publications (i.e., Advanced Materials & Processes) and electronic media.

7.7   Composites Fabricators Association

Composites Fabricators Association
1655 N. Fort Myer Dr., Suite 510
Arlington, VA 22209
Tel: 703·525·0511
Fax: 703·525·0743
E-mail: cfa-info@cfa-hq.org
http://www.cfa-hq.org
(CFA) is the world's largest trade association serving the composites industry. Formed in 1979 to provide education and support for composites fabricators in the successful operation of their businesses, CFA continues to offer leading-edge services that are instrumental in regulatory compliance and formulation, education and training, management, and market expansion. With approximately 800 members including open and closed molders, suppliers, distributors, consultants, academics, and others with a vested interest in the composites market, CFA has earned a reputation as the voice of the industry.

8.   Industry Conferences

8.1   National Design Engineering Show

March
McCormick Place Complex
Chicago, Illinois USA
http://www.manufacturingweek.com/

8.2   Rapid Prototyping & Manufacturing Conference & Exhibition

April
Rosemont, Ill.
Contact Society of Manufacturing Engineers Customer Service,
(800) 733-4763, or (313) 271-1500, Ext. 1600; fax: (313) 271-2861

8.3   Materials Week, International Congress on Advanced Materials, their Processes and Applications

September
Munich, Germany
Werkstoffwoche-Partnerschaft   GBRmbH,
Hamburger Allee 26, 60486
Frankfurt, Germany
Phone: +49 69 79 17 747
Fax: +49 69 79 17 733
E-mail: materialsweek@dgm.de
http://www.materialsweek.org

8.4   IEEE/ACM International Conference on Computer Aided Design

November
Phone: 303 530-4562
Founded in 1982, ICCAD is an annual show focusing on information technology for computer-aided design professionals and engineers.

8.5   IDSA International Design Conference

September
Organized by Industrial Designers Society of America (IDSA)
USA
Tel: +1 703 759 01 00
Fax: +1 703 759 76 79
idsa@erols.com

8.6   Materials Solutions 2000

October
Conference and exhibition organized by ASM International

St. Louis, Missouri, USA
Fax: +1 440 338 46 34

8.7   ICSID Congress and General Assembly (Biannual)

September/October
Congress and exhibition, General Assembly,
Tel: +82 2 708 20 52
Fax: +82 2 36 72 59 71

9.   Key Related Websites

9.1   IBM Patents Website

http://www.patents.IBM.com
The IBM Intellectual Property Network (IPN) has evolved into a premier Website for searching, viewing, and analyzing patent documents. The IPN provides you with free access to a wide variety of data collections and patent information including:

- United States patents
- European patents and patent applications
- PCT application data from the World Intellectual Property Office
- Patent Abstracts of Japan
- IBM Technical Disclosure Bulletins

Searching is fast and easy. Along with a simple keyword search, IPN offers alternative searches by patent number, boolean text, and advanced text that allows for multiple field searching. Browsing provides an organized approach to searching for patents. Through a review of specific classifications, you can identify topics and patents of interest.

9.2   Federal Web Locator

http://www.infoctr.edu/fwl/
The Federal Web Locator is a service provided by the Center for Information Law and Policy and is intended to be the one stop shopping point for federal government information on the World Wide Web. This site is hosted by the Information Center at

Chicago-Kent College of Law, Illinois Institute of Technology.

9.3   Maack Business Services

A Maack & Scheidl Partnership
CH-8804 Au/near Zürich, Switzerland
Tel: +41-1-781 3040
Fax: +41-1-781 1569
http://www.MBSpolymer.com
Plastics technology and marketing business service, which organizes global conferences, and edits a range of reports and studies, which focus on important worldwide aspects of polymer research, development, production, and end uses. Provides updates on plastic costs, pricing, forecast, supply/demand, and analysis. Identified early in the cycle are trends in production, products and market segments.

9.4   Material Safety Data Sheets (MSDS)

9.4.1   MSDSSEARCH.COM, Inc.
http://www.msdssearch.com/
MSDSSEARCH.COM, Inc., is a National MSDS Repository, providing FREE access to over 1,000,000 Material Safety Data Sheets—the largest centralized reference source available on the Internet. MSDSSEARCH.COM is dedicated to providing the most comprehensive single source of information related to the document known as a Material Safety Data Sheet (MSDS). MSDS SEARCH serves as the conduit between users of MSDSs and any reliable supplier. MSDSSEARCH.COM provides access to 350 K MSDSs from over 1600 manufacturers, 700 K MSDSs from public access databases, links to MSDS software, services, training and product providers, links to Government MSDS information, an MSDS discussion forum where you can ask questions, and supplies MSDSs directly from manufacturers via search engine.

9.4.2   The canadian center for occupational health and safety (CCOHS) Promotes a safe and healthy working environment by providing information and advice about occupational health and safety
250 Main Street East
Hamilton ON L8N 1H6 Canada
Tel: 1-800-263-8466 (Toll free in Canada only)/1-905-572-4400
Fax: 1-905-572-4500
http://www.ccohs.ca/products/databases/msds.html
Search MSDS on CCINFOWeb. All databases on CCINFOWeb may be searched for free. The MSDSs are contributed by North American sources, many that are multi-national companies marketing chemical products worldwide. This database meets a growing international requirement for health and safety information on specific chemical products. It helps thousands of users worldwide manage their responsibilities under workplace, environmental and other right-to-know legislation. The MSDS database can be searched quickly and easily for product names and other product identifications, manufacturer or supplier names, dates of MSDSs, or any term used in the text of the MSDS itself.
MSDS records contain information such as:

- Chemical and Physical Properties
- Health Hazards
- First Aid Recommendations
- Personal Protection
- Fire and Reactivity Data
- Spill and Disposal Procedures
- Storage and Handling

10.   Key Corporate Websites

10.1   GE Plastics

http://www.geplastics.com
GE Plastics provides customers with access to a full range of technical services and solutions information available from its webpage Design Solutions Center (http://www.geplastics.com/resins/designsolution/). These include:

10.1.1   GE workspaces: (http://www.geplastics.com/resins/designsolution/workspace/) An internet-based project

collaboration tool. It allows customers to communicate, share, and organize information with GE Plastics project teams in a virtual, secured environment.

10.1.2 GE E-seminars: (http://www.ge-plastics.com/resins/designsolution/seminar/) GE's virtual conference center offers the ability to interact with GE Plastics real-time in "live" on-line conferences. E-Seminar examples include *Material Selection*, which provides the attendee with the knowledge, skills and competencies to determine how application requirements influence the material specification process and *Materials for Single-Use Microwave Food Packaging*, which reviews trends in the growing Freezer-to-Microwave Food Packaging industry.

10.1.3 GE design tools: (http://www.geplastics.com/resins/designsolution/tools/index.html)

- **Datasheets:** Provides password access to GE Plastics product data.
- **Material Selector:** Access to: (1) *GE Select*, a comprehensive database in Microsoft Windows format of the family of GE polymers which allows users to sort for the GE product families and grades of materials that will best meet the specified property ranges, and (2) *CAMPUS*, a worldwide database for plastic materials with uniform global protocol for acquiring and comparing data on competitive plastic materials.
- **Color Selector:** GE Plastics customer color services includes (1) *ColorXpress Services*, dedicated to custom color matching and small lot custom color compounding. This online color match and ordering center includes an extensive online color library and purchasing system. Customers can identify the product and color of choice and fill in an order form. Standards will be shipped in less than 48 hours. Other ColorXpress services offered range from custom color matching to color standards development and maintenance. (2) *MicrolotXpress*, the small lot order center which allows customers to order small quantities of GE

resins (down to 10 pounds) using a standard credit card. Material is shipped in 4 business days.

- **Calculator:** An interactive process wizard that provides quick, effective material, design, processing, and cost solutions. This Engineering Calculator's capabilities include (1) *Material*, to select from a variety of GE Plastics materials; (2) *Design*, which calculates minimum part thickness based upon allowable deflection; (3) *Processing*, which calculates pressure to fill and clamp force; and (4) *Cost*, which calculates estimated material and processing costs for the intended part.

10.1.4 Technical design library: Online Technical Guide Library, provides access to Design Guides, Processing Guides, Secondary Operations Information and Product Literature. Examples include:

- GE Engineering Structural Foam Design and Processing Guide
- Injection Molding Design
- Specific Industry Design Considerations
- Product Design
- Material Selection Guide Table

10.2 Bayer Plastics

http://www.plastics.bayer.com
Application Technology Information (ATIs) from Bayer (http://www.plastics.bayer.com/english/flit.htm) can be downloaded as pdf files. Examples of Bayer literature available for download include:

- Self-tapping screws for thermoplastics;
- Metallized plastics housings to ensure electromagnetic compatibility;
- The mechanical layout of molded parts and molds—ways of achieving optimum results with the Finite Element Method (FEM);
- Data transfer of CAD geometries.

CAMPUS®, the database of Bayer plastic properties can also be downloaded from this web page.

10.3   AlliedSignal-Honeywell

Honeywell Engineered Applications & Solutions
http://www.honeywell-eas.com/
Honeywell Engineered Applications & Solutions is a Honeywell business enterprise that teams Engineering Plastics, Specialty Films and Metal Injection Molding technologies to provide a source for engineered solutions. Honeywell's EAS website supplies a range of technical services and solutions information helpful to the plastics designer.

10.3.1   Troubleshooting tips
http://www.honeywell-plastics.com/
   techinfo/tech-support/trouble.html
A comprehensive guide to symptoms, probable causes and corrections, examples of troubleshooting tips include:

* Extruder Related Problems
* Cast Film Problems
* Tubular Film Problems
* Purging Procedures
* Moisture Considerations
* Drying of "Wet" Nylon 6
* Processing Quality Checklist

10.3.2   Real-time processing tips and tutorials
http://www.honeywell-plastics.com/techinfo/
   audiotips.html
These tips are also available in text only; examples include:

* Snap-Fit Design Applications
* General Snap-Fit Design Guidelines
* Capron Nylon Troubleshooting Guide for Injection Molding
* CAD/CAE Capabilities
* Design Considerations for Injection Molded Parts (Parting lines, draft angles, wall thickness, fillets and radii, bosses, ribs, opening formations, shrinkage, gating, vents, potential knit lines)

10.3.3   The literature shop http://www. honeywell-plastics.com/techinfo/litshop/ litshop.html

View or Download Honeywell Literature such as:

* Product and Technical Support Literature
* Industry Specific Literature
* Rotational Molding with Capron® Nylon Resins
* Injection Molding Processing Guide for Capron® Nylon
* Design Solutions Guide
* Snap-Fit Design Manual
* 1998 Automotive Specifications Guide

10.3.4   Tech information case histories
http://www.honeywell-plastics.com/techinfo/
   snapshot/snapshot.html

10.3.5   Material data   Also accessible through http://www.honeywell-plastics.com are:

* Product Locator: A product-specific search engine which provides datasheets/MSDS sheets.
* CAMPUS® 4.0: A downloadable database program to search for specific resins.

10.4   Montell

Formed   in   1995   from   Royal Dutch/Shell, Montedison polyolefin operations,   and   Himont   polypropylene.
http://www.montell.com/montell/about/
   company.html
To maximise the value of its products to its customers business, Montell has developed a series of services that allow customers to develop products faster and more effectively. These include:

* Co-design—(http://www.montell.com/ montell/products/p-codesign.html) active involvement in product design to optimise the use of material and manufacturing resources.
* CAD and CAE (http://www.montell.com/ montell/products/p-cad.html) computer modelling of product performance and manufacturing processes. Montell's technical centres routinely use sophisticated CAD/CAE systems for customers

developing new applications, achieving often dramatic savings in time and costs by reducing time-consuming prototyping. Using similar techniques to model the future behaviour of the molten polymer while it flows into the mould, Montell can optimise mold design and achieve material economies, as well as reduce development times.

- Piloting (http://www.montell.com/montell/products/p-pilo.html) Montell is able to test new materials and combinations in dedicated laboratory facilities. Montell's development facilities allow testing of new solutions without committing full scale industrial resources.

- Compliance testing (http://www.montell.com/montell/products/p-compil.html) Montell will, on request, provide documentation certifying that the specific Montell polymers to be used by the converter meet the necessary requirements,.ensuring conformity to national and international norms.

10.5   Nypro Inc.

http://www.nypro.com

10.5.1   Nypro Institute:.   Nypro Institute is the corporate university of Nypro Inc., providing educational opportunities to employees, customers, and the general public. Nypro Institute's headquarters are located in Clinton, Massachusetts, and feature a state-of-the-science computer laboratory, a training resource library, and individual multimedia learning centers.

10.5.2   Nypro Online:.   Plastics Education Online Nypro Online is a strategic training initiative of Nypro Inc., the largest multinational custom injection molder in the world. Nypro Online is the first global plastics education provider offering college education and focused plastics training over the internet jointly with its academic partners.

10.6   Equipment Suppliers

10.6.1   Milacron (Milacron Plastics Technologies Group)

http://www.milacron.com/
http://www.milacron.com/PL/PLdefault.html

Milacron is a global leader in plastic processing and metalworking technologies. Milacron's Plastics Technologies Group has the world's broadest line producer of machines, systems, tooling and supplies for the plastics processing industry, with manufacturing facilites in the U.S., Germany and India. The Milacon Group is vertically integrated to produce machines for injection, extrusion and blow molding of plastics and through its D-M-E company, Milacron is also the world's largest manufacturer of basic tooling for the plastics injection molding and die casting industries. D-M-E products include: pre-engineered mold bases, mold design software, mold components, electronic control systems, special tooling and supplies for moldmaking. Other business units of the Milacon Plastics Technologies Group are the Specialty Equipment Group and Contract Services Business Units. SEB is a systems integrator, capable of providing complete processing systems from rail car material unloading to robotic part handling. The CSB Group retrofits, rebuilds, remanufactures and sells used plastics machinery and also creates custom training programs.

10.6.2   Husky

http://www.husky.ca/

Husky is a global supplier of injection molding systems to the plastics industry. Husky designs and manufactures injection molding machines—from 60 to 8000 tonnes, robots, hot runners for a variety of applications, molds for PET preforms, and complete preform molding systems. Customers use Husky's equipment to manufacture a wide range of products in the packaging, automotive and technical industries. The company serves customers in over 100 countries from more than 40 service and sales offices around the world.

10.7   Distribution—General Polymers

10.7.1   Ashland distribution company, a division of Ashland inc.

http://www.ashlandchemical.com/gp.html. The General Polymers Division of Ashland Distribution Company stocks virtually every grade of prime thermoplastic resins and specialties for plastics processors. General Polymers represents and distributes product for the following plastics manufacturing businesses:

- Advanced Elastomer Systems: Thermoplastic Elastomer; Thermoplastic Rubber
- Honeywell Plastics: Nylon 6, 6/6, (Recycled 6); PET (Post Consumer);
- APA Advanced Polymer Alloys: MPR
- AlphaGARY: PVC Rigid, Flexible, Medical, PVC/PUR, PVC Alloy
- Aristech: Polypropylene HPPP, CPPP
- Bayer: ABS; Weatherable ASA&AES; Nylon 6; Polycarbonate; SAN; Polyurethane Elastomers; PC/PET; PC/ABS
- Chem Polymer Corp: Acetal; Nylon 6, 6/6; PBT
- Clariant Masterbatches Division: Color Concentrate
- Dow: ABS; Nylon 6/6; Polycarbonate; Polyethylene, HDPE, LDPE, LLDPE, ULDPE; Polypropylene HPPP, CPPP; Polystyrene HIPS, GPPS, Recycled, Advanced Styrenic Resin; SAN; Polyurethane Elastomers; Polyolefin Plastomer; PC/ABS; Crystalline Polymer; ABS/TPU
- DSM Engineering Plastics, nylon, PBT, Polycarbonate; Thermoplastic Elastomer; PC/ABS; Conductive Resins; Thermoplastics Reinforced and Filled; Thermoplastics Lubricated
- DuPont: Acetal; EVA; Nylon 6, 6/6, 6/12, Mineral Filled 6/6, Industrial; PBT; PET; Polyethylene Modified; Thermoplastic Elastomer; Ionomer; Liquid Crystal Polymer;
- DuPont Dow elastomers: Polyolefin Elastomer
- Equistar: EVA; Polyethylene HDPE, LDPE, LLDPE

- Exxon Chemical; Cross linkable Polyethylene; Purge
- The Geon Company: PVC Rigid, Flexible, High Heat, Composites, Packaging, Static Dissipative, Weatherable
- Huntsman: Polypropylene HPPP, CPPP
- ICI Acrylics: Acrylic
- Montell Polyolefins: Polypropylene HPPP, CPPP, Reinforced Polymers, Aesthetic Polymers, CP, HP, Olefinic Polymer; Engineering Polymers, Elastomeric
- MRC: PET; Polycarbonate; PC/PMMA
- Nova Chemicals: HDPE, LDPE, LLDPE, MDPE, Melt Compounded Black; Polystyrene HIPS, GPPS, Reprocessed, Specialty Polystyrene
- Phillips 66: Butadiene Styrene; PPS
- Prime Source Polymers, Inc.: Custom Compounds
- Spartech Polycom: Filled HDPE; Filled and Reinforced PP; Filled TPO

General Polymers issues a monthly newsletter, CycleTime Tips, which contains technical tips on processing, tool design, etc. The current and past issues of CycleTime Tips may be accessed from the General Polymers webpage (http://www.ashchem.com/home/index.asp? nav_id=4&sub_nav=4)

10.8   PlasticsNet

PlasticsNet.Com, Commerx's flagship is a leading electronic marketplace for the $400 billion domestic plastics industry. PlasticsNet.Com simplifies and streamlines the process of buying and selling, using the speed, access and ease of the internet, to save users time and reduce costs. By enabling plastics processors and suppliers to leverage an advanced e-commerce system, PlasticsNet.Com helps companies capitalize on the benefits of e-commerce without significant investments in resources and technology. PlasticsNet.Com, currently featuring more than 31,000 product SKUs (including resins, materials, equipment and supplies) from companies including General Polymers, MSC Industrial, Van Dorn, Maguire etc.

Registered users count on PlasticsNet. Com for their range of procurement needs. With key strategic alliances including the General Polymers Division of Ashland Distribution and MSC Industrial—PlasticsNet. Com is the one-stop resource for top brand-name resins, machinery and equipment.

In the plastics industry, Commerx delivers e-volved solutions including:

- Procurent solutions for direct and indirect materials
- Supply chain efficiencies
- Web-enabled services

Appendix B

Terminology

A-basis Also called A-allowable. It is the value above which at least 99% of the population of values is expected to fall with a confidence of 95%. See **B-basis; population confidence interval; S-basis; typical basis**.

A-B-C-stages These letters identify the various stages of cure when processing thermoset (TS) plastic that has been treated with a catalyst; basically A-stage is uncured, B-stage is partially cured, and C-stage is fully cured. Typical B-stage are TS molding compounds and prepregs which in turn are processed to produce C-stage fully cured plastic material products; they are relatively insoluble and infusible.

Abscissa The horizontal direction in a diagram.

Accelerator Also called promoter or cocatalyst. It is a chemical substance that accelerates chemical, photochemical, biochemical, etc. reaction during processing, such as crosslinking or degradation of plastics. Action is triggered and/or sustained by another substance, such as a curing agent or catalyst, or environmental condition such as heat, radiation, or a microorganism. It can be used to hasten a chemical reaction with a catalyzed TP or TS plastic. It can be used to reduce the time required for a TS plastic to cure or harden. Often used in room temperature cures. During processing, it undergoes a chemical change.

Additive, slip An additive modifier that acts as an internal lubricant which exudes to the surface of the plastic during and immediately after processing providing the necessary lubricity to reduce or eliminate coefficient of friction in molded parts, film, etc. products.

Adiabatic It is a change in pressure or volume without gain or loss in heat. Describes a process or transformation in which no heat is added to or allowed to escape from the system.

Adiabatic calorimeter Instrument used to study chemical reactions which have a minimum loss of heat.

Adiabatic flame temperature The highest possible temperature of combustion obtained under the conditions that the burning occurs in an adiabatic vessel, that it is complete, and that dissociation does not occur.

Algebra, transfinite A branch of higher mathematics dealing with the algebra of infinity.

Algorithm Also known as a flow chart. It is an abstract description of a procedure or a program. A specified, step-by-step procedure for performing a task that will lead to a correct answer or solving a problem.

Algorithm and artificial intelligence AI programs rely less on algorithms than do conventional programs. Instead they use a procedure which does not guarantee a correct answer.

Algorithm, generic These are a class of machine-learning techniques that gain their name from a similarity to certain processes that occur in the interactions of natural, biological genes. Thus, GA is a method of finding a good answer to a problem, based on feedback received from its repeated attempts to a solution. Each attempt is called a gene.

Algorithm, recognition Computer programs or instruction sets for the recognition of specific phenomena from a processing of data acquired for the system from some external source.

Aluminum foil Al foil is a solid sheet of an appropriate Al alloy, cold rolled very thin, varying from a minimum thickness of about 0.0017 in. (0.00432 mm) to a maximum of about 0.0059 in. (0.1499 mm). In the Al industry, thickness of at least 0.006 in. (0.1524 mm) is sheet material (sheet). After (oil) cold rolling, the foil is annealed to restore its workability. From the standpoint of packaging as well as other applications, one of its most important characteristics is its impermeability to water vapor or gases. Bare foil 1.5 mil (0.0015 in. or 0.0038 mm) and thicker is completely impermeable and used in plastic coating and packaging process systems.

Antioxidant agent Also called aging retardants. AOAs are of major importance to the plastic industry because they extend the plastic's (that are effected by oxygen) useful temperature range and service life during processing and/or product use. The variety of AOAs available and their specific uses are extensive.

Arithmetic It is a branch of mathematics that deals with real numbers and computations with these numbers.

Arrhenius equation Refers to the rates of reaction vs. temperature. It is a rate equation followed by many chemical reactions.

Arrhenius plot A linear Arrhenius plot is extrapolated from the Arrhenius equation to predict the temperature at which failure is to be expected at an arbitrary time that depends on the plastic's heat aging behavior. It is usually 11,000 hours, with a minimum of 5,000 hours. This is the relative thermal index (RTI).

Asbestos It is not the name of a distinct mineral species but is a commercial term applied to fibrous varieties of several silicate minerals such as amosite and crocidolite. These extremely fine fibers are useful as fillers and/or reinforcements in plastics. Property performances include withstanding wear and high temperatures, chemical resistance, and strengths with high modulus of elasticity. When not properly handled or used, like other fibrous materials, they can be hazardous.

Aspect ratio It is the ratio of length to diameter of a material such as a rod or fiber; also the ratio of the major to minor axis lengths of a material such as a product.

A-stage See **A-B-C-stages**.

Asymmetric The opposite to symmetrical. It is irregular to form. Of such form or shape that no point, line, or plane exists about which opposite portions are exactly similar.

Asymptote A straight line connected with a curve such that as a point moves an infinite distance along the curve from the point to the line approaches zero and the slope of the curve at the point approaches the slope of the line.

Attenuation It is the diminution of vibrations or energy over time or distance; it is a decrease in the strength of a signal between two points or between two frequencies.

Bamboo   A grass or plant native to Southeast Asia having rather extremely high performance and having a rather high cellulose content. Use includes specialty papers, light fixtures, fishing rods, building scaffolds, etc. The development of the composite system can be said to be based on the idea of utilizing the growth concept of the natural composite bamboo (which is still being used as a building material in the Asian area). Bamboo stalks receive their high specific strength and modulus of elasticity from unidirectional oriented cellulose fibers that are embedded in a matrix of lignin and silicic acid. A similar situation exists in wood. However, whereas the fibers in wood are usually 1 mm long, in bamboo they reach up to 10 mm. The hallow bamboo stalks are stabilized by evenly spaced flat, strong nodes rectangular to the longitudinal axes.

Barrier plastic   Also called barrier layer. They are materials such as plastic films, sheets, etc. with low or no permeability to different products.

B-basis   The B mechanical property value is the value above which at least 90% of the population of values is expected to fall, with a confidence level of 95%. See **A-basis; population confidence interval; S-basis; typical basis**.

Black-box   A phrase used to describe a device whose method of working is ill-defined or not understood.

British thermal unit   Btu is the energy needed to raise the temperature of 1 lb of water $1°F$ $(0.6°C)$ at sea level. As an example, one lb of solid waste usually contains 4500 to 5000 Btu. Plastic waste contains greater Btu than other materials of waste.

B-stage   See **A-B-C-stages**.

Calculus   It is the mathematical tool used to analyze changes in physical quantities, comprising differential and integral calculations.

It was developed during the 17th century to study four major classes of scientific and mathematical problems of that time. (1) Find the maximum and minimum value of a quantity, such as the distance of a planet from the sun. (2) Given a formula for the distance traveled by a body in any specified amount of time, find the velocity and acceleration of the body at any instant. (3) Find the tangent to a curve at a point. (4) Find the length of a curve, the area of a region, and the volume of a solid. These problems were resolved by the greatest minds of the 17th century, culminating in the crowning achievements of Gottfried Wilhelm (Germany 1646–1727) and Isaac Newton (English 1642–1727). Their information provided useful information for today's space travel.

Cap layer   It is a plastic product that is topped or capped with another plastic.

Catalyst   Basically a phenomenon in which a relatively small amount of substance augments the rate of a chemical reaction without itself being consumed; recovered unaltered in form and amount at the end of the reaction. It generally accelerates the chemical change. The materials ordinarily used to aid the polymerization of most plastics are not catalysts in the strict sense of the word (they are consumed), but common usage during the past century has applied this name to them.

Catalyst, metallocene   Also called single site, Me and m. Metallocene catalysts achieve creativity and exceptional control in polymerization and product design permitting penetration of new markets and expand on of present markets.

Chromatography   A technique for separating a sample material into constituent components and then measuring or identifying the compounds by other methods. As an example separation, especially of closely related compounds, is caused by allowing a solution or mixture to seep through an absorbent such as clay, gel, or paper. Result is that each compound becomes adsorbed in a separate, often colored layer.

Cold flow It is creep at room temperature.

Colorimeter Also called color comparator or photoelectric color comparator. An instrument for matching colors with results about the same as those of visual inspection, but more consistent. Basically the sample is illuminated by light from the three primary color filters and scanned by an electronic detecting system. It is sometimes used in conjunction with a spectrophotometer, which is used for close control of color in production.

Combination mold A mold which has both positive portions or ridges, and cavity portions.

Computer acceptability Information produced via CAD, CAM, CAE, etc. that may require a password.

Computer science and algebra The symbolic system of mathematical logic called Boolean algebra represents relationships between entities; either ideas or objects. George Boole of England formulated the basic rules of the system in 1847. The Boolean algebra eventually became a cornerstone of computer science.

Concentricity Term to describe two circles or cylindrical shapes having a common center and common axis, such as the inside or outside diameters of a barrel or outside diameters of the surface and bearing surfaces of a screw. Deviation from concentricity is referred to as runout. Also refers to the relationship of all inside dimensions to all outside dimensions usually expressed in thousands of inch or millimeter FIM (full indicator movement). Deviation from concentricity is usually referred to as a runout. The concentricity should allow for the maximum part tolerance. The geometry of the part should help indicate the tolerance applied.

Corian DuPont's trade-name for their mineral filled acrylic continuous cast sheet material. This wear resistant and attractive material is used for consumer's kitchen counter tops, bathroom surfaces, fast-food restaurant surfaces, health-care surfaces, etc.

Crystallinity and orientation When crystallites already exist in the amorphous matrix, orientation will make these crystallites parallel. If a plastic crystallizes too far in the melt, it may not contain enough amorphous matrix to permit orientation, and will break during stretching. (Most partially crystalline plastics can be drawn 4 to 5 times.) The degree of crystallinity is influenced by the rate at which the melt is cooled. This is utilized in the fabrication operations to help control the degree of crystallinity. The balance of properties can be slightly altered in this manner, allowing some control over such parameters as container volume, stiffness, warpage, and brittleness. Nucleating agents are available that can promote more rapid crystallization resulting in faster cycle times.

Crystallization The formation of crystallites or groups of plastic molecules in an ordered structure within the plastic as the plastic is cooled from its amorphous state to a temperature below its crystallization temperature.

C-stage See **A-B-C-stages**.

Curing Refers to TS plastics that undergo chemical change (cross-linking) during processing to become permanently insoluble and infusible plastics. TPs do not go through a curing cycle, they go through a cycle of repeatedly softening when heated and hardening when cooled. At times TPs are referred to as curing even though it does not go through a curing stage as in TSs. The term cure was extensively used from the start of the 20th century because practically the only plastics used were phenolics. They are TSs that cure. So the term was an incorrect carryover when TPs became popular. See **Vulcanization**.

Damping The loss of energy, as dissipated heat, that results when a material or material system is subjected to an oscillatory load or displacement. Perfectly elastic materials have no mechanical damping. Damping reduces vibrations (mechanical and acoustical) and prevents resonance vibrations from building up to dangerous amplitudes. However, high damping is generally an indication of reduced dimensional stability, which can be very undesirable in structures carrying loads for long time periods. Many other mechanical properties are intimately related to damping; these include fatigue life, toughness and impact, wear and coefficient of friction, etc.

Deformation   Any part of the total deformation of a body that occurs immediately when the load is applied but that remains permanently when the load is removed.

Denier See Fiber denier

Density, apparent   The weight in air of a unit volume of material including voids usually inherent in the material. Also used is the term bulk density that is commonly used for materials such as molding powders.

Density, bulk   Ratio of weight to volume of a solid material including voids but more often refers to loose form (bulk) material such as pellets, powders, flakes, compounded molding material, etc.

Design allowable   Statistically defined material property allowable strengths, usually referring to stress and/or strain.

Design motion control, mechanical and electronic effects   Selecting a control system is not something that can, or should, be done without proper consideration. Your decision should be guided by a number of parameters. It basically depends on: (1) whether you need a brand-new system or a retrofit, (2) one that is 100% computer controlled or covers select functions, (3) one with leading edge technologies or with just enough high technology to get you up and running, (4) one that designed in-house or by automation specialists, or (5) something in between these choices. It all depends on your specific requirements. In motion controls there are different operating mechanical devices such as actuators that convert the rotation of a motor into linear motion, linear guides, linear

bearings, properly deigned machine structure to ensure rigidity and proper mounting installation, mechanical dampers to isolate the motion system from its environment, ensure control of inertia when components move or cause friction, avoid resonance problems, protect against dirt, etc.

These are a few of the mechanical factors that have much more effect on the electronic design of motion control systems. The electronic engineer must understand the mechanics of motion that are encountered in order for the electronic system to be successful. To decide on electronic and software requirements, it is important factors have to be considered such as product flow and throughput, operator requirements, and maintenance issues.

Deviation   Variation from the a specified dimension or design requirement, usually defining the upper and lower limits. The mean deviation (MD) is the average deviation of a series of numbers from their mean. In averaging the deviations, no account is taken of signs, and all deviations whether plus or minus, are treated as positive. The MD is also called the mean absolute deviation (MAD) or average deviation (AD).

Deviation, root-mean-square   RMS is a measure of the average size of any measurable item (length of bar, film thickness, pipe thickness, coiled molecule, etc.) that relates to the degree of accuracy per standard deviation measurement.

Differential scanning calorimetry DSC is a method in which the energy absorbed or produced is measured by monitoring the difference in energy input (energy changes) into the material and a reference material as a function of temperature. Absorption of energy produces an endothermic reaction; production of energy results in an exothermic reaction. Its use includes studying processing behavior of the melting action, degree of crystallization, degree of cure, applied to processes involving a change in heat capacity such as the glass transition, loss of solvents, etc.

Dilatant Basically a material with the ability to increase the volume when its shape is changed. A rheological flow characteristic evidenced by an increase in viscosity with increasing rate of shear. The dilatant fluid, or inverted pseudoplastic, is one whose apparent viscosity increases simultaneously with increasing rate of shear; for example, the act of stirring creates instantly an increase in resistance to stirring.

Dilatometer Basically it is a pyrometer equipped with instruments to study density as a function of temperature and/or time. It can measure the thermal expansion or contraction of solids or liquids. They also study polymerization reactions; it can measure the contraction in volume of unsaturated compounds. It basically is a technique in which a dimension of a material under negligible load is measured as a function of temperature while it is subjected to a controlled temperature program.

EI theory In all materials (plastics, metals, wood, etc.) elementary mechanical theory demonstrates that some shapes resist deformation from external loads. This phenomenon stems from the basic physical fact that deformation in beam or sheet sections depends upon the mathematical product of the modulus of elasticity (E) and the moment of inertia (I), commonly expressed as EI. This theory has been applied to many different constructions including sandwich panels.

Elastic-plastic transition It is the changes from recoverable elastic behavior to non-recoverable plastic strain which occurs on stressing a material beyond its yield point.

Endotherm A process or change that takes place with absorption of heat and requires high temperature for initiation and maintenance as with using heat to melt plastics and then remove heat; as opposed to endothermic.

Endothermic Also called endoergic. Pertaining to a reaction which absorbs heat.

Endothermic process Processes that absorb heat from the surroundings.

Energy Basically, it is the capacity for doing work or producing change. This term is both general and specific. Generally it refers to the energy absorbed by any material subjected to loading. Specifically it is a measure of toughness or impact strength of a material; as an example, the energy needed to fracture a specimen in an impact test. It is the difference in kinetic energy of the striker before and after impact, expressed as total energy per inch of notch of the test specimen for plastic and electrical insulating material [in-lb (J/m)]. Higher energy absorption indicates a greater toughness. For notched specimens, energy absorption is an indication of the effect of internal multi-axial stress distribution on fracture behavior of the material. It is merely a qualitative index and cannot be used directly in design.

Energy absorption A term that is both general and specific. Generally I refers to the energy absorbed by any material subjected to loading. Specifically it is a measure of toughness or impact strength of a material; the energy needed to fracture a specimen in an impact test.

Energy activation An excess energy that must be added to an atomic or molecular system to allow a process, such as diffusion or chemical reaction, to proceed.

Energy and bottle An interesting historical (1950s) example is the small injection blow molded whiskey bottles that were substituted for glass blown bottles in commercial aircraft; continues to be used in all worldwide flying aircraft. At that time, just in USA, over 500×10^{12} Btu or the amount of energy equivalent to over 80×10^6 barrels of oil was reduced.

Energy and plastic Numerous studies have shown; (1) plastics consume less energy to produce and fabricate products than other materials with glass being the major consumer of energy; (2) their use as a product reduces energy consumption; and (3) more energy can be produced when products are incinerated. Without plastic insulation, major appliances such as refrigerators would use up to 30% more energy. Improvements made in energy efficiency made through the use of plastics in the last decade save more than 53 billion kilowatt-hours of electricity annually in USA. This saves consumers more than $4 billion each year.

Energy conservation   1. Much less energy is used to in the production of plastics then probably any other commercial material. In comparing the ratio of polyethylene (PE) plastic with others, steel requires about three times as much, copper at 18, and aluminum at almost 10. 2. When examining energy consumption or lost, the equipment used in the complete production line as well as the use of plastic products is involved, plastic requires less particularly when compared to glass. Plastics are major contributors to saving energy such as building insulation, reduce weight in automobiles, etc. Upon incineration, plastic provides much more energy that can be put to work.

Energy consumption   The plastics industry consumes about 3% of U.S. total annual oil and gas consumption. This use is more than offset by the savings that plastic products create. Many different studies have substantiated this fact. Worldwide there are areas where the consumption may be lower and possibly greater reaching up to 4%.

Eutectic   It is a mixture of two or more substances that solidifies as a whole when cooled from the liquid state, without changing composition. It is the composition within any system of two or more crystalline phases that melts completely at the minimum temperature.

Eutectic arrest   In a cooling (or heating) curve an approximately exothermal segment corresponding to the time interval during which the heat of transformation from the liquid phase of two or more conjugate solid phases is being evolved (or conversely).

Eutectic composition   It has a minimum melting temperature when two or more liquid solubility curves interact.

Eutectic deformation   The composition within a system of two or more components, which on heating under specified conditions, develops sufficient liquid to cause deformation at the minimum temperature.

Eutectic divorced   Structure where the components of a eutectic mixture appear to be entirely separate.

Eutectic mixture   It is the composition within any system of two or more crystalline phases that melts completely at a relatively low temperature and can be repeatedly solidified and melted. Two or more substances solidify (such as zinc-aluminum, tin, bismuth, etc.) as a whole when cooled from the liquid state without changing composition. They have been used rather extensively since at least the 1940s during certain plastic processing techniques such as injection molding, casting; reinforced plastic bag molding, and compression molding.

Eutectic temperature   Melting temperature of an alloy with a eutectic mixture; it is at the interaction of two or more liquid solubility curves.

Exotherm   It is the temperature vs. time curve of a chemical reaction or a phase change giving off heat, particularly the polymerization of thermoset plastics. Maximum temperature occurs at peak exotherm. Some plastics such as room temperature curing TS polyesters and epoxies will exotherm severely with damaging results if processed incorrectly. As an example, if too much methyl ethyl ketone peroxide (MEK peroxide) catalyst is added to polyester plastic that contains cobolt naphthenate (promoter), the mix can get hot enough to smoke and even catch fire. Thus, an exotherm can be a help or hindrance, depending on the application such as during casting, potting, etc.

Exotherm curve Temperature vs. time curves during a curing cycle. Peak exotherm is the point of highest temperature of a plastic during the cure.

Exotherm heat It is heat given off during a polymerization reaction by the chemical ingredients as they react and the plastic cures.

Exothermic reaction The temperature rise resulting from the liberation of heat by a chemical reaction. It is the opposite of endothermic reaction.

Extensometer An extensometer is an instrument to monitor strain in the linear dimension of a test specimen while a load or force is applied to it. The automatic plotting of load with strain produces stress-strain curves.

Extrudate It is the plastic hot melt as it emerges/discharges/exits from the extruder die's orifice into a desired product form such as film, sheet, etc.

Fabric Any woven, knitted, felted, bonded, knotted, etc., textile material. There are woven and nonwoven fabrics.

Fabric woven, eight-harness satin It is a seven-by-one weave where a filling thread floats over seven warp threads and then under one. Like a crowfoot weave, it looks different on one side than the other side. This weave is more pliable than others and is especially adaptable to forming around the more complex shapes.

Fiber denier It is a unit of weight expressing the size or coarseness but particularly the fineness of a continuous fiber or yarn. The weight in grams of 9000 m (30,000 ft) is one denier. The lower the denier, the finer the fiber, yarn, etc. One denier equals about 40 micron. Sheer women's hosiery usually runs 10 to 15 denier. Commercial work 12 to 15 denier fiber is generated.

Flash It is defined as that portion of the material that flows from or is extruded from the mold [usually where two separate sections (parting line) of the mold meet] during the molding process. When this excess is trimmed off the finished piece, there will usually be a noticeable line around the edge of the product that may or may not distract from the appearance of the finished product. This problem can be avoided by properly designing the product or the molds so as to eliminate the flash problem or place the flash lines where they will not be seen.

Flocked Usually provides a velvet-like layer for decorative products.

Fuzzy logic control Although fuzzy logic control (FLC) may sound exotic, it has been used to control many conveniences of modern life (from elevators to dishwashers) and more recently into industrial process control that include plastic processing such as temperature and pressure. FLC actually outperforms conventional controls because it completely avoids overshooting process limits and dramatically improves the speed of response to process upsets. These controllers accomplish both goals simultaneously, rather than trading one against another as done with proportional-integral-derivative (PID) control. However, FLC is not a cure-all because not all FLCs are not equal; no more than PIDs. FL is not needed in all applications; in fact FLCs used allow them to be switched off so that traditional PID control takes over.

Geomembrane These liners chiefly provide impermeable barriers. They can be characterized as : (1) solid waste containment: hazardous landfill, landfill capping, and sanitary landfill; (2) liquid containment: canal, chemical/brine pond, earthen dam, fish farm, river/coastal bank, waste-water, and recreation; (3) mining, leach pad and tailing ponds; and (4) specialties: floating reservoir caps, secondary containment, tunnel, erosion, vapor barrier, and water purification. Plastics used include medium to very low density PE, PVC, and chlorosulfonated PE (CSPE). (The Romans used in their land and road constructions what we call geomembrane.)

Geometric steradian The solid angle which, having its vertex in the center of a sphere, cuts off an area of the surface of the sphere equal to that of a square with sides of length equal to the radius of the sphere.

Glassy state In amorphous plastics, below the T_g, cooperative molecular chain motions are "frozen", so that only limited local motions are possible. Material behaves mainly elastically since stress causes only limited bond angle deformations and stretching. Thus, it is hard, rigid, and often brittle.

Good manufacturing practice See **Quality system regulation**.

Gough-Joule effect When an elastomer/rubber is stretched adiabatically (without heat entering or leaving the system), heat is evolved. This effect was first reported discovered by Gough in 1805 and rediscovered by Joule in 1859.

Heat sink A device for the absorption or transfer of heat away from a critical element or part. Bulk graphite is often used as a heat sink.

Hooke's Law It is the ratio of normal stress to corresponding strain (straight line) for stresses below the proportional limit of the material.

Inlay or overlay They can be applied to plastic fabricating processes such as moldings during or after molding.

Isothermal 1. Relating to or marked by changes of volume or pressure under conditions of constant temperature. 2. Relating to or marked by constant or equality of temperature.

Kinetic A branch of dynamics concerned with the relations between the movement of bodies and the forces acting upon them.

Kinetic energy dissipated Different plastics provide different degrees and excellencies for producing parts that absorb and dissipate energy, usually from impact.

Kinetic friction It is the friction developed between two bodies in motion.

Kinetic theory A theory of matter based on the mathematical description of the relationship between pressures, volumes, and temperatures of gases (PVT phenomena). This relationship is summarized in the laws of Boyle's law, Charle's law, and Avogadro's law.

Latex Also called emulsion. It is an aqueous dispersion of natural or synthetic elastomeric rubbers and plastics (dispersions of plastic particles in water).

Load The term load means mass or force depending on its use. As an example, a load that produces a vertically downward force because of gravity acting on a mass may be expressed in mass units. Any other load is expressed in force units.

Load amplitude One-half of the algebraic difference between the maximum and minimum loads in the load cycle.

Logarithm The exponent that indicates the power to which a number is raised to produce a given number. Thus, as an example, 1000 to the base of 10 is 3. This type of mathematics is used extensively in computer software.

Lubricity Refers to the load-bearing characteristics of a plastic under conditions of relative motion. Those with good lubricity tend to have low coefficients of friction either with themselves or other materials and have no tendency to gall.

Mathematical dimensional eccentricity The ratio of the difference between maximum and minimum dimensions on a part, such as wall thickness. It is expressed as a percentage to the maximum.

Mathematician Galois Evariste Galois, now recognized as one of the greatest 19th century mathematicians, twice failed the entry exam for the Ecole Polytechnique and a paper he submitted to the French Academy of Sciences was rejected as "incomprehensible." Embittered he turned to political activism and spent six years in prison. In 1832, at the age of 20, he was killed in a duel, reported to have arisen from a lover's quarrel, although their were those who believed that an/agent provocateur of the police was involved.

Mean Arithmetical average of a set of numbers. It provides a value that lies between a range of values and is determined according to a prescribe law.

Mean absolute deviation MAD is a statistical measure of the mean (average) difference between a product's forecast and actual usage (demand). The deviations (differences) are included without regard to whether the forecast was higher than actual or lower.

Mean and standard deviation The statistical normal curve shows a definite relationship among the mean, the standard deviation, and normal curve. The normal curve is fully defined by the mean, that locates the normal curve, and the standard deviation that describes the shape of the normal curve. A relationship exists between the standard deviation and the area under the curve.

Mean, arithmetic More simply called the mean, it is the sum of the values in a distribution divided by the number of values. It is the most common measure of central tendency. The three different techniques commonly used are the raw material or ungrouped, grouped data with a calculator, and grouped data with pencil and paper.

Metallocene See **catalyst, metallocene**.

Mohr's circle A graphical representation of the stresses acting on the various planes at a given point.

Moment of inertia It is the ratio of torque applied to a rigid body free to rotate about a given axis to the angular acceleration thus produced about that axis.

Motion control system MCSs are the major user of electrical power. As with plastic equipment, they can be found in practically every aspect of our lives, performing the task of converting electrical energy to mechanical energy in a series of controlled motion activities such as those in fabricating equipment. There are constant speed, variable speed, and positioning motion control systems. Although there are many good tuning methods and self-tuning algorithms available to properly tune a motor, most of them are keyed to specific brands of motion controllers or specific types of operations. A very popular style of gain algorithm in use is the proportional-integral-derivative.

Necking Also called neck-down or neck-in. (1) It is the localized reduction in cross section that occurs in a material under tensile or compression stress during thermoforming. (2) During fabrication of products necking occurs such as extruding film where the width of extrudate leaving the die is necked-in as it moves downstream.

Non-resonant forced and vibration technique A technique for performing dynamic mechanical measurements in which the sample is oscillated mechanically at a fixed frequency. Storage modulus and damping are calculated from the applied strain and the resultant stress and shift in phase angle.

Oil-canning The property of a panel that flexes past a theoretical equilibrium point, and then returns to the original position. This motion is analogous to the bottom of a metal oilcan when pressed and released. Part flexing can cause stress, fracturing, or undesirable melting of thin-sectioned, flat parts.

Operator's station That position where an operator normally stands to operate or observe the machine.

Ordinate The vertical direction in a diagram.

Orientation and glassy state An important transition occurs in the structure of both crystalline and non-crystalline plastics. This is the point at which they transition out of the so-called glassy state. Rigidity and brittleness characterize the glassy state. This is because the molecules are too close together to allow extensive slipping motion between each other. When the glass transition (T_g) is above the range of the normal temperatures to which the product is expected to be subjected, it is possible to blend in materials that can produce the T_g of the desired mix. This action yields more flexible and tougher plastics.

Orientation and heat-shrinkability There are oriented heat heat-shrinkable plastic products found in flat, tubular film, and tubular sheet. The usual orientation is terminated (frozen) downstream of a stretching operation when a cold enough temperature is achieved. Reversing this operation occurs when the product is subjected to a sufficient high temperature. This reheating results in the product shrinking. Use for these products includes part assemblies, tubular or flat communication cable wraps, furniture webbing, medical devices, wire and pipe fitting connections or joints, and so on.

Orientation and mobility Orientation requires considerable mobility of large segments of the plastic molecules. It cannot occur below the glass transition temperature (T_g). The plastic temperature is taken just above T_g.

Orientation, cold stretching Plastics may be oriented by the so-called cold stretching; that is below its glass transition temperature (T_g). There has to be sufficient internal friction to convert mechanical into thermal energy, thus producing local heating above T_g. This occurs characteristically in the necking of fibers during cold drawing.

Orientation, thermal characteristic These oriented plastics are considered permanent, heat stable materials. However, the stretching decreases dimensional stability at higher temperatures. This situation is not a problem since these type materials are not exposed to the higher temperatures in service. For the

heat-shrink applications, the high heat provides the shrinkage capability.

Orientation, wet stretching For plastics whose glass transition temperature (T_g) is above their decomposition temperature, orientation can be accomplished by swelling them temporarily with plasticizing liquids to lower their T_g of the total mass, particularly in solution processing. As an example, cellulose viscous films can be drawn during coagulation. Final removal of the solvent makes the orientation permanent.

Plasticator A very important component in a melting process is the plasticator with its usual specialty designed screw and barrel used that is used in different machines (extruders, injection molding, blow molding, etc.). If the proper screw design is not used products may not meet or maximize their performance and meet their cost requirements. The hard steel shaft screws have helical flights, which rotates within a barrel to mechanically process and advance (pump) the plastic. There are general purposes and dedicated screws used. The type of screw used is dependent on the plastic material to be processed.

Plasticize It is the mixing action to soften a plastic and make it processable usually through the use of heat.

Plasticizer They are materials that may be added to thermoplastics to increase toughness and flexibility and/or to increase the ease of fabrication. These materials are usually more volatile than the plastics to which they are added.

Plastic, virgin A plastic material in the form of pellets, granules, powder, flock, liquid, etc. that has not been subjected to use or processing other than what was required for its initial manufacture. It is not recycled plastics.

Population confidence interval The limits on either side of a mean value of a group of observations which will, in a stated fraction or percent of the cases, include the expected value. Thus the 95% confidence limits are the layers between which the population mean will be situated in 95 out 100 cases. See **A-basis; B-basis; S-basis; typical basis**.

Positive mold A projecting mold over which the product is formed, usually referred to as a male mold.

Polyolefins Plastics such as polyethylene (PE), polypropylene (PP), and polybutylene (PB) that are derived from unsaturated hydrocarbons (also called olefins).

Primary structure Mainframe of a product is the primary structure, Examples include aircraft main supports, building main beams, and automobile frames. If the primary structure fails, it would be damaging or catastrophic to the product and/or people. See **Secondary structure**.

Processing intelligent What is needed is to cut inefficiency, such as the variables, and in turn cut the costs associated with them. One approach that can overcome these difficulties is called intelligent processing (IP) of materials. This technology utilizes new sensors, expert systems, and process models that control processing conditions as materials are produced and processed without the need for human control or monitoring. Sensors and expert systems are not new in themselves.

What is novel is the manner in which they are tied together. In IP, new nondestructive evaluation sensors are used to monitor the development of a materials microstructure as it evolves during production in real time. These sensors can indicate whether the microstructure is developing properly. Poor microstructure will lead to defects in materials. In essence, the sensors are inspecting the material on-line before the product is produced.

Processing, intelligent communication The information these sensors gather is communicated, along with data from conventional sensors that monitor temperature, pressure, and other variables, to a computerized decision making system. This decision-maker includes an expert system and a mathematical model of the process. The system then makes any changes necessary in the production process to ensure the material's structure is forming properly. These might include changing temperature or pressure, or altering other variables that will lead to a defect-free end product.

Processing, intelligent systematic There are a number of benefits that can be derived from intelligent processing. There is, for instance, a marked improvement in overall product quality and a reduction in the number of rejected parts. And the automation concept that is behind intelligent processing is consistent with the broad, systematic approaches to planning and implementation being undertaken by industries to improve quality. It is important to note that intelligent processing involves building in quality rather than attempting to obtain it by inspecting a product after it is manufactured. Thus, industry can expect to reduce post-manufacturing inspection costs and time. Being able to change manufacturing processes or the types of material being produced is another potential benefit of the technique.

Processing line, downstream The plastic discharge end of the fabricating equipment such as the auxiliary equipment in an extrusion pipe line after the extruder.

Processing line downtime Refers to equipment that cannot be operate when it should be operating. Reason for downtime could be equipment being inoperative, shortage of material, electric power problem, operators not available, and so on. Regardless of reason, downtime is costly.

Processing line, upstream Refers to material movement and auxiliary equipment (dryer, mixer/blender, storage bins, etc.) that exist prior to plastic entering the main fabricating machine such as the extruder.

Qualified products list QPL is a list of commercial products that have been pretested and found to meet the requirements of a specification.

Quality system regulation The past good manufacturing practice (GMP) and process validation (PV) was renamed to quality system regulation (QSR). It is important for the medical device industry (that uses an extensive amount of plastics) and also in other product industries where they want to follow strict processing procedures. It sets up an important procedure for many plastic fabricators to consider that targets to ensure meeting zero defects.

The originator FDA (Food & Drug Administration) defined GMP and PV as a documented program providing a high degree of assurance that a specific process will consistently produce a product meeting its predetermined specifications and quality attributes. Elements of validation are product specification, processing equipment, and process revalidation and documentation. The GMP regulation became effective during 1978. As of October 7, 1996 GMP was revised incorporating many changes; it was renamed to quality system regulation (QSR). The GMP focused almost exclusively on production practices requiring very detailed manufacturing procedures and extremely very detailed documentation.

The QSR major new requirements are in the areas of design, management responsibility, purchasing, and servicing. It encompasses quality system requirements that apply to the entire life cycle of a device.

Radome Also called radiation dome. It is a cover for a microwave antenna used to protect the antenna from the environment on the ground, underwater, and in the air (aircraft nose cone, etc.). The dome is basically transparent to electromagnetic radiation and structurally strong. Different materials have been used such as wood, rubber-coated air-supported fabric, etc. The most popular is the use of glass fiber-TS polyester RPs. The shape of the dome, that is usually spherical, is designed not to interfere with the radiation.

Relative thermal index Section UL 746B provides a basis for selecting high temperature plastics and provides a long-term thermal aging index called the relative thermal index (RTI).

Reynold's number It is a dimensionless number that is significant in the design of any system in which the effect of viscosity is important in controlling the velocities or the flow pattern of a fluid. It is equal to the density of a fluid, times its velocity, times a characteristic length, divided by the fluid viscosity. This value or ratio is used to determine whether the flow of a fluid through a channel or passage, such as in a mold, is laminar (streamlined) or turbulent.

Root-mean-square See **Deviation, root-mean-square**.

Roving The term roving is used to designate a collection of bundles of continuous filaments/fibers, usually glass fibers, either untwisted strands or twisted yarns. Rovings can be lightly twisted; their degree of twisting and format depends on their use. As an example for filament winding they are generally wound as bands or tapes with little twist as possible.

Safety device All machines are equipped (or should be equipped) with applicable electrical, hydraulic, and/or mechanical safety devices. Some of them, such as injection molding machines, have all three modes for safety operations.

Safety interlock A safety device designed to ensure that equipment will not operate until certain precautions are taken and set on the equipment.

Secondary structure A structure that is not critical to the survival of the primary structure. Examples are in aircraft and aerospace vehicles that are not critical to safety of operations such as the interior decorative paneling. See **Primary structure**.

Shrinkage block jig A metal, wood, plastic, etc. shaped block against which parts are held under light or no pressure while cooling to reduce warpage and distortion.

Specific gravity, apparent The ratio of the weight in air of a given volume of the impregnable portion of a permeable material (that is the solid matter including its permeable pores or voids) to the weight in air of an equal volume of distilled water at a stated temperature.

Specific gravity, bulk The weight in air of a given volume of a permeable material (including both permeable and impermeable voids normal to the material) to the weight in air of an equal volume of distilled water at a stated temperature.

Specific heat It is the amount of heat in BTU required to raise one pound of any material one degree Fahrenheit (BTU/lb, 10F).

Stabilizer These are agents or materials present in or added into practically all different plastics to improve their performances that range in the many different requirements needed in the fabricated product to meet performance requirements. They basically inhibit chemical reactions that bring about undesirable chemical degradation.

Standard industrial classification The SIC system published by the U.S. Department of Commerce classifies all manufacturing industries and services produced in USA (transportation, communication, electronic, plastic, etc.). Their digital numbering system follows a pattern that provides input/output detailed information data. Basically the I/O program determines what each of about 470 product level industries consumes from each of the other 370 industries. The manufacturing segments of the plastics industry are in the major group numbers 28 (chemicals and allied products) and 30 (rubber and miscellaneous plastics products). In four-digit listings such as the SIC 2821 (plastics materials), SIC 3081 (unsupported plastics film and sheet), SIC 3084 (plastics bottles), SIC 3086 (plastics foam products), SIC 3088 (plastics plumbing fixtures), and so on.

Statistical benefit Using statistical methods in the design of experiments and data analysis allows designers, etc. to attain benefits that would otherwise be considered unachievable. Benefits include a 20 to 70% reduction in problem solving time; a minimum 50% reduction in costs due to testing, machine processing time, labor, and materials; and a 200 to 300% increase in value, quality, and reliability of the information generated.

Supercooling The rapid cooling of a normally crystalline plastic through its crystallization temperature, so it does not get a chance to crystallize and it remains in the amorphous state.

Synergism Arrangement or mixture of substances in which the total resulting performance is greater than the sum of the effects taken independently such as with alloying/blending.

Thermodynamic It is the scientific principle that deals with the inter-conversion of heat and other forms of energy. Thermodynamics (thermo = heat and dynamic = changes) is the study of these energy transfers. The law of conservation of energy is called the first law of thermodynamics.

Thermodynamic, first law Energy can be converted from one form to another but it cannot be created or destroyed.

Thermodynamic phase transformation In thermodynamic equilibrium a system may be composed of one or several physically distinct macroscopic homogeneous parts called phases, which are separated from one another by well defined interfaces. These phases are determined by several parameters such as temperature, pressure, and electric and magnetic fields. By continuously varying the parameters it is possible to induce the transformation of the system from one phase to another.

Thermodynamic property With the heat exchange that occurs during heat processing, thermodynamics becomes important and useful. It is the heat content of the melts (about 100 cal/g) combined with the low rate of thermal diffusion (10^{-3} cm^2/s) that limits the cycle time of many processes. Also important are density changes, which for crystalline plastics may exceed 25% as melts cool. Melts are highly compressible; a 10% volume change for 10,000 psi (69 MPa) force is typical. Surface tension of about 20 g/cm may be typical for film and fiber processing when there is a large surface-to-volume ratio.

Thermodynamic properties provide a means of working out the flow of energy from one system to another. Any substance of specified chemical composition. perpetually in electrical, magnetic, and gravitational fields, have five fundamental thermodynamic properties, namely pressure, temperature, volume, internal energy, and entropy. All changes in these properties must fulfill the requirements of the first and second law of thermodynamics. The **third law** provides a reference point, the absolute zero temperature, for all these properties although such a reference state is unattainable. The proper modes of applying these laws to the above five fundamental properties of an isolated system constitute the well-established subject of thermodynamics.

Thermodynamic, second law   The entropy of the universe increases in a spontaneous process and remains unchanged in a reversible process. It can never decrease.

Thermodynamic, statistical   This discipline tries to compute macroscopic properties of materials from more basic structures of matter. These properties are not necessarily static properties as in conventional mechanics. The problems in statistical thermodynamics fall into two categories. First it involves the study of the structure of phenomenological frameworks and the interrelations among observable macroscopic quantities. The secondary category involves the calculations of the actual values of phenomenology parameters such as viscosity or phase transition temperatures from more microscopic parameters. With this technique, understanding general relations requires only a model specified by fairly broad and abstract conditions. Realistically detailed models are not needed to understand general properties of a class of materials. Understanding more specific relations requires microscopically detailed models.

Thixotropic   A property of a plastic that is a gel at rest but liquefies upon agitation and losing viscosity under stress. Liquids containing suspended solids are likely to be thixotropic. They have high static shear strength with low dynamic shear strength at the same time. As an example, these materials provide the capability to be applied on a vertical wall and through quick curing action remain in its position during curing.

Tolerance, full indicator movement   FIM is a term used to identify tolerance with respect to concentricity. Terms used in the past were full indicator reading (FIR) and total indicator reading (TIR).

Tooling   Tools include dies, molds, mandrel, jigs, fixtures, punch dies, etc. for shaping and fabricating parts.

Turnkey operation   A complete fabrication line or system, such as an extruder with a thermoformer line with upstream and downstream equipment. Controls interface all the equipment in-line from material delivery to the end of the line handling the product for in-plant storage or shipment out of the plant.

Typical-basis   The typical property value is an average value. No statistical assurance is associated with this basis. See **A-basis; B-basis; C-basis; population confidence interval**.

Vacuum pressure   Gauge pressure in psi (gpsi) is the amount by which pressure exceeds the atmospheric pressure of 14 psi (negative in the case of vacuum). The absolute pressure (psia) is measured with respect to zero absolute vacuum [29.92 in. (101 kPa) Hg]. In a vacuum system it is equal to the negative gage pressure subtracted from the atmospheric pressure. (Gauge pressure + atmospheric pressure = absolute pressure) (1 in. Hg = 0.4912 psi of atmosphere on a product) (1 psi = 2.036 in. Hg).

Variable A quantity to which any of the values in a given set may be assigned.

Variable, deviation The difference between dependent variable and steady state value.

Variable, independent An experimental factor that can be controlled (temperature, pressure, order of test, etc.) or independently measured (hours of sunshine, specimen thickness, etc.). Independent variables may be qualitative (such as a qualitative difference in operating technique) or quantitative (such as temperature, pressure, or duration). Thus, if variable A is a function of variable B, than B is the independent variable.

Variance The mean square of deviations, or errors, of a set of observations; the sum of square deviations, or errors, of individual observations with respect to their arithmetic mean divided by the number of observations less one (degree of freedom); the square of the standard deviation, or standard error.

Virgin plastic See **Plastic, virgin**.

Viscoelasticity A combination of viscous and elastic properties in a plastic with the relative contribution of each being dependent on time, temperature, stress, and strain rate. It relates to the mechanical behavior of plastics in which there is a time and temperature dependent relationship between stress and strain. A material having this property is considered to combine the features of a perfectly elastic solid and a perfect fluid.

Viscoelasticity of metal This subject provides an introduction on the viscoelasticity of metals that has no bearing or relationship with viscoelastic properties of plastic materials. The aim is to have the reader recognize that the complex thermodynamic foundations of the theory of viscoplasticity exist with metals. There have been developments in the thermodynamic approach to combined treatment of rheologic and plastic phenomena and to construct a thermodynamic theory non-linear viscoplastic material that may be used to describe the behavior of metals under dynamic loads.

It has been shown that the thermodynamic foundations of plasticity may be considered within the framework of the continuum mechanics of materials with memory. A nonlinear material with memory is defined by a system of constitutive equations in which some state functions such as the stress tension or the internal energy, the heat flux, etc., are determined as functionals of a function which represents the time history of the local configuration of a material particle.

As a result of simultaneous introduction of elastic, viscous and plastic properties of a material, a description of the actual state functions involves the history of the local configuration expressed as a function of the time and of the path. The restrictions, which impose the second law of thermodynamics and the principle of material objectivity, have been analyzed. Among others, a viscoplastic material of the rate type and a strain-rate sensitive material have been examined.

There are three different approaches to a thermodynamic theory of continuum that can be distinguished. These approaches differ from each other by the fundamental postulates on which the theory is based. All of them are characterized by the same fundamental requirement that the results should be obtained without having recourse to statistical or kinetic theories. None of these approaches is concerned with the atomic structure of the material. Therefore, they represent a pure phenomenological approach. The principal postulates of the first approach, usually called the classical thermodynamics of irreversible processes, are documented. The principle of local state is assumed to be valid. The equation of entropy balance is assumed to involve a term expressing the entropy production which can be represented as a sum of products of fluxes and forces. This term is zero for a state of equilibrium and positive for an irreversible process. The fluxes are function of forces, not necessarily linear. However, the reciprocity relations concern only coefficients of the linear terms of the series expansions. Using methods of this approach, a thermodynamic description of elastic, rheologic and plastic materials was obtained.

The second approach, called the thermodynamic theory of materials with memory.

The fundamental postulates of this approach are as follows: (1) The temperature and entropy functions are assumed, to exist for non-equilibrium states, (2) The principal restriction imposed on the constitutive equations is inequality, and (3) The notion of the thermodynamic state is modified by assuming that the state of a given particle is characterized, in general, by the time history of the local configuration of that particle. It should be emphasized, however, that in particular cases the history of the local configuration of a particle can be determined by giving the actual values of this configuration and its time derivatives. No limitations are introduced for the processes considered. The constitutive equations are in general nonlinear. Within the framework of this approach, thermodynamic foundations of rheologic materials were established. The same was done for plastic materials.

The third approach is called the thermodynamic theory of passive systems. It is based on the following postulates: (1) The introduction of the notion of entropy is avoided for non-equilibrium states and the principle of local state is not assumed, (2) The inequality is replaced by an inequality expressing the fundamental property of passivity. This inequality follows from the second law of thermodynamics and the condition of thermodynamic stability. Further the inequality is known to have sense only for states of equilibrium, (3) The temperature is assumed to exist for non-equilibrium states, (4) As a consequence of the fundamental inequality the class of processes under consideration is limited to processes in which deviations from the equilibrium conditions are small. This enables full linearization of the constitutive equations. An important feature of this approach is the clear physical interpretation of all the quantities introduced.

Each of the three approaches above has its weaknesses and none is commonly accepted. The first is subject to excessive limitations in the form of the assumptions. Its present development does not appear to be promising for the overcoming of the difficulties that are encountered in nonlinear mechanics. The second approach is criticized principally from the point of view of physical foundations. The problem of physical interpretation of quantities such as the temperature or entropy has not found a detailed treatment within the framework of this approach. The advantages of the first approach are the mathematical foundations that are very well developed and offer a possibility of analysis of many interesting processes. They can also be used for the description of nonlinear materials.

The theories of elastic and viscoelastic materials can be obtained as particular cases of the theory of materials with memory. This theory enables the description of many important mechanical phenomena, such as elastic instability and phenomena accompanying wave propagation. The applicability of the methods of the third approach is, on the other hand, limited to linear problems. It does not seem likely that further generalization to nonlinear problems is possible within the framework of the assumptions of this approach. The results obtained concern problems of linear viscoelasticity.

Viscosity Basically it is the property of the resistance of flow exhibited within a body of material. Ordinary viscosity is the internal friction or resistance of a plastic to flow. It is the constant ratio of shearing stress to the rate of shear. Shearing is the motion of a fluid, layer by layer, like a deck of cards. When plastics flow through straight tubes or channels they are sheared and the viscosity expresses their resistance. The melt index (MI) or melt flow index (MFI) is an inverse measure of viscosity. High MI implies low viscosity and low MI means high viscosity. Plastics are shear thinning, which means that their resistance to flow decreases as the shear rate increases. This is due to molecular alignments in the direction of flow and disentanglements.

Viscosity is usually understood to mean Newtonian viscosity in which case the ratio of shearing stress to the shearing strain is constant. In non-Newtonian behavior, which is the usual case for plastics, the ratio varies with the shearing stress. Such ratios are often called the apparent viscosities at the corresponding shearing stresses. Viscosity is measured in terms of flow in Pa·s (P), with water as the base standard (value of 1.0). The higher the number, the less flow.

Viscosity, apparent Defined as the ratio between shear stress and shear rate over a narrow range for a plastic melt. It is a constant for Newtonian materials but a variable for plastics that are non-Newtonian materials.

Viscosity, intrinsic Also called limiting viscosity number. For a plastic, it is the limiting value of an infinite dilution. It is the ratio of the specific viscosity of the plastic solution to its concentration in moles per liter.

Vulcanization A process in which rubber or TS plastic (elastomer) undergoes a change in its chemical structure brought about by the irreversible process of reacting the materials with sulfur and/or other suitable agents. These cross-linking action results in property changes such as decreased plastic flow, reduced surface tackiness, increased elasticity, much greater tensile strength, and considerably less solubility. Similar cross-linking action occurs with thermoset plastics. See **Curing**.

X-axis The axis in the plane of a material used as 0° reference; thus the y-axes is the axes in the plane of the material perpendicular to the x-axis; thus the z-axes is the reference axis normal to the x-y plane. The term plane or direction is also used in place of axis.

Y-axis A line perpendicular to two opposite parallel faces.

Z-axis The reference axis perpendicular to x and y axes.

Appendix C

Abbreviation

AA	acrylic acid	AFPR	Assoc. of Foam Packaging Recyclers
AAE	American Assoc. of Engineers	AFRP	aramid fiber reinforced plastic
AAES	American Assoc. of Engineering Societies	AI	artificial intelligence
AAMI	Association for the advancement of medical Instrumentation	AIA	Aerospace Industries Assoc.
		AIAA	American Institute of Aeronautics & Astronauts
AAMA	American Architectural Manufacturing Assoc.	AIChE	American Institute of Chemical Engineers
AAR	Assoc. of American Railroads		
AAS	acrylate-styrene-acrylonitrile	AIMCAL	Assoc. of Industrial Metallizers, Coaters & Laminators
ABC	acrylonitrile-butadiene-styrene		
ABS	acrylontrile-butadiene-styrene	AIMMPE	American Institute of Mining, Metallurgical & Petroleum
ABR	acrylate-butadiene rubber		
ABC	acrylonitrile-butadiene-styrene	AMBA	American Mold Builders Assoc.
abs.	absolute	AMC	alkyd molding compound
—	acetal (see POM)	AN	acrylonitrile
AC	cellulose acetate	ANSI	American National Standards Institute
ACTC	Advanced Composite Technology Consortium		
		ANTEC	Annual Technical Conference (SPE)
AD	apparent density		
ADC	allyl diglycol carbonate (also see CR-39)	APC	American Plastics Council
		APET	amorphous polyethylene terephthalate
ADCB	asymmetric double cantilever beam		
		APF	Assoc. of Plastics Fabricators
AF	asbestos fiber	API	American Paper Institute
AFBM	Anti-Friction Manufacturer's Assoc.	API	American Petroleum Institute
		APME	Assoc. of Plastics Manufacturers in Europe
AFCMA	Aluminum Foil Container Manufacturer's Assoc.		
		APPR	Assoc. of Post-consumer Plastics Recyclers
AFMA	American Furniture Manufacturer's Assoc.		
		AR	aramid fiber
AFML	Air Force Material Laboratory	AR	aspect ratio

ARP	advanced reinforced plastic	CCPIA	China Plastics Processing Industry Assoc.
ASA	acrylic-styrene-acrylonitrile		
ASCE	American Society of Civil Engineers	CD	compact disk (disc)
		CFC	chlorofluorocarbon
ASM	American Society for Metals	CFR	Code of Federal Regulations
ASME	American Society of Mechanical Engineers	cg	center of gravity
		CLTE	coefficient of linear thermal expansion
ASNT	American Society for Nondestructive Testing		
		cm	centimeter
ASQC	American Society for Quality Control	CMRA	Chemical Marketing Research Assoc.
ASTM	American Society for Testing Materials	CN	cellulose nitrate (celluloid)
		CO	carbon monoxide
atm.	atmosphere or atmospheric pressure	COPE	copolyester
		CP	Canadian Plastics
		CPE	chlorinated polyethylene
bbl	barrel	CPET	chlorinated polyethylene terephthalate
BFRL	Building & Fire Research Laboratory (NIST)		
		CPI	Canadian Plastics Institute
Bhn	Brinell hardness number	CPRR	Center for Plastics Recycling Research (Rutgers Univ.)
BM	blow molding		
BMC	bulk molding compound	CPVC	chlorinated polyvinyl chloride
BO	biaxially-oriented	CR	chloroprene rubber
BOPP	biaxially-oriented polypropylene	CR	compression ratio
		CR-39	diethylene glycol bis-allyl carbonate (see also ADC)
BPF	British Plastics Federation		
BTC	Bottling Technology Council	CRP	carbon reinforced plastics
Btu	British thermal unit	CTFE	chlorotrifluorethylene
Buna	polybutadiene	CV	coefficient of variation
Butyl	butyl rubber		
		d	denier (preferred DEN)
c	centi (10^{-2})	d	density
C	carbon	3-D	three dimension
C	Celsius	D	diameter
C	Centigrade (preference Celsius)	3-D	three-dimensional
C	channel black	DAP	diallyl phthalate
C	composite	dB	decibel
C	stiffness constant	DBM	dip blow molding
Ca	calcium	DBMS	database management system
CA	cellulose acetate (CAc)	DBTT	ductile-to-brittle transition temperature
CAB	cellulose acetate butyrate		
$CaCO_3$	calcium carbonate (lime)	DFA	design for assembly
CAc	cellulose acetate	DFCA	design for competitive advantage
CAMPUS	computer-aided material preselection by uniform standards		
		DFD	design for disassembly
		DFM	design for manufacturability
CAN	cellulose acetate nitrate	DFMA	design for manufacturability/assembly
CAP	cellulose acetate propionate		
CAS	Chemical Abstracts Service	DFQ	design for quality
CBA	chemical blowing agent	DFR	design-for-recycling
CCA	cellular cellulose acetate	DGA	differential gravimetric analysis

DIF	diffusion coefficient	EPS	expandable polystyrene
DMA	dynamic mechanical analysis	ESC	environmental stress cracking
DMC	dough molding compound		
DOX	design of experiments methodology	ESCR	environmental stress cracking resistance
DRC	design rules checking	ETFE	ethylene tetrafluoroethylene
DSC	differential scanning calorimeter	EtO	ethylene oxide
DSD	Duales System Deutschland (German Recycling System)	ETP	engineering thermoplastic
		EU	entropy unit
DSQ	German Society for Quality	EU	European Union
		EUPC	European Assoc. of Plastics Converters
DTA	differential thermal analysis		
DTGA	differential thermogravimetric analysis	EUPE	European Union of Packaging & Environment
DTMA	dynamic thermomechanical analysis	Euro	European currency (started 1 January 1999 and completed 31 December 2001)
DTUL	deflection temperature under load		
DVR	dimensional velocity research	EUROMAP	European Committee of Machine Manufacturers for the Rubber & Plastics Industries (Zurich, Switzerland)
DVR	design value resource		
DVR	Dominick Vincent Rosato		
DVR	Donald Vincent Rosato		
DVR	Druckverformungsrest (compression set/German)	EVA	Environmental Protection Agency
DVR	dynamic velocity ratio	EVA	ethylene-vinyl acetate
DWP	design with plastics	EVAL	ethylene-vinyl alcohol copolymer (or EVOH)
E	elongation	EVOH	ethylene-vinyl alcohol copolymer (or EVAL)
E	modulus of elasticity or Young's modulus		
E_c	modulus, creep (apparent)	F	coefficient of friction
E_r	modulus, relaxation	F	Fahrenheit
E_s	modulus, secant	F	Farad
EBM	extrusion blow molding	F	force
ECTFE	ethylene-chlorotrifluoroethylene	FALLO	Follow ALL Opportunities
ECVM	European Council of Vinyl Manufacturers	FB	fishbone
		FDA	Food & Drug Administration
EDM	engineering development model	FDEMS	frequency-dependent electromagnetic sensor
EEC	European Economic Community		
EI	modulus (times) moment of inertia (or stiffness)	FEA	finite element analysis
		FEM	flexural elastic modulus
EIA	Electronic Industries Assoc.	FEP	fluorinated ethylene propylene
EMI	electromagnetic interference		
EO	ethylene oxide (also EtO)	FLC	fuzzy logic control
EPA	Environmental Protection Agency	FMCT	fusible metal core technology
EPIC	Environmental & Plastic Institute of Canada	fpm	feet per minute
EPP	Expandable polypropylene	FRCA	Fire Retardant Chemicals Assoc.
EPRI	Electric Power Research Institute		

FRP	fiber reinforced plastic	H_2O	water
FRTP	fiber reinforced thermoplastic	H-P	Hagen-Poiseuille
		HPLC	high pressure liquid chromatography
FRTS	fiber reinforced thermoset		
		HPM	hot pressure molding
g	gram	HRc	hardness Rockwell cone
G	giga (10^6)	HRC	high resolution chromatography
G	gravity	HSc	hardness scleroscope number
G	shear modulus (modulus of rigidity)	Hz	Hertz (cycle)
G	torsional modulus	*I*	integral
GAIM	gas assisted injection molding	*I*	moment of inertia
		IAPD	International Assoc. of Plastics Distributors
gal	gallon		
GB	gigabyte (billion bytes)	IB	isobutylene
GD&T	Geometric dimensioning & tolerancing	IBACOS	integrated building & construction solutions
GFRP	glass fiber reinforced plastic	IBM	injection blow molding
GINA	graphical integrated numerical analysis	ICM	injection-compression molding
		ID	internal diameter
GMP	good manufacturing practice (see QSR)	IEC	inelastic energy curve
		IEC	International Electrochemical Commission
GNP	gross national product (GDP replaced GNP in US 1993)		
		IEEE	Institute of electrical & Electronics Engineers
GPa	giga Pascal		
GPC	gel permeation chromatography	IGA	isothermal gravimetric analysis
		IGC	inverse gas chromatography
GPC	graphics performance characterization	IM	injection molding
		IMM	injection molding machine
gpd	grams per denier	in.	inch
GPD	gas phase deposition	I/O	input/output
GPI	Glass Packaging Institute	IOT	initial oxidation temperature
GPPS	general purpose polystyrene	IOY	I owe you
gr	grain	IPE	intelligent processing equipment
GRP	glass fiber-TS polyester reinforced plastic	ips	inch per second
		ISO	International Standardization Organization or International Organization for Standardization
GSC	gas solid chromatography		
h	hour	ISS	ion spectroscopy scattering
H	hysteresis	IT	information technology
H_2	hydrogen	IV	inherent viscosity
HB	Brinell hardness number	IVD	in vitro diagnostic
HCCN	honeycomb core crush number	J	joule
HCFC	hydrochlorofluorocarbon	J_p	polar moment of inertia
HCl	hydrogen chloride	JEOL	Japan Electron Optics Laboratory
HDBK	handbook		
HMC	high strength molding compound	JIS	Japanese Industrial Standard
		JIT	just-in-time
HMW-HDPE	high molecular weight-high density polyethylene	JIT	just-in-tolerance
		JND	just noticeable difference

JSPS	Japan Society for Promotion of Science
JSR	Japanese SBR
JSW	Japan Steel Works
JUSE	Japanese Union of Science & Engineering
JWTE	Japan Weathering Test Center
K	bulk modulus of elasticity
K	coefficient of thermal conductivity
K	Kelvin
K	Kunststoffe (plastic in German)
KB	kilobyte (1000 bytes)
kB	knowledge-based
KBE	knowledge-based engineering
kc	kilocycle
kcal	kilogram calorie
KE	kinetic energy
kg	kilogram
KISS	keep it short & simple
KISS	keep it simple & safe
KISS	keep it simple stupid
KK	thousand
Km	kilometer
KM	Kubelka-Munk theory
km/h	kilometer per hour
KO	knockout
kPa	kilopascal
KRF	Korea (South) Research Foundation
ksi	thousand pounds per square inch ($psi \times 10^3$)
kV	kilovolt
l	length
L	litre (USA liter)
lb	pound
lbf	pound-force
LC	liquid chromatography
L/D	length-to-diameter (ratio)
LDPE	low density polyethylene (also PE-LD)
LDM	light depolarization microscopy
LEED	low energy electron diffraction
LIM	liquid impingement molding (now called RIM)
LIM	liquid injection molding
LLDPE	linear low density polyethylene (also PE-LLD)

LMDPE	linear medium density polyethylene
m	matrix
m	metallocene (catalyst)
m	meter
mg	milligram
mμ	micromillimeter; millicron; 0.000001 mm
μm	micrometer
M	mega
M	million
M_b	bending moment
MAD	molding area diagram
MD	machine direction
MD	mean deviation
MD&DI	Medical Device & Diagnostic Industry
MDD	Medical Devices Directory
Me	metallocene catalyst
mg	milligram
mHDPE	metallocene HDPE (different m/plastics such as mPS, mPP, etc.)
mi	mile
MI	melt index (see MFI)
mike	microinch (10^{-6} in.)
mil	one thousand of inch (10^{-6} in.)
min	minute
min.	minimum
MIPS	medium impact polystyrene
ml	milliliter
mLLDPE	metallocene catalyst LLDPE
MM	billion
MRPMA	Malaysian Rubber Products Manufacturers' Assoc.
Msi	million pounds per square inch ($psi \times 10^6$)
MVD	molding volume diagram
MVT	moisture vapor transmission
MW	molecular weight
MWD	molecular weight distribution
MWR	molding with rotation
Mylar	polyethylene glycol terephthalate
N	Nano (10^{-9})
N	Newton (force)
N	number of cycles

NAAEE	North America Assoc. for Environmental Education		OD	outside diameter
NAAQS	National Ambient Air Quality Standards		OEM	original equipment manufacturer
NACE	National Assoc. of Corrosion Engineers		OPET	oriented polyethylene terephthalate
NACO	National Assoc. of CAD/CAM Operation		OSHA	Occupational Safety & Health Administration
NAGS	North America Geosynthetics Society		oz	ounce
NBS	National Bureau of Standards (since 1980s renamed National Institute of Standards & Technology or NIST)		%vol	percentage by volume (prefer vol%)
			%wt	percentage by weight (prefer wt%)
NCGA	National Computer Graphics Assoc.		P	load
			P	poise
NCP	National Certification in Plastics		P	pressure
NCRC	National Container Recycling Coalition		Pa	Pascal
			PA	polyamide (nylon)
NDE	nondestructive evaluation		PASS	polymer analysis & simulation software
NDI	nondestructive inspection		PBA	physical blowing agent
NDT	nondestructive testing		PC	permeability coefficient
NEAT	nothing else added to it		PC	personal computer
NEMA	National Electrical Manufacturers Assoc.		PC	plastic composite
			PC	plastic compounding
NEN	Dutch standard		PC	plastic-concrete
NFE	non-linear finite element		PC	polycarbonate
NFK	Dutch Plastics Federation		PC	printed circuit
NFPA	National Fire Protection Assoc.		PC	process control
NFPA	National Food Processors Assoc.		PC	programmable circuit
nm	nanometer		PC	programmable controller
NOS	not otherwise specified		pcf	pounds per cubic foot
NOSE	nuisance odor solution evaluator		PDFM	Plastics Distributors & Fabricators Magazine
NPCM	National Plastics Center & Museum		PE	plastics engineer
			PE	polyethylene (UK polythene)
NPE	National Plastics Exhibition (SPI)		PE	professional engineer
NPFC	National Publications & Forms Center (US gov't)		PEI	polyethylene isophthalate
			PEN	polyethylene naphthalate
NPII	National Printing Ink Institute		PET	polyethylene terephthalate
NPRC	National Polystyrene Recycling Co.		PETG	polyethylene terephthalate glycol
NQR	nuclear quadruple resonance		PETS	plastics evaluation & troubleshooting system
NTMA	National Tool & Machining Assoc.		PE-UHMW	ultra-high molecular weight polyethylene (or UHMWPE)
NWPCA	National Wooden Pallet & Container Assoc.		PFA	perfluoroalkoxy alkane
nylon	(see PA)		phr	parts per hundred
			pi	$\pi = 3.141593$
O_2	oxygen		PI	polyimide
O_3	ozone		PI	proportional integral

PIA	Plastics Institute of America	QMC	quick mold change
PID	proportional-integral-derivative	QPL	qualified products list
		QRS	quality system regulation
PIRRG	Plastics Industry Risk Retention Group (insurance)	R	Rankine
PLASTEC	Plastics Technical Evaluation Center (US Army)	R	Reaumur
		R	Reynold's number
PMMA	Plastics Molders & Manufacturers Assoc. (of SME)	R	Rockwell (hardness)
		RA	reduction of area
		radome	radar dome
PMMA	polymethyl methacrylate (acrylic)	RAPRA	Rubber & Plastics Research Assoc.
PMMI	Packaging Machinery Manufacturers Institute	RH	relative humidity
		RMS	root mean square
PN	Plastics News	RP	rapid prototyping
PO	polyolefin	RP	reactive polymer
POM	polyacetal	RP	reinforced plastic
POP	polyolefin plastomer	rpm	revolutions per minute
PP	polypropylene	RT	rapid tooling
ppb	parts per billion	RTP	reinforced thermoplastic
PPFA	Plastics Pipe & Fittings Assoc.	RTS	reinforced thermoset
pph	parts per hour	RTV	room temperature vulcanization
PPO	polyphenylene oxide	Rx	radiation curing
psi	pounds per square inch		
PSI	Polymer Search on the Internet	s	second
		SB	styrene-butadiene
psia	pounds per square inch, absolute	SCT	soluble core technology
		SDM	standard deviation measurement
psid	pounds per square inch, differential	SF	safety factor
psig	pounds per square inch, gauge (above atmospheric pressure)	SF	structural foam
		s.g.	specific gravity (SG)
PT	Plastics Technology magazine	SI	International System of Units
PTFE	polytetrafluoroethylene (TFE)	SI	silicone
PTX	pressure-temperature concentration variables	SI	swelling index
		SIC	Standard Industrial Classification
PUR	polyurethane (also PU, UP)		
p.v.	pore volume	SMCAA	Sheet Molding Compound Automotive Alliance
P-V	pressure-volume (also PV)	SME	Society of Manufacturing Engineers
PV	process validation		
PV	pressure velocity	S-N	stress-number of cycles
PVDC	polyvinylidene chloride	SN	synthetic natural rubber
PVDF	polyvinylidene fluoride	SPC	statistical process control
pVT	pressure-volume-temperature (also P-V-T or pvT)	SPE	Society of the Plastics Engineers
PW	Plastics World magazine (1997 became Molding Systems of SME)	SPI	Society of the Plastics Industry
		sPS	syndiotactic polystyrene
		Spec.	specification
		sp. gr.	specific gravity
QA	quality assurance	sp. vol.	specific volume
QC	quality control	sq.	square

SRI	Standards Research Institute (ASTM)	ULDPE	ultra low density polyethylene (or PE-ULD)
S-S	stress-strain	UR	urethane (also PUR, PU)
STP	Special Technical Publication (ASTM)	USA	United States of America (also USA)
STP	standard temperature & pressure	UV	ultraviolet
t	thickness	V	vacuum
T	temperature	V	velocity
T	time	V	volt
T	torque (or T_t)	VA	value analysis
T_g	glass transition temperature	VCM	vinyl chloride monomer
T_m	melt temperature	VDA	Assoc. of the Automotive
T_s	tensile strength		Industry (Germany)
T&E	test & evaluation	VOC	volatile organic compound
T/C	thermocouple	vol	volume
TCM	technical cost modeling	vol%	percentage by volume
TD	transverse direction	vs.	versus
TF	thermoforming		
TFE	see PTFE	w	width
TGA	thermogravimetric analysis	W	watt
TGI	thermogravimetric index	WP&RT	World Plastics & Rubber
three-D	3-dimensional (3-D)		Technology magazine
TIR	tooling indicator runout	WPC	wood-plastic composite
TIR	total indicator reading	WPC	world product code
TMA	thermomechanical analysis	wt%	percentage by weight
TMA	Tooling & Manufacturing Assoc. (formerly TDI)	WVT	water vapor transmission
		WVTR	water vapor transmission rate
torr	mm mercury (mmHg)		
TP	thermoplastic	WYSIWYG	what you see is what you get
TPE	thermoplastic elastomer		
TPO	thermoplastic olefin (TPE-O)	X	arithmetic mean
TPU	thermoplastic polyurethane	X-axis	axis in plane used as O°
TQC	total quality control		reference
TQM	total quality management	XL	cross-linked
TR	torque rheometer	XLPE	cross-linked polyethylene
TS	thermoset	XPS	expandable polystyrene
TSC	thermal stress cracking		
TSE	thermoset elastomer	Y-axis	axis in the plane perpendicular to X-axis
two-D	2-dimensional (2-D)		
TX	thixotropic	YPE	yield point elongation
Tx	toxic		
		Z-axis	axis normal to the plane of the X-Y axes
UA	urea, unsaturated		
UD	unidirectional	ZDP	zero defect product
UHMW	ultrahigh molecular weight	ZST	zero-strength time
UL	Underwriter's Laboratories	Z-twist	twisting fiber direction

Appendix D

Conversion

The following data uses the decimal point (that is, a dot, as used in the USA) rather than a comma (as widely used in the rest of the world, and eventually to be used in the USA).

a. Alphabetical list of units

Convert from	To	Multiply by
acre (43 560 square US survey feet)	square meter (m²)	4046.873
atmosphere, standard	pascal (Pa)	$1.013\,25 \times 10^5$
	kilopascal (kPa)	101.325
bar	pascal (Pa)	1.0×10^5
	kilopascal (kPa)	100
British thermal unit (Btu) (Int'l Table)	joule (J)	1055.056
British thermal unit (Btu) (thermochem.)	joule (J)	1054.350
Btu per cubic foot (Btu/ft³)	joule per cubic meter (J/m³)	3.7259×10^4
Btu per degree Fahrenheit (Btu/°F)	joule per kelvin (J/K)	1899.101
Btu per hour (Btu/h)	watt (W)	0.293 0711
Btu per pound (Btu/lb)	joule per kilogram (J/kg)	2326
centimeter of water	pascal (Pa)	98.0665
centipoise	pascal second (Pa·s)	0.001
chain (66 USA survey feet)	meter (m)	20.116 84
circular mil	square millimeter (mm²)	5.067×10^{-4}
cubic foot (ft³)	cubic meter (m³)	0.028 317
cubic foot per second (ft³/s)	cubic meter per second (m³/s)	0.028 317
cubic inch (in³)	cubic meter (m³)	$1.638\,706 \times 10^{-5}$
cubic mile	cubic meter (m³)	$4.168\,182 \times 10^9$
	cubic kilometer (km³)	4.168 182
cubic yard (yd³)	cubic meter (m³)	0.764 555
day (mean solar)	second (s)	8.64×10^4
degree	radian (rad)	0.017 453
degree Celsius (°C) (interval)	kelvin (K)	1.0
degree Celsius (°C) (temperature)	kelvin (K)	$t_{°C} + 273.15$
degree Centigrade (interval)	degree Celsius (°C)	1.0
degree Centigrade (temperature)	degree Celsius (°C)	$\approx t_{\text{centigrade}}$

a. (*Continued*)

Convert from	To	Multiply by
degree Fahrenheit (°F) (interval)	kelvin (K)	0.555 5556
	degree Celsius (°C)	0.555 5556
degree Fahrenheit (°F) (temperature)	kelvin (K)	$(t_{°F} + 459.67)/1.8$
	degree Celsius (°C)	$(t_{°F} - 32)/1.8$
degree Fahrenheit hour per Btu (°F · h/Btu)	kelvin per watt (K/W)	1.895 634
degree Rankine (°R) (interval)	kelvin (K)	0.555 5556
degree Rankine (°R) (temperature)	kelvin (K)	$T_{°R}/1.8$
denier	kilogram per meter (kg/m)	1.111×10^{-7}
dyne	newton (N)	1.0×10^{-5}
dyne centimeter	newton meter (N · m)	1.0×10^{-7}
faraday	coulomb (C)	9.649×10^{4}
fathom	meter (m)	1.8288
fermi	meter (m)	1.0×10^{-15}
	femtometer (fm)	1.0
foot	meter (m)	0.3048
foot, USA survey	meter (m)	0.304 8006
foot of water	pascal (Pa)	2989.07
	kilopascal (kPa)	2.989 07
foot pound-force (ft·lbf) (torque)	newton meter (N · m)	1.355 818
foot pound-force (ft·lbf) (energy)	joule (J)	1.355 818
gallon (Imperial)	cubic meter (m³)	$4.546\,09 \times 10^{-3}$
	liter (L)	4.546 09
gallon (USA) (231 in³)	cubic meter (m³)	$3.785\,412 \times 10^{-3}$
	liter (L)	3.785 412
gallon (USA) per day	cubic meter per second (m³/s)	$4.381\,264 \times 10^{-8}$
	liter per second (L/s)	$4.381\,264 \times 10^{-5}$
gallon (USA) per minute (gpm)	cubic meter per second (m³/s)	$6.309\,020 \times 10^{-5}$
	liter per second (L/s)	0.063 090 20
gallon (USA) per horsepower hour	cubic meter per joule (m³/J)	$1.410\,089 \times 10^{-9}$
gamma	tesla (T)	1.0×10^{-9}
hectare	square meter (m²)	1.0×10^{4}
horsepower (550 ft · lbf/s)	watt (W)	745.6999
horsepower (boiler) (≅33470 Btu/h)	watt (W)	9809.50
horsepower (electric)	watt (W)	746
horsepower (metric)	watt (W)	735.4988
horsepower (water)	watt (W)	746.043
hour	second (s)	3600
hour (sidereal)	second (s)	3590.170
hundredweight, long (112 lb)	kilogram (kg)	50.802 35
hundredweight, short (100 lb)	kilogram (kg)	45.359 24
inch	meter (m)	0.0254
inch of mercury	pascal (Pa)	3386.39
	kilopascal (Pa)	3.386 39
inch of water	pascal (Pa)	249.089
kelvin (K) (temperature)	degree Celsius (°C)	$T_K - 273.15$
kilogram-force	newton (N)	9.806 65
kilometer per hour	meter per second (m/s)	0.278
knot (nautical mile per hour)	meter per second (m/s)	0.514 4444
light year	meter (m)	$9.460\,53 \times 10^{15}$
liter	cubic meter (m³)	0.001

a. (*Continued*)

Convert from	To	Multiply by
microinch	meter (m)	2.54×10^{-8}
	micrometer (μm)	0.0254
micron	meter (m)	1.0×10^{-6}
	micrometer (μm)	1.0
mil (0.001 in)	meter (m)	2.54×10^{-5}
	millimeter (mm)	0.0254
mil (angle)	radian (rad)	9.8175×10^{-4}
	degree ($\circ$)	0.056 25
mile, international (5280 ft)	meter (m)	1609.344
mile, nautical	meter (m)	1852
mile, USA statute	meter (m)	1609.347
mile per gallon (US) (mpg)	meter per cubic meter (m/m^3)	4.2514×10^5
	kilometer per liter (km/L)	0.425 1437
mile per hour	meter per second (m/s)	0.447 04
	kilometer per hour (km/h)	1.609 344
mile per minute	meter per second (m/s)	26.8224
millimeter of mercury	pascal (Pa)	133.3224
minute (arc)	radian (rad)	2.9089×10^{-4}
minute	second (s)	60
minute (sidereal)	second (s)	59.836 17
ounce (avoirdupois)	kilogram (kg)	0.028 349 52
	gram (g)	28.349 52
ounce (Imperial fluid)	cubic meter (m^3)	$2.841 31 \times 10^{-5}$
	milliliter (mL)	28.4131
ounce (troy or apothecary)	kilogram	0.031 1348
	gram (g)	31.103 48
ounce (USA fluid)	cubic meter (m^3)	$2.957 35 \times 10^{-5}$
	milliliter (mL)	29.5735
ounce-force	newton (N)	0.278 0139
pica (computer) (1/6 in)	millimeter (mm)	4.233 333
pica (printer's)	millimeter (mm)	4.2175
pint (Imperial)	cubic meter (m^3)	5.6826×10^{-4}
	liter (L)	0.568 26
pint (USA dry)	cubic meter (m^3)	5.5061×10^{-4}
	liter (L)	0.550 61
pint (USA liquid)	cubic meter (m^3)	$4.731 76 \times 10^{-4}$
	liter (L)	0.473 176
point (computer) (1/72 in)	millimeter (mm)	0.352 7778
point (printer's)	millimeter (mm)	0.351 46
poise	pascal second (Pa·s)	0.1
pound (avoirdupois)	kilogram (kg)	0.453 592 37
pound (troy or apothecary)	kilogram (kg)	0.373 2417
pound-force	newton (N)	4.448 222
pound-force foot (lbf · ft) (torque)	newton meter (N · m)	1.355 818
pound-force per foot (lbf/ft)	newton per meter (N/m)	14.593 90
pound-force per pound (lbf/lb)	newton per kilogram (N/kg)	9.8066
pound-force per square inch (lbf/in^2) (psi)	pascal (Pa)	6894.757
	kilopascal (kPa)	6.894 757
pound per cubic foot (lb/ft^3)	kilogram per cubic meter (kg/m^3)	16.018 46
pound per cubic inch (lb/in^3)	kilogram per cubic meter (kg/m^3)	$2.767 990 \times 10^4$
pound per cubic yard (lb/yd^3)	kilogram per cubic meter (kg/m^3)	0.593 2764
pound per foot (lb/ft)	kilogram per meter (kg/m)	1.488 164

a. (*Continued*)

Convert from	To	Multiply by
pound per gallon (US) (lb/gal)	kilogram per cubic meter (kg/m^3)	119.8264
	kilogram per liter (kg/L)	0.119 8264
quart (USA dry)	cubic meter (m^3)	0.001 101 221
	liter (L)	1.101 221
quart (USA liquid)	cubic meter (m^3)	9.463 529 × 10^{-4}
	liter (L)	0.946 3529
rad (absorbed dose)	gray (Gy)	0.01
ream (printing paper)	sheets	500
revolution	radian (rad)	6.283 185
revolution per minute (rpm)	radian per second (rad/s)	0.104 7198
rod (16.5 USA survey feet)	meter (m)	5.029 210
second (angle)	radian (rad)	4.8482 × 10^{-6}
second (sidereal)	second (s)	0.997 2696
square inch (in^2)	square meter (m^2)	6.4516 × 10^{-4}
square mile	square meter (m^2)	2.589 99 × 10^6
square yard (yd^2)	square meter (m^2)	0.836 1274
stokes	square meter per second (m^2/s)	1.0 × 10^{-4}
tex	kilogram per meter (kg/m)	1.0 × 10^{-6}
therm (EEC)	joule (J)	1.0551 × 10^8
therm (USA)	joule (J)	1.0548 × 10^8
ton, assay	gram (g)	29.166 67
ton, long (2240 lb)	kilogram (kg)	1016.047
ton, metric	kilogram (kg)	1000
tonne	kilogram (kg)	1000
ton, register	cubic meter (m^3)	2.831 685
ton, short (2000 lb)	kilogram (kg)	907.1847
ton of refrigeration (12 000 Btu/h)	watt (W)	3516.853
ton (long) per cubic yard	kilogram per cubic meter (kg/m^3)	1328.939
ton (short) per cubic yard	kilogram per cubic meter (kg/m^3)	1186.553
torr	pascal (Pa)	133.322
watt	ergs per second	1 × 10^7
watt hour	joule (J)	3600
watt second	joule (J)	1.0
yard	meter (m)	0.9144
year of 365 days	second (s)	3.1536 × 10^7
year (sidereal)	second (s)	3.1558 × 10^7
year (tropical)	second (s)	3.1558 × 10^7

b. Temperature conversions

TEMPERATURE CONVERSION

−210 to 0

C.	C. or F.	F.
−134	−210	−346
−129	−200	−328
−123	−190	−310
−118	−180	−292
−112	−170	−274
−107	−160	−256
−101	−150	−238
−95.6	−140	−220
−90.0	−130	−202
−84.4	−120	−184
−78.9	−110	−166
−73.3	−100	−148
−67.8	−90	−130
−62.2	−80	−112
−56.7	−70	−94
−51.1	−60	−76
−45.6	−50	−58
−40.0	−40	−40
−34.4	−30	−22
−28.9	−20	−4
−23.3	−10	14
−17.8	0	32

1 to 25

C.	C. or F.	F.
−17.2	1	33.8
−16.7	2	35.6
−16.1	3	37.4
−15.6	4	39.2
−15.0	5	41.0
−14.4	6	42.8
−13.9	7	44.6
−13.3	8	46.4
−12.8	9	48.2
−12.2	10	50.0
−11.7	11	51.8
−11.1	12	53.6
−10.6	13	55.4
−10.0	14	57.2
−9.44	15	59.0
−8.89	16	60.8
−8.33	17	62.6
−7.78	18	64.4
−7.22	19	66.2
−6.67	20	68.0
−6.11	21	69.8
−5.56	22	71.6
−5.00	23	73.4
−4.44	24	75.2
−3.89	25	77.0

26 to 50

C.	C. or F.	F.
−3.33	26	78.8
−2.78	27	80.6
−2.22	28	82.4
−1.67	29	84.2
−1.11	30	86.0
−0.56	31	87.8
0	32	89.6
0.56	33	91.4
1.11	34	93.2
1.67	35	95.0
2.22	36	96.8
2.78	37	98.6
3.33	38	100.4
3.89	39	102.2
4.44	40	104.0
5.00	41	105.8
5.56	42	107.6
6.11	43	109.4
6.67	44	111.2
7.22	45	113.0
7.78	46	114.8
8.33	47	116.6
8.89	48	118.4
9.44	49	120.2
10.0	50	122.0

51 to 75

C.	C. or F.	F.
10.6	51	123.8
11.1	52	125.6
11.7	53	127.4
12.2	54	129.2
12.8	55	131.0
13.3	56	132.8
13.9	57	134.6
14.4	58	136.4
15.0	59	138.2
15.6	60	140.0
16.1	61	141.8
16.7	62	143.6
17.2	63	145.4
17.8	64	147.2
18.3	65	149.0
18.9	66	150.8
19.4	67	152.6
20.0	68	154.4
20.6	69	156.2
21.1	70	158.0
21.7	71	159.8
22.2	72	161.6
22.8	73	163.4
23.3	74	165.2
23.9	75	167.0

76 to 100

C.	C. or F.	F.
24.4	76	168.8
25.0	77	170.6
25.6	78	172.4
26.1	79	174.2
26.7	80	176.0
27.2	81	177.8
27.8	82	179.6
28.3	83	181.4
28.9	84	183.2
29.4	85	185.0
30.0	86	186.8
30.6	87	188.6
31.1	88	190.4
31.7	89	192.2
32.2	90	194.0
32.8	91	195.8
33.3	92	197.6
33.9	93	199.4
34.4	94	201.2
35.0	95	203.0
35.6	96	204.8
36.1	97	206.6
36.7	98	208.4
37.2	99	210.2
37.8	100	212.0

101 to 340

C.	C. or F.	F.
43	110	230
49	120	248
54	130	266
60	140	284
66	150	302
71	160	320
77	170	338
82	180	356
88	190	374
93	200	392
99	210	410
100	212	413
104	220	428
110	230	446
116	240	464
121	250	482
127	260	500
132	270	518
138	280	536
143	290	554
149	300	572
154	310	590
160	320	608
166	330	626
171	340	644

341 to 490

C.	C. or F.	F.
177	350	662
182	360	680
188	370	698
193	380	716
199	390	734
204	400	752
210	410	770
216	420	788
221	430	806
227	440	824
232	450	842
238	460	860
243	470	878
249	480	896
254	490	914

491 to 750

C.	C. or F.	F.
260	500	932
266	510	950
271	520	968
277	530	986
282	540	1004
288	550	1022
293	560	1040
299	570	1058
304	580	1076
310	590	1094
316	600	1112
321	610	1130
327	620	1148
332	630	1166
338	640	1184
343	650	1202
349	660	1220
354	670	1238
360	680	1256
366	690	1274
371	700	1292
377	710	1310
382	720	1328
388	730	1346
393	740	1364
399	750	1382

INTERPOLATION FACTORS

C.	F.		C.	F.	
0.56	1	1.8	3.33	6	10.8
1.11	2	3.6	3.89	7	12.6
1.67	3	5.4	4.44	8	14.4
2.22	4	7.2	6.00	9	16.2
2.78	5	9.0	6.66	10	18.0

NOTE:—The numbers in bold face type refer to the temperature either in degrees Centigrade or Fahrenheit which it is desired to convert into the other scale. If converting from Fahrenheit degrees to Centigrade degrees the equivalent temperature will be found in the left column, while if converting from degrees Centigrade to degrees Fahrenheit, the answer will be found in the column on the right.

$$°F = \frac{9}{5}\,(°C) + 32$$

$$°C = \frac{5}{9}\,(°F - 32)$$

c. SI prefixes

Multiplication factor	Prefix	Symbol
$1\,000\,000\,000\,000\,000\,000 = 10^{18}$	exa	E
$1\,000\,000\,000\,000\,000 = 10^{15}$	peta	P
$1\,000\,000\,000\,000 = 10^{12}$	tera	T
$1\,000\,000\,000 = 10^{9}$	giga	G
$1\,000\,000 = 10^{6}$	mega	M
$1\,000 = 10^{3}$	kilo	k
$100 = 10^{2}$	hecto	h
$10 = 10^{1}$	deka	da
$0.1 = 10^{-1}$	deci	d
$0.01 = 10^{-2}$	centi	c
$0.001 = 10^{-3}$	milli	m
$0.000\,001 = 10^{-6}$	micro	μ
$0.000\,000\,001 = 10^{-9}$	nano	η
$0.000\,000\,000\,001 = 10^{-12}$	pico	ρ
$0.000\,000\,000\,000\,001 = 10^{-15}$	femto	f
$0.000\,000\,000\,000\,000\,001 = 10^{-18}$	atto	a

d. Units in use with SI

Quantity	Unit	Symbol	Definition
Time	Minute	min	$1\ \text{min} = 60\ \text{s}$
	Hour	h	$1\ \text{h} = 60\ \text{min} = 3600\ \text{s}$
	Day	d	$1\ \text{d} = 24\ \text{h} = 86\,400\ \text{s}$
	Week, month, etc.	...	...
Plane angle	Degree	°	$1° = (\pi/180)\ \text{rad}$
	Minute	$'$	$1' = (1/60)°$
			$= (\pi/10\,800)\ \text{rad}$
	Second	$''$	$1'' = (1/60)'$
			$= (\pi/648\,000)\ \text{rad}$
Volume	Litre	L	$1\ \text{L} = 1\ \text{dm}^3 = 10^{-3}\ \text{m}^3$
Mass	Metric ton	t	$1\ \text{t} = 10^{3}\ \text{kg}$
Area	Hectare	ha	$1\ \text{ha} = 1\ \text{hm}^2 = 10^{4}\ \text{m}^2$

e. Recommended pronunciation

Prefix	Pronunciation (USA)[1]	Selected units	Pronunciation
exa	ex' a (*a* as in *a*bout)	candela	can*dell*' a
peta	pet' a (*e* as in p*e*t, *a* as in *a*bout)	joule	rhyme with *tool*
tera	as in *terra* firma	kilometer	*kill*' oh meter
giga	jig' a (*i* as in j*i*g, *a* as in *a*bout)	pascal	rhyme with *rascal*
mega	as in *mega*phone	siemens	same as *seamen's*
kilo	kill' oh		
hecko	heck' toe		
deka	deck' a (*a* as in *a*bout)		
deci	as in *deci*mal		
centi	as in *centi*pede		
milli	as in *milli*tary		
micro	as in *micro*phone		
nano	nan' oh (*an* as in *an*t)		
pico	peek' oh		
femto	fem' toe (*fem* as in *fem*inine)		
atto	as in an*ato*my		

The first syllable of every prefix is accented to assure that the prefix will retain its identity.
Pronunciation of kilometer places the accent on the first syllable, *not* the second.

Appendix E

Mathematical Symbol and Abbreviation

$+$	plus (addition)	a', a''	a-prime, a-second
$-$	minus (subtraction)	a_1, a_2	a-sub one, a-sub two
$\pm$	plus or minus	(), [], {}	parentheses, brackets, braces
$\times$	times, by (multiplication)	$\angle$, $\perp$	angle, perpendicular to
$\div$, /	divided by	a^2, a^3	a-square, a-cube
:	is to (ratio)	a^{-1}, a^{-2}	$1/a$, $1/a^2$
::	equals, as, so is	$\sin^{-1}a$	the angle whose sine is
$\therefore$	therefore	π	pi = 3.141593+
$=$	equals	μ	microns = .001 millimeter
$\sim \approx$	approximately equals	$m\mu$	micromillimeter = .000001.
$>$	greater than	$\sum$	summation of
$<$	less than	ε, e	base of hyperbolic, natural or
$\geq$	greater than or equals		Napierian logs = 2.71828+
$\leq$	less than or equals	$\triangle$	difference
$\neq$	not equal to	g	acceleration due to gravity
$\propto$	varies as		(32.16 feet/sec. Per sec.)
∞	infinity	E	coefficient of elasticity
$\parallel$	parallel to	v	velocity
$\star$	square root	f	coefficient of friction
$\square$	square	P	pressure of load
$\bigcirc$	circle	HP	horsepower
$\circ$	degrees (arc or thermometer)	RPM	revolutions per minute
$'$	minutes or feet		
$''$	seconds or inches		

GREEK ALPHABET

A, α	Alpha	H, η	Eta	N, ν	Nu	T, τ	Tau
B, β	Beta	Θ, θ	Theta	Ξ, ξ	Xi	Y, υ	Upsilon
Γ, γ	Gamma	I, ι	Iota	O, o	Omicron	Φ, φ	Phi
Δ, δ	Delta	K, κ	Kappa	Π, π	Pi	X, χ	Chi
E, ε	Epsilon	Λ, λ	Lambda	P, ρ	Rho	Ψ, ψ	Psi
Z, ζ	Zeta	M, μ	Mu	Σ, σ	Sigma	Ω, ω	Omega

References

1. Rosato, D. V., *PIA Plastics Engineering Manufacturing and Data Handbook*, Kluwer, 2001.

2. Rosato, D. V., *Concise Encyclopedia of Plastics: Fabrication & Industry* (25,000 entries), Kluwer, 2000.

3. Rosato, D. V., *Injection Molding Handbook*, 3ed Edition, Kluwer, 2000.

4. Rosato, D. V., Injection Molding in the 21st Century, *SPE-IMD Newsletter*, Nos. 53 & 54, 2000.

5. Rosato, D. V., Fundamentals of Designing with Plastics, *SPE-IMD Newsletter*, No. 51, 1999.

6. Rosato, D. V., *Extruding Plastics: A Practical Processing Handbook*, Kluwer, 1998.

7. Rosato, D. V., Injection Molding (chapter in R. F. Jones book entitled *Guide to Short Fiber Reinforced Plastics*), Hanser, 1998.

8. Rosato, D. V., Parts Removal With On-Robot Secondary Operations and Simple Sprue Pickers in chapter sections of *Tool & Manufacturing Engineers Handbook, Vol. 9, Materials & Parts Handling*, SME, 1998.

9. Rosato, D. V., *Plastics Processing Data Handbook*, Second Edition, Kluwer, 1997.

10. Rosato, D. V., *Designing with Reinforced Composites*, Hanser, 1997.

11. Rosato, D. V., Design Features that Influence Part Performance, *SPE-IMD Newsletter*, Issue 46, 1997.

12. Rosato, D. V., *Designing with Plastics*, Rhode Island School of Design, Lectures 1987–1997.

13. Rosato, D. V., *Rosato's Plastics Encyclopedia and Dictionary*, Hanser, 1993.

14. Rosato, D. V., *Designing with Plastics and Composites: A Handbook*, Kluwer, 1991.

15. Rosato, D. V., Target for Zero Defects, *SPE-IMD Newsletter*, Issue 26, 1991.

16. Rosato, D. V., Materials Selection, Polymeric Matrix Composites, Chapter in *International Encyclopedia of Composites*, VCH, 1990.

17. Rosato, D. V., Product Design: Plastic Selection Guide, *SPE ANTEC*, May 1990.

18. Rosato, D. V., *Designing with Plastics*, Rhode Island School of Design, Lectures 1987–1990.

19. Rosato, D. V., *Current and Future Trends in the Use of Plastics for Blow Molding*, SME, 1990.

20. Rosato, D. V., *Blow Molding Handbook*, Hanser, 1989.

21. Rosato, D. V., Product Design: Basic Processing Guide for Plastics, *SPE ANTEC*, May 1989.

22. Rosato, D. V., Plastics and Solid Waste, *RISD*. Oct. 1989.

23. Rosato, D. V., Role of Additives in Plastics: Function of Processing Aids, *SPE-IMD Newsletter*, Nov. 1987.

24. Rosato, D. V., Plastic Replaces Aorta Permits Living Normal-Long Life, Newton-Wellesly Hospital-Massachusetts, Mar. 1987.

25. Rosato, D. V., Materials Selection Chapter in *Encyclopedia of Polymer Science and*

Engr., Mark Bikales-Overberger-Menges (eds.): Vol. 9, pp. 337–379, Wiley, 1987.

26. Rosato, D. V., Seminars presented worldwide on different plastic subjects that include Introduction to Plastics, Plastic Markets, Plastic Types and Formulations, Plastic Properties, Designing with Plastics, Testing, Design Parts, Fabrication by different processes, Quality Control, Statistical Control, to Marketing via University of Lowell, Plastics World, ASME, General Motors Institute, SPE. SPI. China National Chemical Construction (Beijing), Hong Kong Production Centre. Singapore Institution, Open University of England, Geneva Development, and Tufts Medical University, 1974 to 1986.

27. Rosato, D. V., *Industrial Plastics in Materials Handling*, International Mgm. Soc., Oct. 1985.

28. Rosato, D. V., Advanced Engineering Design Short Course, ASME Engr. Conference, 1983.

29. Rosato, D. V., Polymers. *Processes and Properties of Medical Plastics in Synthetic Biomedical Polymers*, Szycher-Robinson (eds.), Technomics, 1980.

30. Rosato. D. V., Wealth of Choices: Fasteners for Plastics, *PW*, June 1979.

31. Rosato, D. V., Fatigue Testing, *PW*, May 1977.

32. Rosato, D. V., Uniform Resin Coding, *PW*, May 1977.

33. Rosato, D. V., et al., *Plastics Industry Safety Handbook*, Cahners, l973.

34. Rosato, D. V., et al., *Markets For Plastics*, Kluwer, 1969.

35. Rosato, D. V., *Filament Winding* (in Russian), Russian Publ., 1969.

36. Rosato, D. V., *Environmental Effects on Polymeric Materials, Vol. 1.: Environment* and *Vol. 2: Materials*, Wiley, 1968.

37. Rosato, D. V., *Filament Winding*, Wiley, 1964.

38. Rosato, D. V., and G. Lubin, Application of Reinforced Plastics, 4th International RP Conference. British Plastics Federation, London, UK. Nov. 25–27. 1964.

39. Rosato, D. V., Non-Woven Fibers in Reinforced Plastics, *Ind. Engr. Chem.*, 54,8. 30–37, Sept. 1962.

40. Rosato, D. V., Plastics in Missiles, *British Plastics*, 348–352, Aug. 1960.

41. Rosato, D. V. Capt., All Plastic Military Airplane Successfully Flight Tested, Wright-Patterson AF Base, Ohio, l944.

42. Rosato, D. V., Capt., Theoretical Potential for Polyethylene, USAF Materials Lab., WPAFB, 1944.

43. *Advanced Composites Design Guide*, MIL-HDBK-5C, US Supt. of Documents, GPO, 1980.

44. Allgood, J. R., Structures in Soil Under High Loads, *Journal of Soil Mechanics and Foundations Proceedings*, ASCE, Mar. 1971.

45. Altshuler, T. L., *Fatigue: Life Predictions for Materials Selection & Guide*, ASM Software, 1988.

46. Avallone E., et al., *Mark's Standard Handbook of Mechanical Engineers*, 10th Edition, McGraw Hill, 1996.

47. Bass, M., *Handbook of Optics*, 2ed Edition, McGraw-Hill, 1995.

48. Baumeister, T., and Marks, L. S., *Standard Handbook for Mechanical Engineers*, McGraw-Hill, 1978.

49. Beakley, G. C., *Engineering Design*, Macmillan, 1986.

50. Beck, R. D., *Plastics Product Design*, Kluwer, 1980.

51. Belding, W. E., *Handbook of Engineering Mathematics*, ASM International, 1983.

52. Benjamin, B. S., *Structural Design with Plastics*, Kluwer, 1982.

53. Berkovits, A., Relationship Between Fatigue Life in the Creep Fatigue Region & Stress-Strain Response, NASA, 1988.

54. Binder, K., Non-Isothermal Long Term Creep (17 yrs.) Behavior of Thermoplastics Outdoors, *Kunststoffe*, No. 1988.

55. Bucksbee, J. H., *The Use of Bonded Elastomers for Energy & Motion Control in Construction*, Lord Corp. Industrial Products Division, Erie, PA., 1988.

56. Champoux, R. L., Analytical & Experimental Methods of Residual Stress Effects in Fatigue, ASTM, STP 1004, 1989.

56. Chelapati, C. V. & J. R. Allgood in Feb. 1970 Buckling of Cylinders in a Confined Media, Highway Research Board, Feb. 1970.

57. Crawford, R. J., *Plastics Engineering*, Pergamon, 1987.

58. DellaCorte, C., Static & Dynamic Friction Behavior of Candidate High Temperature Materials, NASA, 1994.

59. *Design & Production of Gears*, Hoechst Celanese, 1984.

60. DiCarlo, J. A., Creep & Stress Relaxation Modeling of Polycrystalline Ceramic Fibers, NASA, 1994.

61. Dorf, R. C., *The Engineering Handbook*, CRC Press, 1996.

62. Dorgham, M., and Rosato, D. V., *Designing with Plastic Composites*, Interscience Enterprises-Geneva, 1986.

63. Dudley, D. W., *Handbook of Practical Gear Design*, McGraw-Hill, 1985.

64. Dym, J. B., *Product Design with Plastics*, Industrial Press, 1983.

65. Ehrenstein, G. W., et al., *Designing with Plastics*, Hanser, 1984.

66. *Engineering Materials Handbook, Vol. 1: Composite*, ASM International, 1987.

67. *Engineering Materials Handbook, Vol. 2: Engineering Plastics*, ASM International, 1988.

68. Ernst, H. A., *Elastic Plastic Fracture Mechanics Methodology for Surface Cracks*, Georgia Inst. of Tech., 1994.

69. Eshbach, O. W., et al., *Handbook of Engineering Fundamentals*, Wiley, 1975.

70. *Fatigue & Tribological Properties-Databank*, DPL, 1998.

71. Felton, L. P., Structural Index Method of Optimum Design, Proceedings, ASME winter annual meeting, New York, 1974.

72. Fischbeck, H. J., *Formulas, Facts, & Constants for Students & Professionals in Engineering, Chemistry, & Physics*, Springer-Verlag, 1987.

73. Freed, A. D., Viscoplastic Model Development, NASA, 1995.

74. French, T. E., et al., *Engineering Drawing & Graphic Technology*, McGraw-Hill, 1986.

75. Ganiac, E. N., et al., *The McGraw-Hill Handbook of Essential Engineering Information & Data*, McGraw-Hill, 1991.

76. Gent, A. N., *Engineering with Rubber*, Hanser, 1992.

77. Glanville, A. B., *Plastics Engineers Data Book*, Machine Publ., 1971.

78. Heger, F. J., *Structural Plastics Design Manual*, ASCE, 1981.

79. Heisler, S. I., *The Wiley Engineer's Desk Reference & Concise Guide*, 2ed Edition, Wiley, 1998.

80. Hengl, R., et al., Influence of Loading History on the Mechanical Properties of Polymeric Materials, *Kunststoffe*, pp. 31–33, Mar. 1989.

74. Hertzberg, R. W., *Deformation & Fracture Mechanics of Engineering Materials*, Wiley, 1976.

81. Hertzberg, R. W., et al., Fatigue Testing-Flaws Makes It Better, *PW*, May 1977.

82. Hertzberg, R. W., et al., *Fatigue of Engineering Plastics*, Academic Press 1980.

83. Hibbeler, R. C., *Engineering Mechanics: Dynamics*, Macmillan, 1974.

84. Hoechst Celanese Corp., *Designing with Plastics*, Engr. Manual, 1989.

85. Horne, M. R., Plastic *Theory of Structures*, MIT Press, 1971.

86. Iida, H., et al., Mechanical Characteristics of Filament Wound Pressure Vessel, NASA, 1987.

87. Jensen, A., et al., *Applied Strength of Materials*, McGraw-Hill, 1975.

88. Kahraman, A., Dynamic Analysis of Geared Rotors, NASA, 1990.

89. Kalamkarov, A. L., et al., *Analysis, Design, & Optimization of Composite Structures*, Wiley, 1997.

90. Kutz, M., *Mechanical Engineers' Handbook*, Wiley, 1998.

91. Lindholm, U. S., *Mechanical Behavior of Materials under Dynamic Loads*, Springer-Verlag, 1968.

92. Lubin, G., *Handbook of Composites*, Kluwer, 1982.

93. *Machinery's Handbook*, 21st Edition, Industrial Press, 1979.

94. Manson, S. S., et al., A linear Time-Temperature Relation of Creep & Stress Rupture Data, NACA, TN 2890, 1953.

95. Marmathy, S., Design of Buildings for Fire Safety, ASTM, STP, 685, 1969.

96. Maxwell, J. C., *Philos. Transactions, Royal Society*, London, UK, Vol. 157, p. 49, 1867.

97. Mazda, F. F., *Electronics Engineer's Reference Book*, 6th Edition, Butterworths, 1989.

98. McCauley, *Guide to the Use of Tables & Formulas in Machinery's Handbook*, Industrial Press, 1996.

99. Miyamoto, Y., et al., *Functionally Graded Materials Design, Processing, & Applications*, Kluwer, 1999.

100. Mobay/Bayer Design Manual: Snap Fit Joints in Plastics, *Mobay Engineering Thermoplastics Bulletin*, Bayer Chemical Corp., 1990.

101. *Optics: Handbook of Plastic Optics*, U.S. Precision Lens Inc.

102. Oberge, F., et al., *Machinery Handbook*, Industrial Press, 1981.

103. Orth, F., et al., Selection of Materials Under Dynamic Loading, *Kunststoffe*, Aug. 1989.

104. Pearson, C. E., *Engineering Mathematics*, Kluwer, 1986.

105. Perry, R. H., *Engineering Manual*, McGraw-Hill, 1976.

106. *Plastics for Aerospace Vehicles*, MIL-HDBK-17A & 17B, US Supt. of Documents, GPO, 1981.

107. Ray, M. S., *Engineering Design*, Prentice Hall, 1987.

108. Roark, R. J., et al., *Formula for Stress & Strain*, McGraw-Hill, 1976.

109. Schwartz, R. T., and D. V. Rosato, *Structural Sandwich Construction in Composite Engr. Materials*, Dietz, A G. H. (ed.), pp. 165–181, MIT Press, 1969.

110. Sims, F., *Engineering Formulas*, Industrial Press, 1999.

111. Spangler, M. G., Structural Design of Flexible Culverts, *Soil-engineering Bulletin* 153, Iowa Engineering Experimentation Station, 1971.

112. Stangle, G. C., *Modeling of Materials Processing*, Kluwer, 1999.

113. *Structural Sandwich Composites*, MIL-HDBK-23A, US Supt. of Documents, GPO, 1974.

114. *Structural Plastics Design Manual*, ASCE No. 63 or No. 023-000-00495-0, US Supt. of Documents, GPO, 1982.

115. *Structural Plastics Selection Manual*, Nos. 63 & 66, ASCE, 1984.

116. *Structural Sandwich Composites*, MIL-HDBK-23A, US Supt. of Documents, GPO, 1974.

117. Timoshenko, S., *Strength of Materials*, Kluwer, 1958.

118. Timoshenko, S., et al., *Elastic Stability*, McGraw-Hill, 1961.

119. Tong, L., et al., *Analysis & Design of Structural Bonded Joints*, Kluwer, 1999.

120. Tuma, J. J., *Engineering Mathematics Handbook*, McGraw-Hill, 1987.

121. Tuma, J. J., *Handbook of Numerical Calculations in Engineering*, McGraw-Hill, 1989.

122. Wetton, R. E., et al., Comparison of Dynamic Mechanical Measurements in Bending, Tension & Torsion, *SPE ANTEC*, pp. 1160–1162, May 1989.

123. Wilson, D., et al., *Composite Design Manual*, University of Delaware, 1980.

124. Young, J. E., et al., *Materials & Processes Optimize Design*, Marcel Dekker, 1985.

125. Young, W. C., *Roark's Formulas for Stress and Strain*, 6th Ed., McGraw-Hill, 1998.

126. *Advances in Industrial Computing Technology*, ISA, 1999.

127. Ali, M., *Applied Intelligence*, Kluwer, 1998.

128. *ASTM Book of Standards*, Annual Issues.

129. *ASTM International Directory of Testing Laboratories*, ASTM, 1999.

130. Berins, M. L., *Plastics Engineering Handbook of The Society of Plastics Industry*, 5th Edition, Kluwer, 1991.

131. Bernhardt, A., Computer Modeling Predicts the Dimensional Tolerances of Molded Parts, *SPE-IMD Newsletter* No. 25, Fall, 1990.

132. Bernhardt, A., et al., Rationalization of Molding Machine Intelligent Setting & Control, *SPE-IMD Newsletter*, No. 54, Summer 2000.

133. Billmeyer, F. W., *Textbook of Polymer Science*, Wiley, 1984.

134. Black, and Decker, *The Reasons for Capron*, World Plastics Technology, 2000.

135. Bottenbruch, L., *Engineering Thermoplastics, Hanser*, 1996.

136. Boyer, T. D., et al., Advances in Thermoplastic Encapsulation of Electrical/Electronic Components, *PE*, Feb. 2000.

137. Bregar, B., Architect's Concept Would Clad Buildings in *Plastic, Plastics News*, p. 3, Feb. 1990.

138. Brostow, W., and Corneliussen, R. D., *Failure of Plastics*. Hanser, 1986.

139. Buckleitner, E. L., et al., *Plastics Mold Engineering Handbook*, 5th Edition, Kluwer, 1995.

140. Bushko, W. C., et al., Estimates for Material Shrinkage in Molded Parts Caused by Time Varying Cavity Pressure, *SPE-ANTEC* 1997.

141. Buzzard, W. S., *Flow Control-Design & Control Systems*, ISA, 1994.

142. Cerwin, N., Your Guide to Mold Steels, *Plastics Auxiliaries*, Mar. 1999.

143. *Chemical Resistance*, PDL, 1994.

144. Coleman, B. D., Thermodynamics of Materials with Memory Treatise, *Arch. Rat. Mech. Anal.*, 1964.

145. Composites at Sea., *Advanced Composites*, Mar.–Apr. 1990.

146. Corripio, A. B., *Design and Application of Process Control Systems*, ISA, 1998.

147. Cushion, R. F., *Engineers Malpractice*, Wiley, 1987.

148. Covas, J. A., et al., *Rheological Fundamentals of Polymer Processing*, Kluwer, 1995.

149. Daido Steel, Tool Steel Selection Software Targets Mold Making Startups, *Modern Mold & Tooling*, Apr. 1999.

150. Dealy, J. M., et al., *Melt Rheology & Its Role in Plastics Processing*, Kluwer, 1990.

151. Defosse, M., Processors focus on Differentiation in Window Profiles, *MP*, Sept. 1999.

152. Delliarciprete, J., et al, Cavity Pressure Transfer Extends Prototype Tool Life, *MP*, Jan. 2000.

153. Denison, B. R., Systems Thinking, Modular Assemblies will Keep Auto Processors, OEMs Profitable, *MP*, Jan. 2000.

154. *Dictionary of Measurement & Control*, 3^{ed} Edition, ISA, 1995.

155. Deming, E. W., *Out of Crisis*, MIT Center for Advancement Engr. Studies, 1986.

156. Earle, J. H., *Engineering Design Graphics*, Addison-Wesley, 1990.

157. Edenbaum, J., *Plastics Additives and Modifiers Handbook*, Kluwer, 1992.

158. *Environmental Briefs*, The Vinyl Institute, Dec. 1999.

159. Eskind, L. G., New Improvements in Intumescence, *PE*, Feb. 2000.

160. Evans, B., et al., Effective Use of Industrial Design in Rapid Product Development, *MD&DI*, Sep. 1999.

161. Ezrin, M., *Plastics Failure Guide, Cause & Prevention*, Hanser, 1996.

162. Finn, G., Applying Six Sigma to Design, *IM*, Dec. 1999.

163. Flick, E. W., *Plastics Additives*, 2^{ed} Edition, PDL, 1993.

164. Fluoroplastics-Non-Melt Processible, PDL, 2000.

165. Fuges, C., Rapid Prototyping-A Journey, Not a Destination, *MoldMaking Tech.*, Feb. 2000.

166. Gee, G. L., New & Emerging Opportunities in Elastomers, *PE*, Jan. 2000.

167. Gegov, A., *Distributed Fuzzy Control of Multivariable Systems*, Kluwer, 1996.

168. Goldfein, S., *General Formula for Creep and Rupture Stresses in Plastics*, MP, Apr. 1960.

169. Goldsworthy, B., Composites for Roads and Bridges, *Plastics Trends*, p. 12, Jan.–Feb. 1990.

170. Gonzalez, M., *Special Rules Govern Design of Tools for Elastomers*, MM&T, Mar. 1999.

171. Gordon, D., Aluminum Molds Go the Distance, *Molding Systems*, Mar. 1999.

172. Gouldson, C., Rapid Tooling Processes and Selection Criteria, *SPE-ANTEC* 1996.

173. Grainger, S., et al., *Engineering Coatings*, PDL, 1998.

174. Grimm, T., Rapid Tooling is not the Future, It is Today, *MoldMaking Tech.*, Feb. 2000.

175. Grolik, F. W., et al., Reinforced PP for Motor Vehicle Interior Fittings. *Kunststoffe*, Dec. 1989.

176. Gruhn, P., et al., *Safety Shutdown systems: Design, Analysis, and Justification*, ISA, 1998.

177. Harris, R. M., *Coloring Technology*, PDL, 1999.

178. *Handbook of Plastics Joining, A Practical Guide*, PDL, 1997.

179. Hauck, C., et al., Auto Modules will Play to Plastics' Strength, *MP*, Jan. 2000.

180. Haut, D., Success by Design, *MDDI*, pp. 48–69, Sept. 1988.

181. Hemmerich, K. J., Polymer Materials Selection for Radiation-Sterilized Products, *MD&DI*, Feb. 2000.

182. Holden, G., et al., *Thermoplastic Elastomers*, 2^{ed} Edition, Hanser, 1996.

183. *ISA Directory of Instrumentation*, ISA, 1999.

184. Janzen, W., and Ehrenstein, G. W., Hysteresis Measurements for Characterizing the Cyclic Strain and Stress Sensitivity of Glass Fiber Reinforced PBT, *SPE ANTEC*, May 1989.

185. Kaufman, M., *The First Century of Plastics*, Plastics Institute, London, UK, 1963.

186. King, R. D., et al., *Future belongs to Biopolymers*, MP, Jan. 2000.

187. Kodak, *Industry Innovator*, World Plastics Technology, 2000.

188. Kraus, H., *Creep Analysis*, Wiley, 1980.

189. Legal Ruling Opens the Door for PAEK in human Implants, *MP*, Sept. 1999.

190. Levy, S., and J. H. DuBois, *Plastics Product Design Engineering Handbook*, Kluwer, 1977.

191. Lowen, R., et al., *Fuzzy Logic*, Kluwer, 1993.

192. Malloy, R. A., *Plastic Part Design for Injection Molding*, Hanser, 1994.

193. Mamzic, C. L., *Statistical Process Control*, ISA, 1995.

194. Mamzic, C. L., *Statistical Process Control*, ISA, 1995.

195. Market Data Book, *Annual Plastics News*, Dec. 1999.

196. McMillan, G., *Dispersing Heat through Conviction: Funnier Side of Process Control*, ISA, 1999.

197. Mennig, G., *Mold-Making Handbook for the Plastics Engineer*, 2^{ed} Edition, Hanser, 1998.

198. Michaeli, W., *Plastics Processing, An Introduction*, Hanser, 1995.

199. Miglierini, M., et al., *Mossbauer Spectroscopy in Materials Science*, Kluwer, 1999.

200. *Modern Plastics Encyclopedia*, Yearly.

201. *Mold Shrinkage and Warpage Handbook*, PDL, 1999.

202. Moore, S., Innovative Reactive Extrusion Technologies bring Added Value to Recycled Plastics, *MP*, Jan. 2000.

203. Murray, C. J., Design with Lives in Mind, *Design News*, Feb. 1990.

204. Murray, C. J., Foam Exhibits Negative Poison's Ratio, *DN*, Dec. 1989.

205. Murrill, P. W., *Fundamentals of Process Control Theory*, 3^{ed} Ed., ISA, 1999.

206. Noller, R., Understanding Tight-Tolerance Design, *PDF*, Mar.–Apr. 1990.

207. *Packaging Encyclopedia*, Packaging, Annual.

208. Peraro, J. S., Limitations of Standardized Tests, *PE*, Jan. 2000.

209. Port, O., In Transportation, One Word Plastics, *Business News*, Mar. 6, 2000.

210. Porter, M. E., *Competitive Advantage: Creating & Sustaining Superior Performance*, Free Press/Macmillan, 1980.

211. Porter, M. E., *Competitive Strategy: Techniques for Analyzing Industries & Competitors*, Free Press/Macmillan, 1980.

212. Portnoy, R. C., *Medical Plastics*, PDL, 1998.

213. *Processing Handbook & Buyers Guide*, PT, Yearly.

214. Rao, N. S., et al., *Design Data for Plastics Engineers*, Hanser, 1991.

215. Rasmussen, S., *Shrink Data and Mold Error*, Camas, Wash., Hewlett-Packard, 1989.

216. Redner, A. S., et al., Measure Stress in Transparent Plastics, ASTM 4093-86.

217. Reinfrank, G. B., Capt. et al., Molded Glass Fiber Sandwich Fuselage (etc.) for BT-15 Airplane, Army A.F., Tech. Report No. 5159, Nov. 8, 1944.

218. Rohsenow, W. M., et al., *Handbook of Heat Transfer*, 3^{ed} Ed., McGraw-Hill, 1998.

219. Rotheiser, J., *Joining of Plastics*, Hanser, 1999.

220. Rubin, I., *Handbook of Plastic Materials and Technology*, Wiley, 1990.

221. Rupprecht, L., *Conductive Polymers & Plastics*, PDL, 1999.

222. Schindler, B. M., Made in Japan, W. Edwards Deming, *ASTM Standardization News*, Feb. 1983.

223. Schmidt, L. B., Opportunities for the Next Decade in Injection Molding, *SPE-IMD Newsletter No. 53*, 2000.

224. Schmidt, L. B., et al., Injection Molding of Polycarbonate Compact Disks: Relationship between Process Conditions, Birefringence, & Block Error Rate, *SPE-ANTEC*, 1992.

225. Schmidt, L. B., A Special Mold & Tracer Technique for Studying Shear & Extensional.

226. Sepe, M., *Dynamic Mechanical Analysis*, PDL, 1998.

227. Sepe, M., Pause for a Brief Word about Thermosets, *IM*, Sept. 1999.

228. Shah, V., *Handbook of Plastics Testing Technology*, 2nd Edition, Wiley, 1998.

229. Shay, R. M., Estimating Linear Shrinkage of Semicrystalline Resins from Pressure-Volume Temperature (pvT) Data, *SPE-IMD Newsletter* 49, Fall 1998.

230. Stangle, G. C., *Modeling of Materials Processing*, Kluwer, 1999.

231. Stephenson, G. M., and Freiherr, G., Cost-Effectiveness: New Yardstick for Medical

Device Evaluation, *MDDI*, pp. 71–73, Mar. 1990.

232. Stoeckhert, K., *Mold Making Handbook*, Hanser, 1983.

233. *The Effect of Creep-600 Graphs*, PDL, 1991.

234. Thompson, J. K., Driving Plastics Use in Auto Design, NASA Tech Briefs, Sep. 1999.

235. Toensmeier, P. A., Lenticular Imaging Adds Perspective to Display, *MP* Jan. 2000.

236. Tres, P. A., *Designing Plastic Parts for Assembly*, Hanser, 1995.

237. Troitzsch, J., *International Plastics Flammability Handbook*, Hanser, 1983.

238. Trzaskoma, P. P., et al., Characteristics of Rigid Polymer Foams as Related to their Use for Corrosion Protection in Enclosed Metal Spaces, NACE, 1999.

239. Trzaskoma, P. P., et al., Corrosion Control in Enclosed Areas of Military Ground Vehicles, *J. of Corrosion*, NACE, 1998,

240. Verbruggen, H. B., et al., *Fuzzy Logarithms for Control*, Kluwer, 1999.

241. Wallenberger, F. T., et al., *Advanced Inorganic Fibers Processes, Structures, Properties, Applications*, Kluwer, 1999.

242. Wille, D., Producing Bubble/Taper Tubing for Medical Applications, *MD&DI*, Jan. 2000.

243. Wright, D. C., *Environmental Stress Cracking of Plastics*, PDL, 1996.

244. Wypych, G., *Handbook of Fillers*, PDL, 1999.

245. Wypych, G., *Weathering of Plastics*, PDL, 2000.

Index